World Sustainability Series

Series Editor

Walter Leal Filho, Wissenschaften Hamburg, Hochschule für Angewandte, Hamburg, Hamburg, Germany

Due to its scope and nature, sustainable development is a matter which is very interdisciplinary, and draws from knowledge and inputs from the social sciences and environmental sciences on the one hand, but also from physical sciences and arts on the other. As such, there is a perceived need to foster integrative approaches, whereby the combination of inputs from various fields may contribute to a better understanding of what sustainability is, and means to people. But despite the need for and the relevance of integrative approaches towards sustainable development, there is a paucity of literature which address matters related to sustainability in an integrated way.

Notes on the quality assurance and peer review of this publication

Prior to publication, the works published in this book are initially assessed and reviewed by an in-house editor. If suitable for publication, manuscripts are sent for further review, which includes a combined effort by the editorial board and appointed subject experts, who provide independent peer-review. The feedback obtained in this way was communicated to authors, and with manuscripts checked upon return before finally accepted. The peer-reviewed nature of the books in the "World Sustainability Series" means that contributions to them have, over many years, been officially accepted for tenure and promotion purposes.

World Sustainability Series

Due to its scope and nature, sustainable development is a matter which is very interdisciplinary, and draws from knowledge and inputs from the social sciences and environmental sciences on the one hand, but also from physical sciences and arts on the other. As such, there is a perceived need to foster integrative approaches, whereby the combination of inputs from various fields may contribute to a better understanding of what sustainability is, and means to people. But despite the need for and the relevance of integrative approaches towards sustainable development, there is a paucity of literature which address matters related to sustainability in an integrated way.

Notes on the quality assurance and peer review of this publication

Prior to publication, the works published in this book are initially assessed and reviewed by an in-house editor. If suitable for publication, manuscripts are sent for further review, which includes a combined effort by the editorial board and appointed subject experts, who provide independent peer-review. The feedback obtained in this way was communicated to authors, and with manuscripts checked upon return before finally accepted. The peer-reviewed nature of the books in the "World Sustainability Series" means that contributions to them have, over many years, been officially accepted for tenure and promotion purposes.

Walter Leal Filho · Dirk Franco ·
Laís Viera Trevisan · Mark Mifsud ·
Kent A. Williams
Editors

Introducing the UN SDGs in Higher Education Institutions

Springer

Editors
Walter Leal Filho
Research and Transfer Centre "Sustainable Development and Climate Change Management" (FTZ NK)
Hamburg University of Applied Sciences
Hamburg, Germany

Laís Viera Trevisan
Hamburg University of Applied Sciences
Hamburg, Germany

Kent A. Williams
Manning School of Business
Acadia University
Wolfville, NS, Canada

Dirk Franco
PXL University of Applied Sciences and Arts
University of Hasselt
Hasselt, Belgium

Mark Mifsud
University of Malta
Msida, Malta

ISSN 2199-7373 ISSN 2199-7381 (electronic)
World Sustainability Series
ISBN 978-3-032-23016-4 ISBN 978-3-032-23017-1 (eBook)
https://doi.org/10.1007/978-3-032-23017-1

This Springer imprint is published by the registered company Springer Nature Switzerland AG
The registered company address is: Gewerbestrasse 11, 6330 Cham, Switzerland

Preface

The threats posed by climate change, increasing inequality, resource degradation, and social injustice threaten our global community and endangers livelihoods in a significant way. In order to meet this unprecedented global challenge, the United Nations released the 2030 Agenda for Sustainable Development in 2015 with 17 Sustainable Development Goals (SDGs) at its center. A visionary, integrative plan for peace and prosperity for people and the planet this universal framework requires a joint and concerted response from every facet of society.

Within this global effort, Higher Education Institutions (HEIs) play a pivotal, multifaceted role. For centuries, universities have been the arbiters of knowledge, the engines of innovation, and the caretakers of tomorrow's leaders. This historical mandate is now transformed into a key contemporary obligation. HEIs are much more than sources of abstract thinking; they are actively involved in the local, national, and global environment. Their core functions—education, research, campus operations, and civic leadership—are potent levers in advancing sustainable development. They are responsible for preparing millions of students with the knowledge, skills, values, and agency needed to navigate and design a sustainable future.

Moreover, their research capabilities are essential to creating the evidence-based solutions needed for attaining the complex, interrelated goals of the SDGs.

Yet, a substantial disparity remains between the acknowledgment of the SDGs and its real and systematic incorporation into the context of higher education in practice. And a number of institutions struggle with very big questions—How do we move beyond ad hoc solutions to whole-institution programs? How can the SDGs be integrated in different areas of our education in the various disciplines (engineering and business, the humanities, the arts)? How do we track our impact and tailor our portfolios to global issues we need to address now? How can campus systems be turned into living labs for sustainability? The necessity of a cohesive, pragmatic, academically compelling guide has never been stronger.

This book, *Introducing the UN SDGs in Higher Education Institutions*, is designed to answer that question very explicitly. As a strategic guide for academia, management, students, and practitioners dedicated to this transformational process, this book shifts from arguing for the sound theoretical and ethical case for the importance of SDG engagement to the frameworks that will help guide actionable actions. It showcases methods of curriculum design, transdisciplinary research methodologies, sustainable campus management, and community engagement. By providing a wide scope of case studies, analytical tools, and reflective thinking, this volume delves on how the process of integrating SDGs can be easier for individuals to understand, and presents a vivid picture of their integration process and makes this process accessible and actionable to individuals in various stages of their sustainability journey.

The academic relevance of this study is immense. In synthesis of multiple strands of research and practice, it adds to the emergent field of Sustainability in Higher Education. It lays out a foundational basis for scholarly discussion about education as a tool for sustainable development itself and because of its pragmatic orientation it serves as a relevant key text for professional development programs and graduate courses. Internationally, the need is universal. The SDGs are not merely an international agenda for the world, but global in scope and the tasks they solve transcend national boundaries. This book is intended for popular worldwide audiences and recognizes the various cultural, economic and institutional settings of different nations as different as countries and points out the responsibilities of HEIs everywhere.

We thank all authors for the very important inputs, and the reviewers for supporting the process. We hope that this publication and its tapestry of understanding and experience will inspire, equip, and embolden the academic community to continue to play a key in implementing the UN SDGs and going beyond it.

Hamburg, Germany — Walter Leal Filho
Hasselt, Belgium — Dirk Franco
Hamburg, Germany — Laís Viera Trevisan
Msida, Malta — Mark Mifsud
Wolfville, Canada — Kent A. Williams

Contents

Curriculum Issues

Assessment Strategy to Embed Key Competencies for Sustainability into the Curriculum: A Case Study of Politics and International Relations

Olga Khrushcheva

Normal people are not accountable.
Stop putting it on them
(2nd year, BA International Relations student)

Abstract According to the World Meteorological Organisation, 2023 was the warmest year on record, with the global average near-surface temperature at 1.45 °C above the pre-industrial baseline (WMO in State of the global climate 2023, 2024a). Scientists predict with an 80% probability that the annual average global temperatures will exceed 1.5 °C for at least one of the years between 2024 and 2028 (WMO in Global temperature is likely to exceed 1.5 C above pre-industrial level temporarily in next 5 years, 2024b). Higher education institutions are responsible for equipping students with the skills they need to engage with sustainability challenges before and after graduation. This chapter examines the experience of integrating key competencies for sustainable development into the assessment strategy of a second-year module delivered to BA (Hons) Politics and BA (Hons) International Relations students at Manchester Metropolitan University. The chapter argues that project-based pedagogy helps students develop self-awareness, collaboration, and strategic and critical thinking needed to create their own agency in addressing sustainability challenges.

Keywords Education for sustainable development · Project-based learning · ESD competencies

O. Khrushcheva (✉)
Manchester Metropolitan University, Manchester, UK
e-mail: o.khrushcheva@mmu.ac.uk

W. Leal Filho et al. (eds.), *Introducing the UN SDGs in Higher Education Institutions*, World Sustainability Series, https://doi.org/10.1007/978-3-032-23017-1_1

1 Introduction

According to the World Meteorological Organisation, 2023 was the warmest year on record, with the global average near-surface temperature at 1.45 °C above the pre-industrial baseline (WMO, 2024a). Scientists predict with an 80% probability that the annual average global temperatures will exceed 1.5 °C for at least one of the years between 2024 and 2028 (WMO, 2024b). The costs of a '1.5 °C warmer world' are high, with the loss of ecosystems, increased frequency of extreme weather events, and impacts on human health (IPCC, 2022). Higher education institutions are responsible for equipping students with the skills they need to engage with sustainability challenges before and after graduation. However, universities haven't yet achieved their full transformative potential. The literature suggests that there is a negative correlation between general awareness of the dangers linked to climate change and personal engagement or action from students (e.g., Barlett et al., 2022; Henderson et al., 2022; Leichenko & O'Brien, 2020). To mitigate the general doom and gloom outlook, Trevion Henderson et al. argue that higher education should aim to convert awareness into behavioral change (Henderson et al., 2022). Educators need to rethink their assessment methods to encourage this shift (Barlett et al., 2022; Elder et al., 2023; Gal, 2023; Henderson et al., 2022; Jimenez & Kabachnik, 2023; Michel, 2022; Nalla et al., 2023; Reza, 2016; Talley & Hull, 2023). Committing to the goals of Education for Sustainable Development (ESD) should assist universities in integrating a solution-oriented approach in discussing sustainability and climate change with students (Alguren, 2021). In 2017, UNESCO defined the key eight ESD competencies as follows: (a) systems thinking, (b) anticipatory, (c) normative, (d) strategic, (e) collaboration, (f) critical thinking, (g) self-awareness, and (h) integrated problem solving (UNESCO, 2017). ESD 'empowers people with the knowledge, skills, values, attitudes, and behaviors to live in a way that is good for the environment, economy, and society' (UNESCO). Education for Sustainable Development puts students at the center of the problem by encouraging the use of self-reflection and self-awareness to connect personal experience with global challenges (Keys & Heck, 2023; Weder, 2022).

This chapter explores how we can integrate the ESD competencies of collaboration, self-awareness, and critical thinking into the undergraduate curriculum. The chapter uses the case study of the second-year undergraduate module 'Politics of Climate Change' delivered to BA (Hons) Politics and International Relations students at Manchester Metropolitan University. The chapter starts with an indicative literature review and proceeds to the analysis of the case study. The chapter concludes that the alignment of the competence-based approach to teaching and learning, together with action-oriented pedagogy, provides a better link between understanding climate change and climate action and encourages students to reflect on their own choices and roles in ensuring a more sustainable future. However, a single module is unlikely to be sufficient to turn climate action into an established practice for students. For this to happen, the ESD competencies should be systematically integrated into the curriculum with the support of university leadership.

2 Indicative Literature Review

There is a consensus in the literature that universities have the strategic potential to transform the way we approach global challenges (Filho et al., 2024; Henderson et al., 2022; Price et al., 2021; Vilmala et al., 2022). With the race against time to achieve the 1.5 °C target, education at all levels is the most strategic way to promote sustainable development values (Vilmala et al., 2022). Stephens and Graham (2009) argue that "universities can be slow to respond when societal needs emerge rapidly," and despite the declarative commitment to sustainability, it is still often integrated as an 'add-on' to a broader portfolio of the university's work rather than an integral part of teaching, learning, and operations (Latter & Capstick, 2021; Sengupta et al., 2020). There is agreement in the literature that the conventional way of teaching topics such as climate change through the prism of a singular discipline and focusing on understanding and raising awareness scares students into inaction (Leichenko & O'Brien, 2020). The problem is presented as too big and complex to process, and conventional pedagogies don't provide the tools students need to translate knowledge into action (Barlett et al., 2022).

To move forward, universities need to focus on a multidisciplinary approach based on competence-building and action-based pedagogies (Rieckmann, 2018). We need to provide the link between global and local and equip students with the skills needed to take action (Elder et al., 2023; Jimenez & Kabachnik, 2023; Keys & Heck, 2023; Lin et al., 2023; McKenzie & Wilson, 2023; Michel, 2022; Nadeem & Nawaz, 2023). Education for Sustainable Development (ESD) offers a mechanism that can bring together these crucial elements: 'subject-relevant learning objectives, key competencies for sustainability, and transformational pedagogies to support sustainable development' (Price et al., 2024). Linking the learning and teaching process with research on sustainability and climate change pedagogies will lead to more effective ways to develop key ESD competencies (Filho et al., 2023). Incorporating the local perspective into a more conventional big-picture approach would help students to not only better understand the environmental problems in their communities but also their role in addressing these problems (Elder et al., 2023; Lin et al., 2023; Nadeem & Nawaz, 2023; Oe et al., 2022; Rodríguez-Zurita et al., 2024), thus helping students to develop agency.

There is agreement in the literature that the successful integration of ESD competencies into the curriculum depends on the correct pedagogy (Cottafava et al., 2019; Martínez Casanovas et al., 2022). Active learning pedagogies are generally seen in the literature as a way to engage students in solving real-world problems (Rodríguez-Zurita et al., 2024). Practice-based work not only introduces students to academic content but also provides them with the opportunity to implement knowledge in practice (Cottafava et al., 2019; Gal, 2023; Jimenez & Kabachnik, 2023; Martínez Casanovas et al., 2022; Michel, 2022; Nalla et al., 2023; Reza, 2016; Talley & Hull, 2023).

This chapter will illustrate how these perspectives can be applied in practice by looking at the case study of the second-year undergraduate module 'Politics of

Climate Change'. The assessment structure for the module was designed in line with the objectives of ESD competence-building and active learning pedagogy.

3 Case Study: Politics of Climate Change Module Delivered to 2nd Year Students at Manchester Metropolitan University

The optional module is delivered to BA (Hons) International Relations and BA (Hons) Politics students in the second year of their course. The module provides an introduction to basic carbon science, the impact of climate change, and the role international organizations, governments, big businesses, not-for-profit sectors, and individuals play in addressing climate change. Initially, the assessment strategy followed a more conventional approach, and students were asked to produce a report on the climate policy of a country of their choice. However, after delivering the module for two years, it became evident that students felt detached from the climate emergency and saw it as something happening to them. They recognized the significance and urgency of the situation but didn't think they had the power to change anything. This observation demonstrated that a basic understanding of climate change and international efforts to combat it is insufficient to encourage climate action. As educators, we need to integrate climate action into the curriculum. The assessment strategy was reviewed to make it more solution-oriented and designed to help students develop key sustainability competencies (Brundiers et al., 2021; Wiek et al., 2011).

Consequently, I revised the assessment strategy from an academic report to a group project with a follow-up reflective report. Students need to work in a group to plan and deliver a project that has the potential to produce a positive impact. After the event, they need to submit a reflective report summarizing their research, reflecting on the planning and delivery of the event, and any lessons they learned from it. In the academic year 2023/2024, students decided to run a clothes swap to promote the importance of reuse and recycling and raise awareness of the damaging impact of fast fashion on the climate. Students selected a clothes swap because of the damaging impact of the apparel industry on the environment and the potential for positive change by promoting more responsible consumption. According to Oxfam, extending the garment life by 9 months of active use can reduce carbon footprint, water consumption, and waste by between 20–30% each (Oxfam, 2019). The clothes swap is also a low-cost event students could easily plan and promote without additional resources. While working on the project, students agreed on the following objectives: to promote the event as widely as possible, to promote the benefits of reusing and recycling clothes in the run-up to and during the event, and to document the clothes donated and swapped during the event in an attempt to calculate the carbon saved. The assessment format was new to students and required a different approach to planning, research, and time management. To help students adjust to the format, the preparation for the final assessment was integrated into the seminar plan. The teaching team provided the advice and guidance necessary for students to succeed, but the bulk of the work was student-led. The students identified the list of actions and assigned responsibility for the actions to different people in the group, including research, promotion, and liaison with local charities. Throughout the semester, the objectives of the assessment were framed and explained to students continuously within the overall academic content and the module learning outcomes to ensure that students understood the reason for the assessment and the intended outcomes.

For this chapter, information about the impact of the teaching and learning experience was collected from the analysis of the 22 reflective reports submitted by students at the end of the module. The reflective reports were analyzed using inductive content analysis (Elo & Kyngäs, 2008). The purpose of the analysis is to identify the key similarities in how students evaluated the project outcomes and analyze if any of these themes indicate that students re-evaluated their place in relation to global challenges.

4 Results and Discussion

At the beginning of the semester, students were asked to list the actors responsible for addressing the climate emergency. Students could list any number of actors. The majority of them named governments (93%), big business (40%), intergovernmental organizations (27%), and the general public (20%). Some responses that did mention the general public sometimes included clarifications, such as 'those with power have greater responsibility'. Others outright rejected the ability of the individual to have any impact: 'Stop blaming regular people!'. In line with the ESD priorities, the syllabus of the module and the assessment strategy are designed to encourage self-reflection and to provide students with the space to consider the impact of their personal choices on the environment (Rieckmann, 2018). The assessment strategy for the module was designed to challenge this view and shift the focus from the global level to the necessity of action on the local level. This chapter uses the analysis of the reflective reports to evaluate the effectiveness of this strategy. In the reflective reports, students were asked to consider the outcomes of the event and lessons learned from taking part in the project. Some students focused more specifically on the impact of fast fashion and ways to mitigate it, but others looked at the outcomes more broadly. Overall, the analysis identified four predominant themes that emerged in the reflective reports: (a) the importance of the circular economy and responsible consumption (Table 1), (b) the importance of community action (Table 2), (c) changed perceptions of individual action (Table 3), and (d) the limited impact of a singular event (Table 4). The tables below illustrate these themes with a selection of quotes from each category.

Table 2 demonstrates how students reflected on the importance of community-building and community action in their reports, starting from the community they created with each other while working on the project to the importance of community-based events in addressing the climate change emergency.

Table 1 The importance of circular economy and responsible consumption

"The project was a brilliant way to introduce people to the benefits of the circular economy"
"It was a good way to start the conversation and encourage action"
"Participation in the project prompted me to re-evaluate my own consumption habits"

Table 2 The importance of community action

"It is important to foster sustainability through community involvement"
"The sense of community emphasises the importance of collective active in addressing climate change"
"I truly feel that the results of the day prove that the individuals can do their part to make a difference"

Table 3 Changed the perception of individual action

"I was inspired to work further on climate issues. I want to create events that attract people's attention and raise awareness about climate"
"My involvement in this event has helped me to educate myself and demonstrate the action I can take on a personal level to reduce the impact of climate change… I realise that small change is important to push a big change"
"This initiative has provided me with a different perspective on how I can participate in a sustainable future… (showed me) the importance of shared responsibility"
"I take pride in contributing to an event that promotes sustainability and promotes change"

Table 4 Limited impact of a singular event

"We need to do more. Individual events can bring some results for the individual carbon footprint, but unless it becomes a habit it is not enough"
"The results of the event are positive, but on its own, it is not enough. We need to do more"

As seen in Table 3, students also reflected on how the project changed their personal position on climate action.

Finally, students also acknowledged that an individual event (see Table 4), however successful, is not enough to bring change unless it develops into systematic action.

The assessment strategy for the module is designed based on research on integrating key ESD competencies into the curriculum: the need to shift focus from the global level to the local level, the importance of project-based and action-based pedagogy, and community-based action. That is why students reflecting on the importance of collective action, community-based events, and the drive for further action is a positive outcome of the project. The themes identified through the analysis of the reflective reports illustrate that the project contributed to the development of such key competencies as self-awareness, collaboration, and critical thinking. The active learning approach encouraged greater engagement from students with the module content not only in the assessment preparation but also during in-class discussions. In their reflective reports, students also mentioned that working on the project helped them to build closer ties with their peers and improved their overall sense of belonging at the university.

If colleagues are planning on revising their assessment strategies and teaching content in line with project-based and active learning pedagogies, it is important to keep in mind the increased workload for students and academic staff alike. This observation is shared by academic staff working on ESD competence-building in their teaching (e.g., Howell, 2021). Student workload can be mitigated by integrating project work into timetable sessions. Academic staff require support from university leadership, including training and required workload allocations (Cottafava et al., 2019).

At the same time, it is important to recognize the limitations of these conclusions. Further study is necessary to evaluate the lasting impact of the project, or more specifically, whether it would lead to long-term behavior change for students who took part in the project. Furthermore, a single module is not sufficient, and a comprehensive review of the curriculum design is essential to ensure that students are equipped with the skills and knowledge required to ensure the transformation to a more sustainable society, together with a robust and comprehensive set of measures taken by universities (Filho, 2011).

5 Conclusions

Assertive action is urgently needed to mitigate the consequences of accelerated climate change on the planet: the loss of biodiversity, the increasing number of extreme weather events, and the economic and human costs of climate change. A 2023 analysis estimated that between 2000 and 2019, the loss and damage cost of climate change globally was approximately $2.8 trillion (UN). And the costs of the climate crisis will continue to grow. Thus, UNICEF predicts that the number of environmental refugees will vary between 25 million and 1 billion by 2050 (UNICEF, 2020). To mitigate the consequences of accelerated climate change, we need to change our socio-economic practices. Education can play a strategic role in the search for sustainable solutions (Alguren, 2021; UNESCO, 2017; Vilmala et al., 2022). This chapter argues that, at the moment, universities struggle to successfully achieve their full transformative potential. Despite a declarative commitment to sustainability, old educational practices and curriculum design provide structural obstacles to a more comprehensive integration of the sustainability agenda. We need to move towards a more multidisciplinary approach to teaching climate change and sustainability with a greater focus on practice-based and competence-building methodologies.

The chapter illustrates how the priorities of ESD competence-building can be integrated into the assessment structure by looking at the case study of the second-year undergraduate module 'Politics of Climate Change'. The module is delivered to undergraduate students enrolled in the BA (Hons) Politics and BA (Hons) International Relations programmes at Manchester Metropolitan University. The learning outcomes of the module are aimed at bridging the gap between understanding the climate emergency and taking action. In particular, the assessment strategy aims to develop the following ESD competencies: self-awareness, collaboration, and critical thinking. Students need to work as a group to deliver a project aimed at creating a positive impact on the environment. The project discussed in this chapter is the clothes swap students organized to promote reuse, recycling, and responsible consumption. This chapter uses the reflective reports students wrote to evaluate the impact of the assessment. Similarly to other publications reflecting on the integration of project-based pedagogy to promote sustainability (Keys & Heck, 2023; Weder, 2022), this project demonstrated positive results: students found the project to be rewarding, they were encouraged to reflect on their own consumption habits, identified the benefits of community engagement, and were inspired to take further action in the future. Despite the positive outcomes, it is important to acknowledge the potential limitations as well. The ESD competencies should be integrated throughout the curriculum structures. University leadership should ensure that academic staff are supported through relevant training and workload allocation.

References

Alguren, B. (2021). How to bring about change—A literature review about education and learning activities for sustainable development. *Discourse and Communication for Sustainable Education, 12*(1), 5–21. https://doi.org/10.2478/dcse-2021-0002

Bartlett, M., Larson, J., & Lee, S. (2022). Environmental justice pedagogies and self-efficacy for climate action. *Sustainability, 14*, 22. https://doi.org/10.3390/su142215086

Brundiers, K., Barth, M., Cebrián, G., et al. (2021). Key competencies in sustainability in higher education—Toward an agreed-upon reference framework. *Sustainability Science, 16*, 13–29. https://doi.org/10.1007/s11625-020-00838-2

Cottafava, D., Cavaglià, G., & Corazza, L. (2019). Education of sustainable development goals through students' active engagement: A transformative learning experience. *Sustainability Accounting, Management and Policy Journal, 10*(3), 521–544. https://doi.org/10.1108/SAMPJ-05-2018-0152

Elder, S., Wittman, H., & Giang, A. (2023). Building sustainability research competencies through scaffolded pathways for undergraduate research experience. *Elementa-Science of the Anthropocene, 1*, 1–12. https://doi.org/10.1525/elementa.2022.00091

Elo, S., & Kyngäs, H. (2008). The qualitative content analysis process. *JAN: Leading Global Nursing Research, 62*(1), 107–115. https://onlinelibrary.wiley.com/doi/full/10.1111/j.1365-2648.2007.04569.x

Filho, W. (2011). About the role of universities and their contribution to sustainable development. *Higher Education Policy, 24*, 427–438. https://doi.org/10.1057/hep.2011.16

Filho, W. L., Abubakar, I. R., Mifsud, M. C., et al. (2023). Governance in the implementation of the UN sustainable development goals in higher education: Global trends. *Environment, Development and Sustainability*. https://doi.org/10.1007/s10668-023-03278-x

Filho, W. L., Sierra, J., Price, E., et al. (2024). The role of universities in accelerating the sustainable development goals in Europe. *Science and Reports, 14*, 15464. https://doi.org/10.1038/s41598-024-65820-9

Gal, A. (2023). From recycling to sustainability principles: The perceptions of undergraduate students studying early childhood education of an education for sustainability course. *International Journal of Sustainability in Higher Education*, 1082–1104.https://doi.org/10.1108/IJSHE-05-2022-0165

Henderson, T. S., Michel, J. O., Bryan, A., Canosa, E., Gamalski, C., Jones, K., & Moghtader, J. (2022). An exploration of the relationship between sustainability-related involvement and learning in higher education. *Sustainability, 14*(9), 5506. https://doi.org/10.3390/su14095506

Howell, R. (2021). Engaging students in education for sustainable development: The benefits of active learning, reflective practices and flipped classroom pedagogies. *Journal of Cleaner Production, 325.*

Jimenez, J., & Kabachnik, P. (2023). Indigenizing environmental sustainability curriculum and pedagogy: Confronting our global ecological crisis via Indigenous sustainabilities. *Teaching in Higher Education, 28*(5), 1095–1107. https://doi.org/10.1080/13562517.2023.2193666

Keys, N., & Heck, D. (2023). Positioning and repositioning in higher education: First-year students engaging with the world. *Studies in Higher Education*, 1014–1027. https://doi.org/10.1080/03075079.2023.2257733

Latter B., & Capstick S. (2021). Climate emergency: UK universities' declarations and their role in responding to climate change. *Frontiers in Sustainability, 2.*

Leichenko, R., O'Brien, K. (2020). Teaching climate change in the Anthropocene: An integrative approach. *Anthropocene, 30.* https://doi.org/10.1016/j.ancene.2020.100241

Lin, J., Fiore, A., Sorensen, E., Gomes, V., Haavik, J., Malik, M., Mok, S. K. J., Scanlon, J., Wanjala, E., & Grigoryeva, A. (2023). Contemplative, holistic eco-justice pedagogies in higher education: From anthropocentrism to fostering deep love and respect for nature. *Teaching in Higher Education, 28*(5), 953–968. https://doi.org/10.1080/13562517.2023.2197110

Martínez Casanovas, M., Ruíz-Munzón, N., & Buil-Fabregá, M. (2022). Higher education: The best practices for fostering competences for sustainable development through the use of active learning methodologies. *International Journal of Sustainability in Higher Education, 23*(3), 703–727. https://doi.org/10.1108/IJSHE-03-2021-0082

McKenzie, M., & Wilson, A. (2023). Sustainability as wild policy: Mobile SDG interventions and land-informed policy in education. *Discourse-Studies in the Cultural Politics of Education, 44*(5), 679–694.

Michel, J. O. (2022). Charting students' exposure to promising practices of teaching about sustainability across the higher education curriculum. *Teaching in Higher Education, 27*(6), 787–813. https://doi.org/10.1080/13562517.2020.1747422

Nadeem, O., & Nawaz, M. (2023). Climate change and sustainable development perceptions of university students in Lahore, Pakistan. *International Research in Geographical and Environmental Education, 32*(3), 181–196. https://doi.org/10.1080/10382046.2022.2154973

Nalla, V., Johnson, C., Ranjit, N., Sen, G., Peddibhotla, A., Anand, M. C., Bhatia, N., & Bazaz, A. B. (2023). Considering curriculum, content, and delivery for adaptive pathways: Higher education and disaster resilient infrastructure in the Indian urban context. *Sustainable and Resilient Infrastructure, 8*(sup1), 143–156. https://doi.org/10.1080/23789689.2022.2134645

Oe, H., Yamaoka, Y., & Ochiai, H. (2022). A qualitative assessment of community learning initiatives for environmental awareness and behaviour change: Applying UNESCO Education for Sustainable Development (ESD) framework. *International Journal of Environmental Research and Public Health, 19*, 3528. https://doi.org/10.3390/ijerph19063528

Oxfam. (2019). *New shocking facts about the impact of fast fashion on our climate.* https://www.oxfam.org.uk/oxfam-in-action/oxfam-blog/new-shocking-facts-about-the-impact-of-fast-fashion-on-our-climate/

Price, E. A. C., White, R. M., Mori, K., Longhurst, J., Baughan, P., Hayles, C. S., Gough, G., & Preist, C. (2021). Supporting the role of universities in leading individual and societal transformation through education for sustainable development. *Discover Sustainability, 2*(1), 49. https://doi.org/10.1007/s43621-021-00058-3

Price, E., Nicholson, D. T., Dunk, T., Lawler, C., Carney, M., Vargas, V. R., Veitch, S., Leigh, S., Singleton, M., & Mottram, S. (2024). Enabling change agents for sustainable development: A whole-institution approach to embedding the UN Sustainable Development Goals in higher education. *International Journal of Sustainability in Higher Education, 25*(7), 1333–1350. https://doi.org/10.1108/IJSHE-02-2024-0112

Reza, M. I. H. (2016). Sustainability in higher education: perspectives of Malaysian higher education system. *Sage Open, 6*(3). https://doi.org/10.1177/2158244016665890

Rieckmann, M. (2018). Learning to transform the world: Key competencies in education for sustainable development. In A. Leicht, J. Heiss, & W. J. Byun (Eds.), *Education on the move* (pp. 39–61). UNESCO Publishing.

Rodríguez-Zurita, D., Jaya-Montalvo, M., Moreira-Arboleda, J., Raya-Diez, E., & Carrión-Mero, P. (2024). Sustainable development through service learning and community engagement in higher education: A systematic literature review. *International Journal of Sustainability in Higher Education*, Vol. ahead-of-print, No. ahead-of-print. https://doi.org/10.1108/IJSHE-10-2023-0461

Sengupta, E., Blessinger, P., & Yamin, T. S. (Eds.). (2020). Integrating sustainable development into the curriculum (1st ed.). In *Innovations in higher education teaching and learning* (Vol. 18, p. 5). Emerald Publishing Limited.

Stephens, J., & Graham, A. (2009). Towards an empirical research agenda for sustainability in higher education: Exploring the transition management framework. *Journal of Cleaner Production, 18*(7), 611–618. https://doi.org/10.1016/j.jclepro.2009.07.009

Talley, E. K., & Hull, R. B. (2023). Systems thinking for systems leadership: Promoting competency development for graduate students in sustainability studies. *International Journal of Sustainability in Higher Education* 1039–1057. https://doi.org/10.1108/IJSHE-11-2021-0489

The IPCC. (2022). *Global warming of 1.5°C: IPCC special report on impacts of global warming of 1.5°C above pre-industrial levels in context of strengthening response to climate change, sustainable development, and efforts to eradicate poverty.* https://dlib.hust.edu.vn/ViewOnline?bitstid=256ba678-fda5-4bb3-ad5e-d2fbe1a7ef08&type=1

UN. *Climate action: Finance and justice.* https://www.un.org/en/climatechange/raising-ambition/climate-finance

UNESCO. *Education for sustainable development.* https://www.unesco.org/en/sustainable-development/education

UNESCO(a). (2017). *Education for sustainable development goals: Learning objectives.* https://www.unesco.org/en/articles/education-sustainable-development-goals-learning-objectives

UNICEF. (2020). *Climate migration and education: are we making our education systems future proof?* https://www.unicef.org.uk/policy/climate-migration-and-education/

Vilmala, B. K., Karniawati, I., Suhandi, A., Permanasari, A., & Khumalo, M. (2022). A literature review of education for sustainable development (ESD) in Science learning: What, why, and how. *Journal of Natural Science and Integration, 5*(1), 35. https://doi.org/10.24014/jnsi.v5i1.15342

Weder, F. (2022). #finaltrashtination. An art-based intervention to collaboratively generate conversations about climate change. *Journal of Science Communication, 21z*, 2. https://doi.org/10.22323/2.21020805

Wiek, A., Withycombe, L., & Redman, C. L. (2011). Key competencies in sustainability: A reference framework for academic program development. *Sustainability Science, 6*, 203–218. https://doi.org/10.1007/s11625-011-0132-6

World Meteorological Organisation. (2024a). *State of the global climate 2023.* https://wmo.int/publication-series/state-of-global-climate-2023

World Meteorological Organisation. (2024b). *Global temperature is likely to exceed 1.5C above pre-industrial level temporarily in next 5 years.* https://wmo.int/news/media-centre/global-temperature-likely-exceed-15degc-above-pre-industrial-level-temporarily-next-5-years

Cultivating Sustainable Futures: Enhancing Study Programs in Africa for Environmental Stewardship

Charlotte Newiadomsky and Lisa Schmitt

Abstract This paper provides insights into the distinctive study programs of two influential projects: GREATER ("Growing Rwanda Energy Awareness Through highER Education") in Rwanda and IIDES-NSC ("Industry-Integrated Dual Engineering Studies in a North South Collaboration") in Côte d'Ivoire. GREATER, a pioneering initiative in Rwanda's energy sector, focuses on transforming sustainability education through innovative curricula and practical experiences. It aims to cultivate a new generation of professionals committed to driving a sustainable energy transition with equal access to electricity for the population in Rwanda. By collaborating closely with local universities and industry partners, GREATER offers a comprehensive range of courses designed to equip students with the knowledge and skills needed to address Rwanda's energy challenges sustainably. Emphasizing renewable energy technologies, energy efficiency, and environmental stewardship, GREATER aims to cultivate a new generation of energy professionals committed to driving sustainable development in Rwanda. In parallel, IIDES-NSC is making significant strides in engineering education in Côte d'Ivoire through its unique industry-integrated dual engineering studies program. By forging partnerships between academic institutions in Europe and Africa, IIDES-NSC offers students in Côte d'Ivoire the opportunity to gain practical experience through internships with leading engineering local companies. This innovative approach ensures that students receive hands-on training and exposure to real-world engineering challenges, preparing them for successful careers in the industry upon graduation. Additionally, IIDES-NSC places a strong emphasis on sustainable engineering practices, instilling principles of environmental responsibility and ethical leadership in future engineers. Both GREATER and IIDES-NSC exemplify pioneering efforts to enhance sustainability education and workforce development in their respective regions. Through their tailored study programs and strategic partnerships, these projects are empowering students in Rwanda and Côte d'Ivoire to become agents of change in

C. Newiadomsky (✉) · L. Schmitt
SWK E^2 - Institute of Energy Technology and Energy Management, Hochschule Niederrhein University of Applied Sciences, Krefeld, Germany
e-mail: charlotte.newiadomsky@hs-niederrhein.de

L. Schmitt
e-mail: lisa.schmitt@hs-niederrhein.de

W. Leal Filho et al. (eds.), *Introducing the UN SDGs in Higher Education Institutions*, World Sustainability Series, https://doi.org/10.1007/978-3-032-23017-1_2

the fields of energy and engineering, driving progress towards a more sustainable future.

Keywords Sustainability education · Energy transition · University-business partnerships · Renewable energies · Practice-oriented learning

1 Introduction

Africa, a continent of diverse geography, cultures, and economies, continues to face significant challenges in the areas of energy and water supply. These two sectors are essential for economic development, public health, and social well-being, yet many African nations struggle to provide reliable access to electricity and clean water for their populations (The World Bank, 2023). Addressing these challenges requires innovative solutions that align with global development priorities, particularly the Sustainable Development Goals (SDGs) of the United Nations. Energy and water supply are primarily addressed through SDG 7 (Affordable and Clean Energy), but the projects also contribute to SDG 4 (Quality Education) and SDG 9 (Industry, Innovation, and Infrastructure), as well as indirectly to SDG 17 (Partnerships for the Goals). Thus, this paper gives answers to the following central research question: How can educational interventions and renewable energy projects address energy poverty and sustainable development? The dual crises of inadequate energy and water infrastructure have far-reaching consequences, impacting everything from education and healthcare to industrial productivity and agricultural output (United Nations, 2024). This introduction will explore the nature of these challenges in the broader African context before narrowing the focus to two countries: Côte d'Ivoire and Rwanda. Both nations have made commendable strides in addressing energy and water issues, yet they still face distinct obstacles that reflect the broader struggles faced by the continent.

Africa's energy landscape is characterized by stark contrasts. On the one hand, the continent possesses abundant natural resources, including vast reserves of oil, gas, and renewable energy sources such as hydro, solar, and wind power (African Energy Chamber, 2022). On the other hand, Africa has some of the lowest rates of access to electricity in the world (UNDP, 2024). According to the International Energy Agency (IEA, 2022), as of 2020, about 600 million people across sub-Saharan Africa lacked access to electricity, with rural areas being particularly underserved. This energy deficit is compounded by unreliable supply systems, with frequent blackouts and voltage instability plaguing both urban and rural regions (IEA, 2022). The lack of electricity is not just a barrier to economic growth but also hinders progress in critical areas like education, healthcare, and gender equality, as women and children are disproportionately affected by energy poverty (IRENA, 2024).

Similarly, the water supply situation is dire in many parts of Africa (UNICEF, WHO, 2022). The United Nations estimates that over 300 million people in sub-Saharan Africa lack access to clean drinking water, and more than 700 million live without basic sanitation facilities (UNICEF, WHO, 2022). This scarcity is exacerbated by poor management of water resources, rapid population growth, and climate change, which is making water availability increasingly unpredictable (United Nations, 2024). The challenges related to water supply are multifaceted, involving not only the physical scarcity of water in some regions but also the lack of infrastructure to store, treat, and distribute it. Moreover, in areas where water is available, it is often contaminated due to pollution and inadequate waste management systems, leading to public health crises such as cholera and other waterborne diseases (UNESCO, 2014).

Against this backdrop, Rwanda and Côte d'Ivoire serve as valuable case studies. Both countries are attempting to reform their energy and water sectors, but the specific challenges they face, differ based on their historical, geographical, and socio-economic contexts.

Rwanda, a small landlocked country in East Africa, has garnered international attention for its impressive development progress in recent years. Following the devastating genocide in 1994, Rwanda has rebuilt itself into one of Africa's fastest-growing economies. The government has prioritized infrastructure development, including energy and water supply, as key components of its Vision 2020 and Vision 2050 strategies. However, despite these efforts, Rwanda still faces significant challenges in achieving universal access to electricity and clean water.

Rwanda's energy sector is smaller and less developed than that of Côte d'Ivoire, but the country has made impressive strides in recent years. In 2000, only 2% of Rwanda's population had access to electricity. By 2022, this figure had increased to over 50% (The World Bank, 2024). Much of this progress is due to the government's aggressive electrification programs, which have focused on expanding the national grid and promoting off-grid solutions like solar home systems. Despite these gains, Rwanda's energy sector faces several challenges. Like Côte d'Ivoire, Rwanda relies heavily on hydropower, which makes it vulnerable to droughts and climate change. The country is also highly dependent on imported fuel for thermal power plants, which raises energy costs and exposes the economy to price fluctuations in the global energy market. Additionally, while urban areas have seen significant improvements in electricity access, rural electrification remains a challenge due to the high costs of expanding the grid to remote areas.

Rwanda has made remarkable progress in improving access to clean water, with around 86% of the population having access to improved water sources as of 2020 (World Bank, 2024). However, challenges remain, particularly in rural areas where water infrastructure is still underdeveloped. Many rural communities rely on natural water sources like rivers and lakes, which are often contaminated, leading to waterborne diseases. The government has launched several initiatives to improve water supply, including the construction of new water treatment plants and the rehabilitation of existing infrastructure. However, like in Côte d'Ivoire, rapid population growth and urbanization are putting pressure on existing water systems. Climate change is also a growing concern, as changes in rainfall patterns are expected to affect water availability in the coming decades.

Côte d'Ivoire, located in West Africa, is one of the region's economic powerhouses. The country has a growing industrial sector and is a major producer of agricultural commodities like cocoa and coffee. Despite this economic growth, Côte d'Ivoire faces significant challenges in providing reliable energy and water services to its population. The country is relatively well-positioned compared to many of its regional neighbors in terms of energy production (World Bank, 2018). The country has invested heavily in its energy sector over the past two decades, particularly in natural gas and hydropower (MPE, 2016). It is one of the few countries in West Africa that generates more electricity than it consumes, enabling it to export energy to neighboring countries like Ghana, Burkina Faso, and Mali. However, despite these achievements, large portions of the Ivorian population still lack reliable access to electricity, particularly in rural areas. As an example, in 2020 around 70% of the total population in Côte d'Ivoire had access to electricity, whereas it was nearly 95% of the urban and only 43% of the rural population (The World Bank, 2024).

The energy sector in Côte d'Ivoire faces several key challenges. First, while the country has increased its energy generation capacity, its transmission and distribution networks are outdated and inefficient. This results in significant energy losses during transmission and frequent power outages, even in major urban centers. Second, the country's heavy reliance on hydropower makes it vulnerable to climate variability (Obahoundje et al., 2024). In years of drought, the reduced water flow can severely limit energy production, leading to blackouts and a reliance on costly fossil fuel imports to make up for the shortfall. Finally, while the government has ambitious plans to increase renewable energy generation, especially solar power, the implementation of these projects has been slow due to financial, technical, and regulatory barriers.

Access to clean water is another major challenge in Côte d'Ivoire. The country's rapid urbanization has outpaced the development of water infrastructure, particularly in cities like Abidjan, where informal settlements have grown rapidly. In these areas, water scarcity is common, and many residents rely on unsafe water sources. In rural areas, the situation is even worse. The lack of reliable water infrastructure means that people often have to travel long distances to fetch water, a task that disproportionately falls on women and girls (Noubactep, 2016). Water management in Côte d'Ivoire is further complicated by pollution and environmental degradation. Industrial waste, agricultural runoff, and untreated sewage are major sources of water contamination, particularly in urban areas. The government has made efforts to improve water access, particularly through partnerships with international organizations and private companies, but progress has been slow. Climate change also poses a growing threat to Côte d'Ivoire's water resources, with increasing variability in rainfall patterns leading to both droughts and floods, which can further strain the country's already fragile water systems (Obahoundje et al., 2024).

In addition to the challenges in energy and water supply, education in many African countries, including Côte d'Ivoire and Rwanda, faces its own set of issues, particularly when it comes to aligning academic programs with the demands of the labor market. One of the key challenges is the lack of practical, hands-on training in many university and technical programs. While both countries have made significant investments in expanding access to education, particularly at the tertiary level, the curricula in many fields remain heavily theoretical, with insufficient emphasis on the practical skills needed for the workforce. As a result, graduates often find themselves ill-prepared to meet the demands of employers, especially in technical sectors such as engineering, energy, and water management.

Moreover, there is a notable lack of collaboration between educational institutions and industries in both Côte d'Ivoire and Rwanda. Universities and vocational training centers are not closely aligned with the needs of local industries, leading to a mismatch between the skills taught in schools and the competencies required by employers. This disconnection not only hampers the ability of graduates to secure jobs but also limits the growth of key sectors that are critical for economic development, such as energy and water infrastructure. The lack of practical training and weak industry linkages exacerbate the issue of high unemployment or underemployment among university graduates. Many young people complete their studies only to face limited job prospects, particularly in fields related to science, technology, engineering, and mathematics (STEM). Without the necessary skills or connections to enter the workforce, graduates often find themselves in informal or low-wage jobs, far removed from their areas of study (Yapo, 2018). This situation creates a vicious cycle, where graduates continue to struggle to find meaningful employment, further straining already fragile economies.

The challenges of energy and water supply in Africa are immense, but so too are the opportunities for progress. Rwanda and Côte d'Ivoire, though different in their historical and economic contexts, both illustrate the complex interplay between infrastructure development, governance, and external factors such as climate change (Newiadomsky & Tietze, 2017). As these countries continue to develop, addressing these challenges will be critical to their long-term economic stability and social well-being. Addressing the educational challenges requires a multi-faceted approach. Educational institutions must redesign curricula to include more hands-on, practical training and internships that provide students with real-world experience. Furthermore, governments and the private sector need to work together to strengthen partnerships between universities and industries, ensuring that academic programs are tailored to meet the evolving needs of the labor market. By doing so, countries like Rwanda and Côte d'Ivoire can not only improve job prospects for graduates but also help drive innovation and growth in key sectors of the economy.

Those were amongst others the reasons for choosing Rwanda and Côte d'Ivoire for the projects that will be presented and discussed in this paper. Their main objectives are to develop sustainable energy solutions in rural regions in Africa and raise awareness on the topics of climate change and renewable energy. To achieve those goals, practice-oriented trainings and study programs are established at the local partner university in Côte d'Ivoire, and energy-related study programs at the partner universities in Rwanda are developed or renewed. The link between universities and the industry is strengthened through the joint development of study programs and collaborations in terms of internships for students and future job perspectives for graduates. By involving other relevant stakeholders like schools or local communities the scope and impact of the project results is increased. In addition, both projects promote staff and student exchange through several mobilities between the participating countries. Both projects GREATER and IIDES-NSC, that are discussed and analyzed in this paper, address the above-mentioned issues through innovative education, technological interventions, and stakeholder collaboration. The projects incorporate collaborative governance processes and employ the quadrupole helix model to integrate academia, industry, government, and civil society in sustainable development efforts.

2 Project GREATER

The project GREATER (Growing Rwanda Energy Awareness Through highER Education) aims to support the Rwandan country in the energy transition needed to sustain its social and economic growth. To fill the energy gap and to enable the Rwandan population to have equal access to electricity, widespread exploitation of renewable energy sources is planned. According to governmental strategies for energy development, low-power, off-grid, distributed power generation will play a key role, calling for both the training of a new generation of skilled professionals and for the widest diffusion of green awareness in the national community. The project aims at tackling both aspects by implementing an inclusive, user-participated design flow. The project has been funded in the ERASMUS+ European framework, under the 2022 CBHE (Capacity Building for Higher Education) call.

The project is participated by four Rwandan Institutions (INES-Ruhengeri, Rwanda Polytechnic, University of Technology and Arts of Byumba and University of Rwanda) and by four European ones (University of Parma, Italy, National Research Council, Italy, Hochschule Niederrhein, Germany and Hochschule Bonn-Rhein-Sieg, Germany). The project started on January 1st, 2023 and has a duration of 3 years, with an expected contribution from the European Union of up to 800,000€. Nine associated partners will support the project, including the Rwandan Ministry of Education and the main national companies active in the energy development market.

The core activity will result in the design of eight new or renewed curricula in the field of smart and green energy. Innovation will come from the introduction of strong digital and Internet-of-things foundations, and a markedly practice-oriented training. Besides new lab facilities at Rwandan Partners Institutions, training will be based also on the innovative introduction of a network of four so-called "Living Labs". In this project, Living Labs are real-life environments equipped with renewable energy systems that provide hands-on training for students and serve as educational resources for local communities. These facilities enable students to gain invaluable practical experience, while communities benefit from increased access to green energy solutions.

The project involves 100 students and 20 university staff members annually, with the specific goal of establishing four fully operational Living Labs by 2026, each powered entirely by renewable energy. These labs are designed to integrate sustainability practices and foster community engagement, ensuring long-term impact.

Through these efforts, the project GREATER directly contributes to SDG 4 by enhancing education quality through curricula revision and design as well as the implementation and integration of Living Labs, SDG 7 by increasing access to clean energy and raising awareness on the topic, and SDG 9 by fostering innovation and partnerships with external stakeholders like industry, local communities and the government. By raising awareness on topics like climate change and renewable energies within the universities and communities, the project promotes lifelong learning through local trainers and reaching out to people already at a young age.

The project adopts different approaches for students and staff. For students, it emphasizes hands-on training and practical internships with industry partners, ensuring they are prepared for real-world challenges. For staff, specialized workshops focus on virtual labs, innovative teaching methods, and strategies for effective collaboration with industry.

A full range of representative stakeholders will be enclosed in all design activities since earlier phases, through a collaborative design thinking procedure. This will ensure project outcomes are consistent with community needs and expectancies, in terms of educational goals, better employability, and socioeconomic benefits coming from the project. Community engagement will therefore be twofold: at the project "input" side, insights and addresses from the community will drive educational design activities, at the "output" side, beyond academic outcomes, transfer activities are planned to tailor specific outcomes targeting, besides industries, different community segments, caring mostly at youth and people with fewer opportunities.

The Rwandan Ministry of Education is involved throughout the whole project to align its objectives with national policies, fostering a supportive environment for implementation. These partnerships further align with SDG 17 by promoting collaboration among academia, industry, and civil society.

The final impact of the project GREATER is expected to enhance educational outcomes by integrating experiential learning and reduce energy consumption through the promotion of renewable solutions. It also seeks to build long-term capacity in the energy sector, aligning with national goals for sustainable development.

3 Project IIDES-NSC

The IIDES-NSC project (Industry-Integrated Dual Engineering Studies in a North-South Collaboration) aims to empower local communities in Côte d'Ivoire to address electricity and water supply challenges independently and sustainably. The project focuses on designing, developing, manufacturing, and operating technologies within the country, alongside creating self-sustaining business models for production and operation tailored to African contexts. Funded by the German Federal Ministry for Economic Cooperation and Development (BMZ) through the DAAD's "Practical Partnerships between Universities and Companies in Germany and Developing

Countries from 2021" program, the project received approximately €720,000 for a three-year duration starting in May 2021.

The core consortium includes Hochschule Niederrhein in Krefeld, Germany, and Université NANGUI ABROGOUA (UNA) in Abidjan, Côte d'Ivoire, with support from the Association of Cameroonian Engineers and Computer Scientists (VKII e.V.) and Toukam Consulting. VKII e.V. and Toukam Consulting provide valuable intercultural and technical expertise. Industry partners include "euregio-solarzentrum GmbH," a German SME specializing in photovoltaic systems, and "Wilo SE," a global leader in pumping systems with branches in Germany and Côte d'Ivoire.

The project strengthens educational institutions by enhancing their capacity for application-oriented research and development. It integrates practical modules into curricula, modeled after Germany's dual study system, which has been successful for decades. Adapted to local conditions, this model is implemented with collaborating companies and students. It aims to equip graduates with the knowledge and confidence to become entrepreneurs, create high-quality jobs, and contribute to their country's development. This initiative also seeks to reduce brain drain by fostering opportunities within the country.

The new master study program developed in the framework of the project IIDES-NSC focuses on courses in energy efficiency, water management, and entrepreneurship, equipping students with the skills needed to develop innovative solutions. With 20 students per cohort, the program provides personalized education that fosters both technical and entrepreneurial capabilities. The project has already seen the graduation of 10 students, 5 female and 5 male, who have gone on to create their own renewable energy businesses, contributing to local innovation and reducing brain drain. A SMART goal of the project is to enable 30% of graduates to establish businesses in the renewable energy sector within three years of completing the program. The final impact is to foster local innovation and create job opportunities in the energy and water sectors, ensuring that local communities can address their challenges sustainably and independently. By addressing water management challenges, the project also supports SDG 6, while its strong focus on collaboration and education reinforces SDG 17.

This project also integrates interdisciplinarity by combining technical education with business and social sciences to address multifaceted challenges. Its alignment with SDG 4 is evident through curriculum enhancements, while contributions to SDG 7 and SDG 9 are demonstrated by renewable energy initiatives and fostering industry-academia partnerships.

4 Governance and Integration

The governance of course selection and participant eligibility is a collaborative process involving university committees, industry advisors, and funding agencies. This ensures that programs remain relevant and aligned with both academic and industry needs. Both projects adopt the quadrupole helix model, which integrates academia, industry, government, and civil society. Academia develops curricula and trains students, while industry provides practical training opportunities and insights into market needs. Government entities offer regulatory support and funding, and civil society engages through Living Labs and community-focused initiatives. This model ensures that solutions are well-supported, sustainable, and inclusive.

5 Results and Future Trends in Sustainability

The GREATER project has made significant progress in achieving its goals of promoting green energy awareness, enhancing higher education study programs, and fostering collaboration between academia and society. A key accomplishment is the development of Living Labs, which serve as hubs for innovation, education, and community engagement.

The first operational Living Lab has been established at INES Ruhengeri in Busengo, Rwanda. This lab includes a photovoltaic system providing off-grid electricity and facilities such as a Community Training Centre. The centre supports activities like a tailoring workshop where women receive training to generate income, alongside spaces for community gatherings and educational workshops. Future Living Labs are being designed to address local needs. At the University of Rwanda in Bugesera, a lab will feature a PV-powered water pump and smart irrigation system. Another lab at the University of Technology and Arts of Byumba will supply electricity to a primary and secondary school, while the lab at Rwanda Polytechnic in Muyanza will provide residential electricity. All Living Labs are managed by partner universities with sustainability ensured through community contracts and ongoing workshops.

Training initiatives have also been a major focus. University staff in Rwanda and Germany participated in workshops on topics like virtual labs, practical teaching methods, and industry collaboration, fostering sustainable education practices. Additionally, two Summer Schools were held in Rwanda, teaching students about renewable energy, climate change, and sustainability, combined with hands-on experience building solar chargers.

Another significant achievement is the development and modernization of curricula. Bachelor's, Master's, and Ph.D. programs in renewable energy are being revised or created to incorporate the latest advancements and practical components like Living Labs. This integration ensures students gain real-world experience and are prepared to address environmental and societal challenges.

Outreach and transfer activities have further extended the project's impact. Surveys, interviews, and events have engaged diverse stakeholders, including students, companies, and community members. Project results were shared during the Global Goal Weeks 2024, and an art contest connected to the Living Labs engaged the local community.

By integrating renewable energy innovation with education and community development, the GREATER project stands out as a model initiative with the potential to inspire similar efforts globally.

The project IIDES-NSC achieved within the project duration the successful development, introduction, and implementation of the new Master's study program, including the set-up and launch of both prototypes, at UNA and the model village Legroskro, respectively.

The planned exchange semester at the Hochschule Niederrhein from the participating students has been successfully completed and all 10 students eventually graduated the Master's study program. The Master's program enabled and encouraged the graduates to develop their own start-up company named "Schlüssel" (German for "key"), which will consult privates and companies in topics of energy efficiency, energy management, renewable energies, water management as well as circular economy and sustainability.

As planned from the project's outset, UNA has continued the Master's program independently, and a new cohort of 20 students began their studies in autumn 2024.

Future initiatives for both the GREATER and IIDES-NSC projects may focus on expanding their concepts to other regions or countries. Additionally, extending the current Living Labs or prototypes could serve as a foundation for follow-up projects.

6 Conclusions

The GREATER and IIDES-NSC projects have addressed critical global challenges, with a focus on climate change, sustainability, and community resilience. By fostering practical, hands-on learning, these projects have empowered students to tackle local challenges and develop sustainable solutions, particularly in rural areas across Africa. Both initiatives have strengthened partnerships between academia and industry, bridging gaps that enhance the employability of graduates and ensure that academic efforts align closely with real-world needs. Furthermore, through outreach and engagement, the projects have raised awareness within communities, encouraging collective action toward sustainability goals.

The GREATER and IIDES-NSC projects have significantly advanced the United Nations' Sustainable Development Goals (SDGs), creating a collaborative model that fosters sustainable development and prepares future leaders to drive meaningful change. Both projects primarily address SDG 4 ("Quality Education"), SDG 7 ("Affordable and Clean Energy"), and SDG 9 ("Industry, Innovation, and Infrastructure"), while also contributing indirectly to SDG 6 ("Clean Water and Sanitation"), SDG 10 ("Reduced Inequalities"), and SDG 17 ("Partnerships for the Goals").

GREATER has made notable contributions to SDG 4 by revising existing curricula and developing new study programs focused on renewable energy. The project enhanced quality education by integrating sustainable practices and hands-on learning through innovative Living Labs, modern classroom laboratories, and targeted staff training. Additionally, the Living Labs, equipped with photovoltaic systems, have supported clean energy access in rural Rwandan communities, contributing to both innovation and infrastructure development.

The project also fostered partnerships between higher education institutions and industry by promoting collaboration and a user-centered design process that engaged stakeholders, including industry and local communities. Through these efforts, GREATER has demonstrated how education and innovation can power sustainable progress and strengthen global partnerships.

The IIDES-NSC project made significant contributions to SDG 4 by developing a practice-oriented Master's program. This program incorporates hands-on projects for planning, developing, and operating prototypes, alongside internships with companies, an exchange semester, and interdisciplinary teaching modules. These modules include English language courses, project management, intercultural seminars, and training on managing international projects across borders.

The prototypes for electricity and water supply at UNA and the model village directly address SDGs 6 and 7 by promoting sustainable access to clean water and energy. The program also advances SDG 9 by equipping students with specialized skills through internships and by introducing an innovative dual Master's program focused on prototype development.

By ensuring gender balance in student selection during the assessment process, the project actively supports SDG 10. Furthermore, SDG 17 is promoted through partnerships between Hochschule Niederrhein and UNA, as well as collaborations between UNA, Ivorian companies, and industrial partners in the project consortium.

References

African Energy Chamber. (2022). *The State of African Energy 2023 Outlook*. Johannesburg, South Africa. Retrieved November 28, 2024, from https://energychamber.org/wp-content/uploads/The-State-of-African-Energy-2023-Report_dg_sp_oct-25.pdf

International Energy Agency. (2022). *Africa Energy Outlook 2022—English Version. World Energy Outlook Special Report*. Retrieved November 29, 2024, from https://iea.blob.core.windows.net/assets/6fa5a6c0-ca73-4a7f-a243-fb5e83ecfb94/AfricaEnergyOutlook2022.pdf

IRENA. (2024). *World Energy Transitions Outlook 2024: 1.5°C pathway*. International Renewable Energy Agency, Abu Dhabi. Retrieved November 28, 2024, from https://www.irena.org/-/media/Files/IRENA/Agency/Publication/2024/Nov/IRENA_World_energy_transitions_outlook_2024.pdf

MPE. (2016). Plan d'Actions National des Energies Renouvelables (PANER) CÔTE D'IVOIRE. Ministère du Pétrole et de l'Energie, Abidjan. Retrieved November 30, 2024, from https://www.se4all-africa.org/fileadmin/uploads/se4all/Documents/Country_PANER/CO%CC%82TE_D%E2%80%99IVOIRE_Plan_d_Actions_National_pour_les_Energies_Renouvelables.pdf

Newiadomsky, C., & Tietze, I. (2017). Beitrag Erneuerbarer Energien zur Verfügbarkeit von Elektrizität und Wasser in Afrika: Ansätze für eine nachhaltige Entwicklung? In W. L. Filho (Hg.), Innovation in der Nachhaltigkeitsforschung – Ein Beitrag zur Umsetzung der UNO Nachhaltigkeitsziele, Springer Spektrum, Wiesbaden, S. 39–59, Print ISBN: 978-3-662-54358-0, Electronic ISBN: 978-3-662-54359-7. https://doi.org/10.1007/978-3-662-54359-7_3

Noubactep, C. (Ed.) (2016). Approvisionnement En Eau Potable En Afrique: Une technologie Appropriée Pour Relever Le Défi. Afrika & Wissenschaft/Africa & Science/Afrique & Science 2. Frankfurt/Main: Africa & Science.

Obahoundje, S., Diedhiou, A., Akpoti, K., Kouassi, K. L., Ofosu, E. A., & Kouame, D. G. M. (2024). Predicting climate-driven changes in reservoir inflows and hydropower in Côte d'Ivoire using machine learning modeling. *Energy, 302*, 131849. ISSN 0360-5442. https://doi.org/10.1016/j.energy.2024.131849

UNDP. (2024). *Annual Report 2023*. One United Nations Plaza, New York, MY 10017. Retrieved November 28, 2024, from https://annualreport.undp.org/assets/Annual-Report-2023.pdf?_gl=1*1kcyvwb*_gcl_au*NjgONjA1MjYxLjE3MzI3OTU1ODk.*_ga*Mzg2NzUzMDI5LjE3MzI3OTU1OTA.*_ga_3W7LPK0WP1*MTczMjc5NzY0OS4yLjEuMTczMjc5Nzc2My40OC4wLjA

UNESCO. (2014). *The United Nations World Water Development Report 2014: Water and energy*. Paris, France. Retrieved November 30, 2024, from https://unesdoc.unesco.org/ark:/48223/pf0000225741

UNICEF, WHO. (2022). *Progress on drinking water, sanitation and hygiene in Africa 2000–2020: Five years into the SDGs*. United Nations Children's Fund (UNICEF) and World Health Organization (WHO). Retrieved November 28, 2024, from https://data.unicef.org/wp-content/uploads/2022/03/Progress-on-drinking-water-sanitation-and-hygiene-in-Africa-JMP-snapshot.pdf?utm_source=chatgpt.com

United Nations. (2024). *The United Nations World Water Development Report 2024: Water for prosperity and peace*. UNESCO, Paris. Retrieved November 28, 2024, from https://unesdoc.unesco.org/ark:/48223/pf0000388948

The World Bank. (2018). *Data bank: Sustainable energy for all*. Retrieved November 29, 2024, from https://databank.worldbank.org/source/sustainable-energy-for-all/Type/TABLE/preview/on

The World Bank. (2023). *Tracking SDG 7: The energy progress report. World Bank, Washington DC*. © World Bank. Retrieved October 28, 2024, from https://databank.worldbank.org/reports.aspx?source=2&series=EG.ELC.ACCS.ZS&country=

The World Bank. (2024). *World development indicators*. Retrieved November 30, 2024, from https://databank.worldbank.org/reports.aspx?source=2&country=CIV

Yapo, A. R. (2018). *Les origines du problème de surqualification en Côte d'Ivoire*. Retrieved October 28, 2024, from https://theconversation.com/les-origines-du-probleme-de-surqualification-en-cote-divoire-99004

Sustainable Minds: Developing an International and Interdisciplinary SDG Course in Higher Education

Frank Joosten

Abstract While the benefits of interdisciplinary, project-based learning in sustainability education are widely acknowledged, there are relatively few concrete examples demonstrating its effective implementation in an international educational setting. This paper presents the development and implementation of *Sustainable Minds*, an international and interdisciplinary initiative aimed at fostering sustainable mindsets among 'green & tech' students at PXL University of Applied Sciences and Arts (Belgium). It explores both a recent 'blended intensive programme' (BIP) funded by the European Union and conducted in March 2024, and the design of a new international joint course on the Sustainable Development Goals (SDGs) using collaborative online international learning (COIL). By providing insights into the design and implementation of these initiatives, this paper aims to contribute to the advancement of sustainable development education in higher education contexts.

Keywords Sustainable development · Higher education · International · Interdisciplinary · Blended learning

1 Introduction

In education for sustainable development (ESD), higher education institutions (HEIs) play a pivotal role in equipping students with the competencies necessary to address complex sustainability challenges. The ESDfor2030 framework (Unesco, 2020) underscores the responsibility of universities in advancing the Sustainable Development Goals (SDGs) through research, education, campus activities, and institutional leadership (Sustainable Development Solutions Network, 2020). These institutions serve as incubators for sustainability innovation, fostering knowledge production, skills development, and interdisciplinary collaboration required for transformative change.

F. Joosten (✉)
PXL-Green & Tech Department, PXL University of Applied Sciences and Arts, Universitaire Campus—Agoralaan—Gebouw H, Diepenbeek, Belgium
e-mail: frank.joosten@pxl.be

W. Leal Filho et al. (eds.), *Introducing the UN SDGs in Higher Education Institutions*, World Sustainability Series, https://doi.org/10.1007/978-3-032-23017-1_3

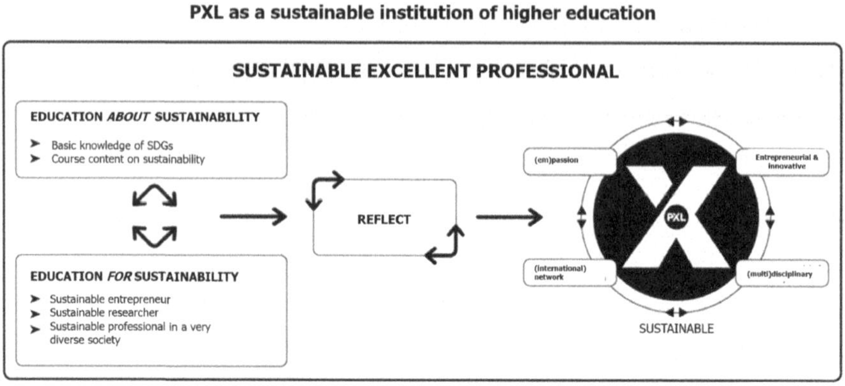

Fig. 1 PXL's framework for educating sustainable excellent professionals

In Belgium, PXL University of Applied Sciences and Arts has taken a leading role in integrating sustainability into higher education. As the first Belgian institution to be awarded the Unitar SDG Champion Certificate, PXL is committed to embedding SDGs across all study programs. Its framework for educating sustainable excellent professionals (see Fig. 1) distinguishes between 'education **about** sustainability'—which focuses on imparting theoretical knowledge about SDGs—and 'education **for** sustainability,' which prioritizes action-based learning, fostering critical thinking, and developing essential competencies for sustainable practice.

The need for competency-based education in sustainability has been well-documented. Wiek et al. (2011) identified key competencies in sustainability education, including systems thinking, anticipatory skills, normative competence, strategic competence, and interpersonal competence. Similarly, Rieckmann (2018) emphasizes the importance of empowering students with transformative competencies that enable them to envision and implement solutions for a sustainable future. These competencies form the foundation of education for sustainability, guiding students to become adaptive, critical, and engaged professionals capable of addressing global challenges.

Integrative approaches to sustainability education are more transformative. Leal Filho et al. (2015) emphasize that integrative approaches, particularly through project-based learning, enhance student engagement and provide real-world contexts for applying sustainability principles. This pedagogical shift moves beyond traditional, siloed academic disciplines and embraces interdisciplinary collaboration as a key driver of sustainability education. Merck and Beermann (2015) further argue that transdisciplinary teaching can have a positive impact on the student motivation and allow for the active transfer of practice-based knowledge.

International collaboration is another important dimension of sustainability education. Given the global nature of sustainability challenges, exposure to different cultural perspectives and problem-solving approaches enriches students' learning experiences. International collaborative programs offer students the opportunity to engage in intercultural dialogue, apply sustainability knowledge in different socio-economic contexts, and develop a global mindset essential for addressing transnational sustainability issues.

Although the benefits of interdisciplinary project-based learning in sustainability education are widely acknowledged, there are relatively few concrete examples of its effective implementation in an international educational setting involving several different nationalities. This paper focuses on the practical implementation of international interdisciplinary project-based sustainability education through two initiatives—both called *Sustainable Minds*—developed within the PXL-Green & Tech department. Section 2 explores an international EU-funded 'blended intensive programme' (BIP) that brought together students from 10 different nationalities and 11 study programs in March 2024. Section 3 presents a newly developed international and interdisciplinary joint course on SDGs in a North–South context, co-organized by PXL and Hochschule Bielefeld (HSBI, Germany). Finally, Sect. 4 presents key lessons learned, reflecting on the challenges and opportunities encountered in the implementation of these initiatives.

2 Blended Intensive Programme

In March 2024, PXL hosted the EU-funded blended intensive programme (BIP) *Sustainable Minds* (2022-1-BE02-KA131-HED-000058832-3). The aim of the programme was to bring together bachelor students (EQF Level 6) from different backgrounds and nationalities, and give them the opportunity to work together in international, interdisciplinary teams for a full week on a real-world case in which the SDGs were a central focus.

Table 1 provides an overview of the partners and study programmes involved.

The **online part** of the blended programme was limited to a short online introduction (before the intensive week) and a short one-hour review and evaluation a few weeks after the physical mobility period. The main focus was on the activities during the **physical week**, from 25 to 29th of March 2024. During that week, the 37 students were divided into eight groups of 4 or 5 students, maximizing group diversity by mixing nationalities and disciplinary backgrounds as much as possible. Staff and lecturers involved in the project took on the role of co-teachers and coaches, complementing each other's expertise.

The **learning objectives** of the programme, defined jointly by all project partners, were the following:

- The student has knowledge of the SDGs and understands their relevance for and impact on (current and future) society.
- The student can work in an international, intercultural, interdisciplinary team.
- The student can think critically, solve problems and elaborate creative, innovative solutions.
- The student can communicate and present ideas in (verbal/written) English.

Table 1 Blended intensive programme *Sustainable minds* 2024: participants

Partner	# Students (# staff)	Bachelor study programmes (EQF Level 6)	Nationalities
PXL University of Applied Sciences and Arts (BE)	11 (3)	Construction (4) Electromechanics (3) Greenery Management (3) Biotechnology (1)	Belgium (8) Netherlands (2) Germany (1)
HSBI Hochschule Bielefeld (DE)	7 (2)	Industrial Engineering (6) Mechanical Engineering (1)	Germany (6) Mexico (1)
SAMK—Satakunta University of Applied Sciences and Arts (FIN)	11 (1)	Mechatronics (5) Artificial Intelligence (3) Energy and Environmental Engineering (3)	Finland (3) India (3) Sri Lanka (2) Myanmar (1) Thailand (1) Romania (1)
Lapin AMK (FIN)	8 (2)	Forestry (5) Agriculture (3)	Finland (8)
Total	37 (8)	11 Different study programmes	10 Nationalities

These learning objectives were assessed at the end of the physical week for all groups and students. All staff were involved in this evaluation, using evaluation rubrics that were agreed upon before the start of the programme. Students also submitted a written peer and self-assessment form in which they commented on their teamwork.

Table 2 lists the variety of **activities** included in the physical week programme.

Following a teambuilding day (day 1), an introduction to the SDGs and an inspiring outdoor visit (day 2), the core of the programme focused on the **real-world SDG case** that the student groups worked on during days 3, 4 and 5. The case was provided by a local entrepreneur who had recently purchased 1.6 hectares of **forest land and a pond** in a nearby town and who was keen to get the students' ideas and feedback on how to transform the area into a sustainable site for the local community, where respect for nature, relaxation, music and leisure are harmoniously integrated. The students learned more about the entrepreneur's **preliminary plans** (none of which have yet been realized):

- a hybrid swimming pond that would not endanger the protected amphibians species currently present;
- an educational trail next to the pond, aimed at primary school children;
- a portable, sustainable video container;
- recycled plastic benches with built-in audio to provide musical impressions of the annual rock festival held nearby;
- a water storage system.

Table 2 Sustainable minds 2024: week programme

	Activities	Lecture/ instruction	Workshop/ exercise	Group work	Outdoor visit	Presentation	Socio-cultural activity
Day 1	Programme outline	x					
	Getting to know each other and teambuilding (1) (Bricks Battle)		x				
	Guest lecture (T. Joye): importance of veteran trees	x					
	Teambuilding (2): team T-shirts and video challenge		x	x		x	
	BBQ & camp fire						x
Day 2	SDG workshop @ PXL Green Office		x	x	x		
	Visit Ecotron UHasselt & Hoge Kempen National Park				x		
Day 3	Sustainable campus tour PXL-Green & Tech				x		
	Actionbound Tour exercise (HSBI)		x	x			
	Introduction SDG case	x					
	Visit Infinity Building & Encon (Bilzen)				x		

(continued)

Table 2 (continued)

	Activities	Lecture/ instruction	Workshop/ exercise	Group work	Outdoor visit	Presentation	Socio-cultural activity
Day 4	Online guest lecture (Merijaan): 100% recyclable surf wear	x					
	Group work: SDG case			x			
	Workshop: sustainable reindeer herding (Lapin AMK)		x				
	Fun and entertainment: bowling and karaoke						x
Day 5	Group work: SDG case			x			
	Final presentations			x		x	
	Ceremony						x

The groups were given GPS coordinates of the area and some concept drawings. They were told to see the entire case as a creative, interdisciplinary 'out-of-the-box' challenge and were asked to deliver the following **outputs**:

- a thorough SDG analysis at target and indicator level, resulting in a comprehensive general overview;
- a basic concept for two specific elements in the plans, targeting different 'personas' (= potential users of the area);
- a basic concept for an additional sustainable element (currently missing in the plans);
- a convincing final presentation

The results exceeded expectations, with most groups conducting an in-depth SDG analysis and proposing some really innovative, well-developed concepts, ranging from sustainable art to pond fountains that respond to musical input, and AR-driven eco-adventure games. All groups agreed that working on a real-world case was highly motivating, although some students expressed a desire for more time to further refine their ideas. A survey conducted among the students at the end of the final day reflected their positive feedback on the programme: 33% rated it as 'excellent' (5 out of 5), 64% as 'good' (4 out of 5) and only 3% as 'satisfactory' (3 out of 5).

3 Joint Course

For two of the partners in the blended intensive programme—PXL (Belgium) and HSBI (Germany)—the international collaboration was the trigger for a new initiative: a **joint, interdisciplinary and international course** *Sustainable Minds*, which was launched in Spring 2025. Table 3 lists the basic course details:

In this course, **collaborative online international learning** (COIL) via the Blackboard Ultra platform enables students from both institutions to work closely on real-world SDG cases. As strong team cohesion significantly enhances student motivation and learning outcomes, and as PXL and HSBI are only a three-hour drive apart, a **physical kick-off event** is organized for all participating students.

After a basic theoretical introduction to sustainable development, students collaborate online for 2 class hours per week over a 10-week period. They work on SDG cases in mixed international and interdisciplinary groups of 4 to 5 students, with lecturers providing **coaching and co-teaching** based on their complementary expertise and backgrounds. Due to significant differences in grading systems between Belgium and Germany, final **grading** is conducted separately by the two institutions. However, both lecturers take into account the collective observations from the international group work in their evaluations.

Thanks to a close connection within the PXL network, the real-world SDG cases in the 2025 course are linked to the **Kakuma Refugee Camp in Kenya**, an international refugee camp overseen by the United Nations High Commissioner for Refugees (UNHCR). The camp is home to around 200,000 people, many of whom are long-term refugees fleeing the war in nearby South Sudan. A local consultant and educator made videos and photos to help PXL and HSBI students gain a deeper understanding of some of the challenges faced in the camp:

- (plastic) waste management & deforestation;
- limited access to affordable power;
- water scarcity and unsanitary conditions;
- overcrowded classrooms;
- etc.

As with the SDG case in the blended intensive programme (see Sect. 3) students are required to conduct a comprehensive SDG analysis and to develop cost-effective solutions to one of the challenges they have investigated in depth. Their proposals

Table 3 Sustainable minds: joint course details

Timing	Spring term (first edition: February–May 2025)
ECTS	3 (= ca. 75 h)
@PXL	Elective course 2nd Bachelor of Construction/2nd Bachelor of Electro-Mechanics
@HSBI	Mandatory course (part of 'Procurement, Production and Logistics') 1st Bachelor of Industrial Engineering—international students

and solutions receive feedback from the local educator. Due to power and infrastructure issues that prevent real-time interactions, feedback is mostly provided asynchronously through recorded video responses.

4 Lessons Learned (and Ongoing)

By providing insights into the design and implementation of both the blended intensive programme (BIP) and the joint course *Sustainable Minds*, this paper aims to contribute to the advancement of sustainable development education in higher education contexts.

However, the path to achieving sustainable development education is (and will continue to be) complex. Just as sustainable development itself involves addressing a wide range of challenges—economic, social, environmental and political—that must be aligned both locally and internationally, so too does the journey of education for sustainable development. Navigating different international academic calendars, establishing an interdisciplinary balance that fosters meaningful international student collaboration, and overcoming logistical or administrative barriers are all part of a trial-and-error process. For a detailed discussion of the potential challenges and bottlenecks inherent in implementing interdisciplinary approaches in higher education, see Logan et al. (2021).

Despite these challenges, the following key lessons—drawn from experience and still evolving—may be of value to for higher education professionals interested in developing similar initiatives:

- **Real-world, community-based cases**: Engaging students with real-world, community-centered challenges significantly increases their motivation, as they see the tangible impact of their work on society. While traditional teaching methods aim to increase students' knowledge of sustainable development (education *about* sustainable development), the most significant societal benefits are likely to come from education *for* sustainable development, i.e. action-oriented learning methods (e.g. project based learning) that cultivate the skills and attitudes necessary for students to become sustainable professionals.
- **Interdisciplinary and international collaboration**: Sustainable development is a global, multifaceted challenge that requires both international and interdisciplinary collaboration. Embedding this international and interdisciplinary cooperation within educational projects provides students with the most authentic learning experience, and preparing them to deal with the complexities of sustainable development now and in the future.
- **Co-teaching**: The collaborative nature of co-teaching brings together complementary expertise from different fields, enriching the educational experience by providing students with a broader and more holistic understanding of sustainable development.

- **Blended learning format**: The combination of physical and online interactions between students is likely to be the most effective approach. Face-to-face meetings foster strong social bonds, which are essential for successful teamwork in any collaborative environment. Meanwhile, online learning allows international collaboration to be extended over time, ensuring that partnerships and exchanges continue when face-to-face meetings are not possible.

By integrating sustainability competencies, embracing interdisciplinary and international collaboration, and adopting transformative learning approaches, higher education institutions can play a vital role in shaping future professionals capable of driving sustainable development.

References

Leal Filho, W., Shiel, C., & Paço, A. (2015). Implementing and operationalising integrative approaches to sustainability in higher education: The role of project-oriented learning. *Journal of Cleaner Production, 133*, 126–135. https://www.sciencedirect.com/science/article/abs/pii/S0959652616305431

Logan, L., Strachan, S., Willison, D., Bain, R., Roberts, J., Mitchell, I., & Yarr, R. (2021). *Reflections on developing a collaborative multi-disciplinary approach to embedding education for sustainable development into higher education curricula*. Emerald Open Research. https://strathprints.strath.ac.uk/77619/1/Logan_etal_EOR_2021_Reflections_on_developing_a_collaborative_multi_disciplinary_approach_to_embedding_education.pdf

Merck, J., & Beermann, M. (2015). The relevance of transdisciplinary teaching and learning for the successful integration of sustainability issues into higher education development. In: W. Leal Filho, L. Brandli, O. Kuznetsova, & A. Paço (Eds.), *Integrative approaches to sustainable development at university level* (pp. 19–25). Springer. https://doi.org/10.1007/978-3-319-10690-8_2

Rieckmann, M. (2018). Learning to transform the world: Key competencies in ESD. In A. Leicht, J. Heiss, & W. J. Byun (Eds.), Issues and trends in education for sustainable development. UNESCO, pp. 39–59. https://unesdoc.unesco.org/ark:/48223/pf0000261802

Sustainable Development Solutions Network. (2020). *Accelerating education for the SDGs in universities: A guide for universities, colleges, and tertiary and higher education institutions*. https://www.unsdsn.org/resources/accelerating-education-for-the-sdgs-in-universities-a-guide-for-universities-colleges-and-tertiary-and-higher-education-institutions/

Unesco. (2020). *Education for sustainable development. A roadmap*. https://unesdoc.unesco.org/ark:/48223/pf0000374802

Wiek, A., Withycombe, L., & Redman, C. L. (2011). Key competencies in sustainability: A reference framework for academic program development. *Sustainability Science, 6*, 203–218. https://link.springer.com/article/10.1007/s11625-011-0132-6

Transatlantic Summer School for Sustainability—Interdisciplinary, Transatlantic and Project-Oriented

Andrea Heilmann, Simone Roth, Nicole Coleman, and Michael Römmich

Abstract The first TSS, held at Harz University of Applied Sciences in 2023, focused on "the future of energy." Student and teacher evaluations indicated that the TSS generated added value in terms of sustainability and the development of future skills. The evaluations also highlighted the need to better integrate research-based input with research-based methodologies to enhance the learning outcomes of future TSS editions.

Keywords Future skills · Sustainability in HEI · Design thinking · Future of energy

1 Introduction

The Transatlantic Summer School (TSS) is a joint teaching–learning program of four partnering institutions: Harz University of Applied Sciences, Ruhr West University of Applied Sciences (Germany), Iowa State University, and Wayne State University (USA). The program offers international experiences and integrates project- and research-based approaches to sustainability education in higher education institutions. Each year, the TSS addresses specific sustainability topics, allowing students from Germany and the USA to deepen their understanding of global challenges

A. Heilmann
Hochschule Harz (University of Applied Sciences), Wernigerode, Germany
e-mail: aheilmann@hs-harz.de

S. Roth (✉) · M. Römmich
University of Applied Sciences Ruhr West, Mülheim an der Ruhr, Germany
e-mail: simone.roth@hs-ruhrwest.de

M. Römmich
e-mail: Michael.Roemmich@hs-ruhrwest.de

N. Coleman
Wayne State University, Detroit, MI, US
e-mail: ncoleman@wayne.edu

W. Leal Filho et al. (eds.), *Introducing the UN SDGs in Higher Education Institutions*, World Sustainability Series, https://doi.org/10.1007/978-3-032-23017-1_4

<table>
<tr><th>Future Skill Categories[1]</th><th>Definition of future skill category</th><th>Learning objectives Online Phase</th><th>Learning objectives Onsite Phase</th></tr>
<tr><td>Classical Skills</td><td>Non-digital key competencies crucial for both individual career success and organizational effectiveness. They include fundamental abilities such as problem-solving and resilience.</td><td>Intercultural communication</td><td>Solution-orientation
Creativity
Entrepreneurship and initiative
Intercultural communication
Resilience</td></tr>
<tr><td>Technological Skills</td><td>Include those skills that are particularly important for the design and efficient use of technologies. These include established technologies that require skills such as software development, as well as new developments that require knowledge of data analytics and artificial intelligence (AI), for example.</td><td>Data analytics

(reflected through different master and bachelor programs by the participating students)</td><td>User-centred design
Data analytics

(reflected through different master and bachelor programs by the participating students)</td></tr>
<tr><td>Digital Skills</td><td>Navigate and actively participate in a digitalized environment. Became essential for professional life and societal engagement (digital Home Offices) during COVID-19.</td><td>Digital collaboration

Digital learning
Digital literacy</td><td>Agile working

Digital learning</td></tr>
<tr><td rowspan="2">Transformative Skills</td><td rowspan="2">Crucial for addressing major societal challenges (climate change, COVID-19). They include mission-driven focus and innovation capability to unite people and unlock new potential.</td><td colspan="2">Ability to judge
Dialogue and conflict orientation</td></tr>
<tr><td></td><td>Innovation competency
Mission oriented
Competence to change</td></tr>
</table>

Fig. 1 Comparison of learning objectives of TSS with the framework of future skills (Stifterverband & McKinsey, 2021)

related to the UN Sustainable Development Goals (SDGs)[1] across diverse scientific fields. The TSS explicitly aims to foster the development of future skills, with a special focus on transformative competencies of students.

Future skills are cross-industry abilities, skills and characteristics that will become increasingly important in all areas of professional and personal life over the next five years. According to the German Stifterverband,[2] which is an "*organization contributed to this effort through several programs funding future skills education, creating networks among various stakeholders in and outside academia, as well as producing numerous analyses and publications*" (Ehlers, 2024, p. 314), future skills consist of 21 future competencies (see Fig. 1). The TSS learning objectives addressed each of the four basic categories and focused on different aspects in the online and onsite phases. A detailed comparison is shown in Fig. 1. This perspective was particularly valuable for the TSS, as there is also a similar framework based on U.S. studies, known as "Competencies for a Career-Ready Workforce," developed by the National Association for Colleges and Employers (National Association of Colleges & Employers, 2024).

Several recent educational approaches aim to develop students' transformative and other future skills in higher education institutions (HEIs). Examples include problem-based learning, research-based learning, experimental learning, foresight learning, action-based learning, long-term learning, and cognitive learning (Redman & Larson, 2011; Trevisan et al., 2024). Specific methods, such as the flipped classroom approach or design thinking, can accompany these frameworks. HEIs incorporate

[1] https://sdgs.un.org/goals; United Nations (2024)

[2] German Stifterverband is a joint initiative started by companies and foundations—the only one in Germany to be devoted entirely to consulting, networking and promoting improvements in the fields of education, science and innovation. Stifterverband für die Deutsche Wissenschaft e.V. (2024); Ehlers (2024, pp. 32–36).

these teaching approaches into their curricula in various ways, such as through specific teaching modules or as part of project- and teamwork.

The innovative approach presented here—the Transatlantic Summer School on Sustainability—is tailored to the curricula of four partner universities, and is explicitly designed to foster future skills. The TSS program addresses a different sustainability-related theme each year, rotating among the four partner universities, and combining digital with onsite teaching–learning components. Based on a comprehensive analysis, the program adopts a methodological approach that integrates research-based and problem-based learning, alongside the design thinking method.

Research-based learning involves participants actively engaging in all phases of the research process: developing research questions, collecting data or conducting experiments, and presenting findings in various formats (Thiem et al., 2023, p. 1040).

Problem-based learning on the other hand, allows learners to collaboratively solve real-world and complex problems (Luna et al., 2024). In the context of the TSS, these real-world problems are directly linked to the UN Sustainability Goals.

Student-centered learning and research/project-based learning foster the students' transformative skills. The theory of "Transformative Learning" has become increasingly popular in education, especially in sustainability studies. This pedagogical approach emphasizes critical thinking and challenges learners to consider alternative and diverse perspectives, thereby enabling them to envision solutions for complex global challenges. Transformative learning is according to Mezirow (1985, p. 22):

> "…the process of becoming critically aware of how and why the structure of our psycho-cultural assumptions has come to constrain the way in which we perceive our world, of reconstituting that structure in a way that allows us to be more inclusive and discriminating in our integrating of experience and to act on these new understandings." Mezirow's theories address emancipatory learning, critical reflection and the discourse of communicative action (Fleming, 2018).

Similarly, to Mezirow's transformative learning theory, the United Nations' concept of Education for Sustainable Development (ESD) focuses on enhancing society's thinking capability, critical judgment, and collaborative skills. Since the UN introduced ESD, numerous universities globally have integrated sustainable-focused practices into curricula. Traditional teaching methods have often been insufficient for effectively addressing the complex sustainability challenges of the twenty-first century. Thus, the shift towards critical thinking and inter-disciplinary collaborations within ESD aims to equip students with the competencies necessary to actively drive societal transformations towards environmental and social sustainability (Filho et al., 2019; UNESCO, 2014).

2 Goals and Concept of the Transatlantic Summer School for Sustainability

When planning the TSS as a joint project between the participating universities, the two primary objectives were:

a. To provide students from various disciplines in Germany and the USA with opportunities to gain international and intercultural experiences.
b. To foster students' development of future skills through the application of research-based learning methods.
c. Students typically acquire intercultural competencies through various opportunities such as semesters abroad, internships or international summer schools. However, many students face significant barriers considering these options. These barriers include financial constraints for longer stays or internships abroad, organizational challenges due to existing commitments at home (e.g., part-time employment), and social or personal factors such as family responsibilities or limited self-confidence. In addition, not all of these opportunities fit into study programs and can lead to postponed graduation. Recognizing these barriers, the four universities jointly developed the TSS as a collaborative international learning concept. The combined online and onsite phases reduce the duration and associated financial and organizational burdens of studying abroad. Through coordination among the universities, participants of the TSS can earn up to 5 ECTS in Germany or 3 credits within the U.S. system, minimizing disruption to their academic progress. Furthermore, social and intercultural activities within the TSS ensures accessibility particularly for students with limited prior intercultural learning experiences.
d. The TSS themes are guided by societal challenges arising from the implementation of the UN SDGs. Successfully addressing these challenges requires students to develop future skills by the end of their studies. With its focus on sustainability and the implementation of the SDGs, the TSS aims to support the development of transformative skills. Transformative skills cannot be taught through conventional methods; rather, they must be acquired and developed through team interactions. The TSS provides an excellent framework for this, as it integrates diverse perspectives from students, experts, and potential users into the decision-making process.

Figure 2 illustrates the structure of the TSS, which specifically aims to strengthen transformative skills. The TSS consists of a 4-week online phase followed by a 2-week onsite phase, each with distinct learning objectives and teaching formats (see as well Fig. 1). While the general structure of the TSS remains consistent each year, each TSS focuses on a specific, highly relevant topic closely linked to the teaching and research activities of the hosting university. Leveraging site-specific resources in the Harz region, the TSS 2023 emphasized sustainable energy supply, defining a clear vision based on megatrends and the SDGs. The goal of the TSS was for

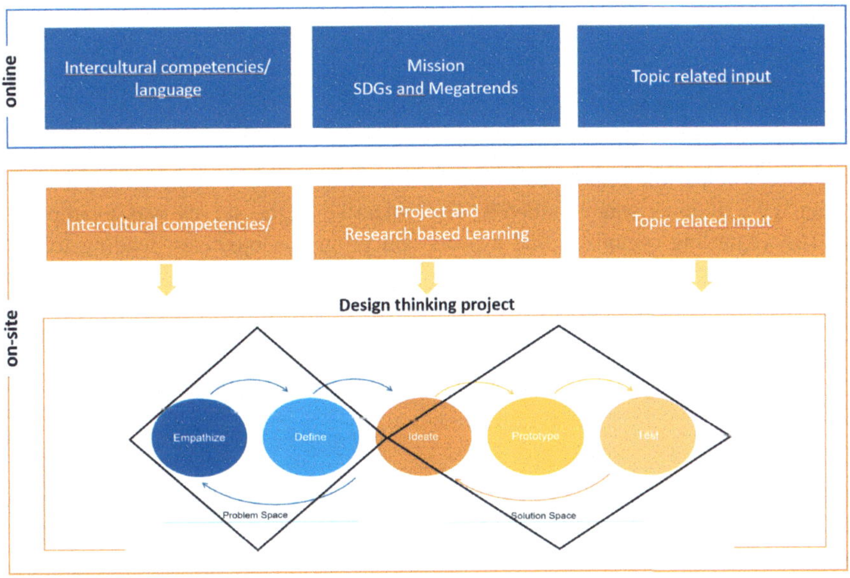

Fig. 2 Structure of the TSS

students, working in intercultural and interdisciplinary teams, to develop solutions aimed at sustainably reducing greenhouse gas emissions.

Experts provided specialist and topic-related input through workshops on energy economics, communication of energy data, smart cities and regions, as well as visits to small power plants. These activities prepared students for the design thinking phase, during which faculty supported the teams with their expertise.

Design thinking, as a method, addresses several future skills. It is a systematic approach used by teams to develop innovative solutions to complex problems across various life contexts, with a strong focus on the needs of people (Daymond & Knight, 2023, p. 2; Lewrick et al., 2018; Lindberg et al., 2011, p. 6; Minet et al., 2024). Therefore, it supports the development of innovation skills and user-centered designs for products. As a team-based approach, design thinking also trains students' ability to engage in dialogue and resolve conflicts. In conclusion, it can be stated that the structure and method mix contribute to the development of future skills across three dimensions (Ehlers & Eigbrecht, 2024, p. 6; Mayer & Schwemmle, 2024, p. 12): future skills related to students´ personal development (learning—individual development-related), the development of specific solutions to selected problems (development—individual object-related), and joint development within student teams (co-creation—individual organization related).

3 Results of the TSS

3.1 Methodology

To assess whether the objectives of the TSS were achieved, a comprehensive evaluation with students was conducted. The objective of the survey was to improve the TSS regarding organizational aspects and program objectives, leading to two research questions:

- How does the TSS contribute to advancing intercultural and international experiences?
- In which areas does the TSS support the development of future skills?

Data collection

Data was collected using structured questionnaires consisting of 28 questions focusing on objectives, organization, and procedures of the TSS, previous participation in study-abroad programs, recommendations for improving the TSS, demographic information (gender, country, age, study program), and consent to be cited anonymously. The self-assessment questionnaire items were adapted from an existing COIL survey of the student evaluation centre, Ruhr West University of Applied Sciences. The questionnaire contained a mix of closed and open questions. A total of 34 students took part in the TSS onsite phase in August 2023. Of these, 26 completed the evaluation questionnaire, resulting in a response rate of 76%. To maximize response rates and representativeness, students filled out the questionnaire at the end of the onsite phase.

Analytical approach and justification

The evaluation employed a mixed-methods analytical approach. Quantitative responses (closed-ended questions) were analyzed using descriptive statistical methods to identify general patterns and insights. Qualitative data from open-ended questions were examined through content analysis (Mayring & Fenzl, 2019). This allowed to systematically identify key themes and deeper insights, enhancing the interpretive depth of the results. The mixed-method approach contributed to methodological robustness.

Limitations and potential biases

Several limitations of the study design were identified. Firstly, the relatively small sample size (26 students) may limit the generalization of the results beyond the specific TSS implementation. Additionally, the student response rate reflects a potential selection bias among student participants. Because qualitative data analysis is inherently subjective, researcher bias was minimized through team discussions and cross-verification during thematic analyses. To address these limitations, future evaluations could incorporate additional qualitative interviews to provide richer contextual information, enhancing overall validity. Selected results from the student survey are presented below, focussing on two core areas: intercultural experience and development of future skills.

3.2 International and Intercultural Experiences

One primary goal of the TSS was to foster intercultural and international competencies among participating students. According to evaluation results, students regarded this aspect as highly valuable, indicating that collaboration within international teams offered unique experiences not typically encountered in their previous educational contexts.

Qualitative feedback from open-ended responses emphasized intercultural experiences as a critical outcome of the program, illustrated by the following student statements:

- "Different cultural backgrounds have different ways of working in groups and on how to approach solving problems."
- "It is so important to overcome our comfort zone and learn about diverse perspectives. Your way is not always the 'only' best way."

Additionally, qualitative data confirmed improvements in competencies related to working in intercultural teams. Students mentioned that they learned "how to work in a diverse group and settle differences in project ideas" as well as "I learned how to be more patient during group work. I think the way Americans work together is much different than the European way, so I had to learn how to be patient."

Quantitative analysis further supported these qualitative insights. Ninety-six percent (96%) of students reported an improvement in their ability to collaborate within the intercultural groups. Additionally, 81% stated that participating in the TSS positively impacted their prospective ability to collaborate effectively in diverse teams in future (see Fig. 3). Over 90% of students also reported feeling better prepared to engage constructively in diverse team settings following their participation. Previous international experiences of students were controlled in the questionnaire (participation to a study abroad program longer than 3 month). Only 10 students had already prior international experience, while 16 students had no previous international experience.

Critical analysis of these findings suggests that while overall outcomes were positive, the extent of intercultural competence development may vary depending on individual backgrounds, a factor that was not systematically controlled in this study. Hence, one limitation of the current approach is the absence of a standardized way to measure intercultural competence before and after the TSS. It is also important to consider the potential for social desirability bias inherent in self-reported data, particularly regarding intercultural skills. While student statements reflect positive experiences, objective measures (e.g., peer- and teacher-assessment of intercultural competencies) could have provided additional validation of theses outcomes.

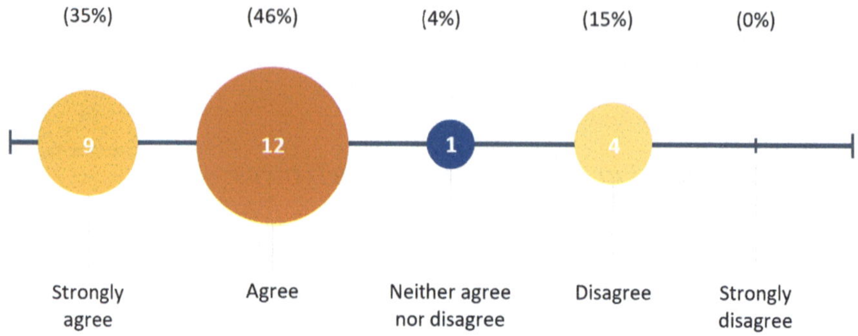

Fig. 3 Ability to work in intercultural teams in the future/ in general (Answers to: "*Participating in this project will have an impact on the way I will work or communicate with people from diverse countries and backgrounds in the future*")

Future TSS iterations might benefit from incorporating such tools alongside subjective evaluations to offer a more comprehensive assessment of intercultural skill development.

Further analytical reflection identifies important implications of theses results for international educational practices: the mixed-format (online followed by onsite) as well as the brief duration of the onsite component (two weeks), were specifically designed to reduce the financial, social, and organizational barriers associated with other international formats (study semester broad etc.). Positive evaluation outcomes suggest that this hybrid approach successfully provided effective international experiences for students typically underrepresented in traditional exchange settings, thereby expanding access and equity for intercultural learning.

3.3 Development of Future Skills Through Research-Based Teaching

The evaluation also examined how effectively the TSS fostered future skills through the application of research-based teaching methods. Questions regarding judgment, innovation competency, and the ability to engage in dialogue and conflict resolution reflect the students' perspectives on the development of future skill.

Developing judgment, critical reflection, and therefore the ability to objectively evaluate sustainability-related date and information is crucial, as decisions in sustainability contexts rarely involve linear processes. Evaluation data indicated that 54% of participating students reported an improvement in their ability to critically evaluate sustainability-related data as a result of the TSS. While this result demonstrates moderate success, it also suggests variability in students' experiences or prior levels of evaluating data. Since these factors were not explicitly controlled, incorporating an explicit baseline assessment of these competencies would have allowed for a more precise measurement of skill development.

The development of innovation skills was strongly supported through structured knowledge-building activities, excursions, expert participation, and systematic introduction to the design thinking methodology during the onsite phase. Students explicitly recognized the significance of design thinking, reflected by frequent references in their open-ended responses (Questions: "Are there specific competencies that you learned during the summer school? What are they?"):

- "*I think my ability to work in groups improved a lot. The design thinking process is a new skill I learned to use and hopefully will help me in the future.*"
- "*I have never reflected on how I work in a group and how tasks get done. This class showed me how to be more deliberate & productive when I lead groups.*"

However, critical reflection on these findings suggests that social desirability bias and students' individual prior experiences before the summer school might also have influenced the results. Introducing structured peer assessment of faculty evaluations during the design thinking phases could further enhance objectivity and reliability when evaluating students' development of innovation skill.

The TSS naturally introduced students to dialogue, collaboration, and conflict management through the intercultural and interdisciplinary composition of teams. This diverse structure was experienced through different working styles of German and U.S.-American students, varied interdisciplinary backgrounds of students and faculty, differences in students' academic progress (Bachelor's or Master's level), and also diverse personal perspectives on sustainability.

Evaluation results indicate substantial improvements in interpersonal skills, with 63% of students reporting that their ability to treat each other respectfully had significantly improved ("greatly" or "very greatly").

While conflict management skills were not explicitly addressed during the TSS, their development emerged implicitly through group interaction. Seventy percent (70%) of students "strongly agreed" or "very strongly agreed" that the summer school improved their ability to identify or address conflicts within a team (see Fig. 4). From an analytical perspective, however, it should be noted, that future TSS iterations could explicitly integrate conflict training to systematically reinforce theses critical skills.

4 Summary and Conclusions

The first TSS, with its focus on "the future of energy" successfully achieved its primary objectives of fostering intercultural experiences and developing future skills among students from Germany and USA. The program´s hybrid structure, combining online and onsite phases, effectively reduced traditional barriers associated with international education programs, such as financial constraints and extended time commitments. This accessible format allowed a diverse group of students to participate, many of whom had limited prior international experience. The program provided

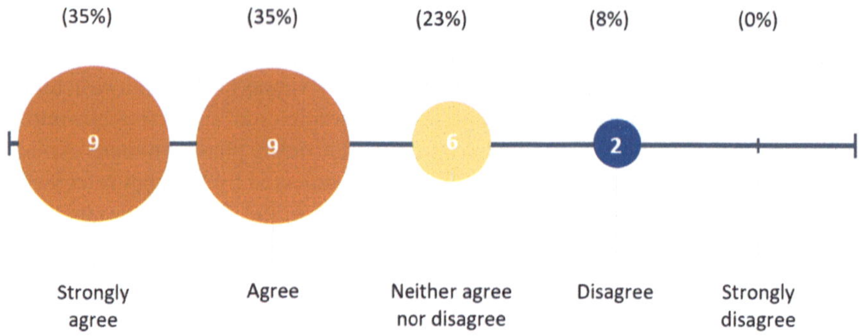

Fig. 4 Ability of conflict management (Answers to: *"The summer school improved my ability to discover conflicts in teams"*)

students with a focused context to apply design thinking methodologies, enhancing their innovation competence. Working in interdisciplinary and intercultural teams, participants addressed real-world energy challenges, fostering critical thinking and problem-solving skills essential for sustainable development.

Development of future skills

Evaluation results demonstrate that the hybrid structure of the TSS effectively facilitated the development of intercultural competencies. Students reported positive experiences collaborating within diverse teams, suggesting that short-term, structured programs can substantially enhance intercultural competencies. However, it needs to be stated that pre-program intercultural competence levels might have also a broader impact and a baseline assessment tool can help show the development of intercultural growth more concretely.

Most of the participating students learned new content and applied this content actively to a sustainability challenge in the field of energy. The increase in knowledge depended as well in this aspect on prior knowledge. Considering the fact that 38% of those surveyed attributed their interest in the TSS to the topic, the focus of the content is important and must continue to be geared towards challenges and innovative solutions in the future. Figure 5 shows that 60% of participants feel better prepared for the transformative process towards a sustainable future (strong and very strong agreement) due to the TSS. This indicates that the second objective was reached.

The use of the design thinking method to strengthen innovation and transformative skills also proved successful. Conflict management, communication and the careful use of information and data is an integral part of the method. In addition, the method also allows each individual student to participate in the solution to the problem that each team sets itself. This increases intrinsic motivation and commitment. Considering the complexity resulting from the implementation of the SDGs, design thinking is a method that is suitable and should be used for research-based education for sustainability in higher education.

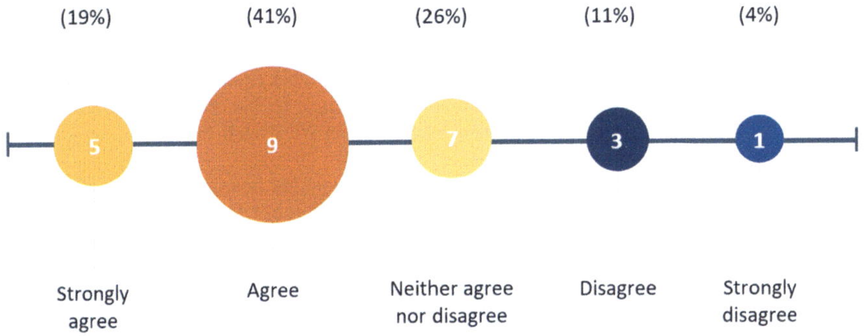

Fig. 5 Improvement of transformative competencies due to the TSS (Answers to: *"I feel better prepared to actively contribute to a solution to sustainability challenges due to the summer school"*)

Limitations and critical reflection

The relatively small sample size (36 students), though sufficient for preliminary conclusions, limits broader generalizability. Moreover, the absence of explicit conflict management training represents another limitation, as skills gained implicitly may vary across student teams. Future TSS implementations could consider explicitly incorporating training modules on conflict management. Analytical reflection on the TSS outcomes reveals valuable insights for broader educational practice in sustainability-focused higher education.

Broader implications and recommendations

The hybrid format of the TSS effectively reduced financial, social, and organizational barriers, facilitating access to intercultural education and contributing positively to equity in international learning opportunities. This outcome provides a replicable model for higher education institutions globally seeking effective and accessible methods for integrating sustainability and transformative skills into curricula.

To enhance the validity and replicability of future programs, subsequent TSS implementations should incorporate longitudinal evaluations and documentation of participants backgrounds and initial skill levels to address methodological limitations.

In conclusion, the TSS exemplifies a practical and effective model for integrating sustainability into higher education curricula through interdisciplinary, transatlantic collaboration, while simultaneously fostering future skills essential for addressing global sustainability challenges.

References

Daymond, J., & Knight, E. (2023). Design thinking in business and management: Research history, themes, and opportunities. In J. Daymond & E. Knight (Eds.), *Oxford research Encyclopedia of business and management*. Oxford University Press. https://doi.org/10.1093/acrefore/9780190224851.013.386

Ehlers, U.-D. (2024). Towards a future skills framework for higher education. In U.-D. Ehlers & L. Eigbrecht (Eds.), *Zukunft der Hochschulbildung—Future Higher Education. Creating the University of the Future* (pp. 21–60). Springer Fachmedien Wiesbaden. https://doi.org/10.1007/978-3-658-42948-5_2

Ehlers, U.-D., & Eigbrecht, L. (2024). Creating the university of the future: A global panorama on future skills. In U.-D. Ehlers & L. Eigbrecht (Eds.), *Zukunft der Hochschulbildung—Future higher education. Creating the University of the Future* (pp. 3–20). Springer Fachmedien Wiesbaden.

Filho, W. L., Shiel, C., Paco, A., & Mifsud, M. (2019). Sustainable development goals and sustainability teaching at universities: Falling behind or getting ahead of the pack? ScienceDirect. https://doi.org/10.1016/j.jclepro.2019.05.309

Fleming, T. (2018). *Mezirow and the theory of transformative learning*. IGI Global (2018). https://doi.org/10.4018/978-1-5225-6086-9

Lewrick, M., Link, P., & Leifer, L. (2018). *The design thinking playbook*. Wiley.

Lindberg, T., Meinel, C., & Wagner, R. (2011). Design thinking: A fruitful concept for IT development? In H. Plattner, C. Meinel, & L. Leifer (Eds.), *Design thinking: Understand—improve—apply* (1st ed., pp. 3–18). Springer Berlin Heidelberg

Luna, P. B., Torne, I. S., & Rio, E. G. (2024). Influence of the problem-based learning methodology on the intrapreneurial intentions of university students. ScienceDirect. https://doi.org/10.1016/j.ijme.2024.101024

Mayer, S., & Schwemmle, M. (2024). The impact of design thinking and its underlying theoretical mechanisms: A review of the literature. *Creativity and Innovation Management*, Article caim.12626. Advance online publication. https://doi.org/10.1111/caim.12626

Mayring, P., & Fenzl, T. (2019). Qualitative inhaltsanalyse. In N. Baur & J. Blasius (Eds.), *Handbuch Methoden der empirischen Sozialforschung* (pp. 633–648). Springer Fachmedien, Wiesbaden. https://doi.org/10.1007/978-3-658-21308-4_42

Mezirow, J. (1985). A critical theory of self-directed learning. In S. Brookfield (Ed.), *Self-directed learning from theory to practice* (pp. 17–30). Jossey-Bass.

Minet, A., Wentzel, D., Raff, S., & Garbas, J. (2024). Design thinking in physical and virtual environments: Conceptual foundations, qualitative analysis, and practical implications. *Technological Forecasting and Social Change, 207*, Article 123596. https://doi.org/10.1016/j.techfore.2024.123596

National Association of Colleges and Employers. (2024). *Career readiness: Competencies for a career-ready workforce*. https://www.naceweb.org/career-readiness/competencies/career-readiness-defined/

Redman, E. F., & Larson, K. L. (2011). *Educating for sustainability: Competencies & practices for transformative action*. ResearchGate. https://www.researchgate.net/publication/228620687_Educating_for_Sustainability_Competencies_Practices_for_Transformative_Action

Stifterverband, & McKinsey. (2021). *Future Skills 2021: 21 Kompetenzen für eine Welt im Wandel*. http://Stifterverband.org/medien/future-skills-welche-Kompetenzen-in-deutschland-fehlen

Stifterverband für die Deutsche Wissenschaft e.V. (2024). *English | Stifterverband*. https://www.stifterverband.org/english

Thiem, J., Preetz, R., & Haberstroh, S. (2023). How research-based learning affects students' self-rated research competences: Evidence from a longitudinal study across disciplines. *Studies in Higher Education, 48*(7), 1039–1051. https://doi.org/10.1080/03075079.2023.2181326

Trevisan, L. V., Filho, W. L., & Pedrozo, E. Á (2024). Transformative organisational learning for sustainability in higher education: A literature review and an international multi-case study. ScienceDirect. https://doi.org/10.1016/j.jclepro.2024.141634

United Nations. (2024, October 16). *The 17 goals | Sustainable development*. https://sdgs.un.org/goals

UNESCO. (2014). *Shaping the future we want: UN Decade of Education for Sustainable Development (2005–2014) final report*. ISBN 978-92-3-100053-9.

Transcending Boundaries: The Fusion of Sustainability Education and Software Engineering Education in a Master's Course

Ashok Tripathi, Jari Porrasd, and Linda Ott

Abstract Humanity's impact, manifested by climate change, shrinking habitats, and depleting natural resources, threatens all life on earth. Indeed, 2023 has proven to be the hottest year on earth in at least a century and a half. The key to maintaining a habitable planet is the broad use of sustainable practices. As pointed out in a 2019 ITiCSE (Innovation and Technology in Computer Science Education) working group report, considering the ubiquity of software which is both a consumer of energy and an enabler of more energy efficient solutions in a vast range of applications, awareness and use of sustainable practices should be considered essential in the development of software. Yet computer science and software engineering education continue to neglect sustainability. Embedding sustainability into the curriculum is essential to ensure that software engineers prioritize the planet in their work. This practice report describes an approach used in a graduate level course titled "Software and Application Innovation" taught at LUT University (Finland), to enhance software engineering students' sustainability competencies. Students were required to design and prototype new software solutions that addressed various sustainability challenges. Through this project, they developed a deeper understanding of global sustainability issues while still meeting the software engineering objectives for the course. Students learned to assess the sustainability impacts of their solutions from different dimensions/perspectives and keep competing demands in balance when designing their software projects. Based on our experiences, we present a practice report that can be used as a starting point when introducing and developing 'sustainability competency' in other software engineering or computer science courses. Key lessons learned are also shared.

A. Tripathi (✉) · J. Porrasd
Software Engineering, LUT University, Lappeenranta, Finland
e-mail: ashok.tripathi@lut.fi

J. Porrasd
e-mail: jari.porras@lut.fi

L. Ott
Computer Science, Michigan Technological University, Houghton, MI, USA
e-mail: linda@mtu.edu

W. Leal Filho et al. (eds.), *Introducing the UN SDGs in Higher Education Institutions*, World Sustainability Series, https://doi.org/10.1007/978-3-032-23017-1_5

Keywords Sustainability education · Software engineering education · SDG · Curriculum intervention · SE4GD

1 Introduction

Human activities are profoundly impacting the planet, upsetting the balance needed for life to thrive. Some argue that we have entered a new geological epoch, the Anthropocene, characterized by climate change and global warming due to human actions (Cook et al., 2016; Oreskes, 2004). Humans have already generated more material than the planet's natural biomass (Elhacam et al., 2020). The relentless use of energy in production and consumption accelerates global warming, causing numerous environmental disruptions (Allen et al., 2018; Masson-Delmotte et al., 2022). In 2022, climate-driven events like Pakistan's massive floods, wildfires across continents, and record-low river levels in Europe and China underscored the crisis, intensifying human suffering, biodiversity loss, and straining societal structures (OCHA, 2023).

Simultaneously, we are in the fourth industrial revolution, Industry 4.0, characterized by the rise of disruptive intelligence and information technologies (ICT), leading to efficiency gains (Bai et al., 2020). Research shows that higher ICT usage can reduce emissions and energy consumption in various sectors (Lin & Huang, 2023). Though ICT consumes about 4–10% of global energy (Ojala & Oksanen, 2021), it has a substantial "handprint" that can lower net greenhouse gas emissions (Andrae, 2021).

Beyond environmental impacts, ICT also influences personal, social, and economic domains, such as education, mental health, communication and commerce (Mansell, 2002; Warschauer, 2004). However, it can also exacerbate disparities, foster social isolation, and impact citizen engagement (Hawi & Samaha, 2017; Selwyn, 2012; Tambini, 2018). ICT professionals must consider these varied effects and their alignment with the UN Sustainable Development Goals (SDGs) (United Nations, n.d.).

Several frameworks exist for sustainability in software engineering (SE) by addressing environmental, social, and economic dimensions. The Karlskrona Manifesto for Sustainability in Software Engineering (Becker et al., 2014) highlights sustainability as a first-class concern, emphasizing long-term thinking across technical, economic, social, and environmental aspects. The Green Software Engineering Principles (Green Software Foundation, 2021) advocate for minimizing energy consumption, optimizing resources, and using sustainable infrastructure. The Triple Bottom Line (TBL) Approach (Elkington & Rowlands, 1999) when applied to software promotes balancing people (social), planet (environmental), and profit (economic) to design sustainable software solutions. Lastly, Lifecycle Analysis in Software Engineering (Penzenstadler & Femmer, 2013) encourages assessing a software system's environmental impact from development through disposal, ensuring long-term sustainability.

The integration of sustainability into SE curricula remains inconsistent, with several emerging trends shaping its adoption. Some universities offer dedicated courses on sustainable SE, covering energy-efficient coding, green computing, and ethical AI, while others embed sustainability topics into core courses, incorporating principles into software design, architecture, and systems engineering (Penzenstadler et al., 2012). Project-based learning through capstone projects and hackathons to foster innovation in sustainable software solutions is another approach (Palacin-Silva et al., 2018). However, a lack of standardization in curricula results in varying levels of graduate preparedness for sustainability-driven roles (Naumann et al., 2011).

Given SE's impact, educating students on sustainability is essential for a sustainable future. This paper explores how sustainability can be integrated into SE course, with the goal of integrating sustainability competency within SE education. It focuses on a course designed to address sustainability challenges via software solutions, moving towards a curriculum where SE and sustainability education are combined.

This research begins with the question: "How can we develop sustainability competency in software engineering students, so they consider sustainability principles when designing software-based ICT products and services?" The ACM-IEEE Computing Curricula 2020 (CC2020) defines SE curriculum guidelines but lacks mechanisms for building sustainability competencies in SE students (Force, 2020). Currently, there are no standard templates to incorporate sustainability into SE education, leaving it largely unaddressed in CC2020.

Sustainability and SE competencies are distinct, and sustainability in SE is a growing field, with foundational literature emerging in 2012 (Penzenstadler et al., 2012). Sustainability in SE can be viewed in two ways: designing sustainable software and creating software that supports sustainability goals (Penzenstadler et al., 2014). Software could thus play a key role in advancing SDGs (Calero et al., 2022). Despite the importance of sustainability in ICT, sustainability education in SE remains limited and underdeveloped.

This paper looks at a mechanism for integrating the teaching of sustainability values and building sustainability competencies within the context of an individual SE course, namely, 'Software and Application Innovation' taught at LUT University, Finland, as part of a Master of Science program in LUT's SE department. From this experience, we explore the following research questions, with the caveat that this work is but the first step of looking at integrating sustainability education into mainstream SE education:

RQ1. How can sustainable thinking be considered as a part of an existing software engineering course to inculcate sustainability competencies in the students?

RQ2. How can developing software-based solutions for local contextual sustainability challenges in a classroom environment increase the perception of global sustainability thinking in the students?

2 Literature Review

2.1 Sustainable Development and the SDGs

The Brundtland Report (WCED, 1987) defines sustainable development as meeting present needs without compromising future generations. This concept emphasizes distributive justice, prioritizing the needs of the world's poor. Sustainability must address global inequalities, ensuring a decent life for all without jeopardizing future well-being (Christen, 2010). The 2023 Agenda for Sustainable Development (United Nations, 2015) provides a global blueprint for a sustainable future, extending beyond human well-being to include all life and ecosystems. Central to this agenda are the 17 Sustainable Development Goals (SDGs), aimed at eradicating poverty, protecting the planet, and ensuring prosperity by 2030. The SDGs tackle key issues such as health, education, inequality, economic growth, climate change, peace, and justice, emphasizing their interconnections and foster a collective action for sustainability.

Sustainable development is closely linked to sustainability, derived from the Latin sustenere, meaning "to endure" or "to uphold." Sustainability refers to an entity or system's ability to endure within its natural lifespan (Venters et al., 2018). Tainter (2006) emphasizes that people sustain what they value, which stems from awareness. His framework (Tainter, 2003) poses four key questions: "Sustain what, for whom, for how long, and at what cost?"—highlighting the core considerations of sustainability.

To develop innovative solutions for sustainability challenges, it is crucial to understand their meaning and connections to visible global issues.

2.2 Sustainability Competencies

Achieving the SDGs and solving sustainability challenges require students to gain effective sustainability education and competencies. Yet, a universally accepted definition of these competencies is lacking (Brundiers et al., 2021). The term "competency" and "competence" (including in their plural forms) appear interchangeably in literature and is used to denote competence, skills, abilities, and capacities (Baartman et al., 2007; Cebrian & Junyent, 2015). Rieckmann (2012) describes "competencies" as personal predispositions to self-organize, encompassing cognitive, affective, and volitional (intentional) aspects as well as motivational elements, thus representing a coming together of knowledge, capacities, and skills reinforced with motives and emotional tendencies. Another similar conceptualization is to consider "competencies" as an application of knowledge, skills, and attitudes toward effective task execution and resolution of a problem (Wiek et al., 2011, 2015).

Specific to the work in our study, we consider the definition of "sustainability competencies" as a comprehensive collection of knowledge, skills, attitudes, and values, such that they enable a practical, tangible actions in addressing the real-world sustainability challenges (UNESCO, 2007; Wiek et al., 2011; Guia, 2020;

Framework	Competencies
Wiek et al., 2011 & 2015	Systems-thinking, Anticipatory/futures-thinking, Normative/values-thinking, Strategic-thinking, Interpersonal/collaborative. *Integrated problem-solving (added in Wiek et al., 2016)*
Brundiers et al., 2021	Wiek et al., 2011 & 2015 + Implementation, Intra-personal/self-awareness.
Redman and Wiek, 2021	Brundiers et al., 2021 + Generic competencies taught in higher education, Disciplinary competencies taught in higher education, Other professional skills.

Fig. 1 Evolution of the sustainability competencies in the scientific literature

Redman & Wiek, 2021). Efforts in sustainability education research have tried to define a framework for these competencies. The earliest, most cited, and highly influential work appeared in 2011 and defines five key competencies (Wiek et al., 2011) and sets the foundation for future similar works. Subsequent research recognizes this work as being pivotal in sustainability education (Grosseck et al., 2019). In 2020, a group of fourteen researchers came together and enhanced the original competencies framework developed by Wiek and his team (2011, 2015), adding two new competencies to the existing five (Brundiers et al., 2021). Subsequently, Redman and Wiek did a literature review focusing on works dealing with the 2011 framework, taking into consideration Brundiers et al.'s (2021) additions, and went on to add additional complementary competencies. Moreover, Redman and Wiek (2021) stated that for achieving sustainable transformations, the key competencies must be inherently interlinked and integrated with each other, as opposed to being a sequential list. Figure 1 presents the competencies, and their evolution as defined in these frameworks.

3 Methodology

This paper presents a case study on integrating sustainability into SE course at LUT University, Finland, in the implementations in spring 2022 and 2023, highlighting key lessons learned. The course, part of the Software Engineers for the Green Deal (SE4GD, 2020) program, emphasizes innovation in addressing sustainability challenges through software solutions. Designed at the intersection of software engineering and sustainability, it brings together students from diverse backgrounds who tackle global and local challenges. Their work extends beyond looking at environmental concerns, addressing social sustainability and broader societal impacts of SE solutions.

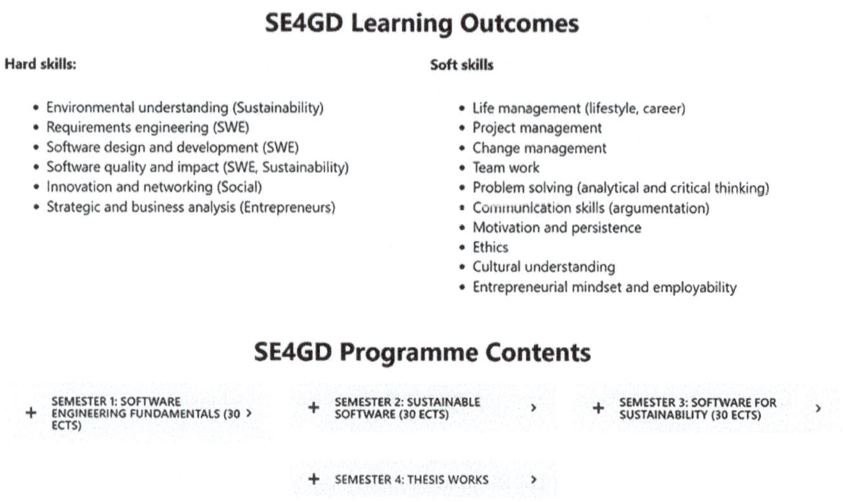

Fig. 2 SE4GD program structure with its learning outcomes

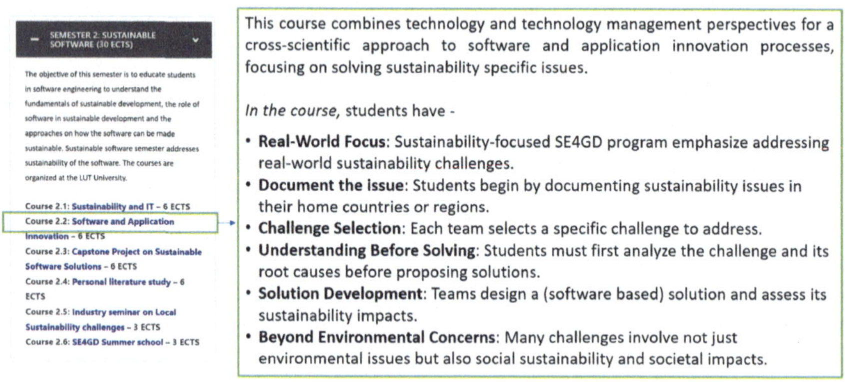

Fig. 3 Learning goals for software and application innovation course

Figure 2 provides an overview of the SE4GD program, while Fig. 3 highlights the selected course—Software and Application Innovation—and its placement within the SE4GD framework.

3.1 Learning Objectives of the Selected Course

In context of SE competencies, this course has the learning objectives and related competencies, as presented in Fig. 4.

At the end of this course, students will be able to:	Software engineering competencies (basis ACM/IEEE, 2014)
Identify and conceptualize an innovation opportunity within the chosen field	Professional knowledge, Design solutions in context
Identify the technical possibilities and limitations within the chosen field	Technical knowledge
Demonstrate technical knowledge, skills, innovation in creating software solutions	Technical knowledge, Design solutions in context, Perform trade-offs
Demonstrate good teamwork in developing and presenting the innovation/solution	Teamwork, Communication

Fig. 4 Learning objectives of the 'Software & Application Innovation' course in context of SE

In addition, this course also focuses on the sustainability specific learning objectives and related competencies. As presented in Fig. 5, what students learn on the completion of the course can be mapped to sustainability competencies as well.

At the end of this course, students will be able to:	Sustainability competencies (basis Wiek et al., 2011 & 2016)
Identify and conceptualize an innovation opportunity for a sustainability challenge in home country/region	Normative/Values-thinking competency
Identify the technical possibilities and limitations in the selected challenge	Anticipatory/Futures-thinking competency, Strategic-thinking competency
Demonstrate technical knowledge, skills, innovation in creating software solutions	Integrated problem-solving competency, Systems-thinking competency
Demonstrate teamwork in developing and presenting the innovation/solution	Interpersonal/Collaborative Competency

Fig. 5 Sustainability-focused learning objectives

3.2 Course Outline

To achieve the intended learning outcomes and competencies both in SE and sustainability, the course has been divided into five phases. Figure 6 encapsulates these phases and the focus area in each.

Phase 1—It starts with finding out meaningful real-world sustainability challenges. As students are from a host of countries, the sustainability challenges tend to have a good coverage in the sense of emphasized SDGs (students are not asked to think of the challenges through SDGs but the SDGs through the identified challenges). Students form into 3–4 person teams, start to define the challenge by thinking of their root causes. After this phase, students develop a clearer understanding of the challenge and the reasons behind it.

Phase 2—Student teams are asked to look at how a similar challenge is seen in other similar countries/regions. The purpose here is not to find solutions for the challenge but to get a deeper understanding of the reasons behind the challenge. For example, the ScholarAid team (affordable education in Indonesia) studied the approaches in France, Italy, India, Pakistan, and China to see how they have achieved more affordable and quality education. After this phase, students a better knowledge on why the challenge(s) persists in their selected region.

Phase 3—Students begin innovating a solution for the identified challenge, building a mock-up of their proposed solution. While developing their solution, students constantly assess the impacts of their solution using the SusAF tool (this is explained

Dividing into 5 distinct phases of implementation (teaching)	Key focus area(s)
Phase 1: Find a challenge	Learn about SDGs. Make 3–4-person team. Pick up a challenge, do root cause analysis
Phase 2: Market and competitor analysis	Look at existing solutions. Competition - did they succeed? Why? Why not? Where? Closer to solution – metrics emerge
Phase 3: Develop a solution and assess the impact	Implement the solution – mock-ups Assess solution -> what is the footprint of the solution -> beyond technology what other effects
Phase 4: Business development	Business case – viability of the solution Business model development Think of pitching the viability of the solution
Phase 5: Dissemination	Spread the word! Adoption and percolation of the solution!

Fig. 6 Course outline in 5 distinct phases

further in Sect. 3.3). The development and assessment are iterative, as bad initial decisions may require students to modify their proposed solution.

Phase 4—Teams develop a business case for their solution, using the Flourishing Business Canvas to demonstrate viability. This integrates with impact assessment, setting a foundation for the business model and pitch.

Phase 5—The last phase of the course is to pitch the solution keeping possible investors or funding organisations in mind. This final phase determines the solution's viability and potential to address a sustainability challenge.

3.3 Implementation Method(s)

The teaching methods and tools for each phase were selected to maximize the learning outcomes. The approaches and tools are summarised in Fig. 6.

Phase 1—Phase 1 starts with a lecture on UN Sustainable Development Goals (SDGs) to help students recognize a range of societal challenges. Students then reflect on major sustainability issues they've encountered in their home countries or regions and select the one they feel is most urgent, based on personal experience rather than metrics. Students then document their chosen challenges on a shared Miro board, including a brief description, supporting evidence, and any linked SDGs. This high-level presentation encourages questions and discussions among classmates.

After all challenges are presented, team building begins. Each student selects a primary and secondary topic of interest, but not from their own country, which fosters learning about different cultural and social norms. Team formation may take a few rounds, with some compromise required; typically, 80% of teams form quickly, while the last 20% may require additional discussion and negotiation. This activity initiates a deeper discussion of the challenges and their underlying causes.

Once topics are finalized, teams prepare a brief presentation from their perspective, covering background, challenge description, root causes, general impacts, and team focus. The student who proposed the challenge acts as an "expert" for the solving team, offering insights on the primary issues identified. If the initial challenge description did not include any SDGs, the team is prompted to link relevant SDGs to it. This alignment is essential as, in later phases, teams will evaluate the impact of their solutions in the context of these SDGs.

Figure 7 presents a sample of work done using the Miro board in phase 1.

Phase 2—At the start of Phase 2, teams have a clear understanding of their challenge and begin analyzing affected stakeholders and comparing similar issues in other countries. First, by examining stakeholders impacted by the challenge in their region, teams gain deeper insights into the problem using stakeholder analysis templates (see Fig. 8). This helps clarify if the identified issue is the actual root cause.

Next, teams compare the same challenge in two or more similar countries, chosen for shared characteristics like size, culture, or development level. For example, teams

Fig. 7 Examples of high-level challenge descriptions on Miro platform

analyzing pollution in the Ganga River compared it to pollution in London's Thames and China's Pearl River. By the end, teams should have a detailed grasp of their challenge.

> If I had an hour to solve a problem, I'd spend 55 min thinking about the problem and 5 min thinking about solutions.
>
> Albert Einstein

Reflecting Einstein's approach to problem-solving, the course emphasizes critical and systems thinking (80%) over solution development (20%), focusing on a thorough understanding of challenges.

Phase 3—In Phase 3, students begin developing solutions for their identified sustainability challenges, progressing in two parallel tracks: mockup development and impact assessment.

Stakeholder Name	**Impact** *How much does the project impact them? (Low, Medium, High)*	**Influence** *How much influence do they have over the project? (Low, Medium, High)*	**What is important to the stakeholder?**	**How could the stakeholder contribute to the project?**	**How could the stakeholder block the project?**	**Strategy for engaging the stakeholder**
Poor People	*High*	*High*	*Sufficient income, Being capable of taking care of basic and social needs*	*Actively searching and trying for available chances*	*Being inactive*	*Advising, informing available chances*
Government	*Medium*	*High*	*Maintaining a better living standard for the people, empowering people economically*	*Implement new rules, reforms, policy, funding, introducing new technologies*	*Unstable political condition, corruption*	*Monthly round-table discussions between government and people representative*
Industries	*High*	*High*	*Providing jobs and opportunities, providing sufficient income for employees*	*CSR (Giving training, giving funds, scholarships), boost local productivity, income redistribution*	*Hire only skilled people*	*Informing and giving them advice on how impactful they are in the project*
Productive Age Workers	*High*	*Medium*	*Work security, financial security, getting a good life standard*	*Sharpen their skill, innovating new things, actively join society*	*Being inactive*	*Inform them the available trainings/jobs and give advice to them door to door*
Children	*High*	*Low*	*Getting education, getting food/nutrition, getting medical cares*	*Actively join class discussion*	*Being inactive*	*Send teachers / volunteers door to door*
Education Institute	*Medium*	*High*	*Sufficient tools for educating students, place to set up the classes*	*Formulate the curricula to capture the required skill set*	*High Tuition Fee*	*Routine discussion with industries to keep update with the needs for industries*
Women	*High*	*Medium*	*Get sufficient aid, medical check, with support and assistance for pregnant women*	*Rising their kids*		*Door to door assistance*

Fig. 8 Example of a stakeholder analysis of one **sustainability c**hallenge

Before creating a mockup, teams identify key features that address root causes of their challenge. For instance, in a problem looking at the Pakistan Flood (due to climate change effect), where delayed information led to loss of life, teams focused on communication as a critical feature. Each feature's potential impact is considered as part of this process.

Teams use various tools for mockup development, including Figma, AdobeXD, MarvelApp, or React. The choice of tools is left open to them.

Impact assessment is conducted using the Sustainability Awareness Framework (SusAF) (Duboc et al., 2020), an open-access, question-based tool available in Zenodo (Betz et al., 2022), that guides teams to evaluate their solution's sustainability impacts. Students summarize main impacts in a SusAD diagram and prioritize features accordingly. Figure 9 presents an example of using SusAF.

The primary learning outcome of Phase 3 is understanding that solution features have varying impacts, which can be intentional, unintentional, positive, or negative. This phase reinforces that all actions carry some level of impact.

Phase 4—A small portion of this course focuses on evaluating how the proposed, impactful solutions could be implemented. While a traditional business canvas helps outline necessary steps for implementation, this course uses the Flourishing Business Canvas to guide business development. This approach provides students with a basic understanding of the economic requirements for their solution, though it remains a minor component of the course overall.

Phase 5—The work dissemination occurred in three ways. First, teams gave a 2-min pitch, using the NABC approach from SRI Institute (Carlson & Wilmot, 2006), to highlight key solution aspects. This pitch was pre-recorded, allowing teams to refine their presentation skills. Second, each team delivered a full 1-h presentation, covering

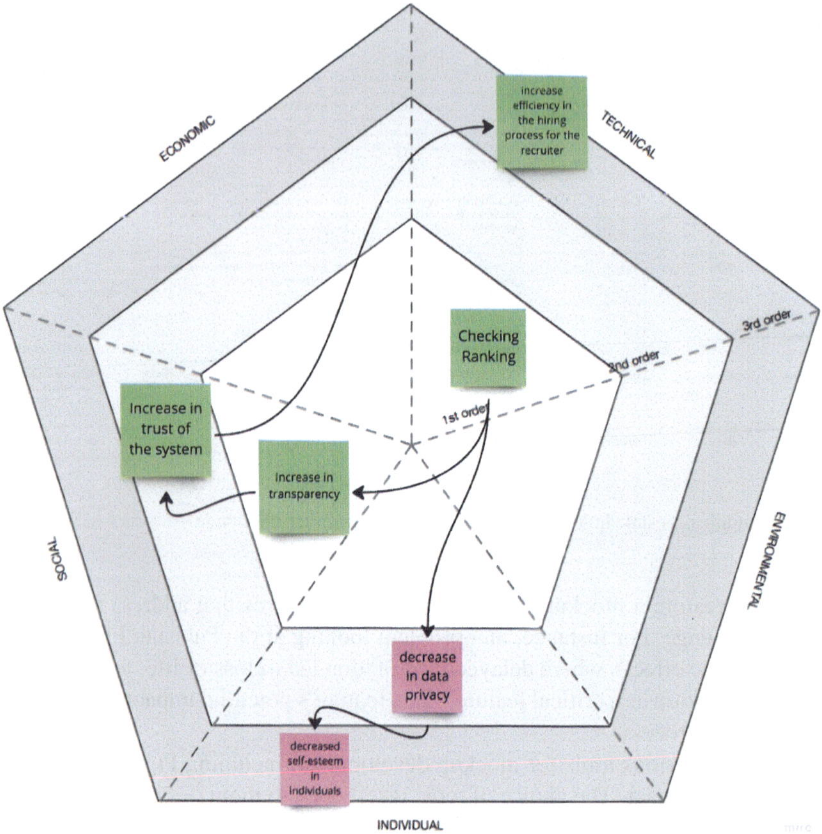

Fig. 9 Example of impact assessment using SusAF—the student team here looks at the impact of their proposed solution to reduce bribery in job selection in Nigeria

all essential elements of their project. Finally, teams submitted their final report as a poster, which emphasizes core insights (see Fig. 10 for a sample). Previously, full reports were used, but posters help students distill their message for greater impact.

At course end, individual students provided feedback on their learning journey via Moodle. Throughout all phases, instructors encouraged and maintained constant engagement to monitor the development of students' sustainability competencies.

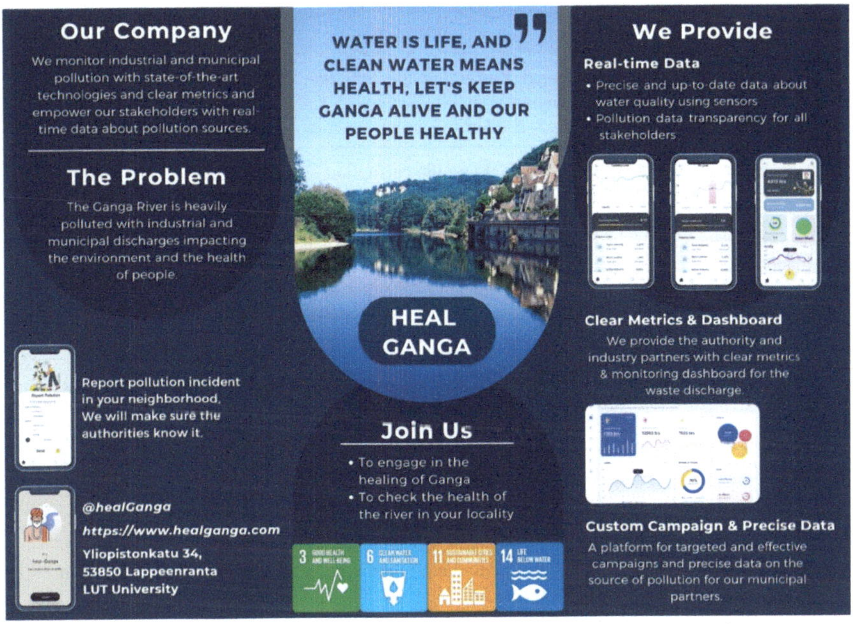

Fig. 10 Example of a poster report from the course. All posters are available online in https://se4gd.lutsoftware.com/programme/swinno/

4 Results

Although this course is running for many implementations, this work reports the two years (2021–2022, 2022–2023) that have had a clear focus on sustainable development and that have been arranged for the students of the Software Engineers for Green Deal programme (SE4GD, 2020).

In 2021–2022, the course had eight groups, and in 2022–2023, five groups. These sustainability challenges are presented in Fig. 11.

A total of 46 students participated (20 in 2022–2023 and 26 in 2021–2022), forming thirteen groups that addressed 16 of the 17 SDGs, excluding SDG 7 (affordable and clean energy). At the start of the course, some students had partial SDG awareness, but none were familiar with all 17 goals. Many were also skeptical about using software to advance SDGs.

By course end, most students reported greater awareness of sustainability issues, SDGs, and competencies, gaining confidence in engaging with sustainability. Some recognized economic potential in their solutions for entrepreneurship. A key insight from using SusAF was identifying both positive and negative impacts, challenging their initial perceptions of purely positive solutions.

Moodle feedback from LUT indicates high engagement and satisfaction. Despite limited free-form input, the question "What was best about this course?" indirectly measured sustainability awareness. In 2021–2022, three of five responses mentioned

2021-2022		
Heal Ganga	Challenge: Preventing the pollution in Indian rivers - Case Ganga *SDGs - 3, 6, 11, 14*	India
Zero corruption Thailand	Challenge: Fighting the corruption in Thailand *SDGs - 16*	Thailand
Waste management in Bangladesh	Challenge: Solving the waste problems in Bangladesh *SDGs - 1, 11, 12, 13, 14*	Bangladesh
Reform your farm	Challenge: Managing the food insecurity in Ethiopia *SDGs - 2*	Ethiopia
My food waste	Challenge: Increasing the awareness of your food waste *SDGs - 12, 13*	Finland
Scholar aid	Challenge: School systems in Indonesia are expensive and lack quality *SDGs - 4*	Indonesia
Plant Go	Challenge: Reducing poverty through improved agriculture SDGs - 1	Nigeria
Affordable housing	Challenge: Managing the housing prices in Indonesia SDGs - 1, 3, 5, 6, 7, 8, 9, 10 , 11, 12, 13, 16, 17	Indonesia
2022-2023		
For the people	Challenge: Extremely high population density in Dhaka due to migration *SDGs - 1, 2, 3, 8, 9, 10, 11, 13, 15*	Bangladesh
AgriAssist	Challenge: Empowering farmers in Ethiopia with knowledge and equipment *SDGs - 1, 2, 3 ,5, 8, 9*	Ethiopia
German e-governance	Challenge: Complex bureaucracy in Germany for small and medium-sized companies *SDGs - 8, 9, 16, 17*	Germany
JobGuard	Challenge: Fighting bribery in Nigeria *SDGs - 1, 4, 8, 10*	Nigeria
FloodFort	Challenge: Managing the effects of floods in Pakistan *SDGs - 9, 11, 13, 16*	Pakistan

Fig. 11 Sustainability challenges and target countries/ regions of different years

"sustainability" or "SusAF," while in 2022–2023, two of eight responses explicitly mentioned "sustainability," with three using descriptive alternatives.

Using insights from the frameworks presented in Fig. 1, we employed the template presented in Fig. 12 as a roadmap for building the sustainability competency in the students participating in the course. Phase 1 covers the elements of eco-literacy and builds awareness among the students. Phase 2, via market analysis, looking at existing working (and failed) solutions, and competitor analysis, covers the elements of awakening. Phase 3, where student teams build their solutions and do impact assessment using SusAF covers the elements of transition. Finally, phases 4 and 5 map onto the action, wherein the students look into aspects of deploying their solutions in an economically viable manner, advocate for the uptake of their solution by end users and create change by solving a sustainability challenge. Students were not exposed to the template in Fig. 12, and yet keywords such as "holistic", "critical thinking", "analytical thinking", "common goal" and "real world problems" are present in their feedback (limited in nature due to the Moodle questionnaire), indicating a convergence of the template intent, and the outcomes of the course.

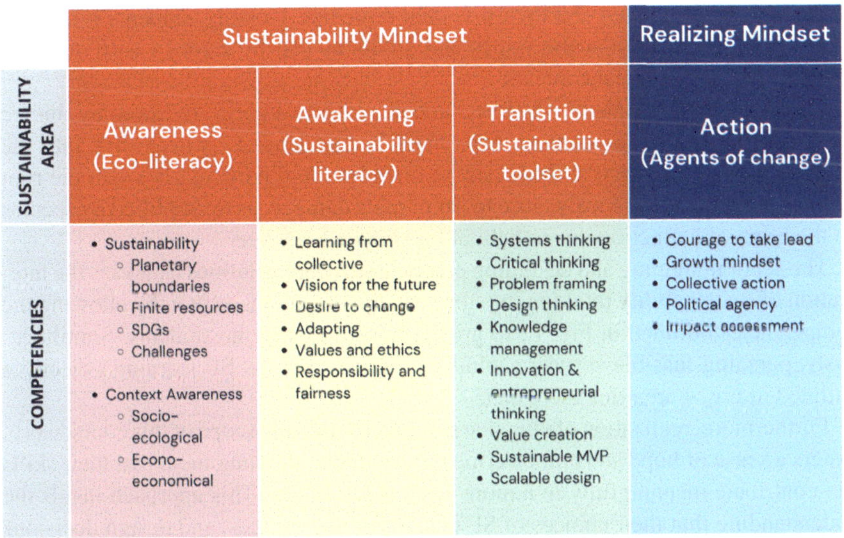

Fig. 12 Template employed for sustainability competency development (First described in the work 'Challenges of integrating sustainability into software engineering education—experiments on building European Software Engineers for the Green Deal programme' by Porras, J., Ashok, T., Shola, O., presented at 1st TSSD @MIT 2023, to be published in World Sustainability Series. North American and European Perspectives on Sustainability in Higher Education)

5 Discussion

The ITiCSE 2019 conference in Aberdeen, UK, featured a working group report by Ian Pollock et al. (2019), emphasizing the need to actively cultivate sustainability competencies in SE students. At LUT University, in the course Software and Application Innovation, we address this imperative by seamlessly integrating sustainability education into core SE curriculum. This approach obviates the necessity for SE students to undertake additional courses solely for sustainability competencies development, although it does require SE educators to receive proper training in delivering sustainability content. Our method revolves around instilling sustainability awareness through an initial exploration of sustainability-related problems, as identified by students within their own countries. Subsequently, we guide these students towards leveraging software technologies to devise potential solutions. Each student identifies problems within their national context and proposes solutions for issues in other nations. This strategy serves a dual purpose: fostering a broader comprehension of sustainability challenges and illustrating the interconnected, interdependent nature of these problems (or at the least the common natures of the underlying core causes of most of the problems) and their potential remedies. Notably, students often find it astonishing how similar the issues and solutions can be across vastly different cultural and geographical contexts.

The paradigm throughout is to foster and increase awareness, followed by sincere deep-thinking toward a solution. Awareness in turn leads to a deep dive into the underlying causes, which at times can be disheartening, particularly when the root causes appear to be perilously close to irreparable damage, as exemplified by species on the brink of extinction due to habitat loss beyond recovery.

The act of imagining and designing practical software solutions catalyses the integration of sustainability thinking into the participants' mindset, thereby allowing the competencies outlined in Fig. 12 to gradually take root in the students. Simultaneously, pursuing feasible software solutions ensures that the SE learning objectives outlined in Fig. 4 are effectively met.

Furthermore, realization of effective solutions within the scope of this coursework, fosters a sense of hope and empowerment in students, showing them that their skills can contribute meaningfully to a more sustainable future. This approach instils the understanding that their choices of SE development practices, and design decisions in SE can lead to direct, positive impacts.

6 Conclusion

This paper presents findings from a case study on incorporating sustainability education into a graduate-level SE course. The primary contribution is using existing sustainability competency framework(s) (Fig. 1) to integrate sustainability competencies into SE coursework, teaching the course, and tracking students' development in sustainability competencies.

Research in sustainability education suggests that sustainability competencies should be foundational to SE competencies, with both sets integrated for meaningful transformation (Mishra & Mishra, 2020; Redman & Wiek, 2021). Our experience supports this.

Addressing the research questions from the introduction.

RQ1—The course should start with teaching sustainability thinking, unencumbered by constraints of sustainability challenges or software (as a technology). Students start with the SDGs, planetary boundaries, and the current state of things in the climate context, placing sustainability at the forefront. With this grounding in place, they can view software as a technical instrument for taking tangible actions for sustainability.

RQ2—In this course, the students define sustainability challenges from their own native countries but do not work on the same challenges. As a rule, they work on challenges present in a different country. This is an effective mechanism to induce awareness and perceptions around global sustainability issues. Moreover, open discussion in the classroom and constant presentations of their work nudge them towards effective cooperation, as well as, in many instances, realizing that the sustainability problems are replicating in more than one locale. They, in turn, can look into accumulated knowledge, learnings, and understandings from cross-locale experiences while working on their locale-specific software solutions for the sustainability challenge under consideration.

Based on realisation of the research questions, we would like to outline the following steps enhance sustainability integration into SE courses, to foster sustainability—awareness to action (Fig. 12) —paradigm. Steps 1–3 are pertinent to the immediate educational context, while step 4 is to be considered as a long-term future orientation to evolve the overall SE ecosystem toward sustainability.

1. Strengthen Sustainability Thinking as a Core Foundation
   a. Build an Introductory Sustainability Module: Begin each course with an intensive module on sustainability thinking, covering SDGs, planetary boundaries, and climate realities.
   b. Encourage Critical Reflection: Implement structured reflections and discussions that allow students to internalize sustainability beyond software engineering constraints.
   c. Interdisciplinary Integration: Invite guest-lecturers/experts, if available, from environmental science, policy, and ethics to broaden students' perspectives.
2. Embedding Sustainability in Software Engineering Practices

a. View Software as a Sustainability Enabler: Teach students to see software as a tool for sustainable solutions rather than focusing only on software engineering principles.
b. Introduce Lifecycle Analysis: Train students in assessing the environmental, social, and economic impact of software throughout its lifecycle.
c. Case-Based Learning: Incorporate case studies of successful and failed sustainable software projects to provide real-world context.

3. Strengthening Global Perspective Through Cross-Context Challenge Solving

a. Structured Cross-Locale Assignments: Require students to work on sustainability challenges outside their native regions, reinforcing global awareness.
b. Expand Knowledge Sharing Mechanisms: Establish formalized cross-team feedback sessions, where students compare challenges and share accumulated insights from different regions.
c. Leverage Open-Source Collaboration: Encourage contributions to open-source sustainability projects to enhance real-world impact beyond the classroom.

4. Encouraging Long-Term Impact Beyond the Course

a. Foster Continuous Engagement: Develop an alumni network where past students mentor new cohorts and share learnings.
b. Encourage Research and Innovation: Provide pathways for students to further develop their sustainability solutions into research projects or startups.
c. Build Industry Partnerships: Collaborate with organizations that focus on sustainability in software engineering to provide students with real-world application opportunities.

Our experience indicates that integrating sustainability education into SE coursework can effectively develop students' sustainability competencies, advancing sustainability education within the SE curriculum. The implementation described here is inherently project-work based and therefore might not be directly translated to all the courses in SE curriculum. For them, alternative approaches may be needed to incorporate sustainability education effectively. For theory-heavy courses, sustainability concepts can be introduced through case studies, discussions, and analytical exercises. In programming-focused courses, sustainability-aware coding practices, energy-efficient algorithms, and software lifecycle considerations can be integrated into assignments. Further research is needed to explore and evaluate these implementation strategies. Additionally, interdisciplinary collaborations with environmental science, ethics, and business studies can provide broader perspectives. Embedding sustainability as a recurring theme across multiple courses, rather than confining it to a single module, can ensure a more comprehensive and cohesive learning experience for software engineering students.

Acknowledgements The research presented here has been conducted within the scope of the European Union-funded Erasmus Mundus Master's program "Educating software engineers with a sustainable mindset to improve the well-being of people through software solutions" (619839).

References

ACM/IEEE. (2014). *Software engineering 2014 curriculum guidelines for undergraduate degree programs in software engineering*. Retrieved October 23, 2024, from https://www.acm.org/binaries/content/assets/education/se2014.pdf

Allen, M., Dube, O. P., Solecki, W., Aragón-Durand, F., Cramer, W., Humphreys, S., & Kainuma, M. (2018). *Special report: Global warming of 1.5 C*. Intergovernmental Panel on Climate Change (IPCC).

Andrae, A. S. (2021). Internet's handprint. *Engineering and Applied Science Letters, 4*(1), 80–97.

Baartman, L. K., Bastiaens, T. J., Kirschner, P. A., & Van der Vleuten, C. P. (2007). Evaluating assessment quality in competence-based education: A qualitative comparison of two frameworks. *Educational Research Review, 2*(2), 114–129.

Bai, C., Dallasega, P., Orzes, G., & Sarkis, J. (2020). Industry 4.0 technologies assessment: A sustainability perspective. *International Journal of Production Economics, 229*, 107776.

Becker, C., Chitchyan, R., Duboc, L., Easterbrook, S., Mahaux, M., Penzenstadler, B., Rodriguez-Navas, G., Salinesi, C., Seyff, N., Venters, C., & Calero, C. (2014). The Karlskrona manifesto for sustainability design. arXiv preprint arXiv:1410.6968

Betz, S., Duboc, L., Penzenstadler, B., Porras, J., Chitchyan, R., Seyff, N., Venters, C. C., & Brooks, I. (2022). SusAF Workbook 6.0 (Version 3). Zenodo. https://doi.org/10.5281/zenodo.7342575

Brundiers, K., Barth, M., Cebrián, G., Cohen, M., Diaz, L., Doucette-Remington, S., ... & Zint, M. (2021). Key competencies in sustainability in higher education—Toward an agreed-upon reference framework. *Sustainability Science, 16*, 13–29.

Button, M., & Cross, C. (2017). *Cyber frauds, scams and their victims*. Taylor & Francis.

Calero, C., Moraga, M. Á., & García, F. (2022). Software, sustainability, and UN sustainable development goals. *IT Professional, 24*(1), 41–48.

Carlson, C. R., & Wilmot, W. W. (2006). Innovation: The five disciplines for creating what customers want. Currency.

Cebrian, G., & Junyent, M. (2015). Competencies in education for sustainable development: Exploring the student teachers' views. *Sustainability, 7*(3), 2768–2786.

Christen, M. (2010, May). A theory of the good for a conception of sustainability. In *The Sixteenth Annual International Sustainable Development Research Conference. Conference Proceedings*, Hong Kong.

Cook, J., Oreskes, N., Doran, P. T., Anderegg, W. R., Verheggen, B., Maibach, E. W., ... & Rice, K. (2016). Consensus on consensus: a synthesis of consensus estimates on human-caused global warming. *Environmental Research Letters, 11*(4), 048002.

Demirci, K., Akgönül, M., & Akpinar, A. (2015). Relationship of smartphone use severity with sleep quality, depression, and anxiety in university students. *Journal of Behavioral Addictions, 4*(2), 85–92.

Duboc, L., Penzenstadler, B., Porras, J., Akinli Kocak, S., Betz, S., Chitchyan, R., ... & Venters, C. C. (2020). Requirements engineering for sustainability: An awareness framework for designing software systems for a better tomorrow. *Requirements Engineering, 25*, 469–492.

Elhacham, E., Ben-Uri, L., Grozovski, J., Bar-On, Y. M., & Milo, R. (2020). Global human-made mass exceeds all living biomass. *Nature, 588*(7838), 442–444.

Elkington, J., & Rowlands, I. H. (1999). Cannibals with forks: The triple bottom line of 21st century business. *Alternatives Journal, 25*(4), 42.

Force, C. T. (2020, November 2020). Computing curricula 2020: Paradigms for global computing education.

Green Software Foundation. (2021). *Green software engineering principles*. Retrieved February 26, 2025, from https://greensoftware.foundation

Grosseck, G., Țîru, L. G., & Bran, R. A. (2019). Education for sustainable development: Evolution and perspectives: A bibliometric review of research, 1992–2018. *Sustainability, 11*(21), 6136.

Guia, B. (2020). *Sustainability competences. A systematic literature review*. Publications Office of the European Union. https://doi.org/10.2760/200956

Hawi, N. S., & Samaha, M. (2017). The relations among social media addiction, self-esteem, and life satisfaction in university students. *Social Science Computer Review, 35*(5), 576–586.

Pollock, I., Alshaigy, B., Bradley, A., Krogstie, B. R., Kumar, V., Ott, L., Peters, A.-K., Riedesel, C., & Wallace, C. (2019). 5 degrees of separation: Computer science education in the age of the anthropocene. In *Proceedings of the Working Group Reports on Innovation and Technology in Computer Science Education* (pp. 1–25).

Lin, B., & Huang, C. (2023). Nonlinear relationship between digitization and energy efficiency: Evidence from transnational panel data. *Energy, 276*, Article 127601.

Mansell, R. (2002). From digital divides to digital entitlements in knowledge societies. *Current Sociology, 50*(3), 407–426.

Masson-Delmotte, V., Zhai, P., Pörtner, H. O., Roberts, D., Skea, J., & Shukla, P. R. (2022). *Global warming of 1.5 C: IPCC special report on impacts of global warming of 1.5 C above pre-industrial levels in context of strengthening response to climate change, sustainable development, and efforts to eradicate poverty*. Cambridge University Press.

Mishra, D., & Mishra, A. (2020). Sustainability inclusion in informatics curriculum development. *Sustainability, 12*(14), 5769.

Naumann, S., Dick, M., Kern, E., & Johann, T. (2011). The GREENSOFT model: A reference model for green and sustainable software and its engineering. *Sustainable Computing: Informatics and Systems, 1*(4), 294–304.

Ojala, T., & Oksanen, P. (2021). Climate and environmental strategy for the ICT sector.

Oreskes, N. (2004). The scientific consensus on climate change. *Science, 306*(5702), 1686–1686.

Palacin-Silva, M. V., Seffah, A., & Porras, J. (2018). Infusing sustainability into software engineering education: Lessons learned from capstone projects. *Journal of Cleaner Production, 172*, 4338–4347.

Penzenstadler, B., Bauer, V., Calero, C., & Franch, X. (2012). Sustainability in software engineering: A systematic literature review.

Penzenstadler, B., & Femmer, H. (2013, March). A generic model for sustainability with process- and product-specific instances. In *Proceedings of the 2013 Workshop on Green in/by Software Engineering* (pp. 3–8).

Penzenstadler, B., Raturi, A., Richardson, D., Calero, C., Femmer, H., & Franch, X. (2014, May). Systematic mapping study on software engineering for sustainability (SE4S). In *Proceedings of the 18th International Conference on Evaluation and Assessment in Software Engineering* (pp. 1–14).

Redman, A., & Wiek, A. (2021, November). Competencies for advancing transformations towards sustainability. In *Frontiers in education* (Vol. 6, p. 785163). Frontiers Media SA.

Rieckmann, M. (2012). Future-oriented higher education: Which key competencies should be fostered through university teaching and learning? *Futures, 44*(2), 127–135.

United Nations Educational, Scientific and Cultural Organization (UNESCO). (2007). *Education for Sustainable Development*. Education for Sustainable Development Unit.

SE4GD. (2020). *Educating software engineers with a sustainability mindset to improve the well-being of people through software solutions*. Retrieved October 28, 2024, from https://erasmus-plus.ec.europa.eu/projects/search/details/619839-EPP-1-2020-1-FI-EPPKA1-JMD-MOB

Selwyn, N. (2012). *Education in a digital world: Global perspectives on technology and education*. Routledge.

Tainter, J. (2003). A framework for sustainability. *World Futures, 59*(3–4), 213–223. https://doi.org/10.1080/02604020310132

Tainter, J. A. (2006). Social complexity and sustainability. *Ecological Complexity, 3*(2), 91–103. https://doi.org/10.1016/j.ecocom.2005.07.004

Tambini, D. (2018). Social media power and election legitimacy.

United Nations (n.d.). *The 17 goals*. Department of Economic and Social Affairs, Sustainable Development. Retrieved October 23, 2024, from https://sdgs.un.org/goals

United Nations Office for the Coordination of Humanitarian Affairs (OCHA). (2023). *2022: A year when disasters compounded and cascaded*. Retrieved October 28, 2024, from https://www.unescap.org/blog/2022-year-when-disasters-compounded-and-cascaded

United Nations. (2015). *Transforming our world: The 2030 Agenda for Sustainable Development*. Department of Economic and Social Affairs, Sustainable Development. Retrieved October 30, 2024, from https://sdgs.un.org/2030agenda

Venters, C. C., Capilla, R., Betz, S., Penzenstadler, B., Crick, T., Crouch, S., … & Carrillo, C. (2018). Software sustainability: Research and practice from a software architecture viewpoint. *Journal of Systems and Software, 138*, 174–188.

Warschauer, M. (2004). *Technology and social inclusion: Rethinking the digital divide*. MIT press.

WCED, S. W. S. (1987). World commission on environment and development. *Our Common Future, 17*(1), 1–91.

Wiek, A., Withycombe, L., & Redman, C. L. (2011). Key competencies in sustainability: A reference framework for academic program development. *Sustainability Science, 6*, 203–218.

Wiek, A., Bernstein, M. J., Foley, R. W., Cohen, M., Forrest, N., Kuzdas, C., … & Keeler, L. W. (2015). Operationalising competencies in higher education for sustainable development. In *Routledge handbook of higher education for sustainable development* (pp. 241–260). Routledge.

Transformative Learning for Sustainability: Bridging Inner Development Goals (IDG) and Sustainable Development Goals (SDG)

Carry Luise Zimmermann

Abstract This article presents a compelling case for a transformative shift in higher education, advocating for the integration of Inner Development Goals (IDGs) to support the achievement of the Sustainable Development Goals (SDGs). As universities worldwide enhance their focus on sustainability education, there is an emergent need for adopting innovative pedagogical approaches that extend beyond conventional academic teaching. This shift involves a holistic educational model that incorporates personal development. By embedding IDGs within academic curriculums, educational institutions can guide students through a profound exploration of personal values, encouraging the development of a sustainable mindset essential for addressing global challenges. The article underscores that cognitive skills, while essential, are insufficient alone for fostering sustainable development; instead, a balanced combination of inner growth and academic learning is required. Incorporating IDGs into educational frameworks equips students with the necessary tools to act with integrity and authenticity, crucial competencies for leading sustainable change. By aligning personal development goals with global sustainability objectives, universities can nurture a generation of responsible, conscious change-makers. This holistic approach not only enhances individual student growth but also aligns with global efforts to achieve the SDGs, illustrating the intertwined nature of individual transformation and global sustainability efforts. The article calls for educators and policymakers to embrace this integrative approach, recognizing its potential to shape a generation committed to sustainable development and societal well-being.

Keywords Transformative learning · Inner development goals · Personal development · Inner growth · Sustainable development

C. L. Zimmermann (✉)
University of Bremen, Sustainable Management, Bremen, Germany
e-mail: c.zimmermann@uni-bremen.de

W. Leal Filho et al. (eds.), *Introducing the UN SDGs in Higher Education Institutions*, World Sustainability Series, https://doi.org/10.1007/978-3-032-23017-1_6

1 Introduction—The Need for Transformative Change in Higher Education

The United Nations introduced the Sustainable Development Goals (SDGs) in 2015 as a comprehensive framework aimed at addressing global challenges such as poverty, environmental protection, and the promotion of peace and prosperity, with a target to achieve these objectives by 2030 (United Nations, 2024). Seventeen goals have been identified and established, each addressing a distinct area of global importance. These include: ending poverty, eradicating hunger, promoting good health and well-being, ensuring quality education, achieving gender equality, providing clean water and sanitation, advancing affordable and clean energy, fostering decent work and economic growth, supporting industry, innovation, and infrastructure, reducing inequalities, creating sustainable cities and communities, encouraging responsible consumption and production, taking urgent climate action, protecting life below water, safeguarding life on land, strengthening peace, justice, and robust institutions, and building partnerships for these goals—all with the overarching commitment to leave no one behind (United Nations, 2024). However, progress towards achieving the goals is being slow and reports highlight that significant efforts are still required. The goals remain far from being fully realised, and in some areas, challenges are intensifying rather than improving. For instance, the 2019 SDG report highlights that despite progress in certain areas, "the natural environment is deteriorating at an alarming rate: sea levels are rising; ocean acidification is accelerating; the past four years have been the warmest on record; one million plant and animal species are at risk of extinction; and land degradation continues unchecked" (United Nations, 2019, p. 2). Furthermore, the global extreme poverty rate increased in 2020 for the first time in over two decades, with an estimated 119 to 124 million individuals falling back into extreme poverty (United Nations, 2021). Similarly, global conflict levels have reached their highest point since the establishment of the United Nations. Reports indicate that approximately 2 billion people now reside in conflict-affected countries, while refugee numbers hit a record high in 2021, and forced displacement continues to rise (United Nations, 2022). Progress in reducing carbon emissions has also stalled or reversed, underscoring the urgency of transitioning to greener and more inclusive economies (United Nations, 2021). The SDG report for 2023 paints a stark picture, stating, "[…] we are leaving more than half of the world behind where progress on more than 50 per cent of the SDGs targets are weak and insufficient and 30% have either stalled or gone into reverse. These include key targets on poverty, hunger, and climate. Unless we act now, the 2030 Agenda could become an epitaph for a world that might have been" (United Nations, 2023, p. 2).

Higher education, as a key multiplier, plays a vital role in achieving the SDGs by educating the leaders, actors, and changemakers of tomorrow. The SDGs advocate that higher education institutions evolve their teaching and learning approaches to better prepare students for addressing the challenges of the twenty-first century and beyond (Ahmad et al., 2023). However, research indicates that insufficient progress has been made in integrating the SDGs into curricula (Leal Filho et al., 2023). Another challenge is that a significant portion of existing teaching formats primarily focus on cognitive and rational learning, thus emphasizing the acquisition of theoretical knowledge rather than fostering a holistic understanding for practical engagement with sustainability issues (Lozano et al., 2013; Nölting et al., 2018; Volk, 2020) that results in a knowledge-action gap (Columbo et al., 2023). Therefore, a fundamental transformation in human values, understanding, and leadership capacities is needed (Ankrah et al., 2023). Furthermore, education should be a catalyst for transformation thus enabling individuals to make thoughtful decisions for both personal and collective actions to enable positive societal change towards safeguarding our planet (Laurie et al., 2016).

2 Implementing the IDG in Higher Education

As stated above, there is an emergent need for adopting innovative pedagogical approaches that extend beyond conventional academic teaching. This shift should include a holistic educational model that incorporates personal development with transformative thinking to help bridge the gap between knowledge and action taking.

One possible approach to address this issue includes the "Inner Development Goals" (IDG)—a global non-profit initiative launched in Stockholm in 2020. This initiative focuses on the inner capabilities needed to achieve the 17 United Nations SDGs. From this movement, the IDG Framework emerged, providing a framework for individual, organizational, and societal change towards a culture of sustainability that resulted in two scientific studies where more than 1000 participants were surveyed (IDG, 2024a, 2024b).

In the first survey, a total of 861 individuals provided complete responses (Growth that matters, 2021). The majority of respondents were women, accounting for approximately 62% of the sample (Growth that matters, 2021). Most participants were between the ages of 30 and 69 (Growth that matters, 2021). Around half of the respondents were Swedish, with an additional 30% coming from other European countries (Growth that matters, 2021). Approximately 12% were from the USA, Canada, Australia, or New Zealand, while only 6% (50 individuals) came from Latin America, Africa, or developing countries in Asia (Growth that matters, 2021). Very few responses were received from Russia, China, and Japan (Growth that matters, 2021). The result was a list of 22 skills and qualities that are needed to successfully build a better world (Growth that matters, 2021).

In the second survey, respondents were asked to review the list of 22 skills and qualities and rank the 10 they considered most important (Growth that matters, 2021). By the time the survey closed, 813 valid responses had been collected, with a participant profile similar to that of the first survey (Growth that matters, 2021). A 23rd item, "sense-making", which had been suggested by several individuals and mentioned frequently in the survey responses, was also added to the list (Growth that matters, 2021). As of 2024, the (continuously evolving) IDG framework consists of five dimensions incorporating the 23 skills summarised above. The skills comprise the abilities and qualities needed to sustainably shape an increasingly complex world through capable action and creativity.

The core assumption of the initiative is that achieving the SDG is only possible if inner development is also addressed. This approach builds on the perspective that inner and outer dimensions, along with individual, collective, and systemic phenomena, are inherently interrelated and co-constructed (Ives et al., 2023). Inner development thus becomes the foundation of sustainable development (IDG, 2024a, 2024b). Wamsler et al. (2024) for example states that more focus is needed on the "[…] connection to ourselves, others and nature." (Wamsler et al., 2024, p. 1), emphasising that a transformation of the inner dimensions, encompassing "[…] individual and collective mindsets, values, beliefs, worldviews and associated cognitive, emotional and relational abilities and capacities […]" (Ives et al., 2023, p. 2778) is needed. Holst et al. (2024) also argues that mental structures need to be considered.

The IDGs are recognized for their novel approach and the importance of personal growth in achieving collective sustainability and systemic change (Ankrah et al., 2023). Unlike traditional goals centred on material progress, the IDGs are also noted for their focus on holistic development, addressing emotional, psychological, social, and spiritual well-being to help navigate the complex challenges of the modern era (Ankrah et al., 2023). The 23 skills categorized within the five dimensions of the IDG Framework are: *1. Being—Relationship to Self, 2. Thinking—Cognitive Skills, 3. Relating—Caring for Others and the World, 4. Collaborating—Social Skills* and *5. Acting—Enabling Change*. The following Fig. 1 gives an overview of the five dimensions and their related skills.

While higher education has increasingly emphasised critical and complex thinking alongside social skills like collaboration and experiential learning through approaches such as service learning, there remains a critical gap in including the

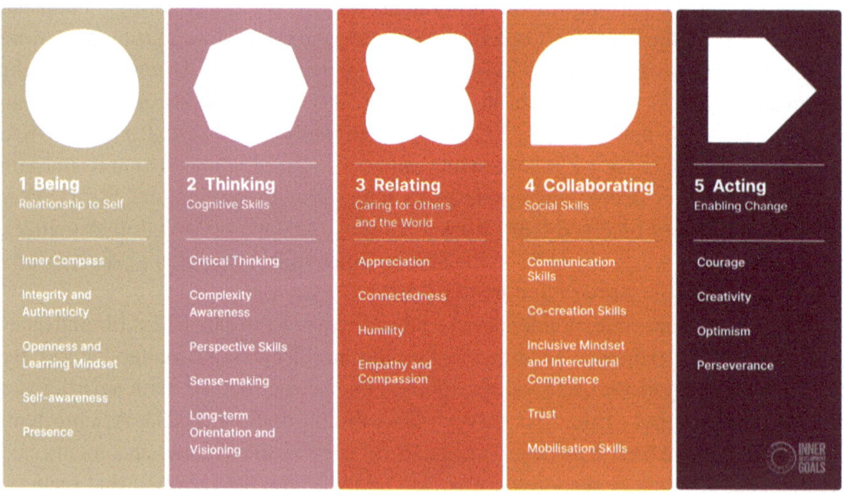

Fig. 1 The five dimensions of the IDG and their related Skill (IDG, 2024a, 2024b)

socio-emotional dimension of learning (Ahmad et al., 2023), fostering self-reflection (Zimmermann, 2025; Zimmermann & Scheering, 2026), and exploring personal values (Zimmermann, 2025). The *Being–Relationship to Self* in the first dimension emphasises the relationship to oneself and remains the least implemented aspect in higher education, limiting the development of inner capacities crucial for aligning personal values with sustainable behaviours. Therefore *Being–Relationship to Self* takes centre stage in this discussion paper, with particular emphasis on cultivating values and fostering a value-based inner compass, and by exploring how this foundational aspect can be implemented in higher education through directly linking it to SDG 12, *Responsible Consumption and Production.* Integrating these concepts, aims to equip students not only with theoretical knowledge but also with the inner tools necessary to make sustainable choices and take meaningful actions. Expanding educational practices to integrate with this dimension, essential for bridging the gap between knowledge and action, is the central focus of this paper.

3 The Intertwined Nature of Individual Transformation and Sustainability

The intertwined nature of individual transformation and sustainability becomes evident after examining the relationship between the IDGs and the SDGs. Individual transformation, as outlined in the IDG framework, focuses on developing inner capacities such as self-awareness, emotional resilience, and a value-based inner compass, which are essential for fostering sustainable behaviours. This aligns with Mezirow's transformative learning theory, where altering one's frame of reference creates the foundation for rethinking actions and engaging with the world in more inclusive and responsible ways (Mezirow, 1997). By linking the IDGs to the SDGs, the process of inner growth and reflection directly contributes to addressing global challenges, underscoring the critical role of personal transformation in driving systemic change for sustainability. The following section provides a brief introduction to the theory and approach of transformative learning.

Transformative Learning

Mezirow, the pioneer of transformative learning theory, describes it as the process of altering one's frame of reference. As adults, we construct a cohesive set of experiences—consisting of associations, concepts, values, emotions, and conditioned responses that collectively shape our worldview (Mezirow, 1997). These frames of reference act as the foundational structures of assumptions, influencing how we interpret and make sense of our experiences (Mezirow, 1997). They selectively shape and limit our expectations, perceptions, thoughts, and emotions, ultimately driving our actions (Mezirow, 1997). Once established, these frames often lead us to transition automatically from one mental or behavioural activity to another without questioning (Mezirow, 1997). Ideas that do not align with these established frames are frequently dismissed as irrelevant, nonsensical, strange, or incorrect (Mezirow, 1997). However, when the conditions are conducive, transformative learners can develop a frame of reference that is more inclusive, nuanced, self-aware, and integrative of their experiences (Mezirow, 1997).

An important aspect of this process is self-reflection, which can be encouraged through reflective tasks. Reflection requires the courage to explore and, at times, question both one's own and other's motives and actions—not as a means of criticism, but as a way to foster wiser and more appropriate future decisions and actions (Sandahl & von Knorring, 2023). Through dialogue and reflection, unconscious response patterns can be reshaped into more constructive coping behaviours, emphasising the transformative potential of reflective practice (Sandahl & von Knorring, 2023).

Drawing on the work of Mezirow (2000), Nohl (2015), and Ball (1999), Grund et al. (2024) outline five key phases of transformative learning. The first phase, called the *novel experience*, serves as the starting point of the transformative learning process. This phase is characterized by a new experience that can challenge learners' existing assumptions, often provoking feelings of discomfort or disorientation—a phenomenon Mezirow describes as a *disorienting dilemma*. Alternatively, the experience may initially appear insignificant but can eventually prompt deeper exploration and engagement (Grund et al., 2024). The second phase, the *reflection phase*, involves learners critically examining their pre-existing assumptions alongside the new experiences they have encountered. This reflection may occur consciously or unconsciously, encouraging learners to reassess their perspectives (Grund et al., 2024). The third phase, known as the *social exchange phase*, highlights the importance of interaction in learning. During this stage, learners share ideas and emotions with others, using collective experiences and feedback to create a sense of security and enhance their understanding (Grund et al., 2024). The fourth phase, *the shift of action*, represents a crucial step in transformative learning. Here, learners experiment with and adopt new roles, behaviours, skills, and competencies, actively integrating what they have learned into practice (Grund et al., 2024). The final phase, the *shift of meaning*, marks the culmination of the transformative process. In this phase, learners experience a profound change in how they view and understand themselves and the world, resulting in a significant transformation from their prior state (Grund et al., 2024).

4 Combining Inner Growth and Academic Learning

To gather insights and gain initial experience in linking the SDGs with the IDGs in higher education, an exploratory study was designed and conducted at the University of Bremen during the summer term of 2024. Established in the early 1970s, the University of Bremen has quickly become known for its interdisciplinary approach, cutting-edge research, and strong emphasis on international collaboration. Key strengths include areas like natural sciences, engineering, environmental studies, and social sciences, with notable research in marine sciences and digital technologies. In the summer term 2024, the University of Bremen enrolled 18,335 students, including approximately 1500 in its Bachelor "Business Administration and Business Economics" program. The following provides a description of a seminar linked to the study.

4.1 Methodology

The seminar was conducted as a guest lecture embedded within the Sustainable Management course of the Bachelor's programs in Business Administration and Economics at the University of Bremen during the summer term 2024. Targeting a large cohort of students, i.e. 196 students, the seminar focused on SDG 12—Responsible Consumption and Production—and the first dimension of the Inner Development Goals (IDGs), namely Being–Relationship to Self. The approach was rooted in transformative learning theory, which posits that lasting change requires both introspection and social interaction. This approach was chosen to bridge the gap between academic sustainability knowledge and personal behavioral transformation. It is an exploratory, qualitative, and action-oriented design that uses narrative elements to capture the nuances of personal transformation in sustainability education.

Given the study's focus on exploring individuals' perceptions and experiences, a qualitative approach was deemed most appropriate. This methodological choice aligns with the constructivist paradigm, which emphasises understanding of how individuals interpret and give meaning to their social realities whilst qualitative research seeks to capture and understand an individuals' perspectives by emphasizing the narratives and verbal expressions they share with researchers (Liamputtong, 2023).

The Study Design: A seminar was developed based on the five stages of transformative learning. It began with a novel experience, followed by a strong emphasis on reflective exercises and social exchange, allowing participants to engage deeply with their learning process. These five stages are discussed in greater detail later and are visually represented in the accompanying figure.

Data Collection and Analysis: While students had opportunities for discussion and idea exchange during the seminar, they were also required to submit written reflections. These 196 reflective papers were later analysed, focusing on key words and phrases that indicated a shift in meaning or behaviour, suggesting a transformation in perspective and decision-making. Expressions such as *"Today I learned," "I realized," "Making more conscious decisions,"* and *"Reflecting"* were particularly examined as indicators of this change.

4.2 Structure of the Seminar

Phase 1—Novel Experience

As described above, the transformative learning process begins with a novel experience. Since the seminar was part of a course attended by 196 students, primarily conducted in a lecture format focusing on the acquisition and deepening of knowledge, the students were given an opportunity to engage in a new experience through the assigned tasks. Before the seminar, students were asked to reflect individually on what sustainable consumption and sustainable production meant to them. Building on this reflection, they were instructed to self-assess how sustainable they considered themselves to be as consumers on a scale from 0 (not at all) to 10 (fully sustainable), along with a justification for their ratings. In addition, students were asked to select an item of clothing or an object they had purchased themselves that held personal significance. They were then asked to bring the item (or a photo in lieu of) to the seminar. The third preparatory task involved reading a scientific paper on the topic of "brand consumption" and summarizing its main findings. During the seminar, students first received a brief introduction to the IDGs and a short presentation on the topic of "brand consumption" to provide thematic context. The "brought in item" and "brand consumption" topics were then integrated to introduce the concept of the inner (value-based) compass, which is a key aspect of the first dimension of the IDGs. This was followed by the active phase of the seminar.

Phase 2—Reflection Phase

In this phase, students were asked to take out the photo of their selected item or piece of clothing and enter a reflective mode. To support their reflection, they were provided with the following questions: *1. What does this item mean to me? (what emotions do I feel when I look at this item (such as joy or pride) and what motivated the purchase?) 2. What brand is the item from? (What values do the brand represent to me personally, do I find these appealing?)* Students were then tasked with connecting the material presented during the seminar to their personal reflections by answering the following question: 3. *How would you describe and explain your inner compass for consumption behaviour? (Sketch your compass and describe and explain it.)* Building on this and the knowledge provided earlier in the seminar, the next exercise included the following reflective questions: *4. What have I learned about consumer behaviour? 5. What have I learned about myself as a person (e.g. about my views, values, or beliefs)? and how has my awareness changed (e.g. have I gained some clarity including about my self-image)?*

Phase 3—Social Exchange Phase

As described earlier, the social exchange phase encourages learners to share their ideas and emotions to foster collective experiences and enhance understanding to form insights. This phase integrated into the reflection phase, allows students to actively engage in discussions and share their emotions, values, and insights throughout the exercises.

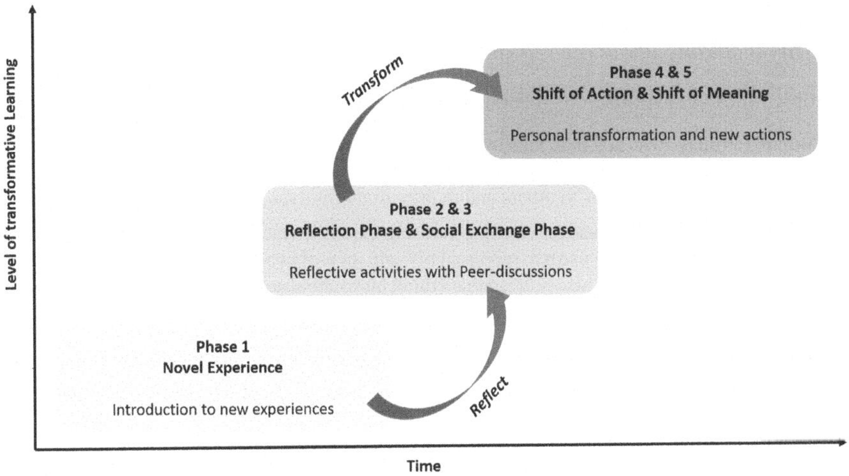

Fig. 2 Transformative learning phases (own illustration, based on Grund et al. (2024)

Phase 4—Shift of Action

Next, students were asked to answer the following transfer questions: *6. How might the knowledge-based insights about myself help shape my life? 7. How might/could my shopping behaviour change?* Thee answers to these two questions examines the potential application of new consumption and behavioural patterns.

Phase 5—Shift of Meaning

As described earlier, this phase is characterized by a fundamental change in how learners perceive and understand themselves and their role in today's world compared to how they felt about this previously (questions 6 and 7). Students submitted their responses as part of a written assignment.

The following graphic provides an overview of the seminar's character and structure (Fig. 2).

5 Results

The key objective of the seminar was to create a connection between SDG 12, Responsible Consumption and Production, and the first dimension of the IDGs, Being–Relationship to Self. This was achieved by initiating a reflection process among students to critically examine their consumption behaviors and develop greater awareness of their values related to consumption (Reflection Phase). To facilitate this introspection, the following questions were employed, as previously outlined: *4. What new content have I learned today? 5. What have I learned about myself as a person today (e.g., about my views, values, or beliefs); has my awareness changed (e.g. an increase in clarity including about my self-image?).*

Example student responses to these questions:

> **Student 1**: Today I learned that consumption is a significant part of identity. The goods we consume influence how we perceive and feel about ourselves. People also use consumption goods to express themselves to others and to present themselves.
>
> **Student 2:** I realized that I use the clothing I wear to express myself and place importance on wearing appropriate attire. I use it to represent my self-image but also to enhance feelings like self-confidence. However, I also noticed that I think more about aesthetics than sustainability, which has made me reflect.

Based on prior reflection through transfer questions and answers, another seminar goal was to examine a potential *Shift of Action* and *Shift of Meaning* by using the following questions.

How can knowledge-based insights help shape my life?

How might/could my shopping behaviour change?

The following are example student answers (out of a total of 196):

> ***Student 1***: The knowledge I gained today will help me make more conscious decisions about the products I buy in the future, considering their impact on the environment and people, as well as reflecting on how often I purchase more than I need. I also hope that in my future career, I can highlight the importance of sustainability to make a positive impact on the environment. Additionally, I want to share my knowledge with my friends and family, not only to shift the direction of my life toward something positive but also to give them the opportunity to do the same.
>
> ***Student 2***: In the future, I want to identify less with my clothing. I want to feel confident and at ease, regardless of my outfit or possessions. My confidence should be rooted in the values I represent. Moreover, knowing that I could help the environment through more sustainable behaviour would make me feel better and prouder about myself.
>
> ***Student 3***: The newly acquired knowledge will also help me by encouraging me to reflect more on the motivations behind my shopping behaviour in the future. I plan to more frequently examine whether my behaviour truly aligns with my personal values. Furthermore, using what I've learned about the Inner Development Goals and Consumer Behaviour Theory, I want to be more mindful of how I can make my shopping habits more sustainable and meaningful. Reflecting on my values and connecting them to my chosen item will help me integrate these values more consistently into my daily life and consciously focus on them whenever possible.

5.1 Key Outcomes, Insights, and Limitations

The exploratory seminar was embedded within the constraints of a traditional lecture course. This introduced a transformative student learning experience that represents a significant milestone in higher education. Designed to bridge the gap between sustainability knowledge (SDGs) and inner development (IDGs), the seminar focused specifically on the first dimension of the IDG framework: *Being–Relationship to Self*.

Initiation of Self-reflection: Students were asked to examine their personal inner, value-based compass, particularly in the context of buying clothing. This reflective process encouraged them to question whether their current consumption choices aligned with what truly made them feel "happy" or fulfilled.

Linking Theory to Practice: By connecting personal values with the broader discourse on sustainable development, the seminar established a critical link between the SDGs and the inner transformative processes advocated by the IDGs.

Preliminary Shifts in Meaning and Action: Evidence from reflective assignments suggests that some students began to experience a potential "conscientiousness-shift" in both their personal meaning-making and their potential for action regarding sustainable behaviours.

5.2 Limitations and Considerations

While the initial results are promising, they are based on a preliminary exploratory examination. As such the author acknowledges several limitations.

Time Constraints: The seminar was limited to a single 90-min session, which only allowed for an initial period of reflective engagement rather than a sustained transformative process.

Result Implications: (i) Although developing an inner (value-based) compass is fundamental, this represents only one facet of the broader competencies required for sustainable transformation. (ii) The brevity of the intervention makes it uncertain whether the observed shifts in meaning and action translate into long-term behavioural change. (iii) The specific context and one-off nature of the seminar limits the generalizability and reach of the findings across other courses/disciplines in academic settings long-term.

Focus on a Single Dimension: By relying exclusively on the "Relationship to Self" dimension due to the time constraint, the seminar did not address the other dimensions of inner development (all 5 dimensions will be needed to foster robust sustainable behaviours).

Contextual Limitations: The exploratory study was carried out during part of an existing lecture course and as such had unknown influence on the students' engagement and the depth of the reflective process.

5.3 Recommendations for Future Research

To address these limitations and better achieve the study's goal of integrating sustainable development with inner transformation in higher education, several improvements should be considered for future iterations.

First, the intervention should be expanded. Extending the reflective seminar across multiple sessions or even an entire semester would allow for deeper engagement and more sustained reflection enabling a strong foundation for transformative learning. Second, the scope should be broadened by incorporating additional dimensions of the IDG framework. This would enable a more holistic approach, explicitly linking various aspects of inner development with the SDGs and fostering a more comprehensive understanding of sustainability. Third, longitudinal studies should be conducted to assess the durability of the observed shifts in meaning and action. Tracking students' reflections, attitudes, and behaviours over time would provide valuable insights into the long-term impact and effectiveness of such interventions.

6 Discussion and Conclusion

The primary aim of this exploratory study was to demonstrate that sustainable development necessitates an inner transformation, a dimension often overlooked in traditional higher education. By using the IDG framework and a transformative learning approach, the seminar indicated a possible personal shift for some students in relation to sustainability. The encouraging outcome, as reflected in student assignments and self-reflection exercises, suggest that even a brief intervention (i.e. within the limitations as described) can catalyse meaningful introspection. While the seminar succeeded in sparking initial reflection on personal values and sustainable consumption, the results should be interpreted as preliminary. Further research and extended interventions are needed to assess whether these reflective processes lead to enduring behavioural changes. Integrating such approaches throughout the curriculum may ultimately foster a more comprehensive and lasting transformation in higher education. In summary, the seminar offered valuable insights into the potential of combining transformative learning with sustainability education whilst also laying the groundwork for further initiatives aimed at embedding inner development within the broader framework of sustainable higher education.

For German language readers seeking more information, including a description of the seminar (emphasising its focus on climate literacy and connection to the concept of emotional competence) a manuscript entitled: *Klimabildung, die bewegt: Die Rolle emotionaler Kompetenz für nachhaltiges Handeln im Kontext des transformativen Lernens* (Zimmermann, 2025) was submitted to the book *Klimawandel und Bildung— Konzepte, Methoden und Perspektiven transformativer Klimabildung*. The book was published in 2025.

References

Ahmad, N., Toro-Troconis, M., Ibahrine, M., Armour, R., Tait, V., Reedy, K., Malevicius, R., Dale, V., Tasler, N., & Inzolia, Y. (2023). CoDesigns education for sustainable development: A framework for embedding education for sustainable development in curriculum design. *Sustainability, 15*, 16460. https://doi.org/10.3390/su152316460

Ankrah, D., Bristow, J., Hires, D., & Henriksson, J. A. (2023). Inner development goals: From inner growth to outer change. *Field Actions Science Reports, 25*, 82–87.

Ball, G. D. S. (1999). Building a sustainable future through transformation. *Futures, 31*, 251–270.

Columbo, S. L., Chiarella, S. G., Lefrancois, C., Fradin, J., & Simione, L. (2023). Why knowing about climate change is not enough to change: A perspective paper on the factors explaining the environmental knowledge-action gap. *Sustainability, 15*, 14856. https://doi.org/10.3390/su152014859

Growth that matters. (2021). *Inner development goals: Background, method and the IDG framework.* https://drive.google.com/file/d/13fcf9xmYrX9wrsh3PC3aeRDs0rWsWCpA/edit

Grund, J., Singer-Brodowski, M., & Büssing, A. G. (2024). Emotions and transformative learning for sustainability: A systematic review. *Sustainability Science, 19*, 307–324. https://doi.org/10.1007/s11625-023-01439-5

Holst, J., Singer-Brodowski, M., Brock, A., & de Haan, G. (2024). Monitoring SDG 4.7: Assessing education for sustainable development in policies, curricula, training of educators and students assessment (input-indicator). *Sustainable Development, 32*(4), 3908–3923

IDG. (2024a). https://innerdevelopmentgoals.org/framework/

IDG. (2024b). https://innerdevelopmentgoals.org/about/resources/

Ives, C. D., Schäpke, N., Woiwode, C., & Wamsler, C. (2023). IMAGINE sustainability: Integrated inner-outer transformation in research, education and practice. *Sustainability Science, 18*, 2777–2786.

Laurie, R., Nonoyama-Tarumi, Y., Mckeown, R., & Hopkins, C. (2016). Contribution of education for sustainable development (ESD) to quality education: A synthesis of research. *Journal of Education for Sustainable Development, 10*(2), 205–318.

Leal Filho, W., Lange Salvia, A., & Pires Eustachio, J. H. P. (2023). An overview of the engagement of higher education institutions in the implementation of the UN sustainable development goals. *Journal of Cleaner Production, 386*. https://doi.org/10.1016/j.jclepro.2022.135694

Liamputtong, P. (2023). Qualitative research in social science: setting the scene. In: P. Liamputtong (Ed.), *How to conduct qualitative research in social science* (pp. 1–13). Edward Elgar Publishing.

Lozano, R., Lukman, R., Lozano, F., Huisingh, D., & Lambrechts, W. (2013). Declarations for sustianability in higher education: Becoming better leaders, through adressing the university system. *Journal of Cleaner Production, 48*, 10–19.

Mezirow, J. (1997). Transformative learning: Theory and practice. *New Directions of Adult and Continuing Education, 74*, 5–12.

Mezirow, J. (2000). Learning to think like an adult: Core concepts of transformation theory. In J. Mezirow (Ed.), *Learning as transformation: Critical perspectives on a theory in progress* (pp. 2–33). Jossey-Bass.

Nohl, A. M. (2015). Typical phases of transformative learning: A practice-based model. *Adult Education Quarterly, 65*(1), 35–49.

Nölting, B., Dembski, N., Pape, J., & Schmuck, P. (2018). Wie bildet man Change Agents aus? Lehr-Lern-Konzepte und Erfahrungen am Beispiel des berufsbegleitenden Masterstudiengangs "Strategisches Nachhaltigkeitsmanagement" an der Hochschuke für nachhaltige Entwicklung Eberswalde. In W. Leal Filho (Ed.), *Nachhaltigkeit in der Lehre - eine Herausforderung für Hochschulen* (pp. 89–106). Springer Spektrum.

Sandahl, C., & von Knorring, M. (2023). *Managing with sense and sensitivity—Professionalism in leadership*. Springer.

United Nations. (2019). *The Sustainable Development Goals Report 2019*. New York. https://unstats.un.org/sdgs/report/2019/The-Sustainable-Development-Goals-Report-2019.pdf

United Nations. (2021). *The Sustainable Development Goals Report 2021.* New York. https://unstats.un.org/sdgs/report/2021/The-Sustainable-Development-Goals-Report-2021.pdf

United Nations. (2022). *The Sustainable Development Goals Report 2022.* New York. https://unstats.un.org/sdgs/report/2022/The-Sustainable-Development-Goals-Report-2022.pdf

United Nations. (2023). *The Sustainable Development Goals Report 2023 special edition—Towards a rescue plan for people and planet.* New York https://unstats.un.org/sdgs/report/2023/The-Sustainable-Development-Goals-Report-2023.pdf

United Nations. (2024). https://www.undp.org/sustainable-development-goals

Volk, B. (2020). Ordnung von Lernzielen - Ordnung des Wissens. Die Bedeutung der Taxonomie von Bloom für die Wissenschaftlichkeit und Praxis der Hochschuldidaktik. In P. Tremp, & B. Eugster (Eds.), *Klassiker der Hochschuldidaktik? Kartografie einer Landschaft* (pp. 219–234). Springer Fachmedien.

Wamsler, C., Osberg, G., Janss, J., & Stephan, L. (2024). Revolutionising sustainability leadership and education: Addressing the human dimension to support flourishing, culture and system transformation. *Climate Change, 177*, 4. https://doi.org/10.1007/s10584-023-03636-8

Zimmermann, C. L. (2025). Klimabildung, die bewegt: Die Rolle emotionaler Kompetenz für nachhaltiges Handeln im Kontext des transformativen Lernens. In D. Pijetlovic, & J. von Au (Eds.), *Klimawandel und Bildung – Konzepte, Methoden und Perspektiven transformativer Klimabildung*, oekom science

Zimmermann, C. L., & Scheering, J. (2026). Navigating climate emotions: A call for emotion sensitive ESD—A scoping review. *International Journal of Sustainability in Higher Education, 27*(10), 256–274. https://doi.org/10.1108/IJSHE-05-2024-0344

Integrating Sustainability into Courses and Programmes—A Transdisciplinary Approach Using Labelling Criteria and SDGs at University of Gothenburg

Johan Boman, Eddi Omrcen, Kassahun Weldemariam, and Fredrik Högberg

Abstract The University of Gothenburg has for long demonstrated a strong commitment to integrating sustainability into education. Various initiatives have been introduced and implemented, including a sustainability labelling system for courses and study programs. The ten criteria used for sustainability labelling were established in 2014 and are based on the system developed by AASHE/STARS. In 2019 the sustainability unit at the university was asked by the university's Board of Education to continue the development of the labelling system by including the UN's 17 Sustainable Development goals (SDGs). A transdisciplinary working group, comprising representatives from the university's eight faculties and other units, was established and met regularly from 2019 to 2022, with a pause during the COVID-19 pandemic. Further developing the labelling system is challenging, as its 10 criteria are well established, and integrating the SDGs into the curriculum and teaching adds complexity due to their interconnected nature. The university's Board of Education mandates that the revised labelling system include support tools for teachers. The working group recommended retaining the existing criteria while allowing courses and programs to be labelled according to the SDGs. To address the challenges of this new approach, various strategies were tested across different faculties. This chapter outlines the development process, and the final outcomes presented to the university's Board of Education. It also reflects on the key challenges and dilemmas faced by the working group, offering insights relevant to any higher education institution seeking to integrate sustainability into education. While the SDGs provide a holistic framework for

J. Boman (✉) · E. Omrcen · K. Weldemariam · F. Högberg
Unit for Property and Sustainable Development, Section for Sustainability, University of Gothenburg, Gothenburg, Sweden
e-mail: johan.boman@gu.se

E. Omrcen
e-mail: eddi.omrcen@gu.se

K. Weldemariam
e-mail: kassahun.weldemariam@gu.se

F. Högberg
e-mail: fredrik.hogberg@gu.se

W. Leal Filho et al. (eds.), *Introducing the UN SDGs in Higher Education Institutions*, World Sustainability Series, https://doi.org/10.1007/978-3-032-23017-1_7

advancing sustainability, their application in curriculum development and defining learning outcomes remains complex.

Keywords Higher education institute · Course labelling · Program labelling · Sustainability · Education assessment · SDGs

1 Introduction

1.1 Background

Sustainability is widely regarded as a positive aspect of societal development. Government institutions and companies often emphasize sustainability in their initiatives as a marker of positive value. Since many civil servants and employees are educated at higher education institutions (HEIs) (United Nations, 2023b), HEIs play a crucial role in developing sustainability competencies (Annelin & Boström, 2022; Corres et al., 2024). For the students to be competitive on the job market HEIs must include sustainability in their education. As a quality mark universities therefore must offer the students a way to know if the university offers courses and programs that can help the student gain these competencies. The competences are needed of both teachers and students to address the complexity of the 17 sustainable development goals (SDGs), set up by the United Nations for global sustainability. Among other issues the SDGs emphases the importance of education: "By 2030, ensure that all learners acquire the knowledge and skills needed to promote sustainable development, including, among others, through education for sustainable development" (United Nations, 2023a).

To avoid "SDG washing" a major challenge for HEI is to "walk the talk" and integrate transformation and institutional changes when implementing sustainable development (SD) in their education (Galán-Muros, 2023). By integration of SD in their curricula and educational programs HEIs can contribute to sustainability awareness of future generations (Alm et al., 2022; Galán-Muros, 2023). Labelling, or mapping, of courses and programs according to set criteria is one way to highlight the sustainable development and the SDGs in the education documentation (Boman et al., 2025a). This labelling can also be used for assessing the inclusion of sustainability in the education of the HEI (Lozano & Barreiro-Gen, 2019; Okanovic et al., 2021; Sánchez-Carracedo et al., 2021).

In the Swedish context, it is important to note that since 2006, Chap. 1, Sect. 5 of the Swedish Higher Education Act (Ministry of Education, 2006) mandates the inclusion of sustainable development in the work of Swedish HEIs:

> In the course of their operations, higher education institution shall promote sustainable development that ensures present and future generations a healthy and good environment, economic and social welfare and justice.

In 2017 the Swedish Higher Education Authority (UKÄ) evaluated the processes used by the Swedish HEI to integrate sustainability in education. Only 12 out of 47 Swedish HEI had systematic processes that support the integration of SD in the education (Finnveden et al., 2020). The University of Gothenburg was one of the 12 HEIs with such documented systematic processes. In the evaluation report the sustainability labelling system for courses and study programs at the University of Gothenburg was suggested to be of national/overall value and interest (UKÄ [the Swedish Higher Education Authority], 2017).

2 Sustainability in Higher Education and the Purpose of the Paper

2.1 Sustainability in Higher Education

Numerous models exist to describe and manage sustainability efforts in higher education institutions (HEIs). One reason for selecting a particular model is to enhance consistency and clarity in the process of integrating sustainability at the institution, often supported by a university strategy. For instance, in 2016, Babeş-Bolyai University (UBB) in Romania adopted the UBB Go Green strategy, and in 2019, UBB, in collaboration with the Department for Sustainable Development of the Romanian Government, organized a sustainable development conference focused on education (Bálint, 2020). A similar sustainability vision strategy is active at the University of Gothenburg (University of Gothenburg, 2020). In some cases a legal framework, like the Higher Education Act in Sweden is in force and set up legislation for sustainability at the HEI (Ministry of education, 2006).

Inclusion of sustainability in the education of the HEI can also be part of a strategic initiative to align the education with the sustainability processes in the society (Blanco-Portela et al., 2017; European commission, 2024). Some universities adopt the strategy of ensuring that current and prospective students can transparently identify and choose study paths that incorporate both their subject area and sustainability, equipping them for their future careers (Janakiraman et al., 2021; Karvinen et al., 2017).

The focus of the content in this chapter is the labelling of courses and programs at the University of Gothenburg. Its roots are to be found in the labelling system that was a part of the STARS system developed by AASHE (2005–2023). The University of Oregon (Mital, 2011), the University of California, Irvine (Reyes, 2013) and Weber State University (Bodine, 2011) in the USA adapted the AASHE/STAR system and implemented it at their universities. The implementation of sustainability labelling systems at the aforementioned three U.S. universities inspired the University of Gothenburg to introduce its own labelling system. While the U.S. systems are no longer active, Gothenburg's remains in use and continues to evolve. This chapter explores the development and impact of this labelling system.

2.2 *Purpose of the Chapter*

This chapter explores the research question: "Can a sustainability labelling system for courses and programs be developed to enhance the integration of sustainability in higher education?" To address this, the study has three key objectives.

First, it aims to describe the process and outcomes of revising the sustainability labelling system for courses and programs at the University of Gothenburg. Second, it critically examines the lessons learned during the development process, including a reflective discussion on the challenges of incorporating the UN Sustainable Development Goals (SDGs). Third, it explores broader challenges in integrating sustainability into higher education curricula.

3 Method

This section outlines our research methodology, grounded in the principles of action research. Additionally, we provide a brief overview of the University of Gothenburg to contextualize the case study.

3.1 *Action Research*

This study is based on the action research methodology, where the authors are active in the development of the university labelling scheme. Action research with its cyclic process of 'plan-act-observe-reflect' is a suitable method with a focus on continuous development, developing new research questions to be part of the ongoing research process (Rönnerman, 2022). The authors have been studying the processes, following up the results, evaluating the process, experiences and lessons learnt. Discussions and reflections have then been part of the process. According to Rönnerman, action research is about development and change of task, process, problem, but also about learning more about the actual change process as it evolves (Rönnerman, 2022). The plans within the labelling change process follow the 'plan-act-observe-reflect' steps. Plans and ambitions that initially were formulated were changed in the development process depending on the outcomes and reflections made during the development of the labelling system.

It is important not only that new knowledge is produced, but also that it is used in the action research process (Rönnerman, 2022). When the participants change their practical work based on the analysis and proved experience, it is possible to see a new and deepened learning. In this case study, the authors can show and reflect on the combination of a 'bottom-up' perspective combined with a 'top-down' perspective, being representatives for different units from the university, and how both perspectives are in some way integrated and intertwined.

3.2 *Short Description of the University of Gothenburg*

With approximately 58,000 students, the University of Gothenburg is one of Sweden's largest universities. It is also one of the country's most comprehensive institutions, encompassing 38 departments across eight faculties. Students can choose from 220 study programs and a diverse selection of 4500 courses. For many years, integrating sustainability into education has been a key objective for the university. However, its size and complexity present significant challenges in achieving this goal.

4 Case Description

4.1 *Background*

The University of Gothenburg's Vision 2030 document (University of Gothenburg, 2020), articulates its vision as follows:

> The vision **A university for the world** expresses the University of Gothenburg's endeavour to be an international higher-education and research institution which assumes responsibility for the development of society while helping to build a sustainable world.

From the vision it is clear that the university will work for a sustainable development of society. There are three overarching commitments that define the development of the university:

- Top-quality education and research
- Sustainable and knowledge-based societal development
- A coherent organization, and an attractive place of work and study

The University of Gothenburg will constantly develop education to meet society's long-term need for competence and knowledge while preparing students for their future working lives. The University will step up its contribution towards the sustainable development goals articulated by the United Nations' General Assembly in Agenda 2030 (United Nations, 2015). In the university's operational plan 2021–2024 it is stated that the University of Gothenburg will strengthen the development of sustainability perspectives in education and research. Part of the goal is further clarified as "The University shall implement a revised sustainability labelling system" with the percentage of labelled courses and programs as indicator.

In the labelling system the inclusion of sustainability in the course or educational program can be graded as related or focused depending on how much sustainability related content the course or program covers (Miljöenheten, 2014). When the grade of sustainability has been established a matching is done between the learning outcomes of the course or program and a set of ten sustainability criteria linked to different aspects of sustainability, Table 1. A course or program where at least one of the learning outcomes clearly shows that the course content meets one of the current sustainability criteria can be labelled. The criteria contain multidisciplinary and interdisciplinary issues and perspectives. They have a clear ambition to include or cover all three dimensions of sustainable development, i.e. social, economic and ecological sustainable development.

4.2 A Revised Labelling System

In 2019, the University of Gothenburg's Board of Education called for the development of a revised labelling system, building on the strengths of the existing criteria-based framework while establishing clearer connections to the UN Sustainable Development Goals (SDGs) and Agenda 2030. The revised system aimed to leverage past experiences and knowledge to enhance support for teachers, provide greater clarity for course and program managers, and improve accessibility for students and other stakeholders. Additionally, it was designed to include structured follow-up mechanisms to ensure effective implementation and continuous development.

4.3 A Revision Working Group

A working group was commissioned to develop the new labelling system and at the same time bringing in as many perspectives as possible. The group should develop a structure for how the labelling system should be evaluated on a regular basis. The working group was led by a project manager at the Gothenburg centre for sustainable development (GMV), in collaboration with students, faculties, the university library, the central university administration and the Unit for Pedagogical Development and Interactive Learning (PIL). The working group met continuously during the years 2019–2021 and developed an overall plan for the revision work and analysed various possibilities to revise the labelling system.

4.4 Meetings and Workshops

The work progress was planned to focus on having workshops as the main method of development, including a piloting phase where the workshop ideas could be tested (Hays & Reinders, 2020). After the workshops and piloting phase, a draft of a revised labelling system should be presented to the board of education. The group foresaw some more planning after an acceptance of the revised labelling system by the board of education, and before implementing the system university wide.

Eight workshops were held where various matters of importance for the revision of the marking system were presented, analysed and discussed, Table 2. It was

Table 1 The 10 criteria for labelling sustainable development in the courses and programs at the University of Gothenburg (Miljöenheten, 2014)

#	Description	Content
1	Sustainability as a concept	The history in a global context of the concept of sustainability and sustainable development and the current study field related to global challenges
2	Analysis from a globalisation perspective	How products, services, or activities in their own lives or in the future professional profession affect the natural environment, social conditions, and the economy in a global perspective, both today and in the future
3	Natural limits	Demographic trends and lifestyle in relation to the exploitation of natural resources, or the finite capacity of natural ecosystems to provide for human needs
4	Maintaining ecosystems	Conservation of natural resources and practices to protect and maintain the integrity of viable ecosystems in the face of rising human demands
5	Human rights and social equity	Distribution, discrimination, health and poverty issues and the mutual interaction between social inequality, poor health, the natural environment and people's opportunities for good living conditions
6	Values, culture and ethics	How norms, culture, religion, ethics and social conditions can shape human behaviour toward the natural world
7	Consumer and customer power	How demands for environmental consideration and social responsibility from private and public clients and consumers affects individuals, policies and corporate strategies and business opportunities
8	Governance and management	How regulations, policies, economic policy instruments and voluntary agreements, and leadership shape human behaviour and the actions of nations and companies in respect of the natural world and social issues
9	Planning and design	How community planning and product and service design can influence human well-being and human impact on the natural environment
10	Stakeholders' work and responsibility	The efforts of various global and local stakeholders, and their monitoring of environmental performance and social and economic responsibility

considered important that different perspectives, points of view, etc. were highlighted rather than discussing ready-made proposals. This was important for the part of the development that concerns learning, knowledge and experience transfer between the participants in the working group (Evans, 2019). The variety of workshops showed also the complexity of how integration of sustainability into education could be done.

In the workshops, an assessment was made regarding the current criteria for labelling, approaches to the SDGs, and several proposals for "tools" were prepared. The initial workshops formed the basis for designing a pilot phase where proposed measures would be tested at each faculty or unit.

Although the development team consisted of a limited number of people, more stakeholders were involved. Between the workshops the ideas and questions brought up at the completed workshop and the ideas to be discussed on the next workshop were brought back to the faculties and units. Here teachers and program committees were involved and contributed with valuable feedback and ideas for further development. All faculties are different and thus a varied response was received. The SDG-IAT tool received mainly positive responses but at the same time the connection between the SDGs and the labelling of courses and programs was considered a problematic barrier by most faculties. Only one faculty accepted the challenge with the SDGs and developed a tool to connect learning outcomes to the university's own criteria and to selected SDGs, see Sect. 5. The students were representatives from the student union and consulted with the student union on a regular basis. The revision of the labelling system was ordered by the Board of Education at the University of Gothenburg and the progress of the development was reported to and discussed in the Board of Education on a regular basis. Since the vision of the university is to contribute to the development of a sustainable society, the support from the Board of Education can be seen an enabler of the work.

To avoid "SGD washing" the results of the revised sustainability labelling system are evaluated by including quality aspects in a systematic way through dialogs with involved stakeholders (Stukalo & Lytvyn, 2021).

Table 2 The implemented workshops that were held during the development of the revised sustainable labelling system at the University of Gothenburg

I	Progression in the education and presentation of the SDG-IAT tool
II	Current criteria and the connection to criteria AASHE /Stars
III	Work with current criteria and the concept of sustainability
IV	Sustainability criteria and connection to the SDGs
V	"Procedure" and "Bridging between criteria and SDGs"
VI	Planning the pilot phase. Tests at the faculties, library, PIL, students
VII	Status at the faculties, PM students, and the situation regarding the web
VIII	Summarizing the pilot tests and planning report writing

5 Results Achieved

Drawing on completed workshops, the pilot phase, in-depth discussions, analyses, and the group's collective expertise, the working group developed the following recommendations for a revised sustainability labelling system:

1. The *university specific criteria.* No changes are needed in terms of content or number. However, faculties or departments may choose to develop their own guidelines on how to implement the labelling system.
2. *The SDGs* should be included in the labelling system. It is suggested that it is a voluntary element in the labelling. As a support in the implementation process, the SDG-Impact assessment tool (Gothenburg Centre for Sustainable Development, 2015) is suggested. It is an online tool to investigate possible links between the course content and the SDGs.
3. *Bridge the criteria and the SDGs.* A description of how the criteria align with the SDGs was developed and tested during the pilot phase, yielding mixed results—it was effective in one faculty but not in another. To support implementation and further development, the recommendation is to facilitate and strengthen connections between the criteria and the SDGs. The concept of bridging is further elaborated in Sect. 5.1.
4. *Procedure for sustainability labelling.* At the university there are working and delegation orders regulating the work at different levels. One faculty wrote a procedural document for the process of labelling courses and programs, based on the working and delegation orders. Since the regulations are similar for all faculties and departments, the procedural document can easily be adopted by others. The procedure is discussed in Sect. 5.2.
5. *Student connection and dialogue:* Faculties and departments should have a dialogue with student representatives about the sustainability label. This will lead to more engagement by the students, more student perspectives, and a better dialog between the student representatives and the teachers.
6. *System support.* As mentioned in Sect. 4.1, several supportive tools have been identified and tested. Discussion forums and workshops are other tools that can be used. The working group saw a need of a support team to ease the continued development of linking sustainability to the education. Tools are further discussed in Sect. 5.3.
7. *Follow-up, review and communication.* A designated support and evaluation team is tasked with discussing and anchoring the labelling with various university stakeholders. Identified communication ways were the learning platform and the student and staff portals of the university web. The support team should also, together with PIL and interested parties, develop and give courses for the teachers' competence development. The support team is investigating the possibilities of a "peer review model" for reviewing the sustainability labelling of courses or programs before the labelling is implemented, and how future environmental audits can act as follow-up of the sustainability labelling.

These seven steps or parts thereof can be seen as "a model" for integrating sustainability into courses and programmes. It gives guidance on what is needed but also on how an integration of sustainability could be managed. Some of the parts, that we believe are of general interest, are further scrutinized in the following sub sections. The so-called bridge between current criteria and SDGs is of value if you have sustainability criteria already implemented. Some kind of procedures are of value to develop and implement a systematic process. We think that "supporting tools" are of value for a systematic process and to guide teachers, course leaders and program managers on how to integrate sustainability and SDGs into the curricula. In the last part the revised labelling system is presented, which summarize the results and outcomes from the workshops and pilot study.

5.1 A Bridge Between Criteria and SDGs

During the development process, strategies for connecting the university's criteria with the SDGs were explored as a way forward. Bridging these frameworks was seen as essential by some faculties, given that both criteria-based and SDG-based labelling needed to be implemented.

As an example of a bridge, we will discuss the bridge adapted by parts of Sahlgrenska Academy (SA), the faculty of medicine, "Support for sustainability labelling of courses and programs at Sahlgrenska Academy". Figure 1 shows the top part of the document. The design is based on the ten university criteria, Table 1, as can be seen in the rows "1". The concept of sustainability development" and "2". Globalization perspective on sustainable development" in Fig. 1. In the first column under each criteria heading the definition of the criterium is given. In the second column the SDGs that the faculty see as relevant for the education at Sahlgrenska Academy and linked to the specific criterium are listed. In the next three columns there are suggestions for concretization and progression for SD elements on three different levels, or themes, introductory, precise, and complex relationships of SA relevance. This content is general for all courses. The content specific for each course follows under heading "Reflection on current/future SD labelling of *[course code and name]* (learning objectives, content and examination elements)", numbered according to the criteria number. Here the teachers working on a course specifies the course code and name, and adds learning objectives, content and examination elements. It is also possible to already at this stage specify potential course literature and add comments.

5.2 Using Procedural Regulations or Action Plans

It is not uncommon that decisions on different changes in the education, for example labelling of sustainability, is handled in a not strictly formal way. The result can be that the action relies of one active person, and the necessary steps are not well documented.

Support for sustainability labeling of courses and programs at Sahlgrenska Academy				
1. The concept of sustainability development				
GU definition	**Potential connection to Agenda 2030**	**Suggestions for concretization and progression for SD elements: possible themes of SA relevance**		
		Introductory themes	**Precise themes**	**Complex relationships**
The concept of sustainable development, its historical development internationally, for the own subject area and the theme of the course.	**Goal 3. Health and well-being** Ensuring healthy lives and promoting well-being for all at all ages	Historical development of the concept of sustainable development including the integration of ecological, economic and social perspectives	Definition, focus, indicators and proposals for measures for Goal 3 as well as critical reflections on their (possible) connections to the own subject area, the theme of the course and the mission of	The relationship between the living conditions of different population groups, their health status and access to good education with a particular focus on the subject area and the theme of the course
	Goal 4. Good education Ensure inclusive and equal education of good quality and promote lifelong learning for all	Historical background to and starting points for Agenda 2030 and current global sustainability goals	Definition, focus, indicators and proposals for measures for Goal 4 as well as critical reflections on their (possible) connections to own studies and SA's work for quality and equal treatment	The connection between critical thinking and systems perspective, and the opportunity to act for sustainable development
		Character and focus of the research field planetary health		
1. Reflection on current/future SD labeling of [course code and name] **(learning objectives, content and examination elements)**				
Learning objectives	**Content**	**Examining part**	**Literature**	**Miscellaneous**
2. Globalization perspective on sustainable development				
GU definition	**Potential connection to Agenda 2030**	**Suggestions for concretization and progression for SD elements: possible themes of SA relevance**		
		Introductory themes	**Precise themes**	**Complex relationships**
Global perspectives on how the natural environment, social conditions and economy together can affect the future professional practice.	**Goal 3. Health and well-being** Ensuring healthy lives and promoting well-being for all at all ages	Historical background to contemporary ecological, economic and social global inequality	Definition, focus and indicators for Objectives 10 and 17 as well as critical reflections on their (possible) connections to the own subject area, the theme of the course and the mission of health care	Connection between the own subject area and the theme of the course, the tasks and activities of the health and medical care, as well as Objectives 10 and 17.'s proposed measures
	CASE. 10 Reduced inequality Reduce inequality within and between countries	Overview of the research fields international relations and global development studies and the concept of globalization	The relationship between the global burden of disease and inequality within and between countries	Analysis of GU/SA's internationalization work based on Objectives 10 and 17
	Goal 17. Implementation and global partnership Strengthen the means of implementation and revitalize the global partnership for sustainable development		The connection between migration and various globalization processes	
2. Reflection on current/future SD labeling of [course code and name] **(learning objectives, content and examination elements)**				
Learning objectives	**Content**	**Examining part**	**Literature**	**Miscellaneous**

Fig. 1 Part of the supportive document at Sahlgrenska Academy (SA), University of Gothenburg, for implementation and evaluation of sustainability labelling of courses and programs

To avoid these pitfalls, the faculty of education adopted a more structured approach to the workflow following the procedural regulations at University of Gothenburg. In their view this was the only way to sustain and secure labelling of courses and programs at the faculty. Their work resulted in two documents, "procedure order for sustainability labelling" and "template procedure order for sustainability labelling of programs", available for others to adopt for their needs. The documents contain clear descriptions of the organizational level for decisions, the need for official registration of the decision, the different actors involved, detailed time plan for preparation and decisions.

All faculties and departments have action plans outlining long- and short-term goals for the coming activities in research and education. Sustainable development (SD) is a compulsory topic in these plans and different faculties and departments have chosen to use the action plan to set up goals for SD in their education. Goals can be on development of sustainability competence, integration of sustainability in courses and educational programs, and support the sustainability labelling process. Since labelling isn't compulsory, the action plan is a possible way to include SD without necessarily labelling courses. In that case there is a problem on the top university level to follow up the development and inclusion of SD in the education, since the actions plans are at faculty and departmental level. On the other hand, the action plans are followed up on the level where they were decided and implemented.

5.3 Supporting Tools

Gothenburg centre for sustainable development[1] (GMV) had previously developed tools that are suitable as support for the teacher when working with the labelling system, such as:

- A toolbox for Sustainable Development for teachers. Developed for each faculty to better suit their framework. Discontinued due to restructuring of the internal web site and to low usage
- SDG Impact Assessment Tool (Gothenburg Centre for Sustainable Development, 2015). An online tool explaining the 17 SDGs and allowing the user to evaluate the impact of the activity, in this case the outcomes of the course or program
- Forums for Sustainability and Sustainable Development in Higher Education. Transdisciplinary forums to enhance discussions on sustainability over the faculty borders
- "GAIA" web system (GAIA stands for Goal, Action, Impact, Assessment) to more clearly and systematically show progression, learning etc. regarding sustainable development in various educational programs (Boman et al., 2025b).

The SDG Impact Assessment Tool (SDGIAT) (Gothenburg Centre for Sustainable Development, 2015) and the UNESCO booklet "Education for sustainable development—learning objectives" (UNESCO, 2017) were the two tools discussed most intensively during the development of the revised sustainability labelling system.

The SDG Impact Assessment Tool (Gothenburg Centre for Sustainable Development, 2015) is a learning and supporting tool to assess the impact of an activity on the SDGs. It can be used on any activity and in this context, it was used for assessing the impact of the assumed knowledge obtained by the students after taking a course or following an educational program at the University of Gothenburg. It is a self-assessment tool where the users can go through the 17 SDGs one by one, learn about the goals, and assess the impact of the activity in focus. The assessment is set on a Likert five step scale between "direct negative impact" and "direct positive impact". Depending on the users' prior knowledge of the SDGs, it can be time consuming. But it is a rewarding exercise since in the end the users have a better knowledge of both the SDGs and the course or program content in relation to sustainability. Of the tools tested during the development process this was the one many of the faculties fund useful. At some departments this tool was later used for staff competence development. The tool is freely available online at https://sdgimpactassessmenttool.org/en-gb (Gothenburg Centre for Sustainable Development, 2015).

The UNESCO booklet "Education for sustainable development – learning objectives" was also considered a useful tool (UNESCO, 2017). It describes how education for sustainable development (ESD) can be seen as a key instrument for reaching the SDGs. The short list of key sustainability competences was used for discussions in the development phase of the labelling and later on as a part of forum discussions on

[1] The centre is a joint initiative between University of Gothenburg and Chalmers University of technology.

how to work with ESD in the education at higher education institutions. The documents also link to the efforts by the Sahlgrenska Academy to bridge the university criteria with the SDG, by discussing and exemplifying the integration of ESD in curricula, Sect. 5.1.

Several other tools for embedding and assessing ESD and SDGs in the education at higher education institutions are presented by other research groups. Sánchez-Carracedo et al. (2021) investigates three different tools to aid inclusion of SDGs in the engineering education. One of the tools the authors found useful is similar to the process of bridging the university's criteria and the SDG created during the labelling development process at University of Gothenburg, Sect. 5.1. Balanced scorecards (BSC) is a method used in many management assessments. Fuchs et al. (2020) used the BSC technique to promote SD development in HEI by green marketing to involve not only the HEI but also other stakeholders.

5.4 *The Suggested Revised Sustainability Labelling System*

The revised sustainability labelling system was required to build on the experience and strengths of the existing criteria-based system while incorporating or aligning with the SDGs and Agenda 2030. It also needed to include proposals for support tools for teachers and provide greater clarity for course and program managers, students, and other stakeholders. Lastly, the system had to include a structured approach for monitoring and follow-up.

The suggested revised sustainability labelling system builds on one system, two processes, four ways to label, and an assessment step, Fig. 2. One system means one labelling system containing all possible options for labelling course curricula and educational programs. A common feature of the revised system is the possibility to label a course or a program as Focused or Related, depending on if sustainability can be said to be the focus of the course or program, or if there are elements of sustainability but sustainability is not the focus.

In this system there are two primary processes: labelling a course or labelling a program. Differentiation is needed since there are different organizational levels

Process	Criteria	Forum	Support tools	Formal decision	Level
Process A: courses	Internal criteria	Collegial discourse	Labelling instructions	Departmental procedure	Dept.
	SDGs		UNESCO ESDG, SDGIAT	Departmental procedure	Dept.
Process B: programs	Internal criteria		Labelling instructions	University's procedure	Faculty
	SDGs		UNESCO ESDG, SDGIAT	University's procedure	Faculty

Fig. 2 Proposed revised system for sustainability labelling of courses and programs. One system (——), two processes (- - -), four ways to label

responsible for the courses and programs, in accordance with the university's work order and the vice-chancellor's delegation order.

In each of the two processes courses or programs can be labelled according to the sustainability criteria or be related to the SDGs. Thus, four ways to label the education at University of Gothenburg. In the suggested revised system, a collegial discourse is included. The reason was to build a collegial knowledge base, to make staff more involved and to facilitate a progression between courses and in programs. Process support is the descriptions of the ten criteria, the SDGIAT, the UNESCO booklet and the support team at the university. Once the suggested revision of the course or program is decided, the decisions on programs are taken on faculty level while course decisions are taken on departmental level.

The revised course curricula or program plan is added to the education database of the university, for later visualization on the web. The database is also feeding the online course and program search tool used by students and other actors interested in the education offered at the university.

For assessment and quality assurance three tracks were suggested.

1. Include the assessment of sustainability as a compulsory part of the normal course or program evaluations.
2. Develop and use a peer-review model for revised course curricula and program plans before they are decided. The process exists for research manuscripts, so it can be developed for educational purposes as well. See for example (UC Berkley Center for Teaching & Learning, 2024).
3. The quality assurance of sustainability labelling can be included as part of the sustainability audits of the university done on a regular basis.

The inclusion of elements of ESD in the European quality assurance (QA) framework is still limited (Galán-Muros, 2023). A possible way forward can be to start by including the ESD or SDG in education in the internal quality assurance, while waiting for the inclusion of an external QA (Stukalo & Lytvyn, 2021).

6 The Decided Labelling System

In October 2021 the suggested new labelling system was presented to the University of Gothenburg's boards of education by the sustainability manager at the University of Gothenburg and the development team leader. The board decided that the suggested revised labelling system should be reworked as a set of guide-lines for sustainability labelling into an informative web-based guide for how the university can work with sustainability labelling, where both the existing criteria and the possibility to label courses and programs according to the SDG are included, as well as to review the possibility of producing a steering document for sustainability labelling at the University of Gothenburg.

In March 2022 the revised labelling system was presented to the board, with the content outlined in the preceding sections of this chapter. Suggestions of more changes to the labelling system were given by the board, and the chairman proposed that each faculty representative discuss the presented revised labelling system to their education committee to discuss if a system containing both the university specific criteria and the SDGs. The need for rules for labelling should also be discussed and weighed against the possibility to have a web-based guide.

In the board of education's meeting in December 2022 the labelling system was once again on the agenda. The board of education decided that a revised sustainability labelling system should be implemented following these principles.

- It should include possibility to use Focused or Related labelling.
- It should use the ten university criteria and/or be related to the SDGs.
- Labelling is not compulsory.
- A web site with examples of labelling should be created.
- Forums for support of ESD discussions should continue.
- If the courses or programs are not labelled, sustainability should be reported in the general quality work of the department or faculty.

This system is now being implemented with a series of supporting forums, The web-based guide is implemented at a slow pace due to the transition of the staff portal to a new web system and lack of necessary staff resources. A more extensive study on experiences and dilemmas of sustainability labelling system for courses and study programs at the University of Gothenburg is presented by Boman et al. (2025a).

7 Lessons Learnt and Limitations of the Study

To be able to take advantage of the knowledge gained in the study at other HEIs, it is important to critically address several aspects of the study.

The most debated aspect of the revised labelling system is the inclusion of the SDGs. Currently, no courses or programs at the University of Gothenburg are labelled based on the SDGs. The decision to incorporate them into the system is ambitious, given that the 17 interlinked SDGs were not originally designed for curriculum development. As suggested by the board of education a web-based collection of examples can be a supportive tool. A drawback of this solution is that it is non-interactive. A better, or complementary, solution can be to form teams of teachers from different educational fields to develop innovative solutions to specific SDGs. This facilitates collaboration and critical thinking. To include other stakeholders, in the academy or in society, one option is to join already existing networks and initiatives like the Higher Education Sustainability Initiative (United Nations, 2024), A Rounder Sense of Purpose (Scherak & Rieckmann, 2022; The RSP Partnership, 2024) or engage in caring communities of practice (Lönngren et al., 2023). A peer review process is also helpful in the development process (UC Berkley Center for Teaching & Learning, 2024).

The student perspective must be considered in the development of sustainability content in courses and programs. Simply linking the curriculum to sustainability criteria and the SDGs is insufficient if students feel disconnected from the material. The syllabus should be clear, learning goals must be explicitly addressed in the classroom, and there should be transparency in how sustainability content is assessed (Drayson & Taylor, 2015; Saleem, 2023).

A key limitation of the sustainability labelling system at the University of Gothenburg is its strong emphasis on learning outcomes. Expanding sustainability integration in education requires a broader focus that also considers course content, assessment methods, and pedagogy. This can in turn lead to further requirements on the teachers motivation and competence to face the challenges of educating for sustainable development (Corres et al., 2024). The University of Gothenburg has used labelling of courses and programs to follow the presence of environmental and sustainability connections in the education for decades. This makes it possible to follow the long-term development compared to limiting project-based approach where the system is initiated as a project and then left on its own when the project ends.

The current limitation of not having the SDGs well included in the labelling system will be addressed in the near future in in the continued efforts of making the education contribute to the sustainable development of society.

8 Conclusions and Outlook

The first aim of the study was to describe the process and results of revising the system for labelling courses and educational programs at the University of Gothenburg, while the third aim focused on discussing the overall challenges of integrating sustainability in courses and programs. Through a detailed account of the development process of the revised labelling system, we have highlighted both the challenges and opportunities of integrating sustainability into higher education. A key advantage of this development was its long-term nature, spanning several years, which allowed for substantial content to be developed, tested, and critically reflected upon—an outcome that would not have been achievable in a short-term, project-based approach.

The second aim of the study was to critically discuss the lessons learned during the development of the labelling system. The SDGs proved challenging to apply, particularly in relation to learning outcomes, due to their relative novelty compared to the university's established criteria and their intricate interconnections. The study explores potential strategies to address these challenges; however, the full implementation of the SDGs at the University of Gothenburg is still ongoing. There are publications available that will be used as the basis for a further discussion of including the SDGs in the education (Adams et al., 2023; Alcántara-Rubio et al., 2022; Fauzi et al., 2023; Hansen et al., 2021; SDSN, 2020; Zhu et al., 2024).

Our final conclusion, after wrestling with the question of whether a sustainability labelling system for courses and programmes can enhance the integration of sustainability in higher education, is that it is indeed possible. However, labelling alone is insufficient. It must be supported by clear procedures, guidelines, tools, follow-up measures, and an ongoing critical discussion to ensure a systematic process rather than one based on ad hoc decisions. Many universities have implemented Green Campus programs, which are valuable initiatives. By incorporating the educational structure into such programs, they can serve as a bridge to implementing a labelling system that integrates sustainability into higher education.

Given the complexity of sustainable development, we conclude that working on the labelling system within a team is highly beneficial. Team discussions are both inspiring and instrumental in fostering a shared understanding of the labelling process and its outcomes at the university. The benefits of labelling courses and programmes include stimulating collegial discussions on education, providing guidance for students, ensuring quality, increasing awareness, and potentially encouraging sustainable behaviours that extend beyond academics.

Despite the lengthy development process, there is still much to learn, both for staff at the University of Gothenburg and at many other higher education institutions. We hope that through shared experiences, we can inspire each other to make meaningful contributions to the sustainability of society.

References

AASHE. (2005–2023). *Association for the advancement of sustainability in higher education*. Retrieved November 28, 2023, from https://www.aashe.org/

Adams, T., Jameel, S. M., & Goggins, J. (2023). Education for sustainable development: Mapping the SDGs to University Curricula. *Sustainability, 15*(10), 8340.

Alcántara-Rubio, L., Valderrama-Hernández, R., Solís-Espallargas, C., & Ruiz-Morales, J. (2022). The implementation of the SDGs in universities: A systematic review. *Environmental Education Research, 28*(11), 1585–1615.

Alm, K., Beery, T. H., Eiblmeier, D., & Fahmy, T. (2022). Students' learning sustainability—Implicit, explicit or non-existent: A case study approach on students' key competencies addressing the SDGs in HEI program. *International Journal of Sustainability in Higher Education, 23*(8), 60–84.

Annelin, A., & Boström, G.-O. (2022). An assessment of key sustainability competencies: A review of scales and propositions for validation. *International Journal of Sustainability in Higher Education, 24*(9), 53–69.

Bálint, M. (2020). *UBB goes green strategy*. https://green.ubbcluj.ro/en/ubb-goes-green-strategy/

Blanco-Portela, N., Benayas, J., Pertierra, L. R., & Lozano, R. (2017). Towards the integration of sustainability in higher education Institutions: A review of drivers of and barriers to organisational change and their comparison against those found of companies. *Journal of Cleaner Production, 166*, 563–578.

Bodine, J. (2011). *Weber State University*. Retrieved November 28, 2023, from https://reports.aashe.org/institutions/weber-state-university-ut/report/2022-04-25/

Boman, J., Omrcen, E., & Weldemariam, K. (2025a). *Sustainability labelling of courses and study programs—Experiences, dilemmas and possible way forward. A case-study from University of Gothenburg*. North American and European Perspectives on Sustainability in Higher Education. W. Leal Filho, Springer Nature.

Boman, J., Omrcen, E., Weldemariam, K., Hanning, A., & Lodin, J. (2025b). Building learning pathways to enhance Education for Sustainable Development (ESD), quality and relevance. *International Journal of Sustainability in Higher Education 26(9),* 353–371

Corres, A., Ruiz-Mallén, I., & Rieckmann, M. (2024). Educators' competences, motivations and teaching challenges faced in education for sustainable development: What are the interlinkages? *Cogent Education, 11*(1), 2302408.

Drayson, R., & Taylor, C. (2015). *The student voice: experiences of student engagement in education for sustainable development* (pp. 627–645). Springer International Publishing.

European Commission. (2024). *Green education initiatives*. https://education.ec.europa.eu/focus-topics/green-education/about-green-education

Evans, T. L. (2019). Competencies and pedagogies for sustainability education: A roadmap for sustainability studies program development in colleges and universities. *Sustainability, 11*(19), 5526.

Fauzi, M. A., Abdul Rahman, A. R., & Lee, C. K. (2023). A systematic bibliometric review of the United Nation's SDGS: Which are the most related to higher education institutions? *International Journal of Sustainability in Higher Education, 24*(3), 637–659.

Finnveden, G., Friman, E., Mogren, A., Palmer, H., Sund, P., Carstedt, G., Lundberg, S., Robertsson, B., Rodhe, H., & Svärd, S. (2020). Evaluation of integration of sustainable development in higher education in Sweden. *International Journal of Sustainability in Higher Education, 21*(4), 685–698.

Fuchs, P., Raulino, C., Conceição, D., Neiva, S., Amorim, W. S. D., Soares, T. C., Andrade de Lima, M., Montenegro De Lima, C. R., Soares, J. C., & Andrade Guerra, J. B. S. O. D. A. (2020). Promoting sustainable development in higher education institutions: The use of the balanced scorecard as a strategic management system in support of green marketing. *International Journal of Sustainability in Higher Education, 21*(7), 1477–1505.

Galán-Muros, V. (2023). *SET4HEI. general guidelines for the implementation of sustainability in higher education institutions*. UNESCO IESALC.

Gothenburg Centre for Sustainable Development. (2015). *SDG impact assessment tool*. Retrieved December 1, 2023, from https://sdgimpactassessmenttool.org/en-gb

Hansen, B., Stiling, P., & Uy, W. F. (2021). Innovations and challenges in SDG integration and reporting in higher education: A case study from the University of South Florida. *International Journal of Sustainability in Higher Education, 22*(5), 1002–1021.

Hays, J., & Reinders, H. (2020). Sustainable learning and education: A curriculum for the future. *International Review of Education, 66*(1), 29–52.

Janakiraman, S., Watson, S. L., Watson, W. R., & Cheng, Z. (2021). Creating environmentally conscious engineering professionals through attitudinal instruction: A mixed methods study. *Journal of Cleaner Production, 291*.

Karvinen, M., Lundgren, U., Mälkki, H., Sorvari, J., Shiel, C., Leal Filho, W., & Pretorius (2017). The implementation of sustainable development in the Nordic Higher Education Institutions (HEIs). In W. Leal Filho, M. Mifsud, C. Shiel, & R. Pretorius (Eds.), *Handbook of theory and practice of sustainable development in higher education* (pp. 169–187). Springer International Publishing AG.

Lönngren, J., Direito, I., Tormey, R., Huff, J. L., & Johri, A. (2023). *Emotions in engineering education* (pp. 156–182). Routledge.

Lozano, R., & Barreiro-Gen, M. (2019). Analysing the factors affecting the incorporation of sustainable development into European Higher Education Institutions' curricula. *Sustainable Development, 27*(5), 965–975.

Miljöenheten. (2014). Hållbarhetsmärkning av kurser och program [Sustainability labelling of courses and programs] (p. 9). Sweden, University of Gothenburg.

Ministry of Education. (2006). *Swedish Higher Education Act* (vol. 1992, p. 1434). M. O. Education.

Mital, S. (2011). *Sustainability course identification*. Retrieved November 28, 2023, from https://reports.aashe.org/institutions/university-of-oregon-or/report/2011-02-25/ER/curriculum/ER-5/

Okanovic, A., Jesic, J., Dakovic, V., Vukadinovic, S., & Andrejevic Panic, A. (2021). Increasing university competitiveness through assessment of green content in curriculum and eco-labeling in higher education. *Sustainability, 13*(2), 1–20.

Reyes, A. (2013). *Sustainability course identification*. Retrieved November 28, 2023, from https://reports.aashe.org/institutions/university-of-california-irvine-ca/report/744/ER/curriculum/ER-5/
Rönnerman, K. (2022). *Aktionsforskning: vad? hur? varför? [ction research. What? How? Why?]*. Studentlitteratur
Saleem, U. (2023). Education for sustainable development in higher education. Students' perspective and experience during a Masters programme in ESD at University of Gothenburg. e. University of Gothenburg/Department of, e. special and s. Göteborgs universitet/Institutionen för pedagogik och.
Sánchez-Carracedo, F., Segalas, J., Bueno, G., Busquets, P., Climent, J., Galofré, V. G., Lazzarini, B., Lopez, D., Martín, C., Miñano, R., de Cámara, E. S., Sureda, B., Tejedor, G., & Vidal, E. (2021). Tools for embedding and assessing sustainable development goals in engineering education. *Sustainability (Basel, Switzerland), 13*(21), 12154.
Scherak, L., & Rieckmann, M. (2022). *Development and assessment of ESD competences: Staff training at the University of Vechta*. Competences in Education for Sustainable Development. Sustainable Development Goals Series (pp. 121–128). Springer International Publishing.
SDSN. (2020). *Accelerating education for the SDGs in universities: A guide for universities, colleges, and tertiary and higher education institutions*. Sustainable Development Solutions Network (SDSN).
Stukalo, N., & Lytvyn, M. (2021). Towards sustainable development through higher education quality assurance. *Education Sciences, 11*(11), 664.
The RSP Partnership. (2024). *The rounder sense of purpose*. https://aroundersenseofpurpose.eu/
UC Berkley Center for Teaching and Learning. (2024). *Peer review of course instruction*. Retrieved December 1, 2024, from https://teaching.berkeley.edu/peer-review-course-instruction
UKÄ [the Swedish Higher Education Authority]. (2017). *Universitets och högskolors arbete med att främja en hållbar utveckling, En Tematisk Utvärdering Del 2, Rapport 2017:12 [The work of universities and colleges in promoting sustainable development, A Thematic Evaluation Part 2]*. Universitetskanslersämbetet.
UNESCO. (2017). *Education for sustainable development goals: Learning objectives*. UNESCO Publishing.
United Nations. (2015). *UN Sustainable development goals*. Retrieved November 30, 2023, from https://www.un.org/sustainabledevelopment/.
United Nations. (2023a). *Transforming our world: The 2030 Agenda for sustainable development*. Retrieved November 22, 2023, from https://sdgs.un.org/2030agenda
United Nations. (2023b). *United Nations Higher Education Sustainability Initiative (HESI)*. Retrieved November 22, 2023, from https://sdgs.un.org/HESI
United Nations. (2024). *Higher education sustainability initiative*. Retrieved December 1, 2024, from https://sdgs.un.org/HESI
University of Gothenburg. (2020). *A university for the world—The University of Gothenburg's vision 2021–2030* (p. 12). University of Gothenburg.
Zhu, B., Wang, T., Liu, G., & Zhou, C. (2024). Revealing dynamic goals for university's sustainable development with a coupling exploration of SDGs. *Scientific Reports, 14*(1), 22715–22799.

Project DITTS: Digital Introduction of the Sustainable Development Goals into Higher Education Teaching Regional Aspects from South Africa and Brazil

Luciana Londero Brandli, Amanda Lange Salvia, Bianca Rebelatto, Gabriela Rodrigues, Rudi Pretorius, Melanie Nicolau, Sanet Carow, Zongho Kom, João Eustachio, Jasmin Röseler, and Walter Leal Filho

Abstract With less than five years for 2030 and to achieve the Sustainable Development Goals (SDGs), initiatives to support their implementation need to be urgently accelerated. Including the goals in higher education teaching is one of these initiatives, considering the role played by universities in promoting the SDGs, offering learning opportunities and training future generations. This study aims to present and reflect on the international project "Digital Introduction of the Sustainable Development Goals into Higher Education Teaching Regional aspects from South Africa and Brazil" (DITTS)—a strong and sustainable partnership between University of Passo Fundo (Brazil), University of South Africa (South Africa) and Hamburg University of Applied Sciences (Germany). The project fosters the creation and dissemination of open educational materials for higher education institutions as a strategy to introduce the SDGs to students from all areas. The collaboration among the partner institutions and countries leverages the expertise and regional insights of each team, thereby enriching each teaching module. The study highlights the impact of the development of the DITTS teaching modules and the project activities. Results indicate that digital learning resources tailored to regional contexts have the potential to enhance student engagement and comprehension of sustainability concepts. This cooperation project and activities contribute to the ongoing discussions on the role of digital education in advancing SDGs and highlight the need for further innovations in sustainability teaching practices, and support further South–South and North–South collaboration and sustainability initiatives in higher education institutions.

L. L. Brandli · A. L. Salvia (✉) · B. Rebelatto · G. Rodrigues
Graduate Program in Civil and Environmental Engineering, University of Passo Fundo, Passo Fundo, RS, Brazil
e-mail: amandasalvia@gmail.com

R. Pretorius · M. Nicolau · S. Carow · Z. Kom
Department of Geography, University of South Africa, Pretoria, South Africa

J. Eustachio · J. Röseler · W. Leal Filho
Research and Transfer Centre "Sustainable Development and Climate Change Management", Hamburg University of Applied Sciences, Hamburg, Germany

W. Leal Filho et al. (eds.), *Introducing the UN SDGs in Higher Education Institutions*, World Sustainability Series, https://doi.org/10.1007/978-3-032-23017-1_8

Keywords Global goals · Sustainability agenda · International collaboration · Open Educational Resources (OER) · Sustainability education

1 Introduction

The increasing urgency to achieve the Sustainable Development Goals (SDGs) by 2030 has intensified efforts to integrate sustainability education within higher education institutions (HEIs). Universities play a fundamental role in fostering sustainability through curriculum development, research, and institutional policies that align with global sustainability targets (Leal Filho et al., 2018). However, despite the widespread recognition of the SDGs, their integration into higher education curricula remains inconsistent across regions (Leal Filho et al., 2015 Leal Filho et al., 2018).

The integration of sustainability in higher education has gained traction in recent years, with numerous studies underscoring its importance in equipping students with the knowledge and skills necessary to address global challenges. Scholars such as Leal Filho et al. (2019a, 2019b) and Rajabifard et al., (2021a, 2021b) emphasise the need for HEIs to adopt a transdisciplinary approach in sustainability education, ensuring that students engage with complex real-world problems. However, challenges persist, including limited faculty training, inadequate teaching resources, and regional disparities in sustainability literacy (Leal Filho et al., 2019a, 2019b, 2024). Previous initiatives, such as the UNESCO sustainability education programs, have laid the foundation for embedding SDGs within academic curricula (UNESCO, 2020), yet their implementation often lacks contextualisation to specific regional challenges (Alcántara-Rubio et al., 2022; Salvia et al., 2019). A significant issue remains the absence of digital learning resources tailored to diverse learning environments, particularly in the Global South (Ketlhoilwe et al., 2020; Krishnamurthy & Sahay, 2023).

Despite existing efforts to integrate sustainability education, there remains a critical lack of structured, digital, and adaptable teaching materials that cater to regional contexts (Ketlhoilwe et al., 2020; Krishnamurthy & Sahay, 2023). The need for digital resources is particularly pressing in the wake of the COVID-19 pandemic, which accelerated the shift toward online learning (Adedoyin & Soykan, 2023; Hermanto & Srimulyani, 2021). However, many institutions in the Global South continue to face challenges in accessing quality digital content that aligns with their specific educational and sustainability needs (Akpan et al., 2024; Langthaler & Bazafkan, 2020; Sareen & Mandal, 2024).

In this perspective, the "Digital Introduction of the Sustainable Development Goals into Higher Education Teaching", a project funded by The German Academic Exchange Service (DAAD), emerges as a strategic initiative aimed at addressing this gap (DITTS, 2025). By leveraging digital learning tools and cross-regional collaborations, the project focuses on incorporating SDG-related content into university teaching, with a particular emphasis on regional perspectives from South Africa, Brazil and Germany. Through an innovative approach that blends open educational resources (OERs) and intercultural academic exchanges, DITTS seeks to strengthen the capacities of faculty members and students, ensuring a more inclusive and globally relevant education framework, especially connected to the SDGs 1, 2, 8–11, 15–17 (DITTS, 2025).

This paper aims to present and reflect on the international project "Digital Introduction of the Sustainable Development Goals into Higher Education Teaching Regional Aspects from South Africa and Brazil" (DITTS) (DITTS, 2025). The significance of research inquiry lies in its potential to inform future sustainability education initiatives and contribute to the broader discourse on inclusive and equitable access to quality learning materials for global sustainability efforts. In order to achieve the goals of this study, this paper is structured as follows: Sect. 2 provides an overview of the DiTTS project, highlighting the outcomes, envisaged outputs, methodology, limitations and exchanges. Section 3, in turn, showcases the impact of the project. Finally, Sect. 4 discusses the lessons learnt in terms of the partnerships and their importance for SDG learning and implementation.

2 DITTS Project

DITTS is an international project partnered by the University of Passo Fundo (UPF—Brazil) and the University of South Africa (Unisa—South Africa), and led by the Hamburg University of Applied Sciences (HAW—Germany). The purpose of the project is to develop sustainable partnerships that would facilitate the joint development of teaching material for HEIs, which could support teaching staff with material to introduce their undergraduate students to the SDGs in general, and in particular to raise the awareness on how these goals can contribute towards achieving Agenda 2030 (Leal Filho et al., 2022). The DITTS project is running for four years (2022–2025) and forms part of the SDGs Partnerships Programme of the German Academic Exchange Service (DAAD), funded through the Federal Ministry for Economic Cooperation and Development (BMZ) of Germany. The DAAD Partnership Programme responds to the growing need to address global challenges through global partnerships from various disciplines, and through adopting successful models of higher education and cooperation. In terms of partnerships, DAAD advocates for the partnership between HEIs in Germany with universities in the Global South (DAAD, 2022, 2023, 2024), which is the foundation of the project described in this chapter.

2.1 Expected Outcomes of the Project

The expected outcomes and outputs of the DiTTs project are summarised in Table 1. The fundamental principles underlying the development of the teaching material and exchanges for the DITTS project are threefold. Firstly, regional challenges, differences and similarities are addressed; secondly, best practices to implement the SDGs are considered; and finally, the development of a rich learning environment for undergraduate students to learn about the SDGs is facilitated. The partners developed the teaching material focused explicitly on the relevance and usefulness of the resources in such a way that it would embrace mutual learning and awareness of the SDGs. In support of the development of the teaching material (OP1; OP2), the project builds on and intends to further strengthen existing informal partnerships between researchers/academics at the three institutions through participation in joint research initiatives, external stakeholders' meetings, publications in journals, reports on sustainability in higher education, and participation in conferences (OP3–5). This has been achieved through various South–South exchange visits by researchers/academics from Unisa and UPF and South-North exchanges with HAW (OP5).

Table 1 Outcomes and envisaged outputs of the DITTS project (DITTS, 2022)

Outcomes (OC)	Envisaged outputs (OP)
OC1: The partner institutions offer study and qualification programs related to the SDGs that are appropriate to the local context and reflect the state of the art in science for SDGs 1, 2, 8, 9, 10, 11, 15, 16, & 17	OP1: Teaching material that relates to the SDGs and suits the local context while reflecting the latest scientific developments
OC2: Established methods are in place for transferring course offerings to other higher education institutions, i.e. their digital Moodle platforms or websites	OP2: Methods for making study opportunities more widely available. By the end of the project, an online resource will be available at the three partners' websites, providing OER teaching materials on SDGs 1, 2, 8, 9, 10, 11, 15, 16, & 17
OC3: Topics related to global circumstances and challenges are established research foci at the institutions involved by means of publications handed in jointly	OP3: Each partner university will hold three stakeholder events over the project period to reach relevant stakeholders outside the university setting
OC4: Methods for the distribution and transfer of knowledge and research findings are in place, i.e. through carrying out events to present the project findings	OP4: Three to four emerging researchers from UPF and Unisa will have attended up to five training sessions each on methodological and sustainability topics
	OP5: By the end of the project, each partner institution will have disseminated the outcomes of the project development and learnings of the SDG materials at least at four national or international events (e.g. conferences, symposia)

2.2 *Framework and Methodology and Limitations Considered*

The initial planning for the DiTTS project (Leal Filho et al., 2022) suggested that the overall structuring of the envisaged learning experience and specifically the teaching materials would build on the methodology followed for a preceding project (2017–2019) by the Research and Transfer Centre "Sustainability and Climate Impact Management" (FTZ-NK), HAW Hamburg. During the latter project, a set of OERs were developed for lecturers to be used flexibly when implementing eight (3–7, 12–14) of the 17 SDGs in their courses. These OERs are hosted as a digital toolbox for SDG leaning on the website of HAW Hamburg and can be accessed at https://dl4sd.org/course/view.php?id=4. This toolbox has a set structure and offers the following resources for each of the eight SDGs: teacher's handbook, PowerPoint presentation, supplementary readings, case studies, links to visualization tools and helpful websites. The teacher's handbook for each of these SDGs is focused on the definition of the SDG, future perspectives, good practices, and exercises, and did not have a regional focus. The DiTTS project aimed to go a step further and include recent developments (e.g. the Covid-19 pandemic, progress made, new threats) into the learning experience and the teaching materials being developed, thus supporting achievement of the UN Agenda 2030 in a direct way. The intended focus was also to include the expertise and regional specificities from the partner universities, thus acknowledging the increased awareness to recognize the importance of regional and localized perspectives in the implementation of the SDGs (Biggeri, 2021; Blicharska et al., 2021; Mohieldin et al., 2023; Moreno Pires et al., 2022; Tan et al., 2019).

The participation of partners from Germany, Brazil, and South Africa in the DiTTS project meant that several of the limitations associated with the preceding set of OERs could be addressed. The first of these links with cultural differences between the three partner regions, meaning that the development of resources and the teachers' handbook need to be culturally relevant and sensitive to the differences in values, traditions, and social norms and that is an aspect that needs to be taken into consideration. This was mitigated by the inclusion of materials related specifically to each region and written by authors from these specific regions to ensure that cultural sensitivity and relevance was addressed. The language barrier between the partner regions could also be considered a limitation and thus the translation of materials into multiple languages relevant to educators in Brazil and South Africa was critical in terms of ensuring that nuances and specific contexts were not lost. A further limitation that had to be accommodated by the methodology was that of the technological disparities between the partner regions. Unequal access to digital tools, the internet, and to electricity supply not only impacted on the development of materials but had to be accommodated within the materials itself as teachers in the three regions would have to mitigate their specific challenges in the application of the resources and the teachers' handbook. Educational standards and differences were also experienced as a limitation and had to be accommodated in the project methodology. Each partner region/country have their unique education standards and curriculum requirements that had to be considered. The development of materials that could be utilised by teachers within all three regions was therefore a challenge but was mitigated by the writing of region-specific materials by authors from each specific region. The final limitation that was experienced and had to be accommodated relates to the differences in available resources of the partner institutions. To develop the required materials implied availability of funding, skills staff and time. In part, the availability of the DAAD grant mitigated this limitation to some extent, however, outside of the grant the three partner institutions had to make significant in-kind contributions to ensure the effective implementation of the project methodology. This specifically relates to making skilled staff available from the academic sector, support departments such as ICT, and language services to ensure that result of the project would include high quality educational materials that could be available as OERs within but not limited to the three partner regions.

Based on the critical evaluation of the preceding OERs, including the limitations identified, the framework for the development of the teaching material in the DiTTS project (explained in Sect. 2.3) had several pedagogical implications on the framework of the new set of teachers' manuals. This included the development of new material that included enabling place-based learning, using case studies to provide more localised content into the curriculum. The new material also includes guidance on active and collaborative learning through the provision of examples of localised projects that learners can be provided to complete as individuals or in teams. The regional approach also allows to enable more enquiry-based learning, through the provision of content and activities encourages critical thinking through the exploration of localised cases studies and activities. The regional approach followed in the presentation of the material provides for differentiated instruction that accommodated all the identified short comings of the previous manuals that would ensure that the cultural and language needs of students are recognised and accommodated.

2.3 Teaching Material

The development of teaching material (OP1) for the DITTS project, took into consideration the reported literature on the importance of including SDGs in the curriculum of HEIs (Kioupi & Voulvoulis, 2020; Leal Filho et al., 2019a, 2019b; Moreno Pires et al., 2022; Rajabifard et al., 2021a, 2021b). The value of the DITTS teaching material is evidenced in the limited research published on the development of similar material (Anyim, 2021; Leal Filho et al., 2019a, 2019b) at a variety of HEIs globally. The most important documented reason for not including SDGs in curricula at HEIs was presented in the research of Leal Filho et al. (2019a, 2019b) and is grounded in the circumstance that stemmed from the lack of sufficient and relevant training (and thus study material) of faculty members. The DITTS project mitigates this reason by providing contextual teaching material that would be available as Open Educational Resources (OERs) to Faculty at all HEIs (OP2).

Thus, the authors of this chapter support Leal Filho et al., (2019a, 2019b) call for HEIs to play crucial roles in developing future leaders, professionals and citizens and become contributors to societal transformations. The research of Rajabifard et al. (2021a, 2021b) that calls upon HEIs to ensure that their graduates become global and regional agents for change is also recognised by the authors. Moreno Pires et al. (2022), argue that to successfully infuse the SDGs into the curricula of HEIs, it is critical that Faculty from a variety of disciplines are consulted. The partnership of DITTS thus addresses the latter, by ensuring that the Faculty involved in the development of the teaching material capitalised on the expertise from Social, Natural, and Applied Sciences.

2.3.1 Content of the Teaching Materials

Information on the SDGs is widely available on various online platforms, for example, UNESCO (https://sdgs.un.org/goals). The overall purpose of these platforms is to make material available to a broader audience and not within a format that can be easily transposed into the curricula of HEIs. Although some researchers, for example, Anon (2020) and SET4HEI (2023), present guidelines on including the SDGs in the HEIs curricula, a more intensive search for such learning material presented limited results. It was noted in the search for relevant teaching material that HAW (https://dl4sd.org/my/) had developed OERs for eight of the 17 SDGs (3–7, 12–14). The DITTS project thus aimed to develop teaching material in the form of teaching handbooks for the remaining nine SDGs, namely 1, 2, 8–11, 15–17, with the aim to support the inclusion of the SDGs into curricula of HEIs and to provide Faculty with resources to compile relevant teaching material.

The approach followed in the handbooks relates to problem-based learning (PBL), which can enhance students' work readiness in sustainability-focused careers. Nguyen et al. (2024) emphasises the importance of PBL, while Silviariza and Handoyo (2021) extend this to spatial problem-based learning (SPBL), which aims to develop students' spatial thinking skills. PBL, and more specifically SPBL (reflected in the regional presentation of each handbook), are much-needed skills required to address the complex nature of implementing and achieving the SDGs.

2.3.2 The Structure of the Teaching Materials

Each teacher's handbook covers a specific SDG and comprises six sections. First section introduces the reader to the SDGs in general, followed by the second section which deals with the specific SDG. The latter explains the significance and the interdependencies with the other SDGs and concludes with a discussion on the advantages and the challenges in implementing the specific SDG. Third section provides a regional (Africa, Europe and Latin America) approach to the impact of three global crises, namely, climate change, COVID-19 and Conflict, on the achievement of the specific SDG. This is followed by the fourth section, which focuses on the progress made towards achieving the specific SDG within each region. Three best practices (per region) of the implementation of the specific SDG are discussed in the fifth section and follow a PBL and SPBL approach. Suggestions of typical assessment questions for each section are provided in the sixth section, which also includes examples of exercises learners can do to incorporate sustainable practices in their lifeworld. Throughout the teacher's handbook, references to supplementary reading materials and OER links on the specific SDG are provided.

Each teaching handbook is accompanied by a PowerPoint presentation that can be used and adapted by educators to incorporate into their teaching material or class. In addition, Story Maps were developed for two of the case studies (one African and one Latin American) per SDG, allowing a more innovative method of engaging with students at a regional level. The manuals were developed in English and then professionally translated into Portuguese and two South African indigenous languages, namely *isiZulu* and *Sesotho sa Leboa* (Northern Sotho). The translation of the teaching handbooks is aligned with research (for example, Mweri, 2020) that indigenous languages are critical in terms of regional cultural heritage and thus by the translation of the material can contribute to the preservation of critical sustainability components such as indigenous knowledge, practices, and values in the Global South.

2.3.3 Incorporating the Teaching Materials into Curricula and Study Materials

The way the teachers' handbooks were developed allows sufficient flexibility for Faculty to incorporate the material into their curricula and connect with the SDGs. This flexibility allows Faculty to make a selection, adapt, or use all the material in new and existing learning material. Furthermore, the handbooks can be used to enhance the Faculty's own knowledge and understanding of a specific SDG. Finally, the handbooks can be adopted by undergraduates and graduates as a good overall orientation of one or more SDGs if they are embarking on a related research project. The regional approach followed in each handbook shows how Faculty can infuse their own regional contexts into their curricula, but at the same time be cognizant of the global impact of the achievement of the SDGs. The case studies, exercises and assessment questions provide examples of good practices that can be used by Faculty through PBL.

2.3.4 Availability of the Developed Material as OERs

The nine teachers' handbooks developed are OERs and can be found on the websites of the three participating universities (Table 2). The license agreement under which these OERs are published, CC-BY-SA 4.0, allows for the materials to be shared and adapted by giving attribution to the developers of the study material.

Table 2 Electronic availability of the teachers' handbooks

Name of the partner university	Link to study material
Hamburg University of Applied Sciences	https://dl4sd.org/course/view.php?id=4
University of Passo Fundo	https://www.upf.br/greenoffice/modulos-ods
University of South Africa	https://www.unisa.ac.za/sites/corporate/default/Colleges/Agriculture-&-Environmental-Sciences/Schools,-departments,-centres-&-units/School-of-Ecological-and-Human-Sustainability/Department-of-Geography

2.4 *Exchanges*

One of the principles underlying the DITTS project is that the educational exchange between HEIs is essential to ensure that the SDGs are effectively integrated into the teaching and learning activities of the participating institutions. The same principle is replicated in similar projects of the DAAD (2022, 2023, 2024) and is further supported by the research of Leal Filho et al. (2023), who indicated that such exchanges were critical to the successful development and integration of teaching material that focuses on sustainable development. Figure 1 reflects the exchanges between the partner universities over the three-year period that helped achieve OP3 (Table 1).

Three important funders meetings arranged by DAAD and attended by project leaders formed the backbone of the successful execution of the DITTS project. The first (April 2022) was the virtual launch of new DAAD projects within the SDG Partnership Programme, while the subsequent project leaders' conferences were held in Costa Rica (April 2023) and Tanzania (July 2024). The important activities at these conferences were the discussions on how SDGs can be implemented at HEIs, as well as important funding information. These conferences were of critical importance as

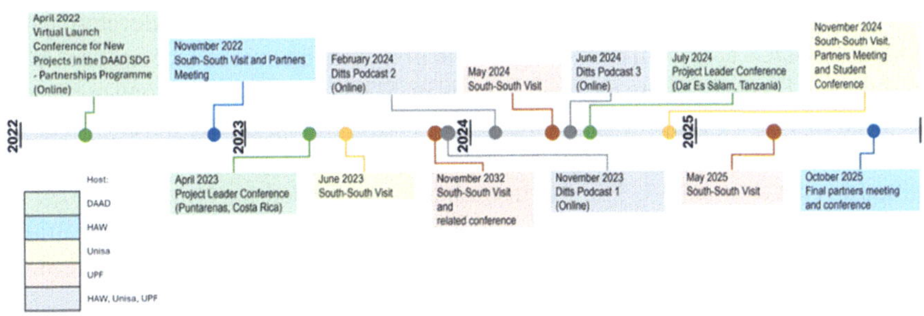

Fig. 1 Project timeframe

project leaders were exposed to similar projects in their respective regions and were provided with the opportunity to network with other project leaders. As is the case with the Reach Alliance (Roll et al., 2024), these meetings allowed for the facilitation of and the sharing of best practices and resources that could enhance the local and global engagement of the SDGs and, thus, the development of relevant teaching material.

Five South–South and South–North visits were scheduled (Fig. 1), and during these visits, the delegates from the partner universities were able to travel to Brazil (Passo Fundo), South Africa (Pretoria) and Germany (Hamburg). The value of these visits is centred on the opportunity to experience the partner countries and institutions' capacity and challenges in meeting the SDGs (Sharma & Varshney, 2023). While these visits were very similar in nature, there was sufficient diversity to ensure that the partners could exchange their experiences and share best practices and technical expertise in a drive to ensure that the teaching material developed by the partner institutions reflected a collective understanding of regional similarities and differences in the teaching of the SDGs.

3 Impact of the Project

The South–South exchange component of the project, with participants from Brazil and South Africa, represents a key aspect of DITTS. These exchanges offer a unique opportunity for researchers to work closely with project partners and also exchange ideas and insights with other relevant stakeholders, sharing knowledge and best practices while addressing regional challenges and lessons learnt. By collaborating with researchers from diverse cultural and academic backgrounds and from different levels, participants gain insights into innovative teaching approaches and strategies that enhance the contextual relevance of SDG education. The exchanges also contribute to capacity building by strengthening research and teaching expertise, empowering participants to reflect on different possibilities and opportunities for sustainability learning in their context. These moments also support collaborative research in terms of academic outputs as scientific publications and are especially useful for creating academic bonds among the team members.

The South–South Exchange within the project has facilitated key developments around the SDGs across different stages. The first exchange took place in June 2023 at UNISA in Pretoria, where the team focused on SDGs 1 and 2. The second exchange occurred in Passo Fundo in November 2023, concentrating on the development of SDGs 8 and 10. A third exchange was held in Brazil in May 2024, where attention shifted to SDGs 9 and 15. The fourth exchange, scheduled for November 2024 at UNISA, focused on SDGs 11 and 16. Finally, the last exchange is planned for Brazil in May 2025, marking the culmination of the project with the development of SDG 17.

The literature points out a series of benefits of academic exchanges. Table 3 presents these benefits and relates them with the impact of the DITTS project.

Table 3 Benefits of academic exchanges and impact in the DITTS context

Benefits international exchange	Impact in the DITTS context
Intercultural proficiency (Clarke et al., 2009)	Ability to foster an environment where participants of the project can effectively interact, communicate, and collaborate with individuals from different cultural backgrounds
Engagement in global issues (Paige et al., 2009)	Discussions on worldwide sustainability challenges, students and educators enhance their ability to address complex global problems through a multicultural lens
Knowledge-sharing (Huang & Chiu, 2020)	By promoting cross-cultural dialogue, educators and students from both regions can share best practices, contextualise the SDGs to their specific realities, and co-develop educational content that is culturally relevant
Cooperation network (Jin & Wang, 2021)	By sharing challenges, enabling intense collaboration, exchanging ideas, and applying collective knowledge

The DITTS project benefits from international exchange by fostering intercultural proficiency, enabling participants to interact, communicate, and collaborate effectively with individuals from diverse cultural backgrounds (Clarke et al., 2009). This engagement enhances the ability of students and educators to address global sustainability challenges through a multicultural lens, promoting discussions on complex global issues (Paige et al., 2009). Furthermore, the project encourages knowledge-sharing by facilitating cross-cultural dialogue, allowing educators and students from different regions to share best practices, contextualise the Sustainable Development Goals (SDGs) to their specific realities, and co develop culturally relevant educational content (Huang & Chiu, 2020). Ultimately, this collaboration creates a strong cooperation network where challenges are shared, ideas are exchanged, and collective knowledge is applied to drive impactful change in sustainability (Jin & Wang, 2021).

One of the activities featured in all the South–South visits included work sessions where partners presented the teaching handbook for each SDG. Intensive workshop sessions followed this to ensure that the development of the final teaching handbooks correctly reflected the realities that each region faces in not only implementing the SDGs but also infusing these realities into teaching material that is relevant to the students and Faculty for each region. In similar projects, albeit related to North–South partnerships (Blicharska et al. 2021), the interactive workshops during the South–South visits for this project ensured that there was equal involvement in the development of the teaching material, and this ensured the balanced representation of the challenges related to the implementation of the relevant SDG in the three regions in each teaching manual.

While not directly related to the writing of the teaching material, the South–South exchanges strategically included various interactions with graduate students from the three partner institutions. The underlying motivation for this type of student interaction was the assumption that graduate students have the potential to play a very important role in the implementation of the SDGs as their research can provide fresh perspectives, insight, and possible innovation in the implementation of the SDGs. Thus, graduate-student interaction formed a major component of the South–South exchanges. At each exchange, graduate students from the partner universities were provided the opportunity to share their SDG research. This culminated in an International Symposium (Sustainable Development Goals (SDGs): *Implications for People and Planet*, 18–19 November 2025), where 70 students were provided the opportunity to interact (in-person and virtually) with their peers across 19 countries with the intent that this interaction would contribute towards a fresh perspective; increase exposure to diverse research themes and methodologies related to the SDGs and that the opportunity would not only enhance their own academic and professional development, that would make a significant contribution to the implementation of SDGs in their respective countries.

Another event organised as part of the project provided a dedicated platform for dialogue, knowledge dissemination, and stakeholder engagement. The Life-Cycle Assessment and Circular Economy Conference, held at the University of Passo Fundo, Brazil, on November 6–7, 2023, brought together experts to discuss the latest developments in Life-Cycle Assessment (LCA) and Circular Economy. The event highlighted how these approaches contribute to sustainable development and climate action, and had the participation of over 50 colleagues from 19 countries. This conference also included an important excursion Be8 (Paso Fundo, Brazil), the largest producer of biodiesel in Brazil, which focuses on different renewable energy sources and the role such energy can play in a more sustainable future in the region. An important outcome of this exchange was a book that is part of the World Sustainability Series (Brandli et al., 2025).

These two events brought together academics, civil society and other stakeholders, facilitating a comprehensive exploration of local, regional and global dimensions of SDG education. The events developed as part of DITTS play a key role in showcasing the outcomes of the project, such as the teaching materials, but they have dedicated topics connected with the project and with the SDGs prepared as part of the teaching modules. They also serve as a bridge between academic research and practical application, ensuring that the project's outcomes align with the needs of various stakeholders, including educators, researchers, and policymakers.

Tourism-related activities are often considered to be a catalyst of economic growth, cultural preservation, and social development and these, in turn, are important building blocks in the achievement of the SDGs (Seraphin & Gowreesunkar, 2021). The exchanges between South Africa and Brazil thus included excursions to provide environmental and cultural tourist experiences that would help the respective partners understand the dynamics and region-specific challenges experienced by each partner. These excursions provided partners with a better depth of understanding in terms of the region-specific aspects related to each SDG as incorporated into each teaching manual.

In addition to the larger events, the project also organised workshops with local stakeholders and the academic communities. These workshops adopted a methodology designed to provide interactive and reflective learning, encouraging active participation from those involved. The goal of these workshops is to test the modules created to be presented by professors in their classes, and foster discussions on how to incorporate the practices and concepts of the SDGs into academic life and the local community. The events usually start with introductory lectures on the SDGs, highlighting the context of each goal. During the presentations, practical examples are discussed on how sustainability can be achieved through public policies and local initiatives, as well as the role of institutions in promoting a more just and sustainable environment.

While the DITTS project has demonstrated significant positive impacts, the process of international collaboration by means of exchanges has also presented challenges. One of the key difficulties encountered relates to the diversity of cultural and national contexts, which influence not only pedagogical approaches but also perceptions of sustainability and the SDGs. Differences in academic practices—e.g., in-person (UPF) or online learning (UNISA), and local realities—e.g., inequality levels and safety concerns, sometimes created the need for additional dialogue and adaptation to ensure that knowledge-sharing and experiences remained equitable and contextually relevant. Additionally, while the use of English across teams largely mitigated language barriers, careful attention was still required to address potential translation challenges. At times, these barriers affected the depth of discussions and the efficiency of collaborative work. However, they also presented valuable opportunities for learning and growth. The immersive nature of the exchanges enabled participants to experience the realities of sustainability implementation in different regions, fostering deeper empathy and understanding. Moreover, experiencing these complexities strengthened team cohesion and communication.

Another strategy used by the DITTS team to connect with a wider audience and in a different format is the project podcast (Strickland et al., 2021; López Valladares & Romero Lora, 2024). The podcast series represents a more dynamic space for sharing insights and experiences of the project team members on sustainability education, teaching about the SDGs, and the project itself. The published episodes are summarised in Table 4.

One of the key benefits of this project podcast was the possibility to reach a wider audience beyond traditional academic circles, making discussions on sustainability education and the SDGs more approachable and dynamic. While the experience of recording and participating in the episodes has been valuable for team members, technical challenges such as ensuring a stable internet connection, maintaining good audio quality across different locations, and dealing with unexpected interruptions during recordings have occasionally represented difficulties. Additionally, the pressure of recording may add another layer of complexity for some project members. Despite these challenges, the podcast has fostered a sense of connection among participants.

Finally, the teaching **modules** developed by the project have the potential to strengthen institutional capacity and promote the systematic inclusion of SDGs in

Table 4 Podcast series

	Episodes/Description	Participants involved
	Episode 1—Introducing the Project Idea. The project leaders in Brazil and South Africa introduce the project and its expected impacts	Luciana Brandli—UPF Rudi Petrorius—UNISA Jennifer Pohlmann—HAW
	Episode 2—Creating the materials. Two researchers from the project share insights and learnings from creating the teaching materials for SDGs	Amanda Salvia—UPF Melanie Nicolau—UNISA Jennifer Pohlmann—HAW
	Episode 3—The student perspective. This episode presents two students talking about their experiences and learnings from participating in the DITTS project	Pedro H.C.Diniz –UPF Lindokuhle Sibiya—UNISA Jennifer Pohlmann—HAW

higher education. The implementation of structured teaching modules can significantly enhance sustainability education by facilitating the integration of SDG-related content into university curricula. These modules, developed by experts in sustainable development, address key challenges such as limited faculty training and insufficient teaching resources, which often hinder the effective incorporation of sustainability topics in higher education. The structured and adaptable materials provided by these modules allow the professors to customise the content according to their specific disciplinary needs and classroom dynamics. Another aspect that strengthens the impact is the transdisciplinary, where the modules are developed to be used across diverse academic fields, from Law to Education, for example.

Additionally, the modules can support a more consistent and evidence-based approach to sustainability education by helping to bridge gaps in faculty expertise and ensuring alignment with global sustainability frameworks (Leal Filho et al., 2021). They also enhance sustainability education in a more accessible and impactful way, as they are readily available in different languages and can be easily downloaded.

Finally, the modules foster student engagement and critical thinking by providing a structured learning pathway (Leal Filho et al., 2024). They begin with basic SDG definitions, then progress to discussions of key challenges, and include exercises designed to help students apply sustainability concepts in their daily lives. This approach strengthens students' capacities by equipping them with the knowledge and skills necessary to address global challenges, ensuring that they actively engage with complex real-world problems.

4 Lessons Learned

The DITTS project has demonstrated the value of international collaboration in enhancing exchange, teaching, and research opportunities related to sustainability education. The South–South and North–South exchanges fostered knowledge-sharing, enabling faculty and students to engage with diverse perspectives on SDG implementation. The development of structured teaching materials as OERs has strengthened institutional capacity, making sustainability education more accessible. Additionally, research collaborations have expanded, resulting in joint publications, stakeholder engagement, and research-policy discussions. These synergies reinforce the role of HEIs as a key player for sustainability learning and highlight the need for continuous partnerships to bridge regional gaps. Even though the diverse expertise and backgrounds of the project partners across the Brazilian, South African, and German institutions can also represent challenges in international cooperation such as this project, they certainly enhanced the quality and applicability of the outputs, ensuring that the teaching materials and research findings reflect a broader range of perspectives and regional contexts.

The main findings of this study highlight the transformative potential of digital learning resources in fostering sustainability education. The results indicate that contextualised SDG teaching materials could significantly enhance student engagement and comprehension of sustainability concepts. Faculty members reported increased confidence in integrating SDGs into their courses, facilitated by the availability of structured, adaptable, and locally relevant content. The study also underscores the importance of South–South and North–South academic collaborations in enriching the quality and applicability of sustainability education materials.

The promotion of collaborative SDG learning via digital and cross-regional approaches is critical in fostering a better global understanding and implementation of the SDGs at all levels. On the one hand, the graduate's presentation of their research to their peers at partner universities allowed for a better cross-regional understanding of the challenges faced with the implementation of the SDGs. The similarities and differences within their local contexts make collaborative learning both challenging and rewarding, and at the same time, it makes a cross-regional approach worthwhile as graduates are more informed and possible catalysts of change. On the other hand, the availability of the teaching handbooks as OERs provide Faculty with introductory material for the nine SDGs in such a way that they can adapt the material to be more specific to their regional contexts. The online availability of teaching handbooks in local languages other than English provides an opportunity for a richer learning environment and, thus, a better understanding of the SDGs.

While digital collaboration has its advantages, especially as, by nature, this type of interaction is more sustainable, the chapter authors advocate that Faculty should not totally rely on online interactions during their collaborations but should seek to advocate a more hybrid approach that includes travel to the regions of partners to foster exchange and collaboration. The value of travel to and exploration of partners' geographical locations helps to understand the differing local contexts, which was so crucial to developing an understanding of the choice, for example, of case studies with a differing focus for the various regions, and also the differing levels of progress and ways in which the crises affect the SDGs in these regions. The collaborative and cross-regional approach leads to richer, more useful teaching materials that provide Faculty with a broader scope and perspective on the matters dealt with in the respective handbooks.

Beyond content development, the partnerships within DITTS have facilitated critical academic exchanges and cross-regional knowledge-sharing. South–South (Brazil and South Africa) and North–South (Brazil, South Africa, and Germany) exchanges have enabled faculty and students to engage with sustainability challenges firsthand, fostering intercultural dialogue and strengthening research collaboration. The joint organisation of workshops, stakeholder meetings, and symposiums has further reinforced the application of SDG-related knowledge, bridging academic research with real-world sustainability practices.

Finally, it is worth considering that partnerships have supported professional development by equipping the beneficiaries (universities and educators) of the outcome of the project with tools and strategies to embed SDGs into teaching. Through collective efforts that emerged from the project, institutions have not only enhanced sustainability education but also established long-term networks for continued collaboration. In sum, the success of DITTS underscores the transformative potential of partnerships in building institutional commitment to sustainability, fostering innovation in teaching methodologies, and empowering students to become global change agents in the pursuit of sustainable development.

References

Adedoyin, O. B., & Soykan, E. (2023). Covid-19 pandemic and online learning: The challenges and opportunities. *Interactive Learning Environments, 31*(2), 863–875. https://doi.org/10.1080/10494820.2020.1813180

Akpan, I. J., Offodile, O. F., Akpanobong, A. C., & Kobara, Y. M. (2024). A comparative analysis of virtual education technology, e-learning systems research advances, and digital divide in the global south. *Informatics, 11*(3). https://doi.org/10.3390/informatics11030053

Alcántara-Rubio, L., Valderrama-Hernández, R., Solís-Espallargas, C., & Ruiz-Morales, J. (2022). The implementation of the SDGs in universities: A systematic review. *Environmental Education Research, 28*(11), 1585–1615. https://doi.org/10.1080/13504622.2022.2063798

Anon. (2020). *Accelerating education for the SDGs in universities: A guide for universities, colleges, and tertiary and higher education institutions—Sustainable development solutions network.* https://www.unsdsn.org

Anyim, W. O. (2021). Sustainable development goal on quality education: A review of e-learning resources and pedagogy in the university system. *Library Philosophy and Practice*, 5578.

Biggeri, M. (2021). Editorial: A "Decade for Action" on SDG localisation. *Journal of Human Development and Capabilities, 22*(4), 706–712

Blicharska, M., Teutschbein, C., & Smithers, R. J. (2021). SDG partnerships may perpetuate the global North–South divide. *Scientific Reports, 11*(1), 22092.

Brandli, L., Dalla Rosa, F., Pretorius R., Avila, L., & Leal Filho, W. (eds.). (2025). *The contribution of life cycle analyses and circular economy to the sustainable development goals*. World Sustainability Series (390 pp.). Springer. https://link.springer.com/book/9783031852992

Clarke, I., III., Flaherty, T. B., Wright, N. D., & McMillen, R. M. (2009). Student intercultural proficiency from study abroad programs. *Journal of Marketing Education, 31*(2), 173–181.

DAAD. (2022, April 25–26). *Virtual Launch Conference for New Projects in the DAAD SDG—Partnerships Programme*. DAAD.

DAAD. (2023, April 24–26). *DAAD—Project leader conference "Higher Education Partnerships as a Key to Sustainable Development"*, Costa Rica. DAAD.

DAAD. (2024, July 22–24). *DAAD—Project leader conference "Educating tomorrow's leaders to pioneer sustainable development—The role of academia"*. Dar Es Salaam. DAAD.

DITTS. (2022). *Project planning summary for submission to the DAAD*. Unpublished Document. HAW, Hamburg

DITTS. (2025). *DITTS. The digital introduction of the Sustainable Development Goals into Higher Education Teaching*. https://www.haw-hamburg.de/forschung/forschungsprojekte-detail/project/project/show/ditts/

Hermanto, Y. B., & Srimulyani, V. A. (2021). The challenges of online learning during the covid-19 pandemic. *Jurnal Pendidikan Dan Pengajaran, 54*(1), Artigo 1.

Huang, M. C., & Chiu, Y. P. (2020). A knowledge tension perspective on management control and performance in international joint ventures. *Journal of International Management, 26*(4), Article 100797.

Jin, J. L., & Wang, L. (2021). Resource complementarity, partner differences, and international joint venture performance. *Journal of Business Research, 130*, 232–246.

Ketlhoilwe, M. J., Silo, N., & Velempini, K. (2020). Enhancing the roles and responsibilities of higher education institutions in implementing the sustainable development goals. In Em G. Nhamo & V. Mjimba (Orgs.), *Sustainable development goals and institutions of higher education* (pp. 121–130). Springer International Publishing. https://doi.org/10.1007/978-3-030-26157-3_10

Kioupi, V., & Voulvoulis, N. (2020). Sustainable Development Goals (SDGs): Assessing the contribution of higher education programmes. *Sustainability, 12*(17), 6701.

Krishnamurthy, R., & Sahay, G. (2023). Higher education for sustainable development goals: bridging the global north and south. IN Em L. Caporarello, P. Kumar, & A. Agrawal (Orgs.), *Higher education for the sustainable development goals: Bridging the global north and south* (pp. 57–75). Emerald Publishing Ltd. https://doi.org/10.1108/978-1-80382-525-020231004

Langthaler, M., & Bazafkan, H. (2020). *Digitalization, education and skills development in the global South: An assessment of the debate with a focus on Sub-Saharan Africa* (Research Report 28). ÖFSE Briefing Paper.

Leal Filho, W., Shiel, C., Paço, A., Mifsud, M., Ávila, L. V., Brandli, L. L., Molthan-Hill, P., Pace, P., Azeiteiro, U. M., Vargas, V. R., & Caeiro, S. (2019a). Sustainable development goals and sustainability teaching at universities: Falling behind or getting ahead of the pack? *Journal of Cleaner Production, 232*, 285–294.

Leal Filho, W., Brandli, L. L., & Pretorius, R. (2022). *Completed project description form for submission to the DAAD*. Unpublished Document.

Leal Filho, W., Dibbern, T., Viera Trevisan, L., Coggo Cristofoletti, E., Dinis, M. A. P., Matandirotya, N., Sierra, J., Shulla, K., Buttriss, G., L'Espoir Decosta, P., & Mbah, M. F. (2023). Mapping universities-communities partnerships in the delivery of the Sustainable Development Goals. *Frontiers in Environmental Science, 11*, 1246875.

Leal Filho, W., Frankenberger, F., Salvia, A. L., Azeiteiro, U., Alves, F., Castro, P.,... & Ávila, L. V. (2021). A framework for the implementation of the Sustainable Development Goals in university programmes. *Journal of Cleaner Production, 299*, 126915.

Leal Filho, W., Pallant, E., Enete, A., Richter, B., & Brandli, L. L. (2018). Planning and implementing sustainability in higher education institutions: An overview of the difficulties and potentials.

International Journal of Sustainable Development & World Ecology, 25(8), 713–721. https://doi.org/10.1080/13504509.2018.1461707

Leal Filho, W., Shiel, C., & do Paço, A. (2015). Integrative approaches to environmental sustainability at universities: An overview of challenges and priorities. *Journal of Integrative Environmental Sciences, 12*(1), 1–14. https://doi.org/10.1080/1943815X.2014.988273

Leal Filho, W., Shiel, C., Paço, A., Mifsud, M., Ávila, L. V., Brandli, L. L., Molthan-Hill, P., Pace, P., Azeiteiro, U. M., Vargas, V. R., & Caeiro, S. (2019b). Sustainable development goals and sustainability teaching at universities: Falling behind or getting ahead of the pack? *Journal of Cleaner Production, 232*, 285–294. https://doi.org/10.1016/j.jclepro.2019.05.309

Leal Filho, W., Trevisan, L. V., Dinis, M. A. P., Ulmer, N., Paço, A., Borsari, B., Sierra, J., & Salvia, A. (2024). Fostering students' participation in the implementation of the sustainable development goals at higher education institutions. *Discover Sustainability, 5*(1), 22. https://doi.org/10.1007/s43621-024-00204-7

López Valladares, H., & Romero Lora, G. (2024). Creating social value through problem-based learning: Using podcasts in business education. In *The Palgrave handbook of social sustainability in business education* (pp. 243–259). Springer Nature Switzerland.

Mohieldin, M., Wahba, S., Gonzalez-Perez, M.A., & Shehata, M. (2023). *Business, government and the SDGs* (162 p.). Springer Nature.

Moreno Pires, S., Mapar, M., Nicolau, M., Patrizi, N., Malandrakis, G., Pulselli, F. M., Nicolau, P. M., Caeiro, S., Niccolucci, V., Theodossiou, N. P., Mancini, M. S., & Galli, A. (2022). Teaching sustainability within the context of everyday Life: Steps toward achieving the sustainable development goals through the EUSTEPs module. *Frontiers in Education, 7*, 639793.

Mweri, J. (2020). Sustainable development goals: Reaching people through their mother tongue. *Linguistics and Literature Studies, 8*(1), 14–25.

Nguyen, L. T. V., Cleveland, D., Nguyen, C. T. M., & Joyce, C. (2024). Problem-based learning and the integration of sustainable development goals. *Journal of Work-Applied Management, 16*(2), 218–234.

Paige, R. M., Fry, G. W., Stallman, E. M., Josić, J., & Jon, J. E. (2009). Study abroad for global engagement: The long-term impact of mobility experiences. *Intercultural Education, 20*(sup1), S29–S44.

Rajabifard, A., Kahalimoghadam, M., Lumantarna, E., Herath, N., Kin Peng Hui, F., & Assarkhaniki, Z. (2021a). Applying SDGs as a systematic approach for incorporating sustainability in higher education. *International Journal of Sustainability in Higher Education, 22*(6), 1266–1284.

Rajabifard, A., Kahalimoghadam, M., Lumantarna, E., Herath, N., Hui, F. K. P., & Assarkhaniki, Z. (2021b). Applying SDGs as a systematic approach for incorporating sustainability in higher education. *International Journal of Sustainability in Higher Education, 22*(6), 1266–1284.

Roll, K., MacLeod, M., Agbodjah, S., & Sánchez Siller, I. M. (2024). Learning in action: Embedding the SDGs through the reach alliance. *Sustainable Earth Reviews, 7*(1), 15.

Salvia, A. L., Leal Filho, W., Brandli, L. L., & Griebeler, J. S. (2019). Assessing research trends related to sustainable development goals: Local and global issues. *Journal of Cleaner Production, 208*, 841–849. https://doi.org/10.1016/j.jclepro.2018.09.242

Sareen, S., & Mandal, S. (2024). Challenges of blended learning in higher education across global north-south: A systematic and integrative literature review. *Social Sciences & Humanities Open, 10*, Article 101011.

Seraphin, H., & Gowreesunkar, V. G. (2021). Tourism: How to achieve the sustainable development goals? *Worldwide Hospitality and Tourism Themes, 13*(1), 3–8.

SET4HEI. (2023). *General guidelines for the implementation of sustainability in higher education institutions*. https://set4hei.org/2023/09/18/guidelines-for-the-implementation-of-sustainability-in-higher-education-institutions/

Sharma, J., & Varshney, S. K. (2023). Role of India in south–south cooperation to achieve sustainable development goals. In *Science, technology and innovation diplomacy in developing countries: Perceptions and practice* (pp. 201–217). Springer Nature Singapore.

Silviariza, W. Y., & Handoyo, B. (2021). Improving critical thinking skills of geography students with spatial problem based learning (SPBL). *International Journal of Instruction, 14*(3), 133.
Strickland, B. K., Brooke, J. M., Zischke, M. T., & Lashley, M. A. (2021). Podcasting as a tool to take conservation education online. *Ecology and Evolution, 11*(8), 3597–3606.
Tan, D. T., Siri, J. G., Gong, Y., Ong, B., Lim, S. C., MacGillivray, B. H., & Marsden, T. (2019). Systems approaches for localising the SDGs: Co-production of place-based case studies. *Globalization and Health, 15*, 1–10.
UNESCO. (n.d.). *The 17 Goals | Sustainable development*. https://sdgs.un.org/goals
UNESCO. (2020). Education for sustainable development: A roadmap. UNESCO. https://doi.org/10.54675/YFRE1448

Evaluation of the Integration of UN SDGs in Architecture Final Undergraduate Projects: Insights from Latin America

Ísis Portolan dos Santos, Elaise Gabriel, Jenifer Godoy Daltrozo, Raiane Pires Tolio, and Debora Cristiele Kummer

Abstract Final undergraduate projects in Architecture provide an opportunity for future professionals to address The United Nations Sustainable Development Goals (SDGs) by integrating them in the design solutions. In Brazil, these projects involve a comprehensive synthesis of knowledge acquired throughout the academic journey. This often culminates in a hypothetical architectural proposal related to some real social and urban challenges. Moreover, federal universities in Brazil are facing a curricular improvement that includes extension projects as a way to increase the connection between academia and society. This study aims to evaluate which and how UN SDGs are integrated across the different stages of the Architecture final projects at the Federal University of Santa Maria (UFSM), located in Southern Latin America. More than 20 designs were evaluated from 2020 to 2024 in five steps of design. The findings revealed the interplay between social requirements and the 17 SDGs, especially the number 3, 7, 10 and 17. Moreover, some SDGs applications increase during the design process, especially those related to water and energy use (7), that can be just added to building, without further integration. The social uses and community development are present in most projects, as a result of the theme chosen by the student to develop their Final Project. In that case, UFSM Undergraduate Architecture Courses can be helping to develop more social consciousness to students rather than just teaching how to apply any sustainable technique.

Í. P. dos Santos (✉) · R. P. Tolio
Architecture Urbanism and Landscape (PPGAUP), Federal University of Santa Maria (UFSM), Santa Maria, Brazil
e-mail: isis.santos@ufsm.br

E. Gabriel
Architecture Urbanism and Landscape (PPGAUP), Federal University of Santa Maria (UFSM), Cachoeira Do Sul, Brazil

J. G. Daltrozo
Architecture (PROPAR), Faculty of Architecture, Federal University of Rio Grande Do Sul (UFRGS), Porto Alegre, Brazil

D. C. Kummer
Architecture and Urbanism(FAUUSP), University of São Paulo (USP), São Paulo, Brazil

W. Leal Filho et al. (eds.), *Introducing the UN SDGs in Higher Education Institutions*, World Sustainability Series, https://doi.org/10.1007/978-3-032-23017-1_9

Keywords UN Sustainable Development Goals · Architectural education · Latin America · Final year projects · Sustainability assessment

1 Introduction

The world's population reached 8 billion in November 2022, with 55% living in urban areas, and the projection is to reach 70% by 2050 (UNSD, 2024). This rapid growth in global population and urbanization is based on the irregular expansion of small cities and intermediate towns without specific planning. Such patterns of expansion caused a variety of complex challenges, including limited access to efficient transportation, the formation of urban heat islands and increased vulnerability to natural disasters such as major floods, access to clean water and sanitation, among others (The World Bank, 2021).

To mitigate these problems, the architecture field rises as a practical instrument to approach environmental and social challenges. Architects play a key role in designing buildings and cities that improve human quality of life, foster sustainability, reduce social poverty, and improve human health and well-being (Thorne & Duran, 2016; Legény et al., 2024). As a result, there is an increasing recognition of the need to integrate sustainable practices into architectural education, equipping the current students and future professionals with the knowledge and skills to approach architecture more responsibly and sustainably (Mohamed & Elias-Ozkan, 2019; Patterson et al., 2021).

The growing emphasis on greener and sustainable cities has been fostered by several global organizations and policy frameworks. Initiatives such as the 17 Sustainable Development Goals, the "Decade of Education for Sustainable Development" both developed by the United Nations (UIA, 2018), highlight the critical role of architectural education in promoting sustainable development. According to several authors, sustainable architecture aligns directly with several SDGs, such as Goals 4, 7, 9, and 11 (Hendawy et al., 2024; Boarin & Martinez-Molina, 2022). These goals relate to quality education, affordable and clean energy, industry, innovation and infrastructure and sustainable cities and communities. Furthermore, there are other SDGs that also can be affected by architectural practices, such as climate action, good health and well-being and clean water and sanitation.

Considering these aspects, many architecture schools have integrated sustainable architecture into their curriculum (Yeretizan, 2020) and have addressed sustainable practices into architecture education by offering opportunities to learn sustainability values and behaviors through practical experiences of collaboration. For example, Trott et al. (2018) assessed how higher education allows the development of achieving SDGs of undergraduate students by providing research approaches through community engagement. This increasing incorporation of sustainable practices into architecture education has been fostered by several authors (Akgun et al., 2023; Ramadan & Abowardah, 2021; Hoskara et al., 2023; Sonetti et al., 2019). However,

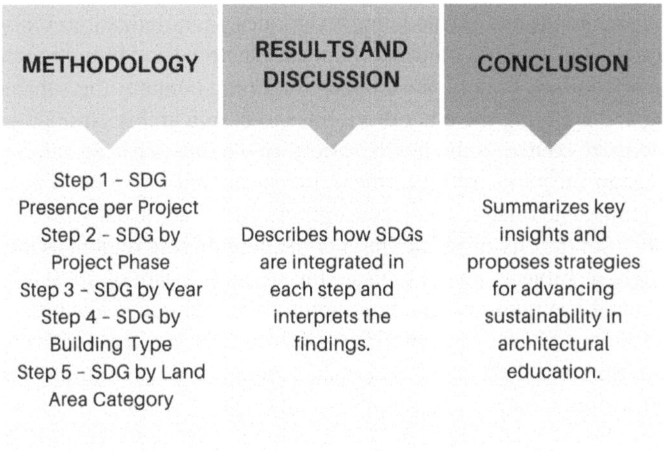

Fig. 1 Roadmap with Study structure and SGD Evaluation Steps

limited research focuses on the extent to which SDGs are integrated into architecture undergraduate projects in Latin America.

Since 2018, the Brazilian Federal Ministry of Education has been requiring outreach actions in the curricula (MEC, 2018), and together with education and research, they formed the three branches of universities. Outreach is how a university extends its expertise to benefit the external community, proposing hypothetical solutions to real problems as a way to learn how to transfer the knowledge made inside the academy. Based on this, this paper aims to investigate if and how graduate students incorporate the United Nations Sustainable Development Goals (UN SDGs) across the different stages of final architecture projects at the Federal University of Santa Maria, and what patterns or gaps can be identified in their implementation. Figure 1 shows the roadmap with study structure applied in this paper.

2 Methodology

The Architecture and Urbanism undergrad courses in Brazil endure 5 years, within 10 periods and diverse courses regarding urbanism planning, landscape design and architecture itself. Moreover, both landscape architecture and urbanism Brazilian careers are part of the same degree. The subjects are usually divided into fields such as drawing, social and economic studies, history, environmental comfort, urban and regional planning, construction technology, and design practices.

At UFSM, the Architecture and Urbanism curriculum is based on four major areas: reflection, representation, materialization and conception. Reflection comprehends the architectural theory and history of art, architecture and society. Representation is a combination of drawing and 3D modeling techniques. Materialization combines technical subjects related to physics and Mechanical, Electrical and Plumbing (MEP) and environmental comfort. Finally, the Conception group contains the subject focused on resuming and applying the other disciplines to design architecture projects.

The undergrad course embodies a variety of themes such as interior design, landscape design for parks, refurbishment, industrial and office buildings, city and regional planning, and others. These conception courses are usually related to social requirements from the city of Santa Maria or the neighboring region, as the outreach curriculum. These subjects intend to make the students aware of all SDG questions presented in local communities, and how architecture, landscape and urban planning can improve them to have a better society. All designs are hypothetical proposals based on external world demands that social groups and associations send to the UFSM Architecture Course. The groups usually call for assistance for some challenges they are facing related to their buildings or land use. Furthermore, the university can not design a formal architecture project because the students do not have legal attribution to it. Alternatively, the university commonly assists these groups by providing field measurement and theoretical studies about the edifice type. The proper architectural designs made by the students are only used to grade and complete the student evaluation.

The Final Design, which is the object of this work, is the last course that students take before graduation. This course also promotes interdisciplinarity, both in its diagnostic phase and in the final product. During the initial stage, students must conduct a comprehensive survey of the context, including physical, social, cultural, environmental, technical, and economic aspects, which informs the entire development of the proposal. These analyses are essential to ensure the project's responsiveness to real and complex conditions. In the final stage, students are required to present a complete architectural solution that includes complementary technical elements, such as structural, hydraulic, electrical, and environmental comfort systems. These components encourage interaction with multiple fields of knowledge, ensuring a more integrated and systemic architectural approach. It is composed of two subjects: in the 9th semester, with a theoretical thesis analyzing the type of building and the urban area with selected site location, and then; in the 10th semester, the architecture design course is composed by a pre-design and the Design Development. The theme and the advisor are chosen by the student, and they decide together about the complexity and range the design will take.

In this study, 21 students designed diverse works from landscape river parks to healthcare institutions. They were all located in South of Brazil, most in Santa Maria, the same city as the UFSM, one in Florianopolis (capital of Santa Catarina State) and other in a small city (less than 100,000 habitants) of Rio Grande do Sul state (Fig. 2). All students were from the same advisor, a professor who currently works with sustainable technologies and SDG. Each project was analyzed through general classification: typology (Architecture or Landscape Architecture), title, year, location, area and Architectural category (Educational, institutional, commercial and offices, healthcare, industrial or cultural buildings).

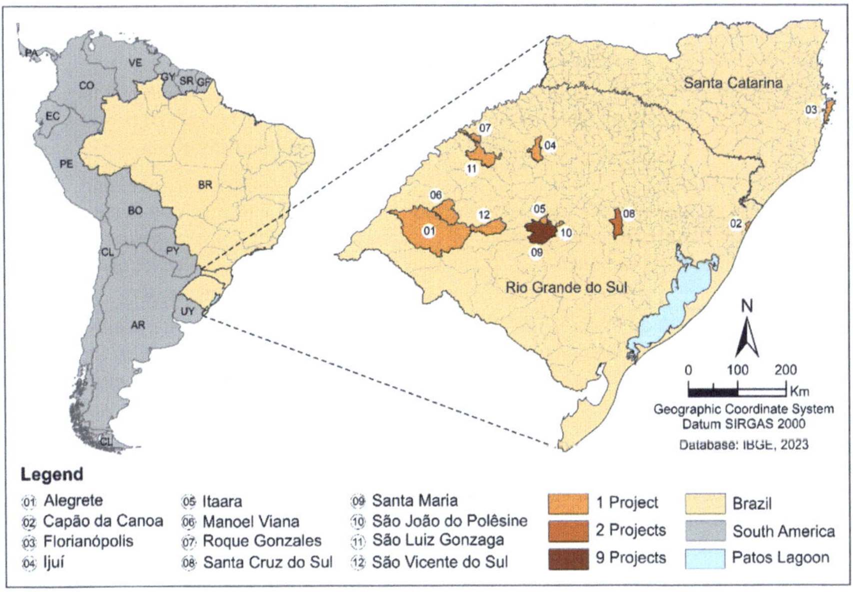

Fig. 2 Location of Architectural Final Design proposals from 2020 to 2024

After that, the SGD solutions were searched for in each step of the design process. In the 9th semester course identifying whether the main goal was a response to a social request or just a business proposition. Social requirement refers to a genuine need or issue within society that requires attention and it is already demanded by some community association or other group of people. As the professors used to choose during the previous courses, at this moment the student has the opportunity for a personal choice this time. The choice for a social requirement can reflect the incentive within outreach actions during the undergrad. The business propositions are made just to create a complexity architecture to a Final Design. Afterwards, the 17 SDGs were converted to architectural applications of each concept based on UIA (2018) to be looked for in each student design proposal (Table 1).

After the SDG classification, all final designs were merged to observe the similarities and prevalence of common features in five steps. At first, the examination evaluated the presence and number of SDGs in the proposals. In the second step, SDGs features were examined in the three phases of the Final design course: precedent research; concept and design phase; and finally, the Design Development. The SDGs presence were identified in each phase to notice if they increase or decrease during architecture planning. In the third step, SDGs presence were ranked by year, to understand if they have been increasing. In the fourth step, projects were classified by type of building and SDG presence, to understand if some of them have more environmental coherence. Finally, the projects were classified by land area use. Sturges rule was applied to determine the 6 categories of 21 samples and De Smith et al. (2024) methodology was used to create the categories threshold combining similar

Table 1 Application of SDG in Architecture. Adapted from UIA (2018)

United Nation Sustainable Development Goals		Architectural features
1	**No Poverty (end poverty in all forms everywhere)**	Affordable, safe, and health buildings and settlement, such as in social housing, co-ops and urban upgrading
2	**Zero Hunger (end hunger, achieve food security and improve nutrition and promote sustainable agriculture)**	Protection existing ecosystems and preservation and expansion of areas for food production, such as urban farming, food-productive cooperatives and regenerative landscape design
3	**Good health and well-being (ensure healthy lives and promote well-being for all at all ages)**	Building design to enable a health, hygiene at indoor air, light, acoustic and thermal comfort and also health care buildings and urban planning and landscapes with public sports place
4	**Quality education (ensure inclusive and equitable quality education and promote lifelong learning opportunities for all**	Schools, Universities, and other educational buildings that enable a learning environment, and even educational spaces in other building categories
5	**Gender equality (achieve gender equality and empower all women and girls)**	Buildings and urban areas that promote security, extensive and safe access to girls, women and LGBTI + citizens
6	**Clean water and sanitation (ensure availability and sustainable management of water and sanitations for all)**	Use of rainwater before been wasted or polluted; protections of freshwater and urban design to withstand climate change related to drought and floods
7	**Affordable and clean energy (ensure access to affordable, reliable, sustainable and modern energy for all)**	Buildings and structures that are energy producers and explore and bioclimatic architecture, and limit their energy consumption and material embodied energy
8	**Decent work and economic growth (promote sustained, inclusive and sustainable economic growth, full and productive employment and decent work for all**	Safe public space and affordable transit routes, buildings that promote marketplace for local producers and handicrafts, or buildings that create a healthy environment for employees
9	**Industry, innovation and infrastructure (build resilient infrastructure, promote inclusive and sustainable industrialization and foster innovation)**	Industries that produce less waste and use less energy or provide solutions safer and healthier than current standards
10	**Reduce inequalities (reduce inequality within and among countries)**	Accessible buildings and urban places that consider physical, mental and cultural diversity and socially responsible

(continued)

Table 1 (continued)

United Nation Sustainable Development Goals		Architectural features
1	**No Poverty (end poverty in all forms everywhere)**	Affordable, safe, and health buildings and settlement, such as in social housing, co-ops and urban upgrading
11	**Sustainable cities and communities (make cities and human settlements inclusive, sofa, resilient and sustainable**	Sustainable design and technologies applied to buildings and cities environment to increase the quality to the human life and to the environment to a more resilient world
12	**Responsible consumption and production (ensure sustainable consumption and production patterns)**	Building Materials and components that produce less waste, reuse of existing buildings and planning new buildings for diverse and long uses
13	**Climate action (take urgent action to combat climate change and its impacts)**	Diminish the CO2 footprint of buildings with efficient energy consumptions, integrating renewable materials and sustainable technologies
14	**Life below water (conserve and use the oceans, seas and marine resources for sustainable development)**	Landscapes and urban planning that reduce cost and add benefits to water-managing infrastructure and also regenerative design to lands close to water
15	**Life on land (protect, restore and promote sustainable use of terrestrial ecosystems, sustainably manage forests, combat desertification, halt and reverse land degradation and halt biodiversity loss)**	Buildings, landscapes and urban planning that preserve ecosystems and biodiversity and also equipment that endure sustainable tourism and environmental education
16	**Peace, justice and strong institutions (promote peaceful and inclusive societies for sustainable development, provide access to justice for all and build effective, accountable and inclusive institutions at all levels**	Local community centers or public buildings that promote inclusion, welcoming, security and non-discrimination
17	**Partnership for the goals (strengthen the means of implementation and revitalize the global partnership for sustainable development)**	Proposals related to projects that involves governments, institutions, research, business and citizens and public–private partnership

areas (Table 2). The second step (presence of SDG in each phase) used absolute number to classify the design phases. All the other phases (first, third, fourth and fifth step) considered the proportion of number of SDG related to the total number of projects.

Each evaluation phase observed the presence of SDGs and tried to elucidate the similarities and motivations of how architectural students applied the UN goals.

Table 2 Categories of designs ranked by area

Category	1	2	3	4	5	6
Inferior threshold (m^2)	2,250	28,000	45,854	91,861	300,000	2,471,000
Superior threshold (m^2)	12,653	28,477	57,213	140,000	300,000	2,471,000
Number of Final Design Projects	11	2	3	3	1	1

Therefore, these results could be used by professors to motivate students to spread the SGD applications in architectural designs.

3 Results

In this section, an analysis of each step presented in the Methodology is shown. At first, the examination evaluated the presence and the quantity of SDGs in each project, considering whether the project goal was a response related to a social requirement or to a business proposition. Examining the participation of the two main goals, it is evident that the majority of proposals provided insights into social requirements (13 projects), such as Institutional and Cultural Buildings, while 8 projects aimed at business propositions, as Commercial and Office Buildings. The preference for addressing real social problems could reflect the influence of outreach initiatives during undergraduate studies.

Along with understanding the main goal of projects, assessing the percentage of SDGs in each goal is essential. The SDGs most frequently cited in the proposals addressing social requirements were SDG 3 (Good Health and Well-being), SDG 7 (Affordable and Clean Energy), SDG 10 (Reduce Inequalities), SDG 12 (Responsible Consumption and Production), and SDG 17 (Partnership for the Goals). These reflect a strong focus on health, sustainability, equality, and collaborative efforts to address critical social challenges. In business propositions, the most cited SDGs were: SDG 3, SDG 7, SDG 13 (Climate Action), and SDG 17. SDG 11, about sustainable cities is usually the most cited in architecture applications, but it did not receive an expressive mark in this work because only sustainable urban applications were considered and not the strategies applied to a singular building.

Contrariwise, SDG 1 (No poverty) and SDG 9 (Industry, Innovation and Infrastructure) did not appear in any project involving business propositions. Notably, the SDGs most frequently cited across both goals showed a significant overlap, with a shared focus on health, energy, and collaboration, and were featured in more than 75% of the proposals (Fig. 3).

Figure 4 shows the 3 projects with more application of SDGs in the Final Design course. Peiter (2024) is a proposal to a Technology Park at Santa Cruz do Sul, a small city in Rio Grande do Sul which has an economy dominated by tobacco farm and factories. This city also has high Human Develompemtn Index (HDI) compared to other cities in the state of Rio Grande do Sul, and holds an enthusiastic group of

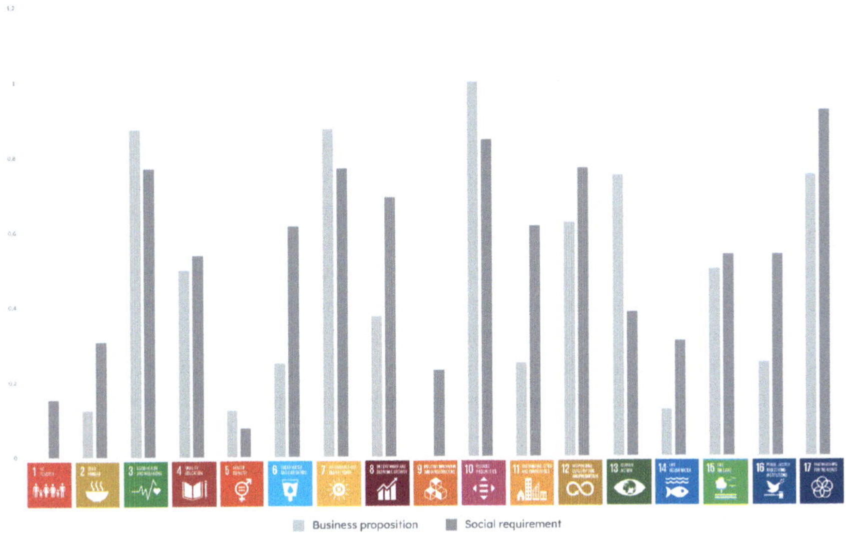

Fig. 3 SDG Presence in business proposition vs. social-requirement

entrepreneurs that already have an association to invest in technology and innovation. The architectural project was designed to fulfill this aim, and incorporate many innovative technologies such as building integrated photovoltaic and water reuse. Also, the park itself was designed to accommodate popular use, as outdoor concerts and trails in the natural presevation area. This project enlists 13 SGDs.

The project from Baier (2024) applied 15 SGDs in the proposal to create a community center for a small river community, impacted many times by Ibicui river floods. This proposal contemplated a multiuse center with space for social gathering, library, ballroom and sports and recreational space for children. This proposal uses architectural features to promote thermal comfort and a sport court that works also as a retention basin in eventual floods.

The Mazzardo (2021) proposal contemplated a neighborhood planning on Carlos Marighella Community, a settlement to landless families living in the Santa Maria skirts between urban and rural areas. This proposal integrates rural–urban necessities

a)  b)  c) 

Fig. 4 Final designs with more application of SDGs in the Final Design course. Source: a) Peiter (2024) b) Baier (2024) and c) Mazzardo (2021)

and provides design for houses, health care centers and urban infrastructure. In addition, each solution considered sustainable techniques based on permaculture design and many partnerships between public institutions, resulting in the application of 10 SGDs.

This integration of the SDGs in the students' proposals reflects a holistic design approach, in which environmental impact, social inclusion, user well-being, and architectural quality are addressed together. Rather than treating sustainability as a technical add-on, the projects demonstrate how diverse and complex demands — including psychological comfort, accessibility, biodiversity, and community needs — can be incorporated from early design stages. This perspective resonates with Vitruvius's classical triad — *firmitas* (solidity), *utilitas* (utility), and *venustas* (beauty) — now expanded to include social and environmental responsibility, reaffirming architecture as a multidimensional discipline with the potential to respond to current societal challenges.

In the second phase, the integration of SDGs was analyzed across the three phases of the Final design project: Precedent research, concept and design phase, and Design Development. During the Precedent research phase, the SDGs most frequently addressed were SDG 17 (Partnerships for the Goals) and SDG 10 (Reduced Inequalities). These appeared in 19 and 18 projects, respectively, reflecting a strong initial focus on collaboration and social equity. SDG 3 (Good Health and Well-being), SDG 4 (Quality Education), and SDG 7 (Affordable and Clean Energy) also featured prominently, with 15 and 10 and 10 mentions, respectively, indicating attention to health, education, and energy concerns early in the process. However, goals like SDG 1 (No Poverty) and SDG 5 (Gender Equality) were significantly less emphasized, appearing in only two projects.

As projects advanced through the Concept and Design Phase and Design Development, SDG 7 showed the most notable increase, rising from 10 mentions in the initial phase to 15 in the final stage, reflecting growing attention to energy solutions (Fig. 5). SDG 15 (Life on Land) also increased slightly, while SDG 10 saw a minor decline, dropping from 18 mentions in the first phase to 16 in the Concept and Design Phase, recovering to 17 in the final stage. Most other SDGs, including SDG 17, remained consistent across all stages, indicating steady attention to partnerships. The findings suggest that while some goals gained prominence, challenges remain in sustaining broader SDG integration, particularly in the later design phases. The most significant increases were observed throughout the process in SDG 7 (Affordable and Clean Energy), which consistently rose across all stages, highlighting the growing attention to energy solutions in architectural design. This trend could indicate that technical aspects of sustainability become more apparent as projects advance toward detailed design. Similarly, SDG 15 (Life on Land) saw a noticeable increase from 9 mentions in the Precedent Research phase to 11 in subsequent stages, underscoring an increased focus on biodiversity and land use. Conversely, SDG 10 (Reduced Inequalities) saw a reduction from 18 mentions in the initial phase to 16 in the Concept and Design Phase and 17 in the Design Development phase, indicating a slight decline in its integration as projects advanced. The prominence of SDG 10 in the early stages aligns with the region's pressing social disparities, likely influencing students' initial project choices. However, its decline in later phases may reflect challenges in maintaining these priorities as designs become more focused on technical and formal aspects.

In the third step, the presence of SDGs was ranked by year, based on the total number of times SDGs were cited each year, divided by the number of projects in that year. Data indicated that the average number of SDGs cited in each project decreased from 9 per project in 2020 to 8 in 2021, then to 7 in 2022 and 2023, before increasing to 10 in 2024. The decrease of SDG citation could be related to the pandemic situation of COVID-19 during 2020 and 2021, when all classes were online. Thus, the students had much less contact with communities and practical activities at real locations in the city, what could explain the minor number during 2022 and 2023, but these would need more research later on.

In the fourth step, the analysis of building types and their alignment with SDGs revealed significant variations in how different typologies addressed sustainability (Fig. 6). While none of the categories covered all 17 SDGs, Cultural Buildings demonstrated the most engagement, addressing 15 goals. SDG 17 and SDG 10 were the most frequently represented, reflecting a focus on collaboration and biodiversity. Some of these goals were demonstrated in the projects through the inclusion of people from different cultural and social backgrounds, landscapes and urban planning, relationships with the natural environment and regenerative design.

However, some goals, such as SDG 5 (Gender Equality) and SDG 9 (Industry, Innovation, and Infrastructure), were not addressed, suggesting areas where the inclusion of social development aspects could be improved. The absence of SDG 5 is particularly relevant, as it indicates a gap in how gender equality is considered in architectural education and practice. Although gender issues are not traditionally

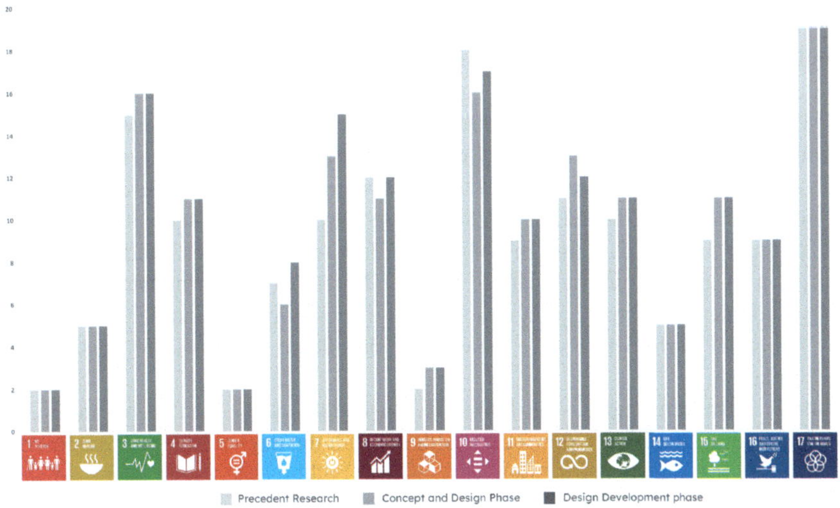

Fig. 5 UN SDGs Quantity X Stage

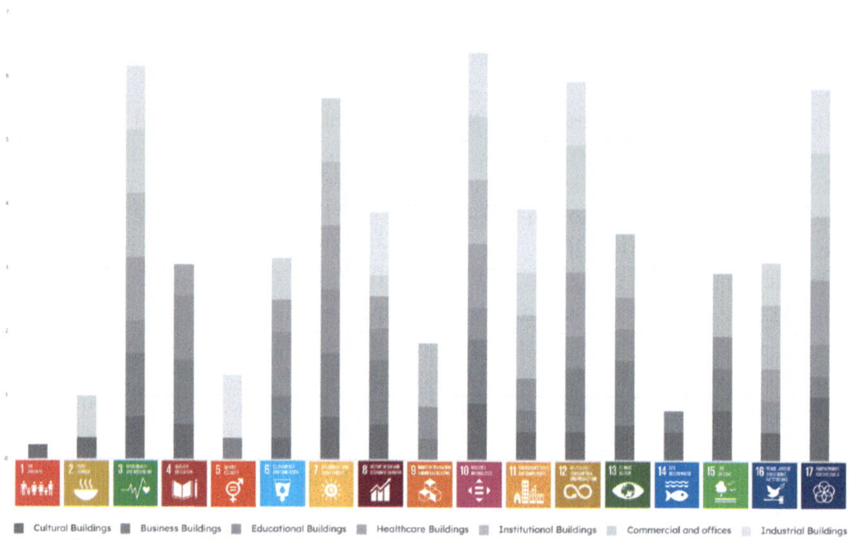

Fig. 6 Distribution of SDGs Across different building typologies

central to architecture, recent studies (Sadeghi & Jangjoo, 2022; Yadav & Jumari, 2024) have begun to highlight how the built environment can be designed to enhance women's safety, mobility, and access to public spaces. The discussion of cities for women, which considers elements such as lighting, accessibility, and the equitable distribution of public spaces, is still emerging but needs to be more explicitly integrated into architectural education. The way cities are planned, and buildings are designed can reinforce or mitigate gender disparities, making it essential to rethink how architecture can contribute to more inclusive and equitable urban spaces.

Industry and innovation may be less cited due to the traditional barrier between architecture and engineering. Although the undergraduate curriculum includes a course on industrial architecture, students often perceive this field as too rigid and lacking creative freedom, which may discourage them from choosing it as a Final Design Project (TFG) theme. The preference for projects that allow greater formal and conceptual experimentation might explain why industrial and infrastructure-related topics are less explored in comparison to other typologies. Addressing this perception in architectural education, perhaps by emphasizing innovation, adaptability, and interdisciplinary collaboration, could encourage more students to engage with industrial architecture as a field with significant potential for sustainable transformation.

Business Buildings addressed 14 SDGs, particularly SDG 3, SDG 10, and SDG 7. These findings indicate a focus on health through the daylighting and ventilation of the buildings, the requalification of a historic building, and solar photovoltaics (BAPV) for example. Equity, energy efficiency, sustainable materials and technologies, landscaping, and areas reserved for biodiversity were present in this group. Educational Building covered 13 SDGs, with SDG 4, SDG 6, and SDG 7 being the most frequently cited. Some projects in this category implemented strategies for conducive learning environments, such as flexible layouts that accommodate diverse teaching methods and collaborative learning. Others support inclusive and ensuring accessibility and addressing various cognitive and social needs.

Healthcare Buildings addressed 10 SDGs, with SDG 3 appearing in all analyzed projects, reflecting their intrinsic focus on health promotion. Several designs incorporated strategies to foster health through sports and wellness facilities, accessibility features, and the integration of public squares to encourage community interaction. Projects often targeted specific populations, such as centers designed for individuals on the autism spectrum, aquatic training centers, and spaces that prioritize inclusivity and adaptability. Moreover, many of these facilities extended their focus beyond individual health to address broader social goals, creating democratic spaces that cater to communities in vulnerable situations.

In the Institutional building case, the project emphasized a strong relationship between landscape design and the natural environment, integrating daylight and ventilation, photovoltaic solar, and biophilic architecture. Sustainable materials and technologies, alongside the design that preserves biodiversity, further demonstrated a focus on environmental stewardship. Additionally, it included proposals for city hall and other government-related facilities, highlighting their role in fostering civic engagement and supporting public administration.

Industrial Buildings exhibited the lowest engagement, addressing only 8 SDGs, with SDG 8 (Decent Work and Economic Growth) being the most frequently represented. Despite their narrower scope, these projects demonstrated innovative approaches by fostering relationships with the natural environment and also promoting women's participation in the workforce, and incorporating accessibility features.

Additionally, some designs included solid resource management systems and partnerships between institutions to enhance operational efficiency and sustainability. The inclusion of extension programs also highlighted the potential of these facilities to contribute to broader educational and social goals. These strategies indicate that, while Industrial Buildings primarily focus on economic growth, they can also integrate environmental and social dimensions when aligned with broader sustainability objectives.

Finally, the projects were classified into six categories based on land area use, and the analysis considered the presence of SDGs by number of projects (Fig. 7). Smaller projects, from category 1 of land area use presented a broader application of SDGs (17 SDGs), addressing multiple sustainable goals simultaneously. The same occurs for category 4 (land area use from 91,861 m^2 to 140,000 m^2), covering 15 SDGs. The land area use with fewer cited SDGs is category 5 (9 SDGs).

Certain SDGs stand out in terms of prominence. SDG 10 (Reduce Inequalities) and SDG 17 (Partnership for the Goals) have the tallest bars, indicating that these goals are the most frequently addressed across all categories of land area use. Conversely, SDG 1, SDG 5 and SDG 14 show relatively fewer land area use variation. The number of projects on each category was also analyzed. Around 52% of projects were classified on category 1 of land area use, followed by 14.3% on categories 3

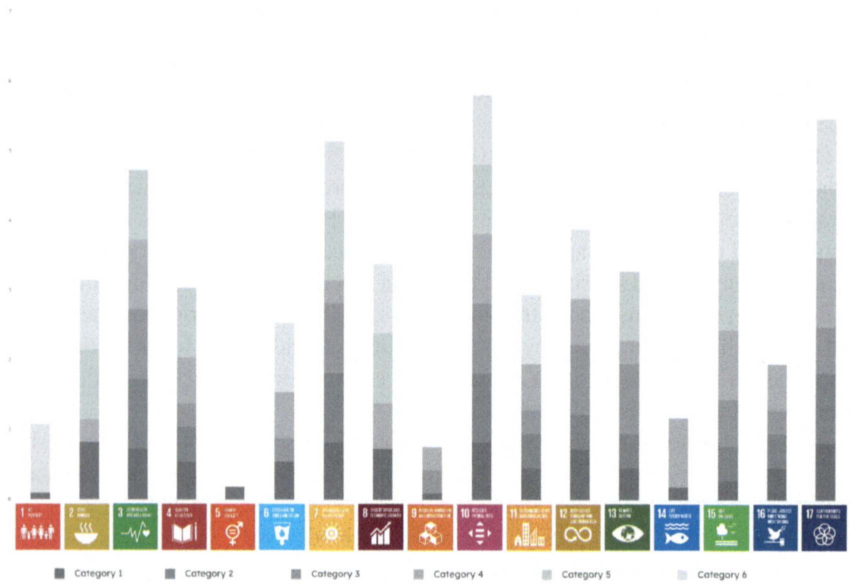

Fig. 7 SDG presence based on land area use

and 4 each. The categories with lesser number of projects were categories 5 and 6, with 1 project each.

The comparison between the frequency and the presence of several SGDs shows they are widely applied to architecture. Nevertheless, these projects are all hypothetical, so publicizing this to governments could be an important action for further studies. This evaluation only shows that Brazilian graduates from UFSM are aware of social and environmental consciousness, but we do not know if they will keep this intention in other designs, nor if the public buildings will be designed in this way with all this SGD consciousness.

4 Conclusion

The study findings underscore the strong relationship between architectural design as a way to implement the Sustainable Development Goals (SDGs). All projects evaluated had a massive presence of SDG as a result of some architectural solutions and the outreach initiatives from the university. This represents the close relation from architecture to social sciences, although it is a course mainly composed of subjects related to mathematics, physics and arts.

As previous research about architecture and SDGs, the Affordable and clean energy goal (SDG 7) has a massive presence in the projects. But the others SDG applications were different in this research. While Hendawy et al. (2024) and Boarin &

Martinez-Molina (2022) cite Education, Infrastructure and Sustainable cities (SDG 4, 9 and 11) as the most related to architectural projects, this work encountered Health, Reduce inequalities and Partnership (3, 10 and 17) as the most mentioned SDG applications at the Final Projects of Federal University of Santa Maria during 2020 and 2024. The previous works could observe Education, infrastructure and sustainable cities as traditional fields related to architecture. Consequently, it would be easier to integrate innovations including SDG with the designs. However, these undergrad students choose their project theme motivated by many other aspects, and in this case, most related to social requirements that they observe during graduation time. As a result, they could find other ways to promote sustainable development in architecture, presenting challenges related to Health, Inequalities and Partnerships. All demands are closely related to social justice regarding community engagement and health infrastructure, which are common social problems in Brazil. It is also endorsed by the high number of projects that answer a social requirement (more than 60%) and the most cited 10 SGD, which aims to reduce inequalities. Moreover, the three projects with more SDG solutions are all outgrowth from established community organizations.

These challenges reinforce the idea that architectural design plays a key role in promoting not only technical performance, but also physical, emotional, and social well-being. In this sense, many of the projects — especially those focused on health and community infrastructure — reflect the concept of a healthy building that supports both individual care and collective integration within the built environment.

About project phases, most SDGs maintain their presence during the whole process. Just some of them had a slight increase. Some related to constructive details do not appear in the theoretical research but appear in the end as a complement (SDG 6 and 7). This could represent more environmental concern when the project is already developed, but this also shows that some strategies are just applied to the design and not considered as fundamental concept and are really integrated to the proposal. This can result in more traditional integration of sustainability in design and not a real composition that considers and embraces sustainability as a concept and promotes new solutions. The water and energy use have this behavior, which can confirm the traditional use of BAPV (building applied photovoltaic) rather than BIPV (building integrated photovoltaic) as cited by Aguacil et. al (2024).

The annual evaluation demonstrates similar measures during the years, with a slight decrease after the COVID-19 pandemic. This outcome may be attributed to online learning (available during isolation times), which limited students' engagement with real communities and practical fieldwork. However, the recovery in SDG application observed in 2024 indicates a gradual reintegration of sustainability-oriented approaches as in-person activities resumed. The building types assessment presents the Cultural Buildings with more adherence to SDGs, and the Industrial buildings as a theme still to be explored by the students. This also shows that SDG 5 is the least cited of any project, which could show a direction to the Professor to understand how Architecture and Urbanism could contribute to this challenge.

The area assessment demonstrates that smaller projects can address more sustainable goals, and may indicate that students in these cases hold more control over all project definitions or have more time to draw and detail each intention proposed.

This appraisal about UFSM Architecture Final courses could reflect the influence of outreach initiatives integrated to curriculum that results in large application of SDGs. Furthermore, even though Architecture has some traditional application techniques to increase environmental sustainability in cities, it also holds a potential to contribute to social development promoting places to integrate communities and the environment.

References

Aguacil, S., Duque, S., Lufkin, S., & Rey, S. (2024). Designing with building-integrated photovoltaics (BIPV): A pathway to decarbonize residential buildings. *Journal of Building Engineering, 110486*, 2352–102. https://doi.org/10.1016/j.jobe.2024.110486

Akgun, Y., Erdoğdu Erkarslan, Ö., & Neşeliler, P. (2023). A guide for a guide: Using UIA publications for an SDG-focused studio. *Archnet-IJAR, 17*, 443–458. https://doi.org/10.1108/ARCH-11-2022-0245

Baier, J. D. O. (2024). Community Center: a space for socializing and support for Balneario Passo Do Umbu—RS. Architecture and Urbanism Final Thesis. Advisor Ísis Portolan dos Santos. Federal University of Santa Maria. Santa Maria.

Boarin, P., & Martinez-Molina, A. (2022). Integration of environmental sustainability considerations within architectural programmes in higher education: A review of teaching and implementation approaches. *Journal of Cleaner Production, 342*, Article 130989. https://doi.org/10.1016/j.jclepro.2022.130989

De Smith, M. J., Goodchild, M. F., & Longley, P. A. (2024). *Classification and clustering. in geospatial analysis: a comprehensive guide to principles, techniques and software tools*, 7th ed. Winchelsea Press. https://www.spatialanalysisonline.com/HTML/index.html?classification_and_clustering.htm

Hendawy, M., Junaid, M., & Amin, A. (2024). Integrating sustainable development goals into the architecture curriculum: Experiences and perspectives. *City and Environment Interactions, 21*, Article 100138. https://doi.org/10.1016/j.cacint.2023.100138

Hoskara, S., Akgun, Y., Erkarslan, O., & Koc, O. (2023). Integration of the UN 2030 Sustainable Development Goals into architectural education in Turkey. In *World Congress of architects* (pp. 815–833). Springer International Publishing.

Legény, J., L·paček, R., Hubinský, T., & Benkovičová, L. (2024). A triangular architectural relation model among sustainability, beauty, and power. *Frontiers of Architectural Research, 13*, 699–711. https://doi.org/10.1016/j.foar.2024.02.011

Mazzardo, R. W. (2021). *Social Center for landless settlements—Carlos Marighella Community*. Architecture and Urbanism Final Thesis. Advisor Ísis Portolan dos Santos. Federal University of Santa Maria. Santa Maria.

MEC. Portary 1350 (2018, December 17). Diario Oificial da União. http://portal.mec.gov.br/index.php?option=com_docman&view=download&alias=102551-pces608-18&category_slug=novembro-2018-pdf&Itemid=30192

Mohamed, K. E., & Elias-Ozkan, S. T. (2019) Incorporating sustainability principles into architectural design education: Results of an experimental design studio. *Journal of Green Building, 214*:143–158. https://meridian.allenpress.com/jgb/article/14/3/143/425983/Incorporating-Sustainability-Principles-Into

Patterson, J., Wyborn, C., Westman, L., Brisbois, M. C., Milkoreit, M., & Jayaram, D. (2021). The political effects of emergency frames in sustainability. *Nature Sustainability, 4*, 841–850. https://doi.org/10.1038/s41893-021-00749-9

Peiter, N. L. (2024). *GAUTEN: Innovation and technology park of Santa Cruz Do Sul—RS*. Architecture and Urbanism Final Thesis. Advisor Ísis Portolan dos Santos. Federal University of Santa Maria. Santa Maria.

Ramadan, M. G., Abowardah, E. S. (2022). Incorporating a sustainability approach in teaching architectural design studio. *The International Journal of Design Education, 17*, 37–64. https://doi.org/10.18848/2325-128X/CGP/v17i01/37-64

Sadeghi, A. R., & Jangjoo, S. (2022). Women's preferences and urban space: Relationship between built environment and women's presence in urban public spaces in Iran. *Cities*. https://doi.org/10.1016/j.cities.2022.103694

Sonetti, G., Brown, M., & Naboni, E. (2019). About the triggering of UN sustainable development goals and regenerative sustainability in higher education. *Sustainability, 11*, 254. https://doi.org/10.3390/su11010254

The World Bank. (2021). *The cities of the future in Latin America: Fewer cars, fewer youth*. Washington, DC, USA. https://www.worldbank.org/cn/news/feature/2017/10/05/ciudades-del-futuro-en-america-latina

Thorne, M., & Duran, P. (2016). *The role that architecture can play in the development agenda*. https://www.devex.com/news/the-role-that-architecture-can-play-in-the-development-agenda-88124

Trott, C. D., Weinberg, A. E., & McMeeking, L. B. S. (2018). Prefiguring sustainability through participatory action research experiences for undergraduates: reflections and recommendations for student development. *Sustainability, 9*, 3332. http://dx.doi.org.ezproxy.aub.edu.lb/10.3390/su10093332

UIA. (2018). *An architecture Guide to the UN 17 sustainable development goals*. Copenhagen. https://www.uia-architectes.org/en/resource/architecture-guide-to-the-un-17-sustainable-development-goals-english/

UNESCO. (2005). *Decade of education for sustainable development: 2005–2014*. Draft International Implementation Scheme.

United Nations Statistics Division (2024) Goal 11. https://unstats.un.org/sdgs/report/2023/goal-11/

Yadav, A., & Kumari, R. (2024). Gender safety perspective in urban planning: The case of pedestrian mobility in Kanpur city. *Cities, 147*. ISSN 0264-2751. https://doi.org/10.1016/j.cities.2024.104845

Yeretzian, A. (2020). Pedagogical methodologies to achieve SDGs in developing economies. In *International Conference on Sustainable Development*. https://ic-sd.org/wp-content/uploads/2019/11/Aram_Yeretzian.pdf

Integrating the SDGs into the Ongoing Activities of Lithuanian Higher Education Institutions

Skaidrė Žičkienė and Zita Tamašauskienė

Abstract Since adopting the United Nations Transforming our World: The 2030 Agenda for Sustainable Development, the SDGs have become an integral part of universities, serving as a flexible and comprehensive framework for institution-wide initiatives. HEIs can contribute to implementing the SDGs by integrating sustainability principles into their curricula, research activities, and daily operations. They also serve as hubs for innovation and solutions to global challenges by collaborating with local communities, businesses, and policymakers. Nevertheless, the extent of HEIs' engagement and the specific SDGs they prioritize remain insufficiently defined. This study addresses this gap by examining four Lithuanian state colleges' most recent Social Responsibility Reports (hereafter SRRs), focusing on their alignment with SD and the SDGs. To date, only two institutions have formally identified priority SDGs and established corresponding targets and indicator frameworks. This research adopts the perspective that SRRs function as an entry point for advancing SDG implementation, as these reports reflect the colleges' activities related to a broad spectrum of SDG objectives. Research indicates that colleges are most engaged in advancing Goal 4 (Quality Education), which is strongly interconnected with Goal 5 (Gender Equality), Goal 10 (Reduced Inequalities), and Goal 3 (Good Health and Well-being). In working towards these objectives, institutions prioritize high-standard education, gender equity, social inclusion, and the well-being of students and staff while extending their activities to community partnerships and fostering responsible citizenship. The SRR prepared by colleges tends to emphasize descriptions of completed actions, highlighting achievements rather than addressing areas for improvement or existing challenges. Nevertheless, these reports provide insights into the colleges' contributions toward the SDGs. Notably, only two of the reviewed colleges have established measurable indicators to evaluate their progress in delivering on the SDGs, which suggests that the systematic integration of the SDGs into institutional strategies and operations remains limited.

S. Žičkienė (✉) · Z. Tamašauskienė
Regional Development Institute, Vilnius University Šiauliai Academy, Šiauliai, Lithuania
e-mail: skaidre.zickiene@sa.vu.lt

Z. Tamašauskienė
e-mail: zita.tamasauskiene@sa.vu.lt

W. Leal Filho et al. (eds.), *Introducing the UN SDGs in Higher Education Institutions*, World Sustainability Series, https://doi.org/10.1007/978-3-032-23017-1_10

Keyword Sustainable development · Sustainable Development Goals · Higher education institutions

1 Introduction

The United Nations has taken a prominent role in coordinating efforts to address critical global challenges affecting humanity. A noteworthy initiative was initiated by adopting the Millennium Development Goals (MDGs) during the United Nations Millennium Declaration in (2000), which outlined eight goals to be achieved by 2015 (UN Millennium Declaration, 2000). Although there was moderate success in realizing these goals, particularly in transforming developing nations, the outcomes prompted the establishment of a broader agenda: "Transforming our world: the 2030 Agenda for Sustainable Development". This agenda introduced a set of 17 SDGs and associated targets designed to stimulate global action over the next fifteen years, focusing on areas of vital importance for both humanity and the planet (United Nations, 2015). The SDGs represent an ambitious framework for creating a better and more sustainable future for all. They specifically address pressing global challenges, including poverty, inequality, climate change, environmental degradation, and issues related to peace and justice. Education for Sustainable Development is essential to achieve all SDGs. It has its dedicated Goal 4, which aims to "ensure inclusive and equitable quality education and promote lifelong learning opportunities for all" (UNESCO, 2018).

Nevertheless, HEIs encounter various challenges in addressing the SDGs. Key barriers include financial constraints that limit funding for developing and implementing sustainability programs. There is often a lack of training opportunities that equip faculty and staff with the necessary skills to engage comprehensively with sustainability principles. Integrating sustainability into existing curricula can also be complex, requiring significant adjustments, time, and human resources.

The paper's research question is "To what extent do Lithuanian state colleges integrate and prioritize the SDGs in their institutional operations, as reflected in their Social Responsibility Reports, and what gaps exist in setting measurable targets and indicators?". The article highlights the interdependence of the 17 SDGs by examining their interlinkages and explores the contribution of HEIs to advancing the SDGs, underlining how HEIs implement specific goals in practice. The paper analyzes the main areas in which HEIs actively engage in realizing the SDGs. Content analysis was applied to the most recent college reports on SD and the SDGs to achieve this. For the purpose of this research, the 17 SDGs were grouped into four SD dimensions—Nature, Economy, Society, and Wellbeing—which served as a framework for systematizing HEI activities reflecting their involvement in SDG implementation. By identifying key categories, the content analysis of SRR enabled the assessment of how colleges contribute to realizing the SDGs.

2 Advancing SDGs Within Higher Education Institutions

Our generation faces pressing social, environmental, and economic challenges, from rising living costs to climate change and inequality. The 2030 Agenda highlights the role of public institutions, particularly HEIs, in successfully achieving the SDGs. Among the 17 SDGs, SDG 4 is directly linked to the activities of HEIs, and responsibility for implementation falls to high schools. The importance of education is also underlined in "Our Vision" (UNESCO, 2015) and in the realization of other goals such as SDG 3 (Target 3.7), SDG 8 (Target 8.6), and SDG 13 (Target 13.3). Adopting the 2030 Agenda and the SDGs in 2015 increased interest in HEIs globally and greatly expanded since 2015 (Toman et al. 2023).

The comprehensive interdependence of the 17 SDGs reflects a holistic approach to SD. Examining the interlinkages between the SDGs, scientists have found a coherence of indicators, where the progress in one indicator was linked to the achievement of another within the same goal, namely SDG 1, SDG 3, SDG 4, SDG 10, SDG 12, and SDG 13 (Pradhan et al., 2017). The specified goals are also interconnected, as SDG 1 aims to eliminate all forms of poverty, and its reduction is closely linked to improved health outcomes (SDG 3). Poverty reduction also relies on SDG 4, as education helps to break the intergenerational cycle of poverty by improving skills, employability, and income and also promotes awareness of sustainable consumption (SDG 12) and climate challenges (SDG 13), encouraging more responsible behavior. SDG 10 is closely linked to SDG 1, as addressing systemic inequalities related to income, gender, ethnicity, or geography is essential to reducing poverty. Growing inequalities between and within countries hinder access to quality education (SDG 4) and healthcare (SDG 3), reinforce unsustainable consumption patterns (SDG 12), and contribute to environmental degradation. SDG 12 addresses the global challenges of overconsumption, waste, and environmental degradation, as unsustainable consumption patterns affect the health (SDG 3) and livelihoods (SDG 1) of vulnerable communities and further deepen inequality (SDG 10). Promoting sustainable consumption is key to reducing poverty and inequality while supporting environmental goals. SDG 13 is a cross-cutting goal that influences and is influenced by all other goals. Climate change has a severe impact on poor and marginalized groups, exacerbating poverty (SDG 1) and inequality (SDG 10), worsening health outcomes (SDG 3), disrupting education systems (SDG 4), and complicating efforts to achieve sustainable consumption and production (SDG 12).

Also, the negative correlations within the same goal were fixed, specifically SDG 7, SDG 8, SDG 9, and SDG 15 (Pradhan et al., 2017), as trade-offs can emerge when these goals are pursued without sufficient integration and environmental safeguards. There is a key tension between SDG 8 and SDG 15. Economic growth and the pursuit of decent work (SDG 8) often drive industrial expansion, infrastructure development, and increased resource extraction. These processes can lead to deforestation, habitat destruction, and biodiversity loss, directly undermining SDG 15, which focuses on protecting, restoring, and sustainable use of terrestrial ecosystems. Similarly, SDG 9 can conflict with SDG 15 if infrastructure projects such as roads, urban sprawl, or large industrial zones disrupt habitats and contribute to environmental degradation. Industrialization often puts significant pressure on natural resources, leading to unsustainable land use and biodiversity loss. There are also trade-offs between SDG 7 and SDG 15. While SDG 7 promotes access to affordable and clean energy, certain renewable energy projects can harm terrestrial ecosystems. At the same time, large-scale bioenergy plantations can lead to land-use change and monocultures, with negative impacts on biodiversity and soil health (SDG 15). In some contexts, SDGs 8 and 9 may also slow progress on SDG 7 if economic growth continues to rely on fossil fuel-based energy systems. This reliance may hinder the transition to renewable energy sources and contribute to further environmental degradation.

The impact of fulfilling one goal does not necessarily generate better outcomes for others, as any organization's contribution to the 2030 Agenda is limited by the existence of positive, negative, and neutral interlinkages among the 17 SDGs (Bebbington & Unerman, 2018). SDG 13 has positive interactions with several goals: SDG 3 due to reduced air pollution, SDG 6 for cutting down water scarcity, SDG 2 as a result of increased food security, SDG 15 for generating responsible land use, SDG 7 enables the utilization of renewable energy (Leal Filho et al., 2023). SDG 1 had the greatest interaction with the other SDGs, meaning that poverty reduction is linked to SDGs 3–6 and SDG 10 progress. However, SDG 12 is linked to trade-offs, particularly in economic progress (Pradhan et al., 2017). It responds to other findings that achieving Goal 1 will have the most synergistic relationships with other SDGs (Kroll et al., 2019).

Practice in implementing the SDGs within the HEIs revealed that priority focus and strong focus was given to SDG 4, SDG 3, SDG 5, SDG 17, SDG 9, SDG 13, SDG 7, and SDG 16, while little (no focus) was captured in the realization SDG 2, SDG 14, SDG 1, SDG 11, SDG 15, SDG 16 (Leal Filho et al., 2021). This suggests that many HEIs prioritize goals directly related to their core missions, while SDG 2 and SDG 14 are often seen as sector-specific, more related to agriculture or marine sciences, which may not be a focus for all universities. Similarly, SDG 1, SDG 11, SDG 15, and SDG 16 often require systemic, policy-level interventions that are beyond the immediate scope of typical HEI activities, making them less central to institutional strategies unless the university specializes in related fields such as social sciences, urban planning, environmental protection or governance. HEIs are working more or less on all SDGs in more than one institutional area, but some SDGs, such as SDG 14 and SDG 2, had the lowest participation rates, while the most often implemented SDGs were SDG 4 and SDG 5 (Toman et al., 2023). Concerning the SDGs disclosure in HEIs published sustainability reports, the most disclosed was SDG 4, the next most important was SDG 10, in the third position was SDG 9; when the least reported were SDG 17, SDG 3, SDG 15, and SDG 2 (Caputo et al., 2021).

HEIs are not officially mandated to adopt or promote the SDGs, but many universities recognize the importance of SD and act voluntarily with different coverages (Serafini et al., 2022; Cleveland et al., 2023). The level of integration of sustainability can range from formulations of policy statements, integration into curricula, and regular curricula updates, offering transdisciplinary courses, 'educate-the-educators' programs, green campus activities, and signing on international charters or declarations (Lozano et al., 2013, 2015). The main areas in which HEIs are engaged in SDGs are given in Table 1.

Education for sustainable development aims to equip individuals with the values, abilities, skills, and knowledge needed to actively contribute to creating a more sustainable society. SDGs are integrated into higher education by applying various learning methods, such as workshops, courses, lectures, degree topics, seminars, and webinars (Molina et al., 2023), and learning strategies: project-oriented learning, simulation games, problem-based learning, service learning, and case studies (Tejedor et al., 2019).

HEIs tend to include the context of the SDGs in their curriculum development and make the SDGs part of the courses (Serafini et al., 2022; Cleveland et al., 2023; Day et al., 2024; Philip et al., 2024) revise teaching content, encourage teaching methods that enable students to develop interdisciplinary thinking, analyzes of multifaceted problems, cooperation in decision-making (HESI, 2021), the interdisciplinary studies improves students' problem-solving capacities and broadens perspectives to create solutions for different challenges (Annan-Diab & Molinari, 2017), while the success of teaching sustainability issues depends on whether sustainability is considered in the core courses (Figueiró & Raufflet, 2015). An analysis of SDG integration within course descriptions and learning outcomes indicated that SDG 4 and 17 were the most frequently addressed goals. This finding highlights that integrating all SDGs in educational programs is neither practical nor beneficial; HEIs can make a more meaningful impact on the SDGs by focusing on providing high-quality education, particularly in alignment with SDG 4 (Prior, 2024). However, there is a position that HEIs should cover all SDGs (Sánchez-Carracedo et al., 2021). The International Association of Universities (hereafter IAU) survey revealed that 65% of universities offer courses specific to SD (Toman et al., 2023).

Teaching staff's education and professional development are vital to advancing sustainability and the SDGs. HEIs offer professional development opportunities to enhance staff and student understanding of SD, but a more structured and systemic approach is required, particularly within teacher education institutions (Gough, 2016). Research revealed that over half of teachers feel inadequately equipped

Table 1 Contribution of higher education institutions to SDGs

Oliveira and Proença (2025)	Leal Filho et al. (2019b, 2021)	Žalėnienė and Pereira (2021)	Toman et al. (2023)
Education	Learning and teaching	Education	Teaching and learning
Research	Research and development	Research	Research
Campus operations and on-campus experiences	Governance and campus operations	Management and governance	Campus operations
Assessment and reporting	Civic engagement and community outreach	Leadership	

to address sustainability issues or teach controversial topics; teachers often lack inspiring examples and expertise in incorporating SD into their teaching (Borg et al., 2012). The follow-up studies disclosed positive changes: the majority of the university staff indicated that they had intermediate (45%) or expert level (36%) knowledge about SDGs, and about 80% of academic staff were involved in SD (Toman et al. 2023).

HEIs are urged to align their research with the SDGs, adopting interdisciplinary approaches to address sustainability challenges and make an institutional impact on global development, and this can be named HEIs contribution to SD (Cottafava et al., 2022). The impact of HEIs should be assessed according to their effectiveness in tackling local challenges through collaboration, civic engagement, and dissemination, and university leaders must find insights into the future evolution of SD and become a place of teaching, research, and society (Corazza & Saluto, 2021). Sustainability researchers have emphasized the vital role of interdisciplinary research in tackling the grand challenges of the twenty-first century (Keynejad et al., 2021); however, interdisciplinarity in research achieving the SDGs remains in infancy (Cottafava et al., 2022), but progress is being made (Wohlgezogen et al., 2020). In reality, HEIS's role goes beyond teaching and research, encompassing the shaping of social norms and influencing policy decisions, thus serving as catalysts for societal change (Etzkowitz et al., 2000).

Cooperation among HEIs includes exchange programs, joint degrees, and joint research, partnerships with external organizations, and community events (Lozano et al., 2015); it promotes growth and development, as collaboration enables resource sharing, curriculum innovation, and reputation enhancement (Chou, 2012), partnerships allow to develop beneficial relations with communities based on social needs (Leal Filho et al., 2021). HEIs partnerships for SD take place at different levels, and NGOs appear to be the first in the partnership line, followed by community organizations and schools (Toman et al. 2023). Studies have also revealed some risky aspects of the relationship between HEIs and the industry. In recent years, fossil fuel companies have increasingly integrated into universities across the U.S., U.K., Canada, and Australia as they fund research, hold positions on governing boards, host recruitment events, and advise on curricula; therefore, such partnerships can lead to biased outcomes favoring the fossil fuel industry, posing risks to academic integrity and potentially obstructing climate action efforts (Hiltner et al., 2024).

Many HEIs integrate SDGs into their strategic plans, including governance, campus operations, teaching, research, and community engagement. Evaluating these efforts is essential, and institutions are encouraged to participate in sustainability assessments. Such evaluations help measure progress, highlight areas for improvement, and strengthen internal collaboration. As an increasing number of universities become involved in the implementation of the SDGs, there is a need to define reliable indicators, which may assess the extent to which specific sub-goals have been reached and the need for a system to monitor and document progress (Leal Filho et al., 2021). The latest IAU report disclosed that 61.2% of HEIs have monitoring tools for SD while comparing with the IAU 2019 survey, (Mallow et al., 2019), the growth was up to 20%. However, internal monitoring was the most common among

HEIs, and external monitoring was also applied but less frequently (Toman et al., (2023).

3 Methodology

The content analysis method was used to examine the most recent college reports related to SD and SDGs. The study aimed to structure the activities carried out by the colleges in relation to the implementation of the SDGs, to identify which SDGs are most and least reported, and which colleges are leading in terms of reported content and completeness of reporting. The reports were analyzed by extracting activities according to the SD dimensions and categorizing them, with activities corresponding to each category. The categorization process included capturing specific words, compound words, phrases, and examples related to the SDGs.

Three generally recognized dimensions of SD are ecological/environmental, social, and economic (Allen et al., 2018; Purvis et al., 2019; Wu et al., 2023; Taylor, 2016). SD is operationalized using more dimensions: environmental, social, economic, individual, and technical (Becker et al., 2016); ecological, economic, social and cultural, good governance (Grindheim et al., 2019); ecological, social, economic, technical, legal and political (Pawłowski, 2008), ecological, social, economic, cultural, and safety (Tomaškinová et al., 2021). The main dimensions of crosscutting consideration are summarized in the so-called "5Ps" of the SDGs: People (SDGs 1–5), Planet (SDG 13–15), Prosperity (SDGs 6–12), Peace (SDG 16), Partnerships (SDG 17) (United Nations, 2015).

We used the Sustainability Compass to group and analyze SDGs in four dimensions: Nature, Economy, Society, and Wellbeing (AtKisson, 2010; Sustainability Accelerator, 2015) (Fig. 1).

The analysis included State Colleges operating in Lithuania, which either produce SRR or have developed indicators to monitor the implementation of the SDGs. For brevity and consistency, the names of the institutions have been standardized by omitting the term "State" as all analyzed colleges are public institutions. The paper uses "Coll." as an abbreviation for "College ". Therefore, the shortened versions such as Vilnius Coll., Kaunas Coll., Klaipėda Coll., Šiauliai Coll., and Panevėžys Coll. are used throughout the text.

Vilnius Coll. is one of Lithuania's largest and leading professional higher education institutions. Professional Bachelor study programs offered at the college are assigned to 12 groups of study fields: engineering, business and public management, arts, social sciences, computer sciences, health sciences, technological sciences, educational sciences, humanitarian studies, veterinary sciences, and agricultural sciences. The study process at the college is structured across 10 faculties, 58 study programs are offered, and about 6647 students were enrolled at the college (2024–25 academic year). Kaunas Coll. is a multi-profile state higher education institution training specialists in technology, informatics, engineering, medicine,

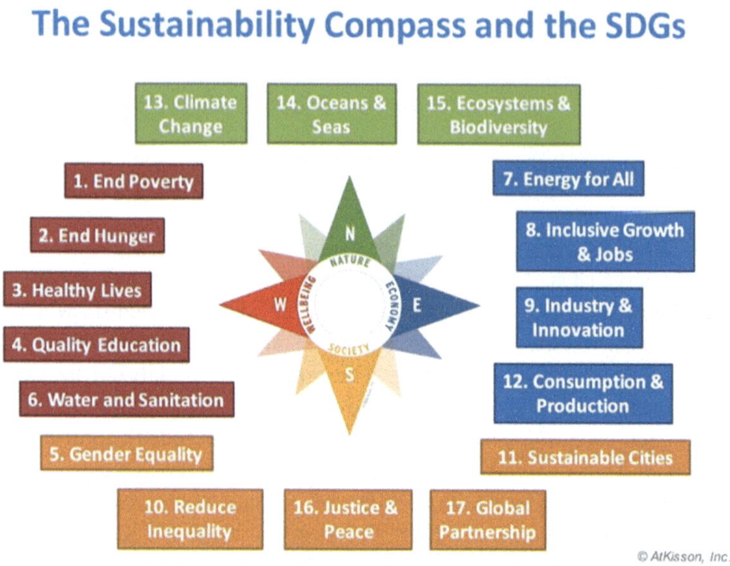

Fig. 1 Sustainability compass and SDGs (AtKisson, 2010; Sustainability Accelerator, 2015)

humanities, social, art, education, business and public management, law, and agricultural sciences. The college organizes its study process through 5 faculties; 52 study programs are available for students, and about 4361 students were enrolled at the college (2024–25 academic year). Šiauliai Coll. is one of Northern Lithuania's leading higher education institutions. It offers 22 study programs in health sciences, social sciences, information technology, engineering, business, and public management, preparing highly qualified specialists for the regional and national labor market. The study programs are offered across two faculties, and about 1726 students were enrolled in the 2024–25 academic year. Panevėžys Coll. is a modern, socially responsible higher education institution that fosters student entrepreneurship and supports sustainable and balanced development. The college has three faculties and offers first-cycle professional bachelor's degree studies from 20 study programs in health sciences, technology, and social sciences. About 1427 students were enrolled at the college in the 2024–25 academic year. Klaipėda Coll. is one of Western Lithuania's largest professional higher education institutions. The college's Bachelor's programs are assigned to 7 groups of study fields: business and public management, health sciences, engineering, computer sciences, technological sciences, social sciences, and educational sciences. The college has three faculties and offers 26 study programs; about 2473 students were enrolled in the college (2024–25 academic year). Key information about the colleges under the research in the context of SDGs is given in Table 2.

Colleges were selected for the study as they are the only ones that produce and publish the SRR and are members of the UN Global Compact. Colleges with the exemption of Klaipėda Coll. submit "The Letter of Commitment" signed by the

Table 2 Colleges under the research

Name	Title of report	Reporting period	Participant in UN global compact	SDGs indicators
Vilniaus College	Social responsibility and SD development report (Vilnius College, 2023)	2023	Since 2018	Set of indicators
Kaunas College	Social responsibility report (Kaunas College, 2022)	2021–2022	Since 2012	Not available
Šiauliai College	Social responsibility report (Šiauliai State College, 2022)	2023	Since 2012	Not available
Panevėžys College	Communication on Engagement (Panevėžys College, 2023)	2021–2022	Since 2011	Not available
Klaipėda College	Not available	Not available	Not available	Set of indicators

directors of the colleges, stating that colleges support the Ten Principles of the UN Global Compact with respect to human rights, labor, environment, and anti-corruption. Vilnius and Klaipėda Coll. have prepared sets of indicators for the SDGs. Vilnius Coll. and universities in Spain and Finland participated in the international "Universities for Sustainable Development" project. The college has developed a set of 48 indicators for measuring the achievements of realizing SDGs: SDG 1–5, SDG 7–13, SDG 15–17. When compiling a set of indicators, individuals from different areas within the university actively participated, i.e., researchers in the sustainability field, lecturers, the sustainability project team, Vice-Rectors, the Head of the International Relations Department, and students. For evaluation purposes, each SDG is analyzed using the following categories: direct positive, indirect positive, no impact, direct negative, and indirect negative. The status of selected indicators is reported for 2021–22 and 2022–23, with a preliminary target set for 2030. Klaipėda Coll. has developed a set of indicators (39) to measure the implementation of the SDGs: SDG 3–4, SDG 9–12, and SDG 17. The SDGs Progress Assessment Report for 2022 and 2023 has been prepared, specifying the selected targets and the indicators for the chosen SDGs. The SDGs Roadmap for 2023–2025 outlines the targets to be achieved by 2025, the expected impact, and the responsible parties.

4 College Activities Implementing the SDGs in Wellbeing Dimension

The well-being of people is important for promoting SD by fostering social stability, economic growth, and environmental resilience. When people have access to adequate nutrition, health care, and education, they are better equipped to improve their lives and break cycles of poverty. Quality education provides individuals with the knowledge and skills they need to make informed decisions, adapt to an ever-changing environment, and be at the forefront of addressing the problems of the modern world (Table 3).

Programs linked to SD encourage interdisciplinary thinking. Students are offered electives on circular economy, sustainable electronics, and sustainable business, and topics such as waste management and business ethics are included in the curricula. SD and social responsibility are addressed in various subjects, and sustainability is integrated into assignments, essays, coursework, and projects, especially students' final theses. The interdisciplinarity approach is also implemented through networking events, forums, cross-departmental initiatives, and interdisciplinary conferences, which actively engage students and faculty members. It is further enhanced through the preparation of interdisciplinary articles and research projects (commissioned research projects) involving specialists from various fields, as well as students, and through the publication of scientific papers on interdisciplinary topics in collaboration with other higher education institutions.

Table 3 College initiatives promoting the SDGs in the Wellbeing dimension

Category	Activities
Learning and teaching	• Lifelong learning • Non-formal education and continuing education • Updating and improvement of study programs and methods • Integration of SD, SDGs, and social responsibility into study programs • Students' final theses focus on sustainability topics • Possibility to choose sustainability-related electives • Counselling for students and schoolchildren • Promotion of information literacy • Permanent feedback on the organization and content of studies • Professional development programs
Financial support and assistance	• Monitoring the competitiveness of teachers' salaries • Diversity of scholarships • Financial support for students with disabilities and from low-income backgrounds • Government-guaranteed soft loans
Healthcare services	• Free psychological support • Free health check-ups • Classrooms for work and rest • Lectures on health and suicide prevention • Promotion of physical activity
Community and business engagement	• Training on well-being topics • Education on the importance of clean water • Courses, including specialized programs to improve professional skills • Well-being campaigns • Volunteering • Involvement of social partners in curriculum development and evaluation • Company visits, internships, joint career events
Research and development	• Writing scientific articles • International conferences • Project activities related to digital education, development of new educational programs, innovative learning and internship modules, promoting student physical activity, technological innovation in the food industry
International cooperation	• Participation in Networks • Academic exchange programs • Collaborative research • International projects • Conferences and workshops

Information literacy, including cyber security, digital innovation, and education, are actively promoted. Food industry-oriented degree programs "Food Technology," "Dietetics," and "Agribusiness Technology," given in Vilnius Coll., contribute to achieving SDG 2.

The financial support depends on the financial capacity of the colleges; there are scholarships based on grades, but also other types of scholarships, such as incentive, one-off, nominal, targeted accessibility, social as well as financial support for students with disabilities and those on low incomes. Significant attention is paid to health and well-being by providing free psychological support in Lithuanian, Russian, and English, organizing lectures on health topics such as nutrition, disease prevention, healthy lifestyles, and suicide prevention (SafeTALK training), multisensory sessions, hippotherapy, art and writing classes. Physical activity for students and staff is encouraged through participation in basketball, volleyball, and football competitions.

The colleges have professional development programs and training for staff, students, and social partners on SD and social entrepreneurship. Lecturers give training on healthy lifestyles, such as "Movement for Health," "Is obesity the norm?" etc. Acts of kindness are performed for the residents of social service homes, volunteering takes place in social actions such as support for children at risk and support for Ukraine, participating in initiatives such as "Peace for Food: Sustainability, Safety for Society and Planet" (Vilnius Coll.).

The lecturers' participation in scientific conferences is linked to public health topics such as "Ageing and Health", "Problems of overweight and obesity in children, solutions", "Current issues in health and social sciences: social partnership for the well-being of society", etc.

The project activities are implemented in various fields of study and training. Among the projects aimed at strengthening education for SD and directly related to the implementation of the SDGs is the project "Universities for Sustainable Development" (Vilnius Coll.), the project "Fostering the Transversal Digital Competences in Higher Education" (Šiauliai Coll.), the project "POWERUP MYHOUSE: Development of innovative learning and practice modules to increase the use of renewable energies for sustainable buildings" (Panevėžys Coll.) is linked to SDG 13 in the Nature dimension, the project "Development of an innovative and cross-sectoral European education proGRam to strEngthEN the training of TECHnicians in the field of renewable energy (GREEN TECH)"(Panevėžys Coll.) is linked to SDG 7, SDG 9 in the Economic dimension, SDG 13 in the Nature dimension. Most projects are interdisciplinary, involving faculty members and researchers from various academic fields. Project activities include international cooperation through conferences, joint events, discussions, and meetings.

The analysis revealed that Klaipėda and Panevėžys Coll. do not directly identify activities related to SDG 1; Vilnius Coll. uses several indicators, e.g., the percentage of students receiving scholarships or low-income student support. Klaipėda, Panevėžys, and Šiauliai Coll. do not name activities directly linked to SDG 2; Vilnius Coll. applies some indicators, e.g., people from the university community participating in initiatives/campaigns promoting a healthy lifestyle. Only some activities concerning SDG 6 are mentioned by Šiauliai Coll. (education on the importance of clean water, etc.). This can be explained by the fact that drinking water in Lithuania is of high quality, as most of it comes from underground sources that are naturally protected from pollution.

Indicators directly related to the realization of the SDGs: 1. Percentage of study programs that integrate the SDGs (Klaipėda Coll.). 2. Number of consolidated research groups that contribute to the SDGs (Vilnius Coll.). 3. Percentage of the university population participating in programs that promote education for SD (Vilnius Coll.).

5 College Activities Implementing the SDGs in the Society Dimension

The social dimension of SD is associated with inclusiveness, equity, and justice. Gender equality empowers individuals, boosts economies, and strengthens communities while reduced inequalities expand opportunities and improve resource access, creating a fairer society. Sustainable cities ensure safe urbanization and better quality of life; in parallel, strong institutions enable governance and partnerships to drive collective commitment to long-term prosperity for all (Table 4).

Special attention is paid to equal opportunity policies to prevent (or neutralize) phenomena such as violence, harassment, and discrimination of all kinds and to ensure equal access to education and employment. The tolerance campaign "Respect and Support" and the Tolerance Day civil initiative "House of Tolerance" have been implemented to achieve these goals.

Emphasis is also placed on maintaining a good psychological working environment by providing access to psychological support and choosing an acceptable form of work and study for staff and students with disabilities. To create equal opportunities for all to participate in common activities, community meetings are translated into sign language (Vilnius Coll.), training, workshops, and initiatives such as taking the Constitution exam help to understand personal rights and duties and promote civic engagement and responsibility.

College communities are involved in public activities and organizing activities for the community, e.g., volunteering at the Red Cross, Childline, blood donation action, language courses to speed up the integration of refugees from Ukraine, etc. Cooperation with the business community is based on mutual trust and benefits through the organization of joint activities, student company visits, and internships. The business community provides scholarships to students of certain study programs (Ignitis Group); also, students prepare their final theses on topics relevant to the business community.

Research on clean transport technologies, energy efficiency, creating a healthy and safe environment, analysis of the energy consumption habits of multi-family

Table 4 College initiatives promoting the SDGs in the Society dimension

Category	Activities
Equality policy	• Violence, harassment, and discrimination prevention policy • Internal rules on equal opportunities • Equal opportunities and diversity programs • Tolerance campaigns
Psychosocial well-being and working conditions	• Risk assessment for psychosocial factors and bullying at work • Psychological support for students and staff • Adaptation of working conditions for staff and students with disabilities • Opportunities for flexible or remote working • Mentoring
Training, workshops, and initiatives to raise consciousness	• Anti-discrimination and gender equality training • Academic integrity Training • Human Rights Day and Anti-Corruption Day celebration • Taking Constitution examination
Research and development	• Writing scientific articles • International conferences • Project activities targeted to students with special needs and socially vulnerable groups • Research on clean transport technologies and energy efficiency, etc.
Community and business engagement	• Mutual social partnerships with universities, schools, educational centers, youth organizations, and NGOs • Non-formal education for children and youth • Organization of summer camps • Volunteer work and social actions • Integration projects for refugees from Ukraine • Humanitarian aid to Ukraine • Community events • Internships in companies • Business scholarships • Study visits • Dissertations relevant to the region and business needs
International cooperation	• Participation in Networks • Participation in alliances • Academic exchange programs • Joint research • International projects • Conferences and seminars

dwellers' inhabitants, and recommendations for renovation and energy efficiency are closely linked to SDG 11.

Project activities focus on students with special needs and vulnerable groups: the project "Promoting the employment of people with disabilities through the development of digital skills and competencies to support a teleworking scheme (DigitAbility)", the project "Educational empowerment and social inclusion of students with disabilities through digital solutions and assistive technologies" overlaps with SDG 1.

The analyzed colleges, except Vilnius Coll. (which counts 6 indicators), do not assess gender equality with specific indicators. Klaipėda Coll. has not included SDG 5 and 16 in the goals to be achieved. Regarding SDG 10, the indicators to measure the achievement are provided by Vilnius Coll., e.g., percentage of students with disabilities/at risk of social exclusion, and by Klaipėda Coll., e.g., facilitation of study and working conditions of disabled students and employees.

SDG 11 by Klaipėda Coll. is measured by several indicators with emphasis on applied research related to environmentally friendly transport, energy efficiency, healthy and safe environments, sustainable food consumption, and zero waste technologies, as well as the accessibility of the college's services to community members, virtual and green mobility, and research related to priorities for an inclusive and creative society; when Vilnius Coll. apply only one indicator "Number of parking spaces for bicycles".

Vilnius Coll. is the only one that measures the realization of Goal 16 using three specific indicators: transparency and accountability reports, participation of the university community in decision-making, and percentage of cases resolved with respect to those reported to the university ombudsman. Some of these indicators are suitable for use in other colleges in the case of reporting practices and creating an ombudsman position.

The 17th SDG plays a vital role in enabling the success of all other goals and targets by promoting collaboration and collective action. In summarizing the international cooperation activities of colleges, which are essential for achieving SDG 17, it is evident that not all activities mentioned in the SRR are directly linked to SDG 17. However, they reflect a clear strategic orientation towards this goal, though several main directions can be identified. Participation in international projects, most of which are interdisciplinary and contribute to several SDGs, e.g., the project "Development of an innovative and cross-sectoral European education program to strengthen the training of technicians in the field of renewable energy (GREEN TECH)" (Panevėžys Coll.) directly supports multiple sustainable development objectives. Project-based activities are closely tied to participation in international conferences, seminars, training, and professional networks. These initiatives strengthen partnerships between participating institutions and foster new connections for future projects and research collaborations. Colleges also actively participate in international interdisciplinary conferences, such as the "Innovative (Eco-) Technology, Entrepreneurship, and Regional Development" conference, the "Health, Environment and Sustainable Development: Interdisciplinary Approach (HESDIA, 2022)" conference (Kaunas Coll.), also joint scientific publications with other HEIs are common and often result from project-based cooperation. Furthermore, colleges are involved in international alliances, e.g., Šiauliai Coll. is a member of the NEOLAiA Alliance (comprising 9 higher education institutions), and Kaunas Coll. signed an agreement in 2023 to establish the "UNINOVIS—Data for L.I.F.E" European University Alliance. Colleges continue to expand the internationalization of their study programs by participating in Blended Intensive Programs (BIP) (Vilnius and Kaunas Coll.) and actively engage in the Erasmus + program (2021–2027), which includes student exchanges, internships/traineeships in foreign companies, teaching visits, staff mobility, and organization of international events such as Erasmus + Weeks, e.g., "Erasmus + Networking" at Šiauliai Coll.

Vilnius and Klaipėda Coll. analyze changes using a full set of indicators. Some indicators are similar and include public–private partnerships, contracts with companies, organizations, and business associations, support received and granted, and participation of students/teachers in development cooperation. Klaipėda Coll., in the context of SDG 17, also rates the "proportion of final theses related to regional development" and the "proportion of students/teachers involved in voluntary activities". At the same time, Vilnius Coll. has added the important measure "participation in networks related to SD" to its set of indicators.

6 College Activities Implementing the SDGs in the Economy Dimension

The economic dimension is important in the context of economic stability and the viability of colleges. Economic stability refers to the financial health of an institution, enabling it to operate effectively without distress, while economic viability ensures sustainable revenue generation to support academic and infrastructural growth. Financially stable colleges can modernize infrastructure, expand research opportunities, and integrate digital learning tools, thereby improving the quality of education. Economic viability facilitates scholarships and financial aid, promoting social equity by reducing financial barriers. Sound institutions can work with governments, businesses, and international organizations to promote sustainability initiatives and global partnerships, implementing sustainability strategies, such as green campuses and carbon reduction plans, attracting international students, and strengthening global knowledge exchange (Table 5).

Activities concerning infrastructure modernization linked to SDG 7 are also linked with SDG 13 in the Nature dimension. Activities include insulation of facades,

Table 5 College initiatives promoting the SDGs in the Economy dimension

Category	Activities
Infrastructure modernization	• Restructuring the energy sector and building renovation • Use of solar energy • Development of electric vehicle infrastructure
Waste reduction and responsible consumption	• Waste sorting • Office paper recycling • Recycling of machinery in accordance with EU directives • Going paperless and using electronic platforms • Secondary use of items • Green purchasing
Education, outreach, and awareness-raising	• Information events and seminars on solar energy/energy efficiency and circular economy • Career guidance events • Environmental protection events • Waste Culture exam • Participation in fairs on green economy ideas • Participation in competitions on entrepreneurship and social entrepreneurship
Research and development	• Writing scientific articles • International conferences • Project activities (entrepreneurship and innovation in the green economy, training of renewable energy professionals, development of social entrepreneurship skills, development of zero-waste technologies, etc.) • Environmental research (development and application of zero-waste technologies and circular economy principles in the construction sector) • Industrial modernization research (development of eco-technologies, green logistics solutions, etc.)
Business engagement	• Organization of conferences • Delivery of lectures • Hosting of students for work placements • Provision of scholarships
International cooperation	• Participation in Networks • Academic exchange programs • Collaborative research • International projects • Conferences and workshops

replacement of windows and doors, renovation of buildings, LED lighting, development of solar power plants on buildings, and installing charging points for electric cars. When green purchases help reduce environmental impact by minimizing waste and pollution, conserving resources, and lowering carbon emissions.

Workshops have been organized, such as the FixLAB, where students repair electronic equipment, and the educational workshops "Worn textiles: pet beds", "Sustainable branding", and "Second life of plastic". Annual environmental and public awareness events are held in and outside the colleges, such as "Inkill Day" and "Christmas Tree—Not Rubbish", students participated in the European Waste Reduction Week and took the "Waste Culture exam". Other important events are employment-related, such as "We've got a job for you!", "Career and study" and "Internship and career days" were organized with social partners. Students participate in entrepreneurship/ social entrepreneurship competitions that promote the development of sustainable business ideas and integration into the labor market, e.g. "Sustainable Business Idea" (Kaunas Coll.), "Future Business Idea" (Panevėžys Coll.).

International conferences and project activities usually focus on several goals; for instance, the project "Three-level Centers of Professional Excellence: Qualification, Entrepreneurship, and Innovation in the Green Economy (3LOE)" (Panevėžys Coll.) meets another goal, namely SDG 4 in Wellbeing dimension.

The social partners are involved in the study process by lecturing on sustainable business and innovation, business processes, and SD practices. Cooperation with companies (e.g., "Teltonika" and "Rimi Lietuva") helps students gain practical sustainability experience. The energy services company "Ignitis Group" has provided scholarships for students who want to contribute to the development of green energy.

Klaipėda Coll. does not have indicators for SDG 7 and 8. At the same time, Vilnius Coll. measures SDG 7 (renewable energy/energy consumption per graduate) and SDG 8 (proportion of students with internships during their studies, student employment rate, and the wage gap. Regarding SDG 9, Klaipėda Coll. counts such indicators: expenditure on research and experimental development; employees working in the field; creation (conversion) of laboratories to promote research in SD and industrial sector technologies. Vilnius Coll. has only one summary indicator: "Increase scientific research and innovation". The fulfillment of SDG 12 in both colleges is assessed by implementing a green procurement policy and paper use/reuse; Klaipėda Coll. also rates the change in hazardous waste generation, purchases from social enterprises or local producers, and the development of electronic library resources.

7 College Activities Implementing the SDGs in the Nature Dimension

The Nature dimension ensures ecological balance, sound livelihoods, and resilience. Healthy ecosystems regulate climate, reduce disaster risks, and provide essential resources when biodiversity conservation supports food security and ecosystem stability, driving sustainable progress. Nature-based solutions like reforestation and conservation aid climate adaptation and resource preservation (Table 6).

All activities concerning infrastructure modernization linked to SDG 13 are also related to SDG 7 in the Economic dimension. These activities include the modernization of electrical installations, installation of solar panels, modernization of heating systems, and renovation of buildings. ISO 14001 environmental management system certification enables continuous environmental assessment, pollution prevention, and green purchases, as pointed out in the dimension Economy. Seminars and lectures on the benefits of solar power plants, the impact of climate change, and the importance of green energy development, together with events and campaigns to promote cycling, walking, and public transport, are all means of raising public awareness and changing behavior.

Studies on the quality of the living environment are being carried out in a variety of directions, including research on water quality and the impact of noise, the study of local flora, projects to assess the impact of green spaces, the valuation of ecosystem services, studies on the efficiency of thermal insulation, and the integration of sustainable technologies into urban infrastructure, among others.

The analysis shows that Klaipėda Coll. does not yet link its activities to the assessment of the goals in the Nature dimension and, therefore, does not provide any information; Vilnius Coll. highlights SDG 13 with several indicators, e.g. a number of studies on climate change and the number of research teams formed on sustainability, SD. The least information is provided on SDG 14. None of the colleges indicate specific activities; only Vilnius Coll. assesses SDG 14 with one indicator: "number of studies or projects on the study and preservation of local flora".

Table 6 College initiatives promoting the SDGs in the Nature dimension

Category	Activities
Infrastructure modernization	• Restructuring the energy sector and building renovation • Use of solar energy • Development of electric vehicle infrastructure • Creation of pedestrian walkways, parking, and recreational areas
Environmental management	ISO 14001 Environmental Management System Certification
Education, outreach, and awareness-raising	• Seminars and lectures on the benefits of solar power plants, the impact of climate change, and green energy development • Climate Week events • Training and public awareness on landscape protection, sustainable urban development • Events and initiatives to promote sustainable mobility, etc.
Research and Development	• Writing scientific articles • International conferences • Project activities related to the training of renewable energy professionals and use of renewable energy for sustainable buildings • Studies on the quality of the living environment
Business engagement	• Participation in conferences • Hosting students for internships • Joint research
International cooperation	• Participation in Networks • Academic exchange programs • Joint research • International projects • Conferences and seminars

8 Conclusions

HEIs are not just centers of learning because they serve as pivotal platforms for shaping future leaders, instilling ethical principles, and fostering the creation of innovative solutions to complex problems. Pursuing sustainability is also linked to reputation management, as commitment to the SDGs demonstrates responsibility, enhances public trust, and strengthens an institution's image.

The colleges are most active in implementing Goal 4 (Quality Education), which is closely linked to Goal 5 (Gender Equality), Goal 10 (Reduced Inequalities) and Goal 3 (Good Health and Well-being). In pursuit of these goals, activities are focused on high-standard education, gender equality and inclusion, the comprehensive well-being of students and teachers, extended activities to cooperation with the community, and acting as responsible citizens.

This is followed by activities related to Goal 7 (Affordable and Clean Energy), Goal 13 (Climate Action), and Goal 12 (Responsible Consumption and Production). Concrete actions and projects are being developed to increase energy efficiency, the use of solar energy, and a continued focus on waste reduction and green purchasing.

Goal 17 (Partnerships for the Goals) combines all activities and is implemented through local and international cooperation and project initiatives. Research and project-based activities may focus on specific areas but are usually linked to achieving several or multiple goals.

Early implementation of SDG-related initiatives and education can strengthen the ability to adapt to future changes and disruptions. It is therefore important to assess sustainability efforts in terms of management, teaching, research, and community engagement, developing measurement instruments. The SRRs produced by colleges are more a description of activities, focusing on what has been achieved rather than what needs to be done to realize improvements or overcome challenges, nevertheless allowing an assessment of the colleges' efforts to act in the context of achieving the SDGs.

Only two of the analyzed colleges have developed measurable outcomes to assess how they deliver on the SDGs, so it cannot be said that colleges successfully integrate the SDGs into institutional policies and practices.

The study has limitations, as it analyzed the activities of colleges that provide SRRs given in the UN Global Compact database or have developed SDG measurement indicators.

References

Annan-Diab, F., & Molinari, C. (2017). Interdisciplinarity: Practical approach to advancing education for sustainability and for the Sustainable Development Goals. The International Journal of Management Education, 15(2), 73-83. https://doi.org/10.1016/j.ijme.2017.03.006

Allen, C., Metternicht, G., & Wiedmann, T. (2018). Prioritising SDG targets: Assessing baselines, gaps and interlinkages. Sustainability Science, 14(2), 421-438. https://doi.org/10.1007/s11625-018-0596-81

AtKisson, A. (2010). *The sustainability transformation. How to accelerate positive change in challenging times* (pp. 344). Routledge.

Bebbington, J., & Unerman, J. (2018). Achieving the United Nations Sustainable Development Goals: An enabling role for accounting research. *J Accounting, Auditing and Accountability, 3*(1), 2–24. https://doi.org/10.1108/AAAJ-05-2017-2929

Becker. C., Betz, S., Chitchyan, R., Duboc, L., Easterbrook, S.M., Penzenstadler, B., Seyff, N., Colin, C., & Venters, C.C. (2016). Requirements: The key to sustainability FOCUS: The future of software engineering. *Published by the IEEE computer society*, 56–65. https://www.cs.toronto.edu/~sme/papers/2016/Beckeretal-IEEESW2016.pdf

Borg, C., Gericke, N., Höglund, H. O., & Bergman, E. (2012). The barriers encountered by teachers implementing education for sustainable development: Discipline bound differences and teaching traditions. *Research in Science & Technological Education, 30*(2), 185–207. https://doi.org/10.1080/02635143.2012.699891

Caputo, F., Ligorio, L., & Pizzi, S. (2021). The contribution of higher education institutions to the SDGs—An evaluation of sustainability reporting practices. *Administrative Sciences, 11*(97), 1–13. https://doi.org/10.3390/admsci11030097

Chou, D. C. (2012). Building a successful partnership in higher education institutions. *International Journal Information Systems and Change Management, 6*(1), 84–97. https://doi.org/10.1504/IJISCM.2012.050350

Cleveland, D., Nguyen, L. T. V., & Nguyen, C. M. (2023). Integrating sustainability in higher education curricula: A transformative approach towards marketing education and brand social responsibility. In *ASCILITE 2023*. https://publications.ascilite.org/index.php/APUB/article/view/543/604

Corazza, L., & Saluto, P. (2021). Universities and multistakeholder engagement for sustainable development: A research and technology perspective. *IEEE Transactions on Engineering Management, 68*(4), 1173–1178. https://doi.org/10.1109/TEM.2020.3020736

Cottafava, D., Ascione, G. S., Corazza, L., & Dhir, A. (2022). Sustainable development goals research in higher education institutions: An interdisciplinarity assessment through an entropy-based indicator. *Journal of Business Research, 151*, 138–155. https://doi.org/10.1016/j.jbusres.2022.06.050

Day, E. L., Petritis, S. J., McFall-Boegeman, H., Starkie, J., Zhang, M., & Cooper, M. M. (2024). A framework for the integration of green and sustainable chemistry into the undergraduate curriculum: Greening our practice with scientific and engineering practices. *Journal of Chemical Education, 101*(5), 1847–1857. https://doi.org/10.1021/acs.jchemed.3c00737

Etzkowitz, H., Webster, A., Gebhardt, C., & Terra, B. R. C. (2000). The future of the university and the university of the future: Evolution of the ivory tower to entrepreneurial paradigm. *Research Policy, 29*, 313–330. https://doi.org/10.1016/S0048-7333(99)00069-4

Figueiró, P. S., & Raufflet, E. (2015). Sustainability in higher education: A systematic review with focus on management education. *Journal of Cleaner Production, 106*(1), 22–33. https://doi.org/10.1016/j.jclepro.2015.04.118

Grindheim, L.T., Bakken, Y., Hauge, K.H., & Heggen, M.P. (2019). Early childhood education for sustainability through contradicting and overlapping dimensions. *ECNU Review of Education, 2*(4):374–395. https://doi.org/10.1177/2096531119893479

Gough, A. (2016). Teacher education for sustainable development: Past, present and future. In W. Leal Filho & P. Pace (Eds.), *Teaching education for sustainable development at university level*. Springer. https://doi.org/10.1007/978-3-319-32928-4_8

Hiltner, S., Eaton, E., Healy, N., Scerri, A., Stephens, J. C., & Supran, G. (2024). Fossil fuel industry influence in higher education: A review and a research agenda. *Wiley Periodicals LLC*. https://doi.org/10.1002/wcc.904

Kaunas College. (2022). Social responsibility report 2021-2022.https://www.kaunokolegija.lt/kk_wp_content/uploads/2024/04/KK_social_responsibility_report_2022.pdf

Keynejad, R. C., Yapa, H. M., & Ganguli, P. (2021). Achieving the sustainable development goals: Investing in early career interdisciplinarity. *Humanities and Social Sciences Communications, 8*(153). https://doi.org/10.1057/s41599-021-00834-6

Kroll, C., Warchold, A., & Pradhan, P. (2019). Sustainable Development Goals (SDGs): Are we successful in turning trade-offs into synergies? *Palgrave Communications, 5*, 140. https://doi.org/10.1057/s41599-019-0335-5

Leal Filho, W., Vargas, V. R., Salvia, A. L., Brandli, L. L., Pallant, E., Klavins, M., Ray, S., Moggi, S., Maruna, M., Conticelli, E., Ayanore, M. A., Radovic, V., Gupta, B., Sen, S., Paço, A., Michalopoulou, E., Saikim, F. H., Koh, H. L., Frankenberger, F., … Vaccari, M. (2019a). The role of higher education institutions in sustainability initiatives at the local level. *Journal of Cleaner Production, 233*, 1004–1015. https://doi.org/10.1016/j.jclepro.2019.06.059

Leal Filho, W., Shiel, C., Paço, A., Mifsud, M., Avila, L. V., Brandli, L. L., Molthan-Hill, P., Pace, P., Azeiteiro, U. M., Vargas, V. R., & Caeiro, S. (2019b). Sustainable Development Goals and sustainability teaching at universities: Falling behind or getting ahead of the pack? *Journal of Cleaner Production, 232*, 285–294. https://doi.org/10.1016/j.jclepro.2019.05.309

Leal Filho, W., Frankenberger, F., Salvia, A. L., Azeiteiro, U., Alves, F., Castro, P., Will, M., Platje, J., Lovren, V. O., Brandli, L., Price, E., Doni, F., Mifsud, M., & Avila, L. V. (2021). A framework for the implementation of the Sustainable Development Goals in university programmes. *Journal of Cleaner Production, 299*, 1–12. https://doi.org/10.1016/j.jclepro.2021.126915

Leal Filho, W. L., Wall, T., Salvia, A. L., et al. (2023). The central role of climate action in achieving the United Nations' Sustainable Development Goals. *Science and Reports, 13*, 20582. https://doi.org/10.1038/s41598-023-47746-w

Lozano, R., Lukman, R., Lozano, F. J., Huisingh, D., & Lambrechts, W. (2013). Declarations for sustainability in higher education: Becoming better leaders, through addressing the university system. *Journal of Cleaner Production, 48*, 10–19. https://doi.org/10.1016/j.jclepro.2011.10.006

Lozano, R., Ceulemans, K., Alonso-Almeida, M., Huisingh, D., Lozano, F. J., Waas, T., Lambrechts, W., Lukman, R., & Hugé, J. (2015). A review of commitment and implementation of sustainable development in higher education: Results from a worldwide survey. *Journal of Cleaner Production, 108*, 1–18. https://doi.org/10.1016/j.jclepro.2014.09.048

Mallow, S., Toman, I., & van't Land, H. (2019). Higher education and the 2030 agenda: Moving into the 'decade of action and delivery for the SDGs'. IAU 2nd global survey report on higher education and research for sustainable development. *International Association of Universities*.https://static1.squarespace.com/static/64ca66e5f086975ce8783d53/t/653bcfbcaf643c254532ba19/1698418623774/iau_hesd_survey_report_final_jan2020.pdf

Millennium Development Goals. (2000). United Nations. https://www.un.org/millenniumgoals/

Molina, Á., Helldén, D., Alfvén, T., Niemi, M., Leander, K., Nordenstedt, H., … Biermann, O. (2023). Integrating the United Nations sustainable development goals into higher education globally: A scoping review. *Global Health Action, 16*(1). https://doi.org/10.1080/16549716.2023.2190649

Oliveira, M. C., & Proença, J. (2025). Sustainable campus operations in higher education institutions: A systematic literature review. *Sustainability, 17*(2), 607. https://doi.org/10.3390/su17020607

Panevėžys College. (2023). Communication on Engagement 2021-2022. https://unglobalcompact.org/participation/report/cop/detail/480118

Pawłowski, A. (2008). How many dimensions does sustainable development have?. *Sustainable Development, 16*(2):81–90. https://doi.org/10.1002/sd.339

Philip, C., Polczynska, M., & Brown, T. (2024). Higher education curriculum design for sustainable development: Towards a transformative approach. *International Journal of Sustainability in Higher Education, 25*(5), 1009–1023. https://doi.org/10.1108/IJSHE-06-2023-0255

Pradhan, P., Costa, L., Rybski, D., Lucht, W., & Kropp, J. P. (2017). A systematic study of Sustainable Development Goal (SDG) interactions. *Earth's Future, 5*(11), 1169–1179. https://doi.org/10.1002/2017EF000632

Prior, D. D., Mysore Seshadrinath, S., & Zhang, M. (Y.), & McCormack, M. (2024). Measuring sustainable development goals (SDGs) in higher education through semantic matching. *Studies in Higher Education*, 1–4. https://doi.org/10.1080/03075079.2024.2386625

Purvis, B., Mao, Y., & Robinson, D. (2019). Three pillars of sustainability: in search of conceptual origins. *Sustainability Science, 14*, 681–695. https://doi.org/10.1007/s11625-018-0627-5

Sánchez-Carracedo, F., Ruiz-Morales, J., Valderrama-Hernández, R., Muñoz-Rodríguez, J. M., & Gomera, A. (2021). Analysis of the presence of sustainability in higher education degrees of the Spanish university system. *Studies in Higher Education, 46* (2), 300–317. https://doi.org/10.1080/03075079.2019.1630811

Serafini, P. G., Moura, J. M., Almeida, M. R., & Rezende, J. F. D. (2022). Sustainable development goals in higher education institutions: A systematic literature review. *Journal of Cleaner Production, 370*, 133473–133493. https://doi.org/10.1016/j.jclepro.2022.133473

Šiauliai State College (2022) Social responsibility report 2022. https://svako.lt/uploads/2022-global-10-25-svako.pdf

Sustainability Accelerator. (2015). *AtKisson Group*. http://atkisson.com/acceleratorlite/

Taylor, S. J. (2016). A review of sustainable development principles: Centre for environmental studies. South Africa: University of Pretoria.

Tejedor, G., Segalàs, J., Barrón, Á., Fernández-Morilla, M., Fuertes, M. T., Ruiz-Morales, J., Gutiérrez, I., García-González, E., Aramburuzabala, P., & Hernández, À. (2019). Didactic strategies to promote competencies in sustainability. *Sustainability, 11*(7), 2086. https://doi.org/10.3390/su11072086

Toman, I., van't, L. H., & Harris, M. (2023). Accelerating action for the SDGs in higher education. In 3rd IAU Global survey report on higher education and research for sustainable development. Paris. https://www.iau-aiu.net/IMG/pdf/iauhesdsurvey2023_accelerating_actions.pdf

Tomaskinova, J., Tomaskin, I., Stastna, M., Theuma, H. (2019). Environmental management challenges in the 21st century: assessment of protected areas management effectiveness in Slovakia. *Journal of Environmental Protection and Ecology 20*(4), 2012–2019.

UNESCO (2015) Education 2030: Incheon declaration and framework for action for the implementation of sustainable development goal 4. Paris: UNESCO.

UNESCO (2018) Education in an interconnected world: ensuring inclusive and equitable development (GEM 2018). https://www.unesco.org/sdg4education2030/en/articles/education-interconnected-world-ensuring-inclusive-and-equitable-development-gem-2018

United Nations Millennium Declaration (Resolution 55/2, adopted by the General Assembly) 2000. New York: United Nations.

United Nations. (2015). Transforming our world: the 2030 agenda for sustainable development. United Nations. https://sdgs.un.org/2030agenda

Vilnius College. (2023). Social responsibility and sustainable development report. https://www.viko.lt/wp-content/uploads/sites/8/2024/04/SOCIALINES-ATSAKOMYBES-IR-DARNIOS-PLETROS-ATASKAITA-2023.pdf

Wohlgezogen, F., McCabe, A., Osegowitsch, T., & Mol, J. (2020). The wicked problem of climate change and interdisciplinary research: Tracking management scholarship's contribution. *Journal of Management & Organization, 26*(6), 1048–1072. https://doi.org/10.1017/jmo.2020.14

Wu, X., Fu, B., Wang, S., Liu, Y., Yao, Y., Li, Y., Xu, Z., Liu, J. (2023) Three main dimensions reflected by national SDG performance. *Innovation (Camb), 4*(6):100507. https://doi.org/10.1016/j.xinn.2023.100507

Žalėnienė, I., & Pereira, P. (2021). Higher education for sustainability: A global perspective. *Geography and Sustainability, 2*, 99–106. https://doi.org/10.1016/j.geosus.2021.05.001

Fostering Sustainability-Related Soft Skills Through Experiential Learning: An Investigation of Interdisciplinary Approaches in Higher Education

Pasquale Marcello Falcone and Ilaria Tutore

Abstract Higher education institutions (HEIs) play a critical role in preparing future leaders to promote sustainable practices. Research underscores the strong link between HEIs and the cultivation of sustainability-oriented attitudes and behaviors, especially within business schools. These institutions are tasked with equipping students with key competencies, such as responsible management, stakeholder engagement, and ethical decision-making, all vital for addressing contemporary sustainability challenges. While progress has been made, questions remain about how sustainability education can effectively nurture essential soft skills like critical thinking, collaboration, and ethical reasoning. Experiential learning approaches—such as service learning, community engagement, action research, and interdisciplinary projects—have shown promise in fostering sustainability competencies. However, further exploration is needed to assess how these methods translate into practical skill development. This study focuses on an experiential learning program conducted in spring 2024, linked to SEA- EU European Universities alliances activities and centered on Circular Economy principles. Through a mixed-methods approach, including surveys and focus groups, the research examines how hands- on, interdisciplinary activities enhance students' sustainability-related soft skills. The findings aim to guide universities in implementing similar initiatives, promoting collaborative projects and real-world problem solving activities. This work contributes to the broader discussion on HEIs' responsibility to prepare future leaders capable of driving sustainable change through responsible management practices.

P. M. Falcone
Department of Business and Economics, University of Naples Parthenope, Naples, Italy

Sustainability Competence Centre, Széchenyi István University, Egyetem Square 1, Győr 9026, Hungary

P. M. Falcone
e-mail: pasquale.falcone@uniparthenope.it

I. Tutore (✉)
Department of Management and Quantitative Studies, University of Naples Parthenope, Naples, Italy
e-mail: ilaria.tutore@uniparthenope.it

W. Leal Filho et al. (eds.), *Introducing the UN SDGs in Higher Education Institutions*, World Sustainability Series, https://doi.org/10.1007/978-3-032-23017-1_11

Keywords Higher Education Institutions (HEIs) · Education for Sustainable Development (ESD) · Experiential learning · Soft skills development

1 Introduction

Higher education institutions (HEIs) play a crucial role in shaping future citizens capable of driving sustainable practices. Research consistently shows a strong relationship between HEIs and the development of sustainable attitudes and behaviors (Greenland et al., 2022; Nguyen et al., 2019; Tetrevova et al., 2021). Universities, especially business schools, are increasingly called upon to equip students with the competencies needed to tackle sustainability issues in their professional lives (Eustachio, Leal Filho, Salvia, Guimarães, et al., 2024a, 2024b; García-Feijoo et al., 2020) and to effectively address and meet Sustainable Development Goals (SDGs) (Sá & Serpa, 2020). These competencies align with broader 21st-century skills, which highlight the need for responsible management, stakeholder engagement, and navigating ethical dilemmas (García-Feijoo et al., 2020). While several studies emphasize the role of sustainability education in shaping future managers (Meyer, 2015; Nguyen et al., 2019; Patel et al., 2017; Sá & Serpa, 2020) and the need to drive to sustainability leadership or transformational leadership, considered an essential aspect of organizational change towards corporate sustainability (Eustachio et al., 2023) there is still the need to provide practical insights on the topic (García-Feijoo et al., 2020). Much of the current literature focuses on the theoretical aspects of sustainability education, exploring its importance but offering few practical insights into how educational interventions can be implemented and measured for effectiveness. However, despite the development of sustainability related soft skills such as critical thinking, collaboration, and ethical reasoning is widely acknowledged as essential, how these skills can be cultivated through sustainability education and their effectiveness remains an area requiring further investigation (García-Feijoo et al., 2020; O'Flaherty & Liddy, 2018).

Among the other pedagogical approaches, previous study already highlighted the role of experiential learning approaches such as Service learning, combined with community engagement and innovative research methods, such as action research, CBPR (Community-Based Participatory Research), interdisciplinary projects or project-based learning, to effectively promote the acquisition of sustainable competencies (Rodríguez-Zurita et al., 2024).In this vein, this chapter aims to investigate the role of Experiential learning into fostering the development of sustainability-related soft skills among university students, focusing on specific case study, the "Circular Spring" program developed during the spring of 2024 for both at bachelor's and master's degrees students at the Interdepartmental School of Economics and Law of the University of Naples Parthenope (Italy). The program, developed as a series of seminars and a student contest, wants to promote Circular Economy (CE) principles through hands-on activities and interdisciplinary collaboration. Employing a mixed-methods approach, integrating both quantitative and qualitative data from questionnaire distributed to participants of the "Circular Spring" program, we assessed the effectiveness of the program to the development of soft skills, including collaboration, ethical decision-making, and critical thinking related to SDGs. The analysis shows that participants in *Circular Spring* reported significantly stronger pro-environmental attitudes and engagement in sustainable behaviors compared to non-participants. Notably, they exhibited higher ecological awareness, stronger support for CE policies, and greater involvement in recycling and responsible consumption practices.

This research contributes to the existing body of knowledge by focusing on the empirical evaluation of a university-led experience designed to promote sustainability. Unlike previous studies, which often lack a practical assessment of intervention outcomes, this study offers insights into how specific experiential and programs can foster soft skills essential for sustainable management. By examining students' experiences and learning outcomes, the study provides evidence-based recommendations for enhancing sustainability education in business schools.

The remainder of this chapter is structured as follows. Section 2 reviews the literature on sustainability education in economics and management universities. Section 3 presents the research methods adopted in the study. Section 4 introduces the Circular Spring case study. Section 5 presents the empirical results and discusses the findings. Finally, Sect. 6 concludes the chapter by outlining the main implications, limitations, and future research directions.

2 Sustainability Education in Economics and Management Universities

In a world grappling with unprecedented environmental, social, and economic challenges, it is increasingly evident that we must teach, learn, and live in fundamentally different ways. The twenty-first century demands a shift from traditional educational models to approaches that equip individuals with the knowledge, skills, and values needed to navigate and address the complexities of our time.

The Environmental Education (EE) has been the first formally used strategy used to involve different levels of society to the awareness and then solving of environmental problems, seeking environmental conservation through knowledge, experience, values and local practices (Acosta Castellanos & Queiruga-Dios, 2022; Ardoin et al., 2020).

This concept was later broadened to include a wider range of definitions. It now encompasses understanding how to be a global citizen and actively engaging in one's society. This involves contributing to the sustainability of the community and seeking solutions to address social and economic issues, alongside natural and environmental challenges (Fekih Zguir et al., 2021).

The current counterpart of EE is education for sustainable development (ESD), a comprehensive and transformative education that focuses on content and learning outcomes, pedagogy, and the learning environment (González-Salamanca et al., 2020).

ESD—considered by UNESCO as a holistic concept "that addresses learning content andoutcomes, pedagogy and the learning environment" (Education for Sustainable Development Goals Learning Objectives, 2017)—is crucial in enabling individuals to rethink their perspectives, actions, and interactions with the world, fostering active engagement in societal transformation. By equipping people with the knowledge, skills, values, and behaviors necessary for shaping a sustainable future, ESD serves as a transformative force driving positive change toward a more sustainable and equitable world (Acosta Castellanos & Queiruga-Dios, 2022; Sá & Serpa, 2020, 2022).

This suggests that individuals must acquire the knowledge, abilities, values, and mindsets necessary to collaborate effectively on sustainable development, calling for a transformative, action-oriented approach to teaching that fosters independent learning, engagement, teamwork, problem-solving, and both interdisciplinary and transdisciplinary thinking (Lozano et al., 2017).

Teaching sustainability is particularly challenging as it requires integrating knowledge from diverse disciplines, bridging formal and informal education. Developing these ambitious profiles of knowledge and skills in students (and in educators) is guided by 21st-century skills, with the goal of preparing them to become future "problem solvers," "agents of change," and "managers of transition"(González-Salamanca et al., 2020; Wiek et al., 2011).

In this process, and specifically in terms of the higher education arena, the notion of Sustainable Higher Education Development (SHED) is all-inclusive and involves, for example, the interaction of higher education with the surrounding environment, economic growth, social equity, equality and the improvement of populations' quality of life (Sá & Serpa, 2020). Higher education is considered important to ensure that all learners can develop the knowledge, attitudes and competencies needed to respond to sustainability challenges throughout their professional and personal lives (Education for Sustainable Development Goals: Learning Objectives, 2017).

Focusing on harmony between the environmental risks and benefits and ensures the provision of natural resources for future generations (Acosta Castellanos & Queiruga-Dios, 2022), ESD in general and SHED take on high centrality in promoting in students both competences to effectively address and meet Sustainable Development Goals (SDGs) and individual sustainability (Sá & Serpa, 2020).

For this purpose, UNESCO published in 2017 a guide that outlines learning objectives, suggests topics, and recommends activities tailored to each SDG. It provides implementation strategies at various levels, ranging from course design to national policies, offering educators adaptable guidance to fit specific learning contexts (Education for Sustainable Development Goals: Learning Objectives, 2017).

For each SDG, the guide defines specific learning objectives across three domains: cognitive, socio-emotional, and behavioral. The cognitive domain focuses on the knowledge and critical thinking skills needed to understand the SDG and the challenges in its achievement. The socio-emotional domain emphasizes the development of social and self-reflection skills, values, attitudes, and motivations necessary for collaboration and advocacy of the SDGs. The behavioral domain highlights competencies required for taking effective action.

Additionally, the guide suggests indicative topics and pedagogical approaches for each SDG, providing a comprehensive framework for integrating sustainability into educational practices.

However, incorporating ESD into university curricula presents a significant challenge for the academic system. In recent years, many universities have taken steps toward its integration, with numerous case studies documenting these implementation efforts (Acosta Castellanos & Queiruga-Dios, 2022; Barth & Rieckmann, 2012; Hinduja et al., 2023).

As distinguished by Lozano et al. (2017), there are four different strategies in integrating Sustainable Development (SD) in curricula: (1) infusing some environmental issues into an existing course; (2) developing a separate SD course; (3) integrating the SD concept into each course; or (4) providing specialization in SD within the faculty curricula (Orlovic Lovren et al., 2020).

Sterling and Thomas (2006) identified four models for integrating sustainability into education. These range from curriculums that remain unchanged, to approaches focused on education about sustainability ("bolt-on" methods), to more integrated strategies for sustainability education ("built-in" methods). The most ambitious model involves redesigning the entire curriculum to embrace sustainable education, a process that demands transforming the university's overarching educational mission. However, more recent studies highlight the importance of teaching issues at the campus, department, or school levels, while generally focusing on all SDGs. This niche debate covers the advantage of adopting cross-curricular approaches (meso-level) rather than stand-alone course initiatives (micro-level) (Fia et al., 2023). Orlovic Lovren et al. (2020) extended the discussion to the department level by introducing four elements for a successful program: (1) interdisciplinary educational courses, (2) improving teachers' skills, (3) enhancing active learning, and (4) setting institutional policies at the faculty level.

Approaches that consider more macro-level teaching strategies toward all 17 SDGs envisages the need to create and develop a wider culture of sustainability that surfaces departments (it has been demonstrated that learning processes which can enable transformative changes largely depend on academic staff and their capabilities and willingness to support such processes (Barth & Rieckmann, 2012) and degree programs, especially for students who are nearing entry to the job market (Albareda-Tiana et al., 2018).

Developing such a mindset has become an essential general skill for all higher education graduates, particularly those from business schools (Chiang & Chen, 2022; Landrum, 2021; Mather et al., 2011).

The call to incorporate the notion of sustainability into education is even more important for economic and business students, considering their future role as the leaders and managers wo will shape tomorrow's companies. It's widely recognized that companies play a pivotal role in driving society's transition toward sustainability, not only through their operations but also by influencing consumer behavior and choices. Equipping business students with sustainability-related skills is essential, as they will not only define future economic policies but also lead organizations in adopting practices that balance economic success with social and environmental responsibility.

This shift stems from the understanding that neglecting sustainability education risks producing graduates driven solely by profit (Hernández-López et al., 2020), potentially resulting in unethical management practices and financial industry crises (Giacalone & Wargo, 2009).

As a result, business education institutions are reevaluating how they integrate sustainability concepts into business-focused curricula (Winfield & Ndlovu, 2019), supporting ESD by adopting Sustainable Development Goals (SDGs) as a central learning objective (Leal Filho et al., 2019). The SDGs serve not only as a strategic framework for embedding sustainability into business education but also as a new language for enterprises (Setó-Pamies & Papaoikonomou, 2020).

This is evidenced by the growing number of business schools becoming a signatory of the Principles for Responsible Management Education (PRME)—a United Nations initiative aiming to "raise the profile of sustainability in schools around the world". This initiative encourages a change in business schools through six principles (purpose, values, method, partnership and dialogue) for implementing responsible management education (RME) aligned with the SDGs and the UN's Global Compact, influencing the implementation of these goals (Chiang & Chen, 2022; Eustachio et al., 2024a, 2024b).

Indeed, the effective implementation of ESD in higher education requires innovative methodologies beyond traditional didactic teaching methods (Barth & Rieckmann, 2012). An alternative methodology that has gained recognition for its potential to drive transformative learning and inclusive knowledge co-generation is Participatory Research (Fernando & Tajan, 2024).

With regards to pedagogical approaches, Rodríguez-Zurita et al. (2024) also demonstrated that project-based learning can effectively promote the acquisition of sustainable competencies. Experiential learning and other innovative research approaches, such as Service learning, work-integrated learning (WIL) projects learning and real-life experiences enable students to participate as change agents in communities, enhancing the relationship between universities and communities in the context of sustainability to develop their interpersonal competencies to serve as ambassadors for SD in their future work life (Alm et al., 2022).

Also, the Education for Sustainable Development Goals Learning Objectives (2017) highlighted the effectiveness of action-oriented learning since it increases knowledge acquisition, competency development and values clarification by linking abstract concepts to personal experience and the learner's life (Kolb, 2014). Despite the vibrant research done so far highlighting the importance of leadership in promoting change towards sustainable development in HEIs, this field is far from being saturated. Current debates focus on how this education should be delivered and the effectiveness of these methodologies.

3 Method

This study adopts a mixed-methods research design to investigate the role of experiential learning in fostering sustainability-related soft skills among university students. The research combines a case study analysis of the "Circular Spring" initiative with quantitative survey data collected from students at the University of Naples Parthenope.

The Circular Spring program was implemented during the spring semester of 2024 at the Interdepartmental School of Economics and Law of the University of Naples Parthenope (Italy). The initiative involved approximately 150 bachelor's and master's students and consisted of seminars, a sustainability contest, and a final dissemination event focused on circular economy principles.

To evaluate the impact of the initiative on students' attitudes and behaviours, a structured questionnaire was administered approximately two months after the conclusion of the program. The survey was distributed to students through email and in-person contact, resulting in a convenience sample of 163 respondents, including both participants and non-participants in the Circular Spring activities. This sample allowed for a comparison between students who participated in the Circular Spring initiative and those who did not.

The questionnaire included 27 items designed to explore different dimensions of sustainability-related attitudes and behaviours, including environmental awareness, sustainable consumption practices, and social participation in environmental initiatives. Responses were measured using Likert-scale questions and treated as continuous variables for the statistical analysis.

To analyse potential differences between students who participated in the initiative and those who did not, the study employed independent sample t-tests. This statistical analysis made it possible to identify whether engagement in experiential learning activities was associated with significant differences in sustainability-related attitudes and behaviours.

4 The Circular Spring Case Study

In order to analyze how can sustainability-related soft skills can be effectively developed in university education through experiential and interdisciplinary learning a mixed-methods approach combining quantitative survey analysis and qualitative insights from anectodical case study developed during the spring of 2024 among around 150 students both at bachelor's and master's degrees level of the Interdepartmental School of Economics and Law of the University of Naples Parthenope (Italy).

The activity, named "Circular Spring", lasted 3 months and hosted a series of events aimed at fostering dialogue and disseminating CE principles and practices within the university community (Fig. 1).

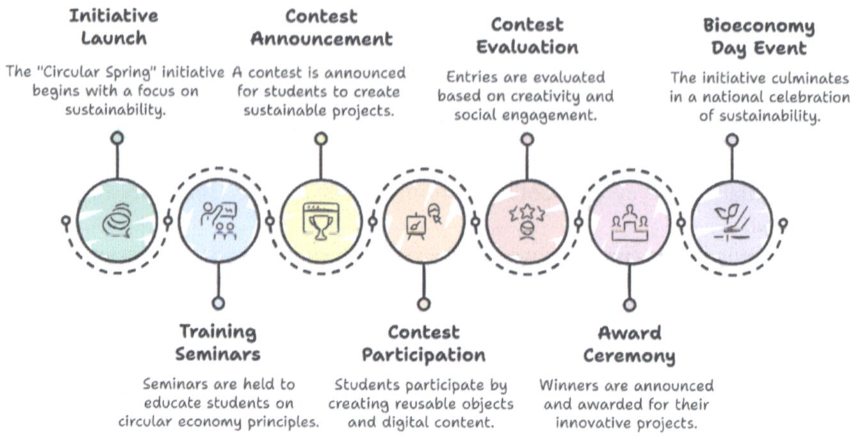

Fig. 1 Timeline of the "Circular Spring"

The initiative included: 1. Two training seminars, with a specific purpose of setting the knowledge and capabilities required to increase the awareness of CE topic on university environment and to enhance skills required to develop the contest: 2. a contest, the winning team will be announced on Bioeconomy Day, and selected members will represent the university at an international event in Gdańsk in June 2024 and 3: a final event, scheduled on the "Bioeconomy Day", officially recognized as a national celebration across Italy.

Prizes were awarded to students who had participated in the "ReCreate & Share" contest, recognizing their innovative contributions to sustainability and creativity. These students represented Parthenope University at the University of Gdansk on June 6–7 2024, where they engaged in discussions, showcased their projects, and exchanged ideas with peers from other SEA-EU institutions.

The SEA-EU Alliance consists of coastal universities that collaborated on key priorities such as ocean literacy, marine research, sustainable development, and intercultural exchange. Through various projects, SEA-EU promotes mobility, interdisciplinary collaboration, and joint research efforts to tackle environmental challenges and drive positive change within European coastal regions. The contest was organized within the broader framework of the Task 4.4 *SEA-EU Goes Greener*, a strategic initiative aimed at fostering sustainability within the SEA-EU Alliance. This initiative sought to create a dynamic sustainability network among SEA-EU universities, facilitating the exchange of best practices, policies, and innovative projects to enhance sustainability impact across campuses.

The promotion of Circular Spring has been ensured by an Instagram dedicated page. During the initiative, the page was populated with contents related to sustainability (Explanation of keywords such as upcycling, SDG, etc.), as well as featuring pictures of the events and work in progress videos created and shared by the students during the development of the project. The 12 Teams uploaded the final promotional video of the upcycling handcrafts: these promotional videos collected more than

4000 likes in 3 days, more than 60 k visualizations and 1 of them become viral on the social network.

In the table 1, the main information about the activities carried out is provided.

5 Results and Discussion

5.1 Sample Characteristics and Empirical Results

To analyze how engagement in sustainability-related activities is correlated with reported improvements in critical thinking, collaboration, and ethical reasoning, we developed several analysis derived from survey conducted to the students of the University just 2 months after the end of the activity. The data were gathered through a structured questionnaire to a convenience sample (n = 163) of students, contacted by mail and in persons.

The dataset provides an overview of participants' demographic, socioeconomic, and political characteristics (Fig. 2). Regarding Circular Spring participation, the majority of respondents (61.3%) did not take part in the event, while 38.7% did.

The gender composition highlights a predominance of female respondents (57.7%), followed by males (41.1%), with a small percentage (1.2%) not specifying their gender. In terms of education level, most participants are enrolled in a bachelor's degree program (57.1%), while 40.5% are at the master's level, and only 2.5% are pursuing a single-cycle degree.

When considering work and study balance, 38.7% of respondents are working students, whereas the remaining 61.3% focus solely on their studies. Family net income data indicate that a considerable proportion (33.1%) belongs to households earning less than 15,000€ annually, while 23.3% fall within the 15,000–25,000€ range. The percentage gradually decreases as income levels rise, with only 6.7% reporting a family income above 70,000€.

Finally, the political orientation of respondents shows that the largest group (33.1%) is uninterested in politics. Among those who do express a preference, the most common affiliation is center-left (22.1%), followed by left (19.6%), center-right (11%), center (8.6%), and right (5.5%). The analysis provides valuable insight into the participants' backgrounds, helping to contextualize their perspectives and behaviours in relation to sustainability and CE initiatives.

The second part of the questionnaire consists of a series of questions aimed at analysing key dimensions of sustainable behaviours and attitudes. Specifically, the 27 questions explore several relevant aspects, such as:

1. Attitudes and Environmental Awareness—Includes personal opinions and ecological awareness.
2. Sustainable Behaviors—Covers responsible purchasing, recycling, reusing, repairing items, sustainable mobility, and the use of eco-friendly transportation.

Table 1 Overview of the "Circular Spring" activities: typology, description, and participant engagement

Title	Typology	Description	Additional information	**Attendees**
Reducing Waste & Creative Design	*Seminar*	The seminar covers research on reducing urban waste impact, creative design for sustainable identities, and community workshops to foster skills in digital manufacturing and entrepreneurship in the green and CE	Guests: *Re-made Community Lab* founders. An organization based in Naples dedicated to the CE, promoting the repair, recycling, and reuse of items to reduce waste and give them a new life. The organization also operates a community workshop where citizens can learn to repair and create objects, fostering a more sustainable lifestyle	~40
From Concept to Creation	*Seminar*	This seminar provides a comprehensive overview of the process of creating audiovisual content. The session outlines key stages, allowing participants to gain a complete understanding of audiovisual production workflow	Guests: *Meleagris firm SRL employe.* A newly established film production company based in Naples, specializing in audiovisual content and covering the entire production process. A co-founder will explain the phases of concept development, production, and finalization, offering students a comprehensive overview of creating an audiovisual product	~50

(continued)

Table 1 (continued)

Title	Typology	Description	Additional information	**Attendees**
ReCreate & Share	*Contest*	Teams consisting of students, faculty, and staff compete to represent the university in an international competition with partner universities. Each team must create: 1. A reusable object: Crafted from recycled or creatively reused materials sourced from the university environment 2. A digital content piece (video/reel): Showcasing the creation process, promoting the object, or highlighting its potential uses	Evaluation criteria: 1. For the Object: Creativity, innovation, quality, functionality, alignment with contest objectives, and positive environmental impact 2. For the Video: • 50% based on originality and alignment with contest goals • 50% based on effectiveness and social engagement (e.g., likes)	60
Bioeconomy day—Designing Sustainability: Strategies, Actions, and Policies for the CE and Bioeconomy	*Conference*	Bioeconomy Day is a nationwide initiative celebrating its sixth edition in 2024 in partnership with *Cluster Spring*. It promotes the use of renewable biological resources and fosters discussions on sustainability Event Highlights: 1. Contest Award Ceremony: Recognition of the winning team 2. Roundtable Discussions: Exploring environmental sustainability strategies, funding opportunities, and project design with representatives from various sectors 3. University Project Presentations: Showcasing of sustainability-focused initiatives developed by university departments	Guests: 1. Manager for the Campania Region in the Division of Energy, Energy Efficiency and Savings, Green Economy, and Bioeconomy 2. Expert in cohesion policy and sustainable development 3. Representative of Legambiente (Italian Environmentalist association) on behalf of the energy community "Comunità Energetica e Solidale di Napoli Est" 4. CEO of "Il Tempio" Agricultural company implementing CE activities	~200

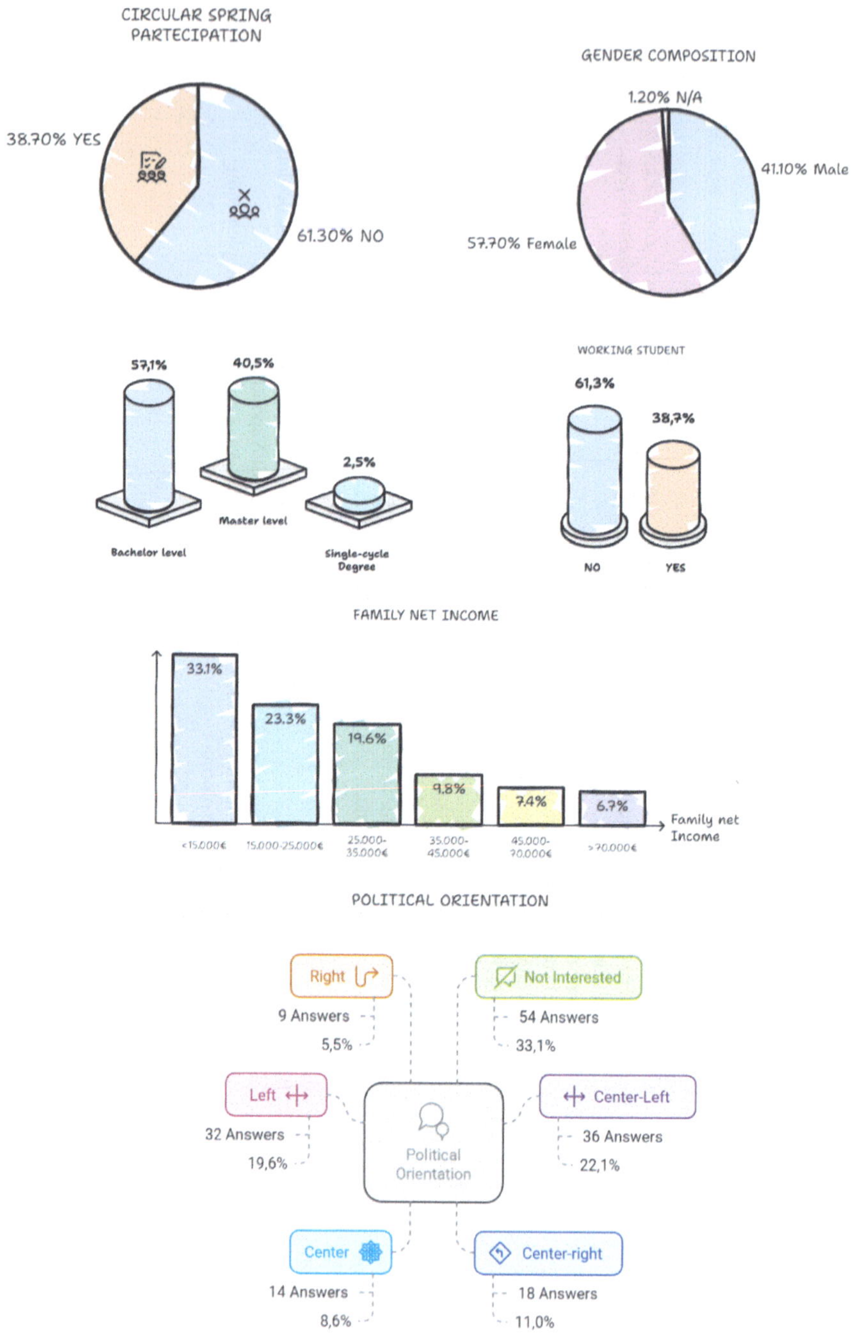

Fig. 2 Composition of the sample

3. Social and Cultural Participation—Relates to volunteering, environmental events, and visits to cultural sites.

Therefore, we conducted a t-test analysis to determine whether the identified dimensions significantly differ between individuals who participated in the activity and those who did not (Appendix 1).

The t-test analysis was conducted to compare differences in the identified dimensions between participants and non-participants in the activity (Fig. 3). The results reveal several significant differences between the two groups.

Regarding environmental attitudes and awareness, participants exhibit a more proactive approach toward sustainability. For instance, they consider environmentally friendly behaviour an important part of their identity ($p < 0.001$, d 1.113) and believe that the government should promote awareness campaigns on the CE ($p < 0.001$, d 1.098). Additionally, compared to non-participants, they are more likely to believe that waste separation influences climate change ($p < 0.001$, d 1.158).

In the domain of sustainable behaviours, participants are more inclined to purchase local and seasonal products ($p < 0.001$, d 1.199) and to choose products with sustainability certifications ($p < 0.001$, $d = 1.162$). They are also more likely to avoid single-use products (p 0.017, d 0.941) and to repair items instead of replacing them (p 0.007, $d = 0.896$).

Finally, in the area of social and cultural participation, significant differences emerge in engagement in environmental activities ($p < 0.001$, $d = 0.344$) and volunteering (p 0.005, $d = 0.348$).

5.2 *Discussion: Implications for Sustainability Education*

While the statistical results highlight clear differences between participants and non-participants, their broader implications for sustainability education require further interpretation.

Statistical analysis confirms the crucial role of HEIs in fostering sustainability-oriented attitudes and behaviours among students, particularly in business schools. The findings reveal that participants in the experiential learning activity exhibit greater environmental awareness and a stronger commitment to sustainable practices compared to non-participants.

In the domain of *Attitudes and Environmental Awareness*, participants demonstrate a stronger sustainability-oriented identity and a heightened recognition of the importance of awareness campaigns. This suggests that hands-on experiences, such as the Circular Spring contest, can reinforce the connection between sustainability education and the development of ethical values and responsible management principles, contributing directly to *SDG 4 (Quality Education).*

Regarding *Sustainable Behaviours*, the fact that participants are more likely to choose sustainable products, reduce single-use materials, and repair objects instead

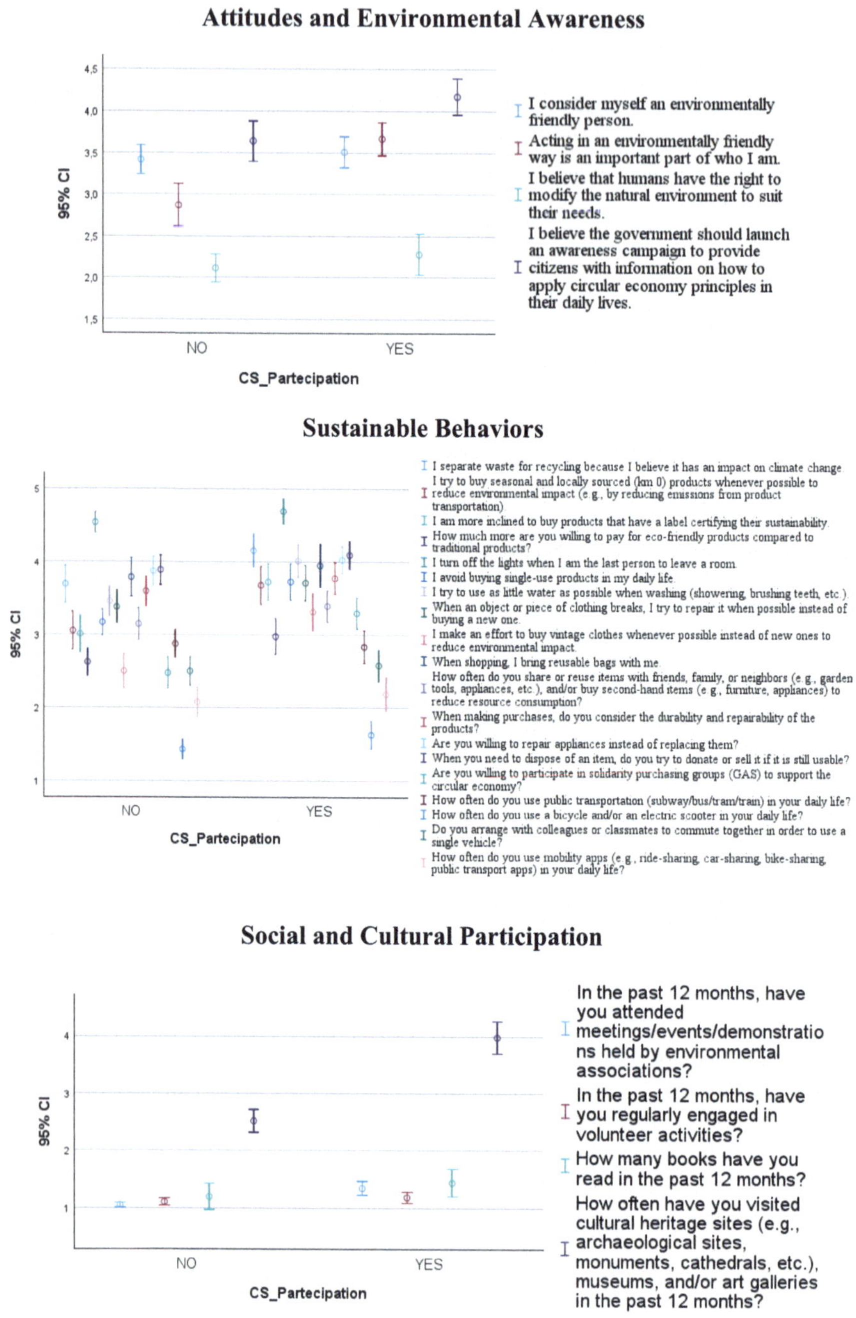

Fig. 3 Error bar chart

of replacing them indicates that practical activities have a tangible impact on translating knowledge into everyday actions. This aligns with *SDG 11 (Sustainable Cities and Communities), SDG 12 (Responsible Consumption and Production) and SDG 13 (Climate Action)* by promoting responsible consumption and reducing environmental impact. Experiential learning fosters the application of sustainability-related competencies in real-world contexts, reinforcing sustainable decision-making in daily life.

For example, the University hosted few months later the "SOS Environment: Commitment and Action for a Sustainable Future" campaign, totally organized by students. The event involved a debate on environmental issues followed by a CleanUp of the venue with the collaboration of Plastic Free NGO. The event- organized by the student volunteer environmental association and supported Italian Network of Universities for Sustainable Development (RUS)—demonstrates how PR campaigns can successfully engage both students and faculty in sustainability efforts.

Another example is the wide participation in the national initiative #miilluminodimeno. The event is celebrated every year on February 16th and serves as a day to honor respect for the environment and the push toward a sustainable future. The theme of 2025, which is gaining increasing awareness and a desire to experiment with new production and consumption practices, focused on fashion and the environmental impact of fast fashion. A group of students, supported by the university, promoted the installation of a clothing collection bin within the university and raised awareness about its use.

Finally, the results in *Social and Cultural Participation* support the idea that interdisciplinary engagement and direct involvement in environmental and volunteer activities can enhance transversal skills such as collaboration and critical thinking. This directly supports *SDG 17 (Partnerships for the Goals)* by encouraging multistakeholder engagement and *SDG 16 (Peace, Justice, and Strong Institutions)* by promoting active citizenship and ethical leadership.

Effective collaboration between multiple stakeholders—such as institutions, businesses, and other universities—is essential for advancing the SDGs. For example, initiatives promoted by the SEA-EU Alliance have significantly contributed to raising awareness, fostering dialogue, and strengthening the commitment to sustainable development. These efforts highlight how engaging a wide range of stakeholders accelerates the realization of sustainability goals by leveraging diverse expertise and resources to drive systemic change.

These findings strengthen the hypothesis that experiential learning approaches effectively cultivate the essential soft skills needed to address sustainability challenges. By integrating hands-on, interdisciplinary activities, HEIs can better prepare future leaders to drive sustainable change through responsible management and stakeholder engagement. This study highlights the importance of implementing similar initiatives to equip students with the competencies necessary for real-world impact.

6 Conclusion

Promoting sustainability is vital to addressing critical global challenges such as climate change, environmental degradation, and social inequality. HEIs play a pivotal role in shaping future leaders who can drive sustainable practices in their personal and professional lives. This study underscores the importance of sustainability education in business schools, where students must be equipped with competencies to navigate the complexities of sustainability issues in their careers. As the research consistently suggests, HEIs are essential in cultivating sustainable attitudes and behaviors among students, helping them develop the critical thinking and ethical decision-making skills required to tackle global challenges (Nguyen et al., 2019; Tetrevova et al., 2021).

One of the key findings from this study is the value of experiential learning in promoting the development of sustainability-related soft skills. The "Circular Spring" program, which focuses on hands-on activities and interdisciplinary collaboration, effectively encourages students to engage with sustainability concepts beyond theoretical learning. These practical experiences allow students to apply their knowledge in real-world contexts, fostering essential skills like collaboration, ethical decision-making, and critical thinking. The study's mixed-methods approach, integrating both quantitative and qualitative data, provides a comprehensive view of how experiential learning activities can enhance sustainability education.

Furthermore, the research highlights the need for interdisciplinary collaboration in addressing sustainability challenges. Considering the specific field in which students in this study are involved, to prepare responsible leaders and manager, it is essential to integrate economic and legal knowledge with broader, interdisciplinary competencies that promote a holistic understanding of sustainability.

Indeed, sustainability cannot be tackled solely from an economic perspective.

In this context, the involvement of external stakeholders—such as businesses, policymakers, and civil society—plays a key role in bridging academic knowledge with real-world applications. The Quadruple Helix Model highlights how universities, industry, government, and society must collaborate to generate sustainable solutions (Franco et al., 2019). Engaging stakeholders in educational initiatives fosters co-creation processes that enhance students' ability to address sustainability challenges through applied knowledge and multi-actor dialogue.

This approach reflects the broader trend in education, where collaboration across disciplines fosters innovation and problem-solving in complex areas like sustainability. By embedding interdisciplinary initiatives into curricula, universities can ensure that future managers develop not only technical expertise in their fields but also the ability to address sustainability challenges from multiple perspectives. This integration enables students to consider diverse viewpoints, develop ethical decision-making skills, and create solutions that balance economic, social, and environmental objectives.

The findings also suggest that sustainability "readiness" may vary according to a student's age and academic level. Younger students may show greater enthusiasm and sensitivity to sustainability issues, suggesting that it could be beneficial to introduce sustainability education at earlier stages of a student's academic journey. This early engagement could foster long-term commitment to sustainable practices, preparing students to be more effective change-makers as they progress through their studies.

From a curriculum design perspective, this research offers several recommendations for universities aiming to integrate sustainability more deeply into business education. The study suggests that universities should encourage collaborative projects across different disciplines, allowing students to work together on sustainability-related challenges. These projects not only promote teamwork but also expose students to the complexities of sustainability issues, encouraging them to think critically and develop innovative solutions.

Moreover, the active participation of external stakeholders in these projects—such as companies engaging in circular economy practices, NGOs advocating for environmental justice, and local policymakers shaping sustainability policies—can enrich the learning experience by providing real-world case studies and mentorship opportunities. This approach aligns with the Quadruple Helix Model, reinforcing the idea that sustainability education should not be confined to academia but rather co-developed with societal actors to maximize its impact.

Additionally, universities should design activities that simulate real-world sustainability challenges, providing students with the opportunity to practice and refine the soft skills necessary for effective sustainable management.

Finally, this study contributes to the ongoing discourse about the role of HEIs in shaping future leaders who can address sustainability challenges through responsible management practices. By focusing on empirical evaluations of educational interventions, it adds valuable insights into the practical application of sustainability education. The findings provide universities with evidence-based recommendations for improving sustainability-related educational programs, ensuring that students are not only knowledgeable but also equipped with the necessary soft skills to drive positive change in a rapidly evolving world.

Overall, these findings highlight several key lessons emerging from the study, particularly regarding the importance of integrating experiential learning approaches and interdisciplinary collaboration into higher education curricula in order to foster sustainability-related competencies among future managers and professionals.

An in-depth analysis of the soft skills developed through experiential learning in HEI significantly reinforces the role of the Inner Development Goals (IDGs), considering fundamental for building the inner capacity to deal with the increasing complexity, in working towards the SDGs (Mubtaker, 2024). Introduced in 2020, this framework is designed to complement and enhance the UN SDGs by developing inner capacities—skills, values, and qualities—that are essential for fostering sustainable behavior and effective leadership in a changing world.

By cultivating internal capacities—such as empathy, critical thinking, and collaboration—experiential learning not only enhances sustainable leadership but also underscores the essential contribution of soft skills to achieving global sustainability objectives.

In conclusion, sustainability education in business schools is not only about imparting knowledge but also about fostering the development of critical soft skills that are essential for navigating the complexities of the twenty-first century.

By implementing experiential learning programs like "Circular Spring," universities can better prepare students to meet the challenges of a sustainable future. The research highlights the need for continued innovation in sustainability education and provides valuable guidance for universities seeking to improve their programs and equip students with the competencies required for sustainable leadership.

While this study provides valuable insights into the role of experiential learning in sustainability education, several limitations must be acknowledged. First, the findings may have limited generalizability, as the sample consists primarily of economics and law students. Future research should explore whether similar outcomes emerge in other academic disciplines.

Second, in order to analyse the effectiveness of the experiential learnings, it could be interesting to assess the long-term impacts on students' sustainability competencies and behaviours. Longitudinal studies would be beneficial in evaluating the persistence of these effects over time.

Additionally, the study relies on self-reported data collected through questionnaires and focus groups, which may introduce social desirability bias. Integrating direct observations or behavioural data could enhance the reliability of the findings.

These considerations also suggest several promising directions for future research in the field of sustainability education.

Future research should consider exploring the relationship between soft skills and the Inner Development Goals (IDGs). Investigating how internal competencies such as self-awareness, empathy, and critical thinking contribute to sustainability leadership may offer valuable insights for SDG-related projects.

Future advancements in sustainability education should also embrace emerging technologies such as AI-driven simulations, gamification, and virtual reality to foster innovative learning experiences that integrate digitalization and sustainability—the twin transition—enhancing students' engagement and problem-solving abilities in real-world sustainability challenges.

Appendix 1: Independent T-Test Result

Variable	Question	Non-participants (n = 100)	Participants (n = 63)	t(df)	p-value	Cohen's d
Attitudes and environmental awareness	I consider myself an environmentally friendly person	3410	3508	–0.737	0.236	0.826

(continued)

(continued)

Variable	Question	Non-participants (n = 100)	Participants (n = 63)	t(df)	p-value	Cohen's d
	Acting in an environmentally friendly way is an important part of who I am	2870	3667	–4.451	0.000	1.113
	I believe that humans have the right to modify the natural environment to suit their needs	2120	2286	–1.140	0.018	0.903
	I believe the government should launch an awareness campaign to provide citizens with information on how to apply circular economy principles in their daily lives	3640	4175	–3.027	0.000	1.098
Sustainable behaviors	I separate waste for recycling because I believe it has an impact on climate change	3690	4159	–2.517	0.000	1.158
	I try to buy seasonal and locally sourced (km 0) products whenever possible to reduce environmental impact (e.g., by reducing emissions from product transportation)	3060	3683	–3.229	0.000	1.199

(continued)

(continued)

Variable	Question	Non-participants (n = 100)	Participants (n = 63)	t(df)	p-value	Cohen's d
	I am more inclined to buy products that have a label certifying their sustainability	3010	3730	–3.853	0.000	1.162
	How much more are you willing to pay for eco-friendly products compared to traditional products?	2630	2984	–2.326	0.700	0.947
	I turn off the lights when I am the last person to leave a room	4540	4698	–1.396	0.046	0.705
	I avoid buying single-use products in my daily life	3170	3730	–3.701	0.017	0.941
	I try to use as little water as possible when washing (showering, brushing teeth, etc.)	3460	4016	–3.582	0.078	0.965
	When an object or piece of clothing breaks, I try to repair it when possible instead of buying a new one	3380	3714	–1.952	0.084	1.064
	I make an effort to buy vintage clothes whenever possible instead of new ones to reduce environmental impact	2510	3317	–4.542	0.014	1.105

(continued)

(continued)

Variable	Question	Non-participants (n = 100)	Participants (n = 63)	t(df)	p-value	Cohen's d
	When shopping, I bring reusable bags with me	3790	3952	–0.795	0.161	1.269
	How often do you share or reuse items with friends, family, or neighbors (e.g., garden tools, appliances, etc.), and/or buy second-hand items (e.g., furniture, appliances) to reduce resource consumption?	3150	3397	–1.531	0.025	1.003
	When making purchases, do you consider the durability and repairability of the products?	3600	3778	–1.142	0.073	0.968
	Are you willing to repair appliances instead of replacing them?	3870	4032	–1.122	0.007	0.896
	When you need to dispose of an item, do you try to donate or sell it if it is still usable?	3890	4095	–1.372	0.019	0.930
	Are you willing to participate in solidarity purchasing groups (GAS) to support the circular economy?	2480	3302	–5.185	0.003	0.985

(continued)

(continued)

Variable	Question	Non-participants (n = 100)	Participants (n = 63)	t(df)	p-value	Cohen's d
	How often do you use public transportation (subway/bus/tram/train) in your daily life?	2880	2841	0.261	0.727	0.922
	How often do you use a bicycle and/or an electric scooter in your daily life?	1440	1635	−1.727	0.091	0.702
	Do you arrange with colleagues or classmates to commute together in order to use a single vehicle?	2510	2587	−0.515	0.050	0.933
	How often do you use mobility apps (e.g., ride-sharing, car-sharing, bike-sharing, public transport apps) in your daily life?	2080	2190	−0.682	0.150	1.006
Social and cultural participation	In the past 12 months, have you attended meetings/events/demonstrations held by environmental associations?	1050	1349	−5.405	0.000	0.344
	In the past 12 months, have you regularly engaged in volunteer activities?	1110	1190	−1.437	0.005	0.348
	How many books have you read in the past 12 months?	1200	1444	−1.416	0.357	1.074

(continued)

(continued)

Variable	Question	Non-participants (n = 100)	Participants (n = 63)	t(df)	p-value	Cohen's d
	How often have you visited cultural heritage sites (e.g., archaeological sites, monuments, cathedrals, etc.), museums, and/or art galleries in the past 12 months?	2520	3984	–8.658	0.587	1.051

References

Acosta Castellanos, P. M., & Queiruga-Dios, A. (2022). From environmental education to education for sustainable development in higher education: A systematic review. *International Journal of Sustainability in Higher Education, 23*(3), 622–644. https://doi.org/10.1108/IJSHE-04-2021-0167

Albareda-Tiana, S., Vidal-Raméntol, S., & Fernández-Morilla, M. (2018). Implementing the sustainable development goals at University level. *International Journal of Sustainability in Higher Education, 19*(3), 473–497. https://doi.org/10.1108/IJSHE-05-2017-0069

Alm, K., Beery, T. H., Eiblmeier, D., & Fahmy, T. (2022). Students' learning sustainability—Implicit, explicit or non-existent: A case study approach on students' key competencies addressing the SDGs in HEI program. *International Journal of Sustainability in Higher Education, 23*(8), 60–84. https://doi.org/10.1108/IJSHE-12-2020-0484

Ardoin, N. M., Bowers, A. W., & Gaillard, E. (2020). Environmental education outcomes for conservation: A systematic review. *Biological Conservation, 241*, Article 108224. https://doi.org/10.1016/J.BIOCON.2019.108224

Barth, M., & Rieckmann, M. (2012). Academic staff development as a catalyst for curriculum change towards education for sustainable development: An output perspective. *Journal of Cleaner Production, 26*, 28–36. https://doi.org/10.1016/J.JCLEPRO.2011.12.011

Chiang, M., & Chen, P. (2022). Education for sustainable development in the business programme to develop international Chinese college students' sustainability in Thailand. *Journal of Cleaner Production, 374*, Article 134045. https://doi.org/10.1016/J.JCLEPRO.2022.134045

Education for Sustainable Development Goals Learning Objectives. (2017). http://www.unesco.org/open-access/terms

Eustachio, J. H. P. P., Caldana, A. C. F., & Leal Filho, W. (2023). Sustainability leadership: Conceptual foundations and research landscape. *Journal of Cleaner Production, 415*. https://doi.org/10.1016/j.jclepro.2023.137761

Eustachio, J. H. P. P., Leal Filho, W., Salvia, A. L., Guimarães, Y. M., Brandli, L. L., Trevisan, L. V., Barbir, J., & Caldana, A. C. F. (2024a). Implementing sustainability in teaching: The role of sustainability leadership and transformational leadership in the context of higher education institutions. *Sustainable Development*. https://doi.org/10.1002/sd.2980

Eustachio, J. H. P. P., Leal Filho, W., Salvia, A. L., Lourenção, M., Guimarães, Y. M., Trevisan, L. V., Barbir, J., & Caldana, A. C. F. (2024b). Responsible management education: The leadership role of PRME business schools. *International Journal of Management Education, 22*(1). https://doi.org/10.1016/j.ijme.2023.100920

Fekih Zguir, M., Dubis, S., & Koç, M. (2021). Embedding Education for Sustainable Development (ESD) and SDGs values in curriculum: A comparative review on Qatar, Singapore and New Zealand. *Journal of Cleaner Production, 319*, Article 128534. https://doi.org/10.1016/J.JCLEPRO.2021.128534

Fernando, A. R. R., & Tajan, G. P. (2024). Education for sustainable development (ESD) through participatory research (PR): A systematic review. *Journal of Cleaner Production, 482*, Article 144237. https://doi.org/10.1016/J.JCLEPRO.2024.144237

Fia, M., Ghasemzadeh, K., & Paletta, A. (2023). How higher education institutions walk their talk on the 2030 agenda: A systematic literature review. *Higher Education Policy, 36*(3), 599–632. https://doi.org/10.1057/s41307-022-00277-x

Franco, D., Vocht, A. De, Kuppens, T., Martens, H., Thewys, T., Vanheusden, B., Schepers, M., & Segers, J. P. (2019). *22 sustainable education: Essential contributions to a 'Quadruple Helix' interaction and sustainable paradigm shift* (pp. 367–393). Brill. https://doi.org/10.1163/9789004396685_022

García-Feijoo, M., Eizaguirre, A., & Rica-Aspiunza, A. (2020). Systematic review of sustainable-development-goal deployment in business schools. *Sustainability (Switzerland), 12*(1). https://doi.org/10.3390/SU12010440

Giacalone, R. A., & Wargo, D. (2009). *The roots of the global financial crisis are in our business schools*. https://www.researchgate.net/publication/280685967

González-Salamanca, J. C., Agudelo, O. L., & Salinas, J. (2020). Key competences, education for sustainable development and strategies for the development of 21st century skills. A Systematic Literature Review. *Sustainability (Switzerland), 12*(24), 1–17. https://doi.org/10.3390/su12241 0366

Greenland, S., Saleem, M., Misra, R., & Mason, J. (2022). Sustainable management education and an empirical five-pillar model of sustainability. *International Journal of Management Education, 20*(3). https://doi.org/10.1016/j.ijme.2022.100658

Hernández-López, L. E., Álamo-Vera, F. R., Ballesteros-Rodríguez, J. L., & De Saá-Pérez, P. (2020). Socialization of business students in ethical issues: The role of individuals' attitude and institutional factors. *International Journal of Management Education, 18*(1). https://doi.org/10.1016/j.ijme.2020.100363

Hinduja, P., Mohammad, R. F., Siddiqui, S., Noor, S., & Hussain, A. (2023). Sustainability in higher education institutions in Pakistan: A systematic review of progress and challenges. *Sustainability (Switzerland), 15* (4). MDPI. https://doi.org/10.3390/su15043406

Kolb, D. A. (2014). *Experiential learning: Experience as the source of learning and development*. FT Press.

Landrum, N. E. (2021). The global goals: Bringing education for sustainable development into US business schools. *International Journal of Sustainability in Higher Education, 22*(6), 1336–1350. https://doi.org/10.1108/IJSHE-10-2020-0395

Leal Filho, W., Shiel, C., Paço, A., Mifsud, M., Ávila, L. V., Brandli, L. L., Molthan-Hill, P., Pace, P., Azeiteiro, U. M., Vargas, V. R., & Caeiro, S. (2019). Sustainable development goals and sustainability teaching at universities: Falling behind or getting ahead of the pack? *Journal of Cleaner Production, 232*, 285–294. https://doi.org/10.1016/j.jclepro.2019.05.309

Lozano, R., Merrill, M. Y., Sammalisto, K., Ceulemans, K., & Lozano, F. J. (2017). Connecting competences and pedagogical approaches for sustainable development in higher education: A literature review and framework proposal. *Sustainability (Switzerland), 9*(10). https://doi.org/10.3390/su9101889

Mather, G., Denby, L., Wood, L. N., & Harrison, B. (2011). Business graduate skills in sustainability. *Journal of Global Responsibility, 2*(2), 188–205. https://doi.org/10.1108/20412561111166049

Meyer, A. (2015). Does education increase pro-environmental behavior? Evidence from Europe. *Ecological Economics, 116*, 108–121. https://doi.org/10.1016/J.ECOLECON.2015.04.018

Mubtaker, Y. (2024). *Inner Development goals: Transformational skills for sustainable development*. Yama Mubtaker. https://books.google.it/books?id=4-JBEQAAQBAJ

Nguyen, N., Greenland, S., Lobo, A., & Nguyen, H. V. (2019). Demographics of sustainable technology consumption in an emerging market: The significance of education to energy efficient appliance adoption. *Social Responsibility Journal, 15*(6), 803–818. https://doi.org/10.1108/SRJ-11-2018-0312

O'Flaherty, J., & Liddy, M. (2018). The impact of development education and education for sustainable development interventions: A synthesis of the research. *Environmental Education Research, 24*(7), 1031–1049. https://doi.org/10.1080/13504622.2017.1392484

Orlovic Lovren, V., Maruna, M., & Stanarevic, S. (2020). Reflections on the learning objectives for sustainable development in the higher education curricula—Three cases from the University of Belgrade. *International Journal of Sustainability in Higher Education, 21*(2), 315–335. https://doi.org/10.1108/IJSHE-09-2019-0260/FULL/PDF

Patel, J., Modi, A., & Paul, J. (2017). Pro-environmental behavior and socio-demographic factors in an emerging market. *Asian Journal of Business Ethics, 6*(2), 189–214. https://doi.org/10.1007/s13520-016-0071-5

Rodríguez-Zurita, D., Jaya-Montalvo, M., Moreira-Arboleda, J., Raya-Diez, E., & Carrión-Mero, P. (2024). Sustainable development through service learning and community engagement in higher education: a systematic literature review. *International Journal of Sustainability in Higher Education*. Emerald Publishing. https://doi.org/10.1108/IJSHE-10-2023-0461

Sá, M. J., & Serpa, S. (2020). Higher education in times of change. In *Higher education in times of change*. Services for Science and Education-United Kingdom. https://doi.org/10.14738/eb.55.2020

Sá, M. J., & Serpa, S. (2022). Higher education as a promoter of soft skills in a sustainable society 5.0. *Journal of Curriculum and Teaching, 11*(4), 1–12. https://doi.org/10.5430/JCT.V11N4P1

Setó-Pamies, D., & Papaoikonomou, E. (2020). Sustainable development goals: A powerful framework for embedding ethics, CSR, and sustainability in management education. *Sustainability (Switzerland), 12*(5). MDPI. https://doi.org/10.3390/su12051762

Sterling, S., & Thomas, I. (2006). Education for sustainability: The role of capabilities in guiding university curricula. *International Journal of Innovation and Sustainable Development, 1*(4), 349–370. https://doi.org/10.1504/IJISD.2006.013735

Tetrevova, L., Vavra, J., & Munzarova, S. (2021). Communication of socially-responsible activities by higher education institutions. *Sustainability (Switzerland), 13*(2), 1–17. https://doi.org/10.3390/su13020483

Wiek, A., Withycombe, L., & Redman, C. L. (2011). Key competencies in sustainability: A reference framework for academic program development. *Sustainability Science, 6*(2), 203–218. https://doi.org/10.1007/s11625-011-0132-6

Winfield, F., & Ndlovu, T. (2019). "Future-proof your Degree": Embedding sustainability and employability at Nottingham Business School (NBS). *International Journal of Sustainability in Higher Education, 20*(8), 1329–1342. https://doi.org/10.1108/IJSHE-10-2018-0196

Integration of the Sustainable Development Goals in Industrial and Environmental Engineering Education

Miren Martínez-Santos and Esther Acha

Abstract This study aims to promote awareness and reflection on the importance of the Sustainable Development Goals (SDGs) among engineering students of the School of Engineering of Bilbao. For this purpose, two projects based on active learning methodologies have been implemented in the subjects of Biotechnology (Degree in Environmental Engineering) and Process Integration and Industrial Chemical Plants (Master in Industrial Engineering). In the Biotechnology course, during three academic years, an activity has been carried out involving the reading of a scientific article related to industrial or environmental biotechnological applications. Students, organized in groups of four, must perform individual and group tasks, culminating in a critical reflection on the impact of the biotechnological application on the SDGs and proposing improvements to reduce its social and environmental impact. Following the experience gained in Biotechnology, a similar analysis was incorporated in the subject of Process Engineering and Chemical Plants. In this subject various industrial processes are studied through active methodologies that include critical analysis of the SDGs. Students in both subjects have reported a significant improvement in their understanding of the SDGs related to energy, climate change, responsible production and sustainable development.

Keywords Engineering students · Active learning methodologies · Sustainability · Degree · Master · Higher education

1 Introduction

The Sustainable Development Goals (SDGs), established by the UN in 2015, represent a crucial global framework for addressing the social, economic and environmental challenges facing the world. The University of the Basque Country (UPV/

M. Martínez-Santos (✉) · E. Acha
Chemical and Environmental Engineering Department, School of Engineering of Bilbao, University of the Basque Country (UPV/EHU), Bilbao, Bizkaia, Spain
e-mail: miren.martinez@ehu.eus

W. Leal Filho et al. (eds.), *Introducing the UN SDGs in Higher Education Institutions*, World Sustainability Series, https://doi.org/10.1007/978-3-032-23017-1_12

EHU) is an example of this reality, as it conducted an internal analysis based on the publication of the UN's 2030 agenda. This process resulted in the creation of its own guide based on its reality: the so-called EHUagenda 2030 for Sustainable Development (2019–2025) (UPV/EHU, 2019a). This EHUagenda 2030 included UPV/EHU's contribution to 12 of the 17 SDGs of Agenda 2030, along with another goal that was defined taking into account the linguistic reality of the Basque Country, on its commitment to linguistic and cultural diversity, the so-called SDG 17 + 1. This agenda was built around UPV/EHU's educational model (IKD, the Basque acronym for "cooperative and dynamic learning"), which was redefined in the latest revision as IKD i3 (teaching x research x sustainability, three words that begin with "i" in Basque).

Higher Education plays a fundamental role in training professionals committed to sustainability, who will be responsible for implementing innovative solutions that contribute to achieving the SDGs. However, for future professionals to deeply understand these goals and their application in the context of their disciplines, it is essential that educational programs effectively integrate them into their training.

Working on the Sustainable Development Goals (SDGs) at university from the early years of undergraduate studies is crucial to train professionals who are aware and committed to sustainability (Leal Filho et al., 2019). Integrating the SDGs into the academic curriculum allows students to develop a critical mind-set and a holistic view of the world, understanding the interconnection between global and local challenges (Torres, 2021). This early training fosters social and environmental responsibility, preparing graduates to face 21st-century challenges with innovative and sustainable solutions (Sáenz, 2020). In addition, education for sustainability promotes transversal skills such as critical thinking, collaboration and problem-solving, essential in any professional field (Pradas & Copelia, 2024).

Universities, as centres of knowledge and innovation, have a fundamental role in disseminating and applying the SDGs (Lozano et al., 2015). By incorporating these goals into their teaching, research and outreach activities, they not only train competent professionals but also citizens dedicated to social and environmental transformation.

Active methodologies emerge as a key pedagogical tool for the integration of SDGs as they promote meaningful learning, critical reflection and active student participation. This pedagogical approach allows not only knowledge acquisition but also skill development for solving complex problems. In this context, these methodologies also play a key role in the development of specific sustainability competencies, such as systems thinking, anticipatory thinking and integrated problem solving. The use of active methodologies in the classroom is essential for integrating SDG concepts. The integration of active learning methodologies, such as project-based learning, cooperative learning, and role-playing, case method, etc., in education not only promotes sustainability competencies but also aligns with established theoretical frameworks in education for sustainable development. According to UNESCO's Education for Sustainable Development (ESD) model, these methodologies foster meaningful and transformative learning, which is essential for addressing current global challenges (UNESCO, 2021). Project-based learning allows students to apply sustainability concepts in real-world contexts, developing critical and practical skills (Freeman et al., 2014). Cooperative learning, on the other hand, promotes collaboration and critical thinking, key elements for solving complex sustainability-related problems (Prince, 2004). Finally, role-playing facilitates the understanding of multiple perspectives and informed decision-making, skills necessary for implementing sustainable solutions (Blanchard & Buchs, 2015). These methodological approaches not only enhance student motivation, participation and engagement but also contribute to the formation of citizens committed to social and environmental transformation (Cebrián, 2020; Martínez Valdivia et al., 2023). Students can connect acquired concepts with social and environmental demands of their professional environment.

Active methodologies also facilitate understanding of interconnections between different SDGs, allowing students to see how their actions can have a positive impact on multiple areas. These methodologies also foster social and environmental responsibility, preparing students to face global challenges with innovative and sustainable solutions. Students who internalize these principles from early academic training are better equipped to integrate sustainable practices into their future careers and professional development, thus contributing to more just and equitable societies. This is important in engineering careers, as well as in many others, due to negative impacts that can arise from engineers' professional actions if they do not consider sustainability globally.

This article explores how implementing active methodologies in Environmental Engineering Degree programs and Industrial Engineering Master's programs can facilitate integrating SDGs into university curricula. Through examples of specific projects in Biotechnology courses and Process Integration & Industrial Chemical Plants courses, we analyse the impacts these methodologies have on students' understanding and implementation of the SDGs while highlighting the importance of transforming university education to effectively and sustainably address global challenges.

2 Methodology

This work shows the implementation of SDGs in the curriculum of undergraduate and master's degree students. These master students differ greatly from undergraduate students in terms of their level of technical knowledge and maturity, so the methodologies used have been different. The methodology used in the undergraduate course (Biotechnology) is described below, followed by the methodology used in the master's course (Process Integration and Industrial Chemical Plants).

2.1 Integrating SDGs in the Biotechnology Course of the Environmental Engineering Degree

The methodology described below was implemented in the Biotechnology course, third year of the Environmental Engineering Degree, in the 2022/23 academic year. This course has a total of 6.0 ECTS taught over 15 weeks. These ECTS credits are distributed as follows: 4.5 for lectures and classroom classes, and 1.5 for laboratory practices. The theoretical-practical contents taught in the lectures and classroom classes are divided into eight major topics, among which is the topic 2: "Characteristics of microorganisms of biotechnological interest", which is the focus of the active methodologies project. This project has a duration of four weeks and the students will have to carry out a series of individual and group activities, for which they will be evaluated. This activity has a weight of 20% of the grade corresponding to the continuous evaluation (35% of the course). One of the advantages of this degree is the low number of students, around ten, which facilitates the implementation and monitoring of active methodologies.

The project uses an active and collaborative methodology aimed at understanding the characteristics and applications of microorganisms in industrial and environmental biotechnology. This project is structured in three stages, designed not only to understand biotechnological applications with microorganisms and their relationship with the SDGs, but also to foster autonomous learning, critical thinking and teamwork, three of the eight transversal competences of the UPV/EHU (UPV/EHU, 2019b). This methodology combines individual work, group discussions and oral presentations, promoting both autonomous learning and teamwork, as well as the development of communication and critical analysis skills.

In the *first stage* of the project, "Understanding and critical reflection on the SDGs", each student must reflect on one or two assigned SDGs. This should consider both the strengths and limitations of the SDGs, as the objective of this first stage is for students to understand that the SDGs are fundamental as guidelines for achieving a more equitable and sustainable world, but their implementation is often conditioned by a series of challenges, such as a lack of resources or coordination between countries, economic disparities between countries, etc. All students will contribute with their opinions to the reflections made by their peers. The teacher also guides the students in their reflection process through comments or questions. To carry out this task, a forum is enabled on the university's virtual platform (eGela-Moodle).

The teacher evaluates this activity (30% of the project grade) using a rubric, in which the following aspects are assessed (scoring each item from 0 to 10): (1) the depth of the critical reflection on the assigned SDG; (2) the quality of the contributions to the reflections of their peers or teacher; and (3) the argumentative soundness and quality of the written expression.

In the *second stage* of the project, "Applications of microorganisms in biotechnology", each group of four students is assigned a scientific article on a biotechnological application for the production of a product of industrial interest (Aimaretti, 2011; González García et al., 2013) or on the resolution of an environmental problem

(Beltrán-Rocha et al., 2017). For the development of this stage, cooperative learning will be used, specifically the 1–2-4 technique:

> Each student reads the assigned article outside the classroom and answers a series of questions about the biotechnological process. Within the group of four, two students share the same questions. In the next session, the pairs who share the questions exchange the information gathered and discuss the answers. Then, the two pairs of each group meet to share all the information. At the end of the session, all members of the group have the same level of knowledge on the same article, regardless of the questions they were initially assigned.
> After the face-to-face session, one forum is created on eGela-Moodle platform per article, where each student publishes a summary of the main conclusions from the worked questions. All students of the group contribute to this forum with information, corrections, and complements. The teacher supervises what has been published in all the forums and raises questions for the students to deepen their explanations. All the groups read also what has been published in the rest of the forums and have the possibility of contributing with their opinion.

Here, the evaluation is done by a rubric, considering: (1) the depth of the information and the contributions made; (2) the quality of the contributions made to the comments of their group mates or other groups members; and (3) the argumentative soundness and quality of the written expression. The evaluation of this stage (30% of the grade) is based on the participations in the forum.

> In a second face-to-face classroom session, new groups are formed. This time three students per group, where there will be one expert per article in each group. Each expert will transfer the most relevant information gathered in the previous work sessions to the other two students, who will concisely write down all the information. At the end of this session, each student should have a complete and deep vision of the three articles.

The task of the teacher in the classroom face-to face sessions is to supervise that the information transmitted by all participants is correct. In addition, he/she clarifies any doubts they may have about the biotechnological process and make the groups more dynamic so that all students participate equally.

The *third stage* of the project, "Impact of biotechnological applications on SDGs", consists of integrating the two previous stages into a scientific poster. This poster must compile all the relevant information of one of the three biotechnological processes and, in addition, answer the following question: Wwhich SDGs would benefit or harm from the implementation of this biotechnological application?". The initial groups of the project (the groups of the second stage) must synthesize all the information gathered from a biotechnological application other than the one in the article they were initially assigned; that is, they make a poster based on the information transmitted orally by their peers. In the elaboration of this poster, students evaluate how biotechnological applications affect the SDGs, positive and/or negative impacts, always arguing critically. Subsequently, the students who are experts on that article review the poster on that topic made by their classmates, providing comments and corrections. The latter have to apply the changes before the oral presentation. Finally, the activity culminates with the oral presentation of the posters.

This last stage, with a weight of 40% of the grade of the project, is divided into two sections: (1) the quality of the poster (20%) and (2) the quality of the oral presentation (20%). Both results are evaluated by means of a rubric. All rubrics are available to students at the beginning of the project.

2.2 *Integrating SDGs in the Process Integration and Industrial Chemical Plants Course of the Master's Degree in Industrial Engineering*

This work was carried out in a subject of the second year of the master's degree in Industrial Engineering. This is the subject Process Integration and Industrial Chemical Plants, in the speciality of Chemical and Process Engineering. It is a 4.5 ECTS course taught over 15 weeks. The students had already worked with flipped learning methodology in another subject from the same area (Design of Chemical Processes, (Requies et al., 2024)), therefore they are used to work with active methodologies. The methodologies described in this work were implemented in the first half of the course, up to week 8. During these 8 weeks 4 topics were described: 1-Introduction, 2-Sulphur industries, 3-Sodium and chlorine industries, and 4-Glass industries. The evaluation of this half of the course was divided into two parts: 60% of the mark was assigned to group or collaborative work and 40% to tests of minimum concepts carried out individually in class for each topic.

In this first half part of the course, the contents were generated by the group of students jointly, being supervised by the teacher to confirm that the minimum technical requirements, necessary to understand each of the industrial sectors studied, were covered. In some cases, the teacher prepared basic material to help students understand more complex aspects of these industries, which was made available to the students at the beginning of the topic. Students were encouraged to complement the information of each sector with sectorial documents available on the official European Union website, BREFs, best available techniques documents, environmental impact assessment, occupational health and safety, etc. In some cases, this content was also complemented with a talk in the classroom by an expert and visits to companies related to the sectors studied in class. The discussions of the SDGs in this subject focused on SDG 7 (Affordable and clean energy), SDG 9 (Industry, innovation and infrastructure), SDG 12 (Responsible production and consumption), and SDG 13 (Climate action).

At the beginning of the course, students were contextualized that part of the knowledge they were going to acquire was not only technical aspects of industrial processes. For this purpose, a group dynamic was carried out on the first day of class. In this session, each group of 2 or 3 students had to discuss a different topic related to the chemical industry: environment, safety, economy, occupational risks, and SEVESO accidents. After a joint discussion, the session ended with a closing by the teacher. In this way, it was explained to them that one of the objectives of this course was to understand the relationship between these aspects in any industry. All these points of view are included within the broad umbrella of the Sustainable Development Goals, which cover everything from energy to more social issues.

In the initial topic of the course, sulphur industries, the SDGs were incorporated through the discussion of a process already known to them, but from an environmental and impact perspective of raw material extraction methods, social environment, etc. In addition, we received a visit in class from an expert in the area of industrial safety, who has been working for 15 years in the chemical industry sector of the Basque Country. One of the nearby companies to the university is a sulphuric acid production plant, which was also visited. This visit allowed students to improve their understanding of the process, visualize the possible impact of raw materials and products, impact on workers, on the environment and occupational health, as well as on energy and access to decent work. This also allowed them to practice what could be a job interview in their future, demonstrating knowledge of the process, asking interesting and pertinent questions, demonstrating how to behave, etc. This topic concluded with a review in class among students of which points of the sulphuric acid production process have the greatest impact on energy and climate, proposing points of improvement, control, research, and innovation. In this way, the topic was concluded by internalizing the SDGs worked on in this subject and applied to a specific industrial process.

In the second topic, the role-playing methodology was used. To achieve a multi-focal study of the industrial processes studied, there were two companies and each student was given one role within one of the companies. These roles were: process engineer, quality engineer, risk engineer, and environmental engineer. In a first session, the concept of the role itself was worked on. They looked for what work each type of engineer does, what type of data they need to collect in their day-to-day work, which parameters are most critical for that type of engineer, what their main objectives are, their difficulties, what type of legislation they have to take into account, what is the impact on the company's environment of their daily work, etc. Once their role was constructed as best as possible, individually and grouping with other students of the same role, they then met in separate groups the students of each company. In each company, all roles were present, resembling real work in any chemical industry. Each company was assigned a different case, which could be a technical problem to solve or a research and development line to design. The groups, each company, worked on their case and had to take into account all the points of view of the different roles. The final dynamic of this topic was a presentation by each group in which they presented their context, initial problem, how they had responded to it including the point of view of each role, and which SDGs they had taken into account in their development. This presentation was done simulating an International Chemical Industry Conference, where the two companies had an oral presentation.

In the last topic (glass Industry), a different active dynamic was used, opting to simulate a trial between two glass companies, to involve students in learning the industrial process under study. In the first session, students were divided into two groups. Each group represented a type of company, one was a traditional glass company and the other was an innovative company producing biodegradable glass. During the second session, students focused on their company, taking into account all the roles worked on in the second topic, including the technical aspects described in the first topic, the SDGs worked on in the subject and any other that each group considered, the strengths and weaknesses of their process (economic, environmental, social, health, etc.). All this was added to the knowledge they had already acquired in class about glass production, which was supported by a visit to a flat glass production company. In the last session, the trial simulation was carried out, with the teacher being the judge, one group being the defence and the other group being the prosecution. Initially, each party explained their main arguments, and then they could rebut the other group, with some students simulating being an expert or a witness, mentioning references to give weight to the data they mentioned, etc. This activity had a limited available time, which forced students to be focused, to synthesize, to distribute tasks, to ensure they addressed key aspects, and to anticipate what the opposing arguments would be, fostering their critical and synthetic analysis. To conclude the activity and the subject, the teacher prepared a generic verdict, emphasizing the importance of taking into account all possible aspects (of the process itself, environmental, social, and health) when dealing with a technical problem in their future as engineers.

3 Results and Discussions

Next, the implications of using active methodologies in the development of transversal competencies and the Sustainable Development Goals (SDGs) are described from the perspective of the two subjects.

3.1 Effect of the Implementation of Active Methodologies on the Development of Transversal Competencies in the Undergraduate Course

Although the teacher involved in this work has been teaching the Biotechnology course since the 2019/20 academic year, it was not until 2022/23 that this collaborative active methodology was implemented. As a tool for evaluating teaching quality at the university, students complete an opinion survey about their teachers' teaching at the end of the course period. This questionnaire includes questions related to planning, methodology, and teaching development, among others. It is worth noting that from 2019/20 to 2023/24, there were two academic years where no statistics were available due to not reaching the minimum required number of enrolled students, one before the implementation of the methodology (2020/21) and one after (2022/

23). Therefore, in the comparison of the results, before versus after the use of active methodologies, we have data from three academic years. To measure the impact of this activity on the teaching–learning process of students, responses to three questions were compared: (1) whether the teaching development promotes a reflective attitude; (2) whether autonomous learning is promoted, and (3) whether teamwork is promoted. These questions are directly related to three of the eight transversal competencies of UPV/EHU: Critical Thinking (CT), Autonomy and Self-regulation (ASR), and Teamwork (TW), respectively.

Figure 1 shows the students' evaluation of these three questions, before (average value of the years 2019/20 and 2021/22) and after (2023/24) the implementation of this activity. The evaluation of the questions related to the competencies CT and TW has significantly improved after the implementation of this methodology, from 4.2 to 4.6 in CT and from 4.4 to 4.7 in TW. However, the improvement in the ASR competency has not been as significant, from 4.3 to 4.4. Although there may be various reasons justifying this low increase, the most plausible one is that this competency, the development of autonomy, was already being worked on in the preparation of laboratory practice reports before the implementation of this activity. On the other hand, another activity that has been worked on in the classroom since the first year, the resolution of numerical problems, also requires autonomous work and critical thinking from students. However, within the framework of this last activity, these competencies were not as clearly defined as in the active methodology, where students are explicitly asked to make a critical reflection on the SDGs and their application in biotechnology. Finally, the teamwork that students did before this methodology was the completion of laboratory practices in pairs. Currently, as explained in the methodology section, teamwork is developed throughout the project.

The implementation of the active methodology contributed to the development of the following competencies, to a greater extent critical thinking and teamwork, and to a lesser extent autonomy and self-regulation. These results are consistent with those found by other authors (Latorre-Cosculluela et al., 2025), who observed that the

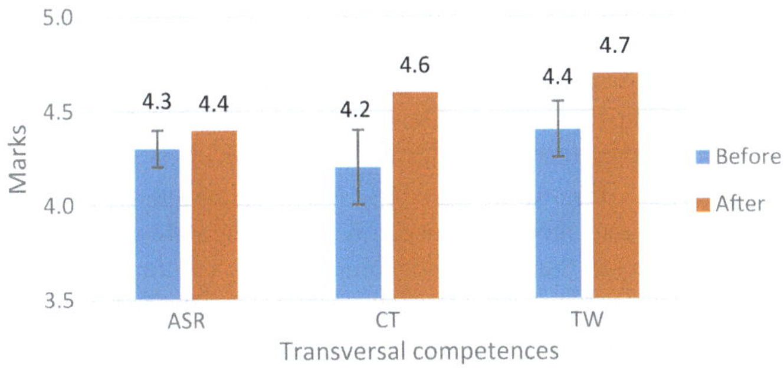

Fig. 1 Evaluation of the transversal competencies in the university official questionnaire done at the end of the course: Autonomy and Self-Regulation (ASR), Critical Thinking (CT), and Teamwork (TW) before (mean (from 0 to 5.0) and standard deviation of the years 2019/20 (N = 5) and 2021/22 (N = 5)), and after (2023/24, N = 10) the application of the active methodology

combination of collaborative learning strategies and active methodologies improved transversal competencies and academic performance.

Furthermore, the use of this methodology has fostered the competence of systems thinking among students, as they have established interconnections between environmental problems and the social and economic effects of each biotechnological application. By way of example, the students who have worked on the biotechnological application for bioethanol production from agroindustrial waste have understood that the production of second-generation bioethanol is a more sustainable alternative to fossil fuels. However, they have also understood that its large-scale implementation also entails environmental, social and economic problems that may limit its viability, such as the production of toxic waste during the production process, the use of large volumes of water, social conflicts over the use of land and resources, high production costs due to high investment in infrastructure and technology, etc. This holistic approach has been applied in the three biotechnological applications worked on the course through this project.

3.2 *Effect of the Implementation of Active Methodologies on the Integration of the SDGs in the Master's Course*

This subject was taught for the first time in the academic year 2024/25. After the development of transversal competencies implemented in recent years in undergraduate courses, the design of this new master's course was focused on using these competencies for a deeper understanding of the chemical processes and the SDGs by students. The results and learnings obtained over the years in the teaching execution of the teacher involved in this design have been implemented as a pilot test in this master's course, whose first half was entirely taught using active methodologies. At the end of the course, a survey was conducted by the teacher among the students participating in the course, with questions related to the methodologies used, the SDGs, and their perception. The most significant results of this questionnaire, especially regarding the SDGs, are described below.

One of the questions in the questionnaire was "How would you rate your level of understanding of the Sustainable Development Goals (SDGs) before starting the course?" and another question referred to their level of understanding after completing the course. Figure 2 shows the results classified according to their responses, with the options they had: very low, low, high, and very high. Half of the students initially answered that their understanding was low or high, in the middle range of response options. However, after taking the course, half of the students rated their knowledge as very high, showing a clear shift towards greater understanding.

Continuing with the idea of knowing the degree of deep understanding of the SDGs achieved by students, to the question "How do you rate your ability to interrelate technical concepts with social, environmental, and economic aspects after the first half of the course?" 100% of the students answered that it was high. This result

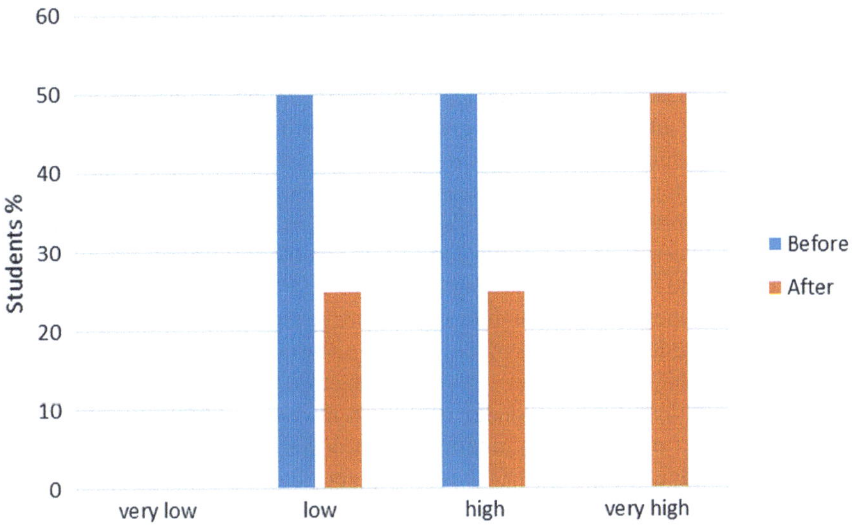

Fig. 2 Students' response to the question "How would you rate your level of understanding of the SDGs before (in blue) and after (in orange) the course"

is very satisfactory, as it demonstrates the clear advantage of using varied active methodologies to work and put oneself in the shoes of a future engineer who takes these transversal aspects into account in the execution of their job, and as part of society with the possibility of altering the impacts of their activity by modifying their way of execution, design, search for solutions, use of technology, etc. When asked "To what extent do you think traditional methodologies have contributed to your understanding of the SDGs?" 50% of the students answered that it was low, showing their perception that learning something as transversal as the SDGs is difficult to acquire if traditional methodologies are used, where the teacher exposes knowledge and students receive information without taking an active part in learning.

In addition, the use of case-learning, cooperative learning, and the role-playing methodology in this project has fostered anticipatory thinking among students, as it forces them to foresee and analyze possible future scenarios within industrial process management. As an example, in the case of students in the role of "risk engineer", they must identify possible safety risks in industrial processes and propose mitigation measures. This develops their ability to visualize future threats and design preventive strategies. The "environmental engineer," on the other hand, must anticipate the ecological impact of operations and evaluate compliance with environmental regulations, strengthening long-term projection. In short, through this methodology students have worked on the integration of multiple perspectives, allowing the student to understand how each decision affects other roles within the company, thus fostering a holistic and prospective vision. This ability is fundamental in sustainability and industrial management, where long-term planning and informed decision making are key.

3.3 Challenges and Lessons Learned in Implementation

The use of active learning methodologies in engineering programs not only improves student learning outcomes, but also contributes significantly to the development of specific competencies in sustainability. These competencies include the systems thinking, anticipatory thinking and integrated problem solving. Systems thinking is fostered through projects that require students to understand and analyze the interconnections between different components of a system (Li et al., 2023), as seen in the Biotechnology and Process Engineering and Chemical Plants projects. Anticipatory thinking is developed through activities that challenge students to envision and plan for future sustainable scenarios, as in role-playing and trial simulations (Robertson & Strachan, 2022). Integrated problem solving is enhanced through team collaboration and practical application of knowledge in real-world contexts, allowing students to approach complex problems from multiple perspectives (Sánchez-Carracedo et al., 2021). To assess the development of these competencies, an assessment framework was used that included detailed rubrics and self-assessment questionnaires, providing a clear and structured view of students' progress in each competency (Cruz et al., 2020).

Finally, the implementation of active learning strategies in the subjects of Biotechnology and Process Engineering and Chemical Plants, although highly beneficial, faced several logistical, institutional and pedagogical challenges that hindered their effective application. At the logistical level, one of the main obstacles lied in the rigidity of academic programs and the limited assignment of hours for practical activities within the current curricula, which restricts the time available for the incorporation of active methodologies, the implementation of which requires a greater time compared to traditional classes. At the institutional level, student resistance to these methodologies was identified, due to the perception of a higher workload than that required for conventional approaches. From a pedagogical perspective, the main challenges include the difficulty in assessing student participation and the challenge of maintaining active engagement throughout the development of the project. Also, absenteeism in class represents an additional problem, as it makes it difficult to meet the objectives of group activities, especially in small groups.

To mitigate these challenges, the teaching team adopted several strategies, among which is the partial reconfiguration of the master classes, integrating practical sessions oriented to the development of the projects. In addition, the weight of these activities in the final grade was increased in order to encourage the active participation of students. Likewise, clear evaluation rubrics were implemented, aligned with the learning objectives, in order to guarantee an objective evaluation of the activities carried out. Finally, the incorporation of peer evaluation allowed the application of self-evaluation and co-evaluation mechanisms, thus favoring a more reflective and collaborative learning process in the students.

4 Conclusions

The objective of this study was to understand the perception of university students regarding sustainability after the implementation of collaborative active methodologies in undergraduate and master's degrees. One of the main achievements of this project was the improvement in students' sustainability training, achieved in parallel with the development of transversal competencies.

In the case of undergraduate students, it was observed that integrating the SDGs into biotechnology not only provided them with a deeper understanding of the application of environmental and sustainable technologies in a real context but also instilled ethical and social principles linked to sustainability values in both personal and professional spheres. Developing a collaborative active methodology around the application of different biotechnological technologies facilitated students' understanding and learning, increased their motivation for the topic in which this activity was framed, and generally provided them with a positive experience. Additionally, the implementation of the active methodology contributed to the development of the following competencies, to a greater extent critical thinking and teamwork, and to a lesser extent autonomy and self-regulation. Finally, students have developed their ability to understand the interconnections between different environmental, social and economic aspects, which gives them a more holistic approach.

Regarding the master's group, students worked more deeply with various active methodologies: flipped learning, case method, role-playing, trial simulation, company visits, and expert visits to the classroom. The first half of the course (2.75 ECTS) was taught using these methodologies, given that they were more mature students with more practice in working with active methodologies and SDGs. In all the methodologies used for the various topics covered in the first half of the course, the SDGs were worked on in parallel, focusing the discussion on SDG 7, SDG 9, SDG 12, and SDG 13, due to their direct relationship with the course content. The result obtained was very satisfactory in terms of understanding of the SDGs, technical knowledge acquired, student involvement, student behaviour during visits, questions asked by students to experts in various fields, etc. From the teacher's point of view, it was an exhausting yet satisfying experience, observing the students' evolution in just a few weeks. In the surveys conducted to the students, it was observed that they also perceived the advantage of using active methodologies to work on the SDGs in a transversal manner, having generated a new way of thinking and executing the technical knowledge they have acquired during their time at university.

The groups with which the initiatives described in this work were launched, both in the undergraduate and master's groups, were small groups, fewer than 10 students. This fact was a key factor for the success achieved, as student involvement was high and the monitoring of the various activities by the teacher was also high and fruitful. This teacher involvement, with quick feedback, in turn, made the students more involved.

The implementation of the Sustainable Development Goals (SDGs) in the engineering curriculum at the University of the Basque Country (UPV/EHU) has proven to be an effective strategy for fostering sustainability competencies. However, for these practices to be applicable in different institutional contexts and scalable to larger classes or other disciplinary fields, certain factors must be considered. Firstly, adapting these methodologies to larger classes can be achieved through the use of educational technologies that facilitate collaboration and student progress tracking, such as online learning platforms and automated assessment tools. Additionally, interdisciplinarity is key to integrating the SDGs across various fields of study. For example, in social science programs, the SDGs can be addressed through case studies and research projects that analyse the social and economic impact of sustainability policies (Dash et al., 2024). In disciplines such as engineering and natural sciences, incorporating problem-based projects and challenge-based learning can facilitate the practical application of the SDGs in real-world contexts. These strategies not only expand the reach of education for sustainability but also promote comprehensive training committed to global challenges.

Finally, integrating the SDGs into engineering education through active learning requires a coordinated effort between university administrators, teachers and curriculum designers. The key to success lies in practice-based teaching, with real projects, innovative technology and assessment aligned with global sustainability goals. Among the main recommendations resulting from this work would be:

(1) Incorporating the SDGs into the institutional vision, establishing formal commitments to integrate sustainability into academic programs.
(2) Designing subjects shared among different disciplines and promoting the integration of interdepartmental projects will allow addressing sustainability issues with a holistic vision.
(3) Providing continuous training to teachers in active learning methodologies and education for sustainability, ensuring that they are equipped with the necessary tools and knowledge to implement these practices effectively.
(4) Incorporating online learning platforms and automated assessment tools to facilitate collaboration and tracking of student progress in large classes.
(5) Implementing a clear and structured evaluation framework that uses detailed rubrics and self- and co-assessment questionnaires to measure the development of sustainability competencies, such as systems thinking and anticipatory thinking, among others.
(6) Designing projects and activities that allow students to apply sustainability concepts in real contexts (service-learning programs), developing practical and critical skills needed to face global challenges.

Acknowledgements The authors wish to thank the University of the Basque Country, UPV-EHU (HBP/PIE i3lab 24-25) for supporting this research.

References

Aimaretti, N. (2011). *Desarrollo de un proceso sostenible para la producción de bioetanol a partir de desechos agroindustriales* (p. 243). Universidad Nacional de Educación a Distancia (UNED), Madrid.

Beltrán-Rocha, J. C., Guajardo-Barbosa, C., Barceló-Quintal, I. D., & López-Chuken, U. J. (2017). Biotratamiento de efluentes secundarios municipales utilizando microalgas: Efecto del pH, nutrientes (C, N y P) y enriquecimiento con CO_2. *Revista De Biología Marina y Oceanografía, 52*(3), 417–427.

Blanchard, O., & Buchs, A. (2015). *Clarifying sustainable development concepts through role playing* (pp. 1–16). SAGE Publications (UK and US).

Cebrián, G. (2020). La educación para el desarrollo sostenible en el currículum universitario: una investigación-acción cooperativa con profesorado. *Revista Iberoamericana de Educación Superior, 11*(30). https://doi.org/10.22201/iisue.20072872e.2020.30.590

Cruz, M. L., Saunders-Smits, G. N., & Groen, P. (2020). Evaluation of competency methods in engineering education: A systematic review. *European Journal of Engineering Education, 45*(5), 729–757. https://doi.org/10.1080/03043797.2019.1671810

Dash, S., Rao, G., & Kumar, P. (2024). *Embedding the SDGs into curricula via an interdisciplinary approach.* Times Higher Education. https://www.timeshighereducation.com/campus/embedding-sdgs-curricula-interdisciplinary-approach

Freeman, S., Eddy, S. L., McDonough, M., Smith, M. K., Okoroafor, N., Jordt, H., & Wenderoth, M. P. (2014). Active learning increases student performance in science, engineering, and mathematics. *Proceedings of the National Academy of Sciences, 111*(23), 8410–8415. https://doi.org/10.1073/pnas.1319030111

González García, Y., Meza Contreras, J. C., González Reynoso, O., & Córdova López, J. A. (2013). Síntesis y biodegradación de polihidroxialcanoatos: Plásticos de origen microbiano. *Revista Internacional De Contaminación Ambiental, 29*(1), 77–115.

Latorre-Cosculluela, C., Sierra-Sánchez, V., & Vázquez-Toledo, S. (2025). Gamification, collaborative learning and transversal competences: Analysis of academic performance and students' perceptions. *Smart Learning Environments, 12*, 2. https://doi.org/10.1186/s40561-024-00361-2

Leal Filho, W., Ruiz Vargas, V., Lange Salvia, A., Londero Brandli, L., Pallant, E., Klavins, M., Ray, S., Moggi, S., Maruna, M., Conticelli, E., Amogre Ayanore, M., Radovic, V., Gupta, B., Sen, S., Paço, A., Michalopoulou, E., Hanisdah Saikim, F., Lye Koh, H., Frankenberger, F., Kanchanamukda, W., Da Cunha D. A., Mohammed Akib, N. A., Clarke, A., Wall, T., & Vaccari, M. (2019). The role of higher education institutions in sustainability initiatives at the local level. *Journal of Cleaner Production, 233*, 1004–1015. https://doi.org/10.1016/j.jclepro.2019.06.059

Li, H., Huang, Z., Yang, T., Zhang, W., Chen, M., Li, Z., & Ren, K. (2023). Individual sustainability competence development in engineering education: Community interaction open-source learning. *PLoS One, 18*(11), Article e0294421. https://doi.org/10.1371/journal.pone.0294421

Lozano, R., Ceulemans, K., Alonso-Almeida, M., Huisingh, D., Lozano, F. J., Waas, T., Lambrechts, W., Lukman, R., & Hugé, J. (2015). A review of commitment and implementation of sustainable development in higher education: results from a worldwide survey. *Journal of Cleaner Production, 108*, 1–18. https://doi.org/10.1016/j.jclepro.2014.09.048

Martínez Valdivia, E., Pegalajar Palomino, M. C., & Burgos García, A. (2023). Metodologías Activas para el Desarrollo Sostenible en la Formación Docente. Revista *Internacional de Educación para la Justicia Social, 12*(2), 191–211. https://doi.org/10.15366/riejs2023.12.2.011

Pradas, F., & Copelia, G. (2024). *Integración de los ODS en Educación Superior para aumentar la motivación y el pensamiento crítico del alumnado universitario* (Integration of the SDGs in Higher Education to increase University Students ' Motivation and Critical Thinking) (Vol. 36, pp. 1–20).

Prince, M. (2004). Does active learning work? A review of the research. *Journal of Engineering Education, 93*(3), 223–231. https://doi.org/10.1002/j.2168-9830.2004.tb00809.x

Requies, J., Laura Barrio, V., Acha, E., Agirre, I., Viar, N., & Gandarias, I. (2024). Integration of sustainable development goals in the field of process engineering through active learning methodologies. *Education for Chemical Engineers, 49*, 26–34. https://doi.org/10.1016/j.ece.2024.08.001

Robertson, E., & Strachan, S. (2022). *Embedding education for sustainable development in the engineering curriculum through challenge-based education.* University of Strathclyde. https://pure.strath.ac.uk/ws/portalfiles/portal/138957071/RobertsonScottISEE2022Embeddingeducationforsustainabledevelopmentintheengineeringcurriculum.pdf

Sáenz, O. (2020, June, 1–16). *Implementación de los ODS en las Instituciones de Educación Superior*. Recomendaciones a partir de la experiencia de una universidad latinoamericana. https://www.researchgate.net/publication/342500766_Implementacion_de_los_ODS_en_las_Instituciones_de_Educacion_Superior_Recomendaciones_a_partir_de_la_experiencia_de_una_universidad_latinoamericana_1

Sánchez-Carracedo, F., Segalas, J., Bueno, G., Busquets, P., Climent, J., Galofré, V. G., Lazzarini, B., Lopez, D., Martín, C., Miñano, R., Sáez de Cámara, E., Sureda, B., Tejedor, G., & Vidal, E. (2021). Tools for embedding and assessing sustainable development goals in engineering education. *Sustainability, 13*(21), 12154. https://doi.org/10.3390/su132112154

Torres, D. I. R. (2021). Contribution of higher education to the sustainable development goals from teaching. *Revista Espanola De Educacion Comparada, 37*(37), 89–110. https://doi.org/10.5944/REEC.37.2021.27763

UNESCO. (2021). *Educación para el Desarrollo Sostenible: hacia la consecución de los Objetivos de Desarrollo Sostenible.* https://unesdoc.unesco.org/ark:/48223/pf0000379748_spa

UPV/EHU. (2019a). *EHUagenda 2030 for sustainable development*. https://www.ehu.eus/documents/4736101/11938005/EHUAgenda-2030-ENG.pdf/487b2c83-51e1-d0e2-dcd1-af419b2b5c26?t=1559656838000

UPV/EHU. (2019b). Universidad del País Vasco/Euskal Herriko Unibertsitatea. Catálogo de competencias transversales de la UPV/EHU. Servicio Editorial de la Universidad del País Vasco (Eds.) (42 pp.). https://www.ehu.eus/documents/1432750/12757375/Cat%C3%A1logo+de+Competencias+trasnversales_cas.pdf

Serious Games as a Starting Point for Sustainability Education: A Behavioral Approach to Teaching the SDGs

André Lucas Novaes, Giselle Cristina Cardoso, Debora Zumkeller Sabonaro, and António Manuel Dias Domingos

Abstract The integration of the Sustainable Development Goals (SDGs) into higher education requires innovative pedagogical strategies that promote student engagement. This study explores the role of serious games as an active learning methodology and as an introductory tool for SDG concepts, emphasizing their connection with the Theory of Planned Behavior (TPB) to strengthen sustainability competencies. The literature shows that serious games foster interdisciplinary learning and strongly engage students with the topics presented in the games. However, despite their recognized potential, gaps remain regarding pedagogical structuring based on behavioral theory, primarily focusing on attitude change rather than merely conveying theoretical concepts. Many games lack a robust theoretical foundation that explores the fundamental principles of sustainable development while integrating a behavioral approach. The research is guided by the following question: How can serious games be used as a strategy to introduce the theme of sustainable development in higher education while simultaneously promoting competencies for sustainable decision-making? By adopting an exploratory and theoretical approach, this study aims to present a structured tool for introducing the teaching of the SDGs in higher education. This work contributes to the debate by synthesizing empirical evidence on

A. L. Novaes (✉) · G. C. Cardoso
Federal Institute of Education, Science, and Technology of the South of Minas Gerais – IFSULDEMINAS, Poços de Caldas, MG, Brazil
e-mail: andre.novaes@ifsuldeminas.edu.br

G. C. Cardoso
e-mail: giselle.cardoso@ifsuldeminas.edu.br

D. Z. Sabonaro
Federal University of Alfenas (UNIFAL), Alfenas, MG, Brazil
e-mail: debora.sabonaro@unifal-mg.edu.br

A. M. D. Domingos
NOVA School of Science and Technology (NOVA FCT), Caparica, Portugal
e-mail: amdd@fct.unl.pt

W. Leal Filho et al. (eds.), *Introducing the UN SDGs in Higher Education Institutions*, World Sustainability Series, https://doi.org/10.1007/978-3-032-23017-1_13

the effectiveness of serious games in sustainability education and proposing fundamental elements of a structured model for serious game design. By integrating behavioral science and game-based learning, this study offers a pathway for embedding SDGs in higher education, fostering engaging learning experiences for students while reducing objections to the interdisciplinary approach to sustainable development.

Keywords Serious games · Education for Sustainable Development (ESD) · Theory of Planned Behaviour (TPB) · Active learning · SDGs

1 Introduction

The integration of Sustainable Development Goals (SDGs) into higher education requires innovative pedagogical approaches to enhance student engagement and promote behavioral change (Albareda-Tiana et al., 2024; Amorós Molina et al., 2023; Filho et al., 2024). Serious games have emerged as effective tools for experiential learning, enabling students to simulate complex decision-making processes related to sustainability (Feltrero et al., 2023; Hsu & Wu, 2023). These games facilitate interdisciplinary learning by aligning gameplay mechanics with SDG concepts, fostering critical thinking, and reinforcing sustainable practices (Andreoni & Richard, 2024; Kilanioti, 2022; Novaes et al., 2023).

By integrating behavioral science principles, particularly the Theory of Planned Behavior (TPB), serious games enhance students' perceived control over sustainable actions, reinforcing their commitment to sustainability beyond the classroom (Ahmadov et al., 2024; Feltrero et al., 2023; Prouty et al., 2018). This gamified approach represents a dynamic entry point for promoting SDG-related awareness and action in educational contexts.

Although recent studies indicate that serious games are effective tools for teaching SDGs (Albareda-Tiana et al., 2024; Damaševičius et al., 2023; Medema et al., 2016; Novaes et al., 2025), structured models guiding their design and pedagogical application are still lacking. Many games are developed without a clear theoretical foundation, making it difficult to assess their educational impact and replicate them across different academic contexts. To address this gap, this study proposes a theoretical model based on the TPB, structuring the game to reinforce sustainable attitudes, social norms, and perceived control in environmental decision-making (Huang et al., 2024). Beyond providing a structured method, serious games stand out as a more engaging and interactive way to introduce SDGs, fostering active learning that encourages experimentation, strategic decision-making, and critical reflection. Unlike descriptive approaches, this innovative framework not only facilitates the pedagogical application of SDGs but also enables an analysis of how games influence student behavior beyond the classroom, encouraging the adoption of sustainable practices in everyday life.

The research is guided by the following question: How can serious games be used as a strategy to introduce the theme of sustainable development in higher education while simultaneously promoting competencies for sustainable decision-making?

By adopting an exploratory and theoretical approach, this study aims to present a structured tool for introducing the teaching of the SDGs in higher education. Grounded in TPB, the proposed framework integrates serious games as an interactive and engaging method to foster sustainability-oriented learning. This model enhances students' understanding of sustainability challenges, strengthens behavioral drivers, and fosters strategic decision-making. By establishing a structured theoretical foundation, this study supports the integration of SDGs into higher education curricula through gamified approaches and lays the groundwork for future empirical research.

2 Serious Game and Active Methodologies

The use of serious games as an active learning methodology for sustainability education has emerged as a promising tool to engage students and foster behavioural changes towards environmental issues. Combining learning and entertainment, serious games offer immersive and interactive experiences that enhance the understanding of complex problems and encourage more sustainable decision-making (Algurén, 2021; Novaes et al., 2025). By integrating cognitive and emotional engagement, these games foster positive attitudes towards sustainability, helping to develop citizens capable of addressing contemporary challenges (Figueiró & Raufflet, 2015; Gatti et al., 2019).

Studies highlight their effectiveness in promoting behavioural change and achieving the SDGs. For example, Ghodsvali et al. (2022) demonstrated that serious games focusing on sustainable consumption improved decision-making in complex scenarios. Similarly, Medema et al. (2016) reported behavioural changes, such as reduced water waste, after participants engaged with an educational game. In healthcare, Damaševičius et al. (2023) found that serious games increased professionals' knowledge of sustainable practices, leading to more environmentally responsible actions.

Serious games are particularly suitable for engaging students, offering opportunities for active participation, collaboration, and critical thinking (Lhamri et al., 2024; Tan et al., 2023). By simulating real-world challenges, these games encourage learners to make decisions, experience their consequences, and develop the skills necessary to address complex socio-environmental issues (Morganti et al., 2017). Thus, serious games provide a dynamic pathway to cultivate sustainability-oriented behaviours while fostering interdisciplinary problem-solving and decision-making skills.

3 Theory of Planned Behaviour

The TPB is a psychosocial framework that explains and predicts human behaviour by focusing on three key components: attitudes towards the behaviour, subjective norms, and perceived behavioural control. These components determine behavioural intention, which is the direct predictor of actual behaviour (Ajzen, 2002).

Attitude towards the Behaviour (ATB): Refers to a person's evaluation of a specific behaviour, shaped by beliefs about its outcomes and associated positive or negative assessments.

Subjective Norms (SN): Relate to the perceived influence of significant others and societal expectations **regarding a behaviour**, as well as the motivation to conform to these expectations.

Perceived Behavioural Control (PBC): Reflects the perceived ease or difficulty of performing a behaviour, shaped by resources, barriers, and facilitating conditions.

The TPB posits that the stronger these three components, the greater the behavioural intention, ultimately leading to action. This framework has been widely applied to understanding pro-environmental behaviours, emphasising the interplay between individual beliefs, social pressures, and perceived control (Carvalho & Neto, 2022). Recent studies demonstrate that serious games can influence pro-environmental intentions and behaviours by incorporating explicit behavioural elements into their mechanics (Figueiró & Raufflet, 2015; Gatti et al., 2019; Novaes et al., 2025). This study advances that discussion by proposing a framework that integrates the TPB elements directly into the serious game mechanics.

When applied to serious games, the TPB provides a robust framework for designing experiences that influence attitudes, reinforce social norms, and empower players to perceive greater control over sustainable actions. By integrating these elements, serious games can effectively motivate learners to adopt sustainability-oriented practices, fostering a deeper understanding of the connections between their decisions and real-world environmental impacts (Ghodsvali et al., 2022).

4 Serious Games and the Integration of SDGs

The integration of the SDGs into courses not directly related to the environment faces significant institutional and cultural obstacles. Institutional barriers include a lack of funding, low support from university administrations, and little institutional priority given to sustainability (Hansen et al., 2021).

Additionally, rigid and bureaucratic curricula often hinder the inclusion of cross-cutting topics such as the SDGs. Faculty members often resist integrating SDGs into curricula due to a lack of specific training and the perception that these topics are incompatible with their traditional disciplines (Browne, 2023). This resistance is amplified by a generational gap, with younger students valuing the SDGs more than older professors (Amorós Molina et al., 2023). To move beyond theoretical approaches and promote effective behavioral changes, it is essential to offer practical, reflective, and contextualized experiences (Saitua-Iribar et al., 2020). Students need to recognize the personal and professional relevance of the SDGs to internalize sustainable concepts (Albareda-Tiana et al., 2024).

In this regard, serious games emerge as effective methodologies aligned with this approach, offering immersive simulations that allow students to explore new perspectives on complex issues such as climate change and sustainable consumption (Gatti et al., 2019; Kilanioti, 2022). In the context of SDGs, they can provide a dynamic and immersive first encounter with sustainability issues, making initial learning more engaging. Studies show a growing interest in using simulation and serious games to address sustainability topics, recognizing the need for innovative approaches to tackle complex global challenges (Feltrero et al., 2023; Hsu & Wu, 2023).

These types of games leverage the interactive and engaging nature of gaming environments to involve users in real-world problems, including sustainability issues (Kilanioti, 2022). For example, a serious game on climate change could place students in the role of decision-makers, challenging them to balance economic development with emission reduction—an experience that activates both cognitive and emotional learning (Huang et al., 2024).

The literature suggests using serious games as a gateway to sustainable development: they introduce SDG concepts playfully, spark student interest and initial awareness, and lay the foundation for more robust methodologies such as applied projects or service-learning (Novaes et al., 2025; Saitua-Iribar et al., 2020). Serious games can sensitize and preliminarily empower students—by teaching basic concepts of sustainable systems or fostering empathy for social and environmental issues—facilitating their later participation in real-world projects (Gatti et al., 2019). After the gaming phase, students tend to be more engaged and contextualized, allowing them to transition into coursework or community action with a better prior understanding (Andreoni & Richard, 2024).

In fact, combined active methods reinforce each other: educators argue that to develop competencies related to SDGs, universities should use methodologies that combine learning with practice (Albareda-Tiana et al., 2024; Saitua-Iribar et al., 2020). Ultimately, the combination of serious games as an initial motivational tool, along with other experience-based approaches, emerges as an effective strategy (Feltrero et al., 2023). This blend progressively and meaningfully engages students—from game to reality—enhancing the chances of attitudinal change and long-term learning about SDGs (Hsu & Wu, 2023).

By integrating behavioral principles and interactive challenges, these games not only facilitate the understanding of complex concepts but also encourage students to reflect on their attitudes toward the challenges of sustainable development. The interdisciplinary approach makes it easier to introduce complex concepts in an accessible manner, preparing students to face the global challenges proposed by the 2030 Agenda (Amorós Molina et al., 2023). Thus, serious games not only increase student engagement but also serve as a powerful gateway to practical SDG comprehension, highlighting the importance of sustainable attitudes in their personal and professional lives.

5 Methodology

This study is a theoretical and exploratory research, focusing on developing a conceptual framework for using serious games as a teaching methodology for Education for Sustainable Development (ESD). The primary goal is to propose a prototype of a serious game, designed based on students' specific educational needs—identified through data collection and literature review—and grounded in the TPB.

This research does not include the practical development or testing of the game, as this phase is beyond the scope of the study. Instead, it aims to establish a structured foundation for future empirical applications, generating hypotheses and insights that can guide further studies. The merit of this study lies in its methodological contribution, proposing a model that integrates serious games and TPB to foster pro-environmental attitudes, intentions, and behaviours.

Although no prototype will be implemented at this stage, the research provides a detailed framework outlining how the serious game could be structured to align with students' learning needs. This includes defining the educational context, learning objectives, and key competencies that the game should develop.

The research methodology follows a structured process divided into three main phases: Requirements Analysis, Conceptual Development, and Serious Game Framework. Figure 1 presents the research framework below.

The first phase of the study focuses on identifying students' learning needs in sustainability education and gathering faculty perspectives on active learning methodologies. This ensures that the proposed serious game aligns with both pedagogical objectives and practical teaching approaches. To achieve this, a survey was conducted with 45 students in their final years of undergraduate studies in Administration and Computer Engineering at the Federal Institute of Education, Science, and Technology of the South of Minas Gerais (IFSULDEMINAS)—Poços de Caldas campus. This research was conducted with ethical approval under protocol number

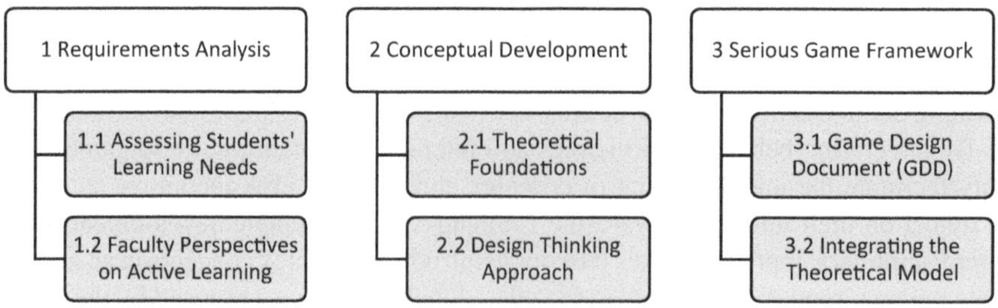

Fig. 1 Research framework

70867723.0.0000.5142, registered on Plataforma Brasil. The findings from this phase provide critical insights for designing a meaningful and engaging serious game experience.

The phase 2 focuses on establishing the theoretical foundations that will guide the game design. The TPB serves as the core theoretical model, explaining how attitudes, subjective norms, and perceived behavioural control influence students' engagement with sustainability-related behaviours. To structure the design process, the Design Thinking approach is employed. This methodology fosters creative problem-solving and ensures that the serious game aligns with students' needs and learning challenges. This phase serves as the bridge between theoretical insights and the practical structuring of the serious game. The process unfolds as follows in results section.

The final phase consists of designing the Game Design Document (GDD), which outlines the game's core mechanics, educational goals, and interactive dynamics. The TPB framework is integrated to ensure that game mechanics are aligned with the three key behavioural components. This phase does not involve developing a functional prototype but instead provides a structured framework for future implementation and empirical testing. The proposed model serves as a roadmap for designing and refining a serious game that fosters sustainability education while grounding behavioural changes in psychosocial theory.

6 Results

The results present students' perceptions regarding the presence of the sustainability theme throughout the course, as well as their views on the influence of games on behaviour. Finally, the findings from interviews conducted during the iterative design thinking process are presented.

Results from the Student Questionnaires

The focus is on understanding students' perceptions and knowledge regarding sustainability, as well as identifying areas for potential improvement in integrating sustainability into the curriculum.

As depicted in Fig. 2, the frequency with which the environmental education theme is addressed varies between the two courses. In, the largest proportion of respondents (30.6%) state that the theme is rarely covered, whereas 2.8% report that it is very frequent. In the Administration Bachelor's degree, 25.0% of students indicate that the topic occasionally appears, while 11.1% believe it has never been covered. Across both courses, only a small proportion report that the theme is frequently discussed.

Bachelor's degree programs in Administration and Computer Engineering, according to student responses. Notably, the frequency of the Sustainable Development theme significantly differs between these Bachelor's degree programs. This is partly due to the fact that Administration students have already taken a discipline

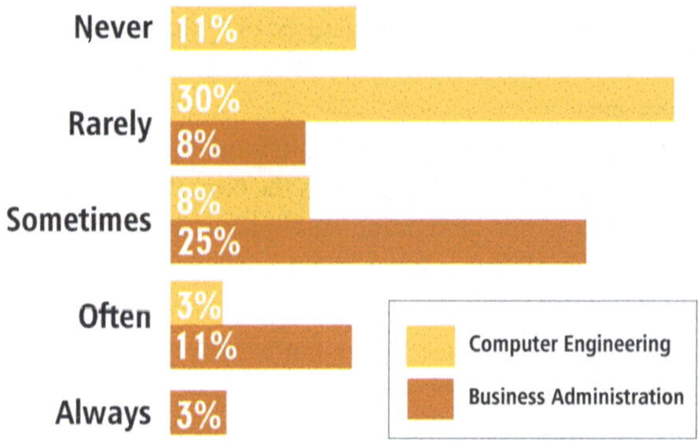

Fig. 2 Frequency of the sustainable development theme in the course

related to Sustainable Development. For these Bachelor's degree students, the presence of the theme is more frequent, with 43.75% indicating that the theme is occasionally addressed and 31.25% frequently. In contrast, the majority of Computer Engineering Bachelor's degree students indicated that the theme is rarely (45%) or never (10%) addressed. However, despite these differences, it's important to note that there is no standardized method regarding this matter. Regardless of these differences, the importance of addressing the Sustainable Development theme should be frequent in any area of knowledge, given the need to tackle global challenges and promote sustainable practices across all spheres of society.

Figure 3 shows the perception of Computer Engineering and Business Administration students regarding the influence of games on behaviour. Business Administration students show greater overall agreement (28% "Agree" and 17% "Strongly Agree"), while Engineering students have lower acceptance but stronger conviction among those who strongly agree (25%). The proportion of neutral responses is higher among Business Administration students (11%) than among Engineering students (6%), suggesting greater uncertainty in this group. This difference may be related to the courses' approaches: Business Administration students, more exposed to organizational behaviour topics, may be more inclined to recognize the influence of games. In contrast, Engineering students, with a more technical perspective, may be more critical when evaluating this relationship. Overall, the data suggest that games are perceived as tools with the potential to influence behaviour, especially when integrated into interactive methodologies. Acceptance varies depending on the academic context and students' familiarity with game-based approaches.

Design Thinking and Serious Game Development

The framework was structured into four phases, following the design thinking methodology.

Empathy—Based on the literature and teachers' perspectives, a lack of student engagement with sustainability concepts was identified, attributed to minimally interactive teaching methods. Students prefer practical, immersive activities that offer autonomy and collaborative learning. STEM-area teachers highlighted educational

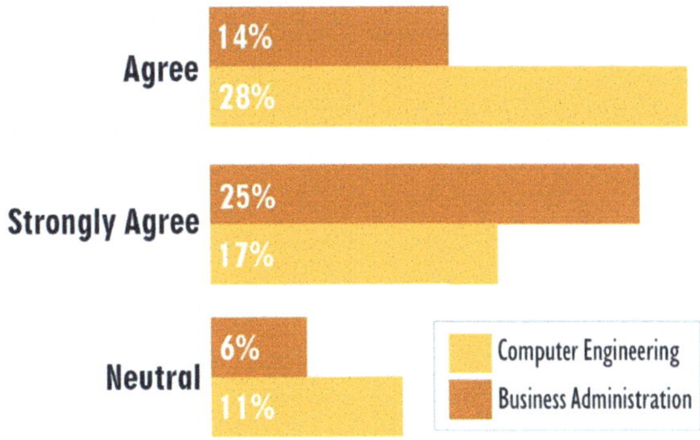

Fig. 3 Can the use of games influence behaviour?

priorities such as the need for simulating real scenarios, strategic decision-making, and interdisciplinary integration.

The qualitative analysis revealed significant gaps in student engagement with sustainable development concepts, again attributed to minimally interactive teaching methods. Data collection through questionnaires and focus groups indicated that students favour practical, immersive activities that provide autonomy in tasks. The primary motivations cited included the practical application of content and the opportunity to learn collaboratively. Table 1 presents a summary of the interviewed teachers' comments and their respective categories.

The interviews revealed four main categories for structuring the serious game: Educational Relevance, Interactive Dynamics and Engagement, Representation of Real Scenarios, and Interdisciplinary Application.

Educational Relevance highlighted the importance of strategic decisions with multidimensional impacts, guiding the design of a sustainable scoring system.

Interactive Dynamics and Engagement underscored the need for collaboration among players and balanced complexity in the mechanics.

Representation of Real Scenarios led to the inclusion of contemporary sustainability challenges through elements such as a modular board and thematic cards.

Interdisciplinary Application ensured that the game connected concepts from different fields and fostered transferable skills, aligning the design with interdisciplinary pedagogical practice.

Definition—Based on the collected data, the core challenge was defined as creating engagement with the theme of sustainable development. The game was designed to simulate strategic decision-making and resource management in sustainable development scenarios. The established educational objectives focused on developing decision-making competencies that encourage students to consider pro-environmental behaviours.

Ideation—Various game concepts were explored, incorporating mechanics such as resource allocation, territory building, and environmental impact management. The final proposal included a modular board, sustainability-themed cards, and a

Table 1 Thematic categories and teachers' statements

Category	Professor	Statement
Educational relevance	Engineering	Students need to understand the systemic impacts of their decisions, especially regarding topics such as energy efficiency and the use of natural resources
	Administration	It is crucial that the game simulates real decision-making scenarios, where students face trade-offs between economic, environmental, and social factors
Interactive dynamics and engagement	Mathematics	Games that involve collaboration among students are more effective, especially when one player's actions have an impact on others
	Physics	The mechanics should challenge students to solve real problems without making them overly complex
Representation of real scenarios	Chemistry	The game needs to depict real issues, such as climate change, water scarcity, and waste management, so that students understand its practical relevance
	Administration	Choices made in the game must have clear and measurable consequences, simulating the impact of corporate actions on the environment
Interdisciplinary application	Biology	Sustainability is inherently interdisciplinary; the game should reflect that by connecting concepts from different fields
	Engineering	The skills developed in the game, such as critical analysis and teamwork, should be transferable to real-world contexts

scoring system based on indicators such as climate change, air quality, and waste treatment. The TPB was integrated into the design to influence players' attitudes, social norms, and perceived control regarding sustainable actions.

Prototyping and Testing—Conceptual prototypes, including wireframes and mock-ups, were developed and validated through an iterative process with an education and game design specialist. Initial feedback resulted in adjustments to the complexity of the mechanics and the clarity of the instructions. The revised prototypes were tested in controlled sessions with STEM-area lecturers. Collaborative decision-making and resource management challenges proved especially effective for promoting active learning. Future iterations will address game balancing and visual improvements based on participant feedback.

Game Design Document

In this board game, students assume the role of leaders of fictional communities, facing challenges related to sustainable development. Each turn, players acquire project cards, with a predetermined number set to ensure interdisciplinarity across STEM subjects, although they can also be prepared in advance by the teacher. These cards represent sustainable actions such as installing solar energy systems, reforestation, or building water-harvesting infrastructure. They are directly related to content previously addressed across various subjects, thus allowing players to apply their existing knowledge and reinforce both interdisciplinary learning and practical application.

The game mechanics involve allocating resources to implement projects, taking into account each decision's environmental impact. Students receive credits to purchase project cards, and at each round they earn additional credits based on their performance, enabling them to fund further projects. Each card can help achieve global indicators. For example, a selective waste collection project card contributes to the waste treatment indicator, reforestation helps meet the climate change indicator, and vehicle electrification benefits air quality. Once all indicators are met, the game ends. Figure 4 below presents the game's indicator scale, the structure of the project cards, and an example.

In Fig. 5, the main board of the game is presented, illustrating a region that must be restored and/or sustainably developed. During the game, players acquire tiles, which are separate pieces representing sustainable achievements. These tiles include vegetation, sustainable cities, and aquifers, and are placed on the designated hexagons, as shown in the bottom-right corner of Fig. 5. Each tile placed advances the students' score, indicated by the numbers around the board.

Additionally, every three rounds players must draw a card from a pile of random events. These event cards—such as droughts or economic crises—simulate uncertainty and require players to adjust their strategies. Environmental impact is monitored throughout the game via the indicators, affected by both the players' actions and the simulated events. This approach enables students to understand the interconnections between natural and social systems, fostering systems thinking skills.

The teacher plays a central role in the game, customising cards and challenges to reflect the content covered in class. Furthermore, the game concludes with a reflection session, during which students discuss their decisions, analyse the impacts, and link the gameplay to the theoretical concepts studied. This stage promotes metacognition, allowing students to evaluate their own performance and recognise the practical relevance of their choices.

Although still a prototype, the game provides an initial structure that offers a dynamic and engaging environment in which students experience, albeit in a simulated manner, the challenges of sustainable decision-making. In addition to enhancing engagement, the game supports the development of key competencies for

Fig. 4 Global game indicators, project card structure, and example

shaping global citizens, such as the ability to solve complex problems, collaborate in multicultural contexts, and think strategically over the long term.

How the Serious Game Introduces SDG Concepts

The game mechanics involve acquiring and implementing sustainable projects, managing resources, and adapting to random events that impact environmental indicators. In this way, participants are encouraged to reflect on real-world challenges, effectively connecting theory and practice. Table 2 presents the main game elements and how each contributes to SDG learning:

The game not only provides a practical introduction to the SDGs but also fosters behavioral change by encouraging the internalization of sustainability values. By allowing students to face challenges and make decisions in a safe and interactive environment, it creates a context conducive to adopting attitudes aligned with sustainable development principles in real life.

Fig. 5 Main game board

In this way, serious games stand out as effective tools not only for transmitting knowledge but also for shaping global citizens who are conscious and engaged with the challenges of the 2030 Agenda.

Theory of Planned Behaviour Applied to the Serious Game

The TPB emphasises the importance of intentions and cognition in determining behaviour, which can be applied in various ways when designing and implementing serious games. Psychological aspects such as attitude, subjective norms, and perceived behavioural control can be mapped onto specific game elements, including objectives, rewards/penalties, rules/mechanics, challenges/obstacles, and social interactivity. This mapping contributes to understanding how game design can influence player behaviour.

The TPB also considers subjective norms, which refer to the perception of relevant social norms. In serious games, it is possible to promote positive social norms by using game mechanics and dynamics that encourage and reward desired behaviours. This can influence players perceived social norms. Table 3 shows how the game elements are related to the TPB components.

The results of this study corroborate previous findings by Ghodsvali et al. (2022), which demonstrated that serious games can foster more sustainable decisions when integrated with active methodologies. However, unlike those studies, this work highlights the importance of structuring the game mechanics based on the TPB, ensuring that players' decisions are guided by theoretically validated principles of behaviour change. In addition studies suggest that incorporating behavioral elements into game mechanics enhances their influence on sustainability actions (Gatti et al., 2019).

Table 2 Game elements and their relationship with the SDGs

Game element	Related SDG(s)	How it contributes
Project cards	SDG 7 (Affordable and Clean Energy), SDG 13 (Climate Action), SDG 15 (Life on Land), SDG 6 (Clean Water and Sanitation) and others	Encourage players to implement sustainable solutions, applying knowledge acquired in related disciplines
Resource allocation system	SDG 12 (Responsible Consumption and Production), SDG 9 (Industry, Innovation, and Infrastructure)	Teaches sustainable resource management and strategic planning for balanced economic development
Global indicators (Waste Treatment, Climate Change, Air Quality)	SDG 11 (Sustainable Cities and Communities), SDG 13 (Climate Action), SDG 14 (Life Below Water)	Players' decisions directly impact these indicators, reinforcing the link between human choices and environmental sustainability
Main board with sustainable tiles (Vegetation, Sustainable Cities, Aquifers)	SDG 15 (Life on Land), SDG 11 (Sustainable Cities and Communities), SDG 6 (Clean Water and Sanitation)	Provides a visual and interactive representation of ecological restoration and sustainable urban planning
Random event cards (Droughts, Economic Crises)	SDG 13 (Climate Action), SDG 1 (No Poverty), SDG 8 (Decent Work and Economic Growth)	Introduces uncertainty and risk management, teaching students the necessity of adaptive strategies when facing socio-environmental challenges
Teacher customization of cards and challenges	Various SDG's	Enables educators to tailor challenges and learning experiences to different educational contexts
Reflection session and metacognition phase	Various SDG's	Encourages critical thinking, discussion, and reflection on sustainability decisions, reinforcing long-term engagement with SDG-related topics

By fostering environmental awareness and sustainable behaviors, serious games contribute to reducing resource consumption, minimizing waste, and mitigating climate change. Their ability to simulate real-world challenges makes them valuable tools for sustainability education, supporting global efforts toward a more sustainable future.

Aligned with UNESCO's ESD framework, the serious game proposed can fosters competencies such as critical thinking, problem-solving, and sustainable decision-making. The ESD should provide practical and participatory learning experiences that enhance environmental awareness and behavioral change (UNESCO, 2014, 2017, 2018).

Table 3 Relationship between TPB components and their application in the serious game

TPB component	Application in the serious game
Attitude towards the Behaviour Students need to have a positive attitude towards the objectives, methods, and assessment processes. It is also crucial to demonstrate the usefulness of sustainable development education	– **Engagement**: Games are naturally captivating and motivating, offering challenges, rewards, and immediate feedback that keep players interested and engaged in learning – **Intrinsic Motivation**: Players are often intrinsically motivated to play, i.e., they do so for enjoyment rather than purely for external rewards. This can boost motivation to learn and explore new concepts – **Utility of the Theme**: By representing real-world situations where sustainable development issues are relevant and directly affect the environment and communities, students see the practical usefulness of these concepts. Game dynamics highlight sustainable behaviours as both attainable and necessary to achieve sustainable development
Perceived Behavioural Control Topics such as education for sustainable development may be perceived as dull. Students can feel discouraged when required to study cross-cutting subjects, so it is vital to provide favourable conditions that encourage engagement during learning	– **Practical Experience**: The serious game offers a chance for students to experience situations and make decisions. This allows them to apply what they have learned in an immediate, practical way, facilitating knowledge retention. Unlike traditional teaching methods, this model encourages exploration, active learning, and nurtures students' interest in learning - **Immediate Feedback and Positive Reinforcement**: The game mechanics provide instant feedback on players' actions, showing positive outcomes of sustainable actions. Additionally, rewards and positive reinforcement—such as victory points and achievement milestones—enhance the sense of control over these actions – **Autonomous Decision-Making**: Players have the freedom to make decisions within the game, giving them a sense of control over their actions and outcomes. This helps them better understand the consequences of their choices and feel in command of them – **Gradual Progression**: The game features a system of incremental progression, where challenges become more difficult as the player advances. This allows players to develop their skills and maintain a sense of control over their progress

(continued)

Table 3 (continued)

TPB component	Application in the serious game
Subjective Norms The classroom context itself promotes subjective norms, given that there is already an implicit sense of social approval when everyone is part of the same course	– **Collectivity**: By using a game-based method, interaction occurs among players and various formed groups. This interaction can help shape students' subjective norms by demonstrating that their peers also value the learning process for sustainability education and are engaged in promoting positive change – **Implicit Social Feedback**: The game's own function, by rewarding sustainable actions and penalising unsustainable ones, implicitly conveys a positive injunctive norm regarding pro-environmental behaviour – **Modelling Collective Behaviour**: Cooperative and competitive interactions among students can showcase the positive impact of collective sustainable behaviours, reinforcing the perception that such behaviours are both achievable when everyone practises and socially approves them, and that an overemphasis on competition over cooperation yields negative results. This may lead to a re-evaluation of descriptive norms, as students recognise the importance and feasibility of certain sustainable behaviours

7 Conclusions

This study explored the potential of serious games as a pedagogical tool for introducing the SDGs in higher education, focusing on active learning methodologies and the TPB. The findings indicate that sustainability is only marginally covered in the curricula of Business Administration and Computer Engineering, with a stronger presence in the former. Additionally, students showed differing perceptions regarding the influence of games on behavior, with greater acceptance among Business Administration students.

Interviews with STEM faculty highlighted the need for interactive and immersive methods that simulate real-world challenges to enhance student engagement. In this context, serious games emerge as a promising approach to bridging theory and practice, providing experiences that stimulate sustainable decision-making and interdisciplinary learning.

This study proposed a board game based on the SDGs, structured according to the TPB to reinforce sustainable attitudes, social norms, and perceived control over environmental decisions. The game simulates sustainable development scenarios, requiring participants to strategically allocate resources and adapt to random events, exposing them to the complexities of sustainability.

However, this research has limitations. First, it presents a theoretical proposal without empirical evaluation of its effectiveness in influencing students' behavior. Additionally, the model was developed from a STEM perspective, which may limit its applicability in social sciences and humanities courses. Future studies should test the game in diverse educational contexts, assessing its ability to foster sustainability competencies and drive long-term behavioral changes.

Despite these challenges, this study contributes by proposing a structured model for the use of serious games in SDG education, reinforcing the importance of innovative approaches in sustainability education.

Acknowledgements This work was supported by the Coordination for the Improvement of Higher Education Personnel -Brazil (CAPES) -Funding Code 001 -Process 88881.981874/2024-01. The authors also thank the support of the Federal Institute of Education, Science and Technology of South of Minas Gerais—IFSULDEMINAS and the Postgraduate Program in Environmental Sciences—PPGCA of the Federal University of Alfenas/UNIFAL.

References

Ahmadov, T., Karimov, A., Durst, S., et al. (2024). A two-phase systematic literature review on the use of serious games for sustainable environmental education. *Interactive Learning Environments*. https://doi.org/10.1080/10494820.2024.2414429

Ajzen, I. (2002). Perceived behavioral control, self-efficacy, locus of control, and the theory of planned behavior. *Journal of Applied Social Psychology, 32*, 665–683. https://doi.org/10.1111/j.1559-1816.2002.tb00236.x

Albareda-Tiana, S., Fernandez-Borsot, G., Berbegal-Mirabent, J., et al. (2024). Enhancing curricular integration of the SDGs: Fostering active methodologies through cross-departmental collaboration in a Spanish university. *International Journal of Sustainability in Higher Education, 25*, 1024–1047. https://doi.org/10.1108/IJSHE-07-2023-0299/FULL/XML

Algurén, B. (2021). How to bring about change—a literature review about education and learning activities for sustainable development. *Discourse Communication Sustainable Education, 12*, 5–21. https://doi.org/10.2478/dcse-2021-0002

Amorós Molina, Á., Helldén, D., Alfvén, T., et al. (2023). Integrating the United Nations sustainable development goals into higher education globally: A scoping review. *Global Health Action, 16*,. https://doi.org/10.1080/16549716.2023.2190649

Andreoni, V., & Richard, A. (2024). Exploring the interconnected nature of the sustainable development goals: The 2030 SDGs Game as a pedagogical tool for interdisciplinary education. *International Journal of Sustainability in Higher Education, 25*, 21–42. https://doi.org/10.1108/IJSHE-11-2022-0378/FULL/XML

Browne, G. R. (2023). The sustainable development goals in a Bachelor of Design course; current integration and benefits, constraints and opportunities for deeper integration. *International Journal of Sustainability in Higher Education, 24*, 1247–1265. https://doi.org/10.1108/IJSHE-02-2022-0043/FULL/XML

Carvalho, L. B., & Neto, J. D. (2022). Serious games may shape the future of accounting education by exploring hybrid skills. *Account Education,* 1–24. https://doi.org/10.1080/09639284.2022.2088241

Damaševičius, R., Maskeliūnas, R., & Blažauskas, T. (2023). Serious games and gamification in healthcare: A meta-review. *Information, 14*,. https://doi.org/10.3390/info14020105

Feltrero, R., Junguitu-Angulo, L., & Osuna-Acedo, S. (2023). Deploying SDG knowledge to foster young people's critical values: A study on social trends about SDGs in an educational online activity. *Sustainability, 15*,. https://doi.org/10.3390/su15086681

Figueiró, P. S., & Raufflet, E. (2015). Sustainability in higher education: A systematic review with focus on management education. *Journal of Cleaner Production, 106*, 22–33. https://doi.org/10.1016/j.jclepro.2015.04.118

Filho, W. L., Trevisan, L. V., Dinis, M. A. P., et al. (2024). Fostering students' participation in the implementation of the sustainable development goals at higher education institutions. *Discover Sustainability, 5*,. https://doi.org/10.1007/s43621-024-00204-7

Gatti, L., Ulrich, M., & Seele, P. (2019). Education for sustainable development through business simulation games: An exploratory study of sustainability gamification and its effects on students' learning outcomes. *Journal of Cleaner Production, 207*, 667–678. https://doi.org/10.1016/j.jclepro.2018.09.130

Ghodsvali, M., Dane, G., & de Vries, B. (2022). An online serious game for decision-making on food-water-energy nexus policy. *Sustainable Cities and Society, 87*, Article 104220. https://doi.org/10.1016/j.scs.2022.104220

Hansen, B., Stiling, P., & Uy, W. F. (2021). Innovations and challenges in SDG integration and reporting in higher education: A case study from the University of South Florida. *International Journal of Sustainability in Higher Education, 22*, 1002–1021. https://doi.org/10.1108/IJSHE-08-2020-0310/FULL/XML

Hsu, C. Y., & Wu, T. T. (2023). Application of business simulation games in flipped classrooms to facilitate student engagement and higher-order thinking skills for sustainable learning practices. *Sustainability, 15*, 16867. https://doi.org/10.3390/SU152416867

Huang, T. C., Ho, S. J., Zheng, W. H., & Shu, Y. (2024). To know, feel and do: An instructional practice of higher education for sustainable development. *International Journal of Sustainability in Higher Education, 25*, 355–374. https://doi.org/10.1108/IJSHE-11-2022-0355/FULL/XML

Kilanioti, I. (2022). Teaching a serious game for the sustainable development goals in the scratch programming tool. *European Journal of Engineering and Technology Research*, 1–9. https://doi.org/10.24018/EJENG.2022.1.CIE.2957

Lhamri, H., Lhamri, I., & Badad, T. (2024). Serious games in education: Transformation of teaching and learning methods through active pedagogy. *World Journal of Advanced Research and Reviews, 22*, 1080–1084. https://doi.org/10.30574/wjarr.2024.22.3.1750

Medema, W., Furber, A., Adamowski, J., et al. (2016). Exploring the potential impact of serious games on social learning and stakeholder collaborations for transboundary watershed management of the St. *Lawrence River Basin. Water (Switzerland), 8*,. https://doi.org/10.3390/w8050175

Morganti, L., Pallavicini, F., Cadel, E., et al. (2017). Gaming for Earth: Serious games and gamification to engage consumers in pro-environmental behaviours for energy efficiency. *Energy Research & Social Science, 29*, 95–102. https://doi.org/10.1016/j.erss.2017.05.001

Novaes, A. L., Cardoso, G. C., & Sabonaro, D. Z. (2023) Proposta de utilização de serious game de sustentabilidade na Educação Ambiental: uma revisão bibliográfica. *Rev Bras Educ Ambient, 18*, 312–328. https://doi.org/10.34024/revbea.2023.v18.15179

Novaes, A. L., Cardoso, G. C., & Sabonaro, D. Z. (2025). Serious game applied as experiential learning for the development behavior of waste reduction and promote education for sustainable development. In W. Leal Filho (Ed.), *The Transformative Power of Higher Education Institutions in Accelerating the Implementation of the UN SDG's*. Springer.

Prouty, C., Mohebbi, S., & Zhang, Q. (2018). Socio-technical strategies and behavior change to increase the adoption and sustainability of wastewater resource recovery systems. *Water Research, 137*, 107–119. https://doi.org/10.1016/j.watres.2018.03.009

Saitua-Iribar, A., Corral-Lage, J., & Peña-Miguel, N. (2020). Improving knowledge about the sustainable development goals through a collaborative learning methodology and serious game. *Sustainabity, 12*, 6169. https://doi.org/10.3390/SU12156169

Tan, C. K. W., Nurul-Asna, H., Campus, M., et al. (2023). Serious games for environmental education 严肃游戏在环境教育中的应用. *Integr Conserv, 2*, 19–42. https://doi.org/10.1002/INC3.18

UNESCO. (2014). Shaping the Future We Want. United Nations Educational, Scientific and Cultural Organization. http://unesdoc.unesco.org/images/0023/002301/230171e.pdf

UNESCO. (2017). Education for Sustainable Development Goals: learning objectives. https://doi.org/10.54675/CGBA9153

UNESCO. (2018). UNESCO Global Action Programme on Education for Sustainable development: Information folder. In: S. 28 (pdf). http://unesdoc.unesco.org/images/0024/002462/246270e.pdf

Latin American Universities Lead the Path to Sustainability: 24 Institutions Participate in the Course on Leadership and Sustainable University Governance

Estibaliz Sáez de Cámara, Yuma Inzolia, Alejandro Velasco, Rosario Romera, Orlando Saenz Zapata, Manrique Arguedas, Rosamaria Moura-Leite, Maria Teresa Holguin, Javier Benayas, and Ángela Maria Plata

Abstract Rectors from 24 universities in Latin America and the Caribbean participated in the Leadership and Sustainable University Governance course, organized by UNESCO IESALC, with the goal of transforming higher education into a driving force for sustainable development. The program helped the reformulation of institutional strategic plans and the design of roadmaps aligned with the Sustainable Development Goals (SDGs) of the 2030 Agenda. It was delivered in a hybrid format,

E. S. de Cámara (✉)
University of the Basque Country (UPV/EHU), Leioa-Bizkaia, Spain
e-mail: estibaliz.saezdecamara@ehu.eus

Y. Inzolia
UNESCO International Institute for Higher Education in Latin America and the Caribbean, Madrid, Spain
e-mail: y.inzolia@unesco.org

A. Velasco
University of Vechta, Vechta, Germany
e-mail: alejandro.velasco@uni-vechta.de

R. Romera
University Carlos III of Madrid, Madrid, Spain
e-mail: rosario.romera@uc3m.es

O. S. Zapata
Latin America and the Caribbean (OSES-ALC), University of Applied and Environmental Sciences (U.D.C.A.), Bogotá, Colombia
e-mail: osaenz@udca.edu.co

M. Arguedas
LATAM Higher Education Observatory Research Team, San José, Costa Rica

R. Moura-Leite
Federal University of Mato Grosso do Sul, Campo Grande, MS, Brazil
e-mail: rosamaria.leite@ufms.br

M. T. Holguin
UNESCO-IESALC, Bogotá, Colombia

W. Leal Filho et al. (eds.), *Introducing the UN SDGs in Higher Education Institutions*, World Sustainability Series, https://doi.org/10.1007/978-3-032-23017-1_14

combining virtual sessions with visits to leading sustainability-focused universities, fostering the exchange of best practices. Supported by international mentors, the course culminated in the signing of a Sustainability Manifesto and the creation of the first Network of Rectors for Sustainability, marking a milestone in university governance towards more sustainable strategies. This chapter presents in detail the characteristics, results and positive impacts of the course with Latin American rectors in the integration of sustainability into their higher education institutions.

Keywords Sustainability · Leadership · Governance · Mentoring · Alliances

1 Introduction. Integration of Sustainability in Higher Education Institutions

In an increasingly interconnected and challenging world, sustainability has become a global priority that transcends national borders and addresses a wide range of social issues, particularly those related to inequality, poverty, lack of education, corruption, violence, and, undoubtedly climate change and widespread environmental degradation. In this context, the 2030 Agenda, with its 17 Sustainable Development Goals (SDGs) and 169 targets, offers an opportunity to promote more equitable and environmentally respectful human development (UN General Assembly, 2015).

For the first time in an international agenda, and as part of its broad scope, the 2030 Agenda expanded the focus beyond primary and secondary education to include Higher Education Institutions (HEIs) in SDG 4- Quality Education. Universities are called to play a key role for SDGs through the dimensions of education, research, outreach, and management (SDSN Australia/Pacific, 2017; Leal Filho et al., 2019; Serafini, 2022; Leal Filho et al., 2024). HEIs play a vital role in training integral, ethical, and conscientious professionals who, through their knowledge and practices, can offer innovative solutions to environmental, social, and economic challenges of the twenty-first century.

The guide prepared by SDSN Australia/Pacific (2017) argues that "it is likely that none of the SDGs can be fulfilled without the involvement of this sector". UN DESA's latest Sustainable Development Goals report (2024) highlights that "Accelerating progress towards SDG4 should be prioritized as it will have a catalytic impact on achieving the overall 2030 Agenda". The Human Development Report 2023/2024 (UNDP, 2024) underscores the critical role of education in reducing inequalities

J. Benayas
Social-Ecological Systems Laboratory, Department of Ecology, Universidad Autónoma de Madrid (UAM), Madrid, Spain
e-mail: javier.benayas@uam.es

Á. M. Plata
Sergio Arboleda University, Bogotá, Colombia
e-mail: angela.plata@usa.edu.co

(SDG 10), fostering social cohesion, and promoting peaceful and inclusive societies (SDG 16).

However, the role of HEIs is more complex than merely contributing to the SDGs, as HEIs are both drivers of the necessary transformations and targets of those transformations. According to Steele and Rickards (2021), universities must ensure that "they are modeling the sort of approaches and impacts they want to engender". This requires leadership and governance structures in HEIs to be approached from a broad and transformative perspective (Chaffin et al., 2016; Galdós et al., 2020). Precisely, this lack of leadership and governance, along with resource constraints and institutional inertia, is one of the main barriers highlighted by Stelle and Richards (2021) to the explicit integration of SDGs into HEIs' strategies. Thus, while many universities recognize the importance of sustainability, the explicit integration of the SDGs into their strategies remains limited.

In this context, and to support this necessary transformation, the UNESCO International Institute for Higher Education in Latin America and the Caribbean (IESALC) designed and developed in 2023/24 the course entitled "Leadership and Sustainable University Governance Program: The Roadmap to the 2030 Agenda". This program embodies a global vision aimed at university rectors across Latin America and the Caribbean, fostering bold leadership, prioritizing strategic alignment with the SDGs, and rethinking traditional structures and practices.

2 Governance for Effective Integration of Sustainability in HEIs

In order to guide and ensure the effective integration of sustainability across all areas of institutional action within universities, it is essential for their leaders to have a comprehensive understanding of the process and a genuine commitment to this purpose. Furthermore, this knowledge and commitment must be shared by other members of the university community, including administrative staff, faculty, researchers, students, workers, and other stakeholders. Institutions must not only possess the capacity to formulate institutional policies but also ensure the participation of all institutional sectors to guarantee effective governance. This perspective is supported by the University Sustainability Evaluation Group (GESU) of the Conference of Rectors of Spanish Universities CRUE-Sustainability, which, in the "Governance" area, includes both the "Environmental Sustainability Policy" and the "Involvement and Awareness of the University Community" dimensions (GESU–CRUE, 2023).

The process of incorporating environmental and sustainability dimensions into HEIs in Latin America and the Caribbean can be identified through various stages, largely linked to the growing interest in specific areas of university action each decade. This process began in the 1970s, primarily focused on educational initiatives (CIFCA, 1978). During the 1980s, interest expanded to include research and university extension activities (UNESCO UNEP and ICFES, 1988; Ángel, 1989). By the 1990s, campus management initiatives gained increasing importance (Sáenz, 2007). However, it was only at the beginning of the twenty-first century that the need to institutionalize HEIs' commitment to environmental and sustainability issues was emphasized at the highest levels of institutional leadership (Benayas et al., 2014; Blanco-Portela et al., 2017; Sáenz et al., 2017). Naturally, the stages of this process and the timing of their development exhibit specific characteristics in each country and institution.

A 2020 study involving 157 HEIs from 13 Latin American countries revealed significant diversity in institutional development levels. Nevertheless, it highlighted that 128 of these institutions reported having a sustainability policy (Leal-Filho et al., 2021), indicating an ongoing process of institutionalization of this topic (Ramísio et al., 2019).

Institutional actors driving the integration of environmental and sustainability dimensions into Latin American universities have varied over time. In the early decades, the majority of these actors were isolated faculty members and researchers within their institutions, with limited influence on policy development (UNESCO, UNEP, and ICFES, 1988). However, their persistent and systematic efforts to promote HEIs' environmental commitments gradually led to the increasing inclusion of sustainability topics in core institutional functions such as education, research, extension, and even campus management (Sáenz & Benayas, 2011).

In recent years, senior university leadership has increasingly recognized the need to institutionalize environmental and social commitments, as reflected in new policies and administrative and academic structures. At the current stage, the leadership of university presidents and senior administrators has played a key role in consolidating and accelerating the integration of sustainability within HEIs (UNESCO IESALC, 2020).

Although the institutionalization of sustainability commitments in Latin American HEIs is relatively recent, governance has emerged as the area with the most significant progress in this regard (Velasco et al., 2024). The OSES-ALC report on the progress of 50 HEIs in terms of knowledge, commitment, and contributions to the SDGs of the 2030 Agenda indicates that "Governance and Participation" is the strongest area of institutional action, with an advancement rate of 82.9% (Sáenz et al., 2023). A similar trend has been observed in the report on 19 Brazilian HEIs, where progress in this area reaches 83.6%. In both Latin America and Brazil, the other five areas of institutional action show progress of 80.0% or less (Moura-Leite et al., 2024).

Clearly, this process is far from being universal, with many institutions still in the early stages of development or implementing isolated initiatives with limited impact. For this reason, programs such as the UNESCO IESALC course -exclusively designed for university presidents- represent a crucial contribution to leadership training needed to consolidate, advance, and enhance the effectiveness of the institutionalization of sustainability commitments in Latin American HEIs.

This study explores how the first edition of the 'Sustainable University Leadership and Governance' course has influenced the transformation of sustainability strategies in Latin American universities through their rectors.

3 Course "Sustainable University Leadership and Governance". Key Aspects

The 60-h course "Sustainable University Leadership and Governance: Leadership and Sustainable University Governance Program: The Roadmap to the 2030 Agenda" is part of the Leadership, Management and Training Program for Sustainable Higher Education developed by the UNESCO International Institute for Higher Education in Latin America and the Caribbean (IESALC).

The primary goal of this course is to build institutional capacities to support the sustainable transformation process of HEI. To achieve this, UNESCO IESALC promotes the development of the educational community's capacities, leveraging its resources and international networks to strengthen urgent efforts toward sustainability and quality education for all.

The course is targeted at rectors of public and private universities in Latin America who are interested in strengthening their Institutional Strategic Plan (ISP) to comply with the commitments of the 2030 Agenda for sustainable development. The specific objectives of the program are:

- To guide university governance teams in understanding the opportunities offered by the SDGs and in structuring a roadmap that enables the contribution to sustainability and aligns with the 2030 Agenda.
- To provide knowledge and solutions that support the implementation of the SDGs at the university.
- To incorporate the principles of the SDGs through governance, management and university culture.
- To explore conceptual and methodological frameworks for promoting the SDGs within the university and advancing the 2030 Agenda.
- To identify and analyze best practices carried out by other universities in training, research, institutional management and community engagement.
- To encourage peer learning, strategic alliances and cooperation on sustainability issues.
- To design an implementation plan as a roadmap with specific activities aimed at achieving the commitments of the 2030 Agenda.

Methodological Framework

The program includes a variety of activities aimed at strengthening the social leadership of rectors. These activities include discovery and immersion sessions, peer learning meetings, and spaces for collaboration and dialogue to exchange best practices and collaborative design strategies. This process strengthens their commitment to the 2030 Agenda and SDGs through the creation of a roadmap for their Institutional Strategic Plan (ISP).

The methodology focuses on knowledge transfer based on best practices, through seminars led by international experts and virtual visits to universities with experience in sustainable governance models. This approach provides first-hand insight into the

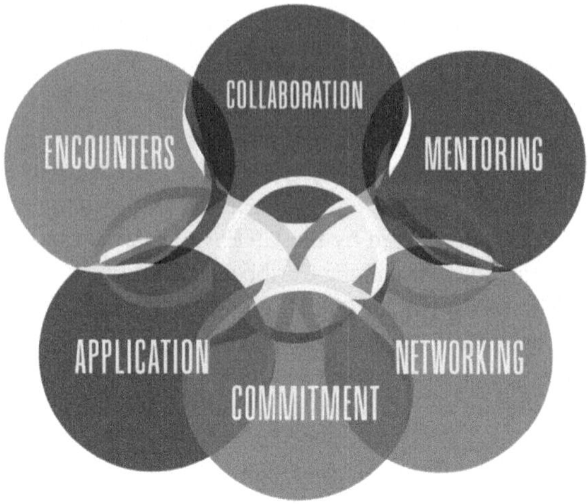

Fig. 1 Key axes of the course "Sustainable University Leadership and Governance"

implementation of sustainable development in HEIs, allowing participants to acquire skills experientially.

This comprehensive training and support model aims to strengthen the integrative project, improving the necessary activities planned for its implementation, as well as for the subsequent measurement and monitoring. The university team, composed of the rector and two other senior university executives, participated in both in person and virtual activities with international experts, to learn first-hand about best practices for the implementation of sustainability within their ISP, using it as a key framework to respond to the 2030 Agenda (Fig. 1).

To achieve this, the following activities were designed to develop the key axes of the program:

1. Encounters: Through face-to-face, synchronous and asynchronous activities, the participants acquire the knowledge needed to lead and implement a strategic plan aimed at complying with the 2030 Agenda.
2. Collaboration: Dialogue and collaborative work spaces essential components of the program, emphasizing the value of participation and the exchange of experiences to promote social learning.
3. Mentoring: University teams receive support from their mentor for six months, during which several tailored sessions are held to guide their efforts.
4. Application: Each team will develop and refine the lines of action in its strategic plan as an integrative project, strengthening their institution's capacity for sustainability.
5. Commitment: Participating rectors sign a commitment to collaborate on the contribution to SDGs of the 2030 Agenda, and form a network of university leaders for sustainability.
6. Networking: Rectors join the Network of Rectors for Sustainability, promoted by UNESCO IESALC.

Course Implementation

The first edition of the course took place in Madrid, Spain, where rectors shared experiences and best practices in sustainability over 2.5 days of work. In collaboration with Casa de América, the sessions were held on September 20, 21 and 22, 2023, with international experts facilitating reflection and guiding activities aimed at strengthening the university's strategic plan. In the face-to-face sessions, two visits were also made to the Polytechnic University of Madrid (UPM) and the Autonomous University of Madrid (UAM). At the end of the face-to-face activities, participants signed a commitment to collaborate on the integration of the 2030 Agenda in their institutions.

During the following 8 weeks, the university team -composed of the rector and two other senior university executives- participated in virtual activities with international experts and received support from specialized mentors through the virtual platform Campus IESALC. These activities were designed to introduce them to best practices for implementing sustainability within their Institutional Strategic Plan (ISP), using it as a key framework to contribute to the SDGs of the 2030 Agenda.

The starting point was a self-diagnosis of the institution's level of maturity in terms of sustainability across its mission areas: teaching and learning, governance, research, and social leadership. Then, four modules (M1-M4) were offered: Sustainability at the Center of the Strategy (M1), Accelerating the SDGs in Universities (M2), Governance and Sustainable Leadership (M3) and Challenges and Paths towards 2050 (M4). These modules included both synchronous and asynchronous training sessions, featuring knowledge transfer, success stories, best practices, and the application of diagnostic, follow-up, and monitoring tools. A description of these modules is provided in Table 1.

Experts led discussions to strengthen the ISP at each university, with a focus on sustainability. Throughout this phase, each university received support from a specialized mentor to help articulate an institutional roadmap and monitor progress over 6 months. The sequence of activities, including their duration and format, is shown in Fig. 2.

4 Results and Discussion

In this first edition of the course, 48 participants from 16 higher education institutions across different countries of Latin America took part as shown in Fig. 3, representing a diverse range of sizes, funding models, and structures, among other characteristics. Participants received support from six expert thematic leaders and ten specialized mentors, who guided them through the thematic modules and the development of their Institutional Strategic Plan.

This section presents participants' evaluations of the program through a quantitative and qualitative analysis of satisfaction surveys and testimonials (Sect. 1). It also includes a selection of Institutional Strategic Plans developed by the participating

Table 1 Description of the modules of the course "Sustainable University Leadership and Governance"

M1.Sustainability at the centre of the strategy	The development of a more just and inclusive world is the central theme of this module. Therefore, it is focused on those aspects that help strengthen the need to cultivate a sustainable culture. This module provides a space for dialogue to explore what sustainability is, what Education for Sustainable Development entails, how the 2030 Agenda is connected to the SDGs, and the relationship with stakeholders and key issues
M2. Accelerating the SDGs in universities	Addressing the challenges of the SDGs of the 2030 Agenda requires new knowledge, innovative approaches, difficult decisions between competing options and, in some cases, profound transformations. This module provides a clear vision of how universities can contribute to implementing the SDGs through their core functions, including teaching–learning, research, governance and, social leadership
M3. Governance and sustainable leadership	A sustainability initiative involves significant transformations that are integral to the activities of higher education institutions. These transformations require a strong institutional commitment, along with governance that enable or that allow the effective participation of various stakeholder groups and the coordination of efforts on sustainability. This module explores the leadership model necessary to strengthen strategic planning processes and, in turn, support sustainability
M4. Challenges and paths towards 2050	Universities are often large institutions with the potential to significantly impact on social, cultural and environmental well-being, both on their campuses and their surrounding communities and regions, sometimes even extending far beyond. These impacts are directly related to all areas of the SDGs and, by acting responsibly, universities can make significant contributions to their achievement. Leveraging their unique position within society, universities -both individually and collectively- can lead, guide, and support local, national, and international efforts towards the SDGs

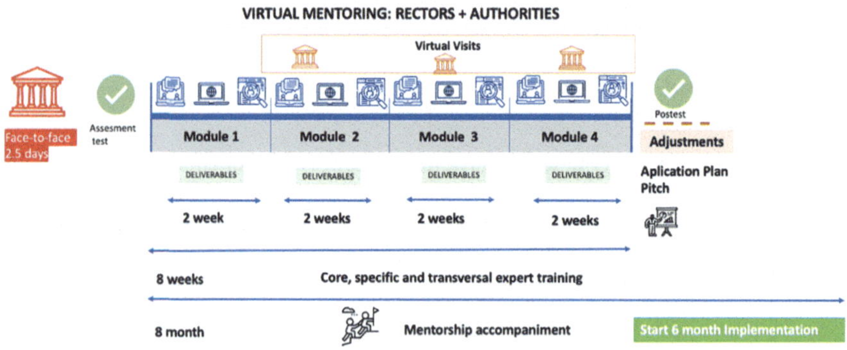

Fig. 2 Sequence of activities of the Course "Sustainable University Leadership and Governance", including duration and format

University	*Country*
Instituto Universitario Hospital Italiano de Buenos Aires	Argentina
Universidad Franz Tamayo	Bolivia
Universidad Cooperativa de Colombia	Colombia
Universidad Nacional Abierta y a Distancia	Colombia
Fundación Universitaria Juan N. Corpas	Colombia
Universidad de Ciencias Aplicadas y Ambientales	Colombia
Escuela Superior Politécnica del Litoral	Ecuador
Universidad De Cuenca	Ecuador
Universidad Católica de Cuenca	Ecuador
Universidad Autónoma de Santa Ana	El Salvador
Universidad Don Bosco	El Salvador
Universidad Pedagógica Nacional Francisco Morazán	Honduras
Universidad Metropolitana de Honduras	Honduras
Universidad Panamericana e IPADE	México
Universidad Nacional de Jaén	Perú
Universidad Peruana de Ciencias Aplicadas	Perú

Fig. 3 Latin American universities by country participating in the first edition of the course

universities (Sect. 2) and highlights an initially unplanned but highly significant outcome: the creation of a Network of Rectors for Sustainability (Sect. 3).

4.1 Surveys and Testimonies of University Team Participants and Mentors

According to the qualitative analysis of the results of the surveys and testimonies from the **university team participants**, the course 'Leadership and Sustainable University Governance' has had a significant positive impact. It enhanced participants' awareness of sustainability in higher education and strengthened the skills of management teams in this area. Participants reported increased recognition of the importance of cross-cutting sustainability, as well as the acquisition of tools and knowledge to implement it within their universities. The course has enabled them to identify areas for improvement and understand the need for a comprehensive, institution-wide approach to sustainability.

Regarding tools and knowledge, participants emphasized the acquisition of concrete tools to drive comprehensive institutional transformation in response to the challenges of sustainability. These tools enable the promotion and evaluation of university progress in relation to SDGs of the 2030 Agenda. In terms of inspiration and motivation, participants expressed feeling inspired to transform their institutions and motivated to advocate policies aligned with the SDGs, developing a sense of responsibility and commitment to sustainability.

In terms of collaborative networks, the creation of networks, such as the 'Network of Rectors for Sustainability' materialized the relevance of the exchange of best practices and the socialization of experiences, strengthening collaboration between universities in the region.

Concerning the impacts of the training process, participants identified several key outcomes. The course helped them move beyond routine practices, enabling them the identification of areas for improvement. The strong focus on the SDGs facilitated the implementation of relevant strategies for their institutions. In relation to the relevance of the course content, the activities aligned with the SDGs prompted reflection on how these goals can be integrated into institutional strategies. In terms of institutional transformation, participants reported changes in both structure and culture, fostering more collaborative work. With respect to knowledge exchange, the opportunity to learn from sustainability initiatives at other universities created a valuable networking environment. In terms of practical methodology, the course provided useful tools, such as the Institutional Strategic Plan (ISP) and matrices for aligning sustainability commitments. Finally, regarding commitment to sustainability, the course reinforced participants' commitment to sustainability, empowering participants to plan and monitor SDG-related activities.

In the matter of evaluating competencies, after the course, all participants reported achieving a sufficient or high level of competency in sustainability. Regarding strategic alliances, partnerships were formed, such as the establishment of a network of rectors for sustainability. Concerning personal and institutional growth, participants reported personal development in their understanding of sustainability and a reinforced institutional commitment to its integration into governance. The testimonies of the rectors highlighted the urgent need to redefine sustainability through

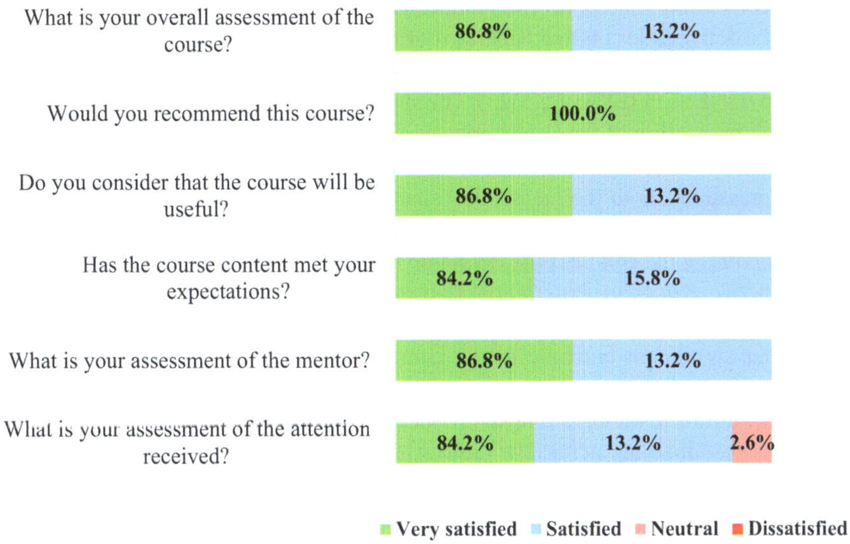

Fig. 4 Results of the main items from the survey answered by the participants of the course

a comprehensive perspective. They called for action to integrate sustainable practices at all levels, including policy reform, the identification of opportunities, and the restructuring of the Institutional Strategic Plan with Education for Sustainable Development (ESD) at its core. Collaboration with other institutions and the consolidation of existing initiatives were also emphasized.

This qualitative analysis has been complemented with a quantitative analysis of a satisfaction survey. It should be noted that, out of 45 participants, 38 responses were collected. It should be noted that the number of participants decreased from 48 to 45 because one university withdrew from the program. The main results of the survey are shown in Fig. 4.

These responses revealed a high level of satisfaction with the course: 86.8% of respondents stated they were very satisfied, and 100% indicated they would recommend the course to others. Additionally, most of the participants were very satisfied with the attention received (84.2%) and their mentor (86.8%). The course met their expectations (84.2%), and they believe the course will be useful for them in their professional field (86.8%). None of the participants were unsatisfied with any of the items in the survey.

There is a high correlation between the data on the attention received and mentorship, considering that one of the closest actors to the course participants was the specialized mentor (See last two items in Fig. 4). The mentor provided technical advice, weekly support to university teams, tutoring, and virtual guidance (both synchronous and asynchronous) during the training, weekly reports on attendance, progress, and risks, management of assigned teams, and communications through social media, email, and platform. Additionally, the mentor supported the achievement of the deliverables outlined in the program, as well as follow-up and monitoring during the six months after the completion of the virtual course on the University's Roadmap (UNESCO IESALC, 2023). The support offered by the mentor contributed to the high promotion of the universities participating in the course (93.75%).

There is also a strong correlation between the quantitative results and the qualitative expressions about the mentorship shared by some of the course participants. They stated "*The exchanges with our mentor have been of great value to us*" and "*The strongest point of the course is the follow-up with the mentors and the opportunity to create a network to share experiences that can enrich each institution in its journey with this topic.*"

Another significant outcome of the course was the endorsement of a sustainability Decalogue by the participating rectors. This commitment underscores a collective dedication to fostering strategic alliances and collaborative networks aimed at promoting sustainability across the Latin American HEIs landscape.

4.2 Best Practices of Institutional Action Plans for the Integration of Sustainability of the Participant Universities

As a key activity of the course, the rectors and participating teams of Latin-American universities, with the support of mentors, have developed 15 Institutional Strategic Plans. This section presents different approaches to integrating the SDGs of the 2030 Agenda in a selection of universities with diverse characteristics. Each institution has developed tailored strategic plans designed to be both viable and effective in addressing their specific institutional and regional challenges related to sustainable development.

Universidad Cooperativa de Colombia

The Universidad Cooperativa de Colombia (UCC) is a private, multi-campus university with a presence in 14 departments across Colombia. It was founded in 1958 to strengthen the solidarity economy and, in particular, cooperativism. UCC stands as evidence of an institution deeply rooted in these principles. As a result, sustainability is a fundamental part of its Statutes and institutional identity.

Through its commitment to University Social Responsibility, UCC embraces the SDGs of the 2030 Agenda and promotes natural, social, healthy, just, and solidarity-driven environments. The university upholds an ethical and responsible academic and professional approach that prioritizes both present and future generations, with the firm belief that a better and more inclusive world is built through collective participation. Additionally, its Critical Educational Model with a Competency-Based Approach provides the curricular framework for developing academic programs that support sustainable development.

When Rector Maritza Rondón and her team joined the UNESCO IESALC course, UCC had established its 2022–2025 Strategic Plan, entitled "Continuing to Navigate Together", which already emphasized sustainability, the SDGs, and the 2030 Agenda. However, participation in the course allowed the university to develop a more cohesive and conscious strategy to advance the SDGs by actively engaging the university community and implementing meaningful and localized institutional changes across its various campuses. This strategy is based on a Sustainability Network, which focuses on two common challenges that are Quality Education (SDG 4) and Climate Action (SDG 13). At the same time, this network creates opportunities to provide tailored solutions to the unique challenges faced in the different territories where UCC operates. The Sustainability Network actively involves the entire university community, integrating all campuses, faculty, students, and external partners within their respective regions of influence (SDG 17- partnerships).

Universidad Metropolitana de Honduras

The Metropolitan University of Honduras (UMH) is a private technical higher education institution located in the city of Tegucigalpa, with a presence in 80% of Honduras' territory. It is currently led by Juana Rosario Duarte Galeas, the rector, who has guided the institution's commitment to sustainability. This leadership enabled its participation in the IESALC course, which, with the commitment of the governance team, contributed to the development of an institutional action plan that acknowledges the institution's progress in relation to the SDGs. Along its journey, the university has made progress in sustainability, such as participating in national and international networks related to the SDGs, like the Network of Honduran Universities and Organizations for Sustainability and the Environment and the Ibero-American Network of Universities for Sustainability and the Environment (ARIUSA). Additionally, as part of its progress in the UNESCO IESALC course, the institution reaffirms the commitment of its university top management and sets as a goal in its action plan to promote the incorporation of ESD in an integrated manner within the University. This aims to raise awareness, understanding, and commitment to addressing the global challenges of sustainable development. To achieve this, the university prioritizes Quality Education (SDG 4), Gender Equity (SDG 5) and Climate action (SDG 13), setting out multiple actions for the short, medium, and long term. These actions include strengthening sustainability on campus, measuring and reducing the carbon footprint, addressing water resource management reinforcing research on climate risks, and enhancing training on gender issues, among others. All of this will allow them to position themselves as a leading university in sustainability in Central America.

Escuela Superior Politécnica del Litoral

The Litoral Polytechnic Higher School (ESPOL) is a public higher education institution based in Guayaquil, Ecuador. Under the leadership of its rector, Cecilia Paredes Verduga, ESPOL has developed a systematic and continuous approach to sustainability in all its areas of work.

To structure this commitment, the Sustainability Program was created in 2018, followed by the establishment of the Sustainability Commission in 2019, which is chaired by the rector and includes the participation of vice-rectors, managers, directors, and other senior officials of the institution. In 2022, the Polytechnic Council, the highest governing body of ESPOL, approved the Sustainability Policy, aimed at fulfilling the SDGs in its functions of teaching, research, social engagement, operational management, and university life. The Department of Sustainability, staffed with dedicated personnel, is responsible for coordinating the implementation of this policy.

That same year, due to its advancements in sustainability, ESPOL ranked first in Ecuador in the UI Green Metric and THE Impact Rankings, a position it has maintained to this day. Additionally, its participation in the UNESCO IESALC Course allowed for the systematization of its experiences and the formulation of an action plan to continue strengthening institutional sustainability.

Universidad Peruana Ciencias Aplicadas

The Peruvian University of Applied Sciences (UPC) is a prominent private educational institution in Peru, founded in 1994. It offers higher education programs in fields such as engineering, health sciences, social sciences, and humanities, among others. As one of its core pillars, UPC has consistently distinguished itself both nationally and internationally as an innovative institution committed to positive societal transformation. Its vision is that by fostering innovation, better solutions can be found for the most significant challenges facing both the country and the planet, thus contributing to sustainable development.

Through the course organized by UNESCO IESALC and under the leadership of Rector Edward Roekaert, UPC was able to analyze and define its strengths and opportunities regarding sustainability management within the university. Tasks and actions necessary to achieve the established objectives were identified. Among these, promoting the training of leaders who acted as agents of change stood out. Following this, the Directorate of Sustainability and Social Responsibility was created, along with the formation of a team to lead the strategies necessary to achieve the outlined objectives. Four strategic pillars were then defined: inclusive and quality education and social impact (SDGs 4, 8, 10), research and technology (SDGs 3, 5, 10, 17), and sustainable campus management (SDGs 12, 13). In pursuit of more inclusive education, a scholarship program has been developed for high-achieving students from vulnerable populations, along with a program that supports the training and personal and professional development of individuals over 50 years of age. Another strategy to raise student awareness of the SDGs involves integrating sustainability into the academic curriculum, as well as its application in projects aimed at addressing issues related to the achievement of the SDGs, such as preventive healthcare campaigns for vulnerable populations, knowledge transfer programs, and skills reinforcement activities, thus, benefiting the community.

During this time, UPC has successfully laid the foundations for managing sustainability and social responsibility within the university. While fully aware that the work is challenging and ongoing, the institution remains confident that this commitment is essential for training future leaders and contributing to the country's development and transformation.

4.3 *Rectors' Network for Sustainability*

As a culmination of the initiative, participants of this first edition signed a Sustainability Decalogue or Manifesto, reaffirming their commitment to the SDGs and establishing the basic principles for the creation of the region's first Network of Rectors for Sustainability. Furthermore, one of the most notable outcomes of this course was the creation of the Ibero-American Rectors' Network for Sustainability (https://www.iesalc.unesco.org/es/rectorsnetwork4sustainability). The network was established during a convention held in Cartagena de Indias (Colombia) in August 2024 that brought together more than 1,200 participants from 19 countries, including rectors, academics and educational leaders.

The Rectors' Network for Sustainability aims to strengthen, from the institutional framework of Ibero-American higher education institutions, the integration of other networks and institutions from various sectors to give greater significance to the commitment to the SDGs of the 2030 Agenda. In this regard, rectors, as public figures, can influence policies, mobilize resources, and lead initiatives that promote sustainability in all aspects of university life. It is worth highlighting that this network, led by the Rector of the Universidad Cooperativa de Colombia Maritza Rondón, represents an unprecedented collective effort in Latin America and the Caribbean. Its commitment to sustainability not only seeks to transform higher education but also to contribute to the development of more equitable and resilient societies.

5 Conclusions

Rectors from 24 universities across Latin America and the Caribbean participated in the Leadership and Sustainable University Governance course. This initiative, led by UNESCO IESALC, aimed at transforming higher education into a driving force for sustainable development. The program focused on reformulating Institutional Strategic Plans and designing actionable roadmaps aligned with the Sustainable Development Goals (SDGs) of the 2030 Agenda.

The course was delivered in a hybrid format, combining virtual sessions with visits to leading universities in sustainability. These visits provided participants with first-hand insights into advanced strategies for strengthening their sustainability units and facilitated the exchange of best practices. The entire process was supported by a team of international mentors specialized in sustainability, governance, and strategic planning, ensuring skilled guidance throughout the program.

Results show that the training process and course have served as catalysts for change, equipping participants with the knowledge and tools necessary to implement sustainable practices within their institutions. The positive feedback and high satisfaction rates indicate that participants now feel more confident in their ability to lead sustainability initiatives and integrate these principles into their institutional frameworks. As they return to their universities, they bring not only enhanced knowledge but also a renewed sense of purpose and commitment to fostering a culture of sustainability.

Moreover, the course has facilitated the establishment of strategic networks among participants and international mentors, promoting the sharing of best practices and experiences that enable various actors and institutions to follow clearer and more accessible pathways. The emphasis on practical methodologies and real-world applications has empowered participants to translate their learning into actionable strategies within their own contexts.

Participating universities, with the support of their mentors, have developed and begun implementing an Institutional Strategic Plan to advance towards sustainability in a comprehensive manner. This transformative approach not only aligns with the SDGs but also fosters innovative and regenerative internal practices in higher education while promoting external collaborations to drive meaningful societal and environmental change, as noted by several participating rectors.

As a culmination of the course, participants of this first edition signed a Sustainability Manifesto and established the region's first Network of Rectors for Sustainability. This milestone underscores the collective commitment of participating universities to leading the transition toward more sustainable educational and institutional models. This course marked a turning point in university governance in the region, strengthening the role of Latin American universities as key drivers in building a sustainable future.

In conclusion, this course has proven to be a crucial driver in advancing sustainability initiatives, fostering collaboration, and strengthening the overall impact of HEIs in the pursuit of a sustainable future. It marks a significant step toward transforming Latin-American higher education into a powerful catalyst for sustainable development, ensuring that future generations are equipped to tackle the complex challenges of our world. The insights gained and the relationships built during this programme will undoubtedly contribute to ongoing efforts to create more sustainable, resilient, and equitable educational environments.

As with all such studies, there are limitations that present opportunities for further research. In this course, rectors were encouraged to adopt governance models that prioritize long-term sustainability, fostering environments where sustainable practices are seamlessly integrated into all university operations and involve all members of the university community. This approach aimed to ensure the continuity of the sustainability commitment beyond each rector's four-year mandate by integrating sustainability into long-term strategic plans and effectively conveying its significance—key objectives of the course. This concerning aspect warrants further investigation, which will be conducted in the coming years, following up on the participating universities, not only with the first edition (2023/23) but also with the second edition (2024/25) of this pioneering course.

References

Ángel, A. (1989). I Seminario Nacional sobre Universidad y Medio Ambiente. Incorporación de la dimensión ambiental en la educación superior. Instituto Colombiano para el Fomento de la Educación Superior y Universidad Nacional de Colombia, 70 pp.

Benayas, J., Alba, D., & Justel, A. (2014). RISU Project. Development of indicators to assess the implementation of sustainability policies in latin american universities. Universidad Autónoma de Madrid and UNEP Regional Office for Latin America and the Caribbean.

Blanco-Portela, N., Benayas, J., Pertierra, L. R., & Lozano, R. (2017). Towards the integration of sustainability in Higher Education Institutions: A review of drivers of and barriers to organisational change and their comparison against those found of companies. *Journal of Cleaner Production, 166* 563–578. https://doi.org/10.1016/j.jclepro.2017.07.252

Chaffin, B. C., Garmestani, A. S., Gunderson, L. H., Benson, M. H., Angeler, D. G., Arnold, C. A. T., Cosens, B., Craig, R. K., Ruhl, J. B., & Allen, C. R. (2016). Transformative environmental governance. *Annual Review of Environment and Resources, 41*, 399–423. https://doi.org/10.1146/annurev-environ-110615-085817

CIFCA. (1978). La formación ambiental en América Latina. Cuadernos del CIFCA No. 8. Centro Internacional de Formación en Ciencias Ambientales, pp 55–139.

Galdós, M., Ramírez, M., & Villalobos, P. (2020). El Rol de las Universidades en la Era de los Objetivos de Desarrollo Sostenible. Instituto de Innovación Basado en Ciencia. Universidad de Talca. https://institutoinnovacion2.utalca.cl/wp-content/uploads/2020/06/WS01-2020_Galdos-Ramirez-Villallobos.pdf

GESU, CRUE-Sostenibilidad. (2023). Diagnóstico de la Sostenibilidad Ambiental en las Universidades Españolas. Informe 2023. Grupo de Evaluación de la Sostenibilidad Universitaria (GESU) de CRUE-Sostenibilidad. https://www.crue.org/wp-content/uploads/2024/05/20240523-InformeSostenibilidadUni2023.pdf

Leal Filho, L., Shiel, C., Paço, A., Mifsud, M., Ávila, L. V., Brandli, L. L., et al. (2019). Sustainable development goals and 693 sustainability teaching at universities: Falling behind or getting ahead of the pack? *Journal of Cleaner Production, 232*(285–294), 694. https://doi.org/10.1016/j.jclepro.2019.05.309695

Leal Filho, W., Amaro, N., Avila, L. V., Brandli, L., Damke, L. I., Vasconcelos, C. R., Hernandez-Diaz, P. M., Frankenberger, F., Fritzen, B., Velazquez, L., & Salvia, A. (2021). Mapping sustainability initiatives in higher education institutions in Latin America. *Journal of Cleaner Production, 315*, Article 128093. https://doi.org/10.1016/j.jclepro.2021.128093

Leal Filho, W., Sierra, J., Price, E., et al. (2024). The role of universities in accelerating the sustainable development goals in Europe. *Science and Reports, 14*, 15464. https://doi.org/10.1038/s41598-024-65820-9

Moura-Leite, R, Sáenz, O., Plata, A, Holguín, M. T., Callejas, M. M., Bravo, M. T., Velasco, A., Arguedas-Camacho, M., López, C., & Parada-Guevara, S. (2024). Avanços no conhecimento, compromisso e contribuições para os Objetivos de Desenvolvimento Sustentável nas instituições de ensino superior da América Latina e do Caribe—Relatório IES brasileiras. https://doi.org/10.13140/RG.2.2.30547.96804. https://oses-alc.net/wp-content/uploads/2024/01/2024-02-22-OSES-ALC-Informe-Brasil.pdf

Ramísio, P. J., Pinto, L. M. C., Gouveia, N., Costa, H., & Arezes, D. (2019). Sustainability strategy in higher education institutions: Lessons learned from a nine-year case study. *Journal of Cleaner Production, 222*, 300–309. https://doi.org/10.1016/j.jclepro.2019.03.054

Sáenz, O. (ed.) (2007). Memorias del Cuarto Seminario Internacional Universidad y Ambiente. Gestión Ambiental Institucional y Ordenamiento de Campus Universitarios. Universidad de Ciencias Aplicadas y Ambientales y Red Colombiana de Formación Ambiental, 289 pp.

Sáenz, O., & Benayas, J. (2011). Higher education, environment and sustainability in Latin America and the Caribbean. In: GUNi, Higher education in the world 4. Higher education's commitment to sustainability: from understanding to action. Global University Network for Innovation (GUNi) and Palgrave MacMillan, pp 161–176.

Sáenz, O., Plata, A., Holguín, M. T., Mora, W. M., & Blanco, N. (2017). Institucionalización del compromiso ambiental de las universidades colombianas. *Civilizar Ciencias Sociales y Humanas, 17*(33), 189–208. https://www.redalyc.org/pdf/1002/100254730012.pdf

SDSN Australia/Pacific. (2017). Getting started with the SDGs in universities: A guide for universities, higher education institutions, and the academic sector. Australia, New Zealand and Pacific Edition. Sustainable Development Solutions Network—Australia/Pacific, Melbourne. https://ap-unsdsn.org/wp-content/uploads/University-SDG-Guide_web.pdf

Serafini, P. G., de Moura, J. M., de Almeida, M. R., & de Rezende, J. F. D. (2022). Sustainable development goals in higher education institutions: A systematic literature review. *Journal of Cleaner Production, 370*, Article 133473. https://doi.org/10.1016/j.jcle-697pro.2022.133473

Steele, W., & Rickards, L. (2021). The Sustainable Development Goals in Higher Education: A Transformative Agenda. Palgrave 722 MacMillan, Switzerland. https://doi.org/10.1007/978-3-030-73575-3

UN DESA. (2024). The Sustainable Development Goals Report 2024—June 2024. New York, USA: UN DESA. UN DESA. https://unstats.un.org/sdgs/report/2024/The-Sustainable-Development-Goals-Report-2024.pdf

UN General Assembly. (2015). Transforming our world: the 2030 Agenda for Sustainable Development, A/RES/70/1 https://www.refworld.org/legal/resolution/unga/2015/en/111816

UNDP (2024). Human Development Report 2023–24: Breaking the gridlock: Reimagining cooperation in a polarized world. New York. https://hdr.undp.org/content/human-development-report-2023-24

UNESCO IESALC. (2020). La contribución de la educación superior a los Objetivos de Desarrollo Sostenible: Marco analítico. Instituto Internacional de la UNESCO para la Educación Superior en América Latina y el Caribe, Caracas, 51 pp. https://unesdoc.unesco.org/ark:/48223/pf0000384239

UNESCO IESALC. (2023). Programa Curso rectores Agenda 2030—Campus IESALCCampus IESALC.https://campus.iesalc.unesco.org › summary (Last Access february 3th 2025)

UNESCO, PNUMA, ICFES. (1988). Diagnóstico de la incorporación de la dimensión ambiental en los estudios superiores en América Latina y el Caribe. In: Universidad y medio ambiente en América Latina. Organización de las Naciones Unidas para la Educación, la Ciencia y la Cultura, Programa de las Naciones Unidas para el Medio Ambiente e Instituto Colombiano para el Fomento de la Educación Superior, pp. 20–27.

Velasco, A., Sáenz, O., Plata, A., Holguin, M. T., Moura-Leite, R., Callejas Restrepo, M. M., Arguedas-Camacho, M., Parada-Guevara, S., López, C., & Bravo Mercado, M. T. (2024). Avances en el conocimiento, compromiso y aportes a los Objetivos de Desarrollo Sostenible en las instituciones de educación superior de América Latina y el Caribe. https://doi.org/10.13140/RG.2.2.19098.00964. https://oses-alc.net/wp-content/uploads/2024/06/2024-06-18-OSES-ALC-informe-a-ministros.pdf

Artificial Intelligence in Universities: Enhancing Teaching, Research, and Extension for Sustainable Development

Tiago Negrão de Andrade, Henrique da Silva Pereira, Rodrigo Augusto Tomba, Bruna Fernanda Damasceno Ramirez, and Fábio Rogério Bueno de Moraes

Abstract The integration of artificial intelligence (AI) into higher education has significantly transformed teaching, research, and community outreach, thereby contributing meaningfully to the advancement of the Sustainable Development Goals (SDGs). This systematic review, conducted in accordance with the PRISMA protocol, investigates how AI has been applied across these three domains, emphasizing its impact on personalized learning, interdisciplinary research collaboration, and the effectiveness of community engagement initiatives. Within teaching, evidence indicates that AI has enhanced student outcomes by up to 30% through curriculum personalization and adaptive learning tools, especially in engineering education and emotionally intelligent systems (Cinar & Bilodeau, 2024; Rahman et al., 2022). In research, AI has bolstered interdisciplinary collaboration by improving data interoperability by 30%, enabling more robust cross-domain knowledge integration (Makwana, 2023). Regarding community outreach, AI has optimized resource distribution, increasing the efficiency of local initiatives by 25% (Borsatto et al., 2024), while simultaneously advancing digital inclusion and innovations in agriculture. Nevertheless, the review also highlights persistent challenges, including institutional resistance, limited infrastructure, and ethical dilemmas—particularly

T. N. de Andrade (✉) · F. R. B. de Moraes
São Paulo State University (UNESP), São Paulo, Brazil
e-mail: tiago.negrao@unesp.br

F. R. B. de Moraes
e-mail: fabio.bueno@unesp.br

H. da Silva Pereira · R. A. Tomba
Centro Universitário Nossa Senhora do Patrocínio (CEUNSP), São Paulo, Brazil
e-mail: henrique.pereira@ceunsp.edu.br

R. A. Tomba
e-mail: rodrigo.tomba@ceunsp.edu.br

B. F. D. Ramirez
University of Campinas (Unicamp), Campinas, Brazil

W. Leal Filho et al. (eds.), *Introducing the UN SDGs in Higher Education Institutions*, World Sustainability Series, https://doi.org/10.1007/978-3-032-23017-1_15

in developing regions. The findings affirm AI's pivotal role in reinforcing the university's core missions and advancing the SDGs, while also underscoring the need to overcome structural and cultural impediments that hinder its full potential.

Keywords Artificial intelligence · Higher education · Sustainable development goals · Teaching innovation · Research collaboration · University extension · Digital inclusion

1 Introduction

Digital transformation has become a central catalyst for technological innovation in recent decades, with artificial intelligence (AI) emerging as one of the most disruptive forces within academic institutions. In the university setting, AI not only enhances the production, dissemination, and application of knowledge but also introduces innovative pedagogical strategies aligned with contemporary societal needs. With increasing urgency to fulfill the United Nations' Sustainable Development Goals (SDGs)—a global agenda addressing issues such as poverty, inequality, and climate change (Gupta et al., 2023)—universities have positioned themselves as critical hubs for technological innovation and social transformation. AI, especially under the paradigm of Augmented Intelligence, is redefining research and learning through tools powered by Artificial General Intelligence (AGI). However, such advancements necessitate a strong foundation in digital literacy to cultivate ethical and critical engagement. Ethical concerns—ranging from plagiarism and data privacy to legitimacy, authorship, and intellectual property—must be critically addressed. AI tools should serve humanity by reflecting our collective intelligence and ethical responsibility as stewards of transformative technologies.

The strategic adoption of AI in higher education represents more than a technological trend; it constitutes a fundamental shift toward long-term solutions with social and environmental impact. As Vázquez-Verdera et al. (2021) assert, AI has the capacity to foster interdisciplinary collaboration and enhance university engagement with the SDGs, positioning academic institutions as agents of transformative societal change. The ongoing integration of AI in universities is thus a direct response to global demands for sustainable innovation.

Current market data reveal continuous AI growth across diverse sectors, including clean energy and climate mitigation efforts, closely aligned with SDGs 7 and 13 (Varriale et al., 2024). This underscores the university's essential role in preparing professionals equipped to navigate and address complex sustainability challenges. As noted by Nerini et al. (2024), the extension of the SDGs to 2050 will depend heavily on the strategic incorporation of digital technologies—an area in which universities are uniquely poised to lead by example.

Consequently, AI adoption in academia should be viewed not merely as an innovation in pedagogy or infrastructure, but as a transformative mechanism for advancing global goals. Understanding how these technologies are implemented and their subsequent impacts remains an urgent area of inquiry.

1.1 Evolution and Current State of Research

The application of AI in education has evolved rapidly. Initial endeavors focused on automated tutoring and computer-assisted instruction. However, advancements in big data analytics and machine learning have since enabled the development of more sophisticated tools, including intelligent chatbots and highly personalized learning environments (Abdalla, 2024; Jo, 2024). These developments have attracted increasing interest from universities aiming to enhance student learning and prepare graduates for the evolving labor market.

Universities have emerged as pivotal institutions for the integrated application of AI across teaching, research, and extension activities, thereby advancing sustainable development. Table 1 outlines the primary domains of AI application within higher education and innovation ecosystems:

Despite this growth, the widespread implementation of AI in universities still encounters barriers, especially in Global South contexts. In Brazil, for instance,

Table 1 Areas of artificial intelligence application in universities

Area of application	Description
Distributed learning	Models operate on distributed or edge-computing frameworks, enhancing privacy, reducing latency, and enabling context-aware decision-making (Yang et al., 2019)
Learning in cognitive architectures	Investigates reinforcement learning in mobile cognitive agents, focusing on efficient training environments and reward optimization (Laird et al., 2017)
Cognitive architectures	Develops systems inspired by human cognition, integrating emotional and social factors within intelligent mobile frameworks
AI for finance	Designs interpretable and fair algorithms for credit risk assessment and fraud detection, with a focus on inclusion and accountability (Sotiropoulos et al., 2021)
AI for marketing	Employs AI for customer personalization, retention analysis, and segmentation, addressing ethical data usage in the Brazilian context (Martin et al., 2020)
Natural language processing	Constructs Portuguese-language NLP models using attention-based neural networks to interpret ambiguity and semantic complexity (Fialho et al., 2021)
Knowledge representation	Explores interpretable models through manifold learning to reduce bias and enhance decision transparency (Ribeiro et al., 2016)

Costa et al. (2022) identify limitations related to infrastructure and institutional readiness, suggesting that technological adoption requires both investment and cultural transformation.

Shabbir et al. (2024) similarly argue that in many developing nations, the transformative capacity of AI remains constrained by academic integrity issues and resistance to change. These persistent obstacles highlight the need for systemic reforms that extend beyond technological capacity to include pedagogical, ethical, and organizational dimensions.

In conclusion, although AI in higher education continues to expand, its full potential will only be realized through strategic investment in infrastructure, institutional change, and human capital development.

1.2 Research Object

This study primarily focuses on exploring the implementation of artificial intelligence (AI) in universities, examining how this technology impacts the personalization of education and the advancement of the Sustainable Development Goals (SDGs). The personalization of teaching through AI is emerging as a recent and promising trend, drawing increasing academic attention. According to Abdalla (2024), personalization in tools such as ChatGPT moderates the relationship between perceived ease of use and the intention to adopt such technologies. Students who perceive the technology as user-friendly are more likely to engage with AI platforms, significantly transforming their educational experience (Abdalla, 2024).

Additionally, the self-learning capacity of AI chatbots, as discussed by Jo (2024), plays a crucial role in knowledge acquisition by enabling learning that is more responsive to individual students' needs. Thus, the study centers on two key components: AI as a tool for personalization in higher education and its integration into the university environment as a strategy to promote the SDGs. Beyond enhancing students' educational experiences, personalized teaching enables universities to prepare professionals equipped to address the global challenges outlined by the SDGs. For instance, Rodriguez and Cobeña (2024) demonstrated that the implementation of personalized AI systems resulted in notable improvements in students' academic performance, with an average final score of 96 out of 100 after using AI in office-based learning. This outcome highlights the positive impact of AI on higher education and its relevance in developing the skills needed to confront global problems.

In today's world, tools powered by Artificial General Intelligence (AGI)—such as ChatGPT, DeepSeek, or Google Gemini—can generate ready-made code, teach complex concepts, and train professionals to act as autonomous agents. However, before the advent of these tools, such capabilities required solid technical knowledge acquired through formal education. Today, universities need professionals in computing, statistics, and mathematics with a strong classical foundation in programming languages to exercise the critical thinking necessary to fully harness the potential of AI tools.

Another crucial technical dimension of this study concerns the impact of AI on knowledge management and the resolution of complex problems, both within academic communities and society at large. Hsieh and You (2021), for example, developed evolutionary AI algorithms that optimized road planning during evacuation scenarios in Taiwan, demonstrating AI's capacity to efficiently address complex community issues. Similarly, Gardeazabal et al. (2021) proposed a knowledge management framework focused on agri-food systems, showing how AI can be effectively integrated to overcome challenges in specific sectors such as agricultural innovation.

Therefore, this study investigates the transformative role of AI in higher education, with a focus on personalized learning and its contribution to the Sustainable Development Goals (SDGs), while also considering the challenges and opportunities associated with the implementation of these technologies. Figure 1 presents a mind map that organizes and synthesizes the review's main findings.

2 Methodology

This study adopts a descriptive qualitative approach, aiming to analyze the implementation of artificial intelligence (AI) in universities and its impact on personalized teaching and the advancement of the Sustainable Development Goals (SDGs). The methodological process was structured into three main stages: bibliographic review, keyword selection, and systematic review design. Each of these stages is detailed below to ensure methodological transparency and scientific rigor.

2.1 Bibliographic Review

The bibliographic review was conducted using high-impact academic databases, including Web of Science, Scopus, and SpringerLink, covering publications from 2018 to 2024. The selection of these databases ensures the inclusion of internationally recognized and peer-reviewed studies of high quality. The literature search focused on identifying articles that explore the application of AI in higher education, with particular emphasis on personalized teaching and its connection to the SDGs. The review process followed strict inclusion and exclusion criteria to ensure the analysis of only highly relevant and methodologically sound studies. Priority was given to publications presenting empirical findings and critically examining the integration of digital technologies in university contexts.

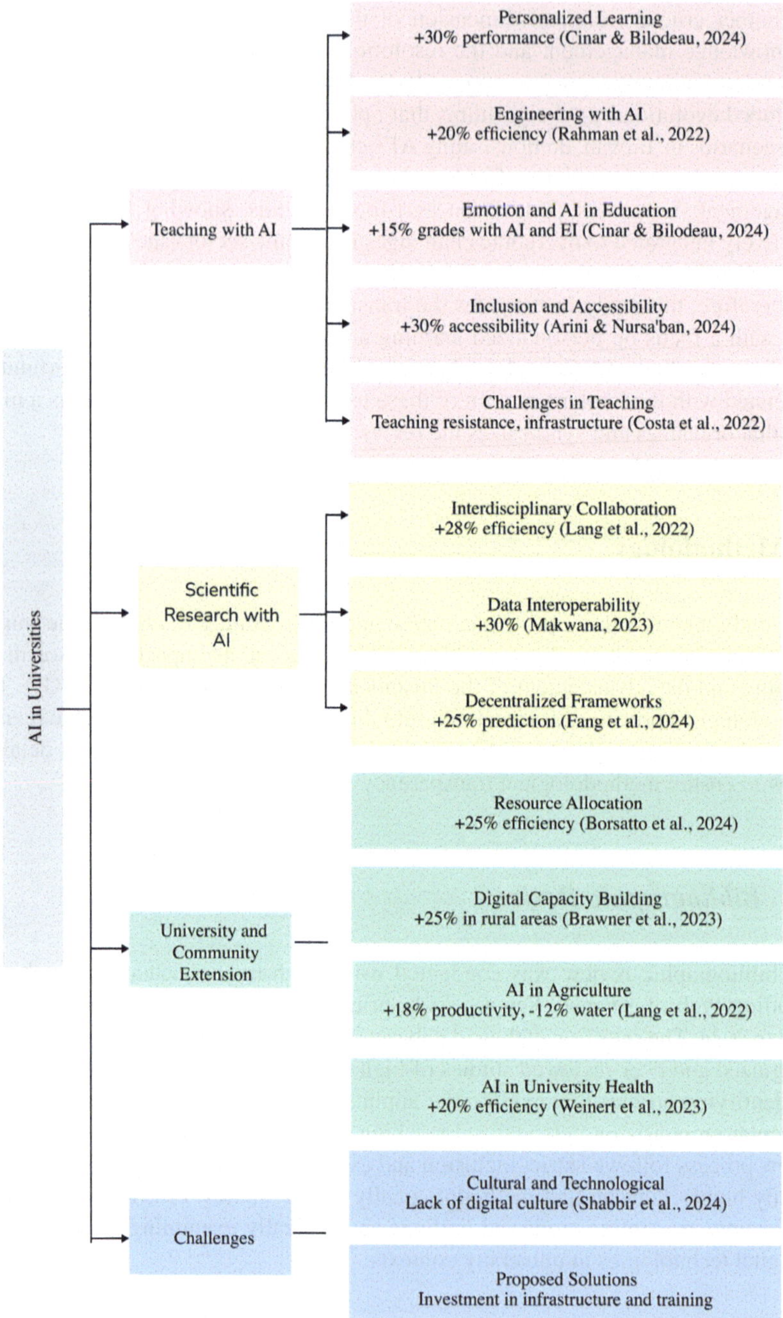

Fig. 1 Impacts of artificial intelligence in different areas: relevant data and results. Designed with Figma

3 Keyword Selection

The selection of keywords was conducted systematically through data mining within the aforementioned databases. The search utilized terms such as "Artificial Intelligence," "Personalized Education," "Sustainable Development Goals," "Higher Education," and "Universities." These keywords were organized into three core thematic categories: (a) emerging technologies in higher education; (b) personalized teaching methodologies; and (c) the impact of AI on the achievement of the SDGs. The selection process was guided by both the frequency of terms in relevant publications and their alignment with the study's objectives. This strategy ensured a comprehensive and targeted bibliographic review, focusing on the most prominent and widely discussed themes in the field.

3.1 Systematic Review Structuring

The systematic review was conducted following the PRISMA (Preferred Reporting Items for Systematic Reviews and Meta-Analyses) protocol to ensure rigor and transparency throughout the study selection process. Widely adopted in systematic reviews, the PRISMA framework guarantees that each phase of the review is executed in a structured and transparent manner. The analysis focused on studies addressing the implementation of AI in higher education and its contributions to achieving the SDGs, considering both the benefits and the associated challenges. Studies were included if they discussed personalized teaching approaches, the promotion of digital inclusion, and the broader process of digital transformation in university settings.

In addition, special attention was given to studies examining institutional and technological barriers to AI integration in universities, particularly those analyzing cultural adaptation and the need for faculty training in the effective use of these tools. The review sought to capture a comprehensive perspective of the literature, identifying not only successful implementations but also the limitations and obstacles faced by higher education institutions worldwide.

4 Results and Discussion

This section is organized into three interrelated parts, each exploring a distinct layer of AI's influence in the academic sphere:

1. **Impact on Teaching and Learning**: This part examines how AI is reshaping personalized education, highlighting academic improvements while also revealing paradoxes related to access inequality, algorithmic bias, and contradictory effects on cognitive development.

2. **Scientific Collaboration and Research**: Here, the role of AI in enhancing interdisciplinary research and data sharing is analyzed, alongside challenges such as institutional constraints, lack of algorithmic transparency, and growing dependence on proprietary models.
3. **University Outreach and Social Impact**: This final part focuses on the use of AI in community-based projects, showing gains in efficiency and digital inclusion, but also addressing issues such as digital exclusion, local mistrust, and sustainability challenges in marginalized contexts.

This tripartite structure enables a progressive analysis—from the classroom to research environments, and ultimately to society—revealing how AI is redefining academia across multiple levels, constantly balancing technological promise with practical and ethical limitations.

From an institutional standpoint, the financial and human resource investments required for sustainable AI integration represent significant challenges. A cost–benefit analysis conducted by Kostoska and Kocarev (2019) on AI curriculum adaptation projects revealed a minimum return-on-investment (ROI) period of three years, with annual maintenance costs consuming 15–20% of the original implementation budget. Such financial realities pose considerable barriers for resource-constrained institutions, potentially reinforcing a new digital divide in global higher education.

The research also points to significant disciplinary variations in AI's effectiveness. While STEM fields consistently report positive outcomes (Rahman et al., 2022; Olubusoye et al., 2023), implementations within the humanities yield more modest results. Kyrpa et al. (2024) found that AI-driven personalization in humanities courses resulted in a 12% improvement in performance, compared to 22% in STEM disciplines—suggesting that different knowledge domains may require context-specific AI strategies.

Moving forward, the path toward responsible AI integration in higher education must address these multidimensional challenges. Evidence from successful cases (Gupta et al., 2020; Sanchez et al., 2023) indicates that hybrid models—combining AI efficiency with human mentorship—yield the most sustainable outcomes. Moreover, the development of culturally aware algorithms and localized training datasets (Arini & Nursa'ban, 2024) may help mitigate algorithmic biases, while robust data governance frameworks (Fang et al., 2024) can address growing concerns around privacy.

Ultimately, while AI offers transformative potential for personalized learning, its implementation demands nuanced, context-sensitive approaches that consider technological limitations, ethical risks, and institutional realities. The future of AI in education will likely be shaped less by technological capabilities alone and more by our collective ability to navigate these complex socio-technical challenges while preserving the humanistic core of higher education.

4.1 Collaborative Research and Data Sharing: The Interdisciplinary Promise and Structural Constraints of AI in Academia

The integration of artificial intelligence (AI) into academic research has significantly enhanced interdisciplinary collaboration, data interoperability, and knowledge production. Empirical studies report that AI-powered platforms have increased research efficiency by 25–30% in fields reliant on large-scale data integration (Fang et al., 2024; Makwana, 2023). Notably, decentralized AI infrastructures have improved predictive modeling capabilities while preserving data privacy—an essential advancement for research involving sensitive information (Fang et al., 2024). Furthermore, the combination of semantic web technologies and AI has increased cross-disciplinary data interoperability by 30%, enabling new connections between disciplines such as engineering, environmental science, and social science (Makwana, 2023). These developments align with SDG 17 (Partnerships for the Goals), as AI promotes global academic cooperation and accelerates progress on complex sustainability challenges.

Despite these advancements, structural and epistemological barriers remain significant. Although AI offers the potential for seamless collaboration, institutional policies frequently inhibit data sharing due to concerns over intellectual property rights, ethical standards, and competitive funding structures (Shu & Gu, 2024). For example, Borsatto et al. (2024) found that although AI optimized resource allocation in over 15,000 university extension projects, only 40% of the institutions involved had adopted standardized data-sharing protocols—thereby limiting scalability. In addition, interdisciplinary AI initiatives often encounter resistance from entrenched academic structures where rigid departmental boundaries hinder integrative efforts (Gaspari et al., 2024). This problem is particularly acute in developing countries, where infrastructure deficiencies further constrain collaborative research capacity (OECD, 2020).

Ethical and Epistemological Challenges in AI-Mediated Research

The growing dependence on AI for data analysis also raises critical ethical concerns, particularly around algorithmic bias and research transparency. Lang et al. (2022) demonstrated that AI models trained on Western-centric datasets produced skewed outcomes when applied to environmental research in Global South contexts, reinforcing longstanding knowledge asymmetries. Similarly, Weinert et al. (2023) observed that AI-enhanced medical research collaborations, while increasing efficiency by 20%, suffered from representational bias in patient data—calling into question the generalizability of their findings.

Moreover, the opaque "black-box" nature of many AI systems complicates peer review and reproducibility—both fundamental principles of academic integrity (Shu & Gu, 2023). Although AI is capable of processing vast and complex datasets, its decision-making logic often lacks transparency, making it difficult for scholars to evaluate the methodological soundness of AI-generated results (Fang et al., 2024).

This issue is particularly pressing in fields such as public health and social sciences, where contextual nuance and interpretability are paramount (Dambha-Miller et al., 2023).

The Role of Startups and Alternative Research Models

Emerging AI startups are increasingly shaping new models of academic collaboration. Santos et al. (2024) reported that AI-powered platforms in urban planning reduced project timelines by 20% by integrating datasets from architecture, economics, and environmental science. Likewise, Brawner et al. (2023) highlighted rural digital inclusion initiatives in which AI-enabled platforms fostered peer-to-peer skill-sharing networks, boosting workforce training efficiency by 25%. These cases illustrate the democratizing potential of AI in expanding participation beyond traditional academic institutions.

However, the growing reliance on private-sector AI tools introduces new challenges, such as data commodification and dependency on proprietary systems. Fırat et al. (2023) warn that such dependence can undermine academic autonomy. Supporting this concern, Silva (2024) found that universities using third-party AI platforms for interdisciplinary research experienced vendor lock-in, ultimately limiting their methodological flexibility and long-term innovation capacity.

Toward a Framework for Sustainable AI Collaboration in Academia

To fully harness the potential of artificial intelligence (AI) in academic research while minimizing associated risks, universities must adopt four interdependent strategies:

- **Standardized Data Governance**: Adoption of the FAIR principles (Findable, Accessible, Interoperable, Reusable) to promote ethical and efficient data sharing practices across institutions (Makwana, 2023).
- **Bias Mitigation Protocols**: Implementation of transparent mechanisms for dataset documentation and algorithmic decision-making processes, aimed at reducing systemic bias (Lang et al., 2022).
- **Hybrid Human-AI Research Models**: Integration of AI tools into research workflows while maintaining critical scholarly oversight to preserve academic rigor and contextual interpretation (Shu & Gu, 2024).
- **Global Equity in AI Access**: Promotion of open-source tools and targeted capacity-building programs to address infrastructural and epistemic disparities in the Global South (OECD, 2020).

While AI has demonstrably enhanced the efficiency of interdisciplinary research, these benefits remain unevenly distributed and are deeply contingent upon addressing ethical, epistemological, and infrastructural barriers. Without systemic reform in data governance, institutional culture, and academic incentive structures, AI-enabled collaboration risks reinforcing existing inequities rather than democratizing knowledge production. Future research must critically assess the long-term implications of AI on scholarly autonomy, inclusivity, and global research parity as the technology becomes increasingly embedded within academic infrastructures.

4.2 Community Outreach and Problem Solving: AI as a Catalyst for Social Impact and Its Operational Constraints

The deployment of AI in university-led community engagement initiatives has shown considerable promise in addressing multifaceted societal challenges—particularly in healthcare delivery, agricultural innovation, and digital literacy. Borsatto et al. (2024) reported that AI-driven resource allocation in 15,000 university extension projects enhanced operational efficiency by 25%, enabling more targeted interventions in underserved areas. These improvements align with SDG 11 (Sustainable Cities and Communities), as demonstrated in urban planning projects by Santos et al. (2024), where AI integration led to a 20% increase in infrastructure management efficiency. Furthermore, the ability of AI systems to process heterogeneous data streams has proven invaluable in rural settings, with Brawner et al. (2023) documenting a 25% improvement in digital skills training through AI-enhanced educational platforms.

Nonetheless, these successes obscure critical implementation challenges. In university hospitals, Weinert et al. (2023) observed a paradoxical effect: while AI improved the efficiency of healthcare interventions by 20%, it simultaneously introduced technological dependencies that undermined long-term sustainability in resource-constrained environments. This finding echoes broader critiques of technological solutionism, where AI-driven interventions overlook local sociocultural dynamics (Semwal et al., 2024). A case in point is the agricultural sector: although Lang et al. (2022) found an 18% increase in crop yields through AI-optimized farming, 62% of smallholder farmers lacked the digital literacy required to sustain these systems independently.

Ethical complexities also arise in AI-mediated community engagement. Shu and Gu (2021) revealed significant disparities in digital inclusion efforts: urban programs experienced 30% growth in participation, whereas rural engagement plateaued at 12%, deepening the digital divide. This asymmetry exemplifies what Silva (2024) defines as "algorithmic peripheralization," wherein AI tools optimized for resource-rich contexts fail to translate effectively in marginal environments. Fırat's (2023) study on blockchain-based community governance further highlighted this issue, finding that 45% of participants distrusted AI-mediated decision-making, raising concerns over transparency and social legitimacy.

Operational barriers compound these ethical concerns. Kans et al. (2020) identified three structural constraints in AI-driven industrial maintenance projects: (1) infrastructure limitations affecting 68% of sites; (2) data scarcity in 55% of cases; and (3) institutional resistance reported by 42% of stakeholders. These challenges are even more pronounced in Global South contexts. Gardeazabal et al. (2021) found that only 28% of communities in agri-food systems possessed the technical capacity to maintain AI interventions without ongoing external support. Moreover, temporal data from Becerra and Paganoni (2023) revealed that while urban development projects experienced immediate 25% efficiency gains with AI adoption, those gains declined over a 24-month period—suggesting limits to the technology's transformative potential in complex and evolving social systems.

Emerging practices point toward more sustainable and participatory models of implementation. Semwal et al. (2024) documented that their "participatory AI" framework—used in rural tourism projects—reduced environmental impacts by 20% while enhancing community ownership through co-design processes. Similarly, Lang et al. (2023) reported that hybrid human-AI systems in waste management achieved 15% higher long-term adoption rates than fully automated solutions. These cases underscore the necessity of context-sensitive adaptation, where AI acts as an augmentative tool rather than a substitute for localized knowledge and community agency.

From these experiences, four guiding principles emerge for the responsible use of AI in community outreach:

1. **Contextual Embeddedness**: Aligning technological interventions with local epistemologies and sociocultural realities (Gardeazabal et al., 2021).
2. **Capacity-Building Infrastructure**: Developing sustainable technical support systems tailored to community contexts (Kans et al., 2020).
3. **Participatory Governance**: Incorporating community stakeholders into the design and deployment of AI systems (Semwal et al., 2024).
4. **Impact Monitoring**: Establishing longitudinal frameworks to evaluate both short- and long-term effects (Becerra & Paganoni, 2023).

Going forward, universities must reimagine their role as intermediaries between technological advancement and community well-being. Rather than positioning AI as a universal solution, institutions should adopt critical, inclusive implementation frameworks that recognize both the promises and limitations of these technologies. This demands a commitment to collaborative engagement throughout the AI lifecycle—from problem definition to post-deployment evaluation—ensuring that digital interventions are not only efficient but also equitable and socially grounded.

5 Conclusion

This study has critically examined the role of artificial intelligence (AI) in reconfiguring the university's core functions—teaching, research, and community outreach—in alignment with the Sustainable Development Goals (SDGs). Rather than presenting AI as a neutral or universally beneficial tool, the analysis underscores its dual nature: a transformative catalyst with the potential to enhance learning, foster interdisciplinary collaboration, and optimize community interventions, but also a vector of new inequalities, dependencies, and ethical dilemmas.

The findings suggest that AI's impact in higher education is neither uniform nor deterministic. Its effectiveness is shaped by institutional preparedness, disciplinary contexts, data governance structures, and the sociotechnical environments in which it is deployed. As such, the promise of AI must be understood not solely through metrics of efficiency, but through its ability to foster inclusive, context-sensitive, and ethically grounded practices.

Looking ahead, universities must take on a proactive role as mediators between technological innovation and social responsibility. This involves not only adopting AI tools but shaping the frameworks that govern their design, implementation, and evaluation. Critical areas for future inquiry include long-term impacts on academic autonomy, participatory algorithm design, and the development of open, equitable AI infrastructures—particularly in underserved regions.

Ultimately, the challenge is not simply to incorporate AI into academic life, but to do so in ways that reinforce the democratic, humanistic, and public-oriented missions of higher education. The trajectory of AI in academia will depend less on technical advances and more on the institutional and ethical choices we make today.

6 Study Limitations

This study presents some limitations, such as reliance on available literature, which may restrict the generalizability of the results. The lack of qualitative data on students' and professors' perceptions limits the understanding of their experiences with AI. Moreover, ethical issues related to data collection and use deserve further investigation. Future studies should explore the effective integration of AI in different curricula and its long-term implications for students' academic and professional training.

Empowering Sustainable Leaders: OAU's Role in SDG Education

P. O. Awoniyi, A. B. Atijosan, and O. C. Olajide

Abstract This study investigates the integration of the United Nations Sustainable Development Goals (SDGs) at Obafemi Awolowo University (OAU) during the 2022/2023 academic year. Recognizing the critical role of higher education in addressing global challenges, the research adopts a mixed-methods approach to examine faculty awareness, curriculum integration, and institutional support for the SDGs. Data were collected through a survey of 87 faculty members, semi-structured interviews with 15 key informants, including university leadership, deans, and department heads, and document analysis of university policies, curricula, research reports, and community engagement initiatives. The findings reveal significant challenges, including limited faculty awareness of the SDGs, weak curriculum integration, and insufficient institutional support, hindering meaningful engagement with the SDGs. This study concludes that well structured faculty development programs, curriculum reform, interdisciplinary collaborations, and strengthened university-community partnerships are crucial to enhance the university's role in advancing sustainable development.

Keywords SDG integration · Higher education · Sustainability · Curriculum integration

1 Introduction

The Sustainable Development Goals (SDGs) serve as a global blueprint for human development, equity, and socio-economic transformation. Comprising 17 goals and 169 targets, they build on the Millennium Development Goals (MDGs) to promote fairness, justice, and sustainability (Griggs et al., 2013). The SDGs focus on three key pillars: social inclusion, economic prosperity, and environmental protection (Veeran et al., 2024). Achieving these goals requires global commitment to poverty eradication (Lampert & Papadongonas, 2016). Achieving these goals requires active

P. O. Awoniyi (✉) · A. B. Atijosan · O. C. Olajide
Centre for Gender and Social Policy Studies, Obafemi Awolowo University, Osun, Nigeria
e-mail: awoniyipo@oauife.edu.ng

W. Leal Filho et al. (eds.), *Introducing the UN SDGs in Higher Education Institutions*, World Sustainability Series, https://doi.org/10.1007/978-3-032-23017-1_16

participation from all stakeholders, including governments, institutions, multilateral organizations, and individuals. Individual actions contribute to broader societal shifts, influencing progress toward SDG targets.

Education, particularly higher education, plays a pivotal role in advancing the SDGs. Goal 4, Quality Education, highlights tertiary education as both a service and an investment that should be equitably accessible regardless of demographic differences. Universities contribute through education, research, and fostering social cohesion (Ashida, 2023). They drive innovation, influence behaviour change, and serve as hubs of critical thinking, stability, and creativity (Shiel et al., 2016; Heleta & Bargus, 2021). Higher education institutions (HEIs) inform the public about SDGs, develop skills essential for achieving them, and facilitate interdisciplinary collaboration among scientists, policymakers, and professionals (Unterhalter & Howell, 2021). Despite perceptions of elitism, universities integrate diverse perspectives, advancing SDG implementation at local, national, and global levels.

The SDGs emphasize universality, indivisibility, and integration, positioning universities as both enablers and partners in their attainment. Universities facilitate knowledge acquisition, research, and skills development essential for SDG progress (Boni et al., 2016; McCowan, 2019). Target 4.7 underscores education's role in embedding SDG principles through pedagogy and curricula. Higher education in Nigeria, like in many developing nations, has faced neglect, limiting its contributions to sustainable development (United Nations, 2015). Heleta and Bargus (2021) argue that despite SDG commitments to inclusivity, African universities have been left behind, weakening their role in sustainable development. However, HEIs remain essential stakeholders in SDG attainment, given their research potential and capacity to shape governance and socio-economic transformation (Aina, 2010; African Union, 2016; Okolie, 2003; UNESCO, 2017; Association of Commonwealth Universities, 2015).

Studies on SDG awareness highlight significant gaps in Nigeria (Omisore et al., 2027; Ejechi, 2018; Shehu et al., 2018) and among university students more broadly (Jati et al., 2019). Beyond awareness, research critiques HEIs' alignment with SDG development outcomes, citing weak systems and ineffective implementation strategies (Unterhalter & Howell, 2021). Despite these challenges, universities have initiated sustainability-driven projects, such as waste management systems and research collaborations (AL-Shamsi et al., 2022; Mori Junior et al., 2019). Global examples include Coventry University's low-carbon projects (Trencher et al., 2013), Australian National University's research contributions to SDGs 4, 7, and 13 (Mawonde & Togo, 2019), and Italian universities incorporating SDGs into curricula (Chaleta et al., 2021).

In Africa, HEIs have made strides despite structural challenges. The University of Cape Town advances SDGs 7 and 11 through carbon footprint research (UCT, 2012), while the University of South Africa implements sustainability initiatives in waste management, energy efficiency, and climate change response (Mawonde & Togo, 2019). Similar efforts have been documented in Kenya's formal higher education sector (Kivati, 2016).

Obafemi Awolowo University (OAU), established in 1961, is a leading Nigerian institution with 13 faculties, 118 departments, and 19 research centers. With a vibrant academic community, OAU is well-positioned to contribute to SDG attainment through its specialized centers, such as the Centre for Gender and Social Policy Studies and the Centre for Energy Research and Development. This study critically examines SDG awareness, research, and engagement at OAU, identifying gaps and strategies for enhancing impact.

This study is primarily guided by two interconnected theoretical frameworks: Institutional Theory and Education for Sustainable Development (ESD). These frameworks are chosen because they offer the most direct and relevant lens for understanding the complexities of SDG integration within a university setting like OAU. Institutional theory helps illuminate the macro-level factors, policies, structures, norms, that shape how the university approaches sustainability, while ESD provides the pedagogical framework for understanding how sustainability is translated into concrete educational practices at the micro-level. While other frameworks like transformative learning, social cognitive theory, and systems thinking offer valuable insights, they are considered secondary to the core focus on institutional influences and educational practice in this particular context.

2 Institutional Theory and Education for Sustainable Development in Higher Education

Institutional Theory provides a useful framework for understanding how universities integrate Sustainable Development Goals (SDGs) into their structures, policies, and practices. According to Scott (2014), institutions shape behavior through regulative, normative, and cultural-cognitive elements, which influence how sustainability initiatives are adopted within higher education. In the context of SDG integration, universities respond to both external pressures, such as global sustainability agendas and national policies and internal motivations, including faculty engagement and student-driven sustainability initiatives (Meyer & Rowan, 2006). This interplay of external and internal factors determines how deeply sustainability principles become embedded within university governance and academic programs.

Education for Sustainable Development (ESD) serves as a critical mechanism for operationalizing sustainability within higher education, embedding SDG principles into curricula, research, and community engagement (Sterling, 2010). Many higher education institutions (HEIs) worldwide have embraced ESD frameworks to align their educational strategies with SDG 4 (Quality Education) and SDG 5 (Gender Equality), reinforcing broader sustainability goals (Leal Filho et al., 2020). For example, the University of Gothenburg in Sweden has integrated sustainability across its curricula and established governance mechanisms to track SDG progress (Ramos et al., 2021). Similarly, the University of Cape Town in South Africa has mainstreamed sustainability education through research initiatives and student-led sustainability projects (Wright & Horst, 2013).

A comparative study by Lozano et al. (2019) highlights that universities in Europe, North America, and Asia have made significant strides in institutionalizing sustainability by embedding SDGs into strategic planning, faculty training, and student-led initiatives. However, challenges remain in developing countries, where limited institutional capacity and funding constraints often hinder effective SDG integration (Bessant et al., 2021). These disparities underscore the need for context-sensitive approaches that consider local institutional realities while aligning with global sustainability goals.

By applying Institutional Theory, we can better understand how HEIs internalize global sustainability norms and translate them into localized institutional practices. At the same time, ESD provides a structured approach to embedding sustainability within academic programs, ensuring that SDG integration moves beyond administrative commitments to practical, measurable educational outcomes. Institutional Theory explains the 'why' and 'how' behind universities' sustainability commitments, while ESD clarifies the 'what' and 'how' of educational implementation. Together, these frameworks offer a robust foundation for examining the complex factors influencing SDG implementation within the university context and provide valuable insights for developing effective strategies to promote sustainability.

2.1 Theoretical Framework

While other theoretical frameworks offer valuable perspectives, Institutional Theory and ESD remain the core focus of this study due to their direct relevance to understanding both the structural and pedagogical dimensions of SDG integration. This dual-framework approach enables a comprehensive analysis of how policies, governance structures, and educational strategies intersect to advance sustainability in higher education.

3 Methodology

This study employed a mixed-methods approach, combining quantitative and qualitative data collection and analysis, to comprehensively understand SDG integration at Obafemi Awolowo University (OAU). This approach was chosen because it allows for a more nuanced exploration of the research problem, capturing both the breadth of faculty awareness and the depth of institutional practices and perspectives (Creswell & Plano Clark, 2018).

4 Research Design

The study adopted a convergent parallel design within the mixed-methods framework (Creswell & Plano Clark, 2018). This design involves collecting both quantitative and qualitative data concurrently, analyzing them separately, and then comparing and integrating the findings to create a more complete understanding of the phenomenon under study. This design was deemed appropriate because it allows for a balanced exploration of both quantitative and qualitative aspects of SDG integration at OAU.

5 Data Collection Methods

Data were collected through three primary methods: document analysis, key informant interviews, and faculty surveys.

6 Document Analysis

A range of university documents was analyzed to identify explicit mentions of the SDGs, integration of sustainability principles, and evidence of SDG-related activities. The documents reviewed included:

- The OAU Strategic Plan (2020–2025): This document was examined to understand the university's overall strategic priorities and whether sustainability and the SDGs were explicitly mentioned or incorporated.
- The University Senate's curriculum guidelines: These guidelines were reviewed to determine if they encouraged or mandated the inclusion of sustainability or SDG-related content in academic programs.
- Departmental course outlines: A stratified random sample of 200 course outlines was selected from five faculties (Sciences, Social Sciences, Arts, Agriculture, and Engineering) to assess the actual integration of SDG-related topics, learning objectives, and assessment methods within specific courses. Course outlines were stratified by faculty and academic level (undergraduate/graduate) before sampling.
- The annual research reports (2021 and 2022) from the Central Office of Research (COR): These reports were reviewed to identify research projects related to the SDGs, funding allocated to such projects, and publications resulting from SDG-related research.
- The 2021 and 2022 reports of Community Engagement activities from the Centre for Gender and Social Policy Studies (CGSPS), OAU: These reports were examined to identify community engagement activities that aligned with the SDGs and the university's efforts to connect research and teaching with community needs.

7 Key Informant Interviews

Semi-structured interviews were conducted with 15 key informants purposively selected to represent diverse perspectives within the university. The interviewees were selected to capture a range of institutional perspectives on SDG integration and included:

- The Vice-Chancellor
- Three Deputy Vice-Chancellors (Academic, Administration, and Research)
- Five Deans (one from each sampled faculty)
- Five Heads of Departments (one from each sampled faculty)
- Two Directors of Centres relevant to sustainability, specifically CGSPS, and Institute of Ecology and Environmental Studies.

The interview guide explored awareness and understanding of the SDGs, current initiatives for SDG integration, perceived challenges and barriers, and recommendations for enhancement. The interviews were semi-structured to allow for flexibility and follow-up questions based on the interviewees' responses (Guest et al., 2017). Interviews lasted approximately 45–60 min, were audio-recorded (with participant consent), and transcribed verbatim.

8 Faculty Surveys

A structured online survey was administered to a representative sample of 87 faculty members from the five selected faculties at OAU. A stratified random sampling technique was used to ensure proportional representation from each faculty and academic rank (e.g., Lecturer, Senior Lecturer, Professor). The target sample size was determined based on a power analysis aiming for 95% confidence and a 5% margin of error. The survey instrument consisted of three sections:

- Demographics (department, faculty, years of experience, academic rank)
- SDG Awareness and Knowledge (multiple-choice and Likert-scale questions assessing familiarity with the 17 SDGs and their relevance to teaching and research)
- SDG Integration in Teaching and Research (open-ended and Likert-scale questions exploring the extent to which SDGs are incorporated into course content, teaching methods, research projects, and student supervision)

The survey was distributed via email using Qualtrics, an online survey platform. Two reminder emails were sent to maximize response rates.

9 Data Analysis

9.1 *Quantitative Data Analysis*

Data from the faculty surveys were imported into R version 4.2.1 for analysis. Descriptive statistics (frequencies, percentages, means, medians, standard deviations, ranges) were calculated for all relevant variables. Chi-square tests of independence were performed to examine relationships between categorical variables (e.g., faculty, department, SDG awareness). Independent samples t-tests or ANOVA (Analysis of Variance), along with post-hoc tests (e.g., TukeyHSD), were used to compare means of continuous variables across different groups (e.g., SDG integration scores by faculty). Correlation analysis was performed to examine relationships between continuous variables (e.g., years of experience, SDG integration scores). Visualizations were created using `ggplot2` to present the quantitative findings.

9.2 *Qualitative Data Analysis*

Interview transcripts and document data were imported into Atlas.ti version 9 for qualitative data analysis. A code list was developed based on the interview guide, literature review, and emergent themes from the data. The data were coded systematically by two researchers independently to ensure inter-coder reliability. Inter-coder reliability was assessed using Cohen's Kappa, resulting in a coefficient of 0.82, indicating strong agreement. Discrepancies in coding were discussed and resolved. Atlas.ti's network views and code co-occurrence tools were used to explore relationships between codes and identify patterns in the data. Memos were used to document interpretations, insights, and connections between the qualitative data and the research questions. Quotations from the interviews and document excerpts were exported from Atlas.ti to illustrate key themes and support the findings. Thematic analysis was employed to identify, analyze, and report patterns (themes) within the data (Braun & Clarke, 2006) (see also Miles et al., 2014 for qualitative data analysis approaches).

9.3 *Data Analysis and Integration*

A mixed-methods approach was employed to ensure a comprehensive understanding of how institutional support affects SDG integration at OAU. This involved triangulating qualitative and quantitative data to validate findings and enhance the reliability of conclusions drawn. According to Creswell and Plano Clark (2018), integrating both qualitative and quantitative data allows researchers to gain deeper insights by corroborating trends identified in numerical data with contextual explanations provided through interviews and focus groups.

Quantitative data were primarily derived from faculty and administrative surveys, measuring perceptions of institutional support, policy effectiveness, and resource allocation for SDG-related initiatives. These numerical data provided patterned trends and statistical associations (Bryman, 2016). However, to complement and deepen these insights, qualitative data from in-depth interviews and focus group discussions were analyzed thematically using Atlas.ti. This qualitative analysis provided contextual narratives explaining why faculty members perceived institutional support as inadequate, offering richer perspectives on barriers such as lack of interdisciplinary collaboration, limited funding, and policy constraints (Leal Filho et al., 2015).

To ensure methodological rigor, both datasets were systematically integrated using data convergence techniques (Plano Clark & Ivankova, 2016). Areas of agreement between qualitative and quantitative findings strengthened the reliability of our conclusions, while discrepancies were further examined to uncover underlying complexities. For instance, while survey results indicated that a majority of faculty perceived institutional policies as inadequate, qualitative responses revealed nuanced challenges such as decentralized budgeting structures and misalignment between university priorities and sustainability goals.

Additionally, potential biases were addressed through several methodological safeguards. First, the use of multiple data sources mitigated the risk of single-method bias (Denzin, 2012). Second, a reflexive approach was employed to acknowledge researcher positionality and its potential impact on data interpretation (Berger, 2015). Third, the member-checking technique was applied, wherein selected participants reviewed preliminary findings to ensure accurate representation of their perspectives (Lincoln & Guba, 1985).

By adopting this systematic integration of qualitative and quantitative findings, this study provides a robust and holistic analysis of SDG integration at OAU, ensuring that institutional challenges are not only quantified but also understood within their broader socio-political and organizational contexts.

9.4 Ethical Considerations

This study was conducted by ethical research principles. Informed consent was obtained from all participants before data collection. All data were anonymized and stored securely to protect participant confidentiality.

10 Results

10.1 Individual Level (Faculty)

10.1.1 Faculty SDG Awareness and Understanding

The survey results, analyzed using R, assessed faculty awareness and understanding of the UN SDGs among the 87 faculty members surveyed. The results (Fig. 1) revealed that while a clear majority (59.8%) reported having heard of the SDGs in general, a substantially smaller proportion (25.3%) could correctly identify at least seven of the 17 goals. Even fewer faculty members (11.5% and 3.4%) could identify 10 or more and 13 or more goals, respectively. This significant difference between general awareness and detailed knowledge highlights a potential barrier to effective SDG integration in teaching and research at OAU. The findings suggest that while faculty are generally familiar with the concept of sustainable development, a deeper understanding of the specific SDGs and their interconnectedness is limited. This knowledge gap underscores the need for targeted interventions, such as faculty development programs, to enhance understanding of the SDGs and their relevance to various disciplines."

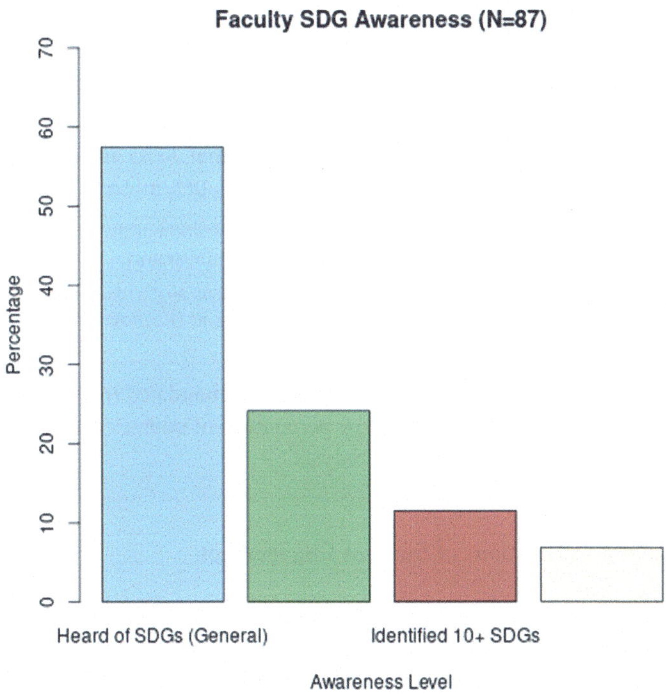

Fig. 1 Faculty SDG awareness

Many faculty members acknowledged a lack of in-depth knowledge about the SDGs.

> I've heard the term 'SDGs,' and I know they're about sustainability, but I couldn't tell you what they all are off the top of my head. (Senior Lecturer/Faculty Member, Sciences)

> I think we need more training on the specific SDGs and how they relate to our disciplines. I'm not sure how I could incorporate them into my teaching. (Professor/Faculty Member, Arts)

These quotes highlight the gap between general awareness and detailed knowledge, which is consistent with the quantitative finding that a smaller proportion of faculty could identify a substantial number of SDGs.

10.1.2 SDG Integration into the Curriculum

This analysis examined the relationship between faculty area and SDG integration into the curriculum. The survey revealed that 72.4% of the 87 faculty members reported no SDG integration, 19.5% reported some integration, and 8.0% reported significant integration. A Chi-squared test was conducted to explore the relationship between SDG integration level and faculty area. The results showed a statistically significant association (χ^2 (4) = 12.7, p =.013). Post-hoc analysis revealed that faculty in the Social Sciences were more likely to report some level of SDG integration compared to faculty in the Sciences and Arts. These findings suggest that SDG integration is not uniform across disciplines and that targeted interventions may be needed to promote integration in specific areas (see Fig. 2).

The document analysis showed that only 18% of sampled course outlines explicitly integrated SDG-related content. This suggests that SDG integration into the curriculum is still in its early stages. The qualitative data reinforces this finding.

> I've looked at a few course outlines, and there's not much mention of the SDGs. I think we need to be more intentional about incorporating them. (Administrator)

> There's no university-wide requirement to include SDG content, so it's really up to individual faculty members. And, as you can see, not many are doing it. (Lecturer/Faculty Member, Sciences)

These interviews point to the lack of systematic or mandated integration of SDGs into the curriculum, which mirrors the low percentage of courses found to explicitly include SDG content in the document analysis.

10.1.3 Faculty Perceptions of Student Engagement

Faculty perceptions of student engagement in SDG-related activities. The survey of 87 faculty members revealed that a minority (25.3%) perceived that they actively involved students in SDG-related projects or activities. A substantial majority (74.7%) of faculty, however, reported not actively involving students in such activities. This disparity suggests a potential disconnect between faculty awareness of the

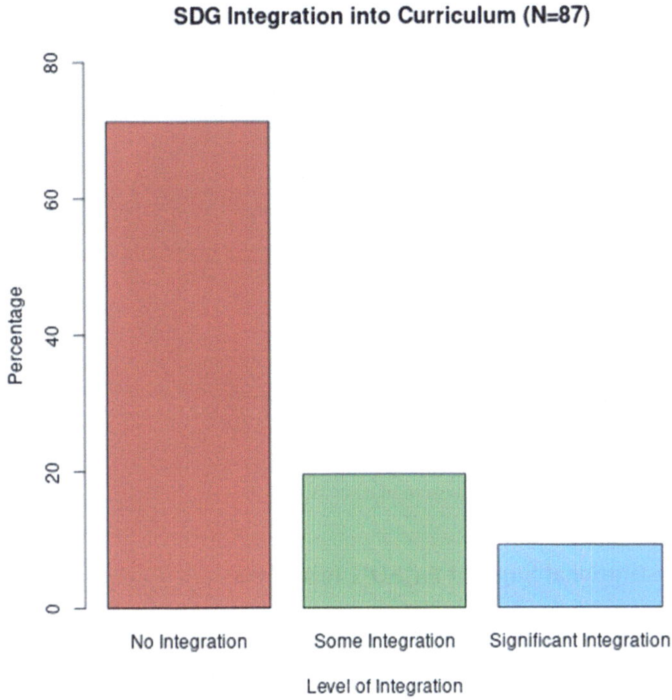

Fig. 2 SDG Integration into Curriculum by Faculty

SDGs and their translation into practical student engagement within the classroom or research settings at OAU. The findings indicate that while some faculty members are actively engaging students with SDG-related work, a large proportion are not, raising questions about the factors that may be contributing to this gap. Further investigation is needed to explore the reasons behind this perceived lack of student engagement and to identify strategies for promoting greater involvement in SDG-related activities."

The qualitative data further illuminates this point.

> I'd love to get students more involved in sustainability projects, but I don't have the resources or the time to develop those kinds of activities. (Senior Lecturer, Engineering)

> Students are interested in these issues, but I haven't found a way to meaningfully integrate them into my courses. (Professor, Social Sciences)

These quotes from faculty members reflect the challenges faculty face in engaging students with SDG-related activities, which aligns with the low percentage of faculty reporting active student involvement in the quantitative data (see Fig. 3).

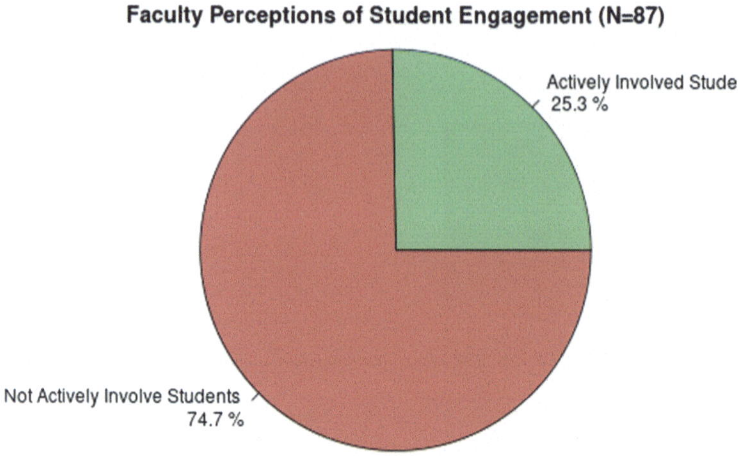

Fig. 3 Faculty involvement of students in SDG activities

10.1.4 Institutional Support for SDG Initiatives

While direct measures of institutional support were not quantified, related indicators suggest the presence of systemic constraints. As shown in Fig. 3, a substantial majority of faculty (74.7%) reported not actively involving students in SDG-related activities, compared to only 25.3% who reported active involvement. This pattern serves as an indirect but important proxy for institutional limitations, particularly in relation to funding, policy guidance, and structural coordination.

These findings suggest that, although there is some level of awareness of the SDGs among faculty, the absence of an enabling institutional environment constrains their effective implementation. Specifically, the lack of dedicated funding, clear institutional frameworks, and coordinated support mechanisms appears to limit the extent to which faculty can translate SDG knowledge into practice. This underscores the critical role of institutional structures in shaping sustainability engagement within higher education contexts. The qualitative data provides further context for these findings.

The qualitative data provides further context for these findings.

> We talk a lot about sustainability, but there's no money to back it up. If we're serious about the SDGs, we need to invest in them. (Senior Lecturer/Faculty Member, Social Sciences)

> The university hasn't provided any guidance on how to integrate the SDGs into our work. We're kind of on our own. (Lecturer/Faculty Member, Arts)

These excerpts illustrate the perceived lack of concrete institutional support, particularly in terms of funding and strategic direction for SDG initiatives. This qualitative evidence reinforces the quantitative pattern observed in Fig. 3, where low levels of faculty engagement in SDG-related activities point to broader structural and institutional constraints.

10.1.5 Challenges and Opportunities for SDG Integration

Challenges:

The most frequently cited challenge was limited resources. Faculty members expressed concerns about the lack of dedicated funding for SDG-related research and teaching initiatives. As one faculty member noted,

> We talk a lot about sustainability, but there's no money to back it up. If we're serious about the SDGs, we need to invest in them. ((Professor/Faculty Member, Arts)

Time constraints were another significant concern, with faculty members reporting heavy teaching loads and administrative responsibilities that made it difficult to dedicate time to SDG-related activities. Inadequate infrastructure, such as limited access to research equipment and technology, was also mentioned as a barrier. A second prominent challenge was lack of awareness about the SDGs among some faculty and staff. Several interviewees acknowledged a need for more training and information on the specific goals and how they relate to different disciplines.

> I've heard the term 'SDGs,' and I know they're about sustainability, but I couldn't tell you what they all are off the top of my head, ((Professor/Faculty Member, Science)

Finally, siloed departments presented a challenge to interdisciplinary collaboration, which is essential for addressing the interconnected nature of the SDGs. "Breaking down the silos between departments is crucial, but it's not easy," commented an administrator.

Opportunities:

Despite these challenges, the study also identified several promising opportunities for advancing SDG integration at OAU. Student enthusiasm for sustainability was a recurring theme. Many faculty members noted the strong interest among students in environmental and social issues, suggesting that tapping into this enthusiasm could be a powerful driver of change.

> Students are really passionate about these issues. We need to tap into that energy, (Administrator/Faculty Member, Arts)
>
> OAU's existing expertise in related fields, such as environmental science and development studies, was also identified as a significant asset.
>
> We have a strong program in environmental science, which provides a great foundation for SDG-related work, ((Lecturer/Faculty Member, Arts))

Finally, the potential for partnerships with NGOs, government agencies, and other universities was seen as a valuable opportunity to leverage external resources and expertise (Holt et al., 2016).

> Collaborating with organizations working on the ground could provide valuable real-world experience for our students and faculty, (Administrator/Faculty Member, Arts)

These findings highlight the complex landscape of SDG integration at OAU, where significant challenges coexist with promising opportunities. Addressing the resource constraints and awareness gaps while leveraging student enthusiasm, existing expertise, and potential partnerships will be crucial for OAU to effectively advance its sustainability agenda. 4.

11 Discussion

This study investigated the integration of the UN Sustainable Development Goals (SDGs) at Obafemi Awolowo University (OAU) using a mixed-methods approach. The findings offer a nuanced understanding of the current state of SDG engagement, highlighting both challenges and opportunities. This discussion section interprets these findings through the lens of Institutional Theory and Education for Sustainable Development (ESD), connecting them to relevant literature and theoretical frameworks.

12 Faculty Awareness and Understanding: Building a Foundation for Action

The finding that a majority of faculty members had heard of the SDGs, but a significantly smaller proportion could correctly identify a substantial number of goals, reveals a crucial gap between awareness and actionable knowledge. This aligns with previous research highlighting the challenge of moving beyond superficial familiarity to a deeper understanding of the SDGs' interconnectedness and specific targets (Hopkins & McKeown, 2012). For example, a study by Sterling (2001) emphasized the importance of transformative learning for fostering genuine understanding of sustainability issues, a concept echoed in our findings. This knowledge deficit can hinder effective integration of SDGs into teaching and research, as faculty may lack the necessary understanding to connect their disciplines to the broader sustainability agenda. As Mezirow's (2000) transformative learning theory suggests, true understanding requires critical reflection and perspective transformation. Faculty need opportunities to critically examine their own assumptions and beliefs about sustainability and explore how their disciplines can contribute to achieving these global goals. The qualitative data reinforces this point, with faculty expressing a need for more training and resources to enhance their SDG literacy. This underscores the importance of targeted faculty development programs that not only provide information about the SDGs but also facilitate critical reflection and perspective transformation, empowering faculty to become agents of change in promoting sustainability within their respective fields (Sterling, 2001). This also connects to the ESD framework (UNESCO, 2017), which emphasizes the development of competencies for critical reflection on sustainability issues. Without a solid foundation of faculty knowledge and understanding, the potential for meaningful SDG integration remains limited.

13 Curriculum Integration: From Policy to Practice

The finding that SDG integration into the curriculum remains limited underscores the challenge of translating institutional rhetoric into concrete action. This echo broader research indicating that SDG integration in higher education often remains fragmented and superficial (Sleurs, 2019). Our findings are consistent with those of Amoros Molina et al. (2023), who reported they found limited information in low- and middle income countries (LMICs) (in which Nigeria is one) on possible methods of integrating the SDGs into higher education. They concluded that this might be as a result of to a lack of funding, documentation, awareness, or recognition associated with the importance of integrating the SDGs in higher education. While OAU's strategic plan may acknowledge the importance of sustainability, the low level of curricular integration suggests a need for more concrete mechanisms to bridge the gap between policy and practice. This aligns with Institutional Theory (DiMaggio & Powell, 1983; Scott, 2008), which emphasizes the influence of institutional structures and norms on organizational behaviour. Without clear institutional direction, support, and incentives, SDG integration is likely to remain ad hoc and dependent on the individual initiatives of a few faculty members. The qualitative data further reinforces this point, with faculty expressing a lack of clear guidelines and support for integrating SDGs into their courses. As Læssøe et al. (2009) argue, ESD requires a holistic, interdisciplinary approach, and OAU should strive to create opportunities for cross-disciplinary collaboration and knowledge sharing (see also Lang et al., 2012 on transdisciplinary research frameworks). The observed variation in integration across faculties suggests that disciplinary differences may play a role, with some disciplines perceived as more readily aligned with SDG themes. However, it is crucial to recognize that all disciplines have a role to play in achieving the SDGs, and OAU should encourage and support SDG integration across the entire curriculum.

14 Institutional Support: Creating an Enabling Environment

The perception among faculty that institutional support for SDG initiatives is inadequate, particularly concerning funding, policies, resources, and interdisciplinary collaboration, presents a major obstacle. This aligns with research indicating that insufficient institutional support is a significant barrier to HEIs seeking to advance sustainability (Læssøe et al., 2009). Our findings corroborate those of Leal Filho et al., 2015, they reported in their study that there is a lack of policies and operational readiness, which shows inadequate institutional support. They further reiterated that due to lack of enabling environment, the obstacles such as lack of resources to support the interdisciplinary approach and system of rewards and academic support, a contrasting culture between each discipline, and procedures and departmental policies and strategies of decentralised budget must be overcome (Leal Filho et al., 2015).

The qualitative data reinforces this point, with faculty expressing concerns about limited resources, lack of clear guidelines, and difficulties in collaborating across departments. As Scott's (2008) institutional theory suggests, universities are influenced by both internal and external pressures. If OAU is to effectively integrate SDGs, it needs to create an internal environment that values and supports sustainability initiatives. Institutional support plays a pivotal role in shaping how universities integrate Sustainable Development Goals (SDGs) into their structures, policies, and academic programs. The presence of clear policies, incentives, and capacity-building mechanisms enhances the sustainability efforts of higher education institutions (HEIs) (Scott, 2014). Universities that prioritize sustainability through top-down institutional strategies create enabling environments where SDG-related initiatives can thrive. These mechanisms include policy incentives, faculty development programs, and funding allocations that support sustainability-related teaching, research, and community engagement (Leal Filho et al., 2020).

One key mechanism of institutional support is policy incentives, which align university governance structures with global sustainability agendas. This can take the form of strategic plans that embed SDG targets into institutional missions, funding schemes that reward SDG-aligned research, and performance metrics that assess faculty contributions to sustainability initiatives (Meyer & Rowan, 2006). For instance, some universities provide internal research grants to encourage interdisciplinary studies on sustainability, while others integrate SDG criteria into faculty promotion and tenure evaluations (Lozano et al., 2019). These strategies incentivize scholars to engage with sustainability themes, ensuring that SDG integration extends beyond administrative commitments to concrete academic contributions.

Another essential institutional mechanism is faculty development, which strengthens educators' capacity to integrate SDG principles into their teaching and research (Sterling, 2010). Faculty training programs, workshops, and interdisciplinary collaboration networks provide educators with the necessary skills and resources to embed sustainability in their curricula (Wright & Horst, 2013). For example, some universities have established sustainability training hubs where faculty members can share best practices and develop new pedagogical approaches for SDG-related education (Bessant et al., 2021). Additionally, mentorship programs and sustainability fellowships enable educators to gain hands-on experience in designing courses that incorporate SDG concepts, thereby institutionalizing sustainability education across disciplines.

Beyond policy incentives and faculty development, universities can also foster SDG integration by allocating dedicated financial and infrastructural support for sustainability programs (Ramos et al., 2021). The establishment of sustainability offices, centers for SDG research, and partnerships with external stakeholders enhances the institutionalization of sustainability efforts (Krizek et al., 2012). The University of Gothenburg, for instance, has successfully embedded SDGs into its academic operations through a combination of governance frameworks, faculty incentives, and student-led sustainability initiatives (Leal Filho et al., 2020). Similarly, the University of Cape Town has implemented policies that integrate sustainability indicators into performance assessments, ensuring accountability in SDG-related initiatives (Wright & Horst, 2013).

By implementing these institutional mechanisms, universities not only strengthen their contributions to the SDGs but also ensure that sustainability becomes a core element of academic culture rather than an isolated initiative. Policy incentives drive structural change, faculty development fosters knowledge dissemination, and financial investments create long-term sustainability pathways. Thus, a comprehensive institutional approach is essential for embedding SDG principles into higher education in a meaningful and impactful way.

15 Faculty Involvement of Students: Engaging the Next Generation

The finding that a relatively small proportion of faculty actively involve students in SDG-related projects or activities suggests a potential disconnect between faculty awareness and action, as well as a missed opportunity to engage students in meaningful ways with sustainability issues. This may be due to a variety of factors, including limited resources, time constraints, or a lack of awareness among faculty about how to effectively engage students in SDG-related work. However, as Brundiers et al. (2021) argue, ESD emphasizes empowering learners to become active participants in creating a sustainable future. Providing students with opportunities to apply their knowledge and skills to real-world problems is essential for fostering a sense of agency and empowering them to become effective agents of change. The qualitative data revealed that many faculty members are keen to involve students in SDG-related projects but face resource constraints and a lack of clear guidance on how to do so. This highlights the need for OAU to provide faculty with resources, training, and support to develop and implement effective student engagement strategies. This engagement is a critical driver of SDG integration in higher education, as active participation fosters a culture of sustainability and empowers students as agents of change (Leal Filho et al., 2020). To strengthen student involvement, universities can adopt three key strategies: curriculum integration, extracurricular initiatives, and institutional incentives. First, embedding SDG-related content into coursework through problem-based learning (PBL) and interdisciplinary collaboration encourages students to apply sustainability principles in real-world contexts

(Sterling, 2010). For example, universities like the University of Gothenburg incorporate sustainability challenges into research projects, enabling students to address pressing societal issues (Ramos et al., 2021).

Second, fostering student-led initiatives through SDG clubs, innovation hubs, and service-learning projects provides experiential learning opportunities. Studies show that participatory approaches, such as campus sustainability campaigns and partnerships with local communities, enhance student motivation and deepen their commitment to sustainable development (Wright & Horst, 2013).

Finally, providing institutional incentives, such as scholarships, awards, and credit recognition for SDG-related projects, encourages greater student engagement. Universities with strong sustainability cultures often establish mentorship programs and funding schemes to support student-driven research and advocacy (Lozano et al., 2019). By implementing these strategies, higher education institutions can create an enabling environment where students actively contribute to advancing the SDGs, ensuring a long-term sustainability impact.

16 Challenges and Opportunities: Navigating the Path Forward

The study identified a range of challenges and opportunities for enhancing SDG integration at OAU. The challenges, limited resources, lack of awareness, and difficulties with interdisciplinary collaboration, are consistent with the findings discussed above and are frequently cited in the literature as barriers to SDG integration in higher education (Hopkins & McKeown, 2012; Sleurs, 2019). These challenges highlight the need for a multi-faceted approach that addresses both individual and institutional factors. However, the study also revealed promising opportunities, including high student interest in sustainability, existing university strengths in related fields, and the potential for partnerships with external organizations. These opportunities suggest that OAU has a strong foundation upon which to build its SDG integration efforts. As DiMaggio and Powell (1983) suggest, organizations often adopt practices and structures that are considered legitimate and appropriate within their institutional field. By positioning itself as a leader in SDG integration, OAU can not only contribute to a more sustainable future but also enhance its reputation and attract students, faculty, and funding opportunities. Connecting research activities directly to sustainable development outcomes is a key strategy for universities (Walshe & Schultz, 2010). This also aligns with the broader societal expectations regarding sustainability, which influence how universities operate (Scott, 2008). By capitalizing on student enthusiasm, leveraging existing expertise, and forging strategic partnerships, OAU can create a synergistic effect, accelerating progress toward a more sustainable university and empowering future generations of leaders who are equipped to address the complex challenges of the twenty-first century.

17 Conclusion

This study has provided valuable insights into the state of SDG integration at Obafemi Awolowo University, highlighting both achievements and persistent challenges. The findings reveal a significant gap between faculty awareness and their understanding of the SDGs, which limits the potential for meaningful integration into the curriculum. The low level of SDG integration in course outlines and teaching practices underscores the need for a deliberate and well-structured institutional approach. Moreover, inadequate faculty engagement of students in SDG-related projects highlights missed opportunities for cultivating sustainability-driven leadership among the next generation. The absence of institutional support, particularly in terms of funding, interdisciplinary collaboration, and administrative guidance, further compounds these challenges. However, these challenges are not insurmountable. By leveraging existing strengths, such as strong student interest, faculty expertise in relevant fields, and potential partnerships with external organizations, OAU can transform its approach to SDG integration. To accelerate progress toward sustainable education, the university must strategically harness these opportunities and strengthen institutional frameworks. These findings contribute to the broader discourse on Education for Sustainable Development and provide critical implications for policy and practice in higher education settings across Nigeria and beyond. With decisive action, OAU has the potential to lead by example, setting a benchmark for SDG integration in Nigerian higher education.

18 Recommendations

To effectively address the challenges identified in this study, a multi-dimensional and strategic approach is essential. Faculty, administrators, and policymakers must work together to foster systemic SDG engagement at OAU through comprehensive interventions in faculty development, institutional support, curriculum reform, interdisciplinary collaboration, and monitoring mechanisms. A key priority is faculty development and capacity building, which requires continuous professional development programs to enhance faculty awareness and understanding of the SDGs across all disciplines. Integrating SDG-related training into faculty induction and existing teaching enhancement programs will ensure that educators are equipped with the necessary knowledge and skills to incorporate sustainability into their teaching. Additionally, developing interdisciplinary teaching modules and resources will facilitate the seamless integration of SDGs into various courses, fostering a more holistic approach to sustainable education.

However, faculty training alone is not enough. Institutional support and clear policies are necessary to sustain and scale these efforts. OAU should formulate policies that mandate and incentivize SDG inclusion in curriculum development and teaching practices. To support this initiative, dedicated funding should be allocated for SDG-related research, teaching, and community engagement activities. Additionally, an SDG Integration Office should be established to track faculty-level implementation and coordinate sustainability efforts across disciplines, while a high-level SDG Task Force within the university's leadership should oversee policy implementation, funding allocation, and long-term institutional sustainability. Together, these structures will ensure that sustainability remains a core priority in OAU's academic and administrative strategies. To enhance curriculum reform and student engagement, SDG themes should be embedded as core components in both general education and discipline-specific courses. Faculty should be encouraged to adopt experiential learning approaches, such as service-learning projects, sustainability challenges, and community-based research, to provide students with hands-on engagement in sustainability issues. Additionally, mentorship programs should be developed, where students can collaborate with faculty on SDG-focused research and initiatives, fostering leadership and critical thinking around sustainable development.

Fostering interdisciplinary collaboration and research is another crucial step toward SDG integration. OAU should promote interdisciplinary research clusters that address SDG-related challenges and encourage cross-departmental collaboration. Strengthening partnerships with external organizations, including government agencies, NGOs, and the private sector, will also support applied research and community-driven solutions. Findings from these interdisciplinary collaborations should directly inform institutional SDG policies, ensuring that research translates into actionable strategies. Finally, monitoring, evaluation, and knowledge sharing mechanisms must be established to track progress and enhance institutional accountability. Periodic assessments, along with faculty and student feedback, will provide valuable insights into the effectiveness of SDG integration efforts. Additionally, OAU should create knowledge-sharing platforms, such as conferences, workshops, and publications, to highlight best practices in SDG education. A publicly accessible repository of SDG-related courses, research outputs, and case studies should also be developed to promote institutional learning and facilitate widespread adoption of sustainability initiatives.

By implementing these strategic interventions, OAU can position itself as a leader in sustainable education and make significant contributions toward the realization of the UN Sustainable Development Goals within the Nigerian higher education sector.

References

African Union. (2016). Continental education strategy for Africa: 2016–2025. Addis Ababa: African Union.

Aina, T. A. (2010). Beyond reforms: The politics of higher education transformation in Africa. *African Studies Review., 53*(1), 21–40.

AL-Shamsi, M., Al-Ali, A. K., & Yi, C. (2022). A review of sustainable campus initiatives in higher education: Motivations, implementation, and assessment. *Journal of Cleaner Production, 370*, 133513.

Amoros Molina, A., Hellden, D., Alfven, T., Niemi, M., Leander, K., Nordenstedt, H.,... & Biermann, O. (2023). Integrating the United Nations sustainable development goals into higher education globally: a scoping review. *Global Health Action, 16*(1), 2190649.

Ashida, A. (2023). The role of higher education in achieving the sustainable development goals. In S. Urata, K. Kuroda, Y. Tonegawa (eds.), *Sustainable Development Disciplines for Humanity*. Sustainable Development Goals Series. Springer, Singapore. https://doi.org/10.1007/978-981-19-4859-6_5

Association of Commonwealth Universities. (2015). Progress and potential: higher education playing its part in the sustainable development goals. Association of Commonwealth Universities.

Berger, R. (2015). Now I see it, now I don't: Researcher's position and reflexivity in qualitative research. *Qualitative Research, 15*(2), 219–234. https://doi.org/10.1177/1468794112468475

Bessant, S., Robinson, Z., & Ormerod, R. (2021). Sustainability in higher education: Institutional challenges and opportunities. *Higher Education Quarterly, 75*(3), 321–340.

Boni, A., Lopez-Fogues, A., & Walker, M. (2016). Higher education and the post-2015 agenda: A contribution from the human development approach. *Journal of Global Ethics, 12*(1), 17–28.

Braun, V., & Clarke, V. (2006). Using thematic analysis in psychology. *Qualitative Research in Psychology, 3*(2), 77–101.

Brundiers, G., Wiek, A., & Redman, A. (2021). Education for sustainable development goals: Transformative potential and implementation challenges. *Sustainability Science, 16*, 831–846.

Bryman, A. (2016). *Social research methods*. Oxford University Press.

Chaleta, E., Saraiva, M., Leal, F., Fialho, I., & Borralho, A. (2021). Higher education and sustainable development goals (SDG)—potential contribution of the undergraduate courses of the school of social sciences of the University of Évora. *Sustainability, 13*(4), 1828. https://doi.org/10.3390/su13041828

Creswell, J. W., & Plano Clark, V. L. (2018). *Designing and conducting mixed methods research*. Sage Publications.

Denzin, N. K. (2012). Triangulation 2.0. *Journal of Mixed Methods Research, 6*(2), 80–88.https://doi.org/10.1177/1558689812437186

DiMaggio, P. J., & Powell, W. W. (1983). The iron cage revisited: Institutional isomorphism and collective rationality in organizational fields. *American Sociological Review, 48*(2), 147–160. https://doi.org/10.2307/2095101

Ejechi, V. (2018). Awareness and Perception of Sustainable Development Goals (SDGs), among Library Personnel in Edo state University Library. https://www.researchgate.net/publication/323391987

Griggs, D., Stafford-Smith, M., Gaffney, O., Shyamsundar, P., Noble, I., Ashford, E.,... & Winkler, H. (2013). Policy: Sustainable development goals for people and planet. *Nature, 495*(7441), 305–307.

Guest, G., Namey, E., & McKenna, K. (2017). How many interviews are enough?: An experiment with data saturation and variability. *Field Methods, 29*(1), 59–82.

Heleta, S., & Bagus, T. (2021). Sustainable development goals and higher education: Leaving many behind. *Higher Education, 81*, 163–177. https://doi.org/10.1007/s10734-020-00573-8

Holt, C. L., Ahern, M., & Lattuca, L. R. (2016). Connecting universities and communities for sustainable development: A review of the literature. *Journal of Higher Education Outreach and Engagement, 20*(1), 1–24.

Hopkins, C. A., & McKeown, R. (2012). *Education for sustainable development: An international perspective*. Routledge.

ICSU & ISSC. (2015). *Review of Target 17.16: Enhance the Global Partnership for Sustainable Development, to complement the efforts of developing countries*. International Council for Science (ICSU) and International Social Science Council (ISSC).

Jati, H. F., Darsono, S. N. A. C., Hermawan, T., Yudhi, W. A. S., & Rahman, F. F. (2019). Awareness and knowledge assessment of sustainable development goals among University students. *Journal Ekonomi & Study Pembangunan, 20*(2), 163–175.

Kivati, G. (2016). The Role of Kenya's formal higher education in sustainable development within the context of globalization. In W. L. Filho, C. Skanavis, A. do Paço, J. Rogers, O. Kuznetsova, & P. Castro (eds.), *Handbook of Theory and Practice of Sustainable Development in Higher Education*, vol. 2, pp. 17–34. https://doi.org/10.1007/978-3-319-47889-0

Krizek, K. J., Newport, D., White, J., & Townsend, A. R. (2012). Higher education's sustainability imperative: How education for sustainable development is evolving. *Journal of Sustainability Education, 4*(1), 56–72.

Læssøe, J., Brundiers, G., Christensen, C. M., & Læssøe, K. (2009). Exploring the transformative potential of education for sustainable development. *Journal of Education for Sustainable Development, 3*(1), 51–62.

Lampert, M., & Papadongonas, P. (2016). *Globality: Towards 2030 without Poverty*. https://utvecklingsarkivet.se/wp-content/uploads/2011/03/Towards-2030-without-poverty-Glocalities2016.pdf

Lang, D. J., Wiek, A., Kearns, S., & Partidário, M. R. (2012). Toward a typology of transdisciplinary research. *Ecology and Society, 17*(2).

Leal Filho, W., Manolas, E., & Pace, P. (2015). The future we want: Key issues on sustainable development in higher education after Rio and the UN decade of education for sustainable development. *International Journal of Sustainability in Higher Education, 16*(1), 112–129.

Leal Filho, W., Shiel, C., & Paço, A. (2020). Implementing sustainable development goals in universities: Challenges and strategies. *International Journal of Sustainability in Higher Education, 21*(5), 763–784.

Lincoln, Y. S., & Guba, E. G. (1985). *Naturalistic inquiry*. Sage Publications.

Lozano, R., Ceulemans, K., & Seatter, C. S. (2019). Institutionalizing sustainability in universities: A global perspective. *Sustainability, 11*(11), 3171.

Mawonde, A., & Togo, M. (2019). Implementation of SDGs at the University of South Africa. *International Journal of Sustainability in Higher Education, 20*(5), 932–950. https://doi.org/10.1108/IJSHE-04-2019-0156

McCowan, T. (2019). *Higher education for and beyond the sustainable development goals*. Palgrave Macmillan.

Meyer, J. W., & Rowan, B. (2006). The new institutionalism in education. *Handbook of Education Policy Research, 2*(3), 121–134.

Mezirow, J. (2000). Learning to think like an adult: Core concepts of transformation theory. In J. Mezirow & Associates (Eds.), *Learning as transformation: Critical perspectives on a theory in progress* (pp. 3–33). Jossey-Bass.

Miles, M. B., Huberman, A. M., & Saldaña, J. (2014). *Qualitative data analysis: A methods sourcebook*. Sage Publications.

Mori Junior, R., Fien, J., & Horne, R. (2019). Implementing the UN SDGs in Universities: Challenges, opportunities, and lessons learned. *Sustainability and Climate Change, 12*(2), 129–133.

Okolie, A. C. (2003). Producing knowledge for sustainable development in Africa: Implications for higher education. *Higher Education, 46*(2), 235–260.

Omisore, A. G., Babarinde, G. M., Bakare, D. P., & Asekun-Olarinmoye, E. O. (2017). Awareness and knowledge of the sustainable development goals in a University community in Southwestern Nigeria. *Ethiopian Journal of Health Sciences, 27*(6), 669. https://doi.org/10.4314/ejhs.v27i6.12

Plano Clark, V. L., & Ivankova, N. V. (2016). *Mixed methods research: A guide to the field*. Sage Publications.

Ramos, T. B., Caeiro, S., & Azeiteiro, U. (2021). Embedding SDGs into higher education institutions: A framework for sustainability governance. *Sustainable Development, 29*(4), 732–745.

Scott, W. R. (2008). Institutions and organizations: Ideas and interests (3rd ed.). Sage Publications.

Scott, W. R. (2014). Institutions and organizations: Ideas, interests, and identities. Sage Publications.

Shehu, M., Shehu, H., & Ode, M. B. (2018). Pattern of smartphone use among clinical medical students of Bingham University Teaching Hospital, Jos. *Central African Journal of Public Health, 4*(5), 137–142. https://doi.org/10.11648/j.cajph.20180405.12

Shiel, C., Leal Filho, W., do Paço, A., & Brandli, L. (2016). Evaluating the engagement of universities in capacity building for sustainable development in local communities. *Evaluation and Program Planning, 54*, pp. 123–134.

Sleurs, W. (2019). Integrating education for sustainable development (ESD) in higher education: A review of challenges and opportunities. *Sustainability, 11*(1), 182.

Sterling, S. (2001). *Sustainable education: Re-visioning learning and change*. Green Books.

Sterling, S. (2010). Transformative learning and sustainability: Sketching the conceptual ground. *Learning and Teaching in Higher Education, 5*(2), 17–31.

Trencher, G. P., Yarime, M., & Kharrazi, A. (2013). Co-creating sustainability: Cross-sector university collaborations for driving sustainable urban transformations. *Journal of Cleaner Production, 50*, 40–55.

UCT. (2012). *ISCN-GULF Sustainable Campus Charter Report*. University of Cape Town.

UN. (2015). *Transforming our world: The 2030 Agenda for Sustainable Development*. Resolution adopted by the General Assembly on 25 September 2015. A/RES/70/1.

UNESCO. (2017). *Education for Sustainable Development Goals: Learning for the world we want*. UNESCO.

United Nations. (2015b). Transforming our world: The 2030 agenda for sustainable development. A/RES/70/1. New York: United Nations.

Unterhalter, E., & Howell, C. (2021). Unaligned connections or enlarging engagements? Tertiary education in developing countries and the implementation of the SDGs. *Higher Education, 81*, 9–29. https://doi.org/10.1007/s10734-020-00651-x

Veeran, S., Duraiswamy, A., Puwanenthiren, P., & Venkatesan, R. (2024). Assessment of students' awareness, knowledge, and accessibility of Sustainable Development Goals (SDGs). *Multidisciplinary Reviews, 7*(7), 2024122. https://doi.org/10.31893/multirev.2024122

Walshe, K., & Schultz, L. (2010). Connecting research and sustainable development: A framework for universities. *Journal of Cleaner Production, 18*(13), 1291–1299.

Wright, T. S. A., & Horst, N. (2013). Exploring faculty perceptions of sustainability in higher education. *International Journal of Sustainability in Higher Education, 14*(2), 159–171.

Global Education and Sustainability in Japanese Universities: A Case Study of Sophia University

Hideki Maruyama and Mikiko Sugiura

Abstract This chapter explores how global education and the Sustainable Development Goals (SDGs) are integrated into Japanese higher education, using Sophia University as a case study. As a mid-sized, globally oriented Catholic university in Tokyo, Sophia has implemented both structural reforms and student-driven initiatives that advance sustainability education. The study analyzes the university's multi-level approach using UNESCO's Whole Institution Approach, focusing on strategic vision, pedagogy, research, campus operations, and community engagement. Institutionally, Sophia prioritized the recruitment of sustainability-focused faculty, redesigned curricula across departments, and leveraged government support through Japan's Top Global University Project, achieving top evaluations among peer institutions. Student initiatives—ranging from project-based learning and campus design to international fieldwork—have transformed the university into a "living lab" for sustainability. The case highlights how the relatively small scale of the university facilitated holistic transformation, allowing for agility in implementation and coherence in vision. Challenges, such as interdisciplinary collaboration and diverse interpretations of the SDGs, are addressed through participatory dialogue and faculty development. Ultimately, the study positions Sophia as a model for embedding sustainability within institutional identity, offering valuable lessons for similarly sized universities worldwide seeking to align global education with transformative sustainability practices.

Keywords Global education · Institutional transformation · Whole Institution Approach (WIA) · Student engagement · Living lab for sustainability

H. Maruyama (✉) · M. Sugiura
Sophia University, Tokyo, Japan
e-mail: hideki@sophia.ac.jp

M. Sugiura
e-mail: sugiura_mikiko@sophia.ac.jp

W. Leal Filho et al. (eds.), *Introducing the UN SDGs in Higher Education Institutions*, World Sustainability Series, https://doi.org/10.1007/978-3-032-23017-1_17

1 Introduction

In Japan's higher education institutions, the integration of global education and the United Nations' Sustainable Development Goals (SDGs) into curricula and practice has become increasingly common. While the implementation of Education for Sustainable Development (ESD)—an advanced concept developed within Japanese universities—was previously criticized for lacking coherence with other educational reforms (Kitamura & Hoshii, 2010), it is now widely promoted as an essential concept in the global discourse on sustainable development (Clark et al., 2020). Notably, Japanese universities had the high representation in the 2019 Times Higher Education Impact Rankings (Ashida, 2023). This development can be attributed not only to universities' role as research institutions dedicated to advancing their respective disciplines but also to the growing expectation that they contribute to their surrounding communities in diverse ways. Moreover, universities are increasingly expected to provide a broad-based education that fosters young individuals with globally relevant competencies, particularly in the context of the global economy (Maruyama, 2025).

This chapter examines how the SDGs have been integrated into global education, using Sophia University—a private university located in the heart of Tokyo—as a case study. Sophia University has the longest-established network of the Catholic Jesuit tradition in Japan and has offered higher education not only to Japanese students but also to foreign residents since it was officially recognized as a university after World War II. As a result, it has gained a reputation within Japan as a university with a highly globalized educational environment (Sugiura et al., 2023).

Ths chapter illustrates how the integration of the SDGs into global education at Sophia University on two distinct levels. On one hand, at the institutional level, it explores how a shared vision for the future has been cultivated within the university and the reforms undertaken to realize this vision, focusing on policy and organizational transformations. While limited in number, the university strategically prioritized the recruitment of faculty members specializing in sustainability and the SDGs, thereby enabling the implementation of the program. On the other hand, the chapter highlights how sustainability education has evolved and deepened as students and faculty actively engage in updating curricula and shaping a sustainable campus environment. The students' activities and commitment have fostered everyday practices of campus sustainability and engagement with the SDGs. While the former represents a top-down approach, in which the university administration implements structural changes and establishes new SDG-related educational programs, the latter demonstrates a bottom-up approach, where experiential learning fosters the practical application of knowledge. This case study provides a unique insight into the organic and effective interplay between these two approaches in advancing sustainability education. Furthermore, this case suggests that the relatively small scale of the university facilitates the implementation of the Whole Institution Approach (WIA) for a sustainable future. WIA to sustainability in education emphasizes the strong connection between the socio-physical environment of an educational institution and its learning processes and outcomes (UNESCO, 2017; Holst et al., 2024).

2 Origin and Organizational Reforms of the University

The history of Sophia University dates back to 1549, when Saint Francis Xavier arrived in Japan and envisioned the establishment of a university in the country (Sophia University, 2023). However, Sophia University officially began functioning as a modern university in 1928 after following the University Ordinance of 1919.[1] Before this ordinance, Japan's higher education landscape was dominated by seven national universities, which were the only officially recognized higher education institutions.[2] With the enactment of the Ordinance, private universities were also granted official recognition. After World War II, Sophia University taught Japanese students and leveraged its English-based higher education programs to provide educational opportunities also for families of U.S. military personnel stationed in Japan. As one of the largest Catholic universities, it has also maintained a strong presence alongside the St. Ignatius Church. Notably, Sophia University is the only Japanese university to have hosted a visit from a sitting pope—Pope John Paul II in 1981 and Pope Francis in 2013.[3]

Like many other institutions, Sophia University has undergone various organizational reforms since the beginning of the twenty-first century. Among these, the publication of the "Grand Layout for the 100th Anniversary[4]" in 2001 marked a significant turning point (Table 1). The subsequent initiatives and accompanying structural reforms have further accelerated the university's commitment to global education. Throughout this process, sustainability has inevitably emerged as an essential agenda, becoming an integral component of institutional development.

The following sections examine the institutional and programmatic developments alongside the initiatives led by students and faculty. However, the distinction between these two levels is not entirely clear-cut, as they do not operate in isolation. Rather, they have mutually reinforced each other, creating a positive cycle of transformation. At the institutional level, as the most important factor for WIA, two notable developments include the establishment of the university-wide Center for Global Education and Discovery and its role as a hub for addressing global challenges (Sect. 2), as well as the launch of the Sustainable Futures Program, designed to promote cross-department collaboration (Sect. 3). Meanwhile, at the student and faculty level, key

[1] Imperial Ordinance No. 388, December 6, 1919 from https://www.mext.go.jp/b_menu/hakusho/html/others/detail/1318056.htm.

[2] Tokyo University(established in 1877); Kyoto University (in 1897); Tohoku University (in 1907); Kyushu University (in 1911); Sapporo Agricultural College (in 1918); Osaka Imperial University (in 1931); and Nagoya Imperial University (in 1939). As of 2024, Japan has 813 universities, approximately 77% of which are private institutions (MEXT, 2024, August 28).

[3] https://www.sophia.ac.jp/jpn/aboutsophia/sophia_spirit/catholic/.

[4] https://www.sophia-sc.jp/disclosure/grand_layout/layout2001.

Table 1 Sophia ground layout initiatives and SDGs/global education

Sophia University's Initiative	*Related SDGs*
Education to uncover issues that lead to humanistic challenges	Raising awareness of issues related to the SDGs
Consideration for the Global Environment (Campus Planning)	Environmental preservation (SDGs 13, 15)
Consideration for the elderly and people with disabilities (Barrier-free)	Inclusive society (SDGs 10,11)
A campus open to society and the local community	Sustainable community (SDG 11)
Improvement of the educational and research life environment, including scholarship programs	Providing quality educational opportunities (SDG 4)
Promoting lifelong education	Providing lifelong learning opportunities (SDG 4)
Sophia University's Initiative	*Global education related*
Promoting prioritized research with international recognition	International research collaboration and knowledge sharing
Promoting academic exchange with domestic and international universities and research institutions	Building a global educational and research network
Training for professionals working in international organizations, NPOs and NGOs	Developing human resources for international cooperation
Implementation of specialized education in foreign languages	Developing global human resources
Improvement of language education (Adaptation to internationalization)	Improving global communication skills
Enhancement of Japanese language education for exchange students and non-Japanese	Support for international students, Multicultural coexistence
Exchange professorship program with domestic and international universities	Internationalization of education, Knowledge Exchange

Source Authors

examples include the engagement of student staff within the restructured Sustainability Office (Sect. 4) and experiential learning programs initiated and led by students and faculty (Sect. 5), followed by discussion.

3 Establishment and Operation of the Global Education Center

As of 2025, the university-wide Center for Global Education and Discovery (CGED) serves as the central coordinating body for global education at Sophia University, overseeing its implementation and ensuring alignment with relevant academic units and administrative organizations. The center originated from Sophia University's global education initiatives, which were established and strengthened within the

Faculty of Foreign Studies after being selected for MEXT's Project for Promotion of Global Human Resource Development in 2012.[5] The primary goal was to cultivate global competencies among students, enabling them to embody the university's educational philosophy: "For Others, With Others" (Hirosato, 2023). In 2015, the university secured additional funding from MEXT's Top Global University Project (2014–2023), which facilitated the transformation of the Global Education Center into an independent, university-wide institution as the CGED, no longer housed under the Faculty of Foreign Studies. This case highlights a distinctive characteristic of Japanese higher education: government funding often serves as a major driving force behind institutional reform.

With the establishment of CGED, functions and roles that had previously been distributed across various departments and faculties were expected to be further enhanced and strategically consolidated under the umbrella of "global education." Specifically, the center took on responsibilities such as planning, developing, managing, and promoting study abroad programs, including traditional exchange programs with more than 800 partner universities and short-term overseas programs. Additionally, it provided study abroad advising, facilitated international partnerships—including the negotiation and conclusion of institutional agreements—developed and managed global education-related courses such as internships, and oversaw applications for and implementation of MEXT-funded projects.

The operational characteristics of the CGED can be summarized in three key aspects. First, the Center consists of six full-time faculty members who engage in research and educational practice. In Japanese universities, educational resources are often limited, and even when full-time faculty are assigned to a similar center rather than a department, their numbers tend to be small. Second, these faculty members collaborate closely with an office known as the Global Education Promotion Office, which covers five main areas:

1. Inter-university agreements, academic exchange, and study abroad promotion
2. Outbound student exchange, risk management, Collaborative Online International Learning (COIL), and the Student Study Abroad Committee
3. Inbound student exchange, Japanese Ministry of Education (MEXT) scholarship recipients, and government-sponsored international students
4. Short-term language programs, short-term training programs, Social Engagement programs, internship programs, and operational meetings
5. Center curriculum and short-term programs

[5] This program provided financial support for university globalization initiatives aimed at enhancing Japan's international industrial competitiveness and strengthening intergovernmental collaboration by fostering young global professionals with advanced language proficiency, communication skills, autonomy, and intercultural understanding. The initiatives eligible for support fell into two categories: (1) the University-Wide Promotion Type, which required institutions to pursue globalization across the entire university while contributing to similar efforts at other domestic universities, and (2) the Distinctive Type, which focused on globalization at the faculty or graduate school level while enhancing the university's overall internationalization. https://www.mext.go.jp/bmenu/houdou/24/09/1326068.htm https://www.mext.go.jp/a_menu/koutou/kaikaku/sekaitenkai/1319659.htm.

This structure allows the implementation of global education to reflect both the perspectives of research and educational practitioners and the administrative perspectives that consider the intentions of the university's upper management and the achievement of institution-wide targets. The full-time faculty members facilitate the dispatch of students who are deemed suitable representatives of the university for study abroad programs, implement related initiatives such as the experiential Social Engagement programs discussed later, and provide educational opportunities for all students by leveraging their respective areas of expertise. Third, effective communication is ensured as a fundamental characteristic of the CGED's operations. Global education initiatives are developed through close interactions between faculty and administrative staff, facilitated by a monthly regular meeting chaired by the CGED's director and attended by the administrative head, along with numerous sub-meetings that support ongoing discussions and coordination.

A key aspect of these activities is that they are made possible through the cooperation and support of all nine faculties and 29 departments across the university, other university administrative units, as well as external institutions. This ensures that global education initiatives align with the university's long-term strategic vision while covering a broad range of global education areas. In this capacity, the CGED serves as a hub for institution-wide coordination (Fig. 1). Additionally, as discussed later, the Diversity and Sustainability Office also plays a role in implementing programs related to the SDGs.

A notable example of how global issues have been integrated into global education through this mechanism is the various Ministry of Education (MEXT)-funded projects in which CGED was involved in both the application and implementation stages. Specifically, these include the Sophia ASEAN International Mobility for Students (SAIMS) Program (2013–2017), which aimed to promote multicultural

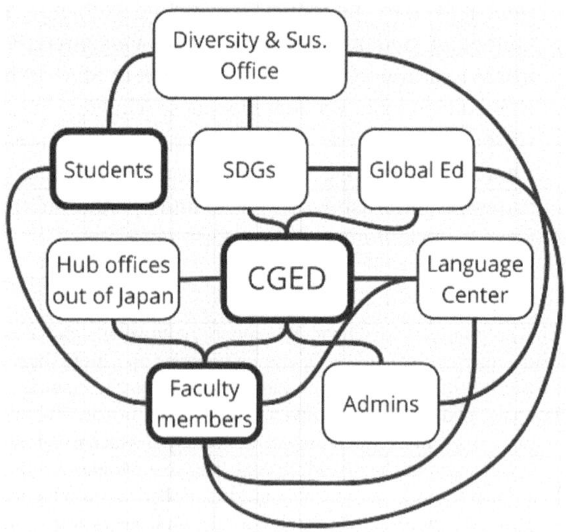

Fig. 1 Hub-functioning CGED collaborates. *Source* Authors

Table 2 Education programs became relevant to the SDGs

	Related SDGs
SAIMS	G4: Quality Education G6: Clean Water and Sanitation G11: Sustainable Cities and Communities G13: Climate Action G14: Life Below Water G15: Life on Land
LAP	G4: Quality Education G10: Reduced Inequalities G17: Partnership for the Goals
COIL	Any issues in align with each learning goals in class with professor's specialities

Source Authors

coexistence and student mobility in the ASEAN region; the Latin America Program (LAP) (2015–2019), a student exchange initiative with six Latin American countries focused on themes of human mobility and coexistence, conducted in collaboration with Nanzan University and Sophia University Junior College Division; and Collaborative Online International Learning (COIL) (2018–2022), a program utilizing ICT to facilitate course collaboration with overseas universities, implemented jointly with Ochanomizu University and the University of Shizuoka.[6]

These programs were designed based on the premise of a globalized socio-economic landscape and aimed to foster discussions among actors from diverse backgrounds and disciplines regarding the implications of globalization for human mobility and resource management. As a result, they facilitated the identification of global issues and their integration into educational programs, fostered new faculty collaborations among those interested in these topics, and enhanced the university-wide curriculum's alignment with the SDGs (Table 2).

Furthermore, even after the completion of the funding periods, the incorporation of experiential fieldwork into courses has become an established practice, reinforcing a hands-on approach to understanding global challenges such as climate change adaptation and biodiversity conservation. A prime example is student participation in *Satoyama* (rural woodland) conservation activities in the Tokyo metropolitan area, which directly aligns with SDG Goal 15: Life on Land (Sugiura, 2024). The high level of student engagement in these initiatives is evident from their proactive involvement in documenting and disseminating their experiences through video production and other creative outputs, further extending their learning (Photo 1).[7]

[6] These initiatives are part of a series of Ministry of Education, Culture, Sports, Science and Technology (MEXT)-funded programs known as the Project for the Promotion of Global Human Resource Development, which began in 2011. The program aimed to foster globally competent human resources and enhanced the global outreach of university education. While ensuring the quality of higher education, it supported efforts to establish international educational collaborations with universities in the target countries and regions.

[7] As an example, a student who visited a *Satoyama* in a SAIMS class called Conservation has sent out a 5-min YouTube video titled Living in balance with the natural world.

Photo 1 Students Engage in a Hands-on Program (Rice Planting). *Source* SACRU Student's YouTube[8]

In this way, the CGED, operated through close collaboration between administrative staff and faculty, has played a leading role in advancing the globalization of Sophia University. At the same time, through the implementation of global education, it has strengthened the integration of SDG goals and targets into educational programs and enhanced stakeholder engagement in these initiatives.

4 Launch of SPSF and Student's Project

As discussed in the previous section, Sophia University applied for the MEXT's Top Global University Project (2014–2023) and had its proposed educational program successfully approved. With a strong advantage in global education compared to other Japanese universities, Sophia University was able to secure valuable information and cooperation from international organizations. The university strategically aligned its educational initiatives with the SDGs and placed particular emphasis on proactive efforts toward sustainable development. This commitment led to the establishment of the Sophia Program for Sustainable Futures (SPSF).

As the institutional structure for the new global education, SPSF is set as a four-year bachelor's program that started in September 2020. Incoming students belong to one of six departments (Economics, Education, Global Studies, Journalism, Management, and Sociology) and receive one of seven degrees (Bachelor in Area Studies,

[8] https://www.youtube.com/watch?v=yg25EvWd6TA.

Economics, Education, International Relations, Journalism, Management, or Sociology) upon graduation (Sophia University, n.d.). This program is Japan's first undergraduate course that directly addresses sustainable futures (Maruyama, 2025). As the strong university initiative, recruitment for SPSF was prioritized in each department.

As the student's initiative, SPSF has unique classes such as the lectures in the first semester and third-year seminar, reflecting on their experiences both on and off campus. In the first-year lectures in and about sustainable futures, all the SPSF students of the six departments are required to take the same lectures to learn sustainability and SDGs with the latest global and local trends. In the third-year seminar for sustainable futures, they all come back together for the seminar to work in groups to define sustainability based on previous research, discuss challenges, create specific projects, and implement these projects through fieldwork and investigations. It becomes essential to design projects in collaboration with students from other departments, making an interdisciplinary approach important. The students first grasp the situation regarding sustainability through Systems Thinking, based on some theories (Meadows, 2008; Stroh, 2015) at the development stage of their project. They finally launch their own project on sustainable futures, aiming at social impact. Projects conducted during the third year of the SPSF are particularly impactful, even influencing university operations. For instance, one student group proposed replacing disposable plastic umbrella bags with water-removal equipment at campus entrances. Though the equipment was installed after the project concluded, the experience gave students a sense of agency in initiating institutional change—something particularly meaningful in Japan, where conformity and deference to authority often dominate educational settings. Through this project, the students realize that they possess the power to change society for their own sustainable futures. Other initiatives include the creation of an on-campus farming site through extracurricular collaboration between faculty and students (Short et al., 2023). Here, the campus itself functioned as a "living lab" (Leal et al., 2020), offering iterative learning and innovation for sustainability.

5 Student's Commitment

The Sophia University Office for the Promotion of Sustainability was established in 2021 to advance research, education, and social engagement contributing to SDGs, as well as to promote diversity initiatives and the realization of an inclusive society. In 2024, it was integrated with the existing Office for the Promotion of Diversity and renamed the Office for the Promotion of Diversity and Sustainability (Diversity & Sus. Office).[9]

[9] The Office for the Promotion of Diversity had been addressing diversity-related issues even before Sophia University formally embraced sustainability. It originated with the "Support Program for Female Researchers in Response to a Global Society" in 2009, followed by the adoption of the "Declaration on the Promotion of Gender Equality" in 2011, led primarily by the Office for the Promotion of Gender Equality. Over time, as the concept of diversity expanded, the office was

Photo 2 Students-lead sustainable campus

Diversity & Sus. Office aims to foster collaboration between students, faculty, and administrative staff to create new values. It oversees interdepartmental projects that promote initiatives such as improving campus energy efficiency and enhancing the campus environment. Student staff play a key role in various activities, including producing instructional videos on waste separation, organizing programs to raise awareness about microaggressions, and compiling the Sophia University SDGs & Sustainability Report, published as the social accountability of the university to the public general. Notably, this office employs student staff, engaging them in event planning, facility improvements, and public relations activities. Through this initiative, students' awareness of sustainability—developed through the university's evolving organizational structure and diverse programs—has translated into concrete outcomes through collaboration with faculty and administrative staff. A notable example is in universal design, where students have taken the lead in planning and implementing projects on campus (Photos 2).[10]

6 Experiential "Sustainability" Study Visit

The Social Engagement Programs (SEP) are hands-on, practice-based programs that allow students to travel abroad and engage in experiential learning on-site. These programs take place during the spring and summer breaks, with most lasting between one to three weeks. The content varies widely, including visits to international organizations, volunteer work, interactions with local universities, and service-learning

restructured into the Office for the Promotion of Diversity, broadening its scope of support and enhancing institutional frameworks. https://diversity-sustainability.sophia.ac.jp/en/en-about/.

[10] https://diversity-sustainability.sophia.ac.jp/en/efforts/4970/.

Photo 3 Students measuring soil acidity in snow

activities. Destinations have included Australia, Estonia, India, South Africa, Thailand, the United States, among others.[11] Sophia University officially grants academic credits to students participating in SEP, which are offered between semesters. Students are allowed to take one SEP per semester, making it theoretically possible to complete up to eight programs out of twelve before graduation. Each SEP carries two academic credits, which contribute toward the 124 credits required for graduation. The program consists of formal lectures before departure, followed by seven to 10 days of practical experience in the field (Maruyama et al., 2020).

Each SEP is designed to reflect the specific conditions of its host location, but one of the most bottom-up and student driven programs is the "Study Visit on Sustainability." By 2024, this program had been conducted eight times annually, except for 2020, when it was canceled due to COVID-19, and 2021, when it was held online. The program provides an exceptional opportunity for participating students to conduct fieldwork in Baltic Sea countries, including the Republic of Estonia, observe natural environments, engage in practical actions, and interview local residents and experts about sustainability. Students are primarily responsible for determining where to visit and what questions to ask, based on their interests in sustainability studies and the SDGs. They receive lectures and training from faculty members on interview techniques and academic background, take the initiative to arrange appointments on-site, and even engage in negotiations when necessary. With these preparations, they travel to the field to pursue their self-defined research themes related to sustainability (Photo 3).

The primary counterparts supporting students on-site are engaged in a UNESCO-affiliated project that has been ongoing since the end of the Cold War in 1989. This long-term environmental education initiative, known as the Baltic Sea Project

[11] https://piloti.sophia.ac.jp/jpn/studyabroad/international/jissengata/jissenn_summer/ and https://piloti.sophia.ac.jp/jpn/studyabroad/international/jissengata/jissen_spring/.

(BSP), involves over 165 secondary schools and several universities in the region. The young local participants from the Baltic nine countries collaborate on field research and environmental measurements, engaging in discussions on sustainability and the SDGs with Sophia students. Interestingly, Sophia students are often more deeply influenced by insights shared by peers of the same generation rather than by comments from teachers or adults in Japan. Through daily reflections conducted among students during the study visit, they gradually become aware of their own learning progress and personal transformation for their own sustainability. Unlike their routine university life, they share living spaces, meals, and daily experiences, which enhances intercultural understanding and adaptability. They also collaborate in forming a community, build trust-based relationships, and deepen their learning by sharing their insights with their peers. The two-week communal experience fosters learning effects similar to those observed in university dormitory life (Blimling, 2014). This type of experiential learning not only enhances cognitive skills, such as knowledge acquisition and methodological understanding, but also cultivates non-cognitive skills, including social competence, independence, and self-emancipatory among the Japanese students (Maruyama, 2022).

7 Discussion

The case of Sophia University illustrates how integrating SDGs into higher education requires a dynamic interplay between top-down structural initiatives and bottom-up student engagement. Such a whole-institution approach (WIA) aims at mainstreaming sustainability into all aspects of the educational institution. It involves rethinking the curriculum, campus operations, organizational culture, student participation, leadership and management, community relationships and research (UNESCO, 2014). In this way, the institution itself functions as a role model for the learners. Sustainable learning environments allow educators and learners to integrate sustainability principles into their daily practices and facilitate capacity-building, competency development and value education in a comprehensive manner (UNESCO, 2017; Washington, 2015).

While early programs such as SAIMS, LAP, and COIL in Sophia were understood in alignment with the SDGs, a significant shift occurred with the establishment of the Promotion of Sustainability Office in 2021. This institutional reform empowered student-led initiatives, which subsequently influenced decision-making processes. Since then, daily sustainability practices and coordinated student efforts have contributed to the institutionalization of sustainability at Sophia University, highlighting the potential of learner-driven change. And meanwhile, education programs such as SPSF aligns closely with UNESCO's WIA, which emphasizes sustainability integration across five dimensions: vision and strategy, learning and teaching, research, campus operations, and community engagement. The university has demonstrated commitment across all these dimensions:

- Strategic Vision: The university embedded the SDGs into its long-term policy document, Grand Layout 3.0, ensuring that sustainability remains a core institutional objective.
- Pedagogical Approach: The SPSF curriculum employs systems thinking and design thinking, encouraging students to engage in project-based learning to address complex sustainability challenges.
- Research and Public Engagement: Interdisciplinary research efforts have gained momentum, fostering collaboration across fields and increasing public relevance.
- Campus Operations: Sustainability practices, co-developed with students, have contributed to reducing environmental impact and improving campus sustainability.
- Community and Global Engagement: Through local and international collaborations, Sophia has extended its sustainability initiatives beyond Japan, positioning itself as a globally engaged institution.

For a lesson from the case, Sophia University's mid-sized scale has played a critical role in the effectiveness of these reforms. Unlike larger institutions, which may face bureaucratic inertia, Sophia was able to implement campus-wide sustainability reforms swiftly and holistically. This suggests that the feasibility of the WIA may, in fact, increase as institutional size decreases. The university's position within the Jesuit network and its designation as the UNESCO Chair on Education for Human Dignity, Peace, and Sustainability (2024) further reinforce its commitment to sustainability education. Sophia University's sustainability initiatives are deeply connected to its institutional identity. Rooted in Catholic and Jesuit traditions, the university's Grand Layout 3.0 reinforces diversity, sustainability, and social justice, aligning with Pope Francis' "Laudato Si',[12]" and the Jesuit Universal Apostolic Preferences (UAPs), which outline four key priorities for Jesuit institutions worldwide.[13] This alignment underscores that sustainability education at Sophia is not merely an external policy adaptation but rather a continuation of the university's foundational values.

Despite these advancements, Sophia has encountered challenges common to Japanese universities, such as generational turnover, declining student populations, and financial constraints. To address these challenges, the university moved away from traditional departmental hiring practices and strategically recruited faculty with expertise in sustainability-related fields. While this decision, led by the president and executive board, enabled the operationalization of programs like the SPSF, it also led resistance among faculty concerned about academic restructuring.

To mitigate these tensions, newly hired faculty participated in professional development programs aimed at deepening their understanding of sustainability education. The university also promoted interdisciplinary research and curriculum reform,

[12] A papal encyclical is a letter issued by the Pope to all bishops, providing guidance on theological or social matters.

[13] Currently, based on this policy, Sophia University is not only continuing and strengthening its existing efforts but also exploring new approaches to further promote university-wide engagement. Through education, research, and social contribution, the university aims to contribute to the creation of a MAGIS ("greater") world—one that strives for continuous improvement and greater justice.

particularly in the SPSF and Green Engineering tracks, where substantial redesigns were implemented over several years. These efforts were facilitated by external funding from MEXT's Top Global University Project, under which Sophia, classified as a Type B (Globalization Leadership) university, consistently resulted the highest evaluation scores across two mid-term and the final evaluations (2017, 2020, and 2025).[14]

The university's transformation can also be analyzed through Giesenbauer and Tegeler's (2020) four-stage model of higher education transformation, which describes the shift from traditional knowledge transmission to a more engaged, participatory learning model. Sophia has moved beyond the initial stages of merely conveying sustainability knowledge and exposing students to disciplinary frontiers. Instead, it has advanced to the later stages of stakeholder dialogue, action research, and the development of a "citizens' university"—a space where students and faculty engage in sustainability discourse while taking responsibility for institutional and societal change.

However, interdisciplinary collaboration remains a challenge. The broad scope of sustainability invites multiple perspectives, which can sometimes lead to conflicts in interpretation. Scholars in development studies and political ecology, for instance, critique the SDGs for reinforcing Western, neoliberal structures (Arora-Jonsson, 2023), while others question the ecological implications of advancing education under current economic paradigms (Komatsu & Rappleye, 2018). Public perceptions are similarly divided, with some viewing SDGs implementation as involving trade-offs, while others emphasize its potential for synergistic benefits (Bain et al., 2019). Given this pluralism, it is essential to frame the SDGs as a platform for dialogue rather than a rigid blueprint, ensuring space for critical engagement and diverse perspectives (Ratner, 2004).

Amid these complexities, student agency shown as their initiatives and actions has emerged as a key driver of institutional transformation. At Sophia, students have taken an active role in shaping sustainability efforts on campus, extending beyond formal classroom learning. Through initiatives such as on-campus farming, material reuse projects, and policy advocacy, students have actively contributed to sustainability practices in both physical and cultural dimensions (Short et al., 2023). Their participation in international workshops and fieldwork abroad highlights the importance of integrating formal and non-formal education in sustainability learning (Maruyama, 2019). These student-led initiatives position Sophia University as a "living lab" (Leal et al., 2020), where sustainability is not only studied but also practiced, negotiated, and internalized through everyday experiences. Such initiatives illustrate that research fosters the co-creation of knowledge, requiring multiple perspectives, critical dialogue on key concepts, and pluralistic, emancipatory approaches to education (Gough & Scott, 2008).

[14] https://www.jsps.go.jp/j-sgu/h26_kekka_saitaku.html.

8 Conclusion

Sophia University exemplifies how sustainability in higher education must be pursued through both systemic institutional transformation and active learner participation. The case highlights the unique advantages of mid-sized universities in implementing holistic sustainability frameworks. As global sustainability challenges intensify, Sophia's experience serves as a valuable model for other institutions seeking to embed sustainability in an authentic, context-sensitive, and community-engaged manner.

In contrast to large-scale universities such as Pennsylvania State University or the University of Colorado in the U.S., or the University of Amsterdam in Europe—where student-led initiatives, community collaborations, and highly specialized research and education in sustainability are more common—Sophia University faces unique challenges as a mid-sized institution with fewer than 60 faculty members directly involved in sustainability-related teaching. In such a context, efficiency across budget, teaching impact, and institutional planning becomes imperative. Rather than replicating models from larger institutions, Sophia has differentiated itself by strengthening sustainability education at the undergraduate level rather than focusing solely on graduate programs.

Moreover, the university's smaller size offers certain advantages: as UNESCO's Whole Institution Approaches (WIAs) suggest, systemic and institution-wide transformation is often more feasible in smaller settings, even for applying for external subsidies. WIAs promote the integration of sustainability across the whole university, enabling both learners and educators to be empowered and enhance sustainable behaviors (Holst et al., 2024; UNESCO, 2017). This model illustrates how relatively smaller universities, like Sophia, can demonstrate agility and cooperation, providing a valuable counter-example for HEIs globally.

However, it is still important to recognize that Sophia University's progress in SDGs education is also rooted in its historical and global foundations. Having embraced internationalism since its inception over a century ago, Sophia benefits from strong networks with both Jesuit institutions worldwide and United Nations agencies. The university's designation as the UNESCO Chair on Education for Human Dignity, Peace, and Sustainability in 2024 further exemplifies this alignment.

References

Arora-Jonsson, S. (2023). The sustainable development goals: A universalist promise for the future. *Futures, 146*, Article 103087. https://doi.org/10.1016/j.futures.2022.103087

Ashida, A. (2023). The Role of Higher Education in Achieving the Sustainable Development Goals. In S. Urata, K. Kuroda, & Y. Tonegawa (Eds.). *Sustainable Development Disciplines for Humanity. Sustainable Development Goals Series*. Springer. https://doi.org/10.1007/978-981-19-4859-6_5

Bain, P. G., Kroonenberg, P. M., Johansson, L. O., Milfont, T. L., Saeri, A. K., Tam, K.-P., Grote, H., Axsen, J., Rus, A., Madsen, L. E., Aavik, T., Sukhov, A., Kõiv, K., Park, J., Pasyk, V., Peixoto, J. F., Zelenski, J. M., Davidov, E., Christodoulou, N., & Harré, N. (2019). Public views of the sustainable development goals across countries. *Nature Sustainability, 2*, 819–825. https://doi.org/10.1038/s41893-019-0365-4

Blimling, G. S. (2014). *Student learning in college residence halls: What works, what doesn't, and why*. Jossey-Bass.

Clark, I., Nae, N. & Arimoto, M. (2020). Education for sustainable development and the "Whole Person" Curriculum in Japan, *Oxford Research Encyclopedia of Education*. https://doi.org/10.1093/acrefore/9780190264093.001.0001/acrefore-9780190264093-e-935.

Giesenbauer, B., & Tegeler, M., et al. (2020). The Transformation of Higher Education Institutions Towards Sustainability from a Systemic Perspective. In F. W. Leal (Ed.), *Universities as Living Labs for Sustainable Development* (pp. 1083–1105). Springer.

Gough, S. R., & Scott, W. A. H. (2008). *Higher education and sustainable development: Paradox and possibility*. Routledge.

Hirosato, Y. (2023). Jochi Daigaku no Gurobaru Kyouiku "100 nen me no Ketsui" to Sono Keishou—SIED kara "Sophia GED" he [Global education at Sophia University: Resolutions for the 100th year and their inheritance—From SIED to "Sophia GED"]. In M. Sugiura & Y. Mizutani (Eds.), *Gurobaru Kyouiku wo Jissen suru—Tayona Ryouiki karano Apurochi [Practice of global education: The approaches from different fields]*. Sophia University Press.

Holst, J., Grund, J., & Brock, A. (2024). Whole Institution Approach: Measurable and highly effective in empowering learners and educators for sustainability. *Sustainability Science, 19*, 1359–1376. https://doi.org/10.1007/s11625-024-01506-5(Accessed31March2025)

Kitamura, Y., & Hoshii, N. (2010). Education for sustainable development at universities in Japan. *International Journal of Sustainability in Higher Education, 11*(3), 202–216.

Komatsu, H., & Rappleye, J. (2018). *Will SDG4 achieve environmental sustainability? (Working Paper No. 4)*. Center for Advanced Studies in Global Education, Arizona State University. https://education.asu.edu/sites/default/files/2022-01/working_paper_4_final.pdf

Leal, F.W., Salvia, A.L., Pretorius, R.W., Brandli, L.L., Manolas, E., Alves, F., Azeiteiro, U., Rogers, J., Shiel, C., & Do Paco, A. (Eds.). (2020). *Universities as living labs for sustainable development*. Springer.

Maruyama, H. (Ed.). (2019). *Cross-bordering dynamics in education and lifelong learning: A perspective from non-formal education*. Routledge.

Maruyama, H. (2022). A deep transformative dimension of ESD in Japanese University: From experiential to emancipatory learning in online and offline environments. *Sustainability, 14*(17). https://doi.org/10.3390/su141710732

Maruyama, H. (2025). Nourishing Change Agents by Undergraduate Program for Sustainable Futures at a Japanese University. In W. Leal, M. Mifsud, L. Trevisan, & P. Pace (Eds.), *Education for Sustainable Development: The Contribution of Universities*. Springer.

Maruyama, H., Saito, Y., & Yoshida, H. (2020). Study tour to wetlands in Estonia. *Wetland Research, 10*, 47–52.

Meadows, D. H. (2008). *Thinking in systems: A primer*. Chelsea Green Publishing.

Ministry of Education, Culture, Sports, Science, and Technology (MEXT). (2024, August 28). *Reiwa 6 nendo gakkou kihon chousa (sokuhouchi) ni tsuite kouhyou shimasu* [Announcing preliminary basic school survey 2024]. Retrieved September 30, 2024, from https://www.mext.go.jp/content/20240821-mxt_chousa01-000037551_001.pdf

Ratner, B. D. (2004). Sustainability as a dialogue of values: Challenges to the sociology of development. *Sociological Inquiry, 74*(1), 50–69.

Short, M., Nagai, G., & Maruyama, H. (2023). Engaging students in co-creating a sustainable campus as a living lab. *Global Concern, Sophia University, 5*, 158–175.

Sophia University. (2023, April 27). *In Sophia University's roots—Together with Francis Xavier's dreams and aspirations*. Retrieved September 30, 2024, from https://www.sophia.ac.jp

Sophia University. (n.d.). *Sophia program for sustainable futures (SPSF)*. Retrieved September 30, 2024, from https://spsf.sophia.ac.jp

Stroh, D. P. (2015). *Systems thinking for social change: A practical guide to solving complex problems, avoiding unintended consequences, and achieving lasting results*. Chelsea Green Publishing.

Sugiura, M. (n.d.). *Satoyama: Showcase of reconciliation ecology*. SACRU Alliance. Retrieved February 11, 2025, from https://www.sacru-alliance.net/satoyama-showcase-of-reconciliation-ecology/

Sugiura, M., & Mizutani, Y. (Eds.). (2023). *Gurobaru Kyouiku wo Jissen suru—Tayouna Ryouiki karano Apurochi* [Practice of global education: The approaches from various fields]. Sophia University Press.

UNESCO. (2014). *UNESCO Roadmap for Implementing the Global Action Programme on Education for Sustainable Development*. http://unesdoc.unesco.org/images/0023/002305/230514e.pdf. Accessed 31 March 2025

UNESCO. (2017). *Education for Sustainable Development Goals: learning objectives*, UNESCO. https://unesdoc.unesco.org/ark:/48223/pf0000247444. Accessed 31 March 2025

UNESCO. (2022). *Knowledge-driven actions: transforming higher education for global sustainability*, UNESCO. https://unesdoc.unesco.org/ark:/48223/pf0000380519. Accessed 31 March 2025

Washington, H. (2015). *Demystifying sustainability: Towards real solutions*. Routledge.

Applied Approaches and Case Studies

A Journey to Professionalism in ESD: A Story from the Mediterranean

Mifsud Mark C., Caruana Mariosa, Tabone Marika, and Vella Maureen

Abstract The importance of education in achieving sustainability is highlighted by the Sustainable Development Goals (SDGs), both specifically as Target 4.7 and also as a key enabler of all the SDGs. Meeting these targets demands a profound shift in attitudes and behaviours from every citizen, achievable through transformative education like Education for Sustainable Development (ESD) (United Nations Educational, Scientific and Cultural Organization, (UNESCO), 2014). In the last decades, there were a number of efforts to promote ESD. At the local level, this included the launch of a Master's degree in ESD by the Centre for Environmental Education Research at the University of Malta. The engagement of qualified ESD professionals is key to fostering this transformation, bridging ESD across sectors to achieve SDG and Green Deal Objectives. Malta currently lacks formal national recognition for ESD specialists, potentially impeding the recognition and credibility of individuals with in-depth expertise in the field. Establishing such recognition through a professional warrant, would enhance development standards while also contributing in achieving sustainability within the community, through accredited ESD professionals. This research employs a mixed-methods approach, combining qualitative and quantitative methodologies to investigate the Maltese ESD landscape. The methods include an initial mapping to identify relevant stakeholders, followed by online questionnaires and in depth interviews to provide deeper insights into stakeholders' perspectives. The outcomes shed light on effective ESD practices and assess the viability for professional recognition of ESD practitioners in Malta. The implications of this study have ramifications that can potentially influence national and international policy development. A number of recommendations are presented.

Keywords Education for sustainable development · Warrant · Mixed-methods · SDGs/Green deal · Malta

M. Mark C.
Centre for Environmental Education and Research, University of Malta, Msida, Malta
e-mail: mark.c.mifsud@um.edu.mt

C. Mariosa · T. Marika · V. Maureen (✉)
Malta Association of Professionals in Education for Sustainable Development, Volunteer Centre 181, Valletta, Malta
e-mail: mapesd.malta@gmail.com

W. Leal Filho et al. (eds.), *Introducing the UN SDGs in Higher Education Institutions*, World Sustainability Series, https://doi.org/10.1007/978-3-032-23017-1_18

1 Introduction

1.1 Global SD and ESD

In the past few decades, Education for Sustainable Development (ESD) has been an integral part in global discourse pertaining to sustainable development (SD) (United Nations Educational, Scientific and Cultural Organization, (UNESCO), 2017). Global efforts highlight the importance of ESD in working towards achieving SD targets such as the Sustainable Development Goals (SDGs), (United Nations, 2015). Meeting these targets demands a profound shift in attitudes and behaviours from every citizen, achievable through transformative education like ESD (UNESCO, 2014). Moreover, proper ESD requires the active participation of various stakeholders, including governments, non-governmental organisations (NGOs), the media, the private sector, as well as educational institutions, to ensure a wide reach (Shaeffer, 2006).

Many European Union (EU) countries, including Malta, are using the SDGs as a base on which to tailor their policies and governance. The EU has adopted two main strategies relating to SD. These are EU Sustainable Development Strategy (Commission of the European Communities, 2009) and the Europe 2020 Strategy—the EU's agenda for growth and jobs (European Commission, 2020). In 2017, the Eurostat developed a reference EU SDG indicator set to monitor progress regarding the SDGs in an EU context.

Also, countries in the Mediterranean region, including Malta, adopted the vision of the Mediterranean Strategy for Sustainable Development (MSSD) 2016- 2025. This is to build "a prosperous and peaceful Mediterranean region in which people enjoy a high quality of life and where sustainable development takes place within the carrying capacity of healthy ecosystems" (UNEP/MAP, 2016, p.4).

Furthermore, the Mediterranean Strategy for Education for Sustainable Development (MSESD) (UNECE, 2005) focused mainly on Mediterranean states and called upon its members to design and implement national strategies. The primary objective of these strategies is to strengthen ESD within the formal, non-formal and informal education, in schools and the wider community through a lifelong learning approach (UfM, 2014; UNECE, 2005).

1.2 Local National Strategies Targeting SD

A National Commission for Sustainable Development was first set up in Malta in 2002 to set up national sustainable development across all sectors in the country. Throughout the process of preparation of the Sustainable Development Strategy (2007–2016), various stakeholders were involved, including the government, civil society and the private sectors (National Commission for Sustainable Development, 2006).

In 2012, The Sustainable Development Act, Cap. 521 was later adopted resulting in a legislative framework mandating Government to mainstream sustainable development in its policies (Government of Malta, 2012). Following that, a new strategy was published for consultation in 2018.

Targets related to SD in Malta are heavily based on global and EU targets. Malta's Sustainable Development Strategy for 2050 (Government of Malta, 2024) outlines a plan of action to achieve strategic goals based on the SDGs in the Maltese context. These strategic goals focus on transition to a climate-neutral green and blue economy, urban planning and culture, health and wellbeing, technology and innovation, and social justice. The document emphasises the importance of education and collaboration between different stakeholders to reach the required targets.

All these national strategies also form part of Malta's contribution to the European Green Deal. Contributions to the European Green Deal extend to the private sector, where companies listed in the Malta Stock Exchange are required to disclose their non-financial information relating to sustainability, specifically done through ESG reporting (Mangion, 2024). Moreover, many private entities engage in Corporate Social Responsibility (CSR), although there seems to be a lack of thorough understanding of the holistic nature of CSR, and most initiatives tend to be limited to awareness-raising (MISCO, 2019).

However, to reach these targets and ensure the success of these strategies, quality ESD is required among all stakeholders (UNESCO, 2020).

1.3 Local ESD Initiatives

In 2004, the Centre for Environmental Education Research (CEER) was set up at the University of Malta (UoM), specialising in coordinating ESD initiatives, promoting and making ESD research more accessible, and facilitating resource transfer and capacity building in Malta and in the Euro-Med region not only through formal education, but also beyond, through collaborations with governmental and non-governmental entities (L-Università ta' Malta, 2020a).

The National Strategy for ESD (NSESD), which was based on the five priority areas identified in the GAP (UNESCO, 2014) was drafted. The NSESD appeals to educate citizens and also to instil "critical reflection, the exploration of alternative lifestyles and choices, good governance and wise management of resources" (NSESD Board of Governors, 2019, p.4). The need for ESD professionals was highlighted in most of its priority action areas. Unfortunately, the NSESD was shelved and is yet to be implemented.

The newly founded Malta Association of Professionals in ESD (mapESD) is another initiative set up by ESD professionals in 2024, aimed to foster a community of ESD professionals and improve the quality of ESD in Malta (Mariosa Caruana, personal communication, September 1, 2024).

1.3.1 Formal, Informal and Non-Formal

Most efforts to address ESD by the Government of Malta target formal education. ESD has been included in the national curriculum framework (Ministry of Education & Employment, 2012) as one of the six cross-curricular themes in compulsory education. The same trend can be noticed in other countries such as the Netherlands, Finland, and Spain (Ametller et al., 2024).

At post-secondary level, all students are required to study Systems of Knowledge, a subject that includes ESD principles like active citizenship, and contains a module dedicated to SD. Environmental Science, another subject, is offered to non-Science students and provides a holistic view of SD. However, packed syllabi, an exam-oriented approach and lack of training makes this challenging (Xerri Hili & Schembri, 2022).

A very powerful catalyst for ESD in local formal education is the collaboration with NGOs, such as Eco-Schools through Nature Trust-FEE, in non-formal educational programmes (Mifsud, 2012). However, participation is voluntary, and most are held during extra-curricular timeslots as there is no time slot allotted for them.

This NGO and others are extending ESD initiatives beyond formal education by targeting different SD pillars, example Koperattiva Internazzjonali (Kopin).

1.4 ESD Certification and Training

The types of training, accreditation and certification related to ESD vary across the globe, however, most efforts are aimed at educators in formal schooling. In fact, 90% of European countries are implementing ESD both in pre-service and in-service training (UNECE, 2016). Teachers in different countries including Bhutan, Finland, Scotland, Belgium, Ethiopia, Sweden and Canada provide pre-service and in-service teacher training on ESD.

ESD training in Cyprus is not only provided as in-service training to teachers but also to principles and directors, targeting a holistic approach. In addition, the concept of training the trainer has been adopted in Cyprus, where a teacher from each school is selected to be trained in becoming an ESD trainer and support educators in their school (Whitby, 2019).

Furthermore, in Indonesia, teachers are given support and professional training through the setting up of ESD centres to provide professional development to educators on ESD (Widjajanti et al., 2014).

Locally both pre-service and in-service teacher training on ESD is offered. The former is mostly carried out by the UoM as part of the teacher training course and the latter is normally provided by governmental educational institutions.

The UoM, also offers accredited study units related to ESD to other courses apart from the teacher training course, both at Bachelor's and Master's level. Moreover, the CEER launched the first local Master's degree in ESD at the UoM in 2011 (Mifsud, 2012). The M.ESD aims to enhance knowledge on all dimensions of SD and equip students with the necessary skills and values to promote a sustainable lifestyle (L-Universita' ta' Malta, 2020b). CEER also offers a shorter course leading to a Certificate in Environmental Education and Interpretation, as well as a Doctor of Philosophy programme.

In the international context, there are also well-renowned entities that also provide ESD certification. For example, the Centre for Environment Education (CEE) in India collaborates with international partners and offers ESD learning material and a diploma programme targeting teachers (Green Teacher: Diploma in Environmental Education, n.d.).

Similarly, the North American Association for Environmental Education (NAAEE) cooperates with national educational entities focusing on the environment and sustainable development to build and review high quality EE programmes (Certification, n.d.).

1.5 Professional Recognition

ESD is a key enabler for SD and is crucial in the pursuit to reach SDG and Green Deal targets. It is thus of utmost importance that those engaged in ESD training possess competences that go beyond the ones required for general quality education (UNECE, 2012). Moreover, quality ESD must not only be limited to formal schooling but must be extended beyond, to the other sectors of society also through informal and nonformal education (UNESCO, 2014). Engaging quality ESD professionals in different sectors of society can help bridge the gap between the different stakeholders and the achievement of their SD objectives (Wong, 2006). The level of national recognition for ESD professionals varies around the world.

In South Africa, an individual with prior educational background can apply for the National Certificate in Environmental Education, Training and Development Practice. This involves an accredited training programme at Diploma level, where educators obtain a formal qualification in Environmental Education to work as "entry-level environmental education practitioners" (South African Qualifications Authority, n.d., para. 2). This goes beyond simple accreditation, but rather satisfies a professional requirement for a specific occupation (South African Qualifications Authority, n.d).

However, to strive for better quality services in different professions, recognition, or regulation, usually in the form of a professional warrant or license, is commonly sought to ensure that individuals with the right amount of expertise and competences are engaged (OECD, 2024).

Presently, to the best knowledge of the researchers, there is no such recognition for ESD professionals in Malta or other countries in the world. This may hinder the acknowledgment and credibility of individuals with extensive expertise in the field.

2 Methodology

This study employed a sequential explanatory mixed-methods design, consisting of two phases where the quantitative findings guided the development of the qualitative component (Creswell, 2014). The mixed-methods approach was selected to achieve both breadth and depth in understanding and validation (Burke Johnson et al., 2007). The primary benefit of this design is that the strengths and limitations of one method are offset by the other (Greene, 2007), allowing for triangulation and complementarity, and resulting in well-supported and well-illustrated findings (Greene et al., 1989).

Firstly, the quantitative data was gathered using random sampling technique by distributing an online questionnaire via diverse social media tools such as Facebook and Instagram, and through an academic platform, targeting adults residing in Malta. There were a total of 107 completed responses (70 Females, 37 Males). The two main age categories represented included the 41 to 50 age cohort with 31.78% and the 31 to 40 age cohort with 28.97% of the total. The majority of respondents had a master degree (53.27%), the second major group being those with a Bachelor Degree at 24.30%.

Through the mapping exercise the stakeholders with specific characteristics and experiences (Diagram 1) that are pertinent for ESD local development, were identified. Using purposeful sampling, key individuals who matched these characteristics and worked in positions that could provide valuable insights into ESD development were selected.

The survey helped identify specific areas for further investigation, which were explored in greater depth through semi-structured interviews as part of the qualitative research. Two different interview formats were developed: one for participants with an ESD background and another for those without. The interviewees received an information letter and consent form prior to the interview, and based on their preference, the interviews were conducted online on a date and time of their convenience. Sixteen participants were interviewed, two from each identified stakeholder, having a mix of people with a background in ESD and not.

For data analysis, the quantitative data was analysed using the statistical Package for social sciences (SPSS). The data was analysed through both descriptive and inferential statistics following checks on the normality of data. With regards to the qualitative data was analysed using Braun and Clarke's thematic analysis framework (Braun & Clarke, 2006). The qualitative data was manually analysed for recurrent themes and sub-themes related to the research's objectives. The results were organised into 6 themes and 10 sub-themes, as shown in Diagram 2.

Triangulation was used to allow for a more comprehensive and in-depth understanding of the research topic. This approach allowed the researchers to not only identify trends and patterns in the data but also to gain insights into the lived experiences and perspectives of individuals on education for sustainable development. Triangulation was used to enhance the reliability and validity of the study findings

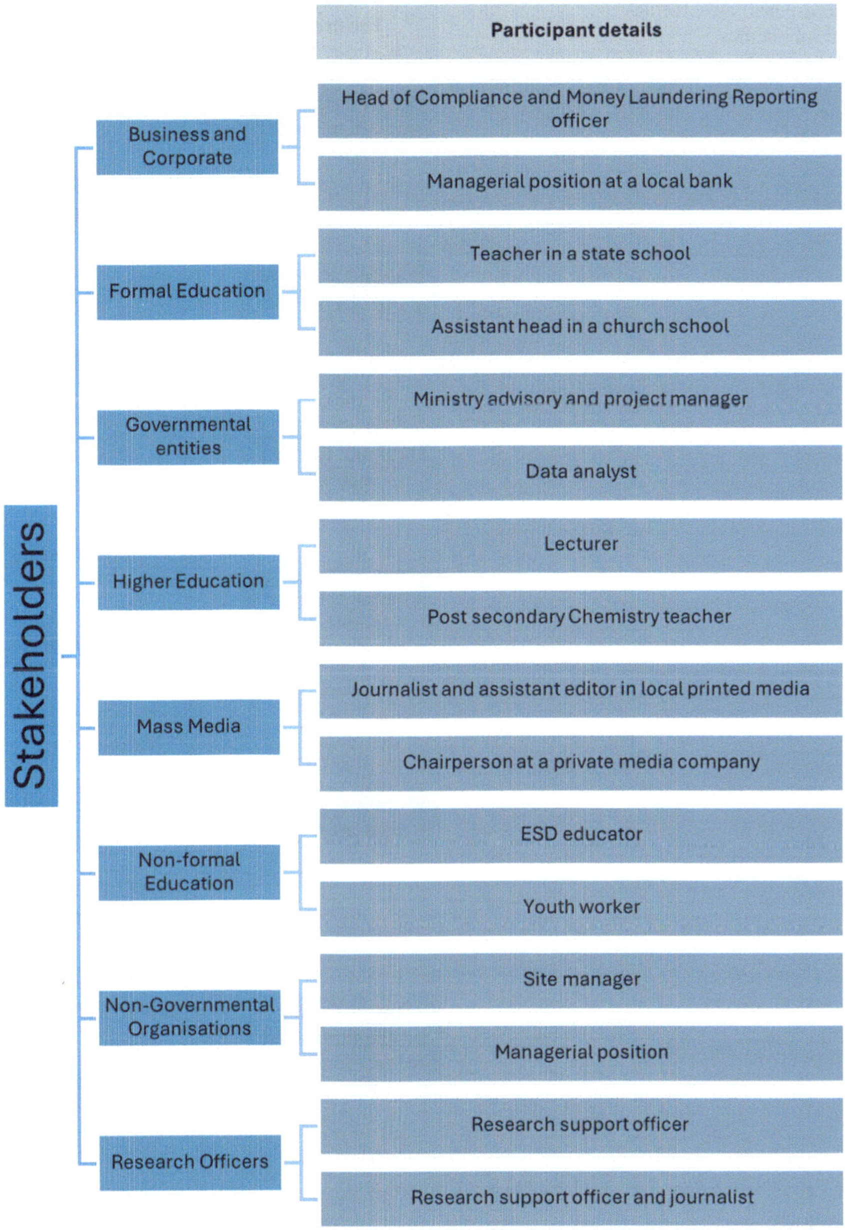

Fig. 1 The identified stakeholders with specific characteristics and experiences

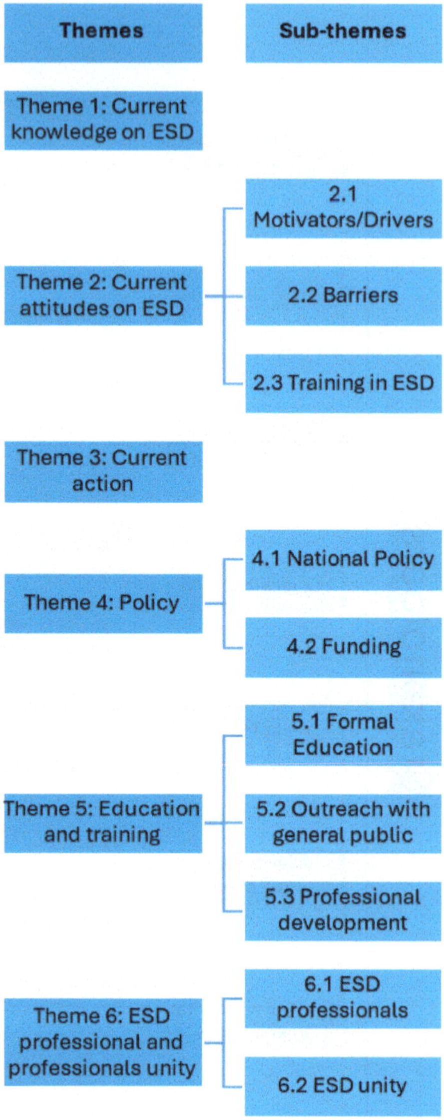

Fig. 2 Qualitative themes and sub-themes

by cross verifying the results obtained through these two different data collection methods.

3 Findings/Results

This section presents the findings of the study, combining both qualitative and quantitative data, under six key themes and their respective sub-themes. The results for each theme are presented using direct quotes from interviews and statistical data from the surveys.

3.1 Theme 1—Current Knowledge on ESD

This theme (refer to Diagram 2 in Sect. 2 for the themes) focused on awareness of SD and ESD.

It is noted that knowledge and awareness of SD is widely recognised with 75.23% of respondents agreeing that SD is a significant concern for Malta. However, there is less understanding of ESD with 69.52% of respondents viewing ESD of significant importance. The findings indicate that the participants that know about ESD are the ones who either have an academic background in ESD, are acquainted with someone that has such a background, or personally participated in ESD training or conferences. Frequently respondents heard of ESD from school and non-formal educational activities. The concept of SD is frequently associated with future generations and environmental concerns, reflecting common textbook definitions. A formal education employee mentioned,

> *I came across it [SD] ages ago in Junior College. We talked about the environment in Environmental Studies. At the time it was called Environmental Education. Then over the years in Social Studies we had to teach SD quite a lot, especially the link with the environment.* (Assistant head in a church school)

While all participants are aware of global and EU SD targets. The UN SDGs are the most frequently mentioned. Those without a background in ESD express familiarity with the SDGs but feel inadequately informed. Conversely, participants with ESD experience demonstrate a stronger understanding of the SDGs. Also, for some participants, the link between SDGs through the environmental aspect is much clearer than through other aspects, such as poverty. This is very prominent in one of the comments of a private business employee:

> *Poverty was always around, and wars ... and poverty has decreased globally. No matter the status, the environment, specifically climate change, is something that you can see and touch. We need to take action and fast. Moreover, everyone should be educated on it.* (Head of compliance and money laundering reporting officer)

Furthermore, when asked to define ESD, most participants without ESD background refer to education in terms of "why" and "how" SD issues were taught. However, it was noted that participants with a background in ESD are more critical of the concept of ESD.

3.2 Theme 2—Current Attitudes on ESD

The attitudes toward ESD initiatives reflect both enthusiasm and scepticism, influenced both by key motivators and barriers. Theme 2—Current Attitudes on ESD was further sub-divided into sub- themes to delve deeper into the different facets of the attitudes on ESD.

3.2.1 Sub- Theme 2.1—Motivators and Drivers

Many participants see the potential of ESD in driving behavioural change, 85.44% of respondents believe ESD could lead to positive behavioural change. However, a lack of a whole-institution approach was also mentioned by different stakeholders. Some participants state that SD and ESD efforts were usually carried out on top of formal duties by passionate members, who do not necessarily have a background in ESD. Also 'passion' was mentioned by many participants as the main perceived motivator with NGOs as the stakeholders with the highest perceived engagement in ESD. As stated by a business organisation employee,

> *NGOs are doing very good work, and they are passionate and committed.* (Managerial position at a local bank)

Other drivers for engaging in ESD initiatives include certain requirements such as funding or financial gain and having an official from top management in the organisation pushing the ESD agenda.

3.2.2 Sub- Theme 2.2—Barriers

Barriers such as a lack of engagement from the government and the business sector were commonly noted, only 15.23% and 16.19% think that that government and business sector are actively involved in ESD initiatives.

Another barrier is the lack of motivation and values from the local community that tend to complain about sustainability issues but take no action to change the situation. In line with this a research officer commented that,

> *It's hard to say whether the people are interested in sustainability ... but I think we also need to put our energy not only in venting our frustrations but also being constructive. We just complain and we stop, or people leave the island. At the end of the day, we need to solve our problems, we cannot run away.* (Research support officer and journalist)

These findings in the qualitative study are mirrored by the results of the quantitative study where 82.86% of the respondents perceive a low level of involvement in ESD initiatives by the community.

The lack of institutional support is perceived to be a challenge to ESD initiatives by 92.92% of the participants. It was also mentioned by the various stakeholders in the interviews. This challenge resulted in existing ESD efforts being undermined. A business organisation employee holds that,

> *Those part of the committee are made up of one top management member, who initiated it, out of his own will. The others, me included, were dragged into it because I was new.* (Head of compliance and money laundering officer)

Other barriers mentioned by the stakeholders are the feeling of "helplessness" and lack of funding. It is also noted that 87% of the survey participants consider funding as either an important or a very important potential challenge to implementing ESD. Another potential challenge mentioned by the stakeholders is that many SD and ESD activities are being carried out either as part of a tick-the-box exercise, or as marketing stunts. This links to the high degree of scepticism among professionals picked up in the survey, with 40.95% agreeing that ESD is met with scepticism in professional circles.

3.2.3 Sub-Theme 2.3—Training in ESD

Despite enthusiasm for ESD, with 85.44% of the survey respondents perceiving ESD as the key to achieving a positive change, there is limited engagement in professional ESD training. There is a consensus for the need for training for personal development, professional development, and the whole organisation advancement. A respondent employed with an NGO remarked,

> *ESD is something lifelong, you cannot say you know enough, but it's important even for professional development. And there are constant updates going on, constant upheavals, so I think to be on top, you must constantly work on yourself.* (Site Manager)

However, some limitations to ESD training are mentioned by different stakeholders, including time and lack of accreditations. Also, although the need for training in ESD is considered important with 85.71% of the respondents thinking that ESD should be made obligatory and 85.72% and 83.81% agree that the government employees and private sector employees, respectively, should have cyclical ESD training. The survey respondents and the stakeholders agree that interest in ESD training locally is still low.

3.3 Theme 3—Current Action

When the participants were asked which stakeholders, they thought were the least engaged in ESD, Businesses and Corporate ranked first by far, followed by Government and mass media. Similarly, the survey respondents perceive the involvement of government (15.23%), private sector (16.19%) and the community (17.14%) as unsatisfactory.

It is noted that there is different level of engagement among local businesses while some lack financial support and do nothing or the bare minimum; others have ESG targets in place and highly value SD and ESD initiatives.

With regards to the governmental sector, participants working within this sector both mentioned how some SD initiatives were being done in practice, but minor applications of ESD were being carried out. Moreover, these did not go beyond sharing of knowledge as stated by a government official,

> *There wasn't a two-way conversation where the staff was informed why these sustainability initiatives were being carried out, for example stickers next to the bathroom faucet, and AC capping initiative.* (Ministry Advisory and project manager)

Moreover, a participant explains how governmental agencies, which are autonomous, carry out ESD initiatives, but they do not refer to them as ESD initiatives.

When asked which stakeholders were perceived to be the most engaged in ESD, NGOs ranked first, followed by formal compulsory and non-formal education. These thoughts are mirrored by the quantitative findings with 22.33% of the respondents involved in ESD projects being engaged in teaching/training programs and 20.39% engaged in community related initiatives.

The NGOs with a specific focus on ESD, having certified ESD educators, comment about the work they do in formal and non-formal education through ESD programmes such as environmental summer schools.

Furthermore, NGOs practise the lifelong learning aspect of ESD by engaging with companies and the public as stated by an NGO employee,

> *Our mission is nature conservation and engaging with the public. Education as the way forward.* (Managerial position)

On the other hand, although there are initiatives in formal compulsory education institutions like Ekoskola, the integration of ESD into the formal education system remains limited to primary and secondary schools with initiatives in the higher educational institutions being sporadic. One participant employed with an NGO mentioned that,

> *Most schools, around 80% of schools in Malta have Eco-schools. The trend is in primary and secondary, not so much in post-secondary and Church.* (Assistant head in a church school).

Moreover, the importance of non-formal education opportunities abroad, to help link local issues with global ones, was also highlighted.

3.4 Theme 4—Policy

3.4.1 Sub- Theme 4.1—National Policy

Participants point to the lack of coherent national policies on ESD. A higher education employee holds that,

> *Globally we have the SDGs to help us achieve SD, and we're behind on many. There are policies at global, European and local levels, but I'm very skeptical about their effectiveness.* (Lecturer)

There is also a general agreement among interviewees that the Government of Malta needs to put in stronger efforts related to SD and ESD. An NGO employee stated,

> *The government impacts other stakeholders quite a lot, easily dictates the agenda, and they have different means to do it. E.g. enacting laws, by ensuring the laws are followed, or simply showing a good example. They have the means to do it.* (Managerial position)

Furthermore, participants working within governmental organisations highlight the need for the government to lead by example and legislate ESD as a requirement. Similarly, 77.45% of the survey respondents indicated that national guidelines to address ESD in the workplace are a main driver to implement ESD in Malta. The stakeholders further comment how ESD duties should not be given as extra duties to existing employees. A government official remarked,

> *They [the government] discussed the need for a person to act as a focal point…however, it should not be an existing employee with extra duties… they do not invest enough to employ an expert on ESD…it has to be a person with this specific role.* (Ministry Advisory and project manager)

Moreover, different stakeholders mention the need for ESD policies, enacted by the government or the entities themselves. There also seems to be mixed opinions on the effectiveness of enforcement, some respondents are in favour of enforcement whereas others suggest the need for instilling motivation and willingness through ESD rather than enforcement.

3.4.2 Sub- Theme 4.2—Funding

The power of the government was put forward again in terms of support through funding, either by providing or funding of an ESD professional to be available for varied stakeholders or to provide a budget for ESD initiatives. An NGO employee stated,

> *Government can provide funding to actually increase ESD…. If there are ESD professionals, these should be funded by the government ….* (Youth worker)

The lack of funding was also identified as a key barrier by the survey respondents with 87% highlighting funding as a significant barrier to implementing ESD initiatives.

Also, the power of the European Union in issuing funding is further highlighted, along with the pressure the EU can put on the national government regarding SD and/or ESD.

3.5 Theme 5—Education and Training

3.5.1 Sub- Theme 5.1—Formal Education

A common suggestion was to focus on education of the younger age groups since the participants deem these as the easiest to work with in terms of ESD and SD. Also, some mention that educating young minds early on ensures ESD becomes ingrained in the individual and this can help government achieve a cultural shift. A business organisation employee stated,

> *You start from school children. When the government wants to instil a cultural change, one has to start with children...Adults find it more difficult to change habits.* (Head of compliance and money laundering officer)

Additionally, the participants focus on the potential of schools for implementing change within the young generation, especially by including ESD across all subject curricula. Moreover, the majority mention the potential of incorporating specific units of ESD within the Higher educational institutions across all courses. Similarly, the survey results suggest that ESD is still not well integrated into higher education institutions, this perception is held by 91.43% of the respondents.

Once again, the holistic approach of educational institution was highlighted to achieve greater outcomes.

3.5.2 Sub- Theme 5.2—Outreach

The participants mention the difficulty in reaching the older generations and suggest methods on how to reach the public. Among the suggestions they mention monetary incentives and through "*giving people the tools that they can use*".

The need for effective ESD outreach, in person or through the media, was mentioned multiple times. A respondent working at a business organisation remarked,

> *... the rest of the community, to reach people you have ...maybe more educational campaigns on media? ...Television and radio are still imp among a certain cohort of people, the other types of media also must be used. Short messages.* (Managerial position in a local bank)

Further to the need for outreach and visibility, several participants emphasise the valuable work being done in academia but express concern about whether this knowledge is reaching the public. There is a need to ensure that research is not confined to academic circles but is effectively communicated to the public using the right communicative tools as noted by a research officer that stated,

> *CEER is the only centre focusing on ESD, there are courses, trainings, research...In theory are training people to do ESD. I'm not sure regarding training outside academia.* (Research officer and journalist)

The lack of outreach activities that engage the public is also mirrored by the survey results with 91.43% of the respondents thinking that there are not enough ESD processes or activities in the community.

3.5.3 Sub- Theme 5.3—Professional Development

There is a desire for more structured professional development in ESD. This is substantiated by feedback from the survey respondents with 85.72% and 83.81% thinking that government employees and private employees, respectively, should have cyclical ESD training. Also, one participant working at a business organisation shared,

> *Professionals need more training in how to implement ESD in their respective fields.* (Head of compliance and money laundering officer)

Professional development training was widely recommended across all stakeholder sectors, with the need to consider specific SD targets for the different sectors and thus the need for quality education through ESD.

3.6 Theme 6—ESD Professionals and Professionals Unity

3.6.1 Sub-theme 6.1—ESD Professionals

Respondents emphasise the importance of having qualified professionals lead ESD initiatives. An NGO employee stated,

> *We need experts in ESD to lead the projects and guide us in achieving sustainability.* (Youth worker)

In fact, a strong majority of participants agree on the need for these professionals, mentioning the skills they can bring about in different sectors. A government employee commented,

> *...a person that sits down with you, and explains what is SD, what happens if it is not achieved, how can we improve our personal lives... Someone who studied ESD keeps the target audience in mind and the things that they are teaching are adapted to the audience.* (Data analyst)

Various stakeholders mention the need for an ESD professional to reach sustainability targets, including in ministries and mass media. Also, 53.33% of the survey respondents strongly agreed that ESD professionals should lead these ESD initiatives. Stakeholders, with a background in ESD further emphasised the need for a professional to have at least a master's level of education.

However, one of the participants, a research officer, was unsure about the need for an ESD professional stating that,

> *I view it as being so cross cutting that I don't see it as a stand-alone profession, but it could be because it is still developing.* (Research support officer)

3.6.2 Sub- Theme 6.2—Professional Unity

There is a positive consensus among participants on the benefits of uniting ESD professionals, including the valuable services they could offer across sectors. A "one-stop shop" of ESD experts providing "priceless skills" and "peace of mind" when needed is highly regarded.

There is a perceived lack of professional unity and collaboration around ESD. A research officer noted,

> *Having an organisation to bring different people together, even if they are all trained by ESD, promoting ESD outside academia is very important.* (Research officer and journalist)

The need for a professional association is also emphasised through the findings of the quantitative study with 55.10% thinking that the absence of a professional association is a barrier to effective ESD. Also, ESD professionals are deemed by 86.54% of respondents as required to facilitate the design and implementation of activities and 78.09% think that warranted ESD professionals should lead ESD initiatives.

Also mentioned by stakeholders that such unity can facilitate conferences, training, and incentives to engage stakeholders in ESD especially if it is recognised on a ministerial level.

3.7 Socio-Demographic Correlations

This section presents the results obtained from the inferential statistics mainly involving Kruskall Wallis or one-way analysis of variance (ANOVA) tests. These were used following a Kolmogorov Smirnov test for normality which indicated that parts of the data were normally distributed but some parts were not.

This statistical test analysed the associations between three socio-demographic indicators of the questionnaire respondents (gender, age, education) and their responses to each statement in the various areas of the questionnaire.

Some of the significant correlations obtained are presented in Table 1. No significant difference was noted between male and female respondents in any of the statements. As displayed, older generations in Malta appear to be more sceptical about ESD effectiveness.

Additionally, there are also some differences between age groups regarding their opinion of cyclical training of employees with less importance afforded by the youngest and oldest age group. Regarding education, a rather bleak picture emerged as a higher level of education is associated with a more futile and helpless view of ESD. It appears that the more highly qualified ones are the more sceptical and the less appreciative ones of the benefits of programmes and education, specifically of an ESD pedagogy. While the less qualified seem to be believers of the change aspect of ESD and are proponents of it.

Table 1 Selection of significant correlations

	Social indicator	P-value meaning result	
There is a lot of skepticism related to ESD among various professions	Age	p = 0,043	Older generation agree more with this statement
Private employees should have cyclical ESD training	Age	p = 0,035	Youngest and oldest groups do not believe in this
An ESD action plan is essential to improve the outcome of our practices	Education	p = 0,040	Higher education up till Master degree qualified, believe less that this is so
ESD is the key to achieving a positive change in behaviour	Education	p = 0,024	Higher education individuals believe that ESD is not the key
Sustainability awareness programmes will lead to behavioural change	Education	p = 0,017	Higher education qualified, believe less that this is so

4 Analysis of Findings

The literature review outlines the importance of knowledge and awareness in fostering sustainable practices, emphasising that education is a pivotal component in shaping perceptions of SD and ESD This is echoed in the findings, where a substantial 75.23% of respondents acknowledge SD and 69.52% ESD as significant concerns for Malta. Participants with academic backgrounds or personal involvement in ESD programs exhibit a deeper comprehension of the SDGs. This is highlighted by the response of a stakeholder that is certified in ESD and that works in Higher Education that states,

> *Globally we have the SDGs to help us achieve SD, and we're behind on many. There are policies at global, European and local levels.* (Lecturer)

This connection between academic background and active involvement in ESD programs highlights the crucial role of both structured education and personal engagement in fostering a nuanced comprehension of sustainability. Recognition of this connection is foundational to initiatives like the Summer Universities led by MEdIES and MIO-ECSDE, which aim to deepen understanding of sustainability through immersive, experiential learning, that goes beyond passive absorption of information by integrating hands- on activities, aligned with the MSESD. By blending theoretical workshops with practical projects in biosphere reserves, these programs address local socio-environmental challenges, fostering skills that reveal the complex dimensions of sustainable development (MIO-ECSDE, 2019).

Another interesting finding emerging from the qualitative data suggests that without ESD training, participants often focus narrowly on environmental aspects of sustainability, underscoring the need for targeted educational interventions to broaden their understanding.

The findings reveal a complex landscape of attitudes toward ESD initiatives, characterised by both enthusiasm for its potential to drive behavioural change (85.44%) and scepticism regarding the commitment of key stakeholders with 40.95% agreeing that ESD is met with scepticism in professional contexts, notably the government and private sector. This duality aligns with the qualitative research with comments supporting the relevance of ESD to promote behavioural change such as the remarks from a respondent working in the non-formal education sector stating,

> *...at 16-18 you will have a lot of effect on society as an individual...So yes if there is someone who is specialised who can do this and help them become better citizens.* (Post-secondary chemistry teacher).

However, despite of the optimism, scepticism among professionals emerged particularly in the government and private sectors. One participant employed in the government sector remarked,

> *Initiatives from the ministries are quite scarce...I would lead by example, showing citizens that I care about SD.* (Ministry Adviser and project manager)

These findings suggest that while there is recognition of ESD's value, systemic barriers impede its broader adoption. The results highlight, a prevalent belief that ESD initiatives often fall on the shoulders of passionate individuals rather than being integrated into organisational frameworks, underscoring the need for a holistic institutional approach.

A respondent employed in the mass media sector states,

> *Usually however it is more that they are passionate rather than they are academically trained.* (Journalist and assistant editor)

Additionally, the high percentage of respondents (92.92%) who perceive a lack of institutional support underscores the findings' assertion that ESD efforts are often undermined. This perception strongly emerged from the comments of the stakeholders interviewed, such as the remark of a government employee that states,

> *There is a director taking care of these things. This person has a lot of other roles, this is just an add-on, as far as I know.* (Data analyst)

The quantitative and qualitative findings reflect a broader trend observed in the literature regarding the challenges of implementing sustainability initiatives in practice (most of the CSR initiatives in private entities locally then to be limited to awareness-raising and depart from thorough understanding of the holistic nature of CSR.

Both the findings and the literature (the National Strategy for ESD drafted in 2016 was shelved and is yet to be implemented) highlight a critical deficiency in coherent national policies related to ESD. Participants express frustration over the weak implementation of existing policies, a sentiment mirrored in the literature (most of the efforts to address ESD by the Government of Malta targets formal education), which advocates for stronger governmental leadership in driving the ESD agenda. The findings indicate that 77.45% of respondents believe national guidelines are crucial for effective ESD implementation, reinforcing the need for government action in providing resources and frameworks for ESD initiatives. The same opinion was mirrored in the interviews with many of the stakeholders emphasising the need for the Government to intervene with policies and mandatory actions, since the government "has the most muscle". However, some of the respondents showed disappointment with the government and the private sector as reflected with comments like "politics destroy the good work" and "they do one thing, more to show off".

Moreover, the findings point to the necessity of dedicated ESD professionals within governmental structures, suggesting that proper investment in human resources is essential for translating policy into practice. A government employee commented,

> *You definitely need a qualified person, in every ministry especially if targets are to be met! It has to be imposed by the government, that at least there is someone part-time.* (Ministry advisory and project manager)

The findings emphasise the pressing need for structured education and training in ESD, with 85.72% of respondents advocating for cyclical training of professionals. To this regard a research officer commented,

> *Since there is the push for SD, I think there should be the interest to educate all stakeholders.* (Research support officer)

Addressing time constraints, lack of accreditation, lack of funding and the feeling of "helplessness" may be the motivation needed for individuals to take part in ESD training programmes, thus fostering a culture of lifelong learning in ESD. There is the perception that a lot of the effort is being made by the few passionate people, but they are overwhelmed by a feeling of helplessness as worded by a research officer,

> *They do their best but may not lead anywhere ... so why bother.* (Research support officer)

The findings strongly support the necessity for qualified ESD professionals to lead initiatives, with 53.33% of respondents agreeing on their importance. Also, as mentioned by a government officer there is the need of,

> *Someone who studied ESD keeps the target audience in mind and the things they are teaching are adapted to the audience.* (Ministry advisory and project manager)

This aligns with the literature, which underscores the value of expertise in effectively guiding ESD efforts across sectors (ESD is a key enabler for SD, engaging quality ESD professionals in different sectors of society can help bridge the gap between the different stakeholders and the achievement of their SD objectives). Furthermore, the findings highlight a perceived lack of professional unity among ESD practitioners, emphasising the need for collaboration and the establishment of a professional association to facilitate knowledge sharing and resource allocation. This sentiment is reflected in the quantitative data, where 55.10% of respondents identify the absence of such unity as a barrier to effective ESD implementation. There is also a positive consensus among participants on the benefits of uniting ESD professionals, including the valuable services they could offer across sectors. A "one-stop shop" of ESD experts providing "priceless skills" and "peace of mind" when needed is highly regarded.

5 Conclusion and Recommendations

The findings and analysis of this study illustrate a complex yet crucial role for ESD in fostering sustainable change in Malta. Despite widespread recognition of ESD's importance, the journey toward effective implementation remains lined with challenges. One of the most significant issues emerging from the data is the feeling of "helplessness" among passionate, qualified ESD practitioners, particularly those within NGOs and grassroots organisations. These individuals often bear the brunt of ESD efforts, investing significant personal energy into initiatives but encountering systemic barriers, lack of institutional support, and limited access to funding. While their passion for ESD fuels their commitment, the overwhelming nature of these challenges has led to frustration and a sense of futility. For many, the lack of formal support translates into concerns about whether their efforts will lead to meaningful, lasting change. As one participant noted, they "do their best, but may not lead anywhere … so why bother," reflecting a broader feeling that passionate efforts alone are insufficient to drive substantial shifts in ESD practices.

This sense of disillusionment underscores an emerging need for qualified ESD professionals who can bridge existing gaps between stakeholders and foster a cohesive network among ESD practitioners. The data highlights that ESD is currently implemented in a fragmented manner, with each sector (e.g., NGOs, government, private sector) working in silos and often lacking a shared vision. A dedicated group of ESD professionals could act as connectors, facilitating collaboration across sectors and fostering a sense of unity among ESD practitioners. Such professionals would bring specialised expertise, capable of aligning grassroots enthusiasm with institutional goals and providing guidance on ESD practices tailored to different audiences.

Moreover, ESD professionals could help overcome the "tick-the-box" approach observed in both government and business sectors, where ESD initiatives are often minimal, performative, or disconnected from broader organisational objectives. Having trained, experienced professionals in ESD roles could ensure that initiatives are both impactful and aligned with genuine sustainability objectives. These professionals would not only facilitate knowledge sharing but could also be instrumental in advocating for ESD policies, guiding organisations in sustainable practices, and offering "one-stop" expertise in ESD for various sectors.

Unity among ESD professionals is also critical to achieving these goals. The study's findings suggest a strong consensus on the need for a professional network or association dedicated to ESD practitioners. Such a network could help address the isolation felt by many stakeholders, provide a platform for knowledge exchange, and advocate for the resources and policy support necessary for effective ESD. With 86.54% of respondents endorsing the idea of a unified professional community, the establishment of an ESD association could serve as a vital step in aligning diverse stakeholders and creating a collective voice for sustainable development.

Recommendations

In summary, the study suggests that the future of ESD in Malta depends on bridging the gap between resource-limited but passionate practitioners and resource-rich institutions with broader reach. A structured approach involving qualified ESD professionals and a dedicated network could foster collaboration, reduce feelings of "helplessness" among practitioners, and establish a sustainable path forward. By fostering unity and drawing on the strengths of all stakeholders, ESD can become the driving force in achieving Malta's sustainable development goals.

Further to these conclusions a set of recommendations are being suggested:

1. **Integrate ESD into Strategic Planning and Employee Objectives**
   Embedding ESD as a core component of organisational strategic plans is essential to ensure long-term commitment and accountability. Organisations should set measurable ESD targets for employees to clarify individual contributions towards sustainability goals and incentivise achievements to foster a culture of sustainable responsibility.
2. **Establish Government Support for ESD Professionals**
   Government funding is critical to supporting ESD professionals across sectors, ensuring organisations have the necessary expertise to meet regulatory and sustainability goals. This support could take the form of subsidies for ESD positions or budget allocations for sector-wide sustainability initiatives, particularly for organisations that lack the resources to hire dedicated ESD staff.
3. **Revise Education Curricula to Embed ESD Competencies**
   To develop a workforce and citizenry equipped with sustainability knowledge, ESD should be integrated into educational curricula at all levels. Appointing ESD leaders in schools and colleges would support this integration, helping to guide curriculum updates and foster an environment that prioritizes sustainability from an early age.
4. **Enhance Community Engagement through Higher Education Institutions (HEIs)**
   HEIs should play a proactive role in promoting ESD by engaging with their communities. By developing community-focused programs, workshops, outreach initiatives and promoting research on ESD; HEIs can empower individuals to take sustainable action and counter feelings of environmental helplessness, positioning ESD as a practical and accessible field.
5. **Establish ESD Professionals as Cross-Sector Resources**
   ESD professionals should be recognised as key resources for various sectors, offering expertise to help organisations meet their ESD goals. Government support for ESD consultations, particularly in under-resourced sectors, would help scale ESD efforts and facilitate knowledge sharing.

6. **Allocate Dedicated Budgets for ESD Implementation**
   Organisations should designate budgets specifically for ESD to support continuous improvement and long-term project sustainability. Budgeting for research and innovation within ESD can also lead to new advancements, keeping organisations at the forefront of sustainable practices.

These recommendations provide a framework for elevating the ESD profession as a cornerstone in the global effort to achieve sustainability, ensuring that organisations, educational institutions, and communities are equipped to address complex environmental challenges effectively whilst contributing to the achievement of the objectives of MSSD and MSESD.

References

Ametller, J., Asikainen, E., Oliva, M. G., & Nemejc, K. (2024). Teacher Training for Education for Sustainable Development (Developing a shared competence framework) (first, pp. 1–67). essay, Czech University of Life Sciences Prague. Retrieved 2024, from https://projects.tuni.fi/uploads/1955/20/880b1ad0-edusta_978-80-213-3324-6.pdf.

Braun, V., & Clarke, V. (2006). Using thematic analysis in psychology. *Qualitative Research in Psychology, 3*(2), 77–101. https://doi.org/10.1191/1478088706qp063oa

Burke Johnson, R., Onwueegbuzie, A. J., & Turner, L. A. (2007). Towards a definition of mixed methods research. *Journal of Mixed Methods Research, 1*(2), 112–133.

Commission of the European Communities. (2009). *Mainstreaming sustainable development into EU policies: 2009 Review of the European Union Strategy for Sustainable Development* 1–16. Retrieved 2024, from https://eur-lex.europa.eu/LexUriServ/LexUriServ.do?uri=COM:2009:0400:FIN:en:PDF.

Creswell, J. W. (2014). Research design: Qualitative, quantitative, and mixed methods approaches (4th ed.). SAGE Publications Inc.

Environment and Resources Authority, & Energy and Water Agency, The 3rd River Basin Management Plan for the Malta Water Catchment District 2021–2027 1–192 (2024). Retrieved 2024, from https://era.org.mt/wp-content/uploads/2024/07/Malta_3rd_River_Basin_Management_Plan.pdf.

European Commission. (2020). *A strategy for smart, sustainable and inclusive growth* 1–35. Retrieved 2024, from https://eur-lex.europa.eu/LexUriServ/LexUriServ.do?uri=COM:2010:2020:FIN:en:PDF

Greene, J. C. (2007). *Mixed methods in social inquiry*. Jossey-Bass.

Greene, J. C., Caracelli, V. J., & Graham, W. F. (1989). Towards a conceptual framework for mixed-method evaluation designs. *Educational Evaluation and Policy Analysis, 11*(3), 255–274.

Green Teacher. (n.d). *Diploma in Environmental Education*. [certification]. Retrieved 2024, from http://www.greenteacher.org/Programme%20Details.html

Government of Malta, Malta's 2030 National Energy and Climate Plan 1–258 (2019). Retrieved 2024, from https://sustainabledevelopment.gov.mt/wp-content/uploads/2021/10/Maltas-National-Energy-and-Climate-Plan.pdf.

L- Universita ta' Malta. (2020a). *Centre for the Environment Education and Research. About Us*. Retrieved from https://www.um.edu.mt/ceer/aboutus

L- Universita ta' Malta. (2020b). Master in Education for Sustainable Development.[courses]. Retrieved from https://www.um.edu.mt/courses/overview/PMESDPET2-2020-1-O

Mangion, G. M. (2024, April 25). ESG takes Roots in Malta. https://maltabusinessweekly.com/esg-takes-root-in-malta/25865/

Mifsud, M. (2012). Environmental education development in Malta: A contextual study of the events that have shaped the development of environmental education in Malta. *Journal of Teacher Education for Sustainability, 14*(1), 52–66. https://doi.org/10.2478/v10099-012-0004-x

Ministry for the Environment, Energy and Regeneration of the Grand Harbour, Sustainable Development Act, Chapter 521 1–7 (2012). Malta; Government of Malta. Retrieved 2024, from https://sustainabledevelopment.gov.mt/wp-content/uploads/2021/10/ACT521-%E2%80%93-Document.pdf.

Ministry for the Environment, Energy and Regeneration of the Grand Harbour, Malta's Sustainable Development Strategy for 2050 1–76 (2024). Government of Malta. Retrieved 2024, from https://sustainabledevelopment.gov.mt/wp-content/uploads/2024/06/Maltas-Sustainable-Development-Strategy-for-2050-1.pdf.

Ministry for the Family and Social Solidarity, National Strategic Policy for Poverty Reduction & for Social Inclusion Malta 2014–2024 1–93 (2015). Government of Malta. Retrieved 2024, from https://familja.gov.mt/wp-content/uploads/2023/04/Poverty-Strategy-14-EN.pdf.

Ministry of Education and Employment. (2012). *A National Curriculum Framework for All*. Floriana: Ministry of Education and Employment. Retrieved from https://curriculum.gov.mt/a-national-curriculum-framework-for-all-2/

MIO-ECSDE. (2019). *2019 Summer University on Protected Areas and Biosphere Reserves takes place in Bulgaria*. https://Mio-Ecsde.org/. Retrieved October 1, 2024, from https://mio-ecsde.org/2019-summer-university-on-protected-areas-and-biosphere-reserves-takes-place-in-bulgaria/

MISCO. (2019). (rep.). *A National Action Plan for CSR* (pp. 1–31). MISCO. Retrieved 2024, from https://www.yumpu.com/en/document/read/63234007/core-national-action-plan-csr-research-2018-converted.

National Commission for Sustainable Development. (2006). (working paper). *A sustainable development strategy for the Maltese Islands 2007–2016* (pp. 1–78). Malta. Retrieved 2024, from https://sustainabledevelopment.gov.mt/wp-content/uploads/2021/10/Sustainable-Development-Strategy-Maltese-Islands-2007-2016-December-2006.pdf.

North American Association for Environmental Education. (n.d.). Certification. [programme]. Retrieved 2024, from https://naaee.org/programs/certification

NSESD Board of Governors. (2019). *Nurturing a Sustainable Society. A National Strategy for Education for Sustainable Development for Malta.*, (working paper), 4–4.

OECD. (2024). (rep.). *Competition and Regulation in Professions and Occupations* (Ser. OECD Roundtables on Competition Policy Papers, pp. 1–58). Retrieved 2024, from https://www.oecd.org/content/dam/oecd/en/publications/reports/2024/05/competition-and-regulation-in-professions-and-occupations_a35fd2e6/218869f5-en.pdf.

Shaeffer, S. (2006). Stakeholders in the decade of Education for Sustainable Development (pp. 34–37). introduction, UNESCO. Retrieved 2024, from https://collections.unu.edu/eserv/UNU:8391/n2005.pdf

South African Qualifications Authority. (n.d.). *National Certificate: Environmental, Training and Development Practice.* [certification]. Retrieved from https://allqs.saqa.org.za/showQualifaction.php?id=22901

UNEP/MAP. (2016). Mediterranean Strategy for Sustainable Development 2016–2025. Valbonne. Plan Bleu, Regional Activity Centre. Retrieved 2024, from https://wedocs.unep.org/bitstream/handle/20.500.11822/7097/mssd_2016_2025_eng.pdf

Union for the Mediterranean. (2014). *Mediterranean Strategy on Education for Sustainable Development (MSESD): Final Draft [Ebook]* (pp. 1,3–4). Retrieved from https://ufmsecretariat.org/wp-content/uploads/2014/05/The-Mediterranean-Strategy-on-Education-for-Sustainable-Development-MSESD.pdf

United Nations. (2015). Transforming our world: The 2030 Agenda for Sustainable Development (A/RES/70/1). *General Assembly, seventieth session.* United Nations.

United Nations Economic Commision for Europe (UNECE). (2005). UNECE Strategy for Education for Sustainable Development. *High-level meeting of Environment and Education Ministries, Vilnius, 17–18 March 2005.* United Nations.

United Nations Economic Commission for Europe (UNECE). (2012). *Learning for the Future. Competencies for Education for Sustainable Development.* UNECE. Retrieved from https://unece.org/DAM/env/esd/ESD_Publications/Competences_Publications.pdf

United Nations Economic Commission for Europe. (2016). Ten Years of the UNECE Strategy for Education for Sustainable Development. UNECE. Retrieved from https://www.unece.org/fileadmin/DAM/env/esd/ESD_Publications/10_years_UNECE_Strategy_for_ESD.pdf

United Nations Educational, Scientific and Cultural Organisation (UNESCO). (2014). *Roadmap for Implementing the Global Action Programme on Education for Sustainable Development.* UNESCO.

United Nations Educational, Scientific and Cultural Organization (UNESCO). (2017). *Education for sustainable development goals: Learning objectives.* UNESCO. Retrieved, 2024 from https://unesdoc.unesco.org/ark:/48223/pf0000247444

United Nations Educational, Scientific and Cultural Organisation (UNESCO). (2020). *Education for Sustainable Development: A roadmap. ESD for 2030.* UNESCO. Retrieved, 2024 from https://unesdoc.unesco.org/ark:/48223/pf0000374802.locale=en

Whitby, A. (2019). *Advancing Education for Sustainable Development Key Success Factors for Policy and Practice.* World Future Council Foundation. 2024, https://www.worldfuturecouncil.org/wp-content/uploads/2019/01/Handbook-ADVANCING-EDUCATION-FOR-SUSTAINABLE-DEVOLOPMENT-by-Alistair-Whitby-WFC_2019.pdf

Widjajanti, D., Matakupan, S., & Didham, R. (2014). Introduction to Education for Sustainable Consumption (ESC) in Indonesia: National Recommendations and Guidelines for Policymakers and Educators.

Wong, L. (2006). Globalization and education for sustainable development: sustaining the future. UNESCO. Retrieved 2024, from https://unesdoc.unesco.org/ark:/48223/pf0000149295?posInSet=1&queryId=N-EXPLORE-32faf5bd-e889-47d8-bd70-7399e2cb9b27

Xerri Hili, C., & Schembri, H. (2022). Cross-curricular teaching within the Learning Outcomes Framework: The experiences of primary teachers in two state schools in Malta. *Malta Journal of Education*, *3*(2), 231–258. https://doi.org/10.62695/jegz4440

Another One Bites the Change—Change Management at Universities for Holistic Sustainability Establishment in Engineering Education

Elisabeth Nierle, Martin Pieper, and Linda Steuer-Dankert

Abstract The importance of sustainability is increasing across global society. For example, within the European Union new, binding standards were established, to ensure a continuous reflection of economic, social and governmental (ESG) sustainability criteria in organizations. By implementing the ESG standards, European Union emphasizes the necessity of competencies like critical thinking, system thinking but also an inter- and transdisciplinary understanding of sustainability, resulting in a continuous change process at educational institutions. To realize this, university structures and strategies are the necessary framework to enable this change and to ensure the provision of save spaces for interdisciplinary discourses, self-development and application-oriented training. As publicly funded educational institutions, universities in Germany have a particular responsibility to reflect society's needs and to assure a continuous reflection of sustainability. Consequently, in North Rhine-Westphalia, universities have committed themselves to sustainability, internationalization and digitalization as part of the Future contract "Strengthening study and teaching" (ZSL). The FH Aachen, as a university of applied sciences located in North Rhine-Westphalia, decided to focus in particular on sustainability in its strategy in 2022. As a result, internal processes were initiated in the ten faculties of FH Aachen to implement sustainability in teaching and research. As part of this paper, the change management process within the Department of Energy Technology will be presented as a lighthouse project at FH Aachen University. A structured approach will be described which, on the one hand, had the objective of establishing subject-specific knowledge of sustainability and, on the other hand, of developing key competencies for sustainable development in teaching. Core elements for achieving the objectives were the qualitative and quantitative analysis of existing concepts for sustainability and, based on this, participatory workshops within the department

E. Nierle (✉) · M. Pieper · L. Steuer-Dankert
Department of Energy Technology, FH Aachen - University of Applied Sciences, Jülich, Germany
e-mail: nierle@fh-aachen.de

M. Pieper
e-mail: pieper@fh-aachen.de

L. Steuer-Dankert
e-mail: steuer-dankert@fh-aachen.de

W. Leal Filho et al. (eds.), *Introducing the UN SDGs in Higher Education Institutions*, World Sustainability Series, https://doi.org/10.1007/978-3-032-23017-1_19

in order to incorporate the various perspectives and embed them effectively in the curriculum.

Keywords Change management · Higher education structure · Strategy · Key competencies · Participatory approach

1 Introduction and Motivation

The global pressure to act on climate change and the ongoing destruction of natural resources has made sustainability a priority in almost all areas of the public and private sector in recent years (Linnér & Wibeck, 2019; Mahardhani, 2023; Rashed & Shah, 2021). As educational institutions, universities in particular have a responsibility to prepare students for the challenges of a sustainable future (Sterling, 2010; UNESCO, 2017). Thus, the integration of sustainability into higher education has shifted from being a voluntary effort to an essential component of university curricula, particularly in disciplines like engineering (Gutierrez-Bucheli et al., 2022; UNESCO, 2022). As global challenges related to climate change, resource depletion, and social inequality grow more pressing, universities have a responsibility to prepare future professionals who are equipped to tackle these issues (Azizi & Sassen, 2023; Moggi, 2023; Sassen & Azizi, 2018). The European Union has recently introduced binding standards that require organizations to adhere to Environmental, Social, and Governance (ESG) criteria, such as the Corporate Sustainability Reporting Directive (CSRD) (Dinh et al., 2023; Hummel & Jobst, 2024). This regulation mandates transparent reporting on non-financial performance, focusing on environmental goals, social responsibilities, and corporate governance. Universities must, therefore, align their educational programs with these evolving expectations to ensure that graduates possess the skills and competencies necessary to contribute meaningfully and sustainable development. Curricular structures that make sustainability a consistent cross-cutting topic play a central role in this. They not only provide students with the necessary subject-specific knowledge, but also promote a critical examination of the ecological, social and economic dimensions of sustainable development (Ruben et al., 2023; Tilbury, 2011).

At the Department of Energy Technology at FH Aachen University of Applied Sciences, the necessity of embedding sustainability in engineering education has been recognized as a strategic priority. This is on the one hand intended to meet the increasing demand from industry, society and politics, all of whom expect graduates to have a comprehensive understanding of sustainability. Furthermore, they should be able to implement these principles in their professional lives. As future employees, students must be trained not only in technical competencies but also in broader concepts such as systems thinking, critical analysis, and an interdisciplinary approach to problem-solving (Ruben et al., 2023; Sanyal, 2024; Tilbury, 2011).

For this reason, a process for the systematic integration of sustainability into Bachelor's degree programs was initiated in 2024 and is still ongoing. A top-down and bottom-up approach was combined for this purpose. As a first step, the aim was to integrate sustainability in various facets into all modules of the curriculum so that the topic of sustainability is not treated in isolation from other disciplines. Such an approach requires the cooperation and support of all teaching staff. Since the freedom of research and teaching is guaranteed by law in Germany, difficulties are to be expected with such serious changes. It is therefore a change process that needs to be prepared and organized accordingly in order to include and involve all faculty members. Furthermore, successful embedding requires a systemic approach that addresses curriculum and module design, mindset change and institutional culture.

Section 2 summarizes the theoretical foundations of change management, with a particular focus on higher education institutions, which form the basis of our process. There are numerous models and frameworks that support implementation (Trevesian, 2024). We use an approach developed at FH Aachen, which is mandatory as part of the regularly conducted accreditations. The Department of Energy Technology was the first faculty to begin implementing this approach. Details and the chronological implementation are described in Sect. 3. Finally Sect. 4 sums up the lessons learned and discusses the transferability to other institutions. The following structured approach guided the transformation (adapted from Trevesian, 2024):

1. **Vision & Leadership Commitment**: The rectorate initiated a comprehensive strategy process for changes in 2022, which includes the universities vision. The essential foundation are the three key pillars: sustainability, internationalization and digitalization, which were also included in the Department's strategies. This institutional priority created the fundamentals for initiating change processes in all departments. However, the specific program-related implementation, including the integration of sustainability, remain the responsibility of the individual departments. To support this decentralized approach, a central working group, including representatives from the rectorate, deans of studies, and the Centre for University Didactics and Quality Development in Teaching and Learning (ZHQ), developed a flexible framework concept for refining study programs. It was deliberately kept open-ended to accommodate the specific requirements and priorities of each department. This balance between top-down strategic direction and bottom-up input from individual departments was critical to the success of the process as it allowed flexibility while ensuring coherence across the institution (Barnat et al., 2024).
2. **Curriculum Integration & Pedagogical Shift**: Working groups were established to identify suitable teaching methods, such as Problem-Based Learning and inter- and transdisciplinary formats, as well as appropriate sustainability content based on relevant literature.
3. **Faculty Development & Institutional Culture**: The appropriate content and teaching methods were subsequently discussed with the teaching staff to explore their integration into the respective modules. This process, supplemented by additional training sessions led by faculty experts, fostered a shared understanding and a common mindset.

4. **Student Empowerment & Engagement**: By law, students are actively involved in the development of teaching. Each faculty has a student dean and student representatives on the faculty council. These representatives, along with other interested students, were engaged throughout the entire process.
5. **Alignment with the UN SDGs**: The SDGs are an essential component and also reflected within the universities sustainability report. During training sessions and discussions with the teaching staff, it was explored how and which SDGs can be addressed in individual modules and how an active connection to teaching can be established.

2 Change Management in Higher Education

Despite the clear relevance of sustainability in the educational context, universities face considerable structural, cultural and organizational challenges in systematically integrating sustainability into their curricula (Barth & Rieckmann, 2012). This is where change management comes into play as an approach to designing and managing change processes (Kotter, 1996). It enables the systematic implementation of change in institutions by taking into account both the organizational structures and the actors involved and their willingness to change (By, 2005; Kotter, 1996). Consequently, to effectively integrate sustainability into curricular structures, universities must take comprehensive and targeted measures that affect not only the teaching content, but also the institutional culture, structures and cooperation with external stakeholders. Sterling (2004) describes this as a "systemic view" which is necessary to achieve long-term change, particularly in the higher education system. Numerous analyses have therefore examined structures, hierarchies, power relations and the distribution of resources that seem to be characteristic for academic institutions and that need to be considered when developing change management measures for the respective institutions. For this reason, the key findings that formed the basis for the approach presented are summarized below.

Birnbaum (2000) distinguishes the **hierarchical structures** and legal authority systems from the non-academic sector as "loosely coupled processes of academic institutions [which] buffer educational from administrative procedures and permit subgroups to operate with significant autonomy. This makes it easier in higher education for an innovation to be publicly 'adopted,' but not actually implemented in a way that affects core institutional processes.". In doing so, Birnbaum (2000) expresses that a superficial implementation of new measures is fundamentally easier, but that it seems difficult to check to what extent implementation is taking place and how long-term it is designed, which he attributes to the non-binding powers of instruction in the academic system. In conclusion, hierarchical control mechanisms seem not appear to be effective or at least questionable from Birnbaum's perspective.

This statement seems surprising as studies show that academic institutions are often divided into clearly defined **hierarchies** (Kezar, 2001; Bleiklie & Byrkjeflot, 2002; Eichhorn & Kubon, 2023) or in Kezars (2001) words "multiple power and authority structures". These hierarchical structures significantly influence decision-making processes and the allocation of resources, which leads to certain power relations. Furthermore, they can lead to a concentration of power in the hands of a few leaders, significantly influencing decisions and priorities (Kezar, 2001). This dynamic has effects on institutional agility and innovation (Birnbaum, 1988).

Hierarchical structures can mean that the opinions and **experiences of employees** at lower levels are not sufficiently incorporated into decision-making processes, which can lead to a feeling of alienation (Bess & Dee, 2008). This is particularly relevant against the background of the necessary acceptance of change initiatives and the support of corresponding measures by employees (Barth & Rieckmann, 2012; Castillo et al., 2018; Kotter, 1996; Straatmann et al., 2016).

Makrakis (2014) sees the transformation of university curricula towards sustainability as a process that integrates the principles of ecological integrity, social and economic justice and human well-being into the content and methods of teaching. This places **teaching staff at the center** of curricular transformation. Studies suggest that transformative learning processes that enable profound change depends largely on the ability and willingness of teachers to facilitate such processes (Barth & Rieckmann, 2012). Embedding sustainability in university curricula is therefore seen as a dynamic process that broadens teachers' perspectives and strengthens their competencies for global responsibility and sustainable action (Makrakis, 2014). Sterling (2004) emphasizes that "[…] sustainability does not simply require an 'add-on' to existing structures and curricula but implies a change of fundamental epistemology in our culture and hence also in our educational thinking and practice".

University employees and their willingness to support change therefore appear to play a central role in change processes. This is supported by Rudhumbus' (2015) study "Managing Curriculum Change from the Middle: How Academic Middle Managers Enact Their Role in Higher Education". Rudhumbu (2015) points out that academic middle managers (AMMs) play a central role in change processes in the higher education system, particularly in the planning and implementation of curriculum changes. The study emphasizes that AMMs are crucial in supporting educational and curricular change and contribute to the successful implementation of change in educational institutions. However, AMMs are often under conflicting demands (role conflict) as they have to perform both academic and administrative tasks while fulfilling expectations from different sides (Rudhumbu, 2015). They act as intermediaries between professors and staff by strategically implementing curriculum adjustments, gathering information and promoting innovation (Amabile & Pratt, 2016). Their role as bridge builders and translators of information is crucial for the effective management of change and the support of employees during the transformation process.

Depending on the hierarchical understanding and authoritarian structures, however, it is questionable to what extent the **employees, as a key position** in change processes in the higher education system, can actively participate in development processes. Birnbaum (2000) emphasizes the role of flatter hierarchies in promoting the innovative capacity of universities. By making the institutional structure more agile, universities can respond better to change and unleash the creative potential of their employees. A dynamic and adaptive organization is able to implement new ideas quickly and adapt to the challenges of the ever-changing academic environment. In a similar vein, Amabile and Pratt (2016) argue that organizational structures, including hierarchies, have a significant impact on creativity and innovation. They call on universities to implement flatter hierarchies in order to increase innovation and promote collaboration between different departments. Such an environment encourages employees to develop creative solutions and pursue innovative and transformative approaches. This is supported by Bess and Dee (2008) which provide important insights into the impact of hierarchies on overall satisfaction and innovativeness in universities. Ezzeddine et al. (2023) add, that clear communication, strategic leadership and the active involvement of all employees seem to be key success factors to increase the agility and innovative capacity of institutions. In summary, **participatory processes** seem to be needed that ensure the integration of different perspectives in change processes across hierarchies and enable active involvement.

In addition to the structurally anchored hierarchies, **bureaucracy** is another special feature of the higher education system. The need to navigate through multiple levels of approval, especially in publicly funded organizations, can significantly delay the implementation of new ideas (Kezar, 2001). For example, Birnbaum (1988) shows that excessive bureaucracy can lead to paralysis of innovation by limiting the flexibility and responsiveness of institutions. Studies show that highly bureaucratized systems have difficulty adapting quickly to change and implementing new ideas, while less bureaucratic structures allow for a faster response to challenges and changes (Birnbaum, 2000).

Implementing sustainability in curricular structures at universities requires a change process critical examining existing hierarchical structures and the creation of a supportive culture that promotes both faculty productivity and employee satisfaction. By reducing bureaucratic hurdles and encouraging open dialog between levels, universities can create an environment that fosters innovation and efficiency. In addition, a supportive and inclusive culture that promotes the involvement of all employees can be critical to maintaining high employee motivation and retention. Universities should therefore develop strategies that actively encourage participation at all levels and ensure that employees' voices are included in change and decision-making processes. In summary, the creation of a contextualized framework for change management in higher education institutions is essential to overcome the challenges mentioned above. Learning from the findings of the scientific literature, a process is therefore presented below which attempts to take these aspects into account in order to be able to implement sustainability in the curricular structures.

3 Change Process at the Department of Energy Technology

The Department of Energy Technology currently has around 1100 students, 950 of whom are studying in the 12 Bachelor's programs and 150 in the three Master's programs. The Bachelor's programs cover the fields of electrical engineering, mechanical engineering and physics engineering, each with a focus on energy technology. The department therefore has great potential for interdisciplinary cooperation.

Sustainability was integrated into the curriculum as part of the reaccreditation of the Bachelor's programs. Thus, the structuring of the change process at the Department of Energy Engineering was dictated by the accreditation elements and deadlines. In total, the department had approximately 6 months to complete the process. The process began with a comprehensive assessment by the faculty of the current integration of sustainability issues into the curriculum, followed by a concerted effort to identify opportunities for improvement and develop new strategies for integrating sustainability into all courses. Figure 1 shows the timeline and different steps of the process, which are explained in more detail in the following.

Step 1: Framework Survey

In December 2023, the first step in the process was a central online survey of all teaching staff in the department. Students from the relevant study programs also took part in the survey in order to gain an additional student perspective. For a better understanding, all participants received an explanation of the terminology so that the questions could be understood in a uniform way. The university-wide survey aimed at

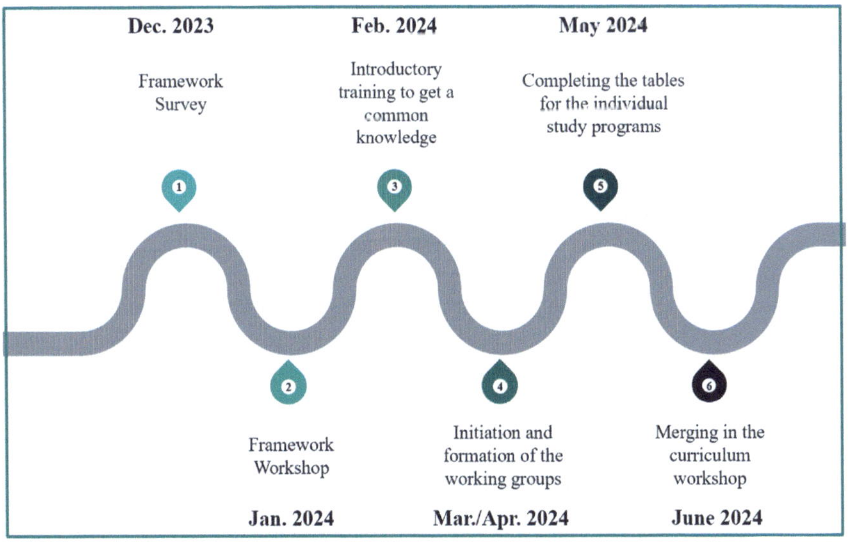

Fig. 1 Road map of the change process in the Department of Energy Technology, FH Aachen

identifying how sustainability was currently being addressed in the study programs and where improvements could be made. It was designed to be comprehensive, covering a range of sustainability issues, from environmental goals to social and economic considerations, reflecting the broad definition of sustainability as outlined by frameworks such as the United Nations Sustainable Development Goals (SDGs). Importantly, the survey was tailored to each department, allowing for discipline-specific insights and making it easier to identify gaps in the curriculum that were relevant to each field of study (Barnat, 2024). The results of the surveys were then compiled and visualized by the ZHQ. This central department of the university is responsible for supporting the curriculum development and reaccreditation processes at FH Aachen.

The first hurdles already became apparent in this first step. A central challenge was the perception of a lack of transparency and external control, which affected the trust of the staff. According to the model of psychological safety (Edmondson & Lei, 2014), resistance arises when participants do not feel they have sufficient control over the process. In hierarchical structures, as implied by the ZHQ's role as the supervising body for curricular processes and reaccreditations, there is the risk of a perception conflict. In addition, some members of the department fear that the results of the survey could be used to evaluate performance or to sanction criticism. Since it was processed centrally, some staff felt that they were merely the objects of the process instead of actively participating in it. Due to this, it appeared that some questions were not answered without bias during the survey.

Step 2: Framework Workshop

As the second step, the results of the survey were discussed with all faculty members, including student representatives, at the Framework workshop in January 2024. This workshop provided a forum for faculty members to review the survey results and collectively decide where improvements were most needed. The workshop was not only a critical step in identifying gaps in the curriculum but also a space for faculty to begin building a shared understanding of what sustainability means in the context of energy technology.

The workshop was led by the ZHQ which resulted in reinforced the power imbalance, as it was primarily conducted by a central office. As a result, some teachers perceived the process as top-down, which in some cases reduced the acceptance and motivation to implement the topics.

Additionally, the workshop revealed that there are significant differences in both knowledge and awareness of sustainability among faculty members. While some lecturers were well-versed in sustainability concepts and eager to integrate them into their teaching, others found the topic abstract and difficult to apply to their specific fields. This discrepancy highlights a key challenge of the change process: translating the general and often theoretical sustainability principles into practical, subject-specific learning outcomes. Thus, it became clear during the workshop that one of the main obstacles was the abstract nature of sustainability.

Recognizing the need for a structured and collaborative approach, the dean took immediate action by organizing a series of information sessions aimed at raising awareness about the importance of sustainability. These sessions were designed to provide the faculty members with a foundational understanding of sustainability issues, helping them see how these topics could be relevant to their individual courses. The kick off was a voluntary two-hour training course on the basics of sustainability for all employees.

Step 3: Ramp up Training—Sustainability

In February 2024, the first internal training on sustainability was conducted by a member of staff from the department who was familiar with the specific requirements, challenges and circumstances of the respective degree programs. In this way, she was able to contextualize the knowledge imparted and adapt it to the needs of the participants, making the training more practical and relevant. In addition, she was perceived as a more credible and trustworthy source than external or central actors such as the staff unit. This approach reduced the perceived power imbalance and created an open learning atmosphere. Employees feel more secure asking questions or expressing uncertainties, which increases the willingness to learn in line with the psychological safety concept (Edmondson & Lei, 2014). Furthermore, the internal departmental specialist was able to provide more sustainable support and maintain contact with the participants, e.g. for further questions or in-depth discussions. It also increased both the participants' trust in the process, relevance and motivation for the transfer of the imparted knowledge into practice. Moreover, a significant increase in networking and discussions on sustainability topics within the department could be observed after the training. Overall, the provided training was a good starting point and encouraged the employees to take responsibility for the process.

Once the described change process has been completed, the various elements must be integrated into all modules. Several training courses are available to teaching staff for this purpose: In general, at FH Aachen, training and support for teaching staff is organized centrally and carried out by ZHQ. They offer ranges from general didactic aspects such as different teaching formats to special consulting on sustainability. Individual training courses are also developed on request. Furthermore, further education offers of an inter-university network (hdw NRW) can be visited.

Step 4: Initiation and Formation of Working Groups

In parallel, three working groups were formed, consisting of faculty members with more experience in sustainability. Additionally student representatives were involved. Each group was tasked with addressing a specific aspect of sustainability integration, and the dean's office took on the role of coordinating and moderating the process. The groups began their work at the start of the semester in March 2024 and completed their work in April.

The first working group focused on identifying the competencies necessary for students to thrive in a sustainable future. Given the abstract nature of competencies like e.g. systems thinking and interdisciplinary collaboration, the group translated these into concrete learning methods that could be applied within the existing study programs. The group's efforts were supported by an extensive literature review, which helped ensure that the proposed learning methods were evidence-based and aligned with current best practices in sustainability education. This also took into account the industry perspective and current political developments, e.g. national, and EU legislation. The final result was a detailed table that mapped specific teaching methods—such as project-based learning, case studies, and inter- and transdisciplinary projects—onto the corresponding modules in the curriculum.

The second working group worked analogously but focused on the specific sustainability topics that needed to be addressed within the study programs. This group identified key concepts such as the definition of sustainability, the United Nations Sustainable Development Goals (SDGs), and the three pillars of sustainability—environmental, social, and economic dimensions. Additionally, they integrated aspects of sustainability reporting, a crucial skill for engineers who will need to comply with evolving regulatory frameworks like the Corporate Sustainability Reporting Directive (CSRD) and Global Reporting Initiative (GRI) standard. Like the first group, they produced a table that linked these topics to all modules, ensuring that sustainability was embedded in both foundational and advanced courses.

The third working group concentrated on technical topics, exploring how sustainability could be integrated into the core engineering subjects of the curriculum. This group examined how traditional engineering disciplines such as mechanics, thermodynamics, and energy systems could be taught through a sustainability lens. For instance, in mechanical engineering courses, students could learn about energy efficiency and sustainable materials, while thermodynamics classes could explore renewable energy sources and their role in reducing greenhouse gas emissions. The group's efforts ensured that sustainability was not treated as an add-on or separate topic but was instead woven into the very fabric of the engineering curriculum.

The challenges of change management described above, such as rigid hierarchies, bureaucratic hurdles and the difficulty of promoting cross-faculty collaboration, were specifically addressed in the working groups. The establishment helped to break down barriers. It allowed for participatory and interdisciplinary collaboration, ensuring that members of the various degree programs within the department were actively involved in the change process. The clear coordination by the dean's office created a framework that minimized bureaucratic delays and promoted exchange between the participants.

Furthermore, the working groups addressed the challenge of translating abstract sustainability skills and topics into concrete teaching methods and selecting sustainability topics in great detail for understandable comprehension. This not only created a shared understanding but also ensured long-term implementation that takes into account both structural and social aspects of change. In doing so, the working groups showed how a systemic approach can promote sustainable change in higher education institutions.

Step 5: Mapping Status quo—sustainability in Existing Teaching Formats

Once each working group had completed its analysis, the results were presented to the entire faculty in a series of meetings held in May 2024. These meetings were critical for ensuring that the proposed changes were accepted and that any remaining concerns could be addressed. Faculty members were asked to review the tables created by the working groups and indicate where they were already incorporating the suggested teaching methods or sustainability topics into their courses. This collaborative approach helped to create a sense of ownership over the process and fostered a shared understanding of the department's sustainability goals. For example, if a faculty member indicated that they were already using a particular teaching method, a mark was made in the table. However, these marks were not added without discussion; lecturers were encouraged to explain how they were implementing the method and whether it aligned with the overall sustainability objectives. This discussion ensures a common understanding of sustainability and the associated issues.

Overall, the meetings were crucial not only to creating acceptance for the proposed changes, but also to addressing the existing structural and cultural challenges. In particular, the hierarchical and bureaucratic structures that often inhibit decision-making processes in higher education institutions were overcome by openly discussing and involving employees in the change process. Working through the tables together helped to reduce resistance and develop a shared vision for integrating sustainability into the curriculum that took into account both institutional goals and the needs of the staff.

Step 6: Merging Results in the Curriculum Workshop

The final step of the change process was the curriculum workshop, held in June 2024. This workshop brought together the results of the working groups and the faculty discussions, allowing for the final adjustments to be made to the curriculum. This phase relied on close and open collaboration between faculty members when discussing the results of the meetings. The structural challenges of change management, such as overcoming hierarchies and promoting participation at all levels, remained present. In particular, balancing academic autonomy and organizational coordination posed a central hurdle.

However, the co-creation curriculum workshop provided again a valuable opportunity to address these tensions through collegial dialogue and the open discussion of concerns. The participatory approach helped to ensure that the staff could identify more clearly with the changes. At the same time, it provided the opportunity to make again specific adjustments in the individual modules and to effectively integrate the sustainability topics. A key aspect was to create incentives for employees to implement these topics in their teaching on a permanent basis, without it being perceived as an additional burden. Here, the change management approach was able to successfully implement a transparent and continuous communication, resulting in a consensus-building and future-oriented curriculum that integrates the basic principles of sustainability into the department.

4 Lessons Learned

The tight schedule—spanning from December to June—was dictated by the accreditation process, which required the new curriculum to be finalized by mid-2024. Despite the time constraints, the department successfully completed all the necessary steps, ensuring that sustainability will not only be included but embedded holistically within the study programs in future. To ensure this, it is necessary to adjust the KPIs from the faculty's objectives based on the finalized implementation in the modules. Possible KPIs include, for example, the frequency of certain teaching methods in the curriculum or the coverage of the SDGs within the curriculum. The structured and transparent approach to the change process, coupled with the collaborative efforts of faculty members, enabled the department to meet its goals on time while fostering a culture of sustainability within the faculty.

The change process at the Department of Energy Technology provided several valuable lessons that can inform future efforts to integrate sustainability in higher education. One of the key insights was the importance of early and transparent communication. Initially, many faculty members experienced shock and resistance when the need for change was first communicated. This reaction highlighted the importance of providing clear and consistent information about the reasons for the change and its anticipated benefits. By addressing concerns early in the process and involving faculty members in decision-making, the department was able to minimize resistance and build engagement.

Another important lesson was the need for flexibility and ongoing support. While some members of the department were already familiar with sustainability concepts and eager to incorporate them into their teaching, others found the topic abstract and difficult to translate into practical learning outcomes. This divergence necessitated a structured approach to building a shared understanding of sustainability and its relevance to engineering education. As the change process progressed, it became clear that faculty members required targeted training and resources to effectively implement sustainability topics in their courses. The department responded by providing workshops and other forms of professional development, ensuring that faculty members will have the tools they needed to succeed. The emotional journey of the change process was non-linear, with periods of frustration and uncertainty interspersed with moments of progress and acceptance. Regular feedback loops and the ability to adapt the process in response to feedback were critical for maintaining momentum and ensuring that the changes were sustainable in the long term.

Finally, the process underscored the importance of creating a structured environment for collaboration. Faculty members needed time and space to exchange ideas and develop a common understanding of sustainability. This was facilitated by the department's decision to hold regular workshops and to moderate discussions externally, ensuring that all voices were heard and that the process remained focused on achieving its objectives. The creation of working groups with specific responsibilities additionally helped to ensure that the change process was manageable and that progress could be tracked over time.

In summary, the following aspects are essential and should be considered when transferring to other higher education institutions:

1. The level of abstraction may be perceived as too high. Training sessions and support from experienced faculty members can help in this regard.
2. Concrete suggestions for sustainability content and appropriate teaching methods are helpful for implementation. Experienced faculty members can also provide support in this regard.
3. Engaging all teaching staff in the concrete implementation process and discussing it collectively is crucial for increasing acceptance and fostering an institutional mindset.
4. Involving faculty members with greater expertise or a strong interest in the topic increases acceptance among the others and can significantly support the process.

References

Amabile, T. M., & Pratt, M. G. (2016). The dynamic componential model of creativity and innovation in organizations: Making progress, making meaning. *Research in Organizational Behavior, 36*, 157–183.

Azizi, L., & Sassen, R. (2023). How universities' social responsibility activities influence students' perceptions of reputation. *Journal of Cleaner Production, 417*, Article 137963.

Barnat, M., Arntz, C., Bernecker, A., Fissabre, A., Franken, N., Goldbach, D., Hüning, F., Jörissen, J., Kirsch, A., Pettrak, J., Rexforth, M., Rosenkranz, J. & Terstegge, A. (2024). Strategische Gestaltung von Studiengängen für die Zukunft: Ein kollaborativ entwickeltes Self-Assessment. In Discussionpaper: Bd. Nr. 33. Hochschulforum Digitalisierung. Online available under: https://hochschulforumdigitalisierung.de/wp-content/uploads/2024/07/HFD_DP_33_Strategische_Gestaltung_von_Studiengaengen.pdf (Last accessed: 12 April 2026)

Barnat, M. (2024). Endlich da: Der Framework Fragebogen, ZHQ Blog. https://zhq-blog.fh-aachen.de/curriculumentwicklung/endlich-da-der-frameworkfragenbogen/ (Last accessed: 12 April 2026)

Barth, M., & Rieckmann, M. (2012). Academic staff development as a catalyst for curriculum change towards education for sustainable development. *Journal of Cleaner Production, 26*(2), 383–391.

Bess, J. L., & Dee, J. R. (2008). Understanding college and university organization: Theories for effective policy and practice. *ASHE Higher Education Report, 33*(3).

Birnbaum, R. (1988). *How colleges work: The cybernetics of academic organization and leadership.* Jossey-Bass.

Birnbaum, R. (2000). The life cycle of academic management fads. *The Journal of Higher Education, 71*(1), 1–16.

Bleiklie, I., & Byrkjeflot, H. (2002). Changing knowledge regimes: Universities in a new research environment. *Higher Education, 44*(3), 519–532.

By, R. T. (2005). Organizational change management: A critical review. *Journal of Change Management, 5*(4), 369–380.

Castillo, C., Fernandez, V., & Sallan, J. M. (2018). The six emotional stages of organizational change. *Journal of Organizational Change Management, 31*(3), 468–493.

Dinh, T., Husmann, A., & Melloni, G. (2023). Corporate sustainability reporting in Europe: A scoping review. *Accounting in Europe, 20*(1), 1–29.

Edmondson, A. C., & Lei, Z. (2014). Psychological safety: The history, renaissance, and future of an interpersonal construct. *The Annual Review of Organizational Psychology and Organizational Behavior, 1*(1), 23–43.
Eichhorn, K., & Kubon, S. (2023). *#Wissenschaftshierarchien* (p. 2023). Hemmnisse im deutschen Wissenschaftssystem.
Ezzeddine, R., Otaki, F., Darwish, S., & AlGurg, R. (2023). Change management in higher education: A sequential mixed methods study exploring employees' perception. *PLOS ONE, 18*(7).
Gutierrez-Bucheli, L., Kidman, G., & Reid, A. (2022). Sustainability in engineering education: A review of learning outcomes. *Journal of Cleaner Production, 330*, Article 129734.
Hummel, K., & Jobst, D. (2024). An overview of corporate sustainability reporting legislation in the European Union. *Accounting in Europe*, 1–36.
Kezar, A. (2001). Understanding and facilitating organizational change in the 21st century: Recent research and conceptualizations. *ASHE-ERIC Higher Education Reports, 28*(4).
Kotter, J. P. (1996). Why transformation efforts fail. Harvard business review.
Linnér, B. & Wibeck, V. (2019). *Sustainability transformations: Agents and drivers across societies.* Cambridge University Press.
Mahardhani, A. J. (2023). The role of public policy in fostering technological innovation and sustainability. *Journal of Contemporary Administration and Management (ADMAN), 1*(2), 47–53.
Makrakis, V. (2014). Transforming University Curricula towards Sustainability: A Euro-Mediterranean Initiative. In K. Thomas & H. Muga (Eds.), *Handbook of Research on Pedagogical Innovations for Sustainable Development* (pp. 619–640). IGI Global.
Moggi, S. (2023). Sustainability reporting, universities and global reporting initiative applicability: A still open issue. *Sustainability Accounting, Management and Policy Journal, 14*(4), 699–742.
Rashed, A. H., & Shah, A. (2021). The role of private sector in the implementation of sustainable development goals. *Environment, Development and Sustainability, 23*(3), 2931–2948.
Ruben, B. D., De Lisi, R., & Gigliotti, R. A. (2023). *A guide for leaders in higher education: Concepts, competencies, and tools.* Taylor & Francis.
Rudhumbu, N. (2015). Managing curriculum change from the middle: How academic middle managers enact their role in higher education. *Higher Education, 4*(1), 106–119.
Sanyal, B. C. (2024). *Higher education and employment: An international comparative analysis.* Taylor & Francis.
Sassen, R., & Azizi, L. (2018). Voluntary disclosure of sustainability reports by Canadian universities. *Journal of Business Economics, 88*, 97–137.
Sterling, S. (2004). Higher education, sustainability, and the role of systemic learning. *Higher education and the challenge of sustainability: Problematics, promise, and practice* (pp. 49–70). Springer, Netherlands.
Sterling, S. (2010). Transformative learning and sustainability: Sketching the conceptual ground. *Learning and Teaching in Higher Education, 5*(3), 17–33.
Straatmann, T., Kohnke, O., Hattrup, K., & Mueller, K. (2016). Assessing employees' reactions to organizational change: An integrative framework of change-specific and psychological factors. *The Journal of Applied Behavioral Science, 52*(3), 265–295.
Tilbury, D. (2011). Education for sustainable development: An expert review of processes and learning. Online available under: https://unesdoc.unesco.org/ark:/48223/pf0000191442 (Last accessed: 12 April 2026)
Trevesian, L. V., Leal Filho, W., & Pedroso, E. A. (2024). Transformative organizational learning for sustainability in higher education: A literature review and an international multi-case study, Journal of Cleaner Production. *Journal of Cleaner Production., 447*, Article 141634. https://doi.org/10.1016/j.jclepro.2024.141634 (Last accessed: 12 April 2026)
UNESCO. (2017). Education for sustainable development goals: Learning objectives. United Nations Educational, Scientific and Cultural Organization. Online available under: https://doi.org/10.54675/CGBA9153 (Last accessed: 12 April 2026)

UNESCO. (2022). Knowledge-driven actions: transforming higher education for global sustainability. United Nations Educational, Scientific and Cultural Organization. Online available under: https://doi.org/10.54675/YBTV1653 (Last accessed: 12 April 2026)

Bridging Knowledge and Action: The Role of SDGs Implementation in Developing Inner Skills Within Higher Education Institutions

Fatma Fourati-Jamoussi, Justas Kažys, Maria-Cristiana Munthiu, and Kimberley Burcke

Abstract Higher education institutions have immense potential concerning the implementation of Sustainable Development Goals (SDGs). The SDGs set by the United Nations for the 2030 Agenda should be a strategic priority for higher education institutions. Although students demonstrate a reasonable level of cognitive understanding (knowledge) of some SDGs, their behavioral commitment to take meaningful action in support of these goals remains relatively low. This gap between knowledge and action highlights the need for innovative approaches to foster deeper student engagement and connection to these goals. Within this context, global education communities are exploring new strategies to accelerate SDGs' progress using inner human resources, such as mindfulness and contemplative practices. These approaches build emotional and psychological resilience and help cultivate the inner capacities to drive sustainable change. To this end, a survey on "Students Challenges, Worries, and Needs Mapping" was conducted across five European countries, gathering 345 responses from higher education students. The results revealed that, on the one hand, most of the respondents, so far, are not satisfied with the level of achievement of SDGs. On the other hand, their knowledge of SDG actions and

F. Fourati-Jamoussi · M.-C. Munthiu
InTerACT Research Unit, The Humanities, Management and Digital Science Department, Institut Polytechnique UniLaSalle, Beauvais, France
e-mail: fatma.fourati@unilasalle.fr

M.-C. Munthiu
e-mail: maria-cristiana.munthiu@unilasalle.fr

J. Kažys (✉)
Hydrometeorological Research Unit, Institute of Geosciences, Vilnius University, Vilnius, Lithuania
e-mail: justas.kazys@gf.vu.lt

M.-C. Munthiu
Laboratoire d'Economie, Finance, Management et Innovation (LEFMI), University of Picardy Jules Verne, Amiens, France

K. Burcke
International Relations Department, Institut Polytechnique UniLaSalle, Beauvais, France
e-mail: kimberley.burcke@unilasalle.fr

W. Leal Filho et al. (eds.), *Introducing the UN SDGs in Higher Education Institutions*, World Sustainability Series, https://doi.org/10.1007/978-3-032-23017-1_20

personal engagement levels were very low. However, most respondents indicated that Inner Development Goals (IDGs) are important for achieving the SDGs. The IDG framework identified the most important inner dimensions (5 categories) and skills (23 subcategories) that could be developed for SDG implementation. This survey provided valuable insights into students' emotional challenges and offered a basis for developing strategies to enhance their engagement with the SDGs.

Keywords MP4s project · Inner Development Goals · Higher education institutions · SDGs

1 Introduction

The United Nations' Sustainable Development Goals (SDGs) represent an urgent call to action for global development across multiple social, environmental, and economic dimensions. As part of the 2030 Agenda, these 17 goals outline a comprehensive blueprint for addressing global challenges like climate change, inequality, and poverty. As stressed by the United Nations Summit on Sustainable Development in 2015, achieving the 2030 Agenda requires the active participation of all stakeholders, including civil society, the private sector, and local communities. While many Higher Education Institutions (HEIs) have embraced the SDGs at various levels, the challenge is to turn students' cognitive understanding of these goals into concrete action. This gap between awareness and engagement highlights the need for innovative approaches that enable students to understand sustainability and to actively participate in it.

This chapter is based on findings from the Erasmus + project "MP4s: Mindfulness Practices for Students in Society," co-funded by the European Union under project ID 2021-1-FR01-KA220-HED-000027504, which aims to help students cope with emotional challenges while developing valuable soft skills through mindfulness and inner development. These skills, such as resilience, self-awareness, and emotional balance, promote personal well-being and encourage a deeper connection with wider goals, such as the Sustainable Development Goals (SDGs). In enabling students to manage stress, engage in mindfulness with thoughtfulness, and develop empathy, MP4s promotes a more holistic type of education that connects personal development with positive social and environmental change. With practical tools and resources for mindfulness, MP4s invites students and educators to integrate well-being into learning, helping students to become mindful and engaged citizens.

The chapter's main question is how HEIs can better implement Sustainable Development Goals (SDGs) by fostering inner skills among students. Specifically, the chapter highlights the gap between students' cognitive awareness of SDGs and their lack of corresponding action (behavioral engagement) toward achieving these goals. The problem to resolve is twofold:

1. **Low engagement with SDG actions**: Although students may understand the SDGs on a theoretical or knowledge level, their personal commitment and active participation in sustainability actions remain low.
2. **Need for inner skills development**: The chapter explores the hypothesis that inner skills development, cultivated through practices like mindfulness (as demonstrated in the MP4s project), might enhance students' pro-environmental behaviors and overall engagement with SDGs through a virtuous cycle.

Thus, the chapter focuses on discovering how inner skills, such as emotional intelligence, self-awareness, and mindfulness, can be developed and aligned with SDG initiatives in HEIs to bridge the gap between knowledge and action. The Inner Development Goals (IDGs), an open-source initiative driven by the nonprofit sector, are proposed as a solution to foster these internal capacities, making students more effective actors of sustainability. The IDGs aim to explore and cultivate the skills and capacities essential for successfully addressing the SDGs (IDG.tools—Inner Development Phase 2 Research Report, 2022).

These inner skills are important not just for developing sustainable competencies but also for supporting mental well-being. As mental health issues among students, from elementary to university levels, become a greater global concern, it is critical to investigate therapies that promote both personal growth and academic performance. Recognizing the importance of addressing these issues, researchers have conducted various studies to explore interventions and challenges related to student mental health. For instance, Flook investigated the impact of mindfulness and yoga interventions on elementary school students and teachers' quality of life (Flook et al., 2010). Through a randomized controlled school-based study, the researchers found that these interventions significantly improved elementary school students' and teachers' quality of life, indicating the potential benefits of incorporating mindfulness and yoga practices into school curricula. Additionally, a study focusing on Spanish university students examined a prevalence of suicidal thoughts and behaviors and their association with distal and proximal factors. The findings revealed the concerning prevalence of suicidal thoughts among university students and highlighted the need for targeted interventions to address underlying risk factors (Blasco et al., 2019). Other studies have shown the impact of mindfulness on pro-environmental behavior (Ballester et al., 2019) and explored the relationship between mindfulness, connectedness to nature, and pro-environmental behavior (Richter & Hunecke, 2022).

This chapter delves into the interplay between the implementation of Sustainable Development Goals (SDGs) in higher education institutions and the development of inner skills through contemplative practices. We begin by exploring the emergence of the SDGs, their historical context, and how global principles are applied within HEIs. This is followed by an analysis of the three core dimensions of SDGs (cognitive, affective, and conative) and by presenting the emergence of Inner Development Goals (IDGs). For each main SDG dimension, we highlight the corresponding Inner Development Goals (IDGs), showing how these inner skills can act as mediators for achieving SDG goals.

In the cognitive dimension, we explore how personal values, norms, and behavior—enhanced by IDGs related to "Being" and "Thinking"—are critical to fostering SDG-related knowledge and action (Leicht et al., 2018; Sterling, 2010). The affective dimension will address the importance of emotional engagement, including empathy and collaboration, emphasizing the IDGs related to "Relating" and "Collaborating" (Ojala, 2013; Riemer et al., 2014). Finally, the conative dimension will focus on action-oriented behaviors, examining how HEIs can empower students to act on SDG initiatives aligned with the IDG of "Acting" (Brundiers et al., 2021; Wiek & Redman, 2022; Wiek et al., 2011).

The chapter draws connections between SDGs and IDGs, providing a theoretical framework that integrates these concepts across the three core SDG. It highlights how inner human development can enhance students' capacity to engage with sustainability efforts. We conclude by discussing potential future research perspectives, including a study on students across several European countries. We propose a methodology that might be implemented in other HEIs in order to better understand how these contemplative practices and inner development skills can further contribute to SDG implementation in accordance with the 2030 Agenda for Sustainable Development.

2 Emergence of SDG Goals

The SDGs were officially adopted in 2015, and their history traces back to earlier international agreements, including the Millennium Development Goals. The SDGs are more widespread in scope, focusing on sustainability and global well-being. Higher Education Institutions are critical stakeholders in the implementation of SDGs as they prepare future leaders, researchers, and professionals who will directly contribute to global development strategies (Berrone et al., 2023).

At the European level, universities are increasingly aligning their strategies with the 2030 Agenda for Sustainable Development. The European University Association (EUA) has been a key advocate in promoting SDG integration in higher education through its Sustainable Development Initiative (also known as Green Transition, s.d.). European universities are part of networks like Erasmus+, which focus on creating cross-border educational programs promoting SDGs such as SDG 4 (Quality Education) and SDG 17 (Partnerships for the Goals). The European Green Deal also plays a significant role, with universities contributing research and innovation to address climate change, biodiversity, and pollution, in line with SDG 13 (Climate Action) (also known as A Green Deal roadmap for universities, s.d.). Additionally, universities have adopted sustainability charters, campus-wide green policies, and student-led initiatives aimed at improving the overall sustainability of academic institutions across Europe (Filho et al., 2024). Programs like the Sustainable Development Solutions Network (SDSN) help universities foster cross-disciplinary collaboration for global SDG implementation (Unsdg | UN Sustainable Development Solutions Network, s.d.).

For instance, at UniLaSalle, a strong emphasis is placed on integrating sustainable development across all disciplines, aligning with the global principles of the SDGs. This French engineering school has adopted sustainability as a core value, embedding it into research, education, and campus life (Fourati-Jamoussi et al., 2015). Programs that focus on sustainable agriculture, climate change, and environmental engineering reflect this commitment. UniLaSalle also actively participates in European and global sustainability initiatives, like Erasmus+ projects and MP4s (Mindfulness Practices for Students in Society), which focus on building emotional resilience and inner skills to support the achievement of SDGs. They also promote SDG awareness campaigns, green campus initiatives, and sustainable engineering practices within their curriculum, ensuring that students are prepared to contribute to global sustainability challenges.

Another example is found at Vilnius University, where SDG principles are integrated through academic programs focused on environmental science, sustainability, and global citizenship. The university fosters interdisciplinary research on sustainability transitions, climate action, and social inclusion. Vilnius University is also a participant in various international sustainability projects, such as the European Green Deal and Horizon Europe programs, which align with several SDGs, including SDG 4 (Quality Education) and SDG 13 (Climate Action) (Sustainable Development Goals—European Commission, s.d.). The university also promotes sustainable campus management practices and engages students in real-world problem-solving initiatives, such as the Green University project, which focuses on reducing the institution's carbon footprint and promoting biodiversity.

Recent critiques, such as Northumbria University, highlight key gaps in the current approach to Education for Sustainable Development (ESD). Although coordinated by UNESCO since 2019, ESD has been criticized for promoting an unsustainable balance between economic growth, social justice, and environmental protection without sufficiently addressing the root causes of environmental crises (Sterling, 2021). Transformative business models based on circular economy should be promoted to move beyond conventional sustainability (Kopnina, 2020).

For achieving SDG goals, the complexity of internal–external transformation requires emergent approaches (Wamsler et al., 2021). IDGs can play a role in helping individuals and communities build the resilience and action they need to create a more positive future (Wood, 2024). Inner development is a collective and cultural process that is ultimately expressed in a community in service of the greater good (Stålne & Greca, 2022). HEIs are tightly interconnected with external stakeholders such as businesses, governments, NGOs, and local communities (Inner Development Goals… 2021; Ankrah et al., 2023). Working with organizations has great leverage to kick-start systemic change at the individual and collective levels. This is the reason why the IDGs focus on working with communities (Stålne & Greca, 2022). One of the main pillars of creating the IDGs communities is "hubs"; more than 600 IDG hubs and networks with different profiles and goals exist around the world (idg.community/). Business and innovation companies have already been successfully integrated into the leadership development frameworks of several corporations, including IKEA, Ericsson, Google, Stena, and Icebug (Bach et al., 2024; Bristow &

Bell, 2023). Another way of integrating IDGs in business is through educational processes (Engel & Janssen, 2024), which comes along with practical examples. According to the SDGs4future project (sdgs4future.org/), which provides innovative education for a more sustainable future, role models play a central role in linking IDGs and SDGs. This is especially important through the educational process of sustainability (Nordén, 2024).

From a political perspective, so far, only the Costa Rican government has officially joined the journey of co-creation and integration of the IDG framework (Stålne & Grecam, 2022). However, despite the growing IDG hubs, research-based information on NGOs and local communities' involvement in SDG activities are still limited. The meso-scale education role (SDG 4) could be a connecting link between the micro-scale (IDGs) and macro-scale (SDGs) worlds. The IDGs are values driven, and therefore, it could stress the good of the group to foster values and competencies which focus on this collective worldview (Wood, 2024). Dijkstra and Mäurer (2024) researched the IDG dissemination and NGOs involvement in South Korea related to gender equality (SDG 5). The Caux Initiatives of Change Foundation organized the forum on Conscious Food Systems (SDG 2) for local communities based on IDGs principles (Dufour, 2024).

3 Dimensions of SDGs

In this section, we first present the level of familiarity European students have with the SDGs, detailing the survey questions and summarizing the results. Next, we explain the three essential dimensions of the SDGs—cognitive, affective, and conative—highlighting their importance in achieving a holistic understanding and implementation. Exploring the SDGs through these core dimensions is critical for assessing their full impact and fostering meaningful engagement among students. These dimensions are integral in promoting both knowledge and action towards sustainability goals.

The MP4s survey "Students Challenges, Worries, and Needs Mapping" collected 345 student responses, with three questions focusing on the SDGs and Agenda 2030: (1) Familiarity with the UN's SDGs and the 2030 Agenda; (2) Perception of progress toward achieving these goals by 2030; and (3) Opinion on whether their country/Europe is moving in the right direction. Responses were measured on a 7-point Likert scale. "Climate Action" (SDG13) was the most frequently selected goal (100 times, except among Italian respondents). Familiarity with the SDGs averaged 4.1 (mode 5), indicating moderate awareness, while perceived achievement of these goals was lower, with an average of 2.8 (mode 3). Responses showed that greater familiarity with the SDGs corresponded to a more critical view of their progress. Most responses regarding the direction of countries/Europe were also less optimistic (average 3.4, mode 4) (MP4s Student Challenges Mapping/PR1 Report, 2023).

Cognitive Dimension

The cognitive dimension emphasizes knowledge and values that inform SDG-related behaviors. Students may be aware of SDGs but struggle with translating this awareness into action. The intersection of cognitive science with learning and behavior has led to numerous studies examining how cognitive processes impact individual engagement, problem-solving, and behavior change. Recent literature explores these cognitive dimensions in various contexts, including cognitive perspectives on learning (Yilmaz, 2011), the influence of cognitive mindsets on environmental behaviors (Spaccatini et al., 2023), and the integration of cognitive-behavioral strategies in teaching (Chipman et al., 2013). These studies provide valuable insights into how cognitive frameworks can shape human interaction, enhance learning outcomes, and promote behavior change.

Affective Dimension

Affective approaches—focused on emotions, values, and attitudes—are increasingly recognized as essential in education for sustainable development (ESD). Researchers have investigated how emotional intelligence, affective engagement, and feedback impact students' understanding and commitment to sustainability, highlighting the role these factors play in supporting the broader implementation of the United Nations Sustainable Development Goals (SDGs). Emotional connections are critical in fostering sustainable behaviors. For instance, a study examined the effects of cooperative learning and project-based learning (PBL) on emotional intelligence in the context of SDG implementation (Lozano et al., 2022). Another study (Buissink-Smith et al., 2011) emphasized that although affective learning is essential for ESD, it remains challenging to assess. They proposed several assessment strategies, including reflective journaling, attitude surveys, and self-assessment tools, as ways to capture affective learning outcomes effectively.

Conative Dimension

The conative dimension, which emphasizes motivation and action, plays a key role in shaping pro-environmental behaviors among students in higher education, often influenced by the institutional environment. This section synthesizes insights from studies exploring transformative educational approaches, particularly within environmental and sustainability contexts, through frameworks like action competence and transformative learning. Action competence, as described by Schnack and Jensen (1997), focuses on developing students' capacity to critically analyze environmental issues, understand societal influences, and make informed decisions that lead to impactful actions. In parallel, transformative learning encourages self-reflection, prompting students to reassess previously held assumptions and adopt more sustainable perspectives. Gravett (2004) highlighted the potential of these methods to inspire significant personal change and align educational goals with sustainable practices. Furthermore, a research study introduced the AACTT framework (Action, Actor, Context, Target, Time) as a model for specifying behaviors in environmental and sustainability contexts, offering a structured approach to behavioral change (Presseau et al., 2019). Molderez extended this idea by examining transformative learning in ecological thinking through the "Students Swap Stuff" project, which promotes sustainability by encouraging students to exchange goods instead of purchasing new items. Her findings indicate that transformative learning experiences like these can foster shifts toward ecological awareness, helping students understand the environmental impacts of their actions and reinforcing a commitment to sustainable lifestyles (Molderez, 2021).

4 Relationship Between SDGs and IDGs

In this section, we define the IDGs and examine their relationship with the SDGs. The Inner Development Goals (IDGs) are a framework of skills and qualities designed to help individuals and organizations, including higher education institutions, address complex societal issues effectively (Inner Development Goals, 2021). Created in 2019 and inspired by the United Nations' Sustainable Development Goals (SDGs), the IDGs promote inner, inter, and outer capacities necessary for global change, in line with what Joanna Macy called "The Great Turning"—a shift to a life-sustaining civilization (Macy, 1998). The framework incorporates insights from psychology, neuroscience, and philosophy and is broken down into five key dimensions with 23 essential skills.

The IDGs emphasize both individual and collective impact, fostering cooperation, empathy, and sustainable action, which are essential for tackling global challenges (Going Deeper, 2023). SDGs, such as Climate Action and Reduced Inequalities, require complex thinking and long-term perspectives—skills that are sometimes underdeveloped in traditional education. Methods that align with the IDGs and strengthen their relevance to the SDGs are therefore favorable in evaluations, as they bridge understanding and action towards sustainable development.

The connection between SDGs and IDGs lies in their shared focus on fostering long-term, systemic change. While SDGs address external global challenges, IDGs focus on developing the internal skills needed to navigate these challenges effectively. By fostering inner competencies such as empathy, resilience, and self-awareness, IDGs complement SDG implementation by equipping students with the pedagogical tools to become sustainability advocates. Finstad-Milion et al. (2021) discussed how engaged research can bridge the gap between theory and practice in sustainability, supporting the integration of IDGs into HEIs. Cooper and Gibson (2022) presented an inner-outer framework for sustainability that underscores the importance of internal skills and mindsets—corresponding to the Inner Development Goals (IDGs)—for achieving sustainable outcomes at broader societal levels. Their model addresses two primary aspects of sustainability: the inner dimension, which includes psychological growth, self-awareness, and mindfulness. The outer dimension encompasses environmental and social targets aligned with SDGs (Cooper & Gibson, 2022).

A systematic approach can guarantee the successful implementation of the Internal Development Goals (IDGs) in higher education institutions (HEIs). Key stakeholders, including teachers-researchers, students, and administrative staff, are engaged after the process starts with an evaluation of the current context, which includes a needs analysis to find gaps in students' internal skills and the institutional culture regarding sustainability (Lozano et al., 2013). The organization should then choose important objectives that are in line with both its mission and the Sustainable Development Goals (SDGs) in order to prioritize IDGs (Issues and trends in education for sustainable development—UNESCO, s.d.). The next step is curriculum integration, which involves creating frameworks to integrate IDGs like collaboration, emotional intelligence, and mindfulness into existing courses or developing new ones (Rieckmann, 2018). Workshops and peer-led projects are examples of training programs that may contribute to skill development. They are, at the same time, measures for monitoring and assessment, as well as surveys, interviews, and reports used in order to make sure that these efforts are successful (Barth, 2014; Tilbury, 2011). Finally, to sustain and expand these efforts, a long-term strategy supported by collaboration and information sharing among European institutions is required (EUA Annual Report, 2023, s.d.). Since 2023, UniLaSalle's Ecological and Societal Transformation team has been integrating the SDGs into their current curriculum. In addition, since 2021, UniLaSalle has integrated IDGs into two optional courses focusing on mindfulness practices for the first and second years of the engineering cycle as part of the MP4s initiative. The pedagogical team provides assistance in presenting the project's findings to interested stakeholders (i.e., professors, researchers, administration, and students) and informs the psychological support department and campus life management concerning the results of IDGs implementation in the curriculum, their evaluation, and possible ways of improvement.

The MP4s survey "Students Challenges, Worries and Needs Mapping" found that most respondents view integrating the IDGs as essential for bridging individual and social responsibilities. In response to the question, "How important do you think developing inner skills is for reaching this goal?"—Likert scale responses were largely positive, with an average of 5.8 and mode of 7. Overall, 67% of respondents emphasized IDG importance, though this varied between 77% for socially inclined students and 63% for environmentally focused ones. The key IDGs identified for achieving the SDGs and Agenda 2030 goals include a sense of responsibility, commitment to the greater good, critical thinking, kindness, active listening, fostering dialogue, and courage to act on values (MP4s Student Challenges Mapping/PR1 Report, 2023).

The alignment of the IDGs with SDGs provides a theoretical foundation for embedding inner skill development into Education for Sustainable Development (see Table 1). The MP4s project incorporates the IDG framework across three dimensions.

Cognitive Dimension: The Inner Development Goals (IDGs) offer a framework to cultivate essential skills for SDG engagement, particularly through "Being" (IDG 1) and "Thinking" (IDG 2). These dimensions foster self-reflection, critical and systems thinking, and a long-term orientation, all critical for addressing global challenges. Some researchers highlighted how mindful learning enhances pro-environmental behaviors by promoting self-expansion and personal growth, thereby deepening one's

Table 1 SDGs dimensions and corresponding IDGs

Theoretical framework	Corresponding IDGs
Cognitive dimension	IDG 1 + 2 = Being and thinking
Affective dimension	IDG 3 + 4 = Relating and collaborating
Conative dimension	IDG 5 = Acting

Source created by the authors

connection to the natural world and encouraging sustainable actions (Tang et al., 2017). Similarly, other authors have explored how a sense of connectedness to nature can drive sustainable behaviors (Barbaro & Pickett, 2016). While Donald et al. (2020) and the Inner Development Goals (2021) underscore the importance of cognitive development within HEIs for cultivating these transformative capacities.

MP4s project's conducted survey revealed that in the sense of "Being" (IDG 1)—*"Openness & Learning-mindset"* skill—has the highest importance and it especially corresponds to "Quality education" (SDG 4) principles (Lane, 2017; Filho et al., 2019; 2023). For the "Thinking" (IDG 2) dimension, the highest recognition was for the *"Long-term orientation & Visioning"* skill. Together with other cognitive skills, it could serve by building starting positions for achieving SDG 1, 2, 3, and 8 (Shtaltovna et al., 2024). Moreover, it could work by connecting human mindset with practical tasks in SDG 9, 10 & 12 (Adshead et al., 2019; Aly et al., 2022).

Affective Dimension: The IDGs framework emphasizes "Relating" (IDG 3) and "Collaborating" (IDG 4), which prioritize empathy, social skills, and care for others to support community-oriented sustainability efforts. Persson, Shrivastava, and Zsolnai explored the role of aesthetics and emotional intelligence in fostering sustainability education, highlighting the emotional component essential in sustainability learning (Persson et al., 2022). Jacob et al. further underscore how mindfulness meditation can enhance pro-environmental behaviors by cultivating empathy and care for the environment (Jacob et al., 2009). Additionally, Blasco et al. (2019) provide insights into the influence of emotional and mental well-being on university students' engagement with SDG-related activities, illustrating how social and emotional competencies are key to global cooperation on sustainability.

"Empathy & Compassion" skill was mostly mentioned in "Relating" (IDG 3). It could be helpful for conscious and connected actions on an international level for SDG 16 and 17 (Grant et al., 2022) and activating education (SDG 4) for sustainable development (Giangrande et al., 2019). Moreover, the development of *"Communication skills"* (IDG 4) is essential and could be addressed to all SDGs. Additionally, it could be seen as "SDG 18: Communication for All" (McLeod & Marshall, 2023).

Conative Dimension: The IDG dimension of "Acting" (IDG 5) promotes transformative change by encouraging practical engagement with the SDGs. Marchi et al. (2021) highlighted the significant impact of individual behavioral changes in reducing household carbon footprints, illustrating the tangible outcomes of personal commitment to sustainability. Similarly, other studies demonstrated how mindfulness practices can enhance individual performance and support sustainable behaviors, aligning well with IDG 5's objectives (Shao & Skarlicki, 2009). Several researchers provide insight into how behavior shifts contribute to sustainability transitions, linking inner development directly to conative actions that advance the SDGs (Tang et al., 2017; Kaufman et al., 2021). Furthermore, recent studies discuss how individual actions post-COVID-19 can align with SDG 7, fostering a shift toward energy sustainability (Madurai Elavarasan et al., 2021).

According to survey results, for environmental goals (SDG 13, 14, 15), *"Perseverance* and *Courage"* were deemed the most crucial conative skills (IDG 5). These skills help to build and secure environmental, social, and economic resilience (Aly et al., 2022; Assarkhaniki et al., 2023). Moreover, perseverance could be one of the most powerful skills shaping long-term orientation for environmental and other SDGs (Sedita et al., 2022).

The MP4s survey "Students Challenges, Worries and Needs Mapping" found that most respondents view integrating the IDGs as essential for bridging individual and social responsibilities. In response to the question, "How important do you think developing inner skills is for reaching this goal?"—Likert scale responses were largely positive, with an average of 5.8 and mode of 7. Overall, 67% of respondents emphasized IDG importance, though this varied between 77% for socially inclined students and 63% for environmentally focused ones. The key IDGs identified for achieving the SDGs and Agenda 2030 goals include a sense of responsibility, commitment to the greater good, critical thinking, kindness, active listening, fostering dialogue, and courage to act on values. For environmental goals (SDG13), perseverance and courage were deemed the most crucial conative skills (MP4s Student Challenges Mapping/PR1 Report, 2023).

5 Perspectives for Future Research

Future research could expand on the integration of IDGs in higher education curricula, investigating their impact on students' engagement with SDGs. Comparative studies across cultural contexts could provide valuable insights into how inner competencies such as emotional intelligence, critical thinking, and ecological awareness influence students' contributions to global sustainability efforts. Studies have shown that frameworks like action competence and transformative learning can shape students' ecological thinking and engagement with sustainable practices, offering a compelling foundation for further research into IDG-driven curricula (Jensen & Schnack, 1997; Molderez, 2021). Furthermore, as other studies have explored cognitive mindsets in pro-environmental behavior, research could examine how IDGs facilitate a proactive sustainability mindset in students, fostering behaviors aligned with SDGs (Spaccatini

Table 2 A methodology for implementing IDGs at UnilaSalle[1]

Stage no.	Stage name	Description
1	Current context evaluation	A needs analysis is needed to find gaps in students' internal skills and the institutional culture regarding sustainability (Lozano et al., 2013)
2	Set objectives	In line with both HEIs mission and the Sustainable Development Goals (SDGs) (Issues and trends in education for sustainable development—UNESCO, s.d.)
3	Curriculum integration	Creating frameworks to integrate IDGs (e.g., collaboration, emotional intelligence, and mindfulness into already-existing courses or developing new ones) (Rieckmann, 2018)
4	Monitoring and assessment	Workshops and peer-led projects such as training programs, surveys, interviews, and reports

Source Created by the authors

et al., 2023). Application-oriented research, such as a longitudinal study across European students' understanding and commitment to SDGs, would deepen our understanding of how IDGs impact both short-term awareness and long-term engagement with sustainability initiatives.

HEIs can use a variety of useful approaches that are adapted to their institutional and curriculum environment in order to successfully implement the IDGs. As demonstrated by UniLaSalle, including IDGs in teaching practices, communicating findings to stakeholders, and working with institutional support structures are part of a methodology we developed (see Table 2) for effectively implementing IDGs, in accordance with a long-term strategy supported by collaboration and information sharing among other European HEI institutions, as suggested by (EUA Annual Report, 2023, s.d.).

Other methods improve emotional intelligence and self-awareness (Brown et al., 2007; Kabat-Zinn et al., 2009). Interdisciplinary implemention of IDGs in HEIs may expose students to real-world problems and experiential learning, such as project-based or service learning, in order to develop social responsibility, empathy, and teamwork (Kolb, 1984; Tilbury, 2011). In addition, reflective journaling and mindfulness training are examples of contemplative and mindfulness techniques, such as case studies and collaborative courses, foster a diversity of viewpoints and connect humanities disciplines to ethical and social justice issues or technology-focused programs to sustainability challenges (Ferrer-Balas et al., 2008). Linking theory and practice through research projects and community partnerships could facilitate students' tackling local and global sustainability issues (Stephens et al., 2008). Curriculum mapping and flexible learning paths enable the integration of IDGs across disciplines, ensuring accessibility for all students. Finally, evaluation and feedback

[1] In accordance with a long-term strategy supported by collaboration and information sharing among European institutions is required (EUA Annual Report 2023, s. d.).

methods, such as continuous self-assessment and iterative feedback loops, enable continual adaptation and improvement (Barth, 2014; Sterling, 2013).

6 Conclusion

The effective implementation of SDGs within Higher Education Institutions (HEIs) can be significantly enhanced by fostering inner competencies through frameworks such as IDGs. The MP4s project illustrates the potential of contemplative and action-based practices—mindfulness, transformative learning, and action competence—as tools for bridging the gap between SDG awareness and concrete action. Studies on the cognitive and affective aspects of education support the notion that nurturing both cognitive and emotional dimensions in students strengthens their commitment to sustainable development (Buissink-Smith et al., 2011; Presseau et al., 2019). This approach aligns with broader findings, which advocate for HEIs as transformative agents in cultivating sustainable behaviors and resilience (Findler et al., 2019; Garrett, 2004). By integrating IDGs into educational strategies, HEIs can foster leaders equipped with the emotional intelligence, cognitive clarity, and resilience necessary for addressing complex sustainability challenges. This integration holds promises for training a generation of ecologically responsible graduates prepared to make substantial contributions to global sustainability.

Beyond individual development, embedding IDGs into educational strategies could reshape institutional practices, promoting a holistic form of education that aligns personal growth with sustainability goals. Practical steps, such as incorporating mindfulness and emotional intelligence programs into curricula, offer HEIs a concrete way to produce well-rounded, sustainability-focused graduates. Future research should examine how IDG-based approaches affect student engagement with the SDGs across different cultural and educational contexts, providing insights into the scalability and effectiveness of IDG-focused curricula globally. Comparative studies could further elucidate how students' emotional intelligence, ecological awareness, and resilience foster sustainable behaviors over time.

Higher education institutions thus have a unique and essential role in advancing the SDGs by championing inner development. Through IDGs, HEIs can empower students to transform knowledge into impactful, sustainable actions, helping to shape a fairer, more resilient, and more sustainable world. This is an urgent call to educators, policymakers, and institutions to integrate these inner development principles, equipping students to become compassionate, engaged global citizens dedicated to meaningful, long-term sustainability.

References

A Green Deal roadmap for universities. (s.d.). https://www.eua.eu/publications/reports/a-green-deal-roadmap-for-universities.html

Adshead, D., Thacker, S., Fuldauer, L. I., & Hall, J. W. (2019). Delivering on the sustainable development goals through long-term infrastructure planning. *Global Environmental Change, 59*, Article 101975. https://doi.org/10.1016/j.gloenvcha.2019.101975

Aly, E., Elsawah, S., & Ryan, M. J. (2022). Aligning the achievement of SDGs with long-term sustainability and resilience: An OOBN modeling approach. *Environmental Modelling & Software, 150*, 105360.https://doi.org/10.1016/j.envsoft.2022.105360

Ankrah, D., Bristow, J., Hires D., & Henriksson, J. A. (2023). Inner Development Goals: From inner growth to outer change. *Field Actions Science Reports*, 25. http://journals.openedition.org/factsreports/7326

Assarkhaniki, Z., Sabri, S., Rajabifard, A., & Kahalimoghadam, M. (2023). Advancing sustainable development goals: Embedding resilience assessment. *Sustainability Science, 18*, 2405–2421. https://doi.org/10.1007/s11625-023-01372-7

Bach, D., Olynec, N., & Cairns-Lee, H. (2024). Leaders need inner development to reach the SDGs. *Sustainability* (the Institute for Management Development.). https://www.imd.org/ibyimd/sustainability/leaders-need-inner-development-to-reach-the-sdgs/

Ballester, L., Alayo, I., Vilagut, G., Almenara, J., Cebrià, A. I., Echeburúa, E., Gabilondo, A., Gili, M., Lagares, C., Piqueras, J. A., Roca, M., Soto-Sanz, V., Blasco, M. J., Castellví, P., Forero, C. G., Bruffaerts, R., Mortier, P., Auerbach, R. P., Nock, M. K., … Group, on behalf of the U. study. (2019). Accuracy of online survey assessment of mental disorders and suicidal thoughts and behaviors in Spanish university students. Results of the WHO World Mental Health-International College Student initiative. *PLOS ONE, 14*(9), e0221529. https://doi.org/10.1371/journal.pone.0221529

Barbaro, N., & Pickett, S. M. (2016). Mindfully green: Examining the effect of connectedness to nature on the relationship between mindfulness and engagement in pro-environmental behavior. *Personality and Individual Differences, 93*, 137–142. https://doi.org/10.1016/j.paid.2015.05.026

Barth, M. (2014). Implementing Sustainability in Higher Education: Learning in an age of transformation. *Routledge*. https://doi.org/10.4324/9780203488355

Berrone, P., Rousseau, H. E., Ricart, J. E., Brito, E., & Giuliodori, A. (2023). How can research contribute to the implementation of sustainable development goals? An interpretive review of SDG literature in management. *International Journal of Management Reviews, 25*(2), 318–339. https://doi.org/10.1111/ijmr.12331

Blasco, M. J., Vilagut, G., Almenara, J., Roca, M., Piqueras, J. A., Gabilondo, A., Lagares, C., Soto-Sanz, V., Alayo, I., Forero, C. G., Echeburúa, E., Gili, M., Cebrià, A. I., Bruffaerts, R., Auerbach, R. P., Nock, M. K., Kessler, R. C., Alonso, J., & Group, the U. study. (2019). Suicidal thoughts and behaviors : Prevalence and association with distal and proximal factors in Spanish University Students. *Suicide and Life-Threatening Behavior, 49*(3), 881–898. https://doi.org/10.1111/sltb.12491

Bristow, J., & Bell, R. (2023). Orienting inner development in organisations. IDG White Paper #1 (focus on corporates), 59 pp. https://cdn.prod.website-files.com/60bf97e4e4d3871f520114d7/65d860817aae264e5d86cbf9_IDG%20Corporate%20White%20Paper%202023.pdf

Brown, K. W., Ryan, R. M., & Creswell, J. D. (2007). Mindfulness: Theoretical foundations and evidence for its salutary effects. *Psychological Inquiry, 18*(4), 211–237. https://doi.org/10.1080/10478400701598298

Brundiers, K., Barth, M., Cebrián, G., Cohen, M., Diaz, L., Doucette-Remington, S., Dripps, W., Habron, G., Harré, N., Jarchow, M., Losch, K., Michel, J., Mochizuki, Y., Rieckmann, M., Parnell, R., Walker, P., & Zint, M. (2021). Key competencies in sustainability in higher education—Toward an agreed-upon reference framework. *Sustainability Science, 16*(1), 13–29. https://doi.org/10.1007/s11625-020-00838-2

Buissink-Smith, N., Mann, S., & Shephard, K. (2011). How do we measure affective learning in higher education? *Journal of Education for Sustainable Development, 5*(1), 101–114. https://doi.org/10.1177/097340821000500113

Chipman, S. F., Segal, J. W., & Glaser, R. (Éds.). (2013). *Teaching Thinking : A Cognitive-Behavioral Perspective* (0 éd., p. 419-438). Routledge. https://doi.org/10.4324/9780203056646-27

Cooper, K. J., & Gibson, R. B. (2022). A novel framework for inner-outer sustainability assessment. *Challenges, 13*(2), 64. https://doi.org/10.3390/challe13020064

Dijkstra, I. J., & Mäurer, J. K. (2024). The Journey and Translation of the Inner Development Goals (IDGs) in South Korea. An investigation of IDG dissemination in Korea as avcontribution to gender equality. Master thesis, Malmö University, 71 pp. https://mau.diva-portal.org/smash/get/diva2:1874762/FULLTEXT03.pdf

Donald, J. N., Bradshaw, E. L., Ryan, R. M., Basarkod, G., Ciarrochi, J., Duineveld, J. J., Guo, J., & Sahdra, B. K. (2020). Mindfulness and its association with varied types of motivation: A systematic review and meta-analysis using self-determination theory. *Personality & Social Psychology Bulletin, 46*(7), 1121–1138. https://doi.org/10.1177/0146167219896136

Dufour, C. (2024). How the Inner Development Goals can help shape Conscious Food Systems: Reflections from the Caux IDG Forum. UNDP Food And Agricultural Commodity Systems. https://www.undp.org/facs/blog/how-inner-development-goals-can-help-shape-conscious-food-systems-reflections-caux-idg-forum

Engel S., & Janssen, C. (2024). The Inner Development Goals as an innovative approach to Sustainable Development: conceptualization, implementation, and evaluation of an experiential university seminar for holistic sustainability education. In: *Engaging in Prosocial Behaviour for an Inclusive Society*, E. Leung (ed.). IntechOpen books, 24 pp. https://doi.org/10.5772/intechopen.1006198

EUA Annual Report 2023. (s.d.). https://www.eua.eu/publications/reports/eua-annual-report-2023.html

Ferrer-Balas, D., Adachi, J., Banas, S., Davidson, C. I., Hoshikoshi, A., Mishra, A., Motodoa, Y., Onga, M., & Ostwald, M. (2008). An international comparative analysis of sustainability transformation across seven universities. *International Journal of Sustainability in Higher Education, 9*(3), 295–316. https://doi.org/10.1108/14676370810885907

Filho, W. L., Sierra, J., Price, E., Eustachio, J. H. P. P., Novikau, A., Kirrane, M., Dinis, M. A. P., & Salvia, A. L. (2024). The role of universities in accelerating the sustainable development goals in Europe. *Scientific Reports, 14*(1), 15464. https://doi.org/10.1038/s41598-024-65820-9

Filho, W. L., Simaens, A., Paço, A., Hernandez-Diaz, P. M., Vasconcelos, C. R. P., Fritzen, B., & Mac-Lean, C. (2023). Integrating the Sustainable Development Goals into the strategy of higher education institutions. *International Journal of Sustainable Development & World Ecology, 30*(5), 564–575. https://doi.org/10.1080/13504509.2023.2167884

Filho, W. L., Skanavis, C., Kounani, A., Brandli, L. L., Shiel, C., Paço, A. D., Pace, P., Mifsud, M., Beynaghi, A., Price, E., et al. (2019). The role of planning in implementing sustainable development in a higher education context. *Journal of Cleaner Production, 235*, 678–687. https://doi.org/10.1016/j.jclepro.2019.06.322

Findler, F., Schönherr, N., Lozano, R., Reider, D., & Martinuzzi, A. (2019). The impacts of higher education institutions on sustainable development: A review and conceptualization. *International Journal of Sustainability in Higher Education, 20*(1), 23–38. https://doi.org/10.1108/IJSHE-07-2017-0114

Finstad-Milion, K., Ceulemans, K., & Avetisyan, E. (2021). Promoting engaged scholarship for sustainability regionally : The case of the PRME France-Benelux Chapter: *Management & Sciences Sociales, N°, 30*(1), 140-154. https://doi.org/10.3917/mss.030.0140

Flook, L., Smalley, S. L., Kitil, M. J., Galla, B. M., Kaiser-Greenland, S., Locke, J., Ishijima, E., & Kasari, C. (2010). Effects of Mindful awareness practices on executive functions in elementary school children. *Journal of Applied School Psychology, 26*(1), 70–95. https://doi.org/10.1080/15377900903379125

Fourati-Jamoussi, F., Agnès, M., Caron, P., Dubois, M. J. F., Leroux, V., Rakotonandraina, N., Kotbi, G., & Sauvée, L. (2015). How to promote, support and experiment sustainability in higher education institutions? The case of LaSalle Beauvais in France. *International Journal of Innovation and Sustainable Development, 9*(3–4), 227–245. https://doi.org/10.1504/IJISD.2015.071856

Giangrande, N., White, R. M., East, M., Jackson, R., Clarke, T., Saloff Coste, M., & Penha-Lopes, G. (2019). A competency framework to assess and activate education for sustainable development: Addressing the UN sustainable development goals 4.7 Challenge. *Sustainability, 11*(10), 2832. https://doi.org/10.3390/su11102832

Going Deeper. (2023). Google Docs. Retrieved 30 October 2024, from https://drive.google.com/file/d/1M3e1ixgNRXXrlFIN93M2WMmApQwDB9MC/edit?usp=embed_facebook

Grant, L., Reida, C., Buesselerb, H., & Addissb, D. (2022). A compassion narrative for the sustainable development goals: conscious and connected action. *The Lancet, 400*(10345), 7–8. https://www.thelancet.com/journals/lancet/article/PIIS0140-6736(22)01061-3/fulltext

Gravett, S. (2004). Action research and transformative learning in teaching development. *Educational Action Research, 12*(2), 259–272. https://doi.org/10.1080/09650790400200248

Green transition. (s.d.). https://www.eua.eu/our-work/topics/green-transition.html

IDG.tools—Inner Development Phase 2 Research Report. (2022). Retrieved 30 October 2024, from https://idg.tools/assets/221215_IDG_Toolkit_v1.pdf

Inner Development Goals: Background, method and the IDG framework. (2021). Growth that Matters AB.

Jacob, J., Jovic, E., & Brinkerhoff, M. B. (2009). Personal and planetary well-being: Mindfulness meditation, pro-environmental behavior and personal quality of life in a survey from the social justice and ecological sustainability movement. *Social Indicators Research, 93*(2), 275–294. https://doi.org/10.1007/s11205-008-9308-6

Jensen, B. B., & Schnack, K. (1997). The action competence approach in environmental education. *Environmental Education Research, 3*(2), 163–178. https://doi.org/10.1080/1350462970030205

Kabat-Zinn, J., & Hanh, T. N. (2009). Full Catastrophe Living: Using the Wisdom of Your Body and Mind to Face Stress, Pain, and Illness. Random House Publishing Group.

Kaufman, S., Saeri, A., Raven, R., Malekpour, S., & Smith, L. (2021). Behaviour in sustainability transitions: A mixed methods literature review. *Environmental Innovation and Societal Transitions, 40*, 586–608. https://doi.org/10.1016/j.eist.2021.10.010

Kolb, D. A. (1984). Experiential learning: Experience as the source of learning and development. Prentice-Hall.

Kopnina, H. (2020). Education for the future: Critical evaluation of education for sustainable development goals. *Journal of Environmental Education, 51*(4), 280–291. https://doi.org/10.1080/00958964.2019.1710444

Lane, A. (2017). Open education and the sustainable development goals: Making change happen. *Journal of Learning for Development, 4*(3), 275–286. https://doi.org/10.56059/jl4d.v4i3.266

Leicht, A., Heiss, J., & Won Jung, B. (2018). Issues and trends in education for sustainable development. UNESCO. https://unesdoc.unesco.org/ark:/48223/pf0000261445?posInSet=4&queryId=06c9baee-0a92-4546-8cb4-d26adbde1a65

Lozano, A., López, R., Pereira, F. J., & Blanco Fontao, C. (2022). Impact of cooperative learning and project-based learning through emotional intelligence : A comparison of methodologies for implementing SDGs. *International Journal of Environmental Research and Public Health, 19*(24), Article 24. https://doi.org/10.3390/ijerph192416977

Lozano, R., Lukman, R., Lozano, F. J., Huisingh, D., & Lambrechts, W. (2013). Declarations for sustainability in higher education: Becoming better leaders, through addressing the university system. *Journal of Cleaner Production, 48*, 10–19. https://doi.org/10.1016/j.jclepro.2011.10.006

Macy, J. (1998). *Coming back to life : Practices to reconnect our lives, our world*. Gabriola Island, BC, Canada ; Stony Creek, CT: New Society Publishers. http://archive.org/details/comingbacktolife0000macy

Madurai Elavarasan, R., Pugazhendhi, R., Jamal, T., Dyduch, J., Arif, M. T., Manoj Kumar, N., Shafiullah, G. M., Chopra, S. S., & Nadarajah, M. (2021). Envisioning the UN Sustainable Development Goals (SDGs) through the lens of energy sustainability (SDG 7) in the post-COVID-19 world. *Applied Energy, 292*. Scopus. https://doi.org/10.1016/j.apenergy.2021.116665

Marchi, L., Vodola, V., Visconti, C., Gaspari, J., & Antonini, E. (2021). Contribution of individual behavioural change on household carbon footprint. In P. Akimov, N. Vatin, A. Tusnin, & A. Doroshenko (Éds.), *E3S Web Conf.* (Vol. 263). EDP Sciences; Scopus. https://doi.org/10.1051/e3sconf/202126305024

McLeod, S., & Marshall, J. (2023). Communication for all and the sustainable development goals. *International Journal of Speech-Language Pathology, 25*(1), 1–8. https://doi.org/10.1080/17549507.2022.2160494

Molderez, I. (2021). How transformative learning nurtures ecological thinking. Evidence from the Students Swap Stuff project. *International Journal of Sustainability in Higher Education, 22*(3), 635-658. https://doi.org/10.1108/IJSHE-05-2020-0174

MP4s Student Challenges Mapping/PR1 Report. (2023). MP4s. Retrieved 30 October 2024, from https://mp4s.eu/mp4s-student-challenges-mapping-pr1-report/

Nordén, B. (2024). Advancing sustainability through higher education: Student teachers integrate inner development goals (IDG) and future-oriented methodologies. *Challenges, 15*(2), 28. https://doi.org/10.3390/challe15020028

Ojala, M. (2013). Emotional awareness: On the importance of including emotional aspects in education for sustainable development (ESD). *Journal of Education for Sustainable Development, 7*(2), 167–182. https://doi.org/10.1177/0973408214526488

Persson, S., Shrivastava, P., & Zsolnai, L. (2022). Art and Aesthetics in Sustainability Education Insights into Beautifying Management Education with François Jullien: *Revue Internationale de Psychosociologie et de Gestion Des Comportements Organisationnels, XXVII*(71), 75–98. https://doi.org/10.3917/rips1.071.0075

Presseau, J., McCleary, N., Lorencatto, F., Patey, A. M., Grimshaw, J. M., & Francis, J. J. (2019). Action, actor, context, target, time (AACTT): A framework for specifying behaviour. *Implementation Science, 14*(1), 102. https://doi.org/10.1186/s13012-019-0951-x

Richter, N., & Hunecke, M. (2022). Mindfulness, connectedness to nature, personal ecological norm and pro-environmental behavior: A daily diary study. *Current Research in Ecological and Social Psychology, 3*, Article 100038. https://doi.org/10.1016/j.cresp.2022.100038

Rieckmann, M. (2018). Chapter 2—Learning to transform the world : Key competencies in ESD (p. 39–59).

Riemer, M., Lynes, J., & Hickman, G. (2014). A model for developing and assessing youth-based environmental engagement programmes. *Environmental Education Research, 20*(4), 552–574. https://doi.org/10.1080/13504622.2013.812721

Sedita, S. R., Blasi, S., & Yang, J. (2022). The cultural dimensions of sustainable development: A cross-country configurational analysis. *Sustainable Development, 30*(6), 1838–1849. https://doi.org/10.1002/sd.2351

Shao, R., & Skarlicki, D. P. (2009). The role of mindfulness in predicting individual performance. *Canadian Journal of Behavioural Science/revue Canadienne des Sciences Du Comportement, 41*(4), 195–201. https://doi.org/10.1037/a0015166

Shtaltovna, Y., Rodriguez Carreon, V., Lindencrona, F., & Donald, W. E. (2024). Cognitive skills within the inner development goals (IDG) framework: Empowering sustainable careers and sustainable development. *GILE Journal of Skills Development, 4*(1), 74–94. https://doi.org/10.52398/gjsd.2024.v4.i1.pp74-94

Spaccatini, F., Riva, P., Richetin, J., Porcelli, E., Pancani, L., Capellini, R., & Sacchi, S. (2023). From past to present (for a better future): The moderating role of cognitive mindset on spillover effects in environmental behaviors. *Current Psychology, 42*(18), 15858–15873. https://doi.org/10.1007/s12144-022-02917-2

Stålne, K., & Greca, S. (2022). *Inner Development Goals: phase 2 research report*. Kävlinge, Sweden and Welver, Germany, 128 pp. https://idg.tools/assets/221215_IDG_Toolkit_v1.pdf

Sterling, S. (2010). Learning for resilience, or the resilient learner? Towards a necessary reconciliation in a paradigm of sustainable education. *Environmental Education Research, 16*(5–6), 511–528. https://doi.org/10.1080/13504622.2010.505427

Sterling, S. (2013). The future fit framework: An introductory guide to teaching and learning for sustainability in HE (Guide). *Journal of Education for Sustainable Development, 7*, 134–135. https://doi.org/10.1177/0973408213495614b

Sterling, S. (2021). Concern, conception, and consequence: re-thinking the paradigm of higher education in dangerous times. *Frontiers in Sustainability, 2*,. https://doi.org/10.3389/frsus.2021.743806

Sustainable Development Goals—European Commission. (s.d.). https://international-partnerships.ec.europa.eu/policies/sustainable-development-goals_en

Tang, Y., Geng, L., Schultz, P. W., Zhou, K., & Xiang, P. (2017). The effects of mindful learning on pro-environmental behavior: A self-expansion perspective. *Consciousness and Cognition, 51*, 140–148. https://doi.org/10.1016/j.concog.2017.03.005

Tilbury, D. (2011). Higher education for sustainability: A global overview of commitment and progress. Higher Education's Commitment to Sustainability: From Understanding to Action, 18-28.

UNESCO. Issues and trends in education for sustainable development—(s.d.). https://unesdoc.unesco.org/ark:/48223/pf0000261445

Unsdg | UN Sustainable Development Solutions Network. (s.d.). https://unsdg.un.org/un-entities/unsdsn

Wamsler, C., Osberg, G., Osika, W., Hendersson, H., & Mundaca, L. (2021). Linking internal and external transformation for sustainability and climate action: Towards a new research and policy agenda. *Global Environmental Change, 71*, Article 102373. https://doi.org/10.1016/j.gloenvcha.2021.102373

Wiek, A., & Redman, A. (2022). What do key competencies in sustainability offer and how to use them. In P. Vare, N. Lausselet, & M. Rieckmann (Éds.), *Competences in Education for Sustainable Development: Critical Perspectives* (p. 27–34). Springer International Publishing. https://doi.org/10.1007/978-3-030-91055-6_4

Wiek, A., Withycombe, L., & Redman, C. L. (2011). Key competencies in sustainability: A reference framework for academic program development. *Sustainability Science, 6*(2), 203–218. https://doi.org/10.1007/s11625-011-0132-6

Wood, P. (2024). The inner development goals: changing educational systems to meet the challenge of human-generated global crises. In Kushnir, K. Sood, M.S.-A. Park, H. Zhong, & N. Serret (Eds.), *Education and Sustainable Development in the Context of Crises: International Case Studies of Transformational Change*, I, Emerald Publishing Limited, Leeds, pp. 125–139. https://doi.org/10.1108/978-1-83797-773-420241008

Yilmaz, K. (2011). The cognitive perspective on learning: its theoretical underpinnings and implications for classroom practices. *The Clearing House: A Journal of Educational Strategies, Issues and Ideas, 84*(5), 204–212. https://doi.org/10.1080/00098655.2011.568989

Change Making in Culinary Arts Education to Promote Sustainability: The Challenges and Opportunities Presented by Participatory Action Research

John Killeen

Abstract A key challenge in education for sustainability is the need for sustainable pedagogical approaches that encompass the learning environment (the classroom), community (students and staff) and assessment (reflection). The core question underpinning this research paper is how can Higher Education (HE) actively assume responsibility as a change-maker committed to producing both sustainable actors and leaders in society? This research addresses the question through a collaborative approach to sustainability in education that focuses on co-creation of knowledge and practice in the classroom. The research will utilise Participatory Action Research (PAR) in culinary arts education to investigate how through growing, cooking and eating food, students and staff can critically engage with their practice and understanding of sustainability. The experience and reflections of the learning community in the kitchen classroom will be used to co-create their own sustainability initiatives in real time. The focus of this paper is an interrogation of the methodological approaches that will be deployed to conduct PAR with a group of culinary arts students over a two-year period. The article is divided into four sections: Firstly, a short autoethnographic reflection on the author's experience of Action Research (AR) in HE is presented. Secondly, the role of PAR in developing sustainable pedagogical approaches and co-creation of sustainable knowledge and practice is critically reviewed. Thirdly, the paper presents an overview of an innovative mixed methods research design underpinning the PAR approach and offers a template of "how to do" primary research on the issue of sustainable practice in HE. Fourthly, the paper addresses power and engagement in the co-creations of knowledge and practice.

Keywords Sustainable education · Culinary · Pedagogy · PAR

J. Killeen (✉)
Department of Hospitality, Tourism & Leisure (HTL), Technological University of the Shannon (TUS Athlone Campus), Westmeath, Ireland
e-mail: John.Killeen@tus.ie; John.Killeen.2023@mumail.ie

Department of Sociology, Maynooth University, Kildare, Ireland

W. Leal Filho et al. (eds.), *Introducing the UN SDGs in Higher Education Institutions*, World Sustainability Series, https://doi.org/10.1007/978-3-032-23017-1_21

1 Introduction

"Those who authentically commit themselves to the people must re-examine themselves constantly" (Freire, 1996:42). In keeping with the spirt of Freire this paper will (re)exam and interrogate the opportunities and challenges presented by Participatory Action Research (PAR) in a Higher Education (HE) setting. The key objective is to explore how HE institutions can fulfil the remit of change-maker, producing sustainability conscious actors and leaders who can effect wider changes in society. The research project is based in a HE institution and focuses on culinary pedagogic practice. Firstly, a short autoethnographic reflection on the author's experience of Action Research (AR) in HE is presented, encompassing chef, educator and researcher praxis. The rationale is to demonstrate the transformative experiences of AR motivating the author to make culinary arts education more sustainable. Adapting such a reflexive approach helps to contextualise the research, and create ideational prompts for the design of the PAR. Secondly, the role of PAR in developing sustainable pedagogical approaches and co-creation of sustainable knowledge and practice is critically reviewed. Thirdly, the paper presents an overview of an innovative, mixed methods protocol underpinning the PAR approach and offers a template of "how to" do primary research on the issue of sustainable practice in HE. The fourth section of the paper takes up the issue of power and engagement in the co-creation of knowledge and practice, and is followed by concluding comments.

2 Reflection of a Chef, Educator and Researcher

Sparkes (2000:21) has characterised auto-ethnographic writing as 'highly personal accounts that draw upon the experience of the author/researcher for the purposes of extending sociological understanding." More specifically, Boyle and Parry (2007) argue that autoethnographic methods can enhance the investigation of organizations and culture. Indeed, they suggest that "increasing use of the first person in the write-up of management research indicates heightened acceptance of the self-narrative as a form of sensemaking within organizational life,' (2007:187). An autoethnographic account is presented here to convey researcher positionality and the multiple perspectives the author brings to the HE research site.

> My time working as a chef fulfilled my inclination towards wanting to use my hands to make something in real time. The appeal to me was the process rather than the finished product. Indeed, I have often described myself as a "technician" rather than a chef. I viewed the chef as a creative and charismatic force, believing these were rare attributes to which only a few could ever aspire. I built my own career on efficiency and continuous improvement. I gravitated towards the problem areas. I felt I could identify the problem and devise a suitable action to allow the kitchen to function better. In hindsight it would appear "actioner researcher" would describe the type of chef I was.

As I transitioned my career into education this is something I wanted to instill in my students. I wanted them to be good technicians first. I began my lecturing career by trying to impress on my students the importance of process and technique. I would demonstrate multiple ways they could do the same thing and test them on where and when they should apply a particular method. I encouraged them to taste each other's food and critique what they had done and how it had impacted the end product. I tried to mimic the situations I felt they would experience in a professional kitchen. Towards the end of my first year of teaching I was frustrated. I could not fathom how a recipe could turn out completely different when followed by twelve students. I sought answers; "why did you do that here when it says this?", "is there a way I can communicate this better?". I was displaying the chef's traditional need for consistency and conformity. A restaurant is judged by how consistent it is in reproducing hundreds of items a night that look and taste the same. In an attempt to prepare them for what lay ahead I assumed what had been so effective for me would work for them. I was wrong. I was trying to prepare them for an industry that is notoriously difficult but in doing so I was actually reinforcing industry norms and perpetuating the problems. I had believed that a professional kitchen functioned as a system of culinary practices in a specific time and place without fully realizing each person in the kitchen brought in their own ways of knowing and meaning.

My views of sustainable food developed along a similar path to my changing approach to culinary education. At first, I had a sense of guilt at the level of food waste that occurs. I told myself that the waste was a necessity of fulfilling the module learning outcomes. There is a long-standing belief amongst faculty that waste is an unfortunate by-product of the teaching and learning process. How could I address this? Why does it happen in the first place? I noted the waste segregation systems in the classrooms were rarely adhered to. I asked students if they adhered to waste segregation at home and at work and the answer was yes. They could not explain why their practice in one space was different to the other. This conundrum led me to engage the students in action research. They were empowered to devise solutions to the waste issues. The root of the problem was wide-ranging but the overriding conclusion was clear. It was not important in their culinary education, they were too busy learning how to cook. The solutions they devised focused on early student engagement, better signage, a clearer system for segregation. They explained that during their first two years in the practical classes the focus was on the cooking. They were aware of sustainability as a concept but it was not made explicit as part of their cookery practice. They highlighted theory-based modules that discussed it but identified a disconnect between theory and practice. They had developed a keen interest in sustainable food but regretted how absent sustainable practice was in their cooking. These insights from the students led me to my current research. In addition to this, several bottom-up sustainability initiatives centred around food began to be emerge in the Hospitality, Tourism and Leisure (HTL) department at my HEI. This inadvertently created moments of tension and contradiction for the community of staff and students that often went unnoticed or were ignored. Could PAR be used to investigate these tensions and contradictions to improve sustainability initiatives?

3 Participatory Action Research in Higher Education

The aim of this section of the paper is to critically interrogate the role of PAR in developing sustainable pedagogical approaches and co-creation of sustainable knowledge and practice. It is not intended to present a simplistic view of AR, in the Northern tradition, vs PAR in the Southern tradition. Rather, it is to provide a re-examination of the transformation of the author's approach to teaching and research arising from ongoing reflection on experience. The AR initially conducted was rooted in Lewin's (1946) "action research" tradition, or the Northern tradition, to which it is sometimes referred (Littman et al., 2021). The focus was on solving a problem through some collective investigation and action. AR begins with the simplest of questions, "how can we improve this situation?"(Reason & Bradbury, 2013:11). Inadvertently, the author positioned himself in the role of 'expert', trying to solve a particular problem that would improve practice in the classroom. Levin and Greenwood (2013) are critical of this approach to AR in HE as it risks presenting a limited worldview to students. The author proposed a binary problem and solution scenario to the students: there is an issue with food waste, how can we reduce it? This limited dialogue and critical insight in defining what the problem was in the first place. Kemmi (2006:459) maintains this type of AR has been widely adopted in education. Instead of being a "vehicle for educational critique" it has become "a vehicle for domesticating students and teachers to conventional forms of schooling". Kemmi (2006) views schooling as the institutional practices and processes that prepare the student to participate in society. There are parallels here with how the author tried to prepare culinary students to best function in kitchens. The student was taught how to participate in the cultural, social and economic life of society, whether that be in university or the kitchen. There was no empowerment of the students to question the dominant narratives within their culinary education. Kemmi (2010) argues that the lack of educational critique in much of AR is due to a tendency to focus on the action phases rather than the reflection. This was clearly evident in the initial undertaking of AR with students: the focus was on action, what could be done differently to reduce food waste. Whenever discussions drifted into wider issues, the author reined it in keeping the focus on the task at hand, and on the technical aspects of the practice (Kemmi, 2006, 2010). This prevented the students from raising broader questions such as; what is sustainability, or why is culinary education based on 100 year old French cookery principles and Escoffier's militaristic kitchen brigade? (Deutsch, 2014; Lewis, 2020; Mac Con Iomaire, 2021). Echoing the author's early teaching practice, the brigade system emphasises discipline, technique, efficiency and conformity to ensure a uniform, predetermined outcome. Deutsch (2014) documents the issues this training has imparted; a hegemony of western cooking, hierarchical culture, discipline and learning by repetition. The author recognises a subconscious inclination to reinforce classical French cookery as a means to demonstrate primary legitimacy as a chef rather than as an educator (Woodhouse, 2015). The author's own role models and teachers had been trained under the doctrine of Escoffier, which espoused the requirement to work up through the ranks from apprentice to master. The master who was once the apprentice (oppressed) becomes the oppressor (Freire, 1996). The challenge here is the "existential duality of the oppressed, who are at the same time themselves and the oppressor whose image they have internalized" (Freire, 1996:43). Recognising the oppressive nature of culinary arts education, the author attempted to work and change things from within without challenging the power structure of the tradition it was based on. In doing so "*systematic education*" was maintained. Real change can only occur by engaging in "*educational projects*" with the oppressed to facilitate their ownership of the change (Freire, 1996:36).

AR provides an opportunity for staff and students to engage in critical reflection as to why traditions in education prevail. This cannot occur if improvement in practice is viewed from the educator perspective only and not genuinely engaging with the perspectives of the students (Kemmi, 2006). Marginalising the students' own experiences and knowledge and assigning them the role of change makers disempowered them (Glassman & Erdem, 2014). The status quo of culinary arts education was maintained by directing their reflection onto their individual practices rather than affording them the opportunity to critique their educational experience. Fals-Borda (1987) outlines how PAR enables the demystification of research so the students can participate and generate their own knowledge. The students become co-researchers and collaborators who identify things they want to change. They create meaning and generate their own knowledge by relating to their own life experience (Freire, 1996). This emancipatory approach to PAR, or the Southern tradition to which it is also referred, centres on producing knowledge amongst the participants (Littman et al., 2021). Levinson (2017) notes that a difficulty in defining PAR is how interchangeably it is used with AR. In the context of this paper, AR is characterized as a broad methodological approach. PAR is differentiated by its commitment to collaboration, co-research, dialogue and critical reflection to represent the collective knowledge generated by all who participate in the research process.

Whilst the criticisms of the hegemonic hold of Escoffier's principles in culinary pedagogy are valid, these programmes sit within the overall HE landscape. It is instructive to provide a brief overview of the opportunity for PAR to challenge such hegemonic narratives in education more generally. Additionally, it provides context for the specific methodological approach presented in this paper. A common thread in much of the research around PAR is an overwhelming criticism of the prevalence of Neoliberalism in HE (Fals Borda, 2006; Huckle & Wals, 2015; Kemmis, 2006, 2010; Kindon & Elwood, 2009; Trott et al., 2018). Levin and Greenwood (2013) argue that while university research remains a purveyor of 'public good' this is increasingly being eroded by neoliberal forces emanating from HE management, government regulatory bodies and private investment funds. Thus, the decline in public funding and the imperative to make social and economic contributions to society has resulted in HE research migrating to the private sector. Kemmi (2006) elaborates on the flaws of an education that is removed from the wider social context within which it is embedded. Limiting education to be understood only in terms of developmental tasks for the individual does not allow for an interpretation and understanding of the world as it has come to be (Kemmi, 2006). In many ways, this echoes the prevailing hierarchy, structure and conformity that is so entrenched in culinary education. The culinary student is taught to understand food and cookery only from the perspective of a useable commodity that is to be mastered by the application of learned skills. Equally, the wider education system has become efficient at educating people in how to best function in a neoliberal economy. Kindon and Elwood (2009) contend that the neoliberal university is responsible to the community in which it exists, through both the research it pursues and the research it omits. As noted above, the author pursued a change in cookery practice but omitted the participants' reflection on the dominant narrative in culinary arts education. Furthermore, the author failed to even consider any dominant narratives in sustainable education.

Fals Borda (2006) contends that the greatest challenge for the university today is to move away from the limiting effects of rigid structures. He suggests the need for education and wider society to put greater emphasis on grassroots groups and the voices of the excluded in dominant systems. One such structure the author assumed in their original AR was his interpretation of the United Nation's Sustainable Development Goals (UNSDGs). AR calls for a flattening of hierarchy in education that can be achieved through the empowerment and involvement of all stakeholders in the community (Kemmi, 2006). The SDGs provide a blueprint for achieving this, for example through greater empowerment of educational communities and increased partnerships with external stakeholders. The value of AR to the SDGs is its ability to facilitate practice among the community rather than just knowledge about action and practice (Kemmi, 2010). The author's original AR used the UNSDGs as a checklist to verify their practice and research, aligning a reduction in food waste to UNSDG target 2.4, to insure a more sustainable food production system. Much like the doctrine of Escoffier, the SDGs were not critically engaged with as part of the educational experience, they were merely used to validate a learning outcome. Huckle and Wals (2015) assert that this sort of Education for Sustainable Development (ESD) ignores the political hegemony of capitalism and encourages people to see sustainability in individual and monetary terms. They go on to suggest this keeps the focus on the behaviour of individuals and distracts from critically reflecting on the dominant political and economic discourse.

Rather than dissect the merits and shortcomings of ESD and the SDGS, the methodological approach proposed here aims to address the gap in dialogical engagement between agents (students and staff) and the dominant narrative of SD in HE. Huckle and Wals (2015) propose that the challenge for sustainable education is to create an environment that enables critical discourse, to participate in sustainable decision-making and develop an understanding of contested meanings for the whole community. The potential for PAR to address these issues in conjunction with the SDGs is being established (see Snapp et al., 2023; Trajber et al., 2019; Trott et al., 2018). In support of this, the PAR approach being presented here empowers the students and staff to collaborate and interrogate the UNSDGs. Instead of assuming a particular target, the research participants collaborative educational experience of growing, eating and cooking sustainable food will facilitate, but not limit, their individual and collective engagement with multiple SDGs such as; SDG 3 Good Health and Well-Being, SDG 4 Quality Education and SDG 15 Life on Land.

Nevertheless, criticisms of PAR being used to reinforce neoliberalism (Fals Borda, 1996; Kemmi, 2006; 2010) remain salient. Snapp et al. (2023) illustrates this by measuring the success of a community agriculture scheme in terms of crop yields. Interestingly, the technologically advanced crops, proposed by experts, did not yield as well as expected so the local farmers adjusted crops using some of their traditional knowledge and practices, achieving better yields (Snapp et al., 2023). In this example, the authors highlight the comprehensive framework provided by multiple UNSDGs, such as food security (SDG 2) and environmental preservation (SDG 15).

The takeaway here is how local knowledge and practice can help inform the SDGs. However, the focus on measuring in terms of crop yields detracts from another point. The success here was in creating a collaborative group of farming communities that shared knowledge to co-create new, improved practice, thus creating a collaborative sustainable community (SDG 11). Rather than only working from the perspective provided by expert organisations and institutions, should the focus not be on the co-researcher's collective knowledge generation and critical interrogation of farming methods? Furthermore, Trott et al. (2018) suggests there is a symbiotic relationship between PAR and SDGs in HE. They note the opportunity for PAR to achieve the SDGs but the challenge is for HE to facilitate PAR by moving beyond traditional measurements of success and performance metrics for staff and students (Trott et al., 2018). There is an inherent tension between facilitating PAR to move beyond the neoliberal university structures for sustainable research, but yet continuing to measure this research against the SDGs. Research funding is more likely when aligned to the SDGs and if the research output can be expressed in measurable economic or environmental terms. Yet the importance of PAR is not having clearly defined outcomes before the research is undertaken. Levinson (2017) proposes true PAR should create an environment for unanticipated pathways. This would not mean discarding the SDGs which have intrinsic value as sustainability principles. Rather, the SDGs provide a broad blueprint for achieving a more sustainable future through dialogue and action. PAR can facilitate applying the SDGs in a local context that enables ongoing discussion, reflection and critique of constitutive discourses and practices.

Glassman and Erdem (2014) note that in order to achieve a cyclical process of discovery and realization (to which Freire (1996) refers), we must accept that we may be working from the perspective of not actually understanding what we think we understand. In culinary arts education, the SDGs and Escoffier's kitchen brigade create an illusion that there is some absolute knowledge to be obtained, "the existing ways of doing things" (Kemmis & McTaggart, 2000). In order to move beyond this "taken-for-grantedness" PAR should facilitate an ongoing dialogue as outlined above. Research communities should be empowered as problem solvers, thinkers and learners to generate "alternative ways of doing things" (Kemmi & McTaggart, 2000). PAR should not be framed by hegemonic power, it should be rooted in the knowledge generation of the participants and how they understand their experience (Fals Borda, 2006). In this way, all participants collaborate in an ongoing process of practice and reflection. Or as Morales (2016:161) puts it, "the researcher is not a facilitator of the process as in action research, but a co-producer of learning."

4 Overview of the PAR Approach

The aim of the study of culinary education transformation is open-ended: to undertake PAR that is collaborative and unconstrained by a clearly defined outcome. The outcomes are essentially unpredictable and the research itself should take unanticipated pathways. This section of the paper presents an overview of an innovative, mixed methods research design (see Fig. 1) underpinning the PAR approach of the culinary education study, offering a template of "how to" do primary research on the issue of sustainable practice in HE. It will also explore some of the challenges and opportunities the current research design may present.

Glassman and Erdem (2014) suggest that the researcher must participate in the experience or find research participants who live the experience as part of their everyday life. The co-creation of knowledge between the researcher and participants (co-researchers) seeks to empower participants to facilitate changes in sustainable practice through their experience in the kitchen classroom setting. The research is longitudinal and will be conducted over a two-year period. Fals Borda (1996) emphasises that PAR requires a long-term persistence and dedication to transform existing systems of power. The two-year cycle (see Fig. 2) is designed to allow such persistence and dedication from the co-researchers. Although a two-year cycle may still not be enough time to implement the desired action of the research group (Bou & Sales, 2024). The outcome of this approach should be dialogue and discussion on sustainability from different co-researchers.

This dialogue is facilitated by multiple perspectives using various methods (see Fig. 1) e.g. staff through semi structured interviews, students through focus groups and creative methods, and the autoethnography, which provides the perspective of researcher, lecturer and chef practitioner. The inclusion of multiple perspectives should provide diverse understandings of sustainability. For example, PAR on teacher activism highlighted the benefit to educators of hearing multiple perspectives beyond their professional colleagues. The latter felt talking to people outside of their day-to-day colleagues helped them to question their practices or ways of thinking and doing (Finefter-Rosenbluh et al., 2023). Kemmi (2013) contends the value of PAR lies in its interrogation of individual and collective perspectives to understand praxis. In doing so it allows for the identification and understanding of any prejudices that are raised in dialogue (Kemmi, 2013). The research design allows for a self-examination by the author of his perspective in comparison with those of colleagues. Furthermore, the student perspective is examined individually through the creative methods and collectively through the focus groups. Finally, a collective perspective is reached among the co-researchers through the PAR process. For Stuart (2022:9), what emerges from theses multiple perspectives is a first-person account of "what have we found out together, and what does it mean"? This collective perspective is not without flaws however. The research design outlined here was arrived at before engaging the co-researchers. As such the research is still somewhat 'on' them rather than 'with' them (Stuart, 2022). This raises the prospect of the research marginalising the voice of the student co-researcher and giving them tokenistic representation (Trajber et al., 2019).

The greatest opportunity for PAR in developing sustainable pedagogical approaches is its ability to represent the student voice and challenge the traditional hierarchy in HE. To address these concerns the role of power and the student voice in PAR must be elaborated (Fig. 2).

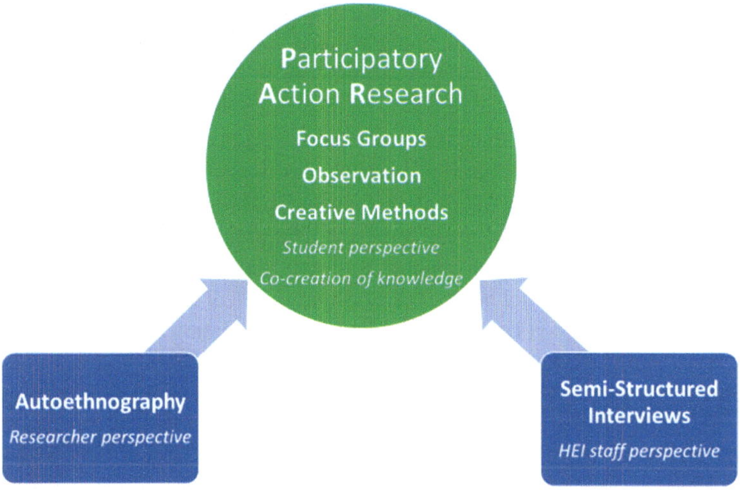

Fig. 1 Participatory action research design

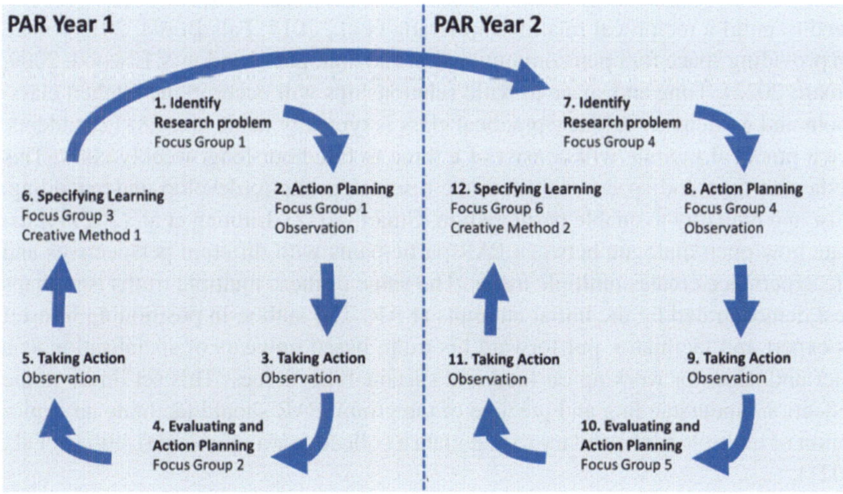

Fig. 2 Two-year participatory action research cycle

5 Power and Engagement in the Co-creation of Sustainable Knowledge and Practice

PAR holds out the prospect of an emancipatory process for participants reducing traditional power dynamics in research (Datta et al., 2015; Fals Borda, 2006). Fals Borda (2013) describes how "authentic participation" in PAR should shorten the distance between the oppressor and the oppressed. This research is actively seeking collaborative practice between staff and students to co-research and collectively combine knowledge and engage in dialogue. It is not about measuring the outcome of the project against pre-ordained metrics. Rather it is about trying to understand the perception of the participants based on real social processes (Fals Borda, 2013). Fals Borda (2013) highlights that PAR is often slow and multidirectional but a fascinating and emancipating experience for all involved. The author's earlier experience has shown how training students to conduct a clearly defined piece of AR results in them learning only how to act better in the current political and social structure, thus maintaining the status quo (Trajber et al., 2019). Renegotiating these traditional roles in PAR is not simple. Students may be intimidated by the active role of researcher (Kindon & Elwood, 2009) or the lack of structure in terms of what they should do (Stuart, 2022). Co-researchers may find it difficult to see themselves as equals and seek guidance from the lead researcher or lecturer (Bou & Sales, 2024). To address this, the focus groups will dedicate time to discussion on the value of everyone's knowledge and experience in the research process.

Lecturers may seek to prevent 'mistakes' or lack the confidence to take on unexpected challenges (Kindon & Elwood, 2009). Therefore, the lecturer and students need to build a reciprocal relationship (Datta et al., 2015; Fals Borda, 2013) based on providing space for open communication and dialogue (Kindon & Elwood, 2009; Stuart, 2022). Time and space to build relationships will occur in the kitchen classroom and gardens. A culinary practical class is typically made up of 8–12 students. Each practical module will consist of a three to five hour-long, weekly class. This dedicated time and space will allow the research group to develop understanding, trust and empathy to enable participation (Stuart, 2022). Littman et al. (2021) illustrate how open dialogue between PAR participants with different perspectives and life experience creates multiple truths. The value of these multiple truths is perhaps best demonstrated by the initial attempts at AR. The author, in positioning himself as expert and facilitator, put forward his truth, based on years of socialisation as a chef and educator working on food and sustainability issues. This set limits to the reflection, understanding and practice of the group. PAR should facilitate an exploration of multiple truths and try to negotiate a collective way forward (Littman et al., 2021).

Stuart (2022) points out the variance in the level of participation a student may engage in. Some may be actively engaged, some may wish to dip in and out, some may wish to dominate discussions and others may see no relevance at all. The level of engagement of a particular participant can be viewed as part of their individual truth and all levels of participation will still inform the collaborative process. Furthermore, participants may wish to participate in the research but may not be willing to engage with predetermined data gathering methods, such as reflective writing (Luguetti et al., 2023). The latter suggests using creative methods such as videos and drawings so the co-researchers can contribute to the research in the way they want. The strategy of deploying creative methods is built into the research design, but unspecified currently to allow the students agree which way they would like to communicate and generate knowledge. This further enhances their empowerment and ownership of the research.

6 Conclusion

PAR has created a pathway for the author to reflect and transition from not trusting students to make the 'right changes' to seeking dialogue and exploring how knowledge and change can be co-created together. This represents an opportunity for sustainable education to transform from "systematic education" that promotes sustainability through a neoliberal lens to sustainable "educational projects" that trusts students and staff to engage in change through critical reflection. These "educational projects" have the potential to incubate bottom-up sustainability initiatives in HE. However, a limitation for these projects is how they can be scaled up across the university as a whole. For example, the ongoing PAR presented in this paper is currently exploring how existing bottom-up sustainable food initiatives are being experienced by the research participants with limited to no funding. The research site itself is situated in a newly formed university navigating the creation and implementation of its own sustainability strategy. Thus, the limitations of these bottom-up initiatives may present an opportunity for further research. How do bottom-up initiatives intersect with top-down sustainability strategy as it emerges in the research site?

PAR presents an approach to pedagogy and research that is transferable to other domains. In its simplest form it is the engagement of students, staff and/or external stakeholders in real time research within their own educational experience. PAR provides a "how to" approach to engage the HEI community in sustainable practice. This may be part of the institute's sustainability strategy and/or may be deployed as a best practice pedagogical model at faculty, department or even modular level.

Elsewhere, the PAR approach is transferable beyond the educational setting. The focus on culinary education relates to cookery as an every day practice. Cookery is an opportunity for multiple stakeholders such as local communities, farmers, and consumers to adopt PAR in our everyday life and engage with the SDGs. A PAR approach to cookery fosters greater consciousness of sustainability through food systems. Much like in the classroom, it identifies food related challenges and allows space to continuously collaborate and devise more sustainable farming, cookery and eating practices.

Finally, PAR offers a pathway to engage educators and students in a collaborative process of teaching and learning to be change-makers in real time. It has the added benefit of fostering a critical and sustainable mindset amongst participants to produce agents of change for the future. This approach puts the social challenges of the SDGs and sustainable education in focus. The co-creation of knowledge should facilitate student and educator led sustainability initiatives that belong to the whole HE community and are representative of not only their disciplines but also their lived experience.

Acknowledgements The author wishes to thank the editorial team of this book, in particular Dr. Dirk Franco for his insightful feedback and suggestions in preparing this manuscript. Thanks are also due to the author's supervisory team Prof. Mary Corcoran and Dr. Delma Byrne for their exceptional support with this paper and the research behind it.

References

Bou, M., & Sales, A. (2024). Shared knowledge construction in participatory action research: Perceptions of researchers and educational community. *Educational Action Research, 32*, 106–126. https://doi.org/10.1080/09650792.2022.2111695

Boyle, M., & Parry, K. (2007). Telling the whole story: The case for organizational autoethnography. *Culture and Organization, 13*, 185–190. https://doi.org/10.1080/14759550701486480

Datta, R., Khyang, N. U., Prue Khyang, H. K., Prue Kheyang, H. A., Ching Khyang, M., & Chapola, J. (2015). Participatory action research and researcher's responsibilities: An experience with an Indigenous community. *International Journal of Social Research Methodology, 18*, 581–599. https://doi.org/10.1080/13645579.2014.927492

Deutsch, J. (2014). *Suppressing desire as culinary discipline: Can culinary education be hedonistic? Should it be?*

Fals Borda, O. (2013). Action research in the convergence of disciplines. *International Journal of Action Research, 9*, 155–167.

Fals Borda, O. (2006). The North-South convergence: A 30-year first-person assessment of PAR. *Action Research, 4*, 351–358. https://doi.org/10.1177/1476750306066806

Fals Borda, O. (1996). Power/knowledge and emancipation. *Systems Practice, 9*, 177–181. https://doi.org/10.1007/BF02172931

Fals-Borda, O. (1987). The application of participatory action-research in Latin America. *International Sociology, 2*, 329–347. https://doi.org/10.1177/026858098700200401

Finefter-Rosenbluh, I., Berry, A., & Ryan, T. (2023). Acting upon student voice-based teaching assessment initiatives: An account of participatory action research for teacher professional learning. *Journal of Teacher Education, 74*, 508–521. https://doi.org/10.1177/00224871231200278

Freire, P. (1996). *Pedagogy of the oppressed* (3rd ed.). Penguin Books.

Glassman, M., & Erdem, G. (2014). Participatory action research and its meanings: Vivencia, praxis, conscientization. *Adult Education Quarterly, 64*, 206–221. https://doi.org/10.1177/0741713614523667

Huckle, J., & Wals, A. E. J. (2015). The UN Decade of Education for Sustainable Development: Business as usual in the end. *Environmental Education Research, 21*, 491–505. https://doi.org/10.1080/13504622.2015.1011084

Kemmis, S. (2010). Research for praxis: Knowing doing. *Pedagogy, Culture & Society, 18*, 9–27. https://doi.org/10.1080/14681360903556756

Kemmis, S. (2006). Participatory action research and the public sphere. *Educational Action Research, 14*, 459–476. https://doi.org/10.1080/09650790600975593

Kemmi, S. (2013). Critical theory and participatory action research. In P. Reason & H. Bradbury (Eds.), *The SAGE handbook of action research: Participative inquiry and practice* (pp. 121–138).

Kemmis, S., & McTaggart, R. (2000). *Handbook of qualitative research* (2nd ed.). Sage.
Kindon, S., & Elwood, S. (2009). Introduction: More than methods—Reflections on participatory action research in geographic teaching, learning and research. *Journal of Geography in Higher Education, 33*, 19–32. https://doi.org/10.1080/03098260802276474
Levin, M., & Greenwood, D. (2013). The future of universities: Action research and the transformation of Higher education. In: P. Reason & H. Bradbury (Eds.), *The SAGE handbook of action research: Participative inquiry and practice* (pp. 211–226).
Levinson, M. (2017). When participants don't wish to participate in participatory action research, and when others participate on their behalf: The representation of communities by real and faux participants. *The Urban Review, 49*, 382–399. https://doi.org/10.1007/s11256-016-0390-9
Lewin, K. (1946). Action research and minority problems. *Journal of Social Issues, 2*, 34–46. https://doi.org/10.1111/j.1540-4560.1946.tb02295.x
Lewis, B. J. (2020). *Culinary sustainability education: A culinary education as sustainability (Ed.D.).* University of Wisconsin - Stevens Point, United States -- Wisconsin, US.
Littman, D. M., Bender, K., Mollica, M., Erangey, J., Lucas, T., & Marvin, C. (2021). Making power explicit: Using values and power mapping to guide power-diverse Participatory Action Research processes. *Journal of Community Psychology, 49*, 266–282. https://doi.org/10.1002/jcop.22456
Luguetti, C., Jice, N., Singehebhuye, L., Singehebhuye, K., Mathieu, A., & Spaaij, R. (2023). 'I know how researchers are […] taking more from you than they give you': Tensions and possibilities of youth participatory action research in sport for development. *Sport, Education and Society, 28*, 755–770. https://doi.org/10.1080/13573322.2022.2084374
Mac Con Iomaire, M. (2021). *From the dark margins to the spotlight: The evolution of gastronomy and food studies in Ireland*. Books/Book Chapters. https://doi.org/10.21427/8ws4-gg21
Morales, M. P. E. (2016). Participatory Action Research (PAR) cum Action Research (AR) in teacher professional development: A literature review. *International Journal of Research in Education and Science, 2*, 156–165.
Reason, P., & Bradbury, H. (2013). *The SAGE handbook of action research: Participative inquiry and practice.* SAGE.
Snapp, S. S., Bezner Kerr, R., Bybee-Finley, A., Chikowo, R., Dakishoni, L., Grabowski, P., Lupafya, E., Mhango, W., Morrone, V. L., Shumba, L., & Kanyama-Phiri, G. (2023). Participatory action research generates knowledge for sustainable development goals. *Frontiers in Ecology and the Environment, 21*, 341–349. https://doi.org/10.1002/fee.2591
Sparkes, A. C. (2000). *Autoethnography and narratives of self: Reflections on criteria in action.* https://doi.org/10.1123/ssj.17.1.21
Stuart, K. (2022). Problematic participation: Reflections on the process and outcomes of participatory action research into educational inequalities. *Educational Action Research, 30*, 342–356. https://doi.org/10.1080/09650792.2020.1851737
Trajber, R., Walker, C., Marchezini, V., Kraftl, P., Olivato, D., Hadfield-Hill, S., Zara, C., & Fernandes Monteiro, S. (2019). Promoting climate change transformation with young people in Brazil: Participatory action research through a looping approach. *Action Research, 17*, 87–107. https://doi.org/10.1177/1476750319829202
Trott, C. D., Weinberg, A. E., & Sample McMeeking, L. B. (2018). Prefiguring sustainability through participatory action research experiences for undergraduates: Reflections and recommendations for student development. *Sustainability, 10*, 3332. https://doi.org/10.3390/su10093332
Woodhouse, A. (2015). *Culinary arts pedagogy: A critical enquiry into its knowledge, power and identity formation.* https://doi.org/10.13140/RG.2.1.4530.6324

Environmental Destructions: A Macroeconomic Approach Leaded by Anthropology

Paul Boucher

Abstract This paper explores the structural causes behind the lack of global sustainability by combining macroeconomic and anthropological perspectives. It examines how historical and contemporary economic systems contribute to environmental degradation and assesses the mechanisms that enable or hinder societal transformation. Through an interdisciplinary approach, this study highlights the interplay between economic interdependence, innovation, and governance. By drawing on case studies of past collapses and technological shifts, it provides a framework for understanding how societies may transition towards sustainability. The findings suggest that achieving systemic change requires a combination of policy reform, technological innovation, and cultural adaptation. Higher education institutions can play a crucial role in fostering this transition by promoting interdisciplinary research, policy engagement, and knowledge dissemination.

Keywords Sustainability development · Macroeconomics · Anthropology · Global interdependence · Higher education

1 Introduction

Throughout Earth's history, climate change and species extinctions have occurred naturally, but the current rapid pace of these events is unprecedented and largely attributable to human activities (IPCC, 2021). Humans extract, transform, and exploit natural resources for food, energy, industry, housing, and infrastructure, among other purposes, with profound and often irreversible impacts on biodiversity (Steffen et al., 2015). The scale of these activities has surged dramatically in the past two centuries, paralleling exponential population growth—from 1 billion people in 1900 to over 8.2 billion today (UN, 2022)—and a sharp increase in greenhouse gas emissions from 4.2 billion tonnes in 1850 to more than 50 billion tonnes in 2023 (Global Carbon Project, 2023).

P. Boucher (✉)
Audencia Business School, Nantes, France
e-mail: paul.boucher@lanseriacompagnie.fr

W. Leal Filho et al. (eds.), *Introducing the UN SDGs in Higher Education Institutions*, World Sustainability Series, https://doi.org/10.1007/978-3-032-23017-1_22

As a result, global temperatures are rising, and environmental degradation is accelerating. This includes the flooding of coastal areas, ocean acidification, desertification, glacier melt, and more frequent extreme weather events. By 2020, artificial masses such as concrete, plastic, and metal exceeded the weight of all living biomass on Earth (Elhacham et al., 2020), and fewer than 5% of natural landscapes remain untouched by human activity (Ellis, 2011). Climate-induced displacement is also on the rise, with over 32.6 million people displaced by environmental disasters in 2022 alone (IDMC, 2023). Projections suggest up to 1.2 billion climate refugees by 2050 (IEP, 2020), exacerbating ongoing conflicts and geopolitical tensions.

Biodiversity loss continues at an alarming rate, with an estimated three species going extinct every hour (Ceballos et al., 2017). Habitat destruction, overexploitation, and environmental degradation are the main drivers. More than half of the world's GDP—approximately $44 trillion—depends on natural resources (WEF, 2020), yet the economic losses from environmental disasters reached $380 billion in 2023 alone. The World Bank (2021) predicts that the collapse of key ecosystem services could shrink global GDP by $2.7 trillion by 2030. Comparatively, the post-9/11 U.S. wars cost $6.4 trillion over 20 years (Crawford, 2021), highlighting the monumental cost of environmental inaction.

Despite growing awareness of environmental issues, there is little evidence of transformative societal change. The world is transitioning towards a multipolar geopolitical landscape, where economic competition continues to prioritize productivity over sustainability (Kagan, 2021). This paper addresses a central research question: What structural and macroeconomic barriers prevent a drastic shift toward sustainability, and what mechanisms could enable such a transformation? By exploring historical and contemporary economic patterns, we aim to identify viable pathways for systemic change. Two approaches are crucial for answering this. First, macroeconomic analysis can explore the structural linkages between human activities and environmental degradation. Second, studying historical societal transformations may offer insights into how fundamental changes could occur in the future (Diamond, 2005). This discourse underscores the urgency of interdisciplinary collaboration to address the intertwined crises of climate change, societal dynamics, and economic dependencies, with a view towards building a sustainable future.

2 International Interdependence and Global Competitiveness

Although some nations are largely self-sufficient in vital resources, the global economy operates as an open market essential for survival, where survival is increasingly conflated with competitiveness. The wealthiest countries rely on a constant supply of raw materials—minerals, food, and especially energy sources such as oil, coal, gas, and uranium—typically sourced from less developed nations in exchange for money (Blas & Farchy, 2021). Notably, the largest GDPs are concentrated in the northern hemisphere, particularly in financial hubs like New York, London, and Tokyo. GDP, however, measures consumption capacity rather than true wealth creation, and these wealthier nations are also the largest consumers of raw materials, driving pollution in direct correlation with GDP size.

Developed economies are deeply dependent on raw material imports from Africa and South America, while for many nations in the Global South, revenue from these exports represents more than 80% of their GDP (Blas & Farchy, 2021). This interdependence is so profound that disruptions can lead to significant consequences for both people and infrastructure, particularly the environment. For instance, even during conflicts such as the war in Ukraine, trade in essential goods like wheat and gas persists between Russia and the EU. Similarly, the Brazilian economy's dependence on deforestation for agriculture illustrates the economic complexity that prevents halting destructive practices.

Furthermore, this system extends beyond material resources to human capital. The migration of labour from the Global South to the Global North, whether skilled or unskilled, exacerbates underdevelopment in the source countries (Blas & Farchy, 2021). Geopolitical competition between states drives overproduction and overconsumption, intensifying environmental degradation. Modern global trade, 90% of which is conducted via maritime transport, further multiplies environmental damage, with issues like oil spills resulting from cost-cutting measures that prioritize profit over safety (Bihouix, 2021; Pitron, 2018). In Australia, for example, coal production has increased due to its profitability, despite being the most polluting energy source.

Environmental legislation and standards, though evolving, are unevenly applied across regions. Some nations, such as Costa Rica and Kenya, have adopted measures to protect biodiversity for ecotourism. In contrast, many developed nations continue to prioritize short-term economic gains over sustainable practices, often resulting in environmentally harmful policies. The European Union's "*Green New Deal*" seeks to address some of these issues by promoting renewable energy and Keynesian-style economic recovery through employment. Morocco's ambitions to become a renewable energy hub exemplify regional efforts, but globally, no country has f"ully transitioned to a sustainable model due to the associated economic challenges.

Additionally, market dynamics exacerbate environmental degradation. Schumpeter's theory of creative destruction (1942) illustrates how profit-driven activities lead to environmental destruction, such as deforestation being followed by agricultural expansion or tourism development. Technological advancements, while having the potential to mitigate environmental impacts, are often delayed for profit maximization. For example, innovations in recycling are underutilized despite their technical feasibility (Bihouix, 2021; Pitron, 2018). The exploitation of labour and environmental resources often goes unchecked, with corruption and social inequalities allowing destructive practices to continue in exchange for economic benefits. Consumer behaviour also plays a critical role in environmental degradation. The demand for cheap, low-quality products—driven by rising inequality—supports unsustainable production methods. The influence of social media and advertising reinforces consumer desires, often leading to unnecessary purchases that further harm the environment. Industries like textiles and electronics exemplify this issue, where brand prestige overshadows environmental and social costs. The audiovisual sector, similarly, contributes to environmental degradation through resource-intensive production processes. Addressing these challenges requires not only innovation but a reimagining of consumption patterns, production methods, and economic structures on a global scale.

3 Innovation and Interstate Competition

Innovation often arises from the need to address challenges more effectively, with major advancements frequently driven by state-led competition. This interstate rivalry, particularly in the realms of technology and defence, has historically led to significant environmental impacts due to the necessity of testing. States are unlikely to abandon such tests, as they are perceived as essential for national security (Buhler, 2017). A historical example is the United States' creation of *Fordlandia* in the Brazilian Amazon during World War II to secure rubber supplies, which devastated local biodiversity (Grandin, 2009). Strategic locations for development have been employed repeatedly, notably in nuclear research, with sites like Los Alamos and the *plutopias* of Richland and Mayak, where environmental and health repercussions, such as cancer and radiation poisoning, continue to be felt (Brown, 2013).

Nuclear testing remains a pivotal issue in state competition. Iconic tests, like the U.S.'s Starfish Prime and the Soviet Union's Tsar Bomba, exemplify the severe, often concealed, environmental damage resulting from such activities (Le Page, 2022). Modern nuclear powers, constrained by satellite surveillance, now turn to virtual simulations to minimize both environmental damage and the costs of physical tests (Schlosser, 2013). Advances in computational power, particularly quantum computing, promise even greater realism in these simulations, reducing the need for real-world testing.

Space innovation mirrors earlier state-driven projects, creating concentrated pollution in areas like launch sites. However, space exploration also provides crucial data on environmental monitoring, particularly via satellites (Logsdon, 2015). Yet, space debris poses a growing risk to these systems, prompting efforts to develop robots capable of recovering orbital waste (Klinkrad, 2006). Although often perceived as distant from everyday life, these innovations have tangible impacts on society, as satellite technology facilitates everything from weather forecasting to communications.

Moreover, technological advancements such as artificial intelligence increase the demand for energy and resources, leading to further environmental concerns (Muehlhauser & Salamon, 2012). While innovation holds the potential to address these issues, abandoning research would limit opportunities for societal transformation. The critical question remains: how can human societies evolve to integrate such innovations sustainably?

4 How Do Societies Change?

Understanding how societies change is a complex process. Societal transformation can be broadly categorized into two types of events: sudden or forced changes, and gradual or structural reconfigurations. The first type, sudden change, often occurs under dire circumstances such as civil war, external interference, or resource shortages. Current examples include Afghanistan, the Central African Republic, Chad, the Democratic Republic of Congo, Ethiopia, Haiti, Libya, Mali, Myanmar, Nigeria, Palestine, Somalia, South Sudan, Sudan, Syria, and Yemen, all of which are considered failed states (Rotberg, 2004). These societies often experience a convergence of multiple crises, including political instability, economic mismanagement, and environmental stress, which leads to systemic upheaval (Cline, 2014; Diamond, 2005, 2020).

Historical examples provide insight into these dynamics without drawing direct parallels. Scholars like Jared Diamond and Eric H. Cline argue that societal collapse is rarely the result of a single cause; rather, it arises from the accumulation of several interrelated factors that disrupt political, economic, and demographic systems (Cline, 2014) (Table 1). As seen with the Hittite or Indus civilizations, societal change may not lead to total disappearance but rather to gradual absorption into other entities (Diamond, 2005). Political scientist James C. Scott posits that the real stability of a society depends on the relationship between its capacity for action and the speed at which it can respond to challenges, rather than simply on territorial size or military power (Scott, 2017).

Indicators of societal decline are often subtle but revealing. For instance, Emmanuel Todd predicted the collapse of the Soviet Union in 1974 by analysing rising infant mortality rates. He has since applied similar metrics to assess the declining educational and social conditions in Western countries (Todd, 2017). While political, cultural, and cognitive dimensions of societies are not inherently irreversible, the loss of knowledge is a serious risk during times of collapse. The Mycenaean Greeks lost literacy after their collapse, as did Britain following the Roman withdrawal. Even the Tasmanians lost the use of fire (Diamond, 2005). Thus, the collapse of modern techno-industrial civilization, while not inevitable, could result in similar losses of vital knowledge. Fortunately, history also provides examples of societal reorganization and renewal (Table 2). These changes, though often unintended, can lead to restructuring and adaptation. Such processes suggest that even in the face of collapse, societies have the potential to evolve and recover, albeit in a transformed state (Tainter, 1988).

Throughout history, most societies have experienced transformative phases, often referred to as developmental periods, which have impacted all levels of society. These phenomena are interconnected, forming a series of cause-and-effect relationships. Notably, certain pivotal events in human history have shaped the trajectory of the entire species, with at least five significant world-altering events identified (Table 3). One key dynamic shaping human history is the relationship between predation and production, which has driven the expansion of human populations and later, states, in

Table 1 Sudden changes examples

Country/civilisation	Period of troubles	Natural sudden events and human sudden events	Political evolution	Territorial evolution	Speed of fall	First outcome after fall	Final outcome of the main political core
USSR	From 1970s to 1991	Economic pressures, ethnic problems	Communist regime proves incapable of maintaining the coherence of the Union	Fallback on the Russian Federation. Appearance of new ethnic states	Quick	Great poverty, several ethnic problems with the Chechens and political weakness	Political reaffirmation. Russia remains a great power
Austrian-Hungary empire	From mid-nineteenth century to 1919	Wars and economic pressures, ethic issues	Fall of the Habsburg monarchy	Fragmentation into several ethnic states. Austria was one of them	Quick	Political instability. Tried to be part of a new great power with the Anschluss	Austria lost the war and became a small and neutral country
Ottoman empire	From the mid-nineteenth century to 1919	Wars and economic pressures, ethnic issues	Fall of the Ottoman monarchy The empire is slowly dismantled by foreign powers	The empire is slowly divided by foreign powers. Turkey becomes a republic based on nationality	Average	Turkey becomes a republic based on the average nationality War of independence and internal power struggles	Turkey becomes a regional power
Mali empire	From the XV to the XVI century	Foreign wars and internal political conflicts	The empire is invaded by external powers	The empire is divided by succession after Mahmud IV	Average	The empire disappears. Populations are dispersed and divided.	The Mali Empire ceases to exist

(continued)

Table 1 (continued)

Country/civilisation	Period of troubles	Natural sudden events and human sudden events	Political evolution	Territorial evolution	Speed of fall	First outcome after fall	Final outcome of the main political core
Mayan kingdoms from classic to postclassic era	Late Classic period (600 to 900 AD)	Drought, internal wars and political instability	The traditional mode of governance changes, the more shared *multepal* system appears	Some cities are slowly abandoned	Slow	Maya culture regains a certain dynamism	Maya kingdoms no longer exist, but Maya culture still exists
Khmer empire	XII-XIV centuries	Drought, foreign wars	Fall of the empire	The empire is divided by external powers, especially the Thai, some cities are abandoned	Average	Cambodia goes into decline until the nineteenth century French colonisation	Khmer culture still exists, and Cambodia is a small independent state
Tiwanaku empire	Around 1000 A. D	Drought, foreign wars,	Fall of the empire	The empire is invaded, the city of Tiwanaku is abandoned	Fast	Population leaves the capital	Tiwanaku no longer exists
Hittite civilisation	Around XIII century BC	Drought, famine, wars	The empire gradually disintegrates and becomes depopulated	Hattusa is abandoned	Slow	The Neo-Hittite Empire rises for a time but can't keep up	Hittite culture no longer exists

Table 2 Provoked changes examples

Local phenomena	New knowledges	Technologies and innovations	New ways of thinking	Way of communication	Demographic evolution	Territorial expansion
American hegemony (XX century from now)	Astrophysics, Bioengineering, Genetic	Electric bulb, computers, space innovation, robotics, nuclear technologies, Internet	*Manifest Destiny, American Way,* imposing democracy worldwide, Global Leadership	Cinema, newspapers, television, internet	From 60 million in 1900 to 333 million in 2024	They have the largest maritime area. Lead the world's most powerful military alliance, spanning the globe
British empire hegemony (1815–1922)	Anthropology and Archaeology, Natural Selection, Electromagnetism, Germ Theory of Disease, Botanical Exploration, Antiseptic Surgery	Telegraph and telephone, Steam engine, railway, Sanitation Innovations	*Rule Britannia,* Utilitarianism, Colonialism: Liberalism, Economic Exploitation, Education and Enlightenment:	Telephone, telegraph, submarine cable	Over 60 million in 1815 to 458 million people at its peak in 1922	Present in all continents, more than 25% of the planet was under British administration
Mongol empire (XIII–XIV centuries AD)	Medical knowledge with surgery, geographical knowledge	Archery, military operations, cartography, improvement of agriculture and gunpowder uses	*Pax Mongolica* Unity and leadership under the Khan, administrative innovations, cultural exchanges, strategic alliances	Nomadic roads, acceleration of communication and exchanges, postal systems	Domination of more than oven 100 million people at its peak during the XVI century	From Poland to the Pacific, Siberia to the Middle East

(continued)

Table 2 (continued)

Local phenomena	New knowledges	Technologies and innovations	New ways of thinking	Way of communication	Demographic evolution	Territorial expansion
Classic Maya era (200–900 AD)	Breakthroughs in astronomy, mathematics and medicine	Development of intensive agriculture, monumental architecture, networks, irrigation	Mayan cosmogony. Development of Mayan arts	Temples, glyphs and codex	From 50 to more than 250 individuals per km^2 in less than 1000 years old	Stretches across the whole of southern Mesoamerica, tropical forest totally anthropized
Greek expansion (V–III centuries B.C.)	Major discoveries in mathematics, medicine and architecture	Military innovations. Navigation	*Apoikia and Hellenism* Democratic experience, birth of Western philosophy and dramaturgy	Academies, agora	Multiplication of Greek cities around the Mediterranean	Expands across the entire eastern Mediterranean basin to India after Alexander the Great
Polynesian expansions (IV–XVIII centuries AD)	Astronomy, Navigation with Etak System	Catamarans, navigation chart	Polynesian myths? Development of tattooing and tapa?	Marae? maritime trading networks	Difficult to estimate how much, but the spread to empty areas of people proves that it took place	Extends over almost the entire habitable surface of Oceania
Egyptian new empire (1550–1077 BC)	Peak of ancient Egyptian art, mathematics, medicine	Military innovation, improvement of constructions techniques	Aton's Experience Diffusion of the book of the dead, the centralized power of the pharaohs led to a bureaucratic state that emphasized loyalty and order	Roads and temples	From 2–3 million to 4–6 million of people between	From Libya to Nubia. Extension to Anatolia

terms of both territory and population. This process has often resulted in widespread destruction of biodiversity. While humans may not have directly exterminated many species, their modes of living have certainly disrupted ecosystems, leading to the extinction of megafauna such as the Denisovans and Neanderthals, and the disappearance of ecological niches like elephants in the Middle East and lions in Asia and Europe (Diamond, 2005). The scale of today's biodiversity loss, however, poses a far greater threat.

This historical predation also extends to cultural diversity. The past two centuries, in particular, have witnessed significant cultural homogenization. The industrial revolution, which facilitated the end of local particularism and the decline of great tribal societies such as the Tuareg and Zulu, was marked by technological innovations like the steam engine and automatic weapons, which allowed European powers to dominate rival populations more efficiently. As a result, not only were vast territories secured, but many languages and traditional knowledge systems were also lost (Singeravelou et al., 2017; Reich, 2018).

Disease played a decisive role in these conquests, with deadly pathogens often accompanying military and colonial expansions. The Black Death, for example, spread by nomadic tribes, devastated societies, while European colonization of the Americas and Oceania similarly led to catastrophic population declines due to disease (Reich, 2018). Modernization, particularly during the Information Age, has coincided with the ongoing sixth mass extinction and the emergence of a globalized, uniform culture.

Innovation, driven by necessity, has always been a hallmark of human progress. James C. Scott argues that agricultural centers were often the birthplaces of early states due to the cultivation of cereals, which enabled early forms of standardization and accounting. This led to the development of writing and more complex social organization. These centers of innovation eventually became spheres of influence, surrounded by peripheral zones that extended their reach (Scott, 2017).

Over the course of history, advances in technology, increased work speed, and especially automation have made life more comfortable. However, all of this has been powered by energy. Scholars like Jean-Baptiste Fressoz and Richard Rhodes have shown that new types of energy do not replace older ones but instead accumulate. As populations grow, political structures become more complex, and organizational and production efficiency improves, energy consumption rises in parallel. This is why the use of fossil fuels since the nineteenth century has been such a pivotal moment (Bonnueil & Fressoz, 2016; Rhodes, 2018).

Is change logical? Not necessarily, and no change is ever instantaneous. Some transformations persist across different epochs. Vere Gordon Childe, a materialist archaeologist and originator of the concept of the Neolithic Revolution, explored these issues in *Man Makes Himself*. He challenged the conventional notion of progress, arguing that innovation is rarely linear. For instance, both oil and solar energy were discovered in the nineteenth century, yet their utilization took place at very different points in time (Rhodes, 2018) Over the past two centuries, literacy rates have risen significantly, and thanks to scientific advancements, our understanding of life's mechanisms and Earth's systems has evolved rapidly. On a human timescale,

Table 3 The five major eras in human history

World event	New knowledges	Technological innovations	Military view	New ways of thinking about the world	Way of diffusion/ way of communications	Global education skill
Late palaeolithic and Mesolithic 45 000 BC–7000 BC	First astronomy, Medicine, early domestication	Bows, Sophistication and diversification of tools (stone, bone, wood)	First evidence of conflicts	Zoomorphic deities, cult of "Venus" Shamanism, totemism, birth of arts and music	Fires, rock painting?	Quite equal
Neolithic and classical eras 7000 BC to XV century	Mathematics, mastery of animal power	Polishing then perfecting tools, pottery, metallurgy, sailing Completed animal domestication, agriculture, metallurgy	Creation of armies, first real wars and conquests	Appearance of anthropomorphic deities, polytheism and monotheism, development of a religious and lineage aristocracy and the concept of warrior, appearance of states	Writing, roads	A small elite
Modern era XV Century to XIX century	Optic and physic, great food exchange and upheaval, standardization of science and measurements	Guns and artillery, print, compass	Start of major tactical and strategic plans	Political affirmation, Mercantilism, humanism and liberalism. First international organisation	University, printing	A small elite
Industrial revolution XIX century to mid XX Century	Biology and microbiology, Knowledge of chemical elements, thermodynamic, fossils energies,	Automatic weapons Steam machines, mechanization of work and armies, railway, chemical industry	Automatization of weapons and machine, colonial wars and world wars	Democracy, Nationalism, Authoritarianism, Internationalisms,	Telegraph, telephone, radio	Beginning of mass education

(continued)

Table 3 (continued)

World event	New knowledges	Technological innovations	Military view	New ways of thinking about the world	Way of diffusion/ way of communications	Global education skill
Information era Since Mid XX Century	Astrophysics and quantic physic, Bioengineering Genetic	Computers and informatic, Internet, contraception, nuclear technology, robotics	Asymmetric wars, nuclear dissuasion	Ecologism, female empowerment, global thinking, transhumanism, emancipation of minorities	Television, Internet and mass media, social media	Almost total literacy of the human species
To Quantum era?						
World event	Inequalities	Transport	Demography growth	Territorial expansions	Energies used	World political-trade hub
Late palaeolithic and Mesolithic 45 000 BC–7000 BC	Quite equal	Boats, canoes	50 000–1 000 000 humans	Homo sapiens spread on earth, end of other human species and megafauna	Fire	None
Neolithic and classical eras 7000 BC to XV century	A small elite	Boats, horse	10 000 000	Sedentary life and cities appear at the expense of nature	Fire, slaves, animals, wind	Middle east
Modern era XV Century to XIX century	A small elite	Ships, horse	500 000 000 around XV century	First real globalization. All human being enters in contact	Fire, slaves, animals, wind	Mediterranean aera
Industrial revolution XIX century to mid XX Century	A small elite	Horses, Ships, iron trail, plane, car	1 billion in 1900	Humans reach all the place on earth	Fire, slaves, animals, wind, coal, oil	North Atlantic area

(continued)

Table 3 (continued)

World event	Inequalities	Transport	Demography growth	Territorial expansions	Energies used	World political-trade hub
Information era Since Mid XX Century	1% control 50% of the wealth	Ships, iron trail, plane, rockets and shuttle	8.2 billion	To space, 6th extinction	Fire, robots, animals, wind, coal, oil, nuclear, gas, hydroelectricity, solar	Southern Asia and Indo-Pacific area
To Quantum era?						

however, societal awareness and decision-making appear agonizingly slow, even though, from a broader perspective, these changes are happening at a fast pace. This underscores how the material and energy failures of a socio-system can be disastrous for entire populations. Technological progress often brings new economic systems, ways of life, and interactions with the world. But it also leads to irreversible destruction. Unlike immaterial dimensions, the material aspects of life (such as consumption and production) rarely revert once certain thresholds are crossed globally. Reversing course, especially when industry monopolizes an object or function, becomes extremely difficult. The manual and artisanal become symbols of prestige, accessible only to a few. Despite resistance or warnings from some, most people opt for what is easiest, quickest, and most efficient. As a result, societies become addicted to technological progress—for better or worse. In other words, there is no retreat without significant damage.

Nikolski (2023) has warned about the potential collapse of women's rights due to environmental issues, which directly impact material conditions. She points out that medical advancements, especially, have reduced maternal mortality and stabilized the number of children per family to around two or three. This, in turn, allowed women to step beyond their traditional roles as parents, accessing greater education, rights, and responsibilities during the nineteenth and twentieth centuries. If energy-saving efforts and a profound restructuring of our consumption and production systems were to occur too abruptly, it could have severe consequences for specific social groups, including democracies themselves.

In summary, contemporary human societies face two options: sudden change triggered by a failure to adapt or induced change with the associated risk of perpetuating current disasters. A purely technological approach to addressing these issues would likely exacerbate the damage. A third, more promising path may emerge—a conscious choice to shape the future deliberately. But time is the deciding factor. We are in a race against time. As history shows, unless societies are forced into change—often at great cost—any transformation will only be accepted if it offers a perceived improvement, despite the inevitable collateral damage. This ideal, of course, is not always realized. Science and engineering will be crucial in navigating this future, and higher education, particularly, can play a leading role in promoting sustainability (Bonnueil & Fressoz, 2016; Kaku, 2011; Rhodes, 2018).

5 Perspectives

While social change often takes time, it is possible to overcome the structural blockages identified. As previously discussed, access to money, food, water, and basic infrastructure (such as energy, schools, hospitals, and leisure facilities) provides stability for regions or communities. However, these essentials are frequently subject to speculation. It is therefore critical to ensure stable wages and supplies. Institutions like the UN and partnerships between nations can support initiatives that address these issues. One crucial aspect is the standardization of laws, taxes, and regulations, which can prevent abuses and environmental degradation. Additionally, bonuses, tax exemptions, and support should be directed to companies and associations working directly towards ecological sustainability. Full profitability will follow later.

These actions must be both transversal and multi-scalar, cutting across different sectors and levels of governance. A holistic perspective is essential in this process, as complex challenges like climate change and resource management cannot be addressed by a single discipline. By adopting a holistic and interdisciplinary approach, we can better understand the intricate relationships between economic, environmental, and social systems. Bringing together expertise from economics, environmental sciences, anthropology, and engineering allows for the design of integrated and adaptive solutions. A true understanding of social and environmental dynamics, coupled with advances in science and technology—particularly in the energy sector—is the key to developing real solutions. Technologies like hydrogen engines derived from water and nuclear fusion hold great promise for creating a prosperous and ecological society. However, these technologies will take years to become fully operational. In the meantime, an approach that balances innovation with controlled energy descent, inspired by the principles of degrowth should be considered. Reducing excessive energy consumption while promoting localized, low-impact alternatives can create resilience and sustainability. Start-up companies often excel in meeting environmental needs. Some, particularly in the maritime sector, have begun offering sail propulsion as an alternative solution while waiting for fuel technologies to improve. These solutions may not yet be as competitive as container ships, but they significantly reduce fuel consumption.

Moreover, protecting and enhancing biodiversity should be a continuous priority. This focus goes beyond individual interests, reminding us that others came before us, and many more will follow. The future is not predetermined. While much of futuristic fiction paints dystopian visions of what is to come, the creation of more positive future narratives can serve as a foundation for real-world progress. Imagining better futures is not naive; it is essential for shaping change that is truly for the better. The role of imagination in shaping the material world is often overlooked. However, the difficulty in implementing widespread change often stems from the need to reassure all stakeholders—a process that precedes any consensus.

As philosopher Bernard Stiegler theorized, life itself is a form of controlled entropy. Stiegler's concept of negentropy[1] suggests that economic systems could learn from this principle. Symbiotic or regenerative economies can be seen as an extension of this idea. Agriculture, agroforestry, and traditional pharmacopoeia are good testing grounds for such approaches. It is worth noting that the Amazon rainforest is partly a human creation, shaped by 8,000 years of human occupation prior to the fifteenth century. Indigenous populations thrived by integrating biodiversity into their economy and society, exemplified by innovations like *terra preta* (dark earth) and the diverse use of species (Boucher, 2018). Human species, therefore,

[1] "Stiegler's concept of neguentropy, notably developed in *Bifurquer: Il n'y a pas d'alternative* (2020), appropriates a term originally formulated in physics and mathematics, first introduced by the mathematician Sadi Carnot.".

has an unprecedented ecological and transformative power that grows with every advance in energy management and technology. If destruction has been the norm, the opposite is also possible, but it requires a deep understanding of life on Earth, including human societies, not just technology.

This deep, long-standing knowledge, embedded in most traditional cultures, must be preserved and reinforced with the support of the scientific community. The ability to cultivate specific plants or sustain various forms of life is invaluable. As previously mentioned, just as some ancient civilizations lost the ability to read and write, most people in today's modern cities can no longer accurately identify birds, trees, or plants. The erosion of this naturalistic knowledge not only threatens the environment but also endangers future generations and the stability of their societies.

Higher education, particularly universities, plays a pivotal role in both the dissemination and application of knowledge. However, today's academic researchers often find themselves isolated and continuously seeking funding, which hinders their ability to fulfill their mission.

Moreover, doctoral supervisors' exploitation of their students, combined with the "publish or perish" mindset, exemplifies the worst aspects of hyper-productivity. These systemic pressures create an environment where researchers struggle to focus on long-term, impactful work. Knowledge dissemination beyond academia is not always guaranteed, further restricting the societal benefits of scientific progress. This gap between academia and society is deepened by a lack of interest or even distrust toward science within civil society, weakening its broader impact.

In response to these challenges, a global strategy can be implemented to strengthen and amplify these impacts through closer collaboration within the Quadrupole Helix model (Carayannis & Campbell, 2009), in which universities, businesses, governments, and civil society work together to develop and implement sustainable solutions. To move beyond academic isolation, universities must actively co-create knowledge with economic and political stakeholders, fostering interdisciplinary collaboration to accelerate the ecological transition and ensure that scientific progress translates into tangible societal benefits.

Despite these challenges, universities possess not only knowledge but also international legitimacy. These institutions could become the catalysts for the profound changes needed to achieve a truly sustainable society. This shift must be interdisciplinary and accompanied by international dialogue. Many interactive events could be organized, bringing together diplomats, business leaders, civil society, and political representatives. From these discussions, action strategies could emerge for every link in the macroeconomic chain. In this process, it is crucial to avoid judgment and instead gather the broadest range of perspectives, even those that are purely pragmatic. Every problem is unique, and there are no one-size-fits-all solutions. To be effective, however, solutions must be considered on a global scale. Imagining positive futures is essential for inspiring action and fostering hope. Optimism can act as a catalyst for practical policies and innovations. By combining idealism with concrete steps, we can create pathways for meaningful social change.

Jean Jaurès once suggested that humanity is more of an ideal than a reality,[2] as we often lack a unified voice. Similarly, the Chinese science fiction author Liu Cixin argues the same idea in his novel *The Dark Forest (2008).* But ecology could serve as common ground for collective action, empowering individuals to foster positive change. Pre-industrial societies often maintained a more sustainable balance between consumption and regeneration, but this balance was disrupted with the rise of mass energy consumption. To restore this balance, a profound societal transformation is needed. We must critically examine our values and practices. This shift in understanding could be as revolutionary as the Copernican, Galilean, and Darwinian revolutions, which reshaped human perception despite initial resistance. Sustainability can be likened to a cathedral; its completion may remain out of sight, yet the significance lies not in the finality of its construction but in the enduring commitment to its ongoing development and the principles it embodies. In this regard, the United Nations' Sustainable Development Goals (SDGs) offer a crucial framework, particularly in linking economic stability (SDG 8), climate action (SDG 13), and biodiversity protection (SDG 15). Integrating these goals into policies and business strategies can provide a structured path toward a more balanced and regenerative economy. By fusing rationality with imagination, we can unlock new possibilities for the future. In the alleged words of physicist Niels Bohr, "*Every great leap in science has changed our perception of the world and thus our way of interacting with it, and every great and deep difficulty carries within itself its own solution. It forces us to change our thinking in order to find it* ".

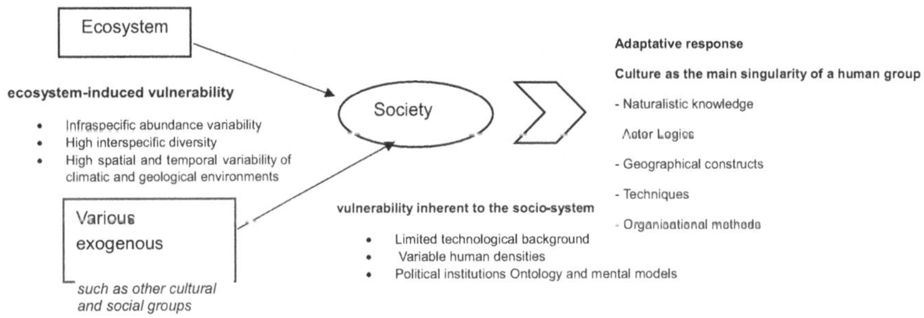

Diagram 1 The mechanisms of emergence: principles of generalities and natural and human logics (Boucher, 2017)

[2] Jaurès' reflections were among the first lines of his editorial, simultaneously marking the launch of his newspaper, *L'Humanité*, on April 18, 1904.

6 Conclusion

This study highlights the structural drivers of environmental degradation, emphasizing the complex interplay between macroeconomic dynamics, global interdependence, and societal transformations. The acceleration of resource exploitation, driven by economic competition and growth imperatives, has placed ecosystems under unprecedented strain. However, history demonstrates that societies can adapt and innovate when institutional and economic structures evolve accordingly. A sustainable transition requires a comprehensive approach that integrates environmental considerations into economic frameworks, fosters technological and organizational innovation, and promotes a cultural shift toward ecological consciousness. These transformations must be supported by academic institutions, which play a crucial role in advancing interdisciplinary research, bridging the gap between science, policy, and industry, and fostering collaborative governance to ensure that sustainability becomes a concrete and actionable priority rather than a theoretical aspiration.

In line with these reflections, this conclusion underscores the necessity of developing sustainable economic models inspired by existing communities, integrating cultural factors to encourage sustainable consumption and fair trade, and fostering ecological economies that prioritize climate justice and biodiversity preservation. These principles must serve as the fundamental pillars of systemic transformation, enabling societies to move beyond the current paradigms and forge a resilient and sustainable future. However, as historical patterns reveal, the temporal dimension of societal change is crucial. The transformations required to achieve sustainability cannot be instantaneous but must be understood in the context of long-term historical dynamics, as illustrated in the comparative table of past societal collapses and restructurations. While these historical insights highlight the necessity of gradual structural evolution, immediate action remains essential to lay the foundation for enduring resilience and a truly sustainable future.

References

Bihouix, P. (2021). *L'âge du low tech, vers une civilisation techniquement soutenable*. Seuil.

Blas, J., & Farchy, J. (2021). *The world for sale: Money, power and the traders who barter the earth's resources*. Penguin Press.

Boucher, P. (2017). *Approches anthropologiques des terres noires dans les Basses Terres Mayas*. Paris 1 Panthéon-Sorbonne.

Boucher, P. (2018). *Amazonie: Une révolution tropicale? Études de la binarité entre présence et absence des terres noires en Amazonie*. Paris 1 Panthéon-Sorbonne.

Bonnueil, C., & Fressoz, J.-B. (2016). *L'événement anthropocène*. Seuil.

Brown, K. (2013). *Plutopia: Nuclear families, atomic cities, and the great soviet and American plutonium disaster*. Oxford University Press.

Buhler, P. (2017). *La puissance au XXIème siècle*. CNRS Éditions.

Carayannis, E. G., & Campbell, D. F. J. (2009). "Mode 3" and "quadruple helix": Toward a 21st-century fractal innovation ecosystem. *International Journal of Technology Management, 46*(3–4), 201–234.

Ceballos, G., Ehrlich, P. R., & Dirzo, R. (2017). Biological annihilation via the ongoing sixth mass extinction signaled by vertebrate population losses and declines. *Proceedings of the National Academy of Sciences, 114*(30), E6089–E6096.

Cline, E. H. 1177. B.C. (2014). *The year civilization collapsed.* Princeton University Press.

Crawford, N. C. (2021). *The costs of war project.* Watson Institute, Brown University.

Diamond, J. (2005). *Collapse: How societies choose to fail or succeed.* Penguin Books.

Diamond, J. M. (2020). *Upheaval: How nations cope with crisis and change.* Penguin Press.

Elhacham, E., Ben-Uri, L., Grozovski, J., Bar-On, Y. M., & Milo, R. (2020). Global human-made mass exceeds all living biomass. *Nature, 588*, 442–444.

Ellis, E. C. (2011). Anthropogenic transformation of the terrestrial biosphere. *Philosophical Transactions of the Royal Society A, 369*(1938), 1010–1035.

Global Carbon Project. (2023). *Global Carbon Budget.*

Grandin, G. (2009). *Fordlandia: The rise and fall of Henry Ford's forgotten Jungle City.* Metropolitan Books.

IDMC. (2023). *Global report on internal displacement 2023.* International Displacement Monitoring Centre.

IEP. (2020). *Ecological threat report 2020.* Institute for Economics & Peace.

IPCC. (2021). *Sixth assessment report.* Intergovernmental Panel on Climate Change.

Kagan, R. (2021). The return of multipolarity: Global politics after the pandemic. *Foreign Affairs.*

Kaku, M. (2011). *Physics of the future: How science will shape human destiny and our daily lives by the year 2100.*

Klinkrad, H. (2006). *Space debris: Models and risk analysis.* Springer.

Le Page, J.-M. (2022). *La menace nucléaire: De Hiroshima à Poutine.* Passés Composés.

Logsdon, J. (2015). *The Penguin history of outer space.* Penguin Books.

Muehlhauser, L., & Salamon, A. (2012). *Artificial intelligence and global risk.* Machine Intelligence Research Institute.

Nikolski, V. (2023). *Féminicène.*

Pitron, G. (2018). *La guerre des métaux rares: La face cachée de la transition énergétique et numérique.* Éditions Les Liens qui Libèrent.

Reich, D. (2018). *Who we are and how we got here.* Oxford University Press.

Rhodes, R. (2018). *Energy: A human history.* Simon & Schuster.

Rotberg, R. I. (2004). *When states fail: Causes and consequences.* Princeton University Press.

Schlossler, E. (2013). *Command and control: Nuclear weapons, the Damascus accident, and the illusion of safety.* Penguin Books.

Scott, J. C. (2017). *Against the grain. A deep history of the earliest states.* Yale University Press.

Schumpeter, J. A. (1942). *Capitalism, socialism, and democracy.* Harper & Brothers.

Singeravelou, P. (dir). (2017) Histoire *du Monde au XIXe siècle.* Fayard.

Stiegler, B. (2020). *Bifurquer: Il n'y a pas d'alternative.* Paris: Les Liens qui Libèrent.

Steffen, W., Richardson, K., Rockström, J., Cornell, S. E., Fetzer, I., Bennett, E. M., & Sörlin, S. (2015). Planetary boundaries: Guiding human development on a changing planet. *Science, 347*(6223), 1259855.

Todd, E. (2017) *Où Allons-nous? Une esquisse de l'histoire humaine.* Seuil.

Tainter, J. A. (1988). *The collapse of complex societies.* Cambridge University Press.

United Nations, Department of Economic and Social Affairs. (2022). *World population prospects 2022: Summary of results.* United Nations.

World Bank. (2021). *The economic case for nature: A global earth-economy model to assess development policy pathways.*

WEF. (2020). *Nature risk rising: Why the crisis engulfing nature matters for business and the economy.*

Higher Education Institutional Sustainability Strategy Requirements: Addressing Strategy at SOAS University London

Ros Taplin and Annabel de Frece

Abstract This chapter reflects on the development of an overarching sustainability strategy for a higher education institution in the United Kingdom, SOAS University of London. The approach was influenced by leading practice in the sector, and the UNESCO *Education for SDGs: Learning objectives.* Drafting of the strategy was informed through participation and partnerships actively involving multiple stakeholders across SOAS as well as review and reflection on current strategies, plans, policy and practice. A clear understanding emerged of the elements that should be incorporated and the importance of a holistic institutional or 'whole of SOAS approach' to sustainability and of the need to facilitate across the institution contributions to the implementation of the SDGs and particularly in the priority areas of education, research and knowledge exchange, collaboration and engagement, social responsibility, and the campus environment. Outcomes established that efforts to pursue sustainability and SDGs implementation by SOAS should be underpinned by four principles, namely, good governance—working collectively aided by credible leadership, with efficient systems and processes for sustainability, resourcing—embedding resources for current and future sustainability, participation and voice—students and staff having a stake in decisions and shaping sustainability, and communication and reporting—effective communication about sustainability across the institution, and impacts and performance measured.

Keywords Sustainability strategy · Sustainable development goals · Higher education institution · Education for sustainable development

R. Taplin (✉) · A. de Frece
Centre for Development, Environment and Policy, Department of Development Studies, SOAS University of London, London, UK
e-mail: rt33@soas.ac.uk

A. de Frece
e-mail: af12@soas.ac.uk

W. Leal Filho et al. (eds.), *Introducing the UN SDGs in Higher Education Institutions*, World Sustainability Series, https://doi.org/10.1007/978-3-032-23017-1_23

1 Introduction

Higher education plays a critical role in responding to the rising concern of global environmental change "…catalyzed by mounting evidence of climate crisis, biodiversity loss, pandemic disease and further systemic issues in unprecedented times of urgency, uncertainty and threat" (Sterling, 2021, p. 1). Many higher education institutions (HEI) globally have adopted institutional strategies and espoused sustainability positions over the last two decades or more.

Research conducted at SOAS University of London in 2022–2023 on teaching and learning approaches for embedding the United Nations Sustainable Development Goals (SDGs) in the curriculum pointed to reforms needed in the institutional policy environment to facilitate better incorporation of the SDGs (Taplin et al., 2025). It was found that whilst SDGs were already embedded in some areas of the curriculum, there was a need to review existing curricula more systematically through the SDGs lens. Without a whole institution approach, including strategic support, and leadership, adapting pedagogy to meet Education for Sustainable Development (ESD) objectives were seen by participants in the study to be challenging. Revision of current SOAS strategies and policies to align with sustainability and drive advancement of the SDGs was seen as a priority for the institution.

In 2024 development of a sustainability strategy for SOAS was initiated. This chapter focuses on research, conducted from June to October 2024 which aimed to contribute towards development of the strategy for implementation in 2025. The objective was to develop a draft strategy based on leading HEI sustainability practice, that addresses the SDGs, ESD and represents the opinions of the SOAS community.

The research aims were to:

1. synthesise aspects of reports and reviews to date on development and implementation of sustainability strategies in HEIs, providing insights into their elements and potential implications of relevance to the development of a sustainability strategy for SOAS,
2. consult student and staff stakeholder opinions on what should be incorporated in the strategy, and
3. develop a draft strategy including provision of recommendations on (a) how to implement the strategy and (b) how to involve different stakeholders across various levels of SOAS in strategy implementation.

The research questions were:

1. How has sustainability been adopted by HEIs from a leading practice perspective?
2. What are the views of students and staff about the current and future sustainability of SOAS University of London?
3. What elements should be incorporated in a sustainability strategy for SOAS University of London?

The following section reviews the literature and aspects of current practice adopted by HEIs related to sustainability. In Sect. 3 the methods used for key informant interviews and survey of staff are outlined. The final outcomes and conclusions sections consider the contributions of the research to the development of the SOAS strategy and broader implications for HEIs.

2 Strategic Approaches to Sustainability in HEIs

Yturzaeta (2023) has identified the drivers for sustainability internally in HEIs as "…the concern for people, planet, and profit in the institution over the short term and pursuit of quality in the HEI's education and operations …sustainability leadership and the desire for institutional sustainability" (p. 104) and externally as "… concerns for the institution's general reputation and for people and the environment beyond its physical boundaries [and] sustainability challenges in society and the desire to contribute to sustainable development goals" (p. 104) (see Fig. 1). While Galán-Muros (2023, p. 9) in the UNESCO SET4HEI (Sustainable Evaluation Tool for HEIs) observes the motivations are social responsibility with goals of aligning with sustainability in curricula, research and administration. Sustainability has been broadly described by EAUC, the leading body for sustainability in the higher education sector in the UK and Republic of Ireland, as relating to: "leadership and governance", "partnerships and engagement", "learning, teaching and research", and "estates and operations" (Appleton, 2017a; EAUC & AUDE, 2022).

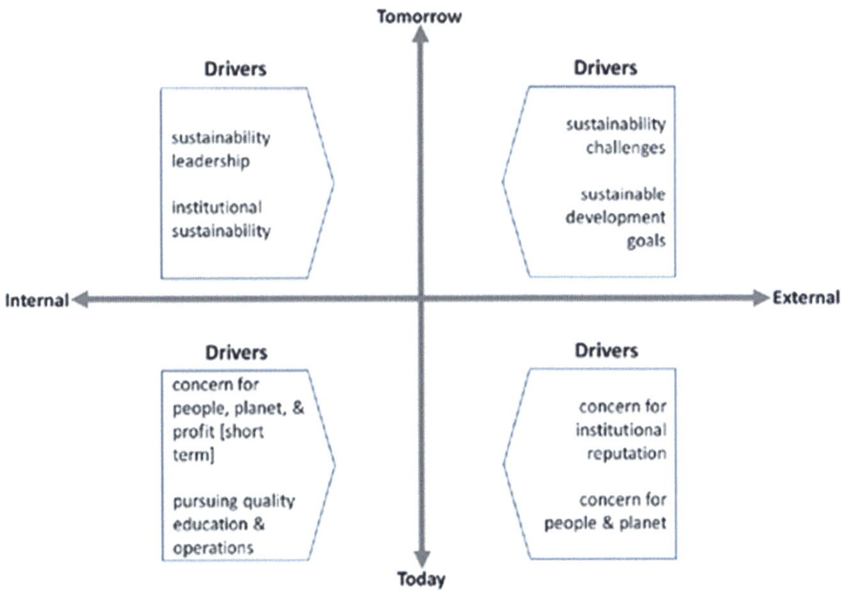

Fig. 1 HEI drivers for sustainability (Yturzaeta, 2023, p. 105)

Learning from the experiences of how sustainability has been adopted by HEIs from a leading practice perspective is the aim of this background assessment of the literature and practice. The areas considered below are defining sustainability from a HEI perspective, overarching HEI sustainability characteristics, and common HEI sustainability processes, and Education for Sustainable Development and Living Labs as particular sustainability foci for HEIs.

How individual HEIs define and communicate their sustainability positions has been examined by many including Ruiz-Mallen and Heras (2020) who emphasise the parallels, differences and complexities, commenting "There is no questioning about the ubiquity and ambiguity of the term sustainability as a critical concept for social change across disciplines and institutions" (p. 3). Pathways taken towards sustainability vary across and within institutions and range from technological solutions, social, economic and political transformation and managerial/risk approaches, reflecting a plethora of values, processes and understandings' (Ruiz-Mallen & Heras, (2020, p3). Harvey et al. (2022) have identified four sustainability themes from a review of the sustainability definitions adopted by HEIs in the United States: three-dimensional (Triple Bottom Line), intergenerational, environmental and ethical. They remark, "because sustainability has many meanings across different contexts, a clear understanding can be a challenge." (p. 182). In contrast, a goal focused approach by HEIs to sustainability is highlighted by Sterling (2013a). Sterling identifies the goals as:

> Integrating actions of conservation and human development...Satisfying basic human needs...Achieving equality and social justice for all... Facilitating social self-determination and cultural diversity... Managing our legacy for future generations... Maintaining ecological integrity...Facilitating social self-determination and cultural diversity... Managing our legacy for future generations...Maintaining ecological integrity...Developing new technologies and product manufacturing processes. (Sterling, 2013a).

The COPERNICUS Charta 2.0 of the European Network on Higher Education for Sustainable Development has goals of: "... contributing to our successful transition towards a sustainable society, which is free, just, equal, solidary and tolerant. A society which is characterized by respect for nature and our fellow humans and by shared responsibility." (Copernicus Alliance, 2011, p.1). The goals and those emphasised by Sterling have a vision of a sustainable society for the future.

HEI sustainability processes

The organisational characteristics needed for HEI sustainability implementation have been evaluated in the *EAUC Next Generation Sustainability Strategy* (Appleton, 2017a, 2017b) where the sustainability structures and approaches of 18 universities (12 in the UK and 6 in other countries: Netherlands, France, United States, Australia and New Zealand) are assessed. The organisational structures that Appleton (2017a, 2017b) found indicated that sustainability is most often led by either a dedicated department or via a decentralised structure but also is in some cases estates-based, elsewhere-based, or student-led. Those HEIs with no dedicated roles were in the process of taking on whole of organisation sustainability (Appleton, 2017a, 2017b,

p. 7). The report advises HEIs adoption of sustainability is best supported by a sustainability strategy, monitored and "...student voice must be listened to and recognised as a driver of change… allowing grassroots initiatives to build support and engagement from the bottom-up…' (Appleton, 2017a, pp. 25–26). Ramiso et al. (2019) emphasise that bottom-up and top-down approaches, that is, multiple-level perspectives are needed for university sustainability. Sterling (2013b) in The Sustainable University: Progress and prospects (Sterling et al., 2013a, 2013b) specifies criteria needed for university sustainability including a sustainability policy, a whole institution action plan and 'culture of organisational learning and improvement' (Sterling, 2013b, pp. 43–44).

More recently, Sanches et al. (2022, 2023) have proposed an organisational approach to move an HEI towards sustainability based on HEI experience reported in the literature and then tested in an HEI in Sao Paulo, Brazil. Their steps for repositioning and structuring a HEI for sustainability include: 'Adhere to international declarations, formalizing HEIs' commitment to sustainable development and the SDGs...Integrate sustainability in the strategic plans and unfold...into action plans... Establish partnerships with other sustainable HEIs (benchmarking; networking)...Implement procedures to sustainability assessment and reporting.' (Sanches et al., 2022, p. 927). Sanches et al. (2022) emphasise the fundamental importance of incorporating in sustainability the strategic plan and strategic planning process of an HEI and cite cases of the lack of success with sustainability implementation across HEIs of universities that adopt sustainability strategies without incorporation of sustainability in their strategic plans. Reviews of HEI sustainability strategies reveal varying approaches (Melles et al., 2022; Sanches et al., 2023).

Bien and Sassen (2020) provide proposals for addressing the changes associated with transformation saying that the aspects of organisational change management needed for sustainability in HEIs are:

- "the need to give meaning to the imprecise concept of sustainability, which happens in a specific context by multiple stakeholders…
- the need to deal with hidden contradictions and tensions that lead to barriers of change due to power perceptions, institutional structures and networks…
- the need to provide interpretational flexibility in order to provide opportunities to discuss sustainability in diverse academic traditions…
- the need to embrace a vision of sustainability that is rooted in plurality and dialogue…
- the need for freedom to explore the idea of sustainability in ways that make contextual sense." (Bien & Sassen, 2020, p. 2)

They also draw on *sensemaking* theory to surface the different conceptualisations of sustainability and the discursive power in institutions to better understand processes of change. In their study, they identified three dimensions that form "the change context of the transition: (1) discourse strategies applied to make and give sense, (2) triggers of resistance and (3) determinants of a collective sensemaking process" (Bien & Sassen, 2020, p. 6). They caution not to ignore the dynamics of change management which can result in pitfalls for sustainability implementation.

Education for Sustainable Development

The strategic adoption of education for sustainability by HEIs is not new. Focus on the role of universities and colleges to "Educate for Environmentally Responsible Citizenship" was part of the 10-point action plan in the Talloires Declaration initiated more than three decades ago in 1990. The Declaration has been signed by over 500 higher education leaders in 50 countries (ULSF, 2024).

UNESCO has been a key driving force as a lead agency promoting ESD since the UN Decade of Education for Sustainable Development (2005–2014). ESD is an ongoing learning process which actively involves multiple stakeholders in change to achieve sustainability. It has had various characterisations, however generally involves five key components:

- "Visioning (imagining a better future)
- Critical thinking and reflection
- Participation in decision-making
- Partnerships and
- Systemic thinking" (Tilbury & Wortman, 2004, p. 11).

In particular, Tilbury and Wortman (2004) emphasise the importance of futures education and understanding, and using processes of visioning for a sustainable future.

Internationally, UNESCO has been influential with leadership and guidance via the ESD for 2030 Global Network programme, *Education for Sustainable Development: A roadmap* (UNESCO, 2020), and Berlin Declaration on Education for Sustainable Development (2020), and *Education for SDGs: Learning objectives* (UNESCO, 2017). In Europe, UNECE has led Capacity-building workshops for facilitating national reporting on implementing the new implementation framework of the UNECE Strategy for ESD (2021–2030) in December 2023 and January 2024.

Many HEIs in the UK and internationally use an ESD approach for their sustainability teaching and learning (e.g. see University of Gloucestershire (2012). AdvanceHE together with the UK Quality Assurance Agency has since advocated a SDG focused approach to implementation of ESD in HEIs since 2021. Recommended approaches to support curriculum design for ESD include *Framework for Education for Sustainable Development* (AdvanceHE, 2024), and *A Handbook for Monitoring and Evaluating Education for Sustainable Development in Higher Education* (White et al., 2023). QAA and AdvanceHE (2021) built on the *Future Fit Framework* of Sterling (2012).

Living labs as sustainability implementation focal points in HEIs

Living labs originated in universities in the early 2000s and have become an effective interdisciplinary mechanism for HEIs to implement collaborative sustainability research and teaching. Hadfield et al. (2023) have examined how living labs are organised in universities internationally through 18 HEI case studies.

They found living labs generally have four interacting characteristics:

- "A research process involving co-design and interdisciplinarity, experimentation and social learning in a real-world setting, and a focus on stimulating innovation and systems change.
- A teaching practice whereby students learn from and participate in applied research and industry engagement.
- Integrated campus management involving experimentation with, and innovative utilisation and management of, campus environments and specific built, natural, and digital assets.
- A partnership and engagement model for multistakeholder industry collaborations to address global challenges on and off campus." (p. 4)

Hadfield et al. (2023) emphasise that:

> "The value proposition for university living labs centres on achieving research impact, enabling experiential learning for students, and integrating stakeholders and activities across institutional silos" (p. 3).

Conclusions about HEI sustainability practice

Many issues influence sustainability in HEIs. This literature and practice review highlights the need for a comprehensive approach to sustainability that encompasses the whole of the HEI. This review also underscores the importance of adoption of an organisational sustainability strategy and acknowledgement by the HEI that sustainability is a dynamic process of change where sensemaking must be addressed (Bien & Sassen, 2020). The SDGs are recognised by Yturzaeta (2023) as an external driver and QAA and Advance HE (2021) has stressed the advantages of using the SDGs as a focus in ESD implementation.

3 SOAS' Current and Future Sustainability

SOAS context

Designing a draft sustainability strategy for SOAS necessarily built on SOAS' ongoing sustainability work, and its current actions and networks. SOAS' sustainability strategy is planned to be the foundation for a new institutional strategic plan as well as all subsidiary policies and strategies, guiding and informing their creation and renewal. These strategies, plans and policies include, but are not limited to: Education Strategy, Research and Knowledge Exchange Strategy, Careers and Employability Strategy, and the SOAS Estates Strategy, which have played an important part in SOAS' sustainability contribution.

Flagship actions by SOAS for sustainability to date have included becoming the first London university to divest in fossil fuels in 2015, many contributions in climate change and sustainable development research, sustainability in learning and teaching, the Decolonising SOAS Vision including research and learning and teaching, the Net Zero Strategy commitment to becoming by 2040, the SOAS Ethical Investment

Policy and the SOAS Supply Chain Code of Conduct and Sustainable Procurement Objectives. In addition, SOAS holds ISO 14001 certification.

Developing the draft strategy

The development of the draft sustainability strategy for SOAS involved processes of scoping, stakeholder engagement and consultation, review and reflection on current SOAS strategies, plans, policy and practice. The key theory or methodological approach that was used was Education for Sustainable Development (ESD) using visioning, systemic thinking, critical analysis and reflection, participation in decision-making and partnerships to actively involve multiple stakeholders across SOAS to achieve change for sustainability. Development of the strategy accordingly has been a process of learning that will continue in implementation, not only about SOAS' operational sustainability and systems thinking related to the curriculum and research, but also of cultural change. A whole of organisational learning process has been initiated with the SOAS community that will be ongoing.

Views of students and staff

The development of the draft strategy very importantly benefitted from the contribution of over 100 of academic and professional staff members and students. Participation was arranged in four ways:

- Sustainability@SOAS survey—students and staff were invited to contribute their visions for the sustainability strategy through email and weekly SOAS Bulletin invitations to do an online survey with over 40 responses.
- Student focus groups—two student focus groups were organised and held with a total of 14 students across the two groups.
- Key informant interviews—over 30 key informants drawn from academics, professional services, library, students, estates and senior leadership were invited to participate in individual online interviews.
- SOAS Sustainability Steering Group—eight staff members were involved as partners giving critical feedback.

The participation process occurred during July and August 2024. Use of online facilities for most of the participation allowed for greater involvement over the university summer break. The 16 questions in the online survey focused on the themes:

- The importance of sustainability for SOAS (Q. 1–6, 13)
- Current efforts by SOAS at promoting sustainability (Q. 7, 9–11)
- Aspects of sustainability that universities should be engaged in (Q. 8)
- What should change or be prioritised regarding sustainability at SOAS (Q. 12, 14, 16)
- Is there a forum to raise sustainability concerns at SOAS (Q. 15)

The key informant interviews and student focus group themes were:

- The most important areas of sustainability or sustainable development initiatives that universities should pursue
- Strategies or initiatives that SOAS has already adopted that contribute towards sustainability
- Other areas of sustainability that SOAS should be actively involved in
- Areas of sustainability that SOAS students could be actively involved in
- Value of incorporating Education for Sustainable Development in the curriculum for all SOAS students
- Benefits and challenges in implementing a unified strategy for sustainability at SOAS
- How important is it that SOAS takes an active stance on sustainability and prioritises sustainability?

The survey was administered online via Microsoft Forms and results of the survey were analysed with Microsoft Forms' built-in reporting and analysis tools. Focus groups and key informant interview qualitative data was thematically analysed using NVivo.

Sustainability@SOAS survey

The Sustainability@SOAS survey was circulated across SOAS in July and August 2024. 48 responses were submitted from students (20), academics (12), professional services (12), academic services (2), and estates (2). Key findings are:

- 85% of respondents said it is 'extremely important' for SOAS to have a Sustainability Strategy.
- 83% of respondents said it is 'extremely important to involve students and staff in SOAS' sustainability initiatives. 15% said it was 'quite important'.
- 42% of respondents were 'aware of any sustainability initiatives currently implemented in SOAS' while the remainder were not aware of any initiatives.
- 13% of respondents said they 'have a forum to raise your concerns around issue of sustainability at SOAS' while 46% said they did not. 41% were unsure.
- Respondents said the 'top priorities in the university's sustainability strategy' should be:

    - Reducing carbon footprint (58%)
    - Supporting social equity and inclusion (37.5%)
    - Conducting impactful research on sustainability (33%)
    - Embedding sustainability in the curriculum (31%)
    - Guaranteeing sustainable investment (27%)
    - Improving waste management and recycling (27%).

The findings of the survey highlight the high priority that students and staff place on sustainability at SOAS, the importance of a sustainability strategy, and significantly, that students and staff wish to be involved in initiatives. It is also clear that students and staff are not always aware of the actions SOAS is currently taking to address sustainability in the institution. Evident in survey responses are issues such as difficulty in making sustainable choices on campus, lack of forums for people to raise concerns about issues relating to sustainability as SOAS and a desire to see sustainability issues embedded in teaching and learning, catering, waste management, and the campus environment. Students and staff have articulated great interest and have made valuable suggestions which will continue to inform sustainability actions as part of an ongoing process and outcome of the SOAS Sustainability Strategy. In addition, communication is key, not only to publicise what action is being taken by SOAS, but also to more clearly identify and articulate how work across the institution connects to key sustainability indicators such as the SDGs.

Student focus groups

The student focus groups of one hour duration discussed the importance of having a sustainability strategy at SOAS and the benefits and challenges of incorporating sustainability into the curriculum. They proposed making sustainability a compulsory part of the program. They expressed concerns about the lack of awareness of students regarding sustainability initiatives at SOAS and the need for better communication and visibility of sustainability initiatives to provide recognition of the positive sustainability work already happening at SOAS. Students in the focus groups suggested various sustainability initiatives, including energy-saving measures, reducing water consumption, and promoting reuse and recycling, composting centres, plant-based food options, improving staff treatment, workshops on consent and sustainability for students, student-led gardens, community volunteering opportunities, and practical projects to increase student involvement in sustainability. Students believe that having a sustainability strategy would enhance their future employability and promote a sense of community and environmental responsibility among SOAS students. The student focus groups also discussed the challenges of fostering a sense of community and responsibility.

Key informant interviews

Open-ended key informant interviews to gain views and insights from within SOAS were held in July and August 2024. Generally, key informants underlined the importance of the role of universities in leading change and setting the agenda for sustainability. They emphasised the need for and benefits of, but also the challenges of implementing a unified sustainability strategy, the incorporation of sustainability into the curriculum, and the importance of and need for better communication and awareness within the institution. Key informants emphasised the importance of allocation of financial resources for sustainability, making trade-offs, changing institutional culture to achieve sustainability goals, and accountability. They also discussed specific areas of focus, such as procurement, waste reduction, and community engagement. They highlighted the value of education for sustainable development, skills development, collaboration, and workplace sustainability. Overall, the conversations stressed the complexity of addressing sustainability in higher education and the importance of taking an active stance on sustainability.

The student key informant perspective was that all areas of sustainability are interconnected and interdependent and that the SOAS focus should be on widening the scope of sustainability and incorporating biodiversity. They discussed the lack of awareness and communication regarding sustainability initiatives at SOAS and the importance of incorporating education for sustainable development across the curriculum. They underlined the need for better communication and recognition of the positive work already happening at SOAS. The financial challenges for SOAS of implementing a sustainability strategy were acknowledged but they emphasised that sustainability is in the interests and identity of SOAS as a place that tackles pressing global issues.

Academic key informants emphasised the need for a comprehensive and integrated approach to sustainability for SOAS, the challenges and benefits of incorporating sustainability into the curriculum, and the importance of addressing areas like climate change and resource use. They also discussed the potential resistance to sustainability initiatives and the need for support and professional development for academics. They also highlighted the importance of workplace sustainability, global cooperation, ethical investment, and interdisciplinary collaboration. Overall, the conversations highlighted the complexity of addressing sustainability in higher education, the surfacing of sustainability as contested field of study and the evident different disciplinary perspectives across the institution, thus highlighting the importance of the need for a collaboration, communication and a clear strategy.

The professional services key informants highlighted importance of sustainability for SOAS and the importance of sustainability as a core value for institutions. They focused on areas such as procurement, cloud computing, reducing travel, waste reduction, education, and utilisation of green spaces. They emphasised the need for a unified sustainability strategy, staff training, effective communication, and collaboration across all departments and areas. Their conversations also touched on the benefits of incorporating sustainability into the curriculum for students. The professional staff key informants also discussed challenges, such as lack of awareness, and highlighted the potential benefits of sustainability initiatives, including improved reputation and funding opportunities.

4 Outcomes

Strategy development and review process

The sustainability strategy approach was drafted for 2025 onwards with a timeline to 2030 for implementation of objectives and actions. Figure 2 depicts the relationship between SOAS' strategy and the quadruple Helix model (König, 2020, p. 10 after Millard, 2018). A detailed operational action plan for sustainability underpins the draft with key performance indicators (KPIs) for measurement of impacts and reporting. The draft strategy was designed as a living document to be updated as required with opportunities for further student and staff participation to be provided regularly throughout the implementation period. Members of the Sustainability Steering Group at SOAS were invited to review the draft in early September

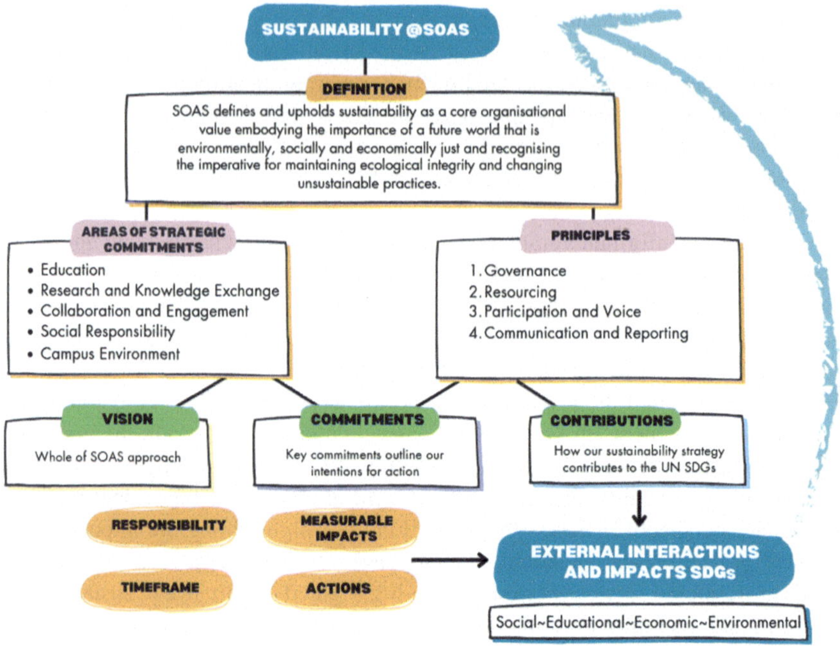

Fig. 2 SOAS' sustainability strategy approach

2024 and give feedback in the strategy visioning and drafting process though meetings and email feedback.

What does Sustainability at SOAS mean?

Consultation with academics, professional services staff, senior management and students highlighted the importance of understanding what sustainability means in the context of SOAS and how a sustainability strategy can best be implemented across the institution. The consultation informed the definition of sustainability for the institution:

> SOAS defines and upholds sustainability as a core organisational value embodying the importance of a future world that is more environmentally, socially and economically just and recognising the imperative for maintaining ecological integrity and changing unsustainable practices.

The SOAS definition and aims for sustainability are in line with the institutional mission to develop local and global knowledge, international perspectives, engage in robust debate, and teach the necessary skills required for students to become global citizens who can make an impact in a changing world. SOAS' aims for research, education, collaboration and partnerships, social responsibility, and campus environment are all underpinned by the objective of attaining more equitable global and local futures.

Contributing to the United Nations Sustainable Development Goals

The draft strategy is designed to achieve a better and more sustainable future through contributions to the attainment of the UN SDGs through teaching and research. It endeavours to ensure that SOAS minimises its footprint on the environment and promotes peace and justice.

A 'whole of SOAS approach'

The draft sustainability strategy embodies a transformative, holistic and enhanced focus on sustainability with defined actions, objectives and targets that will require the active support of all the whole of the SOAS community.

The sustainability strategy's implementation will need a 'whole of SOAS approach'—to be contributed to by collective efforts from across the SOAS community. It will apply to all parts of SOAS activities and campus facilities including colleges, schools, departments, centres, institutes, business and support units.

Implementation will include policy support for practice and collaboration across departments. Examples include: for staff, mandatory training of all teaching staff in ESD run by central academic services, and for students, appointing SOAS student ambassadors for sustainability and facilitating student-led interdisciplinary sustainability projects which contribute and enhance local area social, cultural and environmental spaces.

Students will be at the heart of the sustainability strategy, including student representatives in sustainability decisions and actions as well as supporting the Student Union in developing sustainability initiatives. A survey conducted by student-led education charity 'Students Organising for Sustainability (SOS-UK)' found that 89% of students agreed their place of study should actively incorporate and promote sustainable development, 82% would like to see sustainable development actively incorporated and promoted though all courses (SOS-UK, 2023).

Bringing the outside world into SOAS and engaging with society outside SOAS is integral to the strategy, for example, building research collaborations on sustainability innovation in partnership with other research institutions, civil society, government and business, and acting ethically and with integrity in the supply chain.

Principles

The draft sustainability strategy sets out the commitments and actions for taking forward institutional mission to continue the work achieved so far and to further transform SOAS into a sustainable university. The commitments are underpinned by four principles governance, resourcing, participation and voice, and communication and reporting:

- Governance—Providing visible and credible leadership and working collectively across academic and professional services to adopt processes within the institution which facilitate effective and efficient systems and processes for embedding sustainability practices
- Resourcing—Investing in the future by embedding resources for sustainability within institutional budgets.

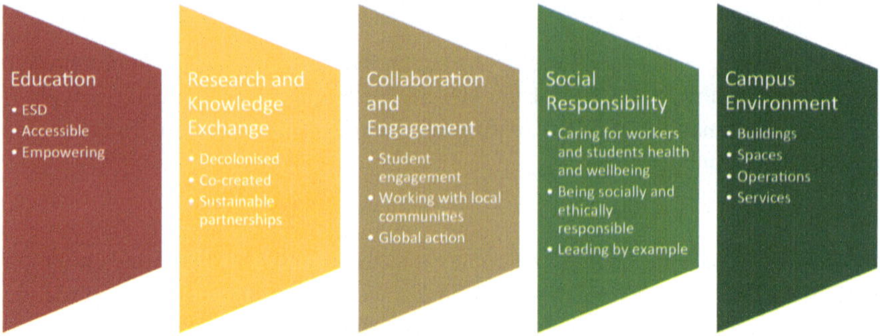

Fig. 3 Achieving our strategic commitments for SOAS' sustainability

- Participation and Voice—Including SOAS students and staff in shaping sustainability initiatives and having a stake in major decisions and implementation.
- Communication and Reporting—Embedding processes of communication that enable exchange of information within the SOAS community concerning sustainability as well as reporting externally on the impact of actions on commitments. Reporting will involve establishment of initial baseline data points to effectively evaluate the impacts measures or key performance indicators (KPIs) of the Strategy.

The strategy design focuses on commitments in five priority areas: education, research and knowledge exchange, collaboration and engagement, social responsibility and the campus environment (see Fig. 3). The strategy provides an overarching approach, sets goals and actions, establishes processes and a culture of practice within the institution to achieve a more sustainable future. SOAS seeks to report on its achievement through university impact rankings, for example, Times Higher Education University SDG Impact Rankings, People and Planet University League, together with commitments such as the Concordat for the Environmental Sustainability of Research and Innovation Practice, and the SDG Accord.

The aims and commitments to principles of sustainability in the five priority areas include:

- Education: To ensure our students gain the skills and competencies to actively contribute towards the building of an equitable and just planetary community and protecting the environment though accessible and empowering education experiences rooted in SOAS' decolonising knowledge approach.
- Research: To undertake research that is co-created through partnerships for sustainability, decolonised and contributes to solving global challenges through change of mindsets and discourses, policies and practices for sustainable development.
- Collaboration and engagement: To facilitate the contribution of students and staff to our local community's sustainability and to support global action on sustainability.

- Social responsibility: To ensure social responsibility through institutional management, teaching, research, services and partnerships that support equality, diversity, inclusion, wellbeing and a healthy sustainable campus environment for all students and staff, and to contribute to social justice and sustainability locally and globally.
- Campus environment: To embed environmental and ethical practices across the whole university and to ensure that all staff, students and visitors can work and study comfortably and make sustainable and ethical choices whilst on campus.

The interconnectedness between each priority area and the interdisciplinary required for achieving institutional sustainability is highlighted in the draft strategy. There is a clear underscoring of the importance of a holistic, interdisciplinary, 'whole of SOAS approach' to sustainability and the need for contributions by all in SOAS to the implementation.

SMART goals (specific, measurable, achievable, relevant, and time-bound) will be pursued throughout SOAS' sustainability strategy's implementation process. Each commitment for the priority areas above has associated actions, KPIs to measure impacts of actions, responsibilities (academic, professional and academic services, Executive Board, student body and groups), and timeframes (short-term 1–2 years, medium 3–4 years, long-term 5+ years).

5 Conclusions

This chapter has shown the journey of one UK HEI, SOAS University of London in developing a draft sustainability strategy. Research for the strategy development addressed three research questions.

The first question tackled how sustainability has been adopted by HEIs from a leading practice perspective. Review of literature and practice highlighted underscores the importance of adoption of an organisational sustainability strategy and the need for a comprehensive approach to sustainability that encompasses the whole of the HEI. Acknowledgement by the HEI that sustainability implementation is a dynamic process of change that should bring in all students and staff is seen as fundamental. This is best implemented by sensemaking to draw out the different conceptualisations of sustainability and the discursive power in the HEI.

The SDGs are recognised in the literature as an external driver for HEI sustainability adoption and they are seen as advantageous as a central focus and especially with ESD implementation in curricula.

The second research question focused on the views of students and staff about the current and future sustainability of SOAS University of London. Thorough consultation with students, academic staff, professional services staff, estates staff and senior management highlighted the importance of understanding what sustainability means in the context of SOAS and how a sustainability strategy can best be implemented across the institution.

This led to the third research question concerning what elements should be incorporated in a sustainability strategy for SOAS University of London. It was established that efforts to pursue sustainability should focus on priority areas of education, research and knowledge exchange, collaboration and engagement, social responsibility and the campus environment. The priority areas and associated commitments importantly need to be underpinned by four principles governance, resourcing, participation and voice, and communication and reporting.

A clear underscoring emerged of the importance of a holistic or 'whole of SOAS approach' to sustainability and contributions by SOAS to the implementation of the SDGs. The interconnectedness between each priority area and the SDGs is highlighted.

Acknowledgements The authors wish to gratefully acknowledge the initiation of and support for this sustainability research by the SOAS Sustainability Steering Group together with contributions by many SOAS colleagues including Elisa Van Waeyenberge, Andrew Goodman, Tom Tanner, Vicki Hart, Tawnya Cheatham, Graeme Were, Emma Keenan, Ola Bankhole and Brian Cronin. We also wish to thank all the SOAS students and staff who generously contributed their time and opinions through participation in focus groups and key informant interviews, and via survey responses.

References

AdvanceHE. (2024). *Framework for education for sustainable development*. https://www.advance-he.ac.uk/knowledge-hub/framework-education-sustainable-development

Appleton, E. (2017a). *Next generation sustainability strategy and structure: Whole-institution approaches to sustainability in universities and colleges*. EAUC. https://www.eauc.org.uk/next_generation_sustainability_strategy_and_str

Appleton, E. (2017b). *Next generation sustainability strategy and structure: Supplementary case studies*. EAUC. https://www.eauc.org.uk/next_generation_sustainability_strategy_and_str

Bien, C., & Sassen, R. (2020). Sensemaking of a sustainability transition by higher education institution leaders. *Journal of Cleaner Production, 256*, Article 120299. https://doi.org/10.1016/j.jclepro.2020.120299

COPERNICUS Alliance. (2011). *COPERNICUS Charta 2.0*. COPERNICUS Alliance—European Network on Higher Education for Sustainable Development. https://www.copernicus-alliance.org/background

EAUC and AUDE. (2022). *Sustainability leadership scorecard user guide*. The Alliance for Sustainability Leadership in Education (EAUC) and Association of University Estates Directors (AUDE). https://www.eauc.org.uk/sustainability_leadership_scorecard

Galán-Muros, V. (2023). General guidelines for the implementation of sustainability in higher education institutions. In *UNESCO International Institute for Higher Education in Latin America and the Caribbean and United Nations Academic Impact*. https://unesdoc.unesco.org/ark:/48223/pf0000387008

Hadfield, P., Sharp, D., Zarea, M., Pigeon, J., Peng, X., Rye, S., & Raven, R. (2023). *Governing university living labs for sustainable development: Lessons from international case studies*. Monash Sustainable Development Institute, Monash University. https://doi.org/10.26180/22138073

Harvey, T., Morales, A., & Middlecamp, C. (2022). Defining sustainability in higher education institutions. *Sustainability and Climate Change, 15*, 182–188. https://doi.org/10.1089/scc.2022.0011

König, J., Suwala, L., & Delargy, C. (2020). Helix models of innovation and sustainable development goals. In W. Leal Filho, L. Brandli, A. Lange Salvia, & T. Wall (Eds.), *Industry, innovation and infrastructure. Encyclopedia of the UN sustainable development goals* (pp. 1–15). Cham: Springer. https://doi.org/10.1007/978-3-319-71059-4_91-1

Melles, G., Lodewyckx, S., & Hariharan, T. S. (2022). Campus sustainability in the Australian higher education sector: Divergence and convergence in planning, reporting and tactics. *International Journal of Sustainability in Higher Education, 23*(1), 87–113. https://doi.org/10.1108/IJSHE-10-2020-0409

Millard, J. (2018). How social innovation underpins sustainable development. In J. Howaldt, C. Kaletka, A. Schröder, & M. Zirngiebl (Eds.) *Atlas of Social innovation* (pp. 41–43). Dortmund: TU Dortmund University. https://www.socialinnovationatlas.net/fileadmin/eBook/Atlas-of-Social-Innovation-2018.epub

QAA, AdvanceHE (2021). *Education for sustainable development guidance.* Quality Assurance Agency for Higher Education and AdvanceHE. https://www.qaa.ac.uk/quality-code/education-for-sustainable-development

Ramísio, P. J., Costa Pinto, L. M., Gouveia, N., Costa, H., & Arezes, D. (2019). Sustainability strategy in higher education institutions: Lessons learned from a nine-year case study. *Journal of Cleaner Production, 222*, 300–309. https://doi.org/10.1016/j.jclepro.2019.02.257

Ruiz-Mallén, I., & Heras, M. (2020). What sustainability? Higher education institutions' pathways to reach the Agenda 2030 Goals. *Sustainability, 12*(4), 1290. https://doi.org/10.3390/su12041290

Sanches, F. E. F., Campos, M. L., Gaio, L. E., & Belli, M. M. (2022). Proposal for sustainability action archetypes for higher education institutions. *International Journal of Sustainability in Higher Education, 23*(4), 915–939. https://doi.org/10.1108/IJSHE-01-2021-0026

Sanches, F.E.F., Souza Junior, M.A.A.d., Massaro Junior, F.R., Povedano, R., Gaio, L.E. (2023). Developing a method for incorporating sustainability into the strategic planning of higher education institutions. *International Journal of Sustainability in Higher Education* 24(4):812–839. https://doi.org/10.1108/IJSHE-10-2021-0439

Sterling, S. (2012). *The future fit framework: An introductory guide to teaching and learning for sustainability in HE.* Higher Education Academy. https://www.advance-he.ac.uk/knowledge-hub/future-fit-framework

Sterling, S. (2013a). *The sustainable university, EAUC Webinar*. Centre for Sustainable Futures/PedRIO Plymouth University, 5 June 2013. https://www.eauc.org.uk/file_uploads/presentation.pdf

Sterling, S. (2013b). The sustainable university—challenge and response. In S. Sterling, L. Maxey, & H. Luna (Eds.), *The sustainable university: Progress and prospects.* Routledge, Abingdon. https://doi.org/10.4324/9780203101780

Sterling, S. (2021). Concern, conception, and consequence: Re-thinking the paradigm of higher education in dangerous times. *Frontiers in Sustainability, 2*, 743806. https://doi.org/10.3389/frsus.2021.743806

Students Organising for Sustainability–UK (SOS-UK) (2023). *Sustainability skills survey 2022–23: Research into students' experiences of teaching and learning on sustainable development.* https://www.sos-uk.org/research/sustainability-skills-survey

Taplin, R., de Frece, A., & Hart, V. (2025). Sustainable development goals in online learning: Action research for enhanced graduate skills at SOAS University of London. In E. J. Schwartz, & D. B. Audretsch (Eds.), *Handbook of teaching and learning in social innovation.* Cheltemham: Edward Elgar.

Tilbury, D., & Wortman, D. (2004). *Engaging People in Sustainability: An IUCN Guide*, IUCN: International Union for Conservation of Nature. IUCN Commission on Education and Communication (CEC). Gland, Switzerland and Cambridge, UK.

ULSF (2024). *Talloires declaration.* https://ulsf.org/talloires-declaration

UNESCO (2017). *Education for SDGs: Learning objectives.* United Nations Educational, Scientific and Cultural Organization. https://doi.org/10.54675/CGBA9153

UNESCO (2020). *Education for sustainable development: A roadmap.* United Nations Educational, Scientific and Cultural Organization. https://doi.org/10.54675/YFRE1448

University of Gloucestershire (2012). *Education for sustainability: A guide for educators on teaching and learning approaches.* https://efsandquality.glos.ac.uk/toolkit/EfS_Educators_Guide.pdf

White, R. M., Möbius, K., King, B., Leask, K., Peres, S., Walker, J., & Higgins, P. (2023). *A handbook for monitoring and evaluating Education for Sustainable Development in higher education.* Quality Assurance Agency for Higher Education. https://www.qaa.ac.uk/membership/collaborative-enhancement-projects/education-for-sustainable-development/monitoring-and-evaluating-education-for-sustainable-development-in-higher-education

Yturzaeta, J. (2023). Assessing sustainability initiatives in higher education institutions. *Journal of Management for Global Sustainability, 8*(1):97–121. https://archium.ateneo.edu/jmgs/vol8/iss1/8

Identification of Knowledge and Skills Gaps Related to Sustainable Development: Two Exploratory Studies with Higher Education Students and Teachers in Europe

Susana Leal, Andriani Piki, João Nascimento, Joke Simons, Luís Barradas, and Sandra Oliveira

Abstract This paper presents the preliminary findings from the Time2Act@SD Erasmus+project, focusing on identifying gaps in knowledge and skills related to sustainable development (SD) and the Sustainable Development Goals (SDGs) in higher education (HE). The study has two main objectives. Firstly, to identify gaps in knowledge and skills related to SD and the SDGs among HE students. This includes analysing students' understanding, attitudes, and behaviours to highlight specific weaknesses, providing a basis for formulating targeted improvements. Secondly, to

S. Leal (✉) · J. Nascimento · L. Barradas · S. Oliveira
School of Management and Technology, Santarém Polytechnic University, Complexo Andaluz, Santarém, Portugal
e-mail: susana.leal@esg.ipsantarem.pt

J. Nascimento
e-mail: joao.nascimento@esg.ipsantarem.pt

L. Barradas
e-mail: claudio.barradas@esg.ipsantarem.pt

S. Oliveira
e-mail: sandra.oliveira@esg.ipsantarem.pt

S. Leal
Life Quality Research Centre (CIEQV), Santarém Polytechnic University, Santarém, Portugal

A. Piki
School of Sciences, University of Central Lancashire Cyprus (UCLan Cyprus), Larnaca, Cyprus
e-mail: APiki@uclan.ac.uk

J. Simons
Centre of Expertise Care and Well-Being, Thomas More University of Applied Sciences, Flanders, Belgium
e-mail: joke.simons@thomasmore.be

S. Oliveira
Centre for Health Studies and Research (CEISUC), Coimbra, Portugal

W. Leal Filho et al. (eds.), *Introducing the UN SDGs in Higher Education Institutions*, World Sustainability Series, https://doi.org/10.1007/978-3-032-23017-1_24

explore prominent gaps in the teaching–learning process within HE, namely the challenges faced by educators trying to incorporate SD and the SDGs in their teaching activities. By understanding what underpins these challenges, this study aims to develop tailored solutions to enhance sustainability-focused education. Two questionnaires were administered, for HE teachers and students, respectively, involving participants from several European countries using a convenience sample. The results will inform the development of innovative tools to improve the teaching–learning process in HE. By engaging both groups of stakeholders, this research aims to provide valuable insights that will guide the development of tools and pedagogical strategies to improve education for sustainability, thereby promoting a deeper integration of sustainability principles into academic curricula.

Keywords Higher education students · Higher education teachers · Knowledge and skills gaps · Sustainable Development · Sustainable Development Goals (SDGs) · Sustainability education

1 Introduction

The introduction of Sustainable Development (SD) into Higher Education Institutions (HEIs) is widely recognised as a critical step towards preparing future professionals to effectively solve global problems. The United Nations' Sustainable Development Goals (SDGs), intended to be implemented by 2030, emphasise the importance of diverse skills, abilities, and attitudes aimed at promoting more equitable and environmentally sustainable societies. Relevant research (Álvarez-Vanegas et al., 2024; Fia et al., 2023; Leal Filho, 2015; Leal Filho et al., 2018) has found significant gaps in the understanding of SD concepts and their application among students and teachers in Higher Education (HE).

This article presents the outcomes from the collective analysis of two transnational studies conducted among HE students and teachers, in the context of Erasmus+Time2Act@SD project, aiming to uncover the gaps in the transfer and acquisition of knowledge linked to sustainability. The study revealed substantial shortcomings in the teaching–learning processes related to SD as well as differences among HE students and teachers in terms of their sustainability knowledge, attitudes, and behaviours. Given the increased recognition of educational institutions' role in addressing social, economic, and environmental challenges, this study highlights the significance of empowering students and teachers, alike, with the necessary sustainability competencies, awareness, and attitudes.

Identifying prominent knowledge and skills gaps is an important first step towards establishing a strong sustainability culture within HE. The research findings can contribute to shaping the creation of novel educational tools, such as Massive Open Online Courses (MOOCs), Virtual Reality (VR) and mobile games, teacher training workshops and co-creation. HEIs can promote social innovation through co-creation with society, industry, and government using collaborative learning tools and a systemic approach (Kumari et al., 2019). HEI's play an important role in promoting social innovation through co-creation with society, business and government using collaborative learning tools and systemic approaches (Franco et al., 2022). Recent research indicates the evolution of collaborative models between universities and industry towards more inclusive approaches, (Galvão et al., 2019; Ketikidis & Solomon, 2018). The quadruple and quintuple helix models highlight this evolution by placing the involvement of diverse stakeholders at the centrer of addressing societal challenges and fostering innovation (Galvão et al., 2019; Ketikidis & Solomon, 2018). To facilitate collaborations within quadruple helix ecosystems, web-based idea management platforms and cultural hubs are increasingly being deployed (Mikelsone et al., 2024; Nørgård & Holflod, 2025). The challenge in the practical implementation of these models is to effectively engage civil society organizations (Roman et al., 2020). The transition to a more open and co-creative research model within higher education requires not only new partnership approaches, but also a fundamental cultural shift (Miller et al., 2018).

Future research should focus on developing coherent frameworks and policies around effective multi-stakeholder collaboration to enable successful technology transfer within universities and the strengthening of regional innovation systems.

By addressing these shortcomings, the study hopes to contribute to the improvement of Education for Sustainable Development (ESD) through a comprehensive and successful integration of SD and SDGs issues in HE.

2 Sustainable Development in Higher Education

Higher Education can help address SD issues through three primary pillars. First, through educating students of all age groups and backgrounds and empowering them to become critical thinkers and responsible problem solvers. Second, by conducting ethical and socially relevant research bringing new knowledge to the community. Finally, the third role entails supporting society as a whole, through the lifelong learning intervention and the promotion of academic citizenship. Consequently, it is crucial that both learners and educators form a clear understanding of SD principles and develop sustainability knowledge, responsible behaviours, and positive attitudes related to sustainability. Identifying knowledge and skills gaps among stakeholders is a crucial initial step in research efforts in this area. However, the current understanding of the dimension of the problem is still limited (Abad-Segura & González-Zamar, 2021; Kopnina, 2020; Zhang et al., 2022). It is crucial for HEIs to explicitly identify the strengths and weaknesses of their current sustainability culture to facilitate the formation of innovative educational pathways. In this way, HEIs can increase their societal impact, and ensure that education, research, and institutional practices contribute to a more sustainable and equitable future. Several sustainability initiatives were identified across the EU landscape ranging from governmental to internal initiatives in universities. These practices take various forms such as changes in the curriculum, offering sustainability-focused units, and interdisciplinary programs. Additionally, universities have implemented community engagement and partnerships, policies and organisational strategies. Some universities have adopted embedded institution-wide sustainability strategies into the campus through environmentally friendly practices. Off-campus requirements and facilities blend environmental criteria in acquisitions and construction investments. Moreover, in recent years, HEIs intensified efforts to promote the SDGs in an effort to understand how they could enhance the agenda for a sustainable future through teaching and learning (Acosta Castellanos & Queiruga-Dios, 2022; Anthony Jnr, 2021; Bautista-Puig & Sanz-Casado, 2021; Leal Filho et al., 2021; Weiss et al., 2021).

Some HEIs documented case studies of their sustainability initiatives. These include various programs ranging from campus sustainability, learning for sustainability, research excellence without boundaries, community partnership activities and a community-focused program. Moreover, through its community network, HEIs are aligning targets in the SDGs: co-benefits in urban planning, circular cities, and reuse and recycling of materials. There is evidence of a growing emphasis towards providing toolkits and multi-year support through facility programs that integrate campus sustainability initiatives with co-curricular opportunities, actions, and outcomes of interest (Brundiers et al., 2021; Kamil et al., 2020; Miranda et al., 2021; Žalėnienė & Pereira, 2021). HEI's must both effectively integrate bottom-up sustainability initiatives into policy alignment, institutional support, and cross-sector collaboration. Additionally, they should create the conditions for the optimal implementation of the SDGs and related innovative approaches in education through top-down strategies.

Despite growing efforts to integrate sustainability into higher education, there are still substantial gaps in teaching approaches and student educational achievements. Numerous teachers lack the proper preparation and resources to efficiently utilise interdisciplinary, project-based, and experiential learning methods, essential for nurturing sustainability skills (Backman et al., 2019; Birdman et al., 2022; Leal Filho et al., 2016; Singha & Singha, 2024). Mokski et al. (2023) propose integrating ESD into existing curricula and creating standalone ESD programs to reach students across disciplines. Sinakou et al. (2019) present a framework combining holistic, pluralistic, and action-oriented teaching for effective ESD. Zinn and Isenmann (2017) describe a "whole institution approach" to incorporate sustainability across all areas of higher education. Martins et al. (2022) introduce an interdisciplinarity-based sustainability framework for management education. Howlett et al. (2016) argue for interdisciplinary curricula and pedagogical practices that build critical and reflective thinking capacities. These studies collectively emphasize the need for innovative, cross-disciplinary approaches to enhance ESD effectiveness in higher education. Theses decrease students struggle to utilise theoretical information in real-life scenarios, impacting their capacity to engage in critical thinking and address sustainability issues (Heinrich et al., 2015; Wyness & Dalton, 2018). In order to address these problems, it is crucial to make systemic changes in educational programs and teaching methods to prepare students more effectively for the intricate issues of sustainability.

3 The Time2Act@SD Project

Time2Act@SD[1] aims at enhancing the knowledge, skills, and attitudes of EU HE students, promoting behavioural change in the area of SD and the SDGs. This aim will be achieved through the development of interactive content (e.g., quizzes, interactive videos, gamified learning experiences with virtual reality games and web/mobile games), digital tools, and innovative teaching methodologies designed to support teachers and engage learners. The project adopts a transversal approach to address environmental issues, climate change, and sustainability, focusing on both HE students and teachers. Learning will encompass knowledge-based resources (Massive Open Online Courses [MOOCs] and webinars) as well as attitude and behaviour changing activities. These will include hands-on practical experiences via a mobile app and an immersive Virtual Reality (VR) game (using gamification strategies and workshops) along with real-world activities (bootcamps) to demonstrate action and ensure transferability to real-life situations.

4 Research Methodology

Two studies were conducted, to gather empirical data from both students and teachers in HE. The methodological steps employed in each study were aligned and involved: (i) a descriptive study with data collected through questionnaires applied to a sample of HE students and HE teachers respectively; (ii) quantitative and descriptive data analysis; (iii) identification of the main gaps; (iv) construction of recommendations for future project tasks.

The Ethics Committee of the Santarém Polytechnic University approved both studies. Informed consent was obtained from all participants, guaranteeing confidentiality and anonymity, and participants were informed that they could withdraw from the study at any time.

4.1 Study 1—HE Students

Initially, an online survey was designed to assess students' familiarity with SD and the SDGs, as well as their perceptions of their own knowledge, attitudes, and behaviours in these areas. Gericke et al.'s 49-item scale was used (Gericke et al., 2019), based on

---

[1] Time2Act@SD is an Erasmus+project led by a consortium of seven organisations: Politécnico de Santarém (Portugal) (coordinator), UCLan Cyprus (Cyprus), Thomas More University of Applied Sciences (Belgium), Gazi Üniversitesi (Turkey), The Language Menu—Learnmera (Finland), Galileo Roma (Italy), and Rosto Solidário (Portugal). For more information about the project, please visit https://time2act.ipsantarem.pt/.

a 5-point Likert scale (from 1 'strongly disagree' to 5 'strongly agree'), for students to express their agreement with the items.

The survey was made available to students from several countries (e.g., Portugal, Belgium, Türkiye, Cyprus, Finland, Italy, Slovenia), between 2 May and 30 June 2023 via the Survey Monkey platform. Descriptive statistics were used to derive meaningful findings.

The sample consisted of 716 HE students, of which 57% (n = 406) were female, 42% (n = 300) were male and 1% did not answer or responded 'other'. About 74% of students (n = 526) were under 26 years old. Most of participants were enrolled in undergraduate courses (68%) and 24% were postgraduate or masters' students, with the remainder enrolled in vocational training or doctoral studies (4% and 3% respectively). The main field of study of the respondents is Social Sciences (48%), followed by Medical and Health Sciences (19%), Engineering and Technology (17%), Natural Sciences (10%) and Humanities and Agricultural Sciences (3%). The student sample includes students studying in Portugal (35%), Türkiye (20%), Cyprus (15%), Slovenia (14%), Belgium (4%), Czech Republic (4%) and Italy (2%), amongst other countries. Because a convenience sampling method was used, it is not possible to report the response rate by country.

4.2 Study 2—HE Teachers

An online survey was designed to gain insights from HE teachers about their experiences in integrating SD and the SDGs into their teaching practices, as well as their perceptions of their own knowledge and degree of awareness in these areas.

The survey included 5-level Likert scale questions, multiple choice questions and open-ended questions to collect quantitative and qualitative data. The survey was made available to teachers across several countries (including and extending the consortium partners, namely Portugal, Belgium, Türkiye, Cyprus, Finland, and Italy). The survey was created using the Survey Monkey platform and was administered between 2 May and 30 June 2023. Descriptive statistics were used to derive meaningful findings.

The final sample includes 209 respondents from HE teachers, with 82% (n = 172) teaching in undergraduate programmes and 59% (n = 124) in Master's programmes, 23% (n = 49) in the vocational education and training (VET) and 24% (n = 50) in doctoral programmes. Teachers were able to choose all the levels of education in which they were teaching.

The sample includes 54% (n = 112) female and 45% (n = 94) male teachers; 39% of teachers are aged in the range 40–49 and 25% are aged in the range 50–59. For 46% of respondents, a doctorate is the highest level of completed formal education; 28% report a postgraduate/Master's degree and 24% of teachers have a postdoctoral degree. Only 2% have bachelor's degree. The main field of study of the participant teachers is Social Sciences (45%), followed by Engineering and Technology (16%), Natural Sciences (15%), Medical and Health Sciences (13%), Humanities (8%) and

Agricultural Sciences (2%). Most teachers are from Portugal (36%), Türkiye (15%), Cyprus (12%) and Belgium (11%). About 36% of the teachers were from Portugal, 15% from Türkiye, 12% from Cyprus, 11% from Belgium, 7% from Italy, 5% from Finland, and 13% from other European countries. Because a convenience sampling method was used, it is not possible to report the response rate by country.

5 Results

The conducted transnational studies provided significant insights on students' and teachers' perceptions with particular focus on their knowledge, attitudes and behaviours, as well as their level of awareness, and preferred learning and teaching methods, respectively. The analysis of their perceptions revealed significant gaps in sustainability education within HE among students and teaching staff, alike. By examining these gaps, a more accurate representation of the challenges HEIs face in preparing students and faculty to address sustainability effectively can be gained. The following section will examine these gaps in depth across three domains: environmental, social, and economic sustainability and explore the implications for teaching and learning practices. Additionally, it will explore potential differences among HE students and teachers in terms of their sustainability knowledge, attitudes, and behaviours.

5.1 Results of Study 1—HE Students

This section presents the main insights gathered among 716 HE students who participated in the study. Only 56% of the students consider themselves familiar (responded with 'agree' or 'disagree') with the term 'Sustainable Development Goals (SDGs)' (Fig. 1). To gain deeper insights, we also explored students' perceptions on their own knowledge, attitudes, and behaviours in relation to SD issues across environmental, social, and economic sustainability, as discussed next.

5.1.1 Students' Perceptions of Their Knowledge of SD Issues

Firstly, we explored the students' perceptions on their SD knowledge. The following three (3) figures illustrate only the responses of students who either 'agree' or 'strongly agree' with the statements focusing on environmental, social, and economic sustainability, respectively. In terms of environmental sustainability, the item which received the lowest rating was K6 which shows that students are less aware of the value of education in protecting themselves against natural disasters (Fig. 2).

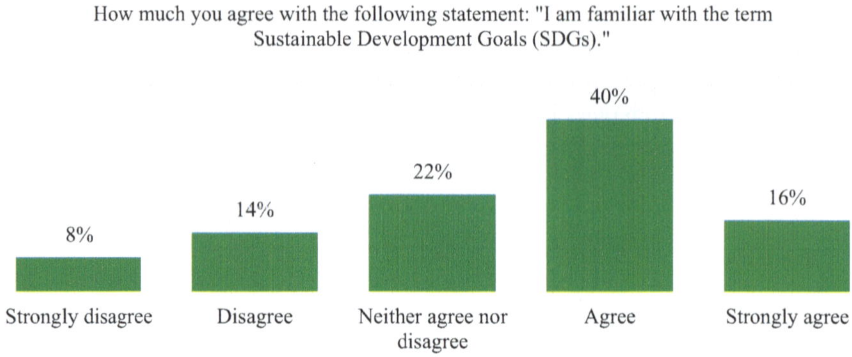

Fig. 1 Level of agreement regarding familiarity with the term SDGs. *Note* N = 716 HE students

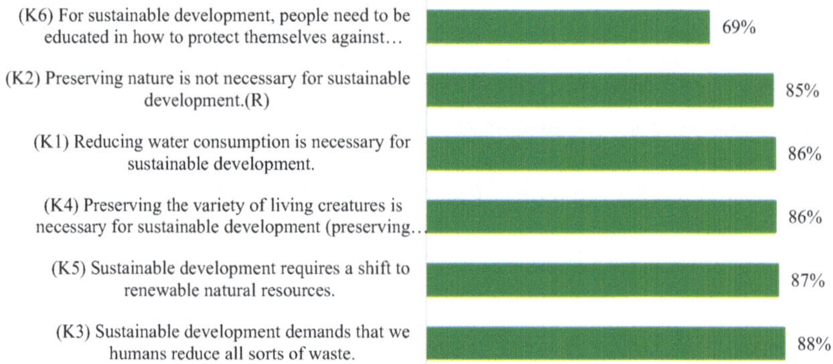

Fig. 2 Level of agreement with environmental sustainability topics (Knowledge). *Note* N = 716 HE students. The percentages represent students who 'agree' or 'strongly agree' with the statements. (R) Reverse score item. For the items with a reverse score, the percentage represents the sum of 'strongly disagree' and 'disagree' responses

Regarding knowledge of social sustainability, students assigned a lower rating to health-related topics such as infectious diseases (K14) and the role of social sustainability for long and healthy life (K7) compared to other items. This indicates lack of knowledge about the importance of stopping infectious diseases such as HIV/AIDS and malaria, as well as ignorance of ways to improve people's chances for health and wellbeing. Specifically, 36% of participating students do not consider that the eradication of these diseases contributes to SD (Fig. 3).

Finally, regarding economic sustainability, students tend to value the importance of corporate social responsibility (K15). However, one in four students (25%) fails to see poverty reduction to achieve SD (K17) and 22% neglect that SD demands people to understand how the economy functions (K18). These two factors, poverty reduction and economic education, are the least valued by students (Fig. 4).

Fig. 3 Level of agreement with social sustainability topics (Knowledge). *Note* N = 716 HE students. The percentages represent students who 'agree' or 'strongly agree' with the statements

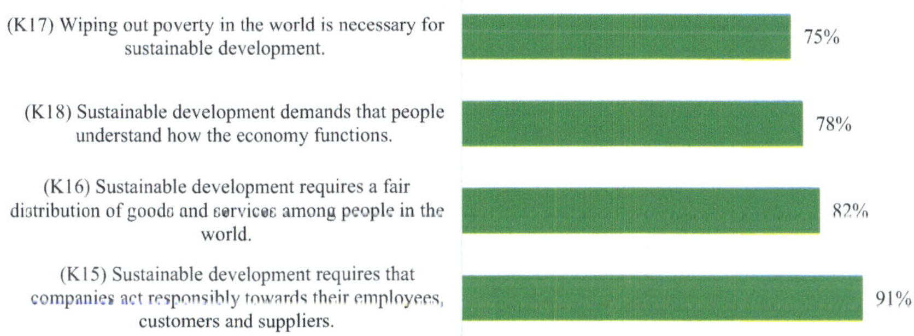

Fig. 4 Level of agreement with economic sustainability topics (Knowledge). *Note* N = 716 HE students. The percentages represent students who 'agree' or 'strongly agree' with the statements

5.1.2 Students' Perceptions of Their Attitudes on SD Issues

Having explored the students' perceptions on their knowledge of SD topics, it is equally important to investigate their attitudes. Students' attitudes towards the use of natural resources and water consumption are two of the areas in the environmental domain that needs the most attention. Only 71% of participants recognise that using more natural resources than necessary will threaten people's health and well-being in the future (71%), while 74% consider it appropriate to use as much water as they want (A4) (Fig. 5).

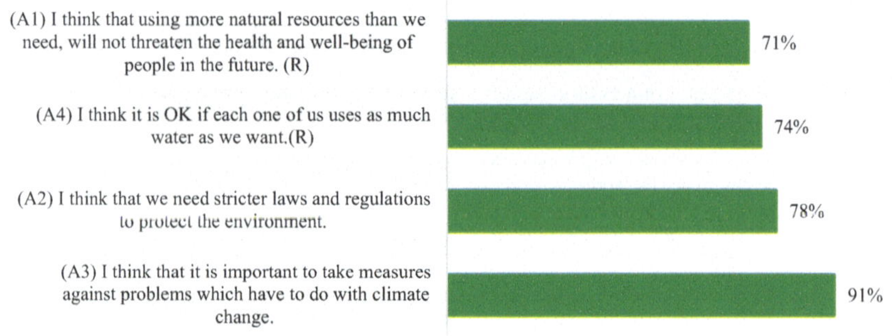

Fig. 5 Level of agreement with environmental sustainability topics (Attitudes). *Note* N = 716 HE students. The percentages represent the percentage of students who 'agree' or 'strongly agree' with the statements. (R) Reverse scored items. For the items with a reverse score, the percentage represents the sum of 'strongly disagree' and 'disagree' responses

When it comes to attitudes towards social sustainability, the role of governments in promoting sustainability received the least support. Only 67% of students think that governments should take SD into account in all their decisions (A8), and only 74% agree that the governments should provide financial support to increase the use of environmentally friendly cars (74%) (Fig. 6).

Overall, students have a positive attitude towards economic sustainability. The question with the lowest level of agreement (76%) is whether wealthy companies should offer the same conditions to employees in poorer countries as those in wealthy countries (A13) (Fig. 7).

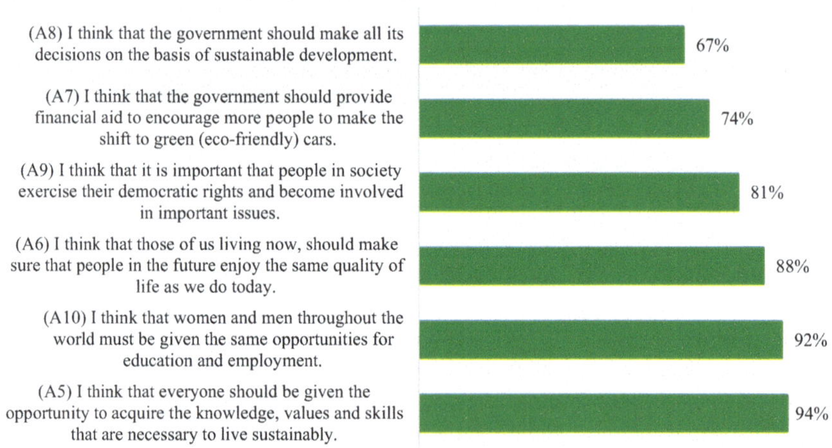

Fig. 6 Level of agreement with social sustainability topics (Attitudes). *Note* N = 716 HE students. The percentages represent the percentage of students who 'agree' or 'strongly agree' with the statements

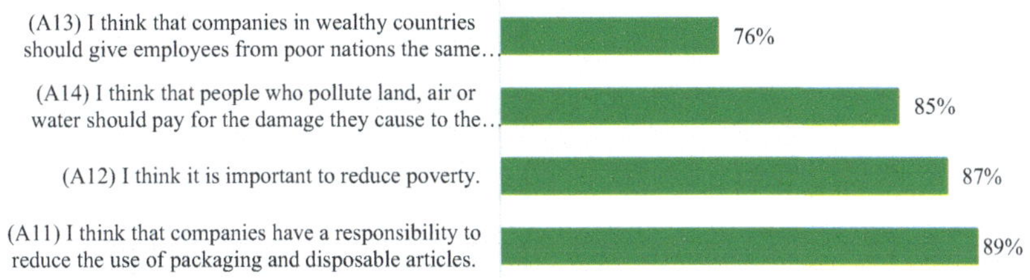

Fig. 7 Level of agreement with economic sustainability topics (Attitudes). *Note* N = 716 HE students. The percentages represent the percentage of students who 'agree' or 'strongly agree' with the statements

5.1.3 Students' Perceptions of Their Behaviour Towards SD Issues

Compared to students' knowledge and attitudes, the study findings show that the greatest gaps exist at the level of students' behaviour towards sustainability. Regarding environmental sustainability, all statements indicate a level of agreement below 75% (Fig. 8).

In the area of social sustainability, two main behaviours were less represented in the student sample (Fig. 9), indicating that only a few students (28%) work on committees (B10) and support an aid organisation/charity or environmental group (45%).

In the economic sustainability domain, the students' actions reflecting their SD behaviours were less under-represented showing levels of agreement below 60% (Fig. 10).

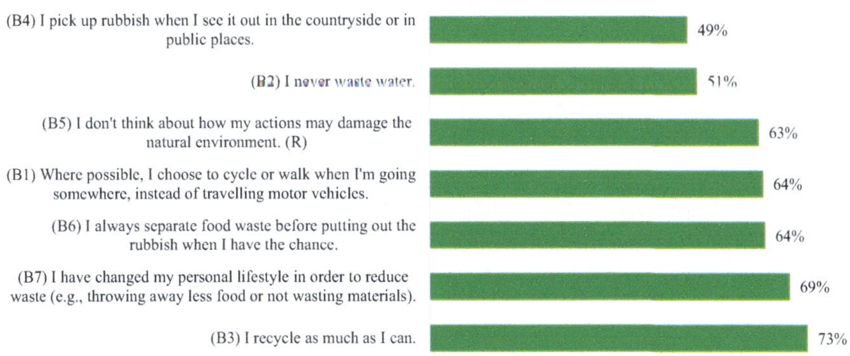

Fig. 8 Level of agreement with environmental sustainability topics (Behaviours). *Note* N = 716 HE students. The percentages represent students who 'agree' or 'strongly agree' with the statements. (R) Reverse score item. For the items with reverse score, the percentage represents the sum of 'strongly disagree' and 'disagree' responses

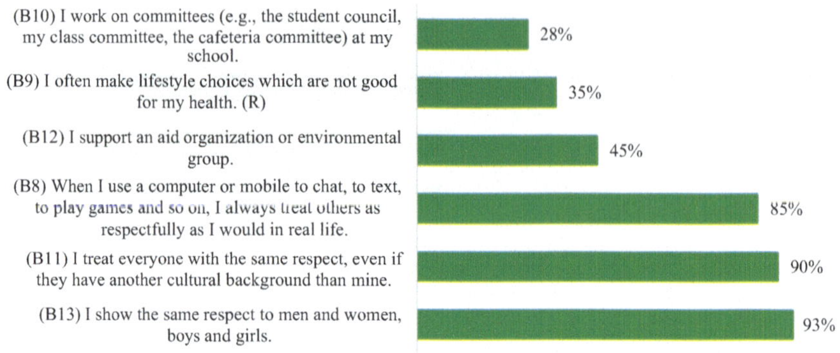

Fig. 9 Level of agreement with social sustainability topics (Behaviours). *Note* N = 716 HE students. The percentages represent students who 'agree' or 'strongly agree' with the statements. (R) Reverse score item. For the items with reverse score, the percentage represents the sum of 'strongly disagree' and 'disagree' responses

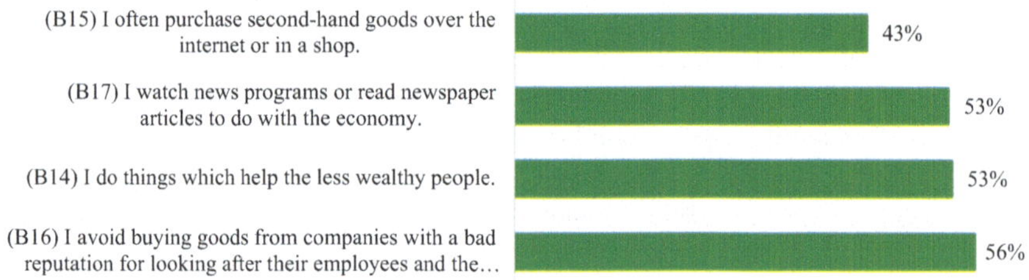

Fig. 10 Level of agreement with economic sustainability topics (Behaviours). *Note* N = 716 HE students. The percentages represent students who 'agree' or 'strongly agree' with the statements

5.1.4 Students' Preferred Methods of Learning About the SDGs

In our study we also wanted to explore the educational methods preferred by students for learning about sustainability (Fig. 11). In order of priority, students reported they prefer to learn via watching documentaries (57%), watching videos (57%), attending seminars (40%), learning through a project (40%), browsing the web for information (33%), playing educational games (31%), and using mobile apps (30%).

5.1.5 Main Gaps Among HE Students Related to SDGs

Following the analysis of students' perceptions, we synthesised the findings to highlight the key gaps in the knowledge, attitudes, and behaviours of HE students in each of the three dimensions of sustainability considering the instrument used), and subsequently we linked these to the relevant SDGs. The outcomes of this synthesis

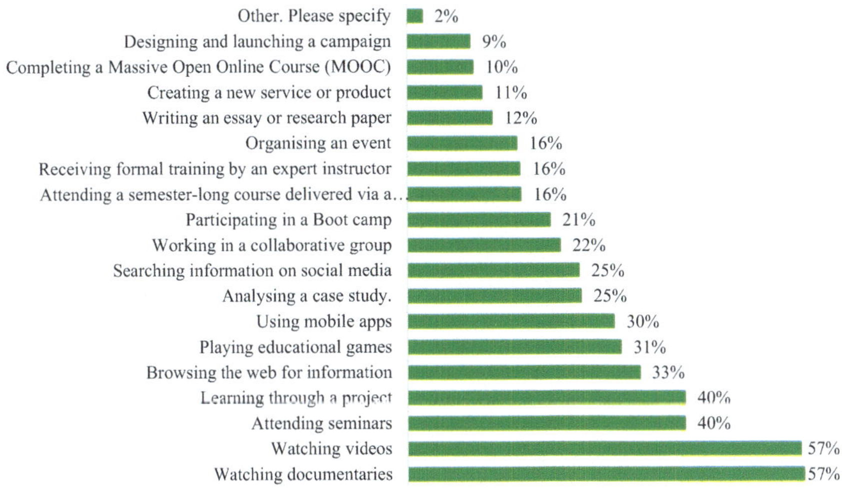

Fig. 11 Students' preferred methods to learn about SDGs. *Note* Optional question. N = 700 students. Each respondent selected the most 6 attractive methods

are illustrated in the following three figures, reflecting environmental, social, and economic sustainability, respectively (Tables 1, 2 and 3).

In summary:

1. **Environmental sustainability gaps**

   Knowledge: Students usually display a lack of awareness about natural resource limitations, pollution impacts, and waste management concepts. Furthermore, there is often a lack of understanding about the importance of saving resources like water and electricity.
   Attitudes: Many students believe that environmental sustainability is the responsibility of the government or industry. Furthermore, the value of biodiversity conservation is frequently underestimated.
   Behaviour: There is a gap between theoretical understanding and practical activity because few students engage in sustainable practices include recycling, water conservation, and environmentally friendly transportation.

2. **Social sustainability gaps**

   Knowledge: Students frequently display a poor understanding of the interrelationships between health and sustainability, as well as a limited awareness of gender and social equality issues.
   Attitudes: The students showed a limited perspective of the government's role on the SD domain.
   Behaviour: Civic engagement and support for social programs are low, indicating a lack of social responsibility and community involvement.

Table 1 Key gaps in knowledge, attitudes, and behaviours of HE students on environmental SD

SD domain	K/A/B	Items with lower level of agreement	Gap identified	SDGs related
Environmental sustainability	Knowledge	For SD, people need to be educated in how to protect themselves against natural disasters	Lack of knowledge about how to protect ourselves against natural disasters	SDG13
	Attitudes	I think that using more natural resources than we need, will not threaten the health and well-being of people in the future	Lack of awareness that the excessive consumption of natural resources is a concern that needs to be addressed to ensure the well-being of future generations	SDG3 SDG12
		I think it is OK if each one of us uses as much water as we want	The prevalence of the idea of unlimited water usage is appealing on an individual level	SDG6
	Behaviours	I pick up rubbish when I see it out in the countryside or in public places	An insufficient number of individuals avoid contributing to pollution by their own actions and help to clean up public and/or natural spaces	SDG6
		I never waste water	Individual behaviours do not reflect the water efficiency use and how this can help mitigate water scarcity	SDG6
		I don't think about how my actions may damage the natural environment	Lack of attention of how their daily activities can have an impact on the degradation of natural resources	SDG12
		Where possible, I choose to cycle or walk when I'm going somewhere, instead of travelling by motor vehicles	Little motivation to replace motorised transport with more health and wellbeing options (e.g. cycling) and less polluting options (clean energy transportation)	SDG3 SDG7 SDG11
		I always separate food waste before putting out the rubbish when I have the chance	Low commitment to prevent and reduce food waste	SDG12

(continued)

Table 1 (continued)

SD domain	K/A/B	Items with lower level of agreement	Gap identified	SDGs related
		I have changed my personal lifestyle in order to reduce waste (e.g., throwing away less food or not wasting materials)	Conscious choices to minimize waste generation is not frequent, particularly by preventing and reducing waste in all materials	SDG12
		I recycle as much as I can	Low motivation to reduce waste generation, especially through recycling and reuse	SDG12

3. **Economic sustainability gaps**

   Knowledge: Students showed lack of economic literacy and insufficient acknowledgment of the role of poverty eradication.
   Attitudes: There is a split in views on corporate social responsibility, as well as a preference for purchasing new goods over used items.
   Behaviour: The lack of participation in economic fairness projects, as well as the inconsistent nature of ethical purchase behaviours.

5.2 *Results of Study 2—HE Teachers*

5.2.1 Level of Familiarity with the SDGs

The level of familiarity with the SDGs, among HE lecturers, varies widely. In fact, 15% of lecturers say they are not at all familiar with the SDGs, and 11% are somewhat familiar. Only 13% feel very familiar with the topic, and 35% feel moderately familiar (Fig. 12). This information underlines that if we want to provide education for SD in HE, lecturers themselves need to be trained in these areas.

5.2.2 SDGs Priorities

Figure 13 shows the SDGs that have been prioritised in the schools or departments where the teachers were lecturing. SDG 4 (Quality Education) is the top priority for teachers, followed by SDG 5 (Gender Equality) and SDG 8 (Decent Work and Economic Growth). In contrast, the least addressed SDGs among teachers in our sample are SDG 14 (Life Below Water), SDG 15 (Life on Land), and SDG 2 (Zero Hunger).

Table 2 Key gaps in knowledge, attitudes, and behaviours of HE students on social SD

SD Domain	K/A/B	Items with lower level of agreement	Gap identified	SDGs related
Social sustainability	Knowledge	For SD, major infectious diseases such as HIV/AIDS and malaria must be stopped	Lack of knowledge about how eliminating infectious diseases such as AIDS or malaria can contribute to sustainable development	SDG3
		Improving people's chances for a long and healthy life contributes to SD	Insufficient knowledge on SDG3 - Ensure healthy lives and promote well-being for all at all ages	SDG3
		Reinforcing girls' and women's rights and increasing equality around the world are necessary for SD	Lack of general knowledge about how to achieve gender equality and empower all women and girls will contribute for SD (SDG5)	SDG5
	Attitudes	I think that the government should make all its decisions on the basis of SD	Lack of understanding of the relevant role that governments can play in sustainable development	SDG16
		I think that the government should provide financial aid to encourage more people to make the shift to green (eco-friendly) cars	Lack of perception of the relevant role governments can play in facilitating access to the use of clean energy equipment (e.g. electric vehicles)	SDG7
	Behaviours	I work on committees (e.g., the student council, my class committee, the cafeteria committee) at my school	Lack of participation in association bodies or other types of committees	SDG4

(continued)

Table 2 (continued)

SD Domain	K/A/B	Items with lower level of agreement	Gap identified	SDGs related
		I often make lifestyle choices which are not good for my health	Lifestyle choices are not prioritising individual's well-being	SDG3
		I support an aid organisation or environmental group	Low levels of individual citizenship	SDG4

Table 3 Key gaps in knowledge, attitudes, and behaviours of HE students on economic SD

SD Domain	K/A/B	Items with lower level of agreement	Gap identified	SDGs related
Economic sustainability	Knowledge	Wiping out poverty in the world is necessary for SD	Low recognition that poverty eradication is a fundamental goal of sustainable development	SDG1
	Attitudes	I think that companies in wealthy countries should give employees from poor nations the same conditions as the employees from wealthy countries	Low awareness of the importance of promoting fairness and equality in the global workforce	SDG8
	Behaviours	I often purchase second-hand goods over the internet or in a shop	Insufficient uptake of buying used goods	SDG12
		I watch news programs or read newspaper articles to do with the economy	Lack of financial literacy skills	SDG4
		I do things which help the less wealthy people	Insufficient support on less wealthy people (e.g. volunteering time, donating resources, offering assistance)	SDG10
		I avoid buying goods from companies with a bad reputation for looking after their employees and the environment	Consumer choices are not grounded on companies' reputation regarding the respect for the employees or environment	SDG12

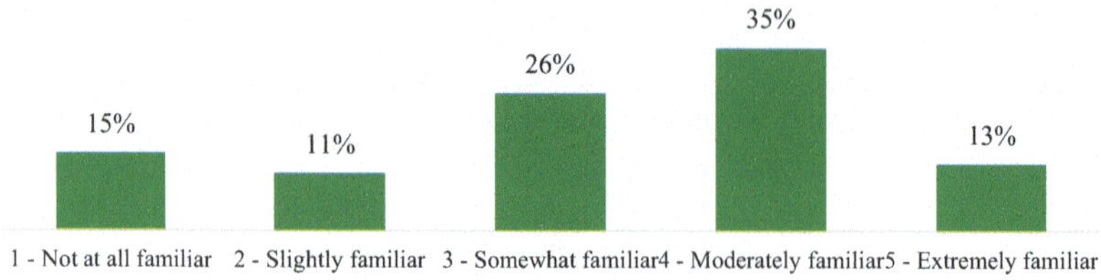

Fig. 12 Teachers' familiarity with the Sustainable Development Goals (SDGs), *Note* N = 209 HE teachers

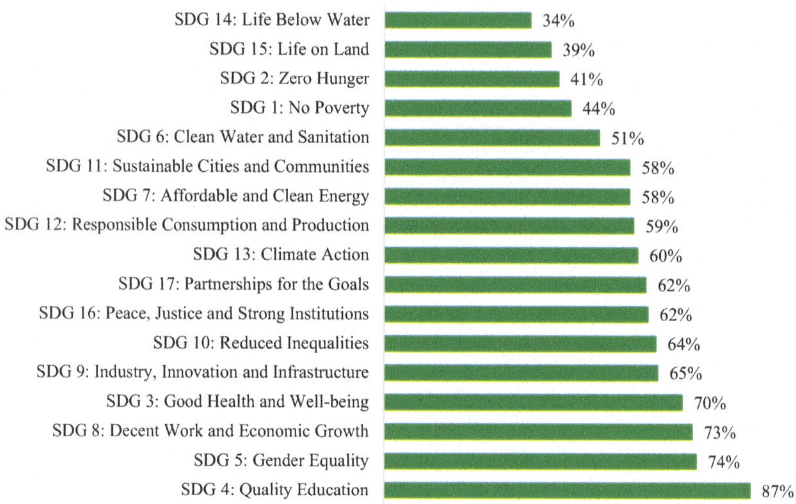

Fig. 13 Priority given to each SDG by higher education departments or schools. *Note* N = 209 HE teachers. Likert scale of 5 points. The percentage shows the sum of the answers '4-moderate priority' and '5-high priority'

5.2.3 SDGs Topics Recently Covered

Figure 14 shows the frequency with which different Education for Sustainable Development (ESD) topics were covered in the last 12 months by the teachers surveyed, in their lectures. Over the past 12 months, teachers have prioritised sustainable development topics related to human health, climate change, and cultural diversity, including intercultural understanding. In contrast, issues such as oceans, land use, and air and atmospheric concerns received less attention.

5.2.4 SDGs Teaching Methods

Lecture-based teaching is the preferred option for teachers to teach SD topics (74% often or always use). Project-based learning comes second, but with only 52% of

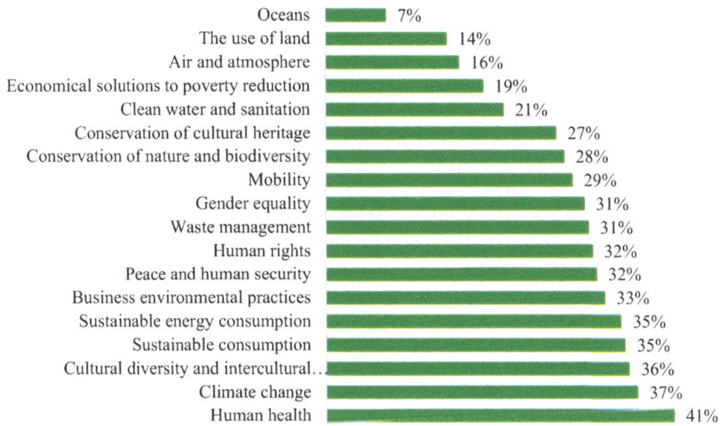

Fig. 14 Regularity of SD issues taught during last 12 months by teachers. *Note* N = 209 HE teachers. Likert scale of 5 points. The percentage shows the sum of the answers '4-often' and '5-always'

fairly frequent use. The least used options are outdoor activities (12%) and service learning (13%) (Fig. 15).

The areas in which most teachers feel the need to develop their knowledge and skills in relation to SD are listed in Fig. 16. Most teachers express a need to enhance their knowledge and skills in areas such as developing project-based

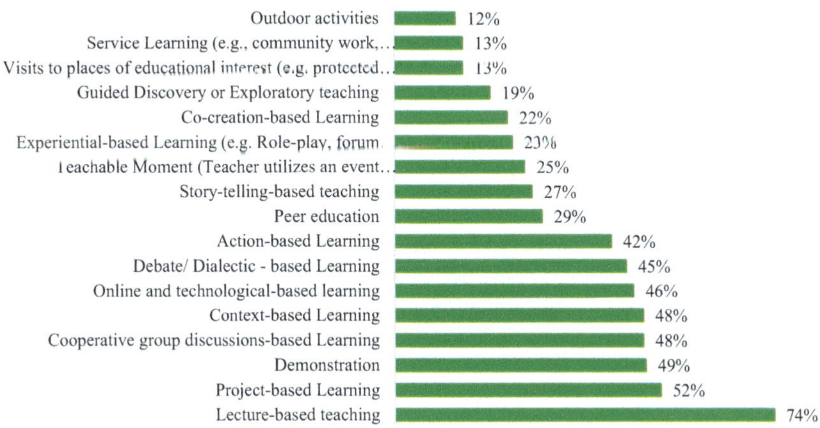

Fig. 15 Regularity of methods used in education for Sustainability Development, during last 12 months by teachers. *Note* N = 209 HE teachers. Likert scale of 5 points. An option of 'I do not know this approach' was also available. The percentage shows the sum of the answers '4-often' and '5-always'

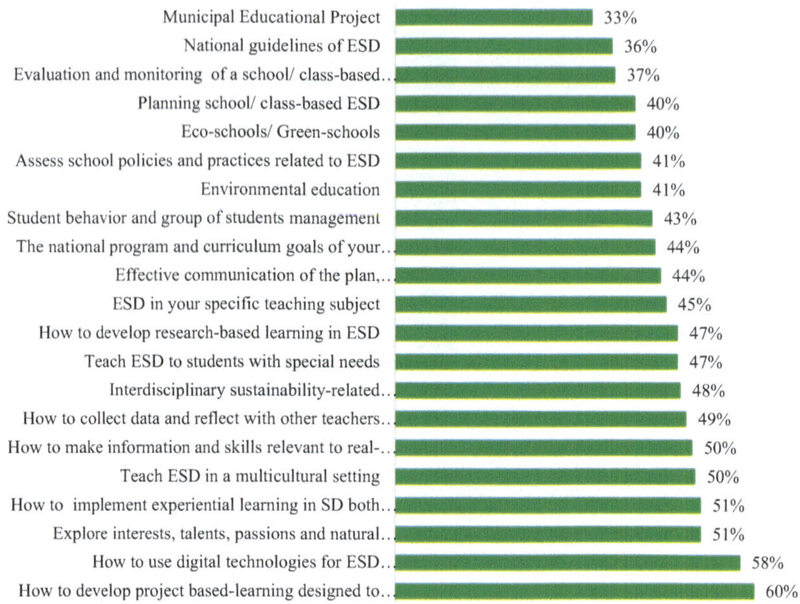

Fig. 16 Necessity of improving the knowledge/competence about ESD by teachers. *Note* N = 209 HE teachers. Likert scale of 5 points. The percentage shows the sum of the answers '4-to a large extent' and '5-to a great extent'

learning, integrating digital technologies into Education for Sustainable Development (ESD), fostering students' interests, talents, passions, and natural curiosity, and implementing experiential learning approaches in sustainability education.

5.2.5 SDGs Competences Developed with Students by Teachers

In the subjects taught by the teachers surveyed, the most developed SD competences are interpersonal competences (75%) and strategic planning competences (72%) (Fig. 17). The least developed is anticipatory competence (48%), system thinking (55%) and intrapersonal competence (55%).

5.2.6 Main Gaps for HE Teachers

Although most of the faculty report that changes have been made to improve sustainability practices at their institution, we can see that 15% are not at all familiar with the SDGs, 11% are slightly familiar, and 26% are somewhat familiar, meaning that more than 50% believe they lack knowledge about the SDGs.

When confronted with the priority given to each SDG in their institution, it is known that SDG4 (quality education) is the more addressed one, followed by SDG5

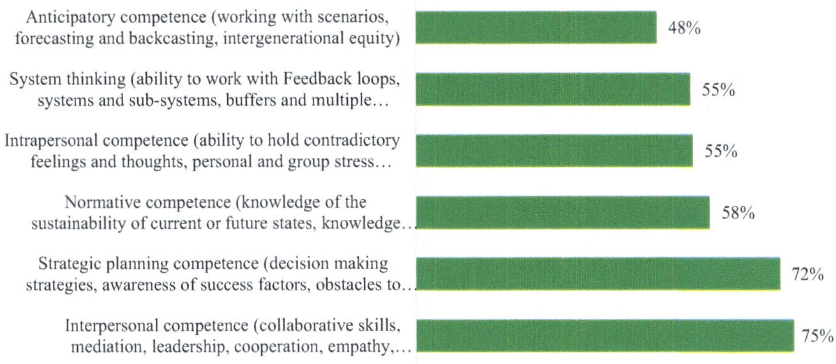

Fig. 17 Competences more often developed with students by teachers. *Note* N = 209 HE teachers. Likert scale of 5 points. The percentage shows the sum of the answers '4-to a large extent' and '5-to a great extent'

(gender equality) and SDG8 (decent work and economic growth). It seems that there is a deeper concern related to social and economic aspects, while environmental issues seem to be the less prioritised ones, with SDG14 and SDG15 at the bottom of the priorities.

In terms of frequency, teachers indicate that the issues or problems they address less in their teaching practices are related to environmental aspects such as oceans, land use and air and atmosphere. Some societal aspects are also low on the list, such as economic solutions to poverty and clean water and sanitation. However, these results should be cross-checked with the area of teaching of the teachers.

The main gaps in the teaching–learning process are related to the use of active and innovative methods for teaching SD and SDGs. In their practice, most teachers use lecture-based teaching (74%), with the more practical and experimental methods being used less.

Teachers may need training to develop their competences in active and innovative methods, as well as to increase their confidence in using active, student-centred practices.

Some approaches in ESD are used by less than 25% of the teachers, in this sense outdoor activities, service learning, field trips, guided discovery or exploratory teaching, co-creation based learning and experiential learning need more investment.

At the same time, teachers also show a large gap in the use of digital technologies to address the topic of SD and SDGs. This may be due to their lack of competence in the use of digital technologies or, if they know how to use technologies, how to use them in these specific topics.

The areas in which teachers feel they need to develop their knowledge and skills related to SD are:

- How to develop project-based learning that addresses real-world problems and issues;

- How digital technologies for ESD can be used purposefully by students to create, develop and share their projects;
- How to explore students' interests, talents, passions and natural curiosity, considering their initial knowledge of SD;
- How to implement experiential learning in SD both inside and outside of the classroom.

In the subjects taught by the teachers surveyed, the SD competences that are less often developed by teachers with students are:

- anticipatory competence (working with scenarios, forecasting and backcasting, intergenerational equity);
- systems thinking (ability to work with feedback loops, systems and subsystems, buffers and multiple variables, nested scales, resilience, and tipping points);
- intrapersonal competence (ability to hold contradictory feelings and thoughts, personal and group stress management cultivating awareness, finding inner peace, compassion, meaning making, experiencing love and connection).

Nevertheless, all SD competencies are equally important and should be equally developed in higher education.

6 Discussion

The study identified critical differences in the sustainability knowledge, attitudes, and behaviours of HE students and teachers, highlighting areas where each group needs support to effectively promote SDGs within academic contexts. Although both groups have a fundamental understanding of sustainability, there are significant differences in their specific levels of familiarity, involvement, and pedagogical approaches, as well as the institutional challenges they face in establishing a sustainable culture.

1. **Understanding of Sustainability and SDGs**

Students: The survey found that while many students reported having a basic understanding of the SDGs, they have considerable gaps in their understanding of specific sustainability challenges, including resource conservation, biodiversity, and ethical economic practices. Students tended to understand the wider principles of sustainability, but struggled to apply this knowledge to specific situations, particularly those concerning environmental and economic sustainability. For example, just a small proportion are aware of water conservation as a key component of sustainable development, indicating the need for more targeted training on sustainability's intricate aspects.

Teachers: The findings indicate that teachers have varying levels of familiarity with the SDGs, with only 13% reporting a good level of awareness and a significant percentage (26%) claiming low knowledge. Although educators are often aware of SDGs relevant to their field, particularly SDGs 4 (Quality Education) and 5 (Gender

Equality), they are often unfamiliar with environmental and economic SDGs such as sustainable land use (SDG 15) and ocean protection (SDG 14). This knowledge gap may hinder their ability to integrate comprehensive sustainability knowledge into their curriculum.

2. **Attitudes Towards Sustainability**

Students: Students generally have positive attitudes regarding sustainability, emphasising the need of environmental stewardship and social equality. However, their comments highlight significant gaps in personal accountability and civic engagement. Many students view sustainability largely as the responsibility of governments or institutions, rather than recognising their own potential contribution. For example, while most students support environmental protection, only about two-thirds believe that government decisions should continuously promote sustainability principles, and even fewer support stricter regulations on water use. This result is markedly different from that obtained by Franco et al. (2022), which found that students favoured stronger government intervention to combat climate change. These findings point to a gap between students' theoretical support for sustainability and their acceptance of individual or community responsibility.

Teachers: In contrast, teachers have positive attitudes towards sustainability and generally support its incorporation into educational frameworks. Despite the positive outlook, teachers cite challenges with limited familiarity with specific sustainability subjects and limited resources make it difficult to provide complete sustainability education. This gap implies that, while educators realise the importance of sustainability, they face structural constraints that prevent successful implementation in within their courses, demonstrating attitudinal alignment with sustainability principles but a lack of institutional support to make them a reality.

3. **Engaging in Sustainable Practices and Behaviours**

Students: The study findings reveal a significant gap between students' understanding of sustainability issues and their engagement in related behaviours. Although students express positive feelings, their actual behaviours—such as waste reduction, recycling, and civic or community involvement—are limited. Only about half of students report consistently conserving water, and even fewer participate in environmental activities or community service. These findings align with Inner Development Goals (IDGs), which emphasize the importance of cultivating personal and interpersonal skills to drive sustainable action (Nordén, 2024). While students theoretically embrace sustainability, they may lack self-awareness, motivation, and the cognitive skills needed to integrate these principles into daily life.

Teachers: The findings also reveal that teachers' sustainability knowledge and actions appear disconnected, particularly in terms of educational techniques. Despite recognising the potential benefits of interactive and experience learning, most teachers continue to use traditional lecture-based methods. Only about half of the surveyed teachers use methods that can actively involve students, such as project-based learning, and less than 15% use service learning, which is critical for real-world sustainability education. This predilection for traditional approaches implies that many educators may lack the resources or experience to successfully switch to more active learning models, which are necessary for nurturing students' systematic thinking and practical problem-solving abilities.

4. **Preferred Learning and Teaching Methods**

Students: According to the survey results, students prefer interactive and visually engaging learning techniques such as documentaries, movies, and project-based learning, which are consistent with their desire for hands-on sustainability experiences. These preferences indicate that students value opportunities to learn about sustainability in practical settings, which can reinforce their knowledge and inspire behaviour change. This preference also aligns with the IDGs, which emphasize the need for cognitive skills, collaboration and action-oriented learning to foster sustainable development. Innovative learning environments also promote critical thinking, experiential learning and emotional engagement and are therefore essential for developing the capacities needed for sustainable action. In turn, the lower preference for traditional teaching methods, such as lectures or theoretical assignments, suggests that students would benefit from a shift to interactive and experiential learning approaches that promote deeper reflection, engagement and a stronger connection to real-world sustainability challenges (Nordén, 2024).

Teachers: On the other hand, teachers indicate a preference for lecture-based instruction, with less use of experiential techniques such as fieldwork or service learning. This result aligns with the findings reported by Lozano et al. (2022). Although some teachers have used project-based learning, their overall teaching methods palette remains consistent with traditional models, suggesting a potential mismatch between instructional practices and student preferences. This discrepancy highlights the need for training and continuous professional growth in interactive teaching practices that better align with students' interests and promote more effective engagement and deeper involvement with environmental, social, and economic SD issues.

7 Conclusions

This study identifies critical gaps in sustainability knowledge, attitudes, and behaviours among higher education students and teachers. Despite widespread awareness of sustainability challenges, there is a striking lack of practical understanding and involvement, highlighting the need for a comprehensive reform of sustainability education. To address these issues, HE should prioritise the creation of specialised curricula that smoothly combine theoretical knowledge with real-world applications. This shift should take place away from traditional lecture-based education and towards experiential, interactive learning approaches that encourage critical thinking, systems analysis, and problem-solving abilities needed to address complex sustainability issues. However, it remains essential to teach students that the SDGs are interconnected, as a positive impact on one SDG can sometimes have a significantly negative effect on another. Therefore, it is crucial to encourage students to adopt sustainable, ethical, and responsible behaviours, enabling them to make informed decisions that balance different sustainability goals and minimize unintended consequences. By investing in teacher training, strategically integrating sustainability knowledge and skills across the curriculum, and launching programmes that promote civic engagement and adherence to the SDGs, HE can play a transformative role in preparing students to demonstrate social intelligence and become active agents and engaged citizens capable of addressing global sustainability challenges.

Funding This work was supported by the European Union Erasmus+programme (grant number 2022-1-PT01-KA220-HED-000087984) and the Life Quality Research Centre by Fundação para a Ciência e Tecnologia (grant number UID/CED/04748/2020).

References

Abad-Segura, E., & González-Zamar, M.-D. (2021). Sustainable economic development in higher education institutions: A global analysis within the SDGs framework. *Journal of Cleaner Production, 294*, 126133. https://doi.org/10.1016/j.jclepro.2021.126133

Acosta Castellanos, P. M., & Queiruga-Dios, A. (2022). From environmental education to education for sustainable development in higher education: A systematic review. *International Journal of Sustainability in Higher Education, 23*(3), 622–644. https://doi.org/10.1108/IJSHE-04-2021-0167

Anthony Jnr, B. (2021). Green campus paradigms for sustainability attainment in higher education institutions—A comparative study. *Journal of Science and Technology Policy Management, 12*(1), 117–148. https://doi.org/10.1108/JSTPM-02-2019-0008

Álvarez-Vanegas, A., Ramani, S. V., & Volante, L. (2024). Service-Learning as a niche innovation in higher education for sustainability. *Frontiers in Education, 9*. https://doi.org/10.3389/feduc.2024.1291669

Backman, M., Pitt, H., Marsden, T., Mehmood, A., & Mathijs, E. (2019). Experiential approaches to sustainability education: Towards learning landscapes. *International Journal of Sustainability in Higher Education, 20*(1), 139–156. https://doi.org/10.1108/ijshe-06-2018-0109

Bautista-Puig, N., & Sanz-Casado, E. (2021). Sustainability practices in Spanish higher education institutions: An overview of status and implementation. *Journal of Cleaner Production, 295*, 126320. https://doi.org/10.1016/j.jclepro.2021.126320

Birdman, J., Wick, A., & Lang, D. J. (2022). Developing key competencies in sustainability through project-based learning in graduate sustainability programs. *International Journal of Sustainability in Higher Education, 23*(5), 1139–1157. https://doi.org/10.1108/ijshe-12-2020-0506

Brundiers, K., Barth, M., Cebrián, G., Cohen, M., Diaz, L., Doucette-Remington, S., & Zint, M. (2021). Key competencies in sustainability in higher education—Toward an agreed-upon reference framework. *Sustainability Science, 16*(1), 13–29. https://doi.org/10.1007/s11625-020-00838-2

Fia, M., Ghasemzadeh, K., & Paletta, A. (2023). How higher education institutions walk their talk on the 2030 agenda: a systematic literature review. *Higher Education Policy, 36*(3), 599–632. https://doi.org/10.1057/s41307-022-00277-x

Franco, D., Macke, J., Cotton, D., Paço, A., Segers, J.-P., & Franco, L. (2022). Student energy-saving in higher education tackling the challenge of decarbonisation. *International Journal of Sustainability in Higher Education, 23*(7), 1648–1666. https://doi.org/10.1108/ijshe-10-2021-0432

Galvão, A., Mascarenhas, C., Marques, C., Ferreira, J., & Ratten, V. (2019). Triple helix and its evolution: A systematic literature review. *Journal of Science and Technology Policy Management, 10*(3), 812–833. https://doi.org/10.1108/JSTPM-10-2018-0103

Gericke, N., Boeve-de Pauw, J., Berglund, T., & Olsson, D. (2019). The sustainability consciousness questionnaire: The theoretical development and empirical validation of an evaluation instrument for stakeholders working with sustainable development. *Sustainable Development, 27*(1), 35–49. https://doi.org/10.1002/sd.1859

Heinrich, W. F., Habron, G. B., Johnson, H. L., & Goralnik, L. (2015). Critical thinking assessment across four sustainability-related experiential learning settings. *Journal of Experiential Education, 38*(4), 373–393.

Howlett, C., Ferreira, J.-A., & Blomfield, J. (2016). Teaching sustainable development in higher education. *International Journal of Sustainability in Higher Education, 17*(3), 305–321. https://doi.org/10.1108/IJSHE-07-2014-0102

Kamil, P. A., Putri, E., Ridha, S., Utaya, S., Sumarmi, & Utomo, D. H. (2020). Promoting environmental literacy through a green project: a case study at adiwiyata school in Banda Aceh City. *IOP Conference Series: Earth and Environmental Science, 485*(1), 012035. https://doi.org/10.1088/1755-1315/485/1/012035

Ketikidis, P., & Solomon, A. (2018). Co-producing knowledge, innovation and growth through university–industry collaboration—Lessons from experience. *Industry and Higher Education, 32*(4), 211–212. https://doi.org/10.1177/0950422218786772

Kopnina, H. (2020). Education for the future? Critical evaluation of education for sustainable development goals. *The Journal of Environmental Education, 51*(4), 280–291. https://doi.org/10.1080/00958964.2019.1710444

Kumari, R., Kwon, K.-S., Lee, B.-H., & Choi, K. (2019). Co-creation for social innovation in the ecosystem context: The role of higher educational institutions. *Sustainability, 12*(1), 307. https://doi.org/10.3390/su12010307

Leal Filho, W., Shiel, C., & Paço, A. D. (2015). Integrative approaches to environmental sustainability at universities: an overview of challenges and priorities. *Journal of Integrative Environmental Sciences, 12*(1), 1–14. https://doi.org/10.1080/1943815x.2014.988273

Leal Filho, W., Amaro, N., Avila, L. V., Brandli, L., Damke, L. I., Vasconcelos, C. R. P., & Salvia, A. (2021). Mapping sustainability initiatives in higher education institutions in Latin America. *Journal of Cleaner Production, 315*, 128093. https://doi.org/10.1016/j.jclepro.2021.128093

Leal Filho, W., Shiel, C., & Paço, A. (2016). Implementing and operationalising integrative approaches to sustainability in higher education: The role of project-oriented learning. *Journal of Cleaner Production, 133*, 126–135. https://doi.org/10.1016/j.jclepro.2016.05.079

Leal Filho, W., Raath, S., Lazzarini, B., Vargas, V. R., de Souza, L., Anholon, R., Quelhas, O. L. G., Haddad, R., Klavins, M., &Orlovic, V. L. (2018). The role of transformation in learning and education for sustainability. *Journal of Cleaner Production, 199*, 286–295. https://doi.org/10.1016/j.jclepro.2018.07.017

Lozano, R., Barreiro-Gen, M., Pietikäinen, J., Gago-Cortes, C., Favi, C., Jimenez Munguia, M. T., & Gladysz, B. (2022). Adopting sustainability competence-based education in academic disciplines: Insights from 13 higher education institutions. *Sustainable Development*, *30*(4), 620–635. https://doi.org/10.1002/sd.2253

Martins, F. P., Cezarino, L. O., Liboni, L. B., Botelho Junior, A. B., & Hunter, T. (2022). Interdisciplinarity-based sustainability framework for management education. *Sustainability, 14*(19), 12289. https://doi.org/10.3390/su141912289

Mikelsone, E., Babris, M., Segers, J.-P., Beitane, A., & Babre, A. M. (2024). Web-based idea management and quadruple helix networking of creative industry cohorts for covid-19. *Business: Theory and Practice*, *25*(1), 362–376. https://doi.org/10.3846/btp.2024.18538

Miller, K., McAdam, R., & McAdam, M. (2018). A systematic literature review of university technology transfer from a quadruple helix perspective: Toward a research agenda. *R&D Management, 48*(1), 7–24. https://doi.org/10.1111/radm.12228

Miranda, J., Navarrete, C., Noguez, J., Molina-Espinosa, J.-M., Ramírez-Montoya, M.-S., Navarro-Tuch, S. A., & Molina, A. (2021). The core components of education 4.0 in higher education: Three case studies in engineering education. *Computers & Electrical Engineering, 93*, 107278. https://doi.org/10.1016/j.compeleceng.2021.107278

Mokski, E., Leal Filho, W., Sehnem, S., & Andrade Guerra, J. B. S. O. D. (2023). Education for sustainable development in higher education institutions: An approach for effective interdisciplinarity. *International Journal of Sustainability in Higher Education, 24*(1), 96–117. https://doi.org/10.1108/ijshe-07-2021-0306

Nordén, B. (2024). Advancing sustainability through higher education: student teachers integrate inner development goals (IDG) and future-oriented methodologies. *Challenges, 15*(2), 28. https://doi.org/10.3390/challe15020028

Nørgård, R. T., & Holflod, K. (2025). Meeting in the middle: Cultural co-creation, transformative partnerships, and ecosystems for public good. *Educational Philosophy and Theory, 57*(2), 112–127. https://doi.org/10.1080/00131857.2024.2384722

Roman, M., Varga, H., Cvijanovic, V., & Reid, A. (2020). Quadruple helix models for sustainable regional innovation: Engaging and facilitating civil society participation. *Economies, 8*(2), 48. https://doi.org/10.3390/economies8020048

Sinakou, E., Donche, V., Boeve-De Pauw, J., & Van Petegem, P. (2019). Designing powerful learning environments in education for sustainable development: A conceptual framework. *Sustainability, 11*(21), 5994. https://doi.org/10.3390/su11215994

Singha, R., & Singha, S. (2024). Application of experiential, inquiry-based, problem-based, and project-based learning in sustainable education. In C. Goi (Ed.), *Teaching and learning for a sustainable future: Innovative strategies and best practices* (pp. 109–128). IGI Global Scientific Publishing. https://doi.org/10.4018/978-1-6684-9859-0.ch006

Weiss, M., Barth, M., Wiek, A., & Von Wehrden, H. (2021). Drivers and Barriers of implementing sustainability curricula in higher education—Assumptions and evidence. *Higher Education Studies, 11*(2), 42. https://doi.org/10.5539/hes.v11n2p42

Wyness, L., & Dalton, F. (2018). The value of problem-based learning in learning for sustainability: Undergraduate accounting student perspectives. *Journal of Accounting Education, 45*, 1–19. https://doi.org/10.1016/j.jaccedu.2018.09.001

Žalėnienė, I., & Pereira, P. (2021). Higher education for sustainability: A global perspective. *Geography and Sustainability,2*(2), 99–106. https://doi.org/10.1016/j.geosus.2021.05.001

Zhang, C., Khan, I., Dagar, V., Saeed, A., & Zafar, M. W. (2022). Environmental impact of information and communication technology: Unveiling the role of education in developing countries. *Technological Forecasting and Social Change, 178*, 121570. https://doi.org/10.1016/j.techfore.2022.121570

Zinn, S., & Isenmann, R. (2017). Munich university of applied sciences towards a "whole institution approach"—Illustrations along x-disciplinary perspectives to education for sustainable development. In W. Leal Filho, C. Skanavis, A. do Paço, J. Rogers, O. Kuznetsova, & P. Castro (Eds.), *Handbook of theory and practice of sustainable development in higher education* (Vol. 2, pp. 215–230). Springer International Publishing. https://doi.org/10.1007/978-3-319-47889-0_16

Insights of the Analysis of the Integration of Education for Sustainable Development with Employability Perspective in Higher Education. A Historical Review

Olatz Zinkunegi-Goitia, Itziar Rekalde-Rodríguez, and Estibaliz Sáez de Cámara

Abstract Universities play a crucial role in preparing students to face the labour market and contribute positively to a society oriented towards sustainability. Integrating Education for Sustainable Development (ESD) with the employability dimension in the curriculum is essential for the development of knowledge, skills and values that enable them access to a work and to face twenty-first century's challenges. This study conducts a historical review of the conceptualization of employability, particularly within the approach of ESD in higher education, to identify the challenges and opportunities arising from this integration. A thorough literature review has been conducted, covering worldwide publications. The data extraction process involved organizing the common themes, followed by analysis and interpretation of the data to determine the findings. The review revealed diverse approaches to conceptualizing employability, with a notable absence of an integrated model that incorporates several perspectives. While the importance of developing employability within ESD and, at the same time, the effectiveness of ESD in the development of crucial skills and values are frequently highlighted, there are very few studies specifically addressing it. Researching employability within ESD in higher education is crucial for developing frameworks, ultimately enhancing graduates' readiness to contribute to sustainable societal progress.

Keywords Education for Sustainable Development (ESD) · Employability · Higher education · Historical review · twenty-first century challenges

O. Zinkunegi-Goitia (✉) · I. Rekalde-Rodríguez
Faculty of Education, Philosophy and Anthropology. Department of Didactics and School Organisation, University of the Basque Country (UPV/EHU), Bilbao, Spain
e-mail: olatz.zinkunegi@ehu.eus

E. S. de Cámara
Director of Sustainability. Vice-Rectorate for Scientific and Social Development and Knowledge Transfer, University of the Basque Country (UPV/EHU), Bilbao, Spain

W. Leal Filho et al. (eds.), *Introducing the UN SDGs in Higher Education Institutions*, World Sustainability Series, https://doi.org/10.1007/978-3-032-23017-1_25

1 Introduction

Higher education institutions have a responsibility to actively contribute to the creation of a more sustainable society, which requires integrating Education for Sustainable Development (ESD) into their curricula and strategic planning (Ferrer-Balas et al., 2010; Sterling, 2021). In this regard, various strategies have been implemented to promote this integration, creating more academic and training environments where ESD is the foundation and central focus. This trend reflects Higher Education Institution's (HEI) commitment to sustainability from a broad perspective and not only from an environmental view (Sedlacek, 2013).

At the same time, HEIs are tasked with preparing their students to face the challenges of their future professional lives, equipping them not only with technical skills but also with soft skills that support their personal and professional development. Employability has gained increasing importance over time. Thus, it has become a strategic goal for public, private, and international organizations (Vázquez-Rodríguez, 2020). In a constantly evolving job market, it is essential that students are ready to actively contribute to society through their respective professional discipline. To achieve this, it is necessary to enhance their employability so that they can advance in the labor market, building professional networks, and accessing the resources necessary for their professional growth (Clarke, 2018; Williams et al., 2016).

Given HEIs' dual commitment—both to sustainability and to professional preparation—it is crucial to investigate the role that ESD approach play in the development of student employability (Jelonek & Urbaniec, 2019). This approach can offer new insights into how to shape professionals who are not only competent but also conscious and committed to social transformation.

2 Objectives and Methodology

The primary aim of this research is to analyze the conceptualization of employability, particularly within the approach of Education for Sustainable Development (ESD) in higher education, to identify the associated challenges and opportunities. To achieve this, the study focuses on the following three specific objectives:

(a) Examine the historical evolution of the concept of employability, providing insights into how it has developed over time.
(b) Analyze the integration of employability within ESD approach as presented in the scientific literature.
(c) Identify the challenges associated with integrating employability into ESD, as well as the opportunities that arise from this integration.

The methodology employed in this study is based on a historical review. As noted by Bögenhold (2020), historical research serves as a methodological tool to deepen our understanding of the processes inherent in the sciences and their evolution. In this context, the review will focus on how the conceptualization of employability has changed over time, which is crucial for contextualizing the present research approach and determining if new perspectives or approaches might be necessary.

To conduct this review, databases of high quality scientific papers such as Web of Science and Scopus were utilized. The search focused on relevant terms including "employability," "higher education," "education for sustainable development," "sustainability," and "sustainable development." This approach ensures a comprehensive examination of the literature, allowing for a thorough understanding of the relationship between employability and ESD in higher education.

3 Results

3.1 The Historical Evolution of the Term "Employability"

At an international level, the concept of employability is not new. Employability appeared in debates about unemployment and the labor market a century ago (De Grip et al., 2004). Despite the existence of various theories, there is still no consensus on the origin of the concept of employability (Barrado, 2023).

Reflections and debates around this concept began a century ago (McQuaid & Lindsay, 2005). At that time, there was a distinction between people who needed help finding work (the unemployed) and those who were actively seeking work (the employable) (Gazier, 1998). However, the concept has gained much more strength in recent decades, leading to a considerable amount of literature on the topic (McQuaid & Lindsay, 2005). Nonetheless, it is accepted that the term originates from the English word "employability," a combination of two words: "employ" (employment) and "ability" (capability) (Barrado, 2023).

Many authors have studied the evolution of the concept of employability, attempting to identify its origin and analyze how its meaning has changed over time (Barrado, 2023). Although the concept of employability has been on the horizon since the 1990s (Lefresne, 1999).The first publications where the term appeared were from the 1950s. At that time, employability responded to economic objectives as there was a high demand for labor, and policies were put in place to help unemployed people obtain full-time jobs (Forrier & Sels, 2003). The concept of full employment was understood as employment for everyone and for life (Creighton, 2007). For example, Feintuch (1955) believed that employability was decisive for securing a job. He also argued that various factors, such as attitude toward work and self-image, had to be considered in developing employability. Public institutions, on a broader scale, took measures to facilitate the integration of workers into the labor market, launching the first policies aimed at promoting employability.

Versloot and his colleagues (1998) also point to the 1950s as the starting point for employability-related publications, but from a different perspective. According to them, the concept arose from concerns about the situation of people without protection in the labor market. Lefresne (1999) places the origin of the concept in medical-social studies conducted in the U.S. during the 1960s, which aimed to distinguish between the "able," the "sick," and the "delinquents." A decade later, the same tests were used to assess the employability of socially disadvantaged groups (Barrado, 2023; Suárez, 2016).

The economic situation changed in the 1970s, and so did the focus on employability, taking on a different perspective. The main goal of policies aimed at increasing employability was to help people get jobs. However, as unemployment grew, the focus shifted to knowledge and skills rather than fixed job-seeking attitudes (Feintuch, 1955). In the 1980s, the attention shifted away from an individual focus on employability to that of organizations and companies. Businesses faced constant challenges and changes during this period, and literature focused on ways to address these challenges. Employability found its place in this discussion, with a new perspective emerging. In this moment, employability was seen as a means of achieving flexibility within organizations, with emphasis on flexibility among workers. Employability was no longer viewed as merely a tool for the labor market but also as a tool for human resources management, helping to optimize employee development within companies (Hiltrop, 1995).

In the 1990s, employability was once again seen as a tool for the labor market, but the focus had shifted since the 1960s and 1970s. Employability was not only useful for the unemployed but also for the entire active population. It was understood as a process that ensured career opportunities both inside and outside of organizations, rather than just a means to access the labor market. Thus, the literature shifted its focus back to the individual, moving away from organizations (Hiltrop, 1995).

As time has passed, especially in the 21st Century, employability has become increasingly important and has been identified as a strategic objective by public, private, and international organizations (Vázquez-Rodríguez, 2020).

Employability was recognized as a key element, especially from the moment the Bologna Process and the European Higher Education Area (EHEA) were established. Consequently, the following points outline the major international declarations and appearances of employability related to the EHEA:

- **Bologna Declaration (1999)**: The goal of the Bologna Process is defined as "promoting the employability of European citizens and the international competitiveness of the higher education system" (European Ministers of Education, 1999).
- **Prague Declaration (2001)**: Ministers acknowledged the contributions made by academic programs to achieving academic quality and sustainable employability and requested higher education institutions to maintain a proactive role (European Ministers of Education, 2001).
- **Berlin Declaration (2003)**: Ministers stressed the need to guarantee a fundamental period of study abroad to help students achieve their European identity, citizenship, and employability (European Ministers of Education, 2003).

- **Bergen Declaration (2005)**: Ministers asked universities to ensure their doctoral programs promoted interdisciplinary training and the development of cross-disciplinary skills to meet the broad needs of the labor market (European Ministers of Education, 2005).
- **London Declaration (2007)**: Ministers affirmed that employability is one of the main goals of the Bologna Process (European Ministers of Education, 2007).
- **Leuven/Louvain-la Neuve Declaration (2009)**: Ministers emphasized that "employability enables individuals to fully seize the opportunities in changing labor markets" (European Ministers of Education, 2009).
- **Budapest/Vienna Declaration (2010)**: Ministers launched the EHEA and committed to implementing the agreed objectives and agenda for the next decade (European Ministers of Education, 2010).
- **Bucharest Declaration (2012)**: Ministers committed to improving graduates' employability and personal and professional development throughout their careers(European Ministers of Education, 2012).
- **Yerevan Declaration (2015)**: Ministers declared employability a priority for the EHEA (European Ministers of Education, 2015).

Employability has increasingly gained importance in academic and political discourses, being identified as a strategic objective by many institutions. However, some experts have criticized the use of employability as a fashionable and "utilitarian" term, especially when it is referenced in discussions without focusing on real solutions to labor problems (Suárez, 2016; Verhaar & Smulders, 1999). Recently, it has been argued that the responsibility for employment is being shifted away from institutions and toward individuals (Brown et al., 2003; Sin et al., 2016; Suárez, 2016).

Gazier (1998) and Weinert et al. (2001) identified different stages in the conceptualization of employability during the twentieth century. According to Suárez (2016), these stages are (Table 1):

Gazier (1998) suggests that the evolution of the concept of employability has passed through three phases. The first phase emerged in the early twentieth century with the dichotomic understanding of employability. The second phase, beginning around the 1960s, saw three different approaches emerge: socio-medical, manpower policy, and flow employability. The third phase, developing in the 1980s and 1990s, focused on labor market outcomes, individual responsibility, and interactive employability. Gazier concludes that the interactive approach has gained the most acceptance, recognizing that employability cannot be solely an individual's responsibility but also involves broader structural factors such as labor market conditions and policies (McQuaid & Lindsay, 2005; Suárez, 2016).

Different ways of understanding employability have emerged over the past decades. At times, the focus has been on the physical or medical skills required to be employed, while at other times, attention has been given to the individual's knowledge or to the demand for labor market and the availability of jobs (Suárez, 2016). Thus, the changes in the conceptualization of employability may be explained by the transformations in employment frameworks and the organization of workers

Table 1 Stages in the conceptualization of employability (during twentieth century)

	Historical moment	Conceptualization of employability
Dichotomic employability	Early 20th century	Distinguished between those who are employable and those who are not
Socio-medical employability	Emerged during World War II	Focused on those unable to work due to social, psychological, or mental deficiencies
Manpower policy employability	1980s (U.S)	Focused on the skills and knowledge required in the labor market
Flow employability	1960 (France)	Focused on the demand for labor in the local and state economy
Labor market performance employability	Late 1970s	Focused on the measurement of the labor market outcomes resulting from specific policies
Initiative employability	1990	Focused on individual initiative and responsibility for career development
Interactive employability	Late 1990	Individual and structural factors affecting employability were balanced

Source adapted from Suárez (2016)

(Moreau & Leathwood, 2006). Today, the initial binary view of being employable or not has been surpassed. It can be said that theories about employability have gained greater complexity; turning it into a multidimensional concept (Williams et al., 2016).

From its historical evolution, it can be concluded that employability is a dynamic term (Forrier & Sels, 2003). As employability is a socially constructed concept composed of social elements, it evolves and changes as society itself changes (Brown et al., 2003; Yorke, 2006). Therefore, when analyzing definitions of employability, it is essential to identify the perspective, discipline, or field of knowledge from which it is approached (Williams et al., 2016). Depending on the viewpoint, employability can be considered an individual trait (Fugate & Kinicki, 2008) the result of labor market conditions (McQuaid, 2006), a measure of job conditions (Forrier & Sels, 2003), or an attitude related to activities that improve employability (Van Dam, 2004). Due to the widespread acceptance of this concept in society, numerous theoretical and empirical studies have been conducted, which have influenced the development of definitions, evaluations, and management strategies (Clarke, 2018; Harvey, 2001; Holmes, 2013; Suleman, 2018).

3.2 Current Conceptualizations of Employability

In the literature, we see that the concept of employability has evolved throughout history. Considering current literature and contemporary conceptualizations, we can also observe various approaches to understanding and analyzing this concept. Considering the literature on the conceptualisation of employability and examining employability research, two main perspectives have emerged in recent decades (see Table 2). On the one hand, Thijssen et al. (2008) identified an individual perspective. This perspective has been predominantly studied in higher education and has often been linked to competencies, specifically those that students in the university community are expected to develop (Dacre-Pool & Sewell, 2007).

For this reason, studies in this sector that identify and evaluate competencies for employability (employability skills) from an individual perspective are particularly noteworthy (Archer & Davison, 2008; Clarke, 2018; Harvey, 2001; Holmes, 2013; Osmani et al., 2019; Rothwell et al., 2008; Suleman, 2018). However, there are also conceptualizations within this sector that go beyond competencies, though they still remain focused on the individual level.

On the other hand, a broader perspective can be identified in this conceptualization. Thijssen and his colleagues (2008) referred to as the "organizational perspective" and the "social perspective." This perspective asserts that, in addition to individual characteristics, other dimensions also influence employability. Research from this perspective has been conducted in various fields, such as economics, business organization, human resource management, sociology, psychology, and even education.

Nevertheless, as stated previously, individual perspectives have dominated research, particularly in higher education, with a specific focus on competencies. As a result, many frameworks have been created to capture these competencies. García-Aracil and Van der Velden (2008) noted the emergence of several frameworks for employability competencies and, recognizing the need for a unified framework, conducted a factor analysis. They identified six types of competencies for employability: organizational (related to the ability to organize work); specialized (more specific competencies required by disciplines or specific jobs); methodological (problem-solving and addressing issues); generic (critical thinking

Table 2 Perspectives on the conceptualisation of employability

Perspective	Field	Research topics
Individual perspective	Higher education	Individual characteristics: competencies, social capital, identity
Broad perspective	Sociology psychology economy human resources education	Beyond individual characteristics (such as: signal management, contextual components)

Source Own elaboration

and communication skills); participative (planning, decision-making, responsibility); and socio-emotional (interpersonal skills and teamwork).

Additionally, the European Union (2016), in a study conducted within the Erasmus+ program, defined a catalogue of transversal competencies considered key for employability. In this catalogue, they identified competencies outlined in the aforementioned reference frameworks, distinguishing a total of 12 competencies: intercultural skills and global awareness; flexibility and adaptability; strategic and innovative thinking; organization and time management; decision-making; teamwork; empathy; problem-solving; learning orientation; negotiation skills; leadership; and collecting and processing information.

To conclude the conceptualization of employability, it is noteworthy to mention authors who have understood individual competencies within a contextual framework. Williams and his colleagues (2016) conducted a systematic review of the literature on employability, recognizing the need for a unified analysis due to the diverse perspectives from which this concept has been approached. They identified and analysed 16 conceptualizations with the goal of building a common framework. In this framework, three main dimensions have been defined: capital, professional development, and context. The capital dimension is subdivided into three categories: human capital, social capital, and psychological capital. In terms of professional development, they distinguish between organizational management and self-management. Finally, in the context dimension, they highlight demand factors and culture.

3.3 The Integration of Employability Within ESD Approach

While some authors highlight the importance of developing employability within ESD approach, there are very few studies that thoroughly examine this integration. The research that does exist in higher education focuses primarily on competency development (Cebrian et al., 2020; Jelonek & Urbaniec, 2019; Leal Filho et al., 2015; Perez-Salgado et al., 2018).

The paper of Jelonek and Urbaniec (2019) aims to identify key sustainability competencies that enhance the employability of higher education graduates in Poland. Through a comprehensive literature review, the authors develop a framework of sustainability competencies. This framework is analysed using empirical data from in-depth interviews and focus group discussions regarding the employability competencies needed in the Polish labor market. The study addresses critical gaps in understanding the sustainability competencies of higher education graduates in relation to current and future labor market's demands. The findings contribute to the literature on sustainable employability by exploring the connection between sustainability competencies and employability, offering insights for HEI in designing and refining educational programs to promote the development of these competencies.

Perez-Salgado et al. (2018) explore sustainability competencies from the perspective of professional practitioners, using an action research approach. The study involved two workshops where participants reflected on the knowledge, skills, attitudes, and behaviors needed to lead successful change processes in various business and professional contexts. The focus is on "intervention competence". This includes the ability to facilitate decision-making and interventions, learn from practice and connect it to scientific knowledge, engage in strategic thinking and actions, act goal-oriented and ethically, and manage complexity and diverse stakeholder interests. The competence is practiced in environments with competing values and power dynamics. The paper concludes with suggestions for further research and practice.

Cebrian et al. (2020) paper presents recent research on sustainability and ESD competencies. It covers various topics, including literature reviews, curriculum development, pedagogical methods (e.g., action research, serious games, augmented reality), and the assessment of sustainability competencies linked to employability and quality assurance. The paper highlights the growing focus on ESD competencies and calls for further research to operationalize and assess these competencies in students and educators.

Leal Filho et al. (2015) present case studies from Hamburg University of Applied Sciences and Bournemouth University that demonstrate the benefits of integrating sustainability into research and teaching. They state that to address the challenges faced by higher education institutions (HEIs), it is essential to provide faculty with the flexibility to develop both formal and extracurricular resources focused on sustainability and employability skills. Collaborating with employers and policymakers can help aligning these initiatives with the needs of the green economy, ensuring that sustainability remains a central theme in teaching and learning. They conclude emphasizing that higher education should focus on developing competencies related to employability and sustainability.

Alimehmeti et al. (2024) study examines the impact of Sustainable Development Goals (SDGs) in higher education curricula on the employability of master's graduates. Analyzing data from 5,784 graduates of the University of Bologna, the researchers used logistic regression to assess job acquisition, job coherence, and employment quality three years post-graduation. Results indicate that graduates with SDG-related coursework achieve better employment outcomes, with a particular emphasis on the economy category leading to higher job acquisition rates and salaries. The findings underscore the significance of integrating SDGs into curricula to enhance graduates' employability.

4 Conclusions

The conclusions of this research highlight the importance of integrating Education for Sustainable Development (ESD) into higher education to promote sustainability and enhance student's employability. The historical analysis reveals that, despite the growing attention to both concepts, the direct link between ESD and employability has been insufficiently explored, underscoring the need for a more integrated framework.

(a) The concept of employability has evolved significantly over time. The historical context and the fields of study related to this concept have influenced its understanding. On one hand, there is a more individual perspective that particularly considers personal characteristics, developed competencies, and social capital. On the other hand, there is a broader perspective that acknowledges additional factors influencing employability, such as familial and social contexts. The authors emphasize the importance of creating a joint framework for conceptualizing employability that integrates both perspectives.
(b) The literature review reveals that, while the importance of developing sustainability competencies is acknowledged, there is a lack of studies that explore how HEIs can effectively integrate ESD into their programs to improve employability. This gap highlights the need for future research to examine how higher education institutions can design academic programs that explicitly connect sustainability with employability in today's and future job markets.
(c) As the integration of ESD into higher education continues to gain traction, it is essential to explore both the opportunities and obstacles that this process presents. While challenges exist in aligning ESD with employability, the potential benefits for students and institutions are significant. By addressing these challenges, HEIs can better prepare their graduates for a changing workforce and the demands of sustainable development.

    One of the primary obstacles to integrating employability into ESD is the lack of a unified conceptual framework. While both concepts have been extensively studied separately, the literature reveals a gap in terms of a coherent and structured model that effectively combines ESD and employability. This absence of a comprehensive approach complicates the understanding of how these two important areas can be integrated in higher education. Furthermore, although the importance of developing employability within ESD is widely acknowledged, there are very few studies that specifically address this issue. Most research focuses on competency development, but there is a notable lack of broader empirical studies. This limits a deeper understanding of how ESD can enhance graduates' employability.

    In light of this literature review, several opportunities can be highlighted. The growing global focus on sustainability, both in business and society, presents a major opportunity for HEIs to position themselves as leaders in educating professionals who are prepared to meet the challenges of sustainable development. In addition, this focus opens up a wide area for research and scientific inquiry to advance the field. Moreover, researching employability within ESD in higher education is essential for creating frameworks that can enhance graduates' readiness to contribute to sustainable societal progress. By exploring how ESD can support the development of employability, HEIs can better prepare students for the demands of the future workforce while also equipping them to drive positive social and environmental change.

This study emphasizes that, to develop professionals who are not only competent in their fields but also conscious and committed to sustainability, HEIs must advance in creating educational frameworks that integrate ESD with employability development, and this requires further research.

References

Alimehmeti, G., Fia, M., & Paletta, A. (2024). The sustainability-to-employment pipeline: The impact of SDG-related curricula on graduates' employability. *Studies in Higher Education, 1–15*. https://doi.org/10.1080/03075079.2023.2299328

Archer, W., & Davison, J. (2008). *Graduate employability: The views of employers*. The Council for Industry and Higher Education.

Barrado, A. R. (2023). *Los mapas de empleabilidad como herramientas fundamentales de los servicios públicos de empleo para diseñar las políticas activas de empleo*. Universidad de Extremadura.

Bögenhold, D. (2020). History of economic thought as an analytic tool: Why past intellectual ideas must be acknowledged as lighthouses for the future. *International Advances in Economic Research, 26*(1), 73–87. https://doi.org/10.1007/978-3-030-79193-3_8

Brown, P., Hesketh, A., & Wiliams, S. (2003). Employability in a knowledge-driven economy. *Journal of Education and Work, 16*(2), 107–126. https://doi.org/10.1080/1363908032000070648

Cebrian, G., Junyent, M., & Mula, I. (2020). Competencies in education for sustainable development: Emerging teaching and research developments. *Sustainability, 12*(2). https://doi.org/10.3390/su12020579

Clarke, M. (2018). Rethinking graduate employability: The role of capital, individual attributes and context. *Studies in Higher Education, 43*(11), 1923–1937. https://doi.org/10.1080/03075079.2017.1294152

Creighton, M. (2007). *Employability: What is it and is it important?* [University of Glasgow]. https://doi.org/10.5525/gla.thesis.71129

Dacre-Pool, L., & Sewell, P. (2007). The key to employability: Developing a practical model of graduate employability. *Education and Training, 49*(4), 277–289. https://doi.org/10.1108/00400910710754435

De Grip, A., Van Loo, J., & Sanders, J. (2004). The industry employability index: Taking account of supply and demand characteristics. *International Labour Review, 143*(3), 211–233. https://doi.org/10.1111/j.1564-913x.2004.tb00269.x

European Union. (2016). *Catalogue of transversal competences key for employability*. Erasmus+.

European Ministers of Education. (1999). *The Bologna declaration of 19 June 1999: Joint declaration of the European Ministers of Education*.

European Ministers of Education. (2001). *Prague Communiqué: Towards the European higher education area*. https://ehea.info/media.ehea.info/file/2001_Prague/44/2/2001_Prague_Communique_English_553442.pdf

European Ministers of Education. (2003). *Berlin Communiqué: Realising the European higher education area*. https://ehea.info/media.ehea.info/file/2003_Berlin/28/4/2003_Berlin_Communique_English_577284.pdf

European Ministers of Education. (2005). *Bergen Communiqué: The European higher education area—A shared responsibility*. https://ehea.info/media.ehea.info/file/2005_Bergen/53/8/2005_Bergen_Communique_Spanish_580538.pdf

European Ministers of Education. (2007). *London Communiqué: Towards the European higher education area: Responding to challenges in a globalised world.* https://ehea.info/media.ehea.info/file/2007_London/69/7/2007_London_Communique_English_588697.pdf
European Ministers of Education. (2009). *Leuven/louvain-la-Neuve Communiqué: The Bologna process 2020: The future of the european higher education area.* https://ehea.info/media.ehea.info/file/2009_Leuven_Louvain-la-Neuve/06/1/Leuven_Louvain-la-Neuve_Communique_April_2009_595061.pdf
European Ministers of Education. (2010). *Budapest-Vienna declaration: The European higher education area: A new stage of cooperation.* https://ehea.info/media.ehea.info/file/2010_Budapest_Vienna/64/0/Budapest-Vienna_Declaration_598640.pdf
European Ministers of Education. (2012). *Bucharest Communiqué: The European higher education area in the 21st century.* https://ehea.info/Upload/document/ministerial_declarations/Bucharest_Communique_2012_610673.pdf
European Ministers of Education. (2015). *Yerevan Communiqué.* https://ehea.info/media.ehea.info/file/2015_Yerevan/70/7/YerevanCommuniqueFinal_613707.pdf
Feintuch, A. (1955). Improving the employability and attitudes of "difficult to place" persons'. *Psychological Monographs, 69*, 392–397.
Ferrer-Balas, D., Lozano, R., Huisingh, D., Buckland, H., Ysern, P., & Zilahy, G. (2010). Going beyond the rhetoric: System-wide changes in universities for sustainable societies. *Journal of Cleaner Production, 18*(7), 607–610. https://doi.org/10.1016/j.jclepro.2009.12.009
Forrier, A., & Sels, L. (2003). The concept employability: A complex mosaic. *International Journal of Human Resources Development and Management, 3*(2), 102–124. https://doi.org/10.1504/IJHRDM.2003.002414
Fugate, M., & Kinicki, A. J. (2008). A dispositional approach to employability: Development of a measure and test of implications for employee reactions to organizational change. *Journal of Occupational and Organizational Psychology, 81*(3), 503–527. https://doi.org/10.1348/096317907X241579
García-Aracil, A., & Van der Velden, R. (2008). Competencies for young European higher education graduates: Labor market mismatches and their payoffs. *Higher Education, 55*(2), 219–239.
Gazier, B. (1998). Employability-definitions and trends. In *Employability: Concepts and policies* (pp. 37–71). European Employment Observatory.
Harvey, L. (2001). Quality in higher education defining and measuring employability. *Quality in Higher Education, 7*(2), 97–109. https://doi.org/10.1080/1353832012005999
Hiltrop, J. M. (1995). The changing psychological contract: The human resource challenge of the 1990's'. *European Management Journal, 13*, 286–294.
Holmes, L. (2013). Studies in Higher Education Competing perspectives on graduate employability : Possession, position or process ? *Studies in Higher Education, 38*(4), 538–554. https://doi.org/10.1080/03075079.2011.587140
Jelonek, M., & Urbaniec, M. (2019). Development of sustainability competencies for the labour market: An exploratory qualitative study. *Sustainability, 11*(20). https://doi.org/10.3390/su11205716
Leal Filho, W., Shiel, C., & do Paco, A. (2015). Integrative approaches to environmental sustainability at universities: an overview of challenges and priorities. *Journal of Integrative Environmental Sciences, 12*(1), 1–14. https://doi.org/10.1080/1943815X.2014.988273
Lefresne, F. (1999). Employability in the heart of the European employment strategy. *Transfer: European Review of Labour and Research, 5*(4), 460–480.
McQuaid, R. W. (2006). Job search succes and employability in local labour markets. *The Annals of Regional Science, 40*(2), 407–421.
McQuaid, R. W., & Lindsay, C. (2005). The concept of employability. *Urban Studies, 42*(2), 197–219. https://doi.org/10.1080/0042098042000316100
Moreau, M. P., & Leathwood, C. (2006). Graduates' employment and the discourse of employability: A critical analysis. *Journal of Education and Work, 19*(4), 305–324. https://doi.org/10.1080/13639080600867083

Osmani, M., Weerakkody, V., Hindi, N., & Eldabi, T. (2019). Graduates employability skills: A review of literature against market demand. *Journal of Education for Business, 94*(7), 423–432. https://doi.org/10.1080/08832323.2018.1545629

Perez-Salgado, F., Abbott, D., & Wilson, G. (2018). Dimensions of professional competences for interventions towards sustainability. *Sustainability Science, 13*(1), 163–177. https://doi.org/10.1007/s11625-017-0439-z

Rothwell, A., Herbert, I., & Rothwell, F. (2008). Self-perceived employability: Construction and initial validation of a scale for university students. *Journal of Vocational Behavior, 73*(1), 1–12. https://doi.org/10.1016/j.jvb.2007.12.001

Sedlacek, S. (2013). The role of universities in fostering sustainable development at the regional level. *Journal of Cleaner Production, 48*, 74–84. https://doi.org/10.1016/j.jclepro.2013.01.029

Sin, C., Tavares, O., & Amaral, A. (2016). Who is responsible for employability? Student perceptions and practices. *Tertiary Education and Management, 22*(1), 65–81. https://doi.org/10.1080/13583883.2015.1134634

Sterling, S. (2021). Concern, conception, and consequence: Re-thinking the paradigm of higher education in dangerous times. *Frontiers in Sustainability, 2*(December), 1–13. https://doi.org/10.3389/frsus.2021.743806

Suárez, B. (2016). Empleabilidad: Análisis del concepto. *Revista De Investigación En Educación, 14*(1), 67–84.

Suleman, F. (2018). The employability skills of higher education graduates: Insights into conceptual frameworks and methodological options. *Higher Education, 76*(2), 263–278. https://doi.org/10.1007/s10734-017-0207-0

Thijssen, J. G. L., Van Der Heijden, B. I. J. M., & Rocco, T. S. (2008). Toward the employability-link model: Current employment transition to future employment perspectives. *Human Resource Development Review, 7*(2), 165–183. https://doi.org/10.1177/1534484308314955

Van Dam, K. (2004). Antecedents and consequences of employability orientation. *European Journal of Work and Organizational Psychology, 13*(1), 29–51.

Vázquez-Rodríguez, A. (2020). *Tesis De Doctorado Educación No Formal Y Empleabilidad De La Juventud*. Universidade de Santiago de Compostela.

Verhaar, C. H. A., & Smulders, H. R. M. (1999). Employability in practice. *Journal of European Industrial Training, 23*(6), 268–274. https://doi.org/10.1108/03090599910284605

Versloot, A. M., Glaudé, M., & Thijssen, J. G. L. (1998). *Employability: Een pluriform arbeidsmarktfenomeen*. Max Goote Kenniscentrum voor Beroepsonderwijs en Volwasseneneducatie.

Weinert, P., Baukens, M., Bollérot, P., Pineschi-Gapénne, M., & Walwei, U. (2001). *Employability: From theory to practice*. Transaction Publishers.

Williams, S., Dodd, L. J., Steele, C., & Randall, R. (2016). A systematic review of current understandings of employability. *Journal of Education and Work, 29*(8), 877–901. https://doi.org/10.1080/13639080.2015.1102210

Yorke, M. (2006). *Employability in higher education: What it is, What it is not. Learning & employability series*. Higher Education Academy.

Shaping Future-Oriented Managers: Integrating Sustainability Competencies into the Business Economics Curriculum

Tom Emile Kuppens, Cathy Macharis, Leen Audenaert, and Tom Janssen

Abstract As global challenges intensify, the need for business economists to embrace sustainable approaches becomes crucial. They must consider the impact of their decisions on the ecological and social environment. This necessitates integrating sustainability competencies into their skillset, aligning with the requirements of the circular economy and the European Green Deal. Therefore, the Faculty of Social Sciences and Solvay Business School of Vrije Universiteit Brussel (VUB) has begun a process of structurally implementing sustainability into their curricula. This process began with a survey revealing fragmented and superficial knowledge of sustainability among students. In response, the Faculty committed to introducing sustainability into their strategic plan and established a working group to implement the plan. They proposed a new obligatory course, 'Sustainability Economics', to fill curriculum gaps. In 2022, the process was guided by the Sustainable Education Point of the Flemish Government to define a new profile for a future-oriented business economist. A writing group, including students, staff, alumni, business representatives, NGOs, government, and sustainability experts, was formed to determine this profile. Based on their input, twelve characteristics of a future-oriented

T. E. Kuppens (✉)
Vrije Universiteit Brussel, Multidisciplinair Instituut Lerarenopleiding—Brussels Research Institute for Teacher Education, Brussels, Belgium
e-mail: tom.emile.kuppens@vub.be

School for Educational Studies, research group U-RISE & Centre for Environmental Sciences, research group Environmental Economics, UHasselt—Hasselt University, Agoralaan, Diepenbeek, Belgium

COPERNICUS Alliance, Lüneburg, Germany

C. Macharis
Vrije Universiteit Brussel, House of Sustainable Transitions, Brussels, Belgium
e-mail: cathy.macharis@vub.be

L. Audenaert · T. Janssen
Duurzaam Educatiepunt (Departement Omgeving), Brussels, Belgium
e-mail: leen.audenaert@vlaanderen.be

T. Janssen
e-mail: tom.janssen@vlaanderen.be

W. Leal Filho et al. (eds.), *Introducing the UN SDGs in Higher Education Institutions*, World Sustainability Series, https://doi.org/10.1007/978-3-032-23017-1_26

economist were defined. These include using business economics knowledge for sustainable development, following sustainability frameworks and exploring new business models in the circular and bio-based economy. The aim is to gradually implement this new profile into the curriculum, ingraining the twelve characteristics throughout the programme. Finally, the curriculum was evaluated against these characteristics, and a new learning line for sustainability was defined.

Keywords Sustainability competencies · Business economics · Green deal · Circular economy · Learning line for sustainability

1 Introduction

In the rapidly evolving landscape of economic studies, the integration of sustainability has emerged as a critical and pressing concern (Cifuentes-Faura & Noguera-Méndez, 2023; de Muijnck & Tieleman, 2021). As global challenges such as climate change, resource depletion, and social inequality intensify, economists are faced with the imperative to re-evaluate traditional economic paradigms and embrace a more sustainable approach (Bäuerle, 2020).

Companies are increasingly searching for people who have this knowledge and competencies (Janssens et al., 2021). The European Green Deal will require businesses to work in a more holistic way (Lamenta & Grzybowska, 2023). However, many companies are not ready for the change and they do not have the right people in house to support them in the transition. Lifelong learning programmes are necessary to get people back to university for the necessary knowledge and competencies (Bezjak & Štremfel, 2024), but students should be prepared to play their role in the transition (EUA, 2022).

This chapter is based on the experiences in introducing sustainability in the economic programmes at the Faculty of Social Sciences and Solvay Business School of VUB. The chapter describes the process and provides the output of the profile created as a lighthouse for 'future-oriented business economists'. 'Business economist' refers to anyone with an academic bachelor's or master's degree in the fields of (1) economics and business economics and (2) commercial sciences and business administration. They play an influential role in society, making important investment and policy decisions as managers or executives in companies or government institutions (Ceulemans & De Prins, 2010). These decisions are made now, while the results manifest in the long term necessitating a future-oriented perspective.

Such a business economist is responsible for the survival of the organisation. A future-oriented business economist can guide organisations through a crisis situation. Whether it is a pandemic, rising energy prices, or the climate crisis: responsible business economists lead organisations that have sufficient resilience to survive such crises.

The future is uncertain, but that such crises will continue to occur is a near certainty. Raw materials will run out, climate change will strike, and consumers will become more critical. To face future challenges, business economists must master a set of sustainability competencies (Lambrechts et al., 2013; Ploum et al., 2018). Future-oriented business economists go beyond reducing negative impacts on people and the environment. They think regeneratively, for example, improving the environment through their business practices.

This chapter presents the characteristics that a future-oriented business economist should possess to devise solutions to the sustainability challenges in the near and distant future. These traits assume a thorough integration of traditional business competencies with sustainability competencies. Future-oriented business economists use their insights into market forces and consumer behaviour, for example, to help consumers make sustainable purchasing decisions.

2 The Curriculum Design Process

The set of characteristics for a future-oriented business economist is the result of a curriculum design process (see Fig. 1) that took place at the Faculty of Social Sciences and Solvay Business School of VUB. A survey was executed in 2020 to get an idea on how much knowledge and competencies were brought to the students (both business economics and business engineer). The result of the survey was that the students' knowledge about sustainability was very fragmented and superficial. However, the Faculty aims to thoroughly prepare students of Business Economics and Business Engineering for the sustainability challenges they will face in their careers. After all, these are becoming increasingly prominent within the business world. Therefore, in 2021 the Faculty decided to put the introduction of sustainability in their strategic plan. This formal commitment helped to finally work towards a clear goal of introducing sustainability in the curriculum. A very important aspect of the process, was the start of an internal working group on sustainability. This informal group of concerned colleagues helped to co-create and support the further development of the rest of the process. In this group, the idea of a new course 'Sustainability Economics' was formed that could become the cornerstone course in the learning pathway. This new course would fill the gaps in the curriculum, be obligatory and positioned in the third bachelor year.

As a next step, the Faculty started to collaborate with the Sustainable Education Point (translation of *Duurzaam Educatiepunt*) of the Flemish Government. This collaboration consisted of three phases. The initial phase (May–December 2022) focused on assessing the current state of sustainability integration within the existing curricula. The curriculum at VUB was compared with other economic programmes in Flanders (Belgium). This was achieved through interviews among 16 professors and 5 alumni to identify where and how sustainability concepts were already being incorporated. Based on these insights, a comprehensive vision for sustainability in the Business Economics and Business Engineering programmes was developed.

This phase also involved defining the profile of the future-oriented business economist. The profile text for a sustainable entrepreneur (Vekeman et al., 2022)

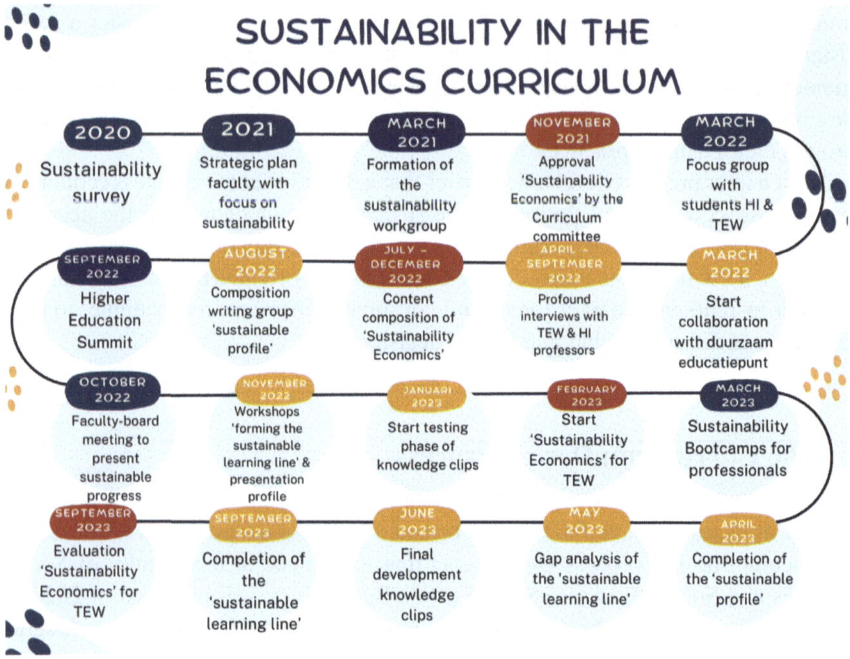

Fig. 1 Curriculum design process for a future-oriented business economist

was taken as a starting point. That text was based on a similar curriculum design process guided by the Sustainable Education Point, but then in a university of applied sciences. Put differently, the profile of a sustainable entrepreneur is developed to shape the curricula of professional bachelor's degrees of Business Sciences and Business Administration in universities of applied sciences. In contrast, the profile of a future-oriented manager is targeted at curriculum designers of academic bachelor's and master's degrees of Business Economics and Business Engineering in universities. A writing group, including students, staff, alumni, business representatives, NGOs, government, and sustainability experts, was formed to determine this profile. The writing group consisted of 11 members, including the authors of this chapter. The writing group met twice: once to determine the ambition level and the goals of the profile text, and a second time to review its first draft. In order to prepare the deliberation of the ambitions during the first meeting, the members of the writing group were asked to fill out an open-ended survey. This questionnaire consisted of the following questions:

1. How would you describe a sustainable graduate business economist or business engineer? What (sustainability) competencies should they have? What must they be able to know and do, and what skills do they need?
2. How should the profile of the current graduate in business economics and business engineering differ from that of a future, sustainable graduate?

3. Do you know a good example of a curriculum of the future in business and economics? Why, what is its essence?
4. What publications (reports, articles, books, …) should be read to build a profile? Why, what is their essence?
5. Which professors/experts/inspiring entrepreneurs on the matter should be talked to when creating the profile? Why, what is their essence?
6. Would you like to add anything or do you have any comments?

Based on their input, eleven characteristics were defined, outlining the future competencies in sustainability that students should acquire. Following on from the first draft, the writing group suggested some revisions. For instance, they proposed a twofold strategy for formulating the characteristics: on the one hand they should be sufficiently ambitious, whilst on the other hand they should not be too radical that faculty member support is lost. For this purpose, a twelfth characteristic (see characteristic 1) was added at the start of the profile text. Consequently, the first characteristic recognises the importance of understanding how businesses and the economy currently work, even when one wants to foster sustainable development. In order to increase uptake and understanding, each characteristic was summarised with keywords, some persona/job descriptions and a complementary didactical section were added, and finally an explanatory glossary was annexed. The second draft was then presented to the faculty members for feedback. One suggestion made was to explicitly include the circular and biobased economy as this might attract the appeal of the more technology oriented business engineers (see characteristic 10). Subsequently, a dedicated learning pathway (see Fig. 2) on sustainability was created, culminating in the design of a new curriculum.

During the second phase (academic year 2022–2023), the newly developed sustainability learning pathway was embedded into various subjects across the curriculum. A key focus was placed on further developing the 'Sustainability Economics' course, which was positioned as the cornerstone course of the learning pathway. 'Sustainability Economics' is a 6 ECTS credits course. Initially it was offered only in the third year of the bachelor's degree of Business Economics, but later it was opened to third year students of the bachelor's degree of Business Engineering as well.

The third and final phase (academic year 2023–2024 and beyond) aimed at deepening the integration of the sustainability learning pathway within the Business Economics programme. The third phase also included the introduction of the 'Sustainability Economics' course across all other faculty programmes, ensuring a comprehensive and cohesive approach to sustainability education.

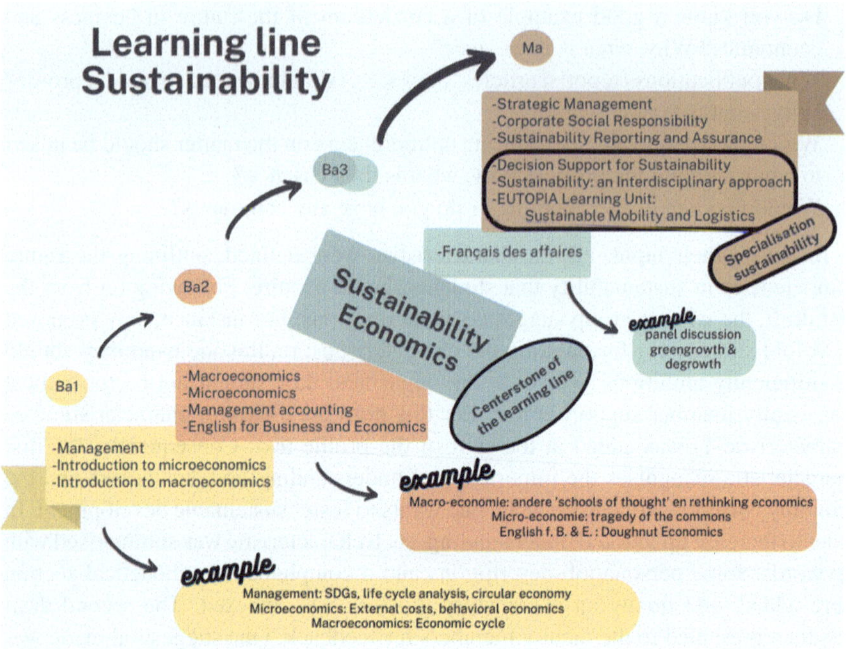

Fig. 2 Sustainability learning pathway in the business economics curriculum

3 Sustainability Competencies

A lot of reference frameworks on sustainability competencies exist. In the Dutch language, Roorda (2009) developed his VESTIA+D model, a set of professional competencies for sustainable development described at four proficiency levels within the context of Dutch higher education. Internationally, the publication of Wiek et al. (2011) is considered a primer for relevant sustainability competencies in higher education. Led by Rieckmann (2017), UNESCO published learning objectives for each of the sustainable development goals somewhat later. Later, Ploum et al. (2018) added a competency framework for sustainable entrepreneurs. Brundiers et al. (2021) updated the work of Wiek et al. (2011) with a view to a widely accepted reference framework for sustainability competencies in higher education. Besides competency frameworks focused on learning for sustainability, there are also initiatives focused on learning for regeneration and thrivability. This involves teaching people how they can improve the economic system with a view to nature restoration on the one hand (regeneration), and how they can create a new future in which human life flourishes in harmony with nature on the other (thrivability). For example, the NGO r3.0 published a blueprint for a transformation of education based on seven transformative learning perspectives aimed at regeneration and thrivability (Smitsman et al., 2021).

In this publication, we start from GreenComp (Bianchi et al., 2022), a competency framework developed by the European Commission—similar to EntreComp and DigiComp for entrepreneurship and digital literacy, respectively. It was developed to establish a common basis within Europe for sustainability competencies within all forms of education. GreenComp consists of a set of twelve sustainability competencies grouped into four competency domains:

- embodying sustainability values;
- embracing complexity in sustainability;
- envisioning sustainable futures;
- acting for sustainability.

Although GreenComp primarily provides a framework for learning for ecological sustainability, the four competency domains are sufficiently generical to include social sustainability.

4 A Future-Oriented Business Economist …

The writing group proposed a set of twelve characteristics for a future-oriented business economist to prepare them for the sustainability challenges they might face throughout their professional career. The Faculty Board formally approved these twelve characteristics and agreed to use them as a beacon to shape their business economics and business engineering curricula. The table on the next page shows how each of the competency areas of GreenComp are translated to the context of higher business economics education in terms of the twelve characteristics. The characteristics have been numbered and are visually summarised in Fig. 3. Each of the numbered characteristics is explained in more detail after the picture. Table 1 shows the link between the characteristics and GreenComp's competency domains.

(1) **deploys business knowledge in a targeted way for sustainable development**

Future-oriented business economists possess a strong foundation in business economics, encompassing general economic knowledge (e.g. macro-, meso- and micro-economics) and business sciences (e.g. accountancy, finance, marketing, strategic management, labour and organisational psychology, operational management, logistics, business law, information technology). They apply this knowledge for sustainable development, addressing both current and future societal and environmental challenges. They respond to issues such as raw materials supply, new technologies, geopolitical shifts, and cost inflation. They prepare companies for changing consumer preferences towards sustainability (Reichheld et al., 2023). They realise in the 'war on talent' that the future working force (including millennials and Generation Z) is willing to give up salary to work for a sustainable company (Yamane & Kaneko, 2021). Finally, they take into account that socially responsible companies usually perform better financially (Busch & Friede, 2018).

Fig. 3 Twelve characteristics of a future-oriented business economist (illustration made by Bart Bruneel)

(2) **uses the planetary and social boundaries as a compass**

Future-oriented business economists use planetary and social boundaries of the doughnut economy (Raworth, 2018) as a compass in both their professional and personal actions, respecting the earth's carrying capacity in their decisions and behaviour. They focus on both ecological and social sustainability making it a basic attitude and automated 'reflex' in their decision-making. They are accountable for any violation of these boundaries.

They develop their own vision of sustainability in which they see it as a dynamic and continuous improvement process, guiding business and economic activities towards a safe and just space for humanity.

A future-oriented business economist starts from the idea of the 'embedded economy' (Raworth, 2018) and realises that the economy is part of society, while the latter is part of the earth ecosystem. Without a healthy earth, humanity is at risk; without a well-organised society (including volunteering, 'commons', democracy), a functioning economy is not possible.

Table 1 Sustainability competencies contextualised to future-oriented business economics

Characteristic: a future-oriented business economist	Embodying sustainability values	Embracing complexity in sustainability	Envisioning sustainable futures	Acting for sustainability
(1) deploys business knowledge in a targeted way for sustainable development	x			
(2) uses the planetary and social boundaries as a compass			x	x
(3) follows (inter)national sustainability frameworks		x		
(4) is a systems thinker		x		
(5) thinks of the long run			x	
(6) exhibits regenerative leadership				x
(7) takes a pluralistic view of the economy	x			
(8) reports transparently on sustainability efforts		x		
(9) is an entrepreneur with a multiple mission	x			x
(10) explores new business models in the circular and biobased economy			x	
(11) is an inclusive entrepreneur	x			x
(12) is able to develop policy instruments to embed sustainability within organisations			x	

(3) **follows (inter)national sustainability frameworks**

Future-oriented business economists consider historical perspectives to understand current and future (inter)national policy frameworks for sustainable development. Future-oriented business economists understand why concepts such as sustainability and sustainable development have emerged, and are aware of key actors and milestones in the development of these frameworks, such as "Silent Spring" (Carson, 1962), the report "Limits to Growth" of the Club of Rome (Meadows et al., 1972), "Our Common Future" (Brundtland, 1987), the triple bottom line (Elkington, 1997), the 2015 Paris Climate Agreement, the 2021 Corporate Sustainability Reporting Directive (CSRD) and most recently the 2024 Corporate Sustainability Due Diligence Directive (CSDD) of the European Commission.

Future-oriented business economists are also expected to be able to reflect critically on these policy frameworks and regulations. They conduct an informed debate on concepts such as economic growth, green growth and degrowth.

Future-oriented business economists are expected to be able to position themselves in this spectrum of sustainability, to participate in the debate on these policy frameworks and to translate them into their own business context. They are also aware of the shifting baseline syndrome, which affects perceptions of environmental quality over time (Soga & Gaston, 2018).

(4) **is a systems thinker**

Future-oriented business economists think in systems (Ploum et al., 2018). They can deal with the complexity and uncertainty inherent in an economic system embedded in a well-functioning society and a safe planet.

This implies that business economists assess the impact of business activities, investment and policy decisions and the current economic system (including the functioning of the money market) on the environment. Conversely, business economists are expected to understand the impact of societal challenges (such as climate change, income inequality and resource scarcity) on those business activities and the firm's prospects of survival.

Future-oriented business economists are thus aware of the linkages that exist between economic, social and ecological systems. They can describe these linkages in the form of mutually reinforcing (positive) or mutually counteracting (negative) feedback loops. This resonates with the case made for integrating system dynamics within economics education (Pavlov et al., 2024).

(5) **thinks of the long run**

Future-oriented business economists prioritise long-term survival over short-term profit maximisation, understanding that the latter often comes at the expense of the environment and subsequently jeopardises a company's long-term survival. Future-oriented business economists therefore focus on value creation and maximisation, interpreting value more broadly than the purely economic concept of added value, but creating sustainable value (Manninen et al., 2024), for instance, at social, financial and environmental levels. In other words, future-oriented business economists strive for sustainability maximisation instead of profit maximisation.

Future-oriented business economists realise that focussing exclusively on short-term shareholder value maximisation puts a company at risk (Kurznack et al., 2021). Therefore, they look for both short-term investments for incremental change towards sustainability together with business opportunities that are more disruptive in nature and bring benefit only in the long term. They are sometimes willing to forgo short-term profits in favour of the long-term survival of the company.

(6) **exhibits regenerative leadership**

Future-oriented business economists think in terms of regeneration. They go beyond reducing negative environmental impact, but instead, they are creating positive contributions to environment and society (Das & Bocken, 2024).

They lead change processes inside and outside their own company, promoting both behavioural and systemic change. They choose sustainability in their private and professional lives. They have the so-called 'inner development goals' that help accelerate the transition to a sustainable society through business and the economy. This is a set of 23 science-based, inner skills and qualities divided into 5 dimensions that help accelerate the achievement of the 17 sustainable development goals (Jordan et al., 2021).

Regenerative leaders also understand human behaviour because of the importance of behavioural change in the sustainable transition. The sub-discipline of behavioural economics, which brings together knowledge from psychology, economics, biology and neuroscience, seeks explanations for human behaviour. Within this perspective, the concept of 'nudging' can be of special interest.

(7) **takes a pluralistic view of the economy**

Future-oriented business economists have a pluralistic view of the economy, integrating heterodox schools of thought within the economic sciences (e.g. ecological economics, behavioural economics, the degrowth movement, feminist economics, new monetary models) to deal with the current polycrisis (Schulz et al., 2024).

In light of the current energy crisis, it is not trivial for future-oriented business economists to understand how energy markets work. They have an understanding of energy and environmental economics, but they also know the limits of market instruments thanks to their understanding of behavioural and ecological economics. Their education emphasises the origin and history of diverse economic theories, preparing them to address wicked problems.

(8) **reports transparently on sustainability efforts**

Future-oriented business economists report transparently on their company's environmental, social and governance (ESG) aspects (Simmons et al., 2024). They understand how the environmental impact of a product can be measured over its entire life cycle, for example using life cycle analysis (LCA), and they can use the data to comply with ESG regulations.

They use standardised procedures to avoid greenwashing and distinguish the latter from truly sustainable practices. They are aware of the landscape of legislation (NFRD, CSRD, ESG) indicators (GRI) and labels/standards (such as B Corp, Science Based Targets). They use various tools to help companies become more sustainable such as Future Fit Business Benchmarks or Doughnut Design for Business, or Flemish-inspired tools such as Sustatool.

They leverage data and digitalisation for instance for the purpose of materials passports and monitoring the environmental impact of products during the use phase.

(9) **is an entrepreneur with a multiple mission**

Future-oriented business economists are purpose-driven entrepreneurs who create value on multiple levels (i.e. economically, but also ecologically and socially). To this end, they not only create financial value for their own company or shareholders, but additionally create value for its stakeholders, of which the environment is certainly one.

They understand the complexity and dynamics of the business ecosystem in the context of a rapidly changing environment. Their education includes traditional business concepts and tools developed specifically for purpose-driven entrepreneurship.

(10) **explores new business models in the circular and biobased economy**

Education for the Circular Economy (ECE) is a rising body of literature within the field of (higher) education for sustainable development (Kirchherr & Piscicelli, 2019). Future-oriented business economists embrace circular business models, value preservation and service design, aligning with the European Green Deal. They are familiar with the principle of cradle-to-cradle prioritising business activities from a hierarchy of R-strategies. They design business processes, products, technologies and business models for circularity, incorporating 'design for disassembly' and 'reverse logistics'. Moreover, dismantling and repairing products requires many hands, so future-oriented business economists seek added value through intensive collaboration with the social economy.

They spot interesting business opportunities in these new business activities for sustainable innovations, while evaluating technologies through the application of techno-economic and life-cycle analysis. They choose an appropriate business model from the possible menu of choices such as product service systems, platform models, sharing economy, using templates such as the flourishing business canvas, the regenerative business model canvas, or the triple layered business model canvas. They set up extensive networks across the supply chain but also with a wide set of players from the innovation ecosystem such as governments, universities, communities, and users. Integrating these new economic frameworks (such as circular economy and cradle-to-cradle) contributes to reflections about the economic growth paradigm and promotes sustainable futures (Kowasch, 2022).

(11) **is an inclusive entrepreneur**

Future-oriented business economists are inclusive entrepreneurs. They not only consider the company's shareholders, but also have empathy for and consider the other stakeholders. They pursue a diverse employee pool and monitor a healthy work-life balance for their employees. Factors of production are remunerated in a fair manner and socio-cultural differences are taken into account. Integrating inclusive entrepreneurship in higher education can help reducing social inequalities within the region (O'Brien & Cooney, 2025). For the programmes involved, it is therefore important to consider decolonising curricula due to its strong relevance in the economics curriculum (Bayer & Rouse, 2016).

(12) **is able to develop policy instruments to embed sustainability within organisations**

Future-oriented business economists develop appropriate policy instruments to achieve sustainability in organisations. These can be internal policy instruments in human resource management and performance management, where sustainability performance management points them towards the right direction. They assist in developing appropriate tools within government institutions to move society towards greater sustainability (see also earlier).

5 Lessons Learnt and Limitations

One of the lessons learnt by drafting the profile of a future-oriented business economist, was that the original content of the 'Sustainability Economics' needed to be revised. The new course set-up is based on several perspectives, including a global, macroeconomic, socio-political, firm, financial and personal perspective to a sustainable economy. Each perspective includes themes that altogether represent the final attainment levels of the twelve aforementioned characteristics. For instance, the global perspective includes global crises (such as the climate crisis) and pluralistic thinking. The macroeconomic perspective discusses GDP and alternative indicators, geoeconomics and the interplay between the economy and the environment. The socio-political perspective deals with existential economics, north–south justice and the European Green Deal. The intrinsic value of sustainability for firms, the identification of greenwashing, the role of technology and of firms in the circular economy make up the firm perspective. Within the financial perspective students learn about sustainable finance and the role of ESG criteria. For the personal perspective students are encouraged to take up sustainable leadership and inner development. Finally, some themes such as external costs and valuing the environment, or the Corporate Sustainability Reporting Directive (CSRD) extend across multiple perspectives.

A key limitation is that the final profile is the result of a curriculum design process and not of an academic research project. In order to increase international uptake, the authors suggest conducting a systematic review followed by a Delphi study to properly validate this first exploratory version of the profile of a future-oriented business economist.

6 Conclusion

This chapter presented twelve characteristics that a future-oriented business economist should hold. These include that a future-oriented business economist (1) deploys business knowledge in a targeted way for sustainable development, (2) uses the planetary and social boundaries as a compass, (3) follows (inter)national sustainability frameworks, (4) is a systems thinker, (5) thinks of the long run, (6) exhibits regenerative leadership, (7) takes a pluralistic view of the economy, (8) reports transparently on sustainability efforts, (9) is an entrepreneur with a multiple mission, (10) explores new business models in the circular and biobased economy, (11) is an inclusive entrepreneur, and (12) is able to develop policy instruments to embed sustainability within organisations. These twelve characteristics can be used by higher education institutions that want to implement the European GreenComp framework into business economics and management education.

Acknowledgements The curriculum design process was supported by the Sustainable Education Point of the Department of Environment and Spatial Development of the Flemish Government, which resulted in the publication in Dutch of the profile of future-oriented business economist (Kuppens, 2023). This chapter translates, summarises and complements the information contained in that publication. The authors also wish to acknowledge the contribution of the members of the writing group: Thomas Crispeels, Arne Daneels, Bruno Delepierre, Sam de Pape, Yolan Gielen, Luc Hens, Maja Lardot, Sebastiaan Michiels, Koen Miseur, Alexandre Thys and Tom Vermeiren. A final acknowledgement goes to Bartel Bruneel, who illustrated the visual representing the twelve characteristics of the future-oriented business economist. The authors thank Dr Alex Mifsud, who is assistant professor in education for sustainability at the Northumbria University (United Kingdom), for her kindness to proofread this book chapter as a native speaker.

References

Bäuerle, L. (2020). Reproduction, deconstruction, imagination on three possible modi operandi of economic education. *Journal of Social Science Education, 19*(3), 22–39. https://doi.org/10.4119/jsse-3378

Bayer, A., & Rouse, C. E. (2016). Diversity in the economics profession: A new attack on an old problem. *Journal of Economic Perspectives, 30*(4), 221–242. https://doi.org/10.1257/jep.30.4.221

Bezjak, S., & Štremfel, U. (2024). Lifelong learning for green skills: priorities, implementation framework, and challenges. In *Lifelong learning for green skills and sustainable development* (pp. 19–30). Springer Nature Switzerland. https://doi.org/10.1007/978-3-031-66947-7_3

Bianchi, G., Pisiotis, U., Cabrera, M., Punie, Y., & Bacigalupo, M. (2022). *The European Sustainability Competence Framework*. https://doi.org/10.2760/13286

Brundiers, K., Barth, M., Cebrián, G., Cohen, M., Diaz, L., Doucette-Remington, S., Dripps, W., Habron, G., Harré, N., Jarchow, M., Losch, K., Michel, J., Mochizuki, Y., Rieckmann, M., Parnell, R., Walker, P., & Zint, M. (2021). Key competencies in sustainability in higher education—Toward an agreed-upon reference framework. *Sustainability Science, 16*(1), 13–29. https://doi.org/10.1007/s11625-020-00838-2

Brundtland, G. H. (1987). *Our common future: Report of the world commission on environment and development.*

Busch, T., & Friede, G. (2018). The robustness of the corporate social and financial performance relation: A second-order meta-analysis. *Corporate Social Responsibility and Environmental Management, 25*(4), 583–608. https://doi.org/10.1002/csr.1480

Carson, R. (1962). *Silent spring*. Houghton Mifflin.

Ceulemans, K., & De Prins, M. (2010). Teacher's manual and method for SD integration in curricula. *Journal of Cleaner Production, 18*(7), 645–651. https://doi.org/10.1016/j.jclepro.2009.09.014

Cifuentes-Faura, J., & Noguera-Méndez, P. (2023). What is the role of economics and business studies in the development of attitudes in favour of sustainability? *International Journal of Sustainability in Higher Education, 24*(7), 1430–1451. https://doi.org/10.1108/IJSHE-10-2022-0324

Das, A., & Bocken, N. (2024). Regenerative business strategies: A database and typology to inspire business experimentation towards sustainability. *Sustainable Production and Consumption, 49*, 529–544. https://doi.org/10.1016/j.spc.2024.06.024

de Muijnck, S., & Tieleman, J. (2021). *Economy studies a guide to rethinking economics education*. https://www.economystudies.com/wp-content/uploads/2021/10/Economy-Studies-pdf-versie.pdf

Elkington, J. (1997). *Cannibals with Forks: The triple bottom line of 21st century business*. New Society Publishers.

EUA. (2022). *A university vision for the European Green Deal*.

Janssens, L., Kuppens, T., & Van Schoubroeck, S. (2021). Competences of the professional of the future in the circular economy: Evidence from the case of Limburg, Belgium. *Journal of Cleaner Production, 281*, Article 125365. https://doi.org/10.1016/j.jclepro.2020.125365

Jordan, T., Reams, J., Stålne, K., Greca, S., Henriksson, J. A., Björkman, T., & Dawson, T. (2021). *Inner development goals: Background, method and the IDG framework*.

Kirchherr, J., & Piscicelli, L. (2019). Towards an Education for the Circular Economy (ECE): Five teaching principles and a case study. *Resources, Conservation and Recycling, 150*, Article 104406. https://doi.org/10.1016/j.resconrec.2019.104406

Kowasch, M. (2022). Circular economy, cradle to cradle and zero waste frameworks in teacher education for sustainability. *International Journal of Sustainability in Higher Education, 23*(6), 1404–1425. https://doi.org/10.1108/IJSHE-10-2021-0428

Kuppens, T. (2023). *De eigenschappen van de toekomstgerichte bedrijfseconoom vlaanderen.be/duurzaameducatiepunt*. https://publicaties.vlaanderen.be/view-file/62206

Kurznack, L., Schoenmaker, D., & Schramade, W. (2021). A model of long-term value creation. *Journal of Sustainable Finance & Investment, 1–19*,. https://doi.org/10.1080/20430795.2021.1920231

Lambrechts, W., Mulà, I., Ceulemans, K., Molderez, I., & Gaeremynck, V. (2013). The integration of competences for sustainable development in higher education: An analysis of bachelor programs in management. *Journal of Cleaner Production, 48*, 65–73. https://doi.org/10.1016/j.jclepro.2011.12.034

Lamenta, Z. A., & Grzybowska, K. (2023). Impact of the European green deal on business operations—preliminary benchmarking. *Sustainability, 15*(10), 7780. https://doi.org/10.3390/su15107780

Manninen, K., Laukkanen, M., & Huiskonen, J. (2024). Framework for sustainable value creation: A synthesis of fragmented sustainable business model literature. *Management Research Review, 47*(1), 99–122. https://doi.org/10.1108/MRR-03-2022-0177

Meadows, D. H., Meadows, D. L., Randers, J., & Behrens, W. (1972). *Limits to growth*.

O'Brien, E., & Cooney, T. M. (2025). Enhancing inclusive entrepreneurial activity through community engagement led by higher education institutions. *Journal of Enterprising Communities: People and Places in the Global Economy, 19*(2), 177–201. https://doi.org/10.1108/JEC-10-2023-0189

Pavlov, O. V., Cavana, R. Y., Wheat, I. D., Saeed, K., Radzicki, M., & Dangerfield, B. (2024). Introducing systems thinking and system dynamics models in economics education. In *The 42nd International System Dynamics Conference*.

Ploum, L., Blok, V., Lans, T., & Omta, O. (2018). Toward a validated competence framework for sustainable entrepreneurship. *Organization & Environment, 31*(2), 113–132. https://doi.org/10.1177/1086026617697039

Raworth, K. (2018). *Doughnut economics—Seven ways to think like a 21st-century economist.* Cornerstone.

Reichheld, A., Peto, J., & Ritthaler, C. (2023, September 18). Research: Consumers' sustainability demands are rising. *Harvard Business Review.*

Rieckmann. (2017). *Education for sustainable development goals: Learning objectives.* UNESCO. https://doi.org/10.54675/CGBA9153

Roorda, N. (2009). *Competentiekaarten voor duurzame ontwikkeling.*

Schulz, J., Hötte, K., & Mayerhoffer, D. M. (2024). Pluralist economics in an era of polycrisis. *Review of Evolutionary Political Economy, 5*(2), 201–218. https://doi.org/10.1007/s43253-024-00128-y

Simmons, V., Serafin, A., Stampone, A., & Rayeski, L. A. (2024). Integrating ESG into the accounting curriculum: Insights from accounting educators. *Issues in Accounting Education, 39*(2), 85–106. https://doi.org/10.2308/ISSUES-2022-080

Smitsman, A., Baue, B., & Thurm, R. (2021). *r3-0_Blueprint_9_V3.*

Soga, M., & Gaston, K. J. (2018). Shifting baseline syndrome: Causes, consequences, and implications. *Frontiers in Ecology and the Environment, 16*(4), 222–230. https://doi.org/10.1002/fee.1794

Vekeman, J., Audenaert, L., & Schildermans, P. (2022). *Hoe ziet het profiel van de Duurzame Ondernemer eruit?* https://www.omgeving.vlaanderen.be/sites/default/files/2022-11/Brochure%20Profiel%20Duurzame%20Ondernemer.pdf

Wiek, A., Withycombe, L., & Redman, C. L. (2011). Key competencies in sustainability: A reference framework for academic program development. *Sustainability Science, 6*(2), 203–218. https://doi.org/10.1007/s11625-011-0132-6

Yamane, T., & Kaneko, S. (2021). Is the younger generation a driving force toward achieving the sustainable development goals? Survey experiments. *Journal of Cleaner Production, 292*, Article 125932. https://doi.org/10.1016/j.jclepro.2021.125932

Strategic Advancement: Intentionally Aligning University Library Practices with the UN SDGs

Sarah K. Tribelhorn

Abstract This study examines the critical role of higher education institutions and academic libraries in advancing the United Nations' Sustainable Development Goals (SDGs). Integrating the SDGs into university operations, research, and curricula is essential for developing future leaders capable of addressing global challenges. The interconnected SDGs provide a framework for tackling poverty, inequality, and climate change, with academic libraries playing a vital role in providing access to relevant information and fostering information literacy. Drawing on established methodologies and practical strategies, this research explores effective approaches to SDG integration within academic libraries, using the San Diego State University (SDSU) Library as a local case study. By assessing existing initiatives, building internal capacity, prioritizing key areas, integrating SDGs into strategic plans, and monitoring progress, libraries can demonstrate their commitment to sustainability and social responsibility. This study highlights the importance of aligning library operations with the SDGs to contribute to a more sustainable and equitable future.

Keywords Academic libraries · Benchmarking · Sustainable development · Sustainability · UN SDGs

1 Introduction

Higher education institutions globally are tasked with preparing future leaders who can address complex global challenges. Integrating the United Nations' (UN) Sustainable Development Goals (SDGs) into the operations of higher education institutions, research, and curricula ensures students develop the knowledge, skills, and values necessary for sustainable development (Abo-Khalil, 2024). Sustainable development can be defined as "development that meets the needs of the present without compromising the ability of future generations to meet their own needs" (UN, 1987). The SDGs have been developed and found to promote critical thinking,

S. K. Tribelhorn (✉)
San Diego State University Library, San Diego, CA, USA
e-mail: stribelhorn@sdsu.edu

W. Leal Filho et al. (eds.), *Introducing the UN SDGs in Higher Education Institutions*, World Sustainability Series, https://doi.org/10.1007/978-3-032-23017-1_27

problem-solving, and interdisciplinary collaboration, all essential skills for global graduates entering a rapidly changing world (Abo-Khalil, 2024). The interconnected SDGs represent a comprehensive framework for tackling pressing global issues like poverty, inequality, and climate change. Higher education and libraries, as centers of knowledge and learning, have a crucial role in contributing to these solutions (Abo-Khalil, 2024; Molina et al., 2023). As determined by Molina et al. (2023) the SDGs are aimed at ending poverty; developing strategies for improving health, social, and economic inequalities; promoting economic growth; improving education; and enhancing environmental health globally. Therefore, institutions of higher learning have a role to play in these goals. Academic libraries play a vital role in providing access to information and resources in higher education institutions related to the SDGs, fostering information literacy, and supporting research that addresses these goals (Beasley, 2024). Higher education institutions and libraries have a social responsibility to contribute to the well-being of their communities and the world at large. Aligning with the SDGs demonstrates a commitment to this responsibility. Academic libraries have a social role that extends beyond supporting research and teaching/learning. They benefit from and contribute to society's networks of information resources (Beasley, 2024). Ultimately, aligning with the SDGs is about building a more sustainable and equitable future for all. Higher education and libraries globally have a key role to play in this effort.

To effectively integrate the SDGs into university and library operations, several key steps are essential, as has been documented for several Australian academic libraries (Thorpe & Gunton, 2022). Firstly, assessing existing initiatives and strategies that align with the SDGs provides a solid foundation. Secondly, building internal capacity and ownership of the SDGs among faculty, staff, and students is crucial for successful implementation. Thirdly, prioritizing key areas of focus, identifying opportunities for improvement, and addressing potential challenges is necessary. Fourthly, integrating the SDGs into strategic plans, policies, and operational procedures ensures their systematic implementation. Finally, monitoring progress, evaluating impact, and sharing results promotes transparency, accountability, and continuous improvement (Thorpe & Gunton, 2022). Beasley (2024) supports this approach, suggesting practical strategies for integrating SDGs into library operations. For example, the existing strengths within library collections that align with the SDGs can be highlighted and utilized. This could be carried out through the creation of a comprehensive library resource guide to showcase all SDG-related content, making it easily accessible to researchers and the wider community. Additionally, outreach to relevant campus and community members could be carried out to collaborate on curating and publicizing these collections, fostering a broader understanding of the SDGs and their significance (Beasley, 2024).

In addition to these international perspectives, understanding the significance of the SDGs is equally crucial in local settings, as demonstrated by the San Diego State University (SDSU) Library's initiatives. Understanding the importance of aligning academic library work in the SDSU Library with SDGs was underscored by several key factors. Firstly, the International Federation of Library Associations and Institutions (IFLA) emphasizes the vital role libraries play in achieving the SDGs and powering sustainable development (IFLA, 2020). Secondly, the SDSU Library's membership in the American Library Association (ALA), which adopted sustainability as a core value in 2019 (ALA, 2019), reinforces the library's commitment to sustainability. Lastly, sustainability is a core component of both SDSU's (2020) and the SDSU Library's strategic plans (SDSU Library, 2022a).

Recognizing the global importance of sustainable development, the SDSU library mapped its work to the SDGs in 2022–2023 as a benchmarking process. The initial goal of this project was to create an exhibit that would be part of a larger library display showcasing campus research related to the SDGs. This mapping provided a baseline to articulate the work being carried out toward sustainable development and each of the 17 goals. This "map" provided insight into the challenges and opportunities regarding sustainability in our library, and it aligned with the strategic plan of the university and library, as previously mentioned. This mapping activity provided a broad overview of sustainability in our library. However, although work toward the SDGs was being done, it was incidental and not intentionally being carried out to meet the SDGs and was also not articulated in this context. The initial benchmarking process was documented (Tribelhorn, 2023a) and the results have been regularly communicated to our library and greater community in various ways to illustrate our continual efforts to align with the SDGs.

The initial mapping process had several positive consequences. The first was realizing the need within the library to have a dedicated team working on sustainability. As such, a library sustainability committee was formed. The sustainability committee is dedicated to promoting sustainability in the library community by reducing its environmental impact, fostering social equity, and encouraging economic feasibility. The committee aims to educate the community about sustainable practices, reduce the library's carbon footprint, and serve as a model for environmentally sound operations (SDSU Library, 2022b).

To achieve the vision and mission of the sustainability committee, a Library Sustainability Policy (SDSU Library, 2024a) was developed to hold the library accountable for achieving the goals. One of the goals was to become accredited as "sustainable" by the benchmarking certification program of the Sustainable Libraries Initiative (SLI, 2018). The process of this certification has resulted in our library being recognized as a key player in the development of the new campus Climate Action Plan.

It was previously ascertained that certification and benchmarking processes could be a way of improving sustainability efforts in academic libraries (Tribelhorn, 2023b). Therefore, in addition to this certification, the Sustainability Committee is strategically building on the initial benchmarking study to focus on the strengths and challenges that were highlighted. In addition, the committee has identified specific targets within each of the SDGs and focus efforts on these. This chapter focuses on four of the SDG targets. The aim is to illustrate the relevance and impact of the SDSU library in context of the SDGs, and to educate students, faculty, staff and our community about the SDGs and sustainability.

2 Methods

2.1 SDG Target Benchmarking Process

To strategically align the work the SDSU Library is doing to achieve the SDG targets, a robust and comprehensive analysis, drawing upon employee roles, library project documentation, and website content, was performed following the methodological framework of Thorpe and Gunton (2022).

A survey was sent via email to all library employees through the internal Listserv to better understand their job functions and current projects. The email outlined the purpose of the study and encouraged participation. Based on the survey responses, a spreadsheet was created, and the work being carried out by all library employees was linked either directly or indirectly to each of the 17 SDGs. The mapping was based on the original analysis carried out by Tribelhorn (2023a), which was originally based on mapping suggested by the ALA Task Force on United Nations 2030 Sustainable Development Goals (ALA, 2021) for academic libraries. The methodological framework involved a targeted approach that allowed for a thorough examination of the library's contributions to specific goals, while acknowledging the valuable work being done in relation to the broader SDG framework.

Once this mapping was completed, specific targets were strategically identified for each goal depending on the work being carried out. Thus, the library was able to articulate the work being carried out toward sustainable development clearly and then intentionally focus on the SDG targets. The process was documented, and the findings have been communicated within the library and greater community.

The survey did not constitute research as defined by 45 CFR 46.102(e)(1) (National Archives & Records Administration, 2018), and was not required to be reviewed by the Independent Review Board (IRB) since it did not generate identifiable private information.

3 Results

Although the SDSU Library's efforts encompass all 17 SDGs, this study focused specifically on four goals: Quality Education, Reduced Inequalities, Sustainable Cities and Communities, and Partnerships for the Goals. These goals were prioritized due to their close alignment with the library's current strategic initiatives, allowing for a more focused and in-depth analysis within the constraints of this project's scope and available resources. This targeted approach enabled a thorough examination of the library's contributions to these specific goals, while acknowledging the valuable work being done in relation to the broader SDG framework. Future research may expand to encompass a wider range of SDGs, providing a more comprehensive view of the library's overall contributions to sustainable development.

3.1 SDG 1: No Poverty

For SDG 1: No Poverty Target 1.4 was identified: "By 2030, ensure that all men and women, in particular the poor and vulnerable, have equal rights to economic resources, as well as access to basic services, ownership and control over land and other forms of property, inheritance, natural resources, appropriate new technology and financial services, including microfinance" (UN Department of Economic and Social Affairs, n.d.).

As a library with a sustainability focus, it is essential to always consider ways to reduce costs and financial burdens to our students. One way to do this is through open access resources (OER), such as textbooks and e-books (Boczar & Pascual, 2017). In addition, given that OER often match the quality of traditional textbooks without compromising student learning (Cheung et al., 2012), they could be provided by the library to make these resources more affordable to students. Other research indicated considerable cost-savings to students through the adoption of OER (Hilton et al., 2014).

At SDSU, the library is working with our Instructional Technology Services, the Book Store, and our publishers, to provide first day open access to prescribed course textbooks. The aim of this is to increase access, affordability, and participation for students. The library has a dedicated Scholarly Communication and Open Initiatives librarian that sources these materials and makes them digitally available through library subscriptions with a budget provided by the bookstore. Since the inception of this program and partnerships in 2022, 151,299 students have enrolled in it, with 46,198 OER resources being provided through the library, resulting in considerable savings for the students: total savings $18,980,652—equivalent of about $127 per student on campus (ShopAztecs, 2024).

3.2 SGD 10: Reduced Inequalities

For SDG 10: Reduced Inequalities Target 10.3 was identified: "Ensure equal opportunity and reduce inequalities of outcome, including by eliminating discriminatory laws, policies and practices and promoting appropriate legislation, policies and action in this regard" (UN Department of Economic and Social Affairs, n.d.).

One of the ways that the SDSU Library is working toward reducing inequalities, and specifically by ensuring equal opportunities is through the support of Open Access (OA) publishing. Open Access (OA) aims to make scholarly research accessible, shareable, and reusable by removing barriers (Tennant et al., 2016). This approach promotes rapid dissemination of research findings, enabling broader participation in scholarly discourse and fostering innovation. Key motivations for OA include enhancing equity within and between research communities (Veletsianos & Kimmons, 2012) and mitigating the global North–South research divide (Adcock &

Fottrell, 2008) by breaking down financial barriers to accessing research, particularly for researchers in low-income countries or institutions with limited budgets. In addition, it also facilitates global collaboration by making research accessible to a diverse range of researchers, regardless of their geographical location or institutional affiliation. OA also promotes transparency and accountability of the research process (Björk B-C & Korkeamäki, 2017).

In the SDSU Library, the Scholarly Communication and Open Initiatives librarian has secured current open access publishing agreements in line with the university's research focus with five publishers: American Chemical Society, Cambridge Journals, Elsevier, IOP Journals, Wiley Journals. These agreements allow the library to pay a premium to publishers, enabling our researchers to publish OA articles in participating journals. This means SDSU researchers can publish in Gold OA or Hybrid OA journals without incurring Article Processing Charges (APCs), or at a significantly reduced cost (SDSU Library, 2024b). With these agreements SDSU authors have published more than 350 articles resulting in a savings of over $1,000,000. This has allowed many first-time authors or new researchers to publish without having to secure funding beforehand, ultimately making their research more accessible to more people.

3.3 SDG 13: Climate Action

For SDG 13: Climate Action we identified Target 13.3 "Improve education, awareness-raising and human and institutional capacity on climate change mitigation, adaptation, impact reduction and early warning" (UN Department of Economic and Social Affairs, n.d.).

The most fundamental reason why this was selected is based on the quote by Jamie Beck Alexander, Director of Drawdown Labs at Project Drawdown: "Every job is a climate job" (Rissman, 2023). Given the academic library's primary role in supporting learning, teaching, and research, it may seem counterintuitive to consider its potential role in addressing the climate crisis. However, libraries also serve a broader scholarly function by sharing resources and knowledge. By leveraging this role, libraries can actively contribute to global efforts to combat climate change, particularly through generating, providing, and disseminating relevant research and information (Appleton & Woolley, 2023).

Based on the initial project of mapping the work being carried out to the SDGs, it was determined that some structure was needed to accomplish the sustainability goals of the library. Therefore, as previously mentioned, a library sustainability committee was formed. This committee motivated for enrollment in the first official academic cohort of libraries in the Sustainable Libraries Initiative Certification Program (SLI, 2018). This aligns with the SDG work being carried out in the library and is a two-year benchmarking program analyzing all aspects of sustainability in the library, including environmental practices, economic feasibility, and social equity. Based on the outcomes of this certification program, the SDSU Library is better able to support campus sustainability initiatives, with library representatives on the SDSU Senate Sustainability Committee, and the SDSU President's Sustainability Advisory Council. In addition to the benchmarking being carried out, the sustainability committee is committed to providing educational programs drawing attention to climate issues and encouraging more sustainable lifestyles. The work of the sustainability committee is being carried out with no budget, so low-cost programming with related education has been developed, for example a T-shirt-to-Tote workshop around Earth Day, and a Sustainable Gift-Wrapping workshop for the holidays. Other events include citizen science programming focusing on climate action, and a digital spring clean to highlight the importance of reducing our carbon footprint.

3.4 SDG 17: Partnerships for the Goals

For Goal 17: Partnerships for the Goals, Target 17.17 was identified: "Encourage and promote effective public, public–private and civil society partnerships, building on the experience and resourcing strategies of partnerships" (UN Department of Economic and Social Affairs, n.d.).

Embree and Gilman (2020) recognized the role of libraries, often as the center of their campuses and communities and pursued the development of their academic library as a "Sustainability Hub". With the evolution of this Sustainability Hub, they recognized the importance of community collaborations and partnerships. Similarly, Tans (2017) noted the importance of collaborations between campus units to advance sustainability goals in academic libraries for their communities. These partnerships and collaborations are fundamental for sustainable development, including community and stakeholder participation (Tans, 2017).

Some of the partnerships identified in the SDSU Library mapping process can also be included in other goals. However, some of the partnerships related to target 17.17 include the work being carried out by the San Diego Circuit Information Issues for Social Impact Committee. San Diego Circuit started out as a lending consortium between the academic and public libraries in San Diego County. This committee was formed during COVID-19 as the Health Advisory Committee, with representatives from each of the collaborating libraries, and was based on the county's need to provide accurate health information to the public, focusing primarily on underrepresented community members, to combat health misinformation. To ensure the underrepresented communities were reached and received the support they needed, town halls in different communities were hosted to understand their requirements. The areas with most need were identified through indexes developed by our local government. The projects included the development of a comprehensive toolkit and resources (San Diego Circuit Libraries, 2022), advertising campaign, and author talks, based on community input. The toolkit for this has had 8,525 views since 2022. The work this group achieved was possible through a collaborative grant and was so successful that the programming was further developed for climate resiliency, also resulting in a toolkit (San Diego Circuit Libraries, 2024), author talks, and educational programming. The toolkit for this has had 3,719 views since January 2024.

Also resulting from this San Diego Circuit collaboration were campus, and specifically, student partnerships with the SDSU Library. For one of the above-mentioned sustainability related author talks, the libraries collaborated with two SDSU student committees: the sustainability committee and mental health committee. These student bodies had input into what they wanted the event to look like, what topic they would like to have highlighted, the author, and they also hosted the event.

In addition to these partnerships, the SDSU library launched citizen (participatory) science programming, a direct result of the first San Diego Circuit collaboration to highlight equity and diversity in health care. The library also developed citizen science programs to support campus research and initiatives focused on plastic pollution control and biodiversity, collaborating with the San Diego State University Biodiversity Museum for Earth Day and the City Nature Challenge.

The SDSU library has many other partnerships, some of them extend to our cross-border communities in Mexico, particularly in Tijuana, and Oaxaca. These are digitization projects, where the SDSU library is helping to digitize important documents with the institutions involved. This is helping to preserve their history and provides access to previously inaccessible material. In these partnerships we are ensuring that we do not control what happens to the digitized products, but that the partner institutions are making these decisions. In addition, the library also has many campus department collaborations, working together with faculty to develop programs to support student success.

4 Limitations

Despite peer review, the study acknowledged the potential for subjectivity in mapping employee roles to the SDGs. To enhance transparency in future analyses, a rubric with clear criteria could be implemented. The study also relied on self-reported survey data, which may introduce biases or inaccuracies. Although all employees were invited to participate, the 20% response rate, supplemented by website information and follow-up interviews, resulted in potential data gaps due to non-response. Time constraints limited further data collection. Additionally, the scope of the study focused on employee roles, library projects, and the website, potentially overlooking areas such as collection development and community outreach. Future analyses will expand the scope for more comprehensive results and incorporate feedback from external stakeholders to validate SDG alignment.

5 Conclusion

In conclusion, this research demonstrates that academic libraries possess the potential to make a substantial contribution towards addressing global challenges, even in a local context, and fostering a more sustainable and equitable future. Through strategic efforts, the SDSU Library has successfully contextualized, articulated, and intentionally aligned its work with the SDG targets. Increased communication of these efforts leads to greater integration of the SDGs within the library's operations. Recognizing that sustainable development is an ongoing and evolving process, the SDSU Library will continue to map its activities to the targets of all 17 SDGs. This study reveals that progress towards sustainability does not require perfection; rather, it emphasizes the importance of initiating action, even through small-scale efforts. These incremental gains generate momentum, ultimately fostering broader societal change. The research has identified a relatively simple process that enables all libraries to contextualize their contributions. By adopting this approach, academic libraries can effectively demonstrate their sustained commitment to sustainability and social responsibility, thereby assuming leadership roles within their respective campuses.

References

Abo-Khalil, A. G. (2024, April 18). Integrating sustainability into higher education challenges and opportunities for universities worldwide. *Heliyon, 10*(9), e29946. https://doi.org/10.1016/j.heliyon.2024.e29946

Adcock, J., & Fottrell, E. (2008). The North-South information highway: case studies of publication access among health researchers in resource-poor countries. *Glob Health Action, 1*(1). https://doi.org/10.3402/gha.v1i0.1865

American Library Association. (2019). *Core values of librarianship*. https://www.ala.org/advocacy/advocacy/intfreedom/corevalues

American Libraries Association. (2021). *ALA task force on United Nations 2030 sustainable development goals*. https://www.ala.org/aboutala/ala-task-force-united-nations-2030-sustainable-development-goals

Appleton, L., & Woolley, N. (2023). The role of academic libraries in climate action (Editorial). *New Review of Academic Librarianship, 29*(2), 117–122. https://doi.org/10.1080/13614533.2023.2230652

Beasley, G. (2024, December 20). *What is the role of academic libraries in sustainable development?* Katina: Librarianship Elevated. https://katinamagazine.org/content/article/future-of-work/2024/role-of-academic-libraries-in-sustainable-development

Björk, B.-C., & Korkeamäki, T. (2020). Adoption of the open access business model in scientific journal publishing: A cross-disciplinary study. *College & Research Libraries, 81*(7), 1080. https://doi.org/10.5860/crl.81.7.1080

Boczar, J., & Pascual, L. (2017). E-books for the classroom and open access textbooks: Two ways to help students save money on textbooks. *The Serials Librarian, 72*(1–4), 95–101. https://doi.org/10.1080/0361526X.2017.1309830

Cheung, S. K. S., Yuen, K. S., Li, K. C., Tsang, E. Y. M., & Wong, A. (2012). Open access textbooks: Opportunities and challenges. In K. C. Li, F. L. Wang, K. S. Yuen, S. K. S. Cheung & R. Kwan (Eds.), Engaging learners through emerging technologies. In *ICT 2012. Communications in*

Computer and Information Science (Vol. 302). Berlin, Heidelberg: Springer. https://doi.org/10.1007/978-3-642-31398-1_18

Embree, J., & Gilman, N. (2020). The library as a campus sustainability hub: a case study in increasing community engagement and collaboration in sustainability through academic libraries. *International Journal of Librarianship, 5*(2):26–44. https://doi.org/10.23974/ijol.2020.vol5.2.172

Hilton III, J., Robinson, T., Wiley, D., & Ackerman, J. (2014). Cost-savings achieved in two semesters through the adoption of open educational resources. *International Review of Research in Open and Distributed Learning, 15*(2):67–84. https://doi.org/10.19173/irrodl.v15i2.1700

International Federation of Library Associations and Institution. (2020). *Powering sustainable development*. https://www.ifla.org/units/sustainable-development/

Molina, A. A., Hellden, D., Alfven, T., Niemi, M., Leander, K., Nordenstedt, H., Rehn, C., Ndejjo, R., Wanyenze, R., & Biermann, O. (2023). Integrating the United Nations sustainable development goals into higher education globally: A scoping review. *Global Health Action, 16*, 2190649. https://doi.org/10.1080/16549716.2023.2190649

National Archives and Records Administration (2018) *Code of federal regulations: Part 46—Protection of human subjects*. https://www.ecfr.gov/on/2018-07-19/title-45/subtitle-A/subchapter-A/part-46

Rissman J (2023) Every job is a climate job with Project Drawdown [Podcast]. *Invested in Climate, 37*. https://investedinclimate.com/podcast/every-job-is-a-climate-job-with-project-drawdown-ep-37/

San Diego Circuit Libraries. (2022). *Library toolkit for addressing health misinformation*. https://libguides.sdsu.edu/library-toolkit-addressing-health-misinformation

San Diego Circuit Libraries. (2024). *Sustainability and climate change resources for San Diego*. https://libguides.sdsu.edu/Circuit-Sustainability

San Diego State University. (2020). *We rise. We defy. Transcending borders, transforming lives. San Diego State University Strategic Plan, 2020–2025*. https://www.sdsu.edu/strategic-plan/pdfs/sdsu-strategic-plan-2020-2025-report-082620.pdf

San Diego State University Library. (2024a). *SDSU University library sustainability policy*. https://library.sdsu.edu/about/policies-guidelines/_files/sdsu-university-library-sustainability-policy-draft.pdf

San Diego State University Library. (2024b). *Open access support @ SDSU*. https://libguides.sdsu.edu/oa_support

San Diego State University Library. (2022a). *Transcending borders: The San Diego State University Library Strategic Plan, 2022–2025*. https://library.sdsu.edu/about/strategic-plan/priorities

San Diego State University Library. (2022b). *Sustainability in the SDSU library*. https://libguides.sdsu.edu/c.php?g=1361366&p=10142617

ShopAztecs. (2024). *day1ready—Formerly equitable access*. https://www.shopaztecs.com/day1ready?srsltid=AfmBOoogB_ZcOyrbcEqTBm9Q4GDjEu0Qz25XXwqu90eD4bDyFaZYxRy8

Sustainable Libraries Initiative. (2018). *What is SCLP?* https://sustainablelibrariesinitiative.org/about-us/program-faq

Tans, E. D. (2017). Sustainable academic libraries: A campus partnership at Michigan State University. In W. Leal Filho, M. Mifsud, C. Shiel & R. Pretorius (Eds.), *Handbook of theory and practice of sustainable development in higher education. World sustainability series*. Cham: Springer. https://doi.org/10.1007/978-3-319-47895-1_6

Tennant, J. P., Waldner, F., Jacques, D. C., Masuzzo, P., Collister, L. B., & Hartgerink, C. H. (2016). The academic, economic and societal impacts of Open Access: An evidence-based review. *F1000Res, 5*, 632. https://doi.org/10.12688/f1000research.8460.3

Thorpe, C., & Gunton, L. (2022). Assessing the United Nation's Sustainable Development Goals in academic libraries. *Journal of Librarianship and Information Science, 54*(2), 208–215. https://doi.org/10.1177/09610006211005528

Tribelhorn, S. K. (2023a). Planning for sustainability: Articulating efforts toward the United Nations Sustainable Development Goals. *College & Research Libraries News, 84*(7). https://crln.acrl.org/index.php/crlnews/article/view/25960/33896

Tribelhorn, S. K. (2023b). Preliminary investigation of sustainability awareness and activities among academic libraries in the United States. *Journal of Academic Librarianship, 49*(3), Article 102661. https://doi.org/10.1016/j.acalib.2022.102661

United Nations. (1987). *Our common guture: The world commission on environment and development*. OUP. http://www.un-documents.net/our-common-future.pdf

United Nations Department of Economic and Social Affairs. (n.d.). *The 17 goals*. https://sdgs.un.org/goals

Veletsianos, G., & Kimmons, R. (2012). Assumptions and challenges of open scholarship. *The International Review of Research in Open and Distributed Learning, 13*(4), 166–189. https://www.oerknowledgecloud.org/archive/1313-10780-1-PB.pdf

Transforming Cities, Empowering Universities: The City-University Partnerships Model at the Universitat Politècnica de València for SDGs

Oksana Udovyk, Guillermo Palau Salvador, Ana Escario Chust, and Jordi Peris-Blanes

Abstract City-University Partnerships (CUPs) hold transformative potential in addressing urban sustainability challenges by bridging academic knowledge with real-world urban governance. These partnerships serve as crucial mechanisms for fostering systemic change by integrating research, policy, and practice to advance the Sustainable Development Goals (SDGs). However, the extent to which CUPs contribute to deep sustainability transformations depends on their governance structures, strategic orientation, and ability to institutionalize long-term collaboration. This chapter examines the transformative potential of CUPs by analyzing the evolution, governance, and impact of the Valencia-UPV CUP. It highlights key enabling factors such as trust-building, alignment with international sustainability frameworks, and strategic relationship development. Additionally, it addresses the challenges faced, including political volatility and the necessity of embedding collaborative practices within institutional frameworks. Drawing on this case study, the chapter provides practical insights and theoretical contributions to strengthening CUPs as vehicles for systemic change, emphasizing their role in advancing SDGs and fostering resilient, inclusive urban governance.

Keywords Urban sustainability · Climate neutrality · Transformation · Systemic change · Political volatility

1 Introduction

The United Nations' Sustainable Development Goals (SDGs) have become pivotal in shaping global sustainability agendas, with Higher Education Institutions (HEIs) increasingly recognized as key players in their implementation. Through education,

O. Udovyk (✉) · G. Palau Salvador · A. Escario Chust · J. Peris-Blanes
INGENIO (CSIC-UPV), Universitat Politècnica de València, Valencia, Spain
e-mail: oudovyk@upv.edu.es

W. Leal Filho et al. (eds.), *Introducing the UN SDGs in Higher Education Institutions*, World Sustainability Series, https://doi.org/10.1007/978-3-032-23017-1_28

research, and community engagement, HEIs have the potential to drive sustainability efforts at both local and global scales. One particularly effective mechanism for realizing the SDGs at the local level is the development of City-University Partnerships (CUPs), where universities collaborate with local governments and various stakeholders to tackle urban sustainability challenges. These partnerships align local initiatives with the broader goals of the SDGs, creating powerful synergies for change.

CUPs can be instrumental in driving urban transformation toward sustainability (Keeler et al., 2019). They move beyond traditional university-community engagement by fostering co-creation of knowledge, resources, and strategies to address complex urban problems. A growing body of literature highlights the potential of CUPs to catalyze systemic change by integrating research, education, and practice to address interconnected challenges such as climate change, social equity, and economic development (e.g. Guenduez et al., 2024; Torfing et al., 2024). Transformative CUPs can address historic and systemic issues, foster shared values across organizations, and establish specific, measurable sustainability goals (Keeler et al., 2018). This holistic approach aligns with the broader understanding of sustainability transitions, which require integrated solutions across multiple sectors and scales (Loorbach & Wittmayer, 2024).

For example, the Viable Cities program in Sweden exemplifies the impact of CUPs on urban sustainability. Viable Cities is a strategic innovation program that brings together universities, municipalities, industry, and civil society to accelerate the transition toward climate-neutral cities by 2030, all led by the University. By fostering innovation and aligning local actions with national and international sustainability frameworks, the program demonstrates the potential of collaborative efforts between city and University in achieving ambitious climate objectives (Moreno-Serna et al., 2024).

What such a case shows us is that achieving true sustainability transformation demands long-term collaboration between universities and cities, with a shared commitment to developing a transformative portfolio. As argued by Keeler et al. (2023), this type of partnership must be built on enduring relationships, underpinned by sustained engagement and deep relationship-building among organizations. The process takes time and dedication but is crucial for creating the kind of systemic change required for lasting sustainability. For transformative CUPs to be successful, they must address historic and systemic issues, foster shared values across organizations, and establish specific, measurable sustainability goals.

However, while CUPs are becoming more common, many still focus narrowly on short-term, project-based collaborations focused on specific areas such as research, development, or workforce training (Caughman et al., 2020). Although these projects may incorporate elements of sustainability, few CUPs are designed explicitly to prioritize and advance **transformational** sustainability goals. These goals often involve addressing systemic issues such elimination of energy poverty (Keeler et al., 2018; Trencher et al., 2014), which require the reconfiguration of socio-technical systems through a combination of technical solutions, policy reforms, and behavioral changes.

This chapter aims to address this gap by examining the Valencia City—Universitat Politècnica de València (Valencia-UPV) CUP, a partnership that has evolved to prioritize long-term, transformational sustainability goals. By exploring the dynamics of such partnership, we aim to identify the **key enabling factors** and **supporting structures** that facilitate this transformation. Additionally, we will investigate the **challenges** the Valencia-UPV CUP encounters in its pursuit of transformative goals. Understanding these factors and challenges will provide valuable insights for other CUPs seeking to transition into transformative partnerships and build the capacity needed to address systemic urban challenges. In Peris el al. (2024) and Udovyk et al. (2024) we showed that the Valencia-UPV CUP was moving from routine or strategic modes of operation into a transformative partnership with the focus on net-zero city transformation. Here we will explore **the enabling factors and supporting structures** that facilitated this partnership. Additionally, we will investigate the **challenges** the CUP has encountered in its pursuit of transformative goals.

This chapter will therefore analyze the process of defining the Valencia-UPV CUP agreement and examine its implementation within UPV.

To do so, the conceptual framework for understanding City-University Partnerships (CUPs) is outlined, plus their potential for transformative impact. An in-depth case study of the Valencia-UPV CUP is hten presented to illustrate real-world applications, followed by an analysis of enabling factors, support structures, and challenges. The chapter concludes with key lessons learned, successes and practical recommendations for universities and cities seeking to build long-term, transformative partnerships for sustainable urban development, adding valuable insights to the ongoing discussions on CUPs, urban transition studies, and mission-oriented governance.

2 City-University Partnerships: Conceptual Framework

2.1 City-University Partnerships (CUP)

City-University Partnerships (CUPs) are strategic, long-term collaborations between academic institutions and municipalities, created to address complex urban challenges and promote sustainable development. These partnerships operate within a broad ecosystem involving various stakeholders, including local governments, businesses, community groups, and academic institutions. By leveraging their combined resources and expertise, CUPs generate innovative solutions to pressing urban issues, particularly those related to sustainability (Keeler et al., 2019).

CUPs are instrumental in aligning local sustainability efforts with broader global frameworks, such as the United Nations Sustainable Development Goals (SDGs) and the European Union's climate objectives. These partnerships not only contribute to policy development and urban planning but also transform university campuses into "living labs" where SDG-aligned innovations can be tested and scaled (Keeler et al., 2018). Universities play a pivotal role by integrating sustainability into their curricula and producing research directly addressing SDG-related challenges. Additionally, CUPs enhance community engagement through participatory research, ensuring local stakeholders' voices are an integral part of sustainability efforts (Caughman et al., 2020).

2.2 Modes of Operation in City-University Partnerships

CUPs share common structural elements, such as formal agreements between universities and city administrators, but their operational modes differ significantly. These modes can be categorized into three types of partnerships based on the complexity of the challenges they address: routine, strategic, and transformative partnerships (Caughman et al., 2020):

Routine partnerships are transactional, focusing on straightforward, short-term projects with minimal collaboration. An example is the mapping of street trees in Portland (Keeler et al., 2019), which required limited joint effort and addressed a simple problem. **Strategic partnerships** involve more co-creation between university and city, where both parties contribute to shaping goals and solutions for more complex, value-driven challenges. An example is the collaboration between Tempe, Arizona, and Arizona State University (ASU) (Keeler et al., 2019) to create a climate action plan, addressing multifaceted issues such as climate resilience. Finally, **transformative partnerships** are deeply collaborative, aiming for systemic change by tackling complex, "deep-rooted" problems. These partnerships involve long-term planning, ongoing adaptation, and mission alignment between institutions. A notable example is the partnership between the University of British Columbia and the City of Vancouver, which focuses on holistic climate adaptation strategies (Keeler et al., 2019).

Given the increasing complexity of current sustainability challenges, there is a growing need for transformative partnerships. These partnerships need to focus on long-term solutions that involve continuous learning and adaptation, making them particularly suitable for addressing issues such as those defined by SDGs.

2.3 Transformative CUPs: Building Capacity for Systemic Change

A central feature of transformative City-University Partnerships (CUPs) is their focus on building transformative capacity (Wolfram, 2016), within both the university and the city institutions involved. Transformative capacity refers to the ability of organizations and individuals to initiate, support, and sustain systemic changes toward sustainability (Wolfram, 2016). Unlike routine or strategic partnerships, transformative CUPs aim to foster deep, long-lasting change by integrating sustainability principles into the core operations of both institutions (Keeler et al., 2018). This requires more than just collaboration on specific projects—it involves creating an ongoing process of mutual learning, co-creation, and structural adaptation that can address the complex, interconnected challenges cities face, such as delivering SDGs.

Keeler et al. (2018) suggest that for CUPs to develop transformative capacities, they must foreground at the outset relationship development and maintain an orientation toward shared sustainability values, goals, and missions. These shared goals must then be supported by structures that ensure the partnership's long-term health and functionality. To do so, **there has to exist**:

Motivation	Sustained drive from stakeholders to maintain long-term engagement
Reflexivity	Feedback mechanisms to adjust and evolve partnership goals over time
Resources	Availability of financial, political, and human capital to support growth
Process	Formal and informal structures that enhance partnership effectiveness
Participation	Engagement across different levels of the organizations involved
Understanding	Shared comprehension of the values, cultures, and histories of the partners

Furthermore, Keeler et al. (2023), identify number of **enabling** factors for CUPs **transformative capacities development**:

Building Trust	Establishing trust through understanding and addressing past conflicts between partners
Focus on Transformational Goals	Expanding partnership objectives to tackle systemic issues and root causes of sustainability problems
Invest in Relationships	Dedicating time and effort to developing strong relationships among stakeholders
Supportive Norms and Structures	Creating organizational norms and structures that support long-term collaboration
Design Motivating Actions	Developing collaborative activities that have meaningful, mutual benefits for all parties involved

These factors are crucial in ensuring that transformative partnerships do not become transactional or strategic, but instead remain focused on long-term systemic change. By addressing these elements, CUPs can build the transformative capacity needed to tackle sustainability challenges effectively.

3 Methodology

This study employs a qualitative case study methodology to analyze the Valencia-UPV City-University Partnership (CUP). The case study method allows for an in-depth exploration of the CUP's development, governance structures, and also project implementation, making it well-suited to investigating the nuances of city-university transformative partnerships. Data was gathered through a combination of primary and secondary sources, offering a comprehensive understanding of the CUP's dynamics.

For the primary sources, participant observation was used since two of the authors served as ctive participants in the CUP's creation and steering processes. This method provided a unique insider perspective that enriched the analysis with firsthand observations of decision-making processes, stakeholder interactions, and the iterative nature of the CUP's development. The authors maintained reflexive journals throughout this period, documenting real-time insights into the governance challenges and informal mechanisms that are often underrepresented in formal documents. This ethnographic component captured the lived experiences of those involved, offering deeper insights into the dynamics of trust-building, conflict resolution, and institutional change.

In addition, a collaborative workshop was facilitated with around 30 participants from both the UPV and the City Council of Valencia. This workshop played a key role in sharing ideas for the configuration of the CUP, enabling stakeholders to collectively define the agreement and governance structure. The participatory approach of the workshop fostered a collaborative vision and ensured that the partnership was aligned with the needs and priorities of both the university and the city.

Following Flyvbjerg's (2006) emphasis on context-based insights, informal conversations were used to gather nuanced data within the Valencia-UPV CUP. Conducted organically during interactions, these conversations offered unique perspectives that complemented formal data sources. Reflective notes ensured accuracy, while anonymity and ethical guidelines were maintained throughout. These interactions were meticulously noted and anonymized to maintain ethical standards and ensure the integrity of the data.

For the secondary sources, the analysis included key documents such as the CUP agreement, progress reports, strategic plans, and related policy documents from both the Universitat Politècnica de València (UPV) and the City of Valencia. These documents provided formal insights into the governance structure, project goals, and alignment with broader sustainability frameworks such as the European Union's climate mission.

We acknowledge the potential for bias inherent in participant observation, especially given the authors' dual role as both researcher and practitioner in the CUP's development. This insider perspective provided privileged access and deeper insight into the partnership's operations but also risked subjective interpretation of successes and challenges. To mitigate this potential bias, we employed triangulation, cross-referencing ethnographic observations with data from document analysis (Flyvbjerg, 2006). This ensured that findings were grounded in multiple sources of evidence, providing a more balanced and holistic understanding of the CUP's dynamics., Reflexive practices were also adopted, which involved internal conversations to critically assess our own observations against documented evidence, enabling us to address potential subjective biases effectively.

3.1 *Case Study: The València-UPV CUP on Climate Neutrality*

The City-University Partnership (CUP) between the Universitat Politècnica de València (UPV) and the City of Valencia exemplifies a strategic and dynamic collaboration aimed at advancing urban sustainability and climate action.

The broader context for this collaboration is shaped by the EU's push for climate neutrality, which has prompted cities across Europe to implement innovative and systemic sustainability strategies to become ¨net-zero¨. The EU has followed the mission-oriented approaches, as advocated by, emphasizing the setting of ambitious, measurable targets while mobilizing diverse stakeholders and resources. The EU's climate mission operates within this framework, providing cities with the necessary structure and incentives to accelerate their transitions towards climate neutrality.

Valencia was an early adopter of this approach, integrating mission-oriented governance into its urban policies as early as 2019, pre-empting its selection in 2022 as one of the 100 cities participating in the EU Climate-Neutral and Smart Cities Mission. This selection further reinforced the city's commitment to a participatory and collaborative approach, involving a broad array of stakeholders in shaping Valencia's sustainable future. Among these actors, UPV emerged as a particularly influential partner, actively contributing research, technological expertise, and institutional support to the city's climate transition efforts.

The formation of the CUP between UPV and the City of Valencia was not an isolated event but rather the outcome of a structured and evolving collaboration. Over time, the partnership evolved from isolated academic contributions to a more integrated, co-governance model in which the university played a key role in policy development, project implementation, and citizen engagement. This transition was facilitated through various institutional mechanisms, including joint research initiatives, funding programs, and policy co-design processes. The collaboration has fostered a synergistic exchange of knowledge and expertise, enabling Valencia to leverage the university's scientific and technological capabilities while offering UPV a real-world testing ground for sustainability solutions. The effectiveness of such a partnership is increasingly recognized in urban governance literature, as universities serve as anchor institutions that can drive systemic change by bridging the gap between policy, research, and community action.

By closely examining the timeline of the CUP's development, it becomes clear that this partnership has undergone a progressive evolution, shaped by key policy decisions, institutional commitments, and external funding opportunities.

This timeline highlights the key moments in the creation of the City-University Partnership (CUP) between the Universitat Politècnica de València (UPV) and the municipality of Valencia. Each step reflects the progressive development of this collaborative effort (see Fig. 1).

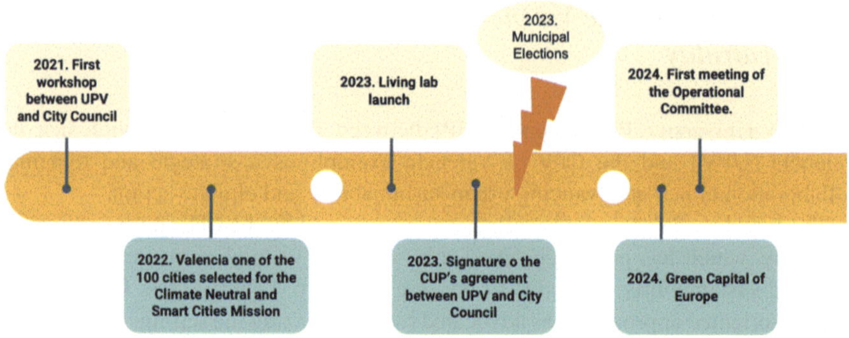

Fig. 1 Temporality of the actions for Valencia's CUP. *Source* Authors

- 2021: Workshop between UPV and the City Council, laying the foundation for future cooperation, defining the strategic lines for the agreement and the governance of the CUP.
- 2022: Valencia's selection as one of the 100 cities in the EU's Climate Neutral and Smart Cities Mission positioned the city as a leader in climate action.
- 2023: Marked by the launch of a living lab and the formal signing of the CUP agreement, this year saw major advancements. The municipal elections also influenced the direction of local governance and sustainability priorities.
- 2024: The CUP's Operational Committee held its first meeting, formalizing the structure for collaboration. In the same year, Valencia was named Green Capital of Europe, celebrating its achievements in sustainability.

3.2 *Starting Point*

The journey of the City-University Partnership (CUP) between the Universitat Politècnica de València (UPV) and the City of Valencia began with Valencia's early commitment to a mission-oriented approach in 2019. Inspired by the European Union's mission-driven framework, Valencia city started to develop its Mission Framework, later identifying **Climate** as its first city mission, setting an ambitious goal to address urban sustainability challenges through an integrated, strategic approach. This decision marked a significant shift for the city, moving from isolated sustainability projects to a comprehensive, long-term vision aimed at achieving climate neutrality in cooperation with different actors, including University.

In parallel, UPV reinforced its commitment to sustainability by establishing the **Alliance of UPV Chairs for Sustainability** in 2020. This alliance, a coalition of twelve specialized *cátedras* (chairs), was foundational in positioning UPV as a key player in Valencia's sustainability transition. Each chair focused on a distinct area of sustainability, ranging from **energy transitions** and **urban resilience** to **climate governance** and **innovation for sustainable cities**. Funded by the City of Valencia and the regional government, the Alliance acted as a bridge between academia, policy, and the community, fostering interdisciplinary research and collaboration.

In 2020, the Vice Rectorate of Sustainability at UPV reached out to Sweden's **KTH Royal Institute of Technology** to explore its model for urban sustainability and its partnerships with Swedish cities through the **Viable Cities initiative**. Viable Cities is a collaborative national program in Sweden focused on achieving climate neutrality by supporting cities with knowledge, technology, and research. Through this initiative, KTH collaborates with various Swedish cities, offering expertise in sustainable urban planning and technology-driven solutions to meet climate targets. This early engagement provided UPV valuable insights into structuring city-university collaborations for climate action.

A year later, in 2021, the UPV organized a **first workshop between UPV and the City Council**, which set the stage for collaboration. This workshop served as a comprehensive reflection and collaborative brainstorming session that was instrumental in shaping the final agreement. Discussions, facilitated by an expert team (see Fig. 2), encompassed not only strategic directions but also governance structures, participant roles, and both potential and desired future scenarios. This holistic approach allowed participants to align on key priorities and establish a shared framework for the partnership's long-term objectives.

In addition, 8 strategic lines of action were identified by analyzing the information gathered during the participatory workshop and trying to define clear activities for the operationalization of the CUP.

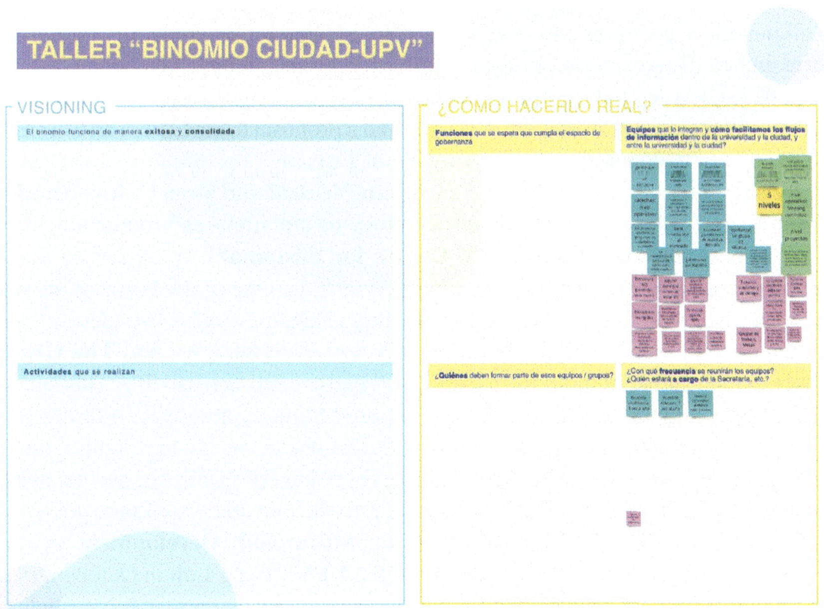

Fig. 2 Example of one of the visual supports that the groups used for reflection. *Source* Authors

L1: Develop an innovative way of working together.
L2: Share strategic insights.
L3: Generate and share valuable knowledge to contribute to the Climate Mission.
L4: Develop projects with transformative capacity.
L5: Offer experimentation spaces and controlled test beds (sandbox).
L6: Work on strategic alliances and collaborations that act as leverage.
L7: Create and manage forums for dialogue and multi-stakeholder reflection.
L8: Communicate the social value of urban transformation with the university community.

3.3 CUP Development—Pivotal Initial Actions

A pivotal moment in the CUP's trajectory occurred in 2022, when **Valencia was selected as one of the 100 cities in the EU's Climate Neutral and Smart Cities Mission**. This selection positioned Valencia as a leader in climate action, catalyzing its commitment to innovative, sustainable urban development. This EU initiative was instrumental in framing the city's ambitions, aligning them with European targets, and setting the stage for a deeper collaboration with Universitat Politècnica de València (UPV) to meet climate neutrality goals.

In response to this momentum, the municipality started informal **Mission Friday's meetings**, bringing together different departments of the City of Valencia and also UPV. These regular meetings facilitated strategic dialogue and co-creation, enabling the alignment of priorities and fostering a shared vision for sustainability initiatives, addressing challenges in real-time, and reinforcing trust among stakeholders.

In 2022, UPV and the City of Valencia, together with other stakeholders from the city ecosystem, jointly developed and submitted a proposal to the European Commission to position Valencia as a **Green Capital of Europe**. This initiative aligned with the city's recent designation in the EU's **Climate Neutral and Smart Cities Mission** and showcased Valencia's commitment to achieving ambitious environmental goals.

In response, the **Alliance of UPV Chairs for Sustainability** played an active role in advancing UPV's internal sustainability efforts. One of the first actions was implementing foresight exercises and conducting extensive surveys to understand the community's attitudes toward climate action. A notable initiative was **The Grand Debate**, which gathered input from over 2,000 participants across UPV. This survey assessed both support and resistance to proposed climate initiatives, offering critical insights into areas of alignment and potential obstacles. By highlighting areas needing additional engagement, The Great Survey helped UPV tailor its sustainability strategies, fostering campus-wide commitment to the university's climate goals.

Simultaneously, UPV's **Vice-Rectorate for Sustainable Development**, central to implementing climate strategies, launched the **UPV Living Lab** in October 2022. The Living Lab was conceived as an experimental, co-creative research space where students, faculty, and community stakeholders could collaborate on projects aimed at achieving carbon neutrality and broader environmental sustainability. This hub not only served as a practical space for innovation but also as a model for how academic institutions can drive applied climate solutions that directly impact the university's operations and, by extension, the wider city.

The Vice-Rectorate reinforced this commitment by issuing calls for innovative, sustainability-focused projects aligned with the **València 2030 Climate Mission**, concentrating on campus decarbonization. Selected projects tackled renewable energy, sustainable water management, and the reduction of carbon emissions in daily campus operations. These projects were designed not only to improve the university's sustainability practices but also to serve as case studies for larger urban initiatives.

In an effort to further embed sustainability into academic programs, UPV launched an initiative to integrate **Final Bachelor's (TFG) and Master's (TFM) Thesis projects** with the decarbonization goals of both the campus and the city. This initiative, aligned with the **UniCities2030 project**, created new pathways for collaboration between UPV faculty and city officials, promoting a multidisciplinary approach to sustainability. By focusing student research on tangible urban challenges, the TFM and TFG projects enabled students to contribute to real-world sustainability efforts. Joint supervision by professors and municipal experts ensured that these projects bridged academic research and practical application, directly contributing to sustainable solutions on campus and within Valencia.

A significant milestone in the formalization of this partnership was the creation of a **joint commission** between UPV, the City Council of Valencia, and the Las Naves Foundation. This commission was tasked with drafting iterations of the City-University Partnership (CUP) agreement, each version acting as a foundation for discussion and refinement. Through collaborative meetings, stakeholders worked to align the university's sustainability initiatives with the city's overarching climate objectives. The final CUP agreement emerged from these discussions, solidifying a structured framework for coordinated action on carbon neutrality. This agreement encapsulated the shared commitment of UPV and the City of Valencia to the **València 2030 Climate Mission**, fostering a cohesive approach to urban sustainability and resilience.

3.4 Signing of the Valencia-UPV CUP Agreement

In 2023, a landmark agreement between the Universitat Politècnica de València (UPV) and the Valencia City Council formally cemented their collaborative efforts towards sustainability. This **agreement** signaled a unified commitment to embed science, research, and innovation into the city's transformation process, strategically aligning these efforts with global sustainability frameworks. The **Valencia-UPV City-University Partnership (CUP)** creates a platform for co-governance, fostering dialogue and collaboration, which benefits both the city and the university across multiple dimensions. The formalization of this agreement has provided

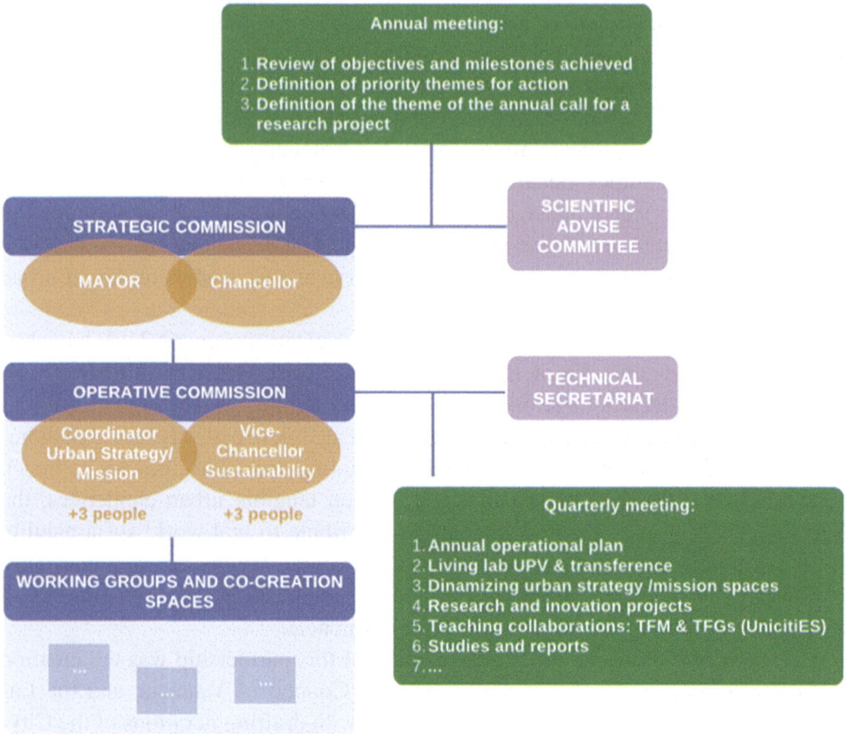

Fig. 3 Governance structure of the Valencia-UPV City-University Partnership (CUP). *Source* Peris-Blanes et al. (2024)

a clear structure for collaboration, outlining strategic priorities and establishing a governance framework, as depicted in Fig. 3.

The governance model is designed to facilitate innovation, collaboration, and strategic decision-making. At its core is a **strategic commission**, co-chaired by the Mayor of Valencia and the Rector of UPV, which defines the partnership's key research and action priorities. This commission is guided by input from a scientific committee composed of distinguished experts. The **operative commission** is jointly led by the UPV Vice-Rectorate for Sustainable Development, the UPV Vice-Rectorate for Innovation and Transfer and the City Council's Urban Strategy Coordinator, supported by three additional key members. Together, they define the annual operating plan and oversee a range of tasks, including the management of the UPV Living Lab, the promotion of the València 2030 Urban Strategy, the Missions València 2030 initiatives, research and innovation projects, teaching collaborations, and the development of specialized reports and studies.

This **dual leadership** ensures a balanced approach, blending operational management with long-term strategic vision. Additionally, working groups and co-creation processes further strengthen the relationship between the university and the city, encouraging joint participation and input from both sides.

Furthermore, to support the development of the City-University Partnership (CUP) Agreement, a Scientific Commission was established, composed exclusively of academics affiliated with the Universitat Politècnica de València. This commission, made up of researchers and faculty with expertise in sustainability, climate science, and urban planning, provided critical insight and technical guidance to ensure that the agreement's initiatives were research-driven and impactful. Meeting regularly, the Commission evaluated proposed actions, provided scientific recommendations, and worked to align the agreement with UPV's sustainability goals. Their expertise was instrumental in shaping an evidence-based framework for the CUP agreement, advancing the university's commitment to carbon neutrality and environmental resilience.

This comprehensive governance structure fosters the dynamism of the Valencia-UPV CUP by integrating research, innovation, educational initiatives, and the Living Lab, all unified under a shared commitment to sustainable urban development and alignment with the EU Missions.

To enhance this structured governance, specific strategic actions (see above) were linked to practical initiatives. For example, calls for research and social innovation projects support knowledge generation (L3, L4), while the Living Lab Climate and Ciutat Lab foster real-world testing environments (L5). Teaching, thesis projects, and internships embed sustainability into academia, aligning students and faculty with urban goals (L2, L6), while spaces for dialogue, such as the Grand Debate, promote inclusive community engagement (L7, L8).

3.5 Changing Political Landscape and Implications for Future CUP Direction

The 2023 municipal elections in Valencia brought a conservative shift in political leadership. The election results brought the **Partido Popular (PP)**, a conservative party, into power, signaling a potential shift in the city's sustainability priorities. The new administration has expressed an interest in emphasizing **technological innovation** and economic competitiveness, which, while not inherently at odds with sustainability, marks a departure from the comprehensive climate action and social justice orientation of the previous administration.

A year later, in 2024, Valencia finally became the **Green Capital of Europe**, a recognition that has since become a centerpiece of the newly elected administration's agenda, overshadowing the previously ambitious **València 2030 Climate Mission**.

This change has had a notable impact on the city's sustainability policies and the trajectory of collaborative initiatives like the City-University Partnership (CUP). Under the València **2030 Climate Mission**, CUP initiatives thrived, with a focus on community-based sustainability efforts, participatory governance, and systemic urban transformations that integrated both environmental and social aspects of sustainability. Changes have raised concerns among stakeholders within the CUP and broader sustainability communities in Valencia, as it introduces uncertainty regarding the continuity of climate neutrality efforts and the València 2030 Climate Mission.

Although the new political leadership has not explicitly discarded Valencia's sustainability goals, there are growing indications that the current administration may prioritize a market-driven, technology-focused approach. This strategy aligns closely with another branch of the UPV's interests, particularly the Vice-Rectorate of Innovation and Transfer, which emphasizes smart technologies and digitalization. While this direction supports innovation and economic growth, it also shifts attention away from the social justice and climate mission toward projects like the Green Capital designation and recent approvals, such as the sandbox initiative. This focus, however, may risk sidelining the participatory and community-centered approaches that were central to the city's prior climate strategies, potentially diminishing community engagement and the integrated social, environmental, and economic considerations critical for broad, transformative sustainability efforts.

The first meeting of the CUP Operational Committee in 2024 highlighted this political shift. Initially intended to review and advance agreed-upon sustainability actions, the meeting instead reflected a more strategic focus on initiatives aligned with the Green Capital designation. The discussions centered on using the CUP to operationalize collaboration between UPV and the city council on this theme, rather than exclusively supporting the Climate Mission.

The political transition has also led to significant changes in **personnel** involved in the CUP, both from the city's administration and within the university. Personnel changes on both sides of the partnership have implications for continuity and institutional memory. On the city's side, new appointments in key positions, particularly within the departments responsible for the Innovation and Climate Change areas, have introduced a different set of priorities and perspectives on the CUP's role. These new actors may prioritize different aspects of the partnership, potentially shifting focus away from the previously established long-term sustainability goals.

4 Analysis and Discussion

This analysis delves into the enabling factors and support structures that have driven the development of the **Valencia-UPV City-University Partnership (CUP)**. By examining these alongside key challenges, this section contextualizes the Valencia-UPV CUP within the broader literature on sustainability partnerships and city-university collaborations.

4.1 Building Trust

Trust has proven to be foundational to the CUP's success, allowing for effective, long-term collaboration between Universitat Politècnica de València (UPV) and the City of Valencia. Research on collaborative partnerships consistently underscores the importance of trust in fostering open communication, reducing conflict, and facilitating joint decision-making (Emerson et al., 2012; Healey, 1998). In the Valencia-UPV CUP, trust developed through years of shared projects and aligned goals has minimized friction, allowing for ambitious climate initiatives that align with the **València 2030 Climate Mission**.

However, as with many partnerships, the CUP's trust has primarily been built at the individual level, reliant on personal connections rather than embedded institutional processes. This reliance on individual-based trust, while effective in the short term, exposes the partnership to potential fragility as personnel changes occur. Trencher et al. (2014) caution against the vulnerability of CUPS. Immature partnerships can be overly dependent on individuals, and continuity in institutional memory and organizational stability may waver with staff turnover. For the CUP to sustain itself as a transformative partnership, it must institutionalize trust-building practices through formal agreements, embedded processes, and organizational frameworks that transcend individual actors, fostering a more resilient partnership structure (Goddard & Vallance, 2013). Although all these elements were considered in the city-university partnership, the initial period from inception to the time of political change in 2022 was too short to enable the partnership demonstrate its benefits in practice. Consequently, not all the political and technical actors within the City Council perceived the real value of what a partnership with the university could imply.

4.2 Focus on Transformational Goals

The CUP's alignment with the **European Union's climate mission** has established a transformative framework that situates Valencia's local climate efforts within a global context. The partnership's explicit commitment to transformational goals, such as climate neutrality and resilience, underscores the importance of integrating local actions with broader international targets, an approach reinforced by studies on transnational sustainability partnerships (Keeler et al., 2019). This alignment has initially allowed the CUP to move beyond isolated projects toward a cohesive vision for long-term systemic change, making Valencia a leader in urban climate action.

However, such alignment depends heavily on political stability and continuity, factors that have proven challenging in Valencia's shifting political landscape. In 2023, the election of a more conservative leadership signaled potential changes in Valencia's sustainability trajectory, placing greater emphasis on technological innovation and economic growth over comprehensive climate action. This shift aligns with observations from other studies, where **political volatility** has disrupted long-term sustainability projects and compromised CUP goals (Evans et al., 2015; Trencher et al., 2014). Sustaining transformational goals within the CUP framework will require adaptability to navigate political changes while maintaining climate neutrality as a core component of the city's identity and development. As Keeler et al. (2018) suggest, partnerships like the CUP must cultivate cross-party support and establish enduring commitments to prevent shifts in political will from derailing long-term goals. However, no matter how well-intentioned these ambitions may be, they come up against an increasingly radicalized political reality in which climate change denialist parties confronting the Agenda 2030 are gaining significant power influence, as in the case of València City. Consequently, the València experience clearly shows the limitation of these institutional agreements to address transformational goals in a context of political radicalization.

4.3 Investment in Relationships

A key factor driving the CUP's success has been its substantial investment in relationship-building and fostering a collaborative culture, a practice that aligns with successful CUPs globally (Evans et al., 2015). Through initiatives like Mission Fridays, strategic conferences, and regular collaborative meetings, the CUP has nurtured a resilient, synergistic partnership model that goes beyond transactional relationships. This approach aligns with Emerson et al. (2012), highlighting the role of sustained engagement in fostering joint ownership and trust among diverse stakeholders.

However, again, the CUP's reliance on personal relationships raises questions about the longevity of its collaborative culture. As with the establishment of trust, relationship-building within CUPs must transition from individual-level interactions to institutionalized practices. This shift would help the CUP avoid potential disruptions from personnel changes and enable a more consistent, enduring collaborative culture. Studies on city-university partnerships, such as those by Trencher et al. (2014) and Healey (1998), underscore the importance of formalized, embedded collaboration mechanisms that ensure continuity, such as regular co-governance structures, joint assessments, and strategic project alignment processes. For the Valencia-UPV CUP, embedding these practices within its governance framework will help institutionalize collaboration, preserving its transformative potential even as individual actors change.

4.4 Supportive Norms and Structures

The formal governance structure established by the **Valencia-UPV CUP agreement** in 2023 has been instrumental in providing a stable foundation for effective coordination, accountability, and adaptability. The governance model—centered on a **dual**

leadership structure co-chaired by Valencia's Mayor and UPV's Rector—facilitates a balanced approach to day-to-day management and long-term strategic planning. This structured governance mirrors successful models seen in other transformative CUPs, such as those in Amsterdam and Vancouver, where formalized governance has allowed city-university partnerships to remain agile in responding to changing urban challenges (Keeler et al., 2018).

Yet, the efficacy of governance structures is contingent on adequate resources and support mechanisms. Without dedicated funding and cross-sectoral support, structured governance risks becoming a "paper framework" that is unable to fulfill its objectives effectively. Studies on CUPs highlight the importance of stable funding and strong political backing to empower governance structures, ensuring that the partnership can meet evolving demands and scale initiatives (Goddard & Vallance, 2013; Bulkeley et al., 2014). For the CUP, enhancing its governance will require the establishment of sustainable financial support and adaptive mechanisms, such as contingency funding and flexible governance roles that can respond dynamically to external pressures.

Furthermore, while initial funding, including budget resources for technical office to assist the partnership in its daily work, enabled the CUP to launch critical projects like the UPV Living Lab, political change affected the further financing and the availability of human and budget resources. This ambiguity created doubts about the continuity and long-term impact of the partnership, despite the relative availability of necessary resources. Studies on CUPs across Europe emphasize that while financial stability is crucial, consistent political backing is equally essential for transformative capacity, allowing partnerships to prioritize sustained initiatives over fragmented, short-term projects (Goddard & Vallance, 2013; Trencher et al., 2014).

4.5 Designing Motivational Actions

Developing **collaborative activities with meaningful, mutual benefits** for all parties involved has been a defining feature of the Valencia-UPV CUP. The partnership has rolled out a variety of initiatives aimed at aligning university research, city policies, and community interests, which has been instrumental in maintaining stakeholder motivation and engagement over time. Activities such as Mission Fridays, collaborative workshops, and the **UPV Living Lab** serve as focal points for hands-on sustainability projects, offering tangible benefits to both UPV and the City of Valencia. These actions embody the essence of **motivational actions** as described by Keeler et al. (2023), where mutual benefit fuels long-term engagement and deepens collaboration between CUP partners.

Moreover, these collaborative initiatives have fostered a participatory culture within the CUP, involving students, faculty, city officials, and community stakeholders. This inclusive approach to project design and execution ensures that each initiative addresses a variety of interests, which not only strengthens commitment to CUP goals but also builds community trust. Engaging the broader community through public events and participatory research has enhanced the relevance of CUP projects, as local residents see direct impacts on their urban environment.

The CUP's ability to maintain sustained motivation through these initiatives reinforces Keeler et al.'s (2018) recommendation that CUPs develop meaningful, impactful actions to support long-term goals. For the Valencia-UPV CUP, continuing to expand these motivating activities while adapting them to evolving community needs will be essential. Although the political context doesn't seem to favor it, by doing so, the CUP can aspire to remain as a dynamic and transformative partnership, capable of responding to both urban challenges and the needs of its stakeholders.

5 Conclusions

City-University Partnerships (CUPs) are increasingly recognized not only as engines for driving urban sustainability and delivering on global goals such as the UN SDGs and the EU Climate-Neutral and Smart Cities Mission, but also as catalysts for institutional transformation within higher education institutions (HEIs). The case of the Valencia-UPV CUP demonstrates that such partnerships do more than facilitate knowledge transfer to cities—they also reshape the universities themselves. Through sustained co-governance, participatory foresight, and interdisciplinary engagement, the CUP has embedded sustainability into UPV's internal operations, strategic priorities, and academic culture.

This partnership has catalyzed institutional innovation at UPV, including the creation of the Alliance of Chairs for Sustainability, the launch of the UPV Living Lab, and the integration of climate neutrality into student theses, curricula, and research agendas. These initiatives reflect a structural shift in how the university envisions and enacts its third mission—not merely as a service provider to society, but as a co-creator of transformative futures. In this sense, the CUP has become a driver of reflexive learning, enabling UPV to experiment with new governance forms and pedagogical approaches while aligning its institutional identity with the SDGs.

The Valencia case further illustrates the mutual reinforcement between internal university transformation and external urban impact. As UPV adapted its institutional structures and priorities to meet the city's climate neutrality goals, it also deepened its own sustainability commitment—becoming not just a partner in Valencia's transition, but a living testbed for systemic change. Even in the face of political shifts and uncertainty, the CUP has helped institutionalize sustainability within UPV, strengthening its capacity to remain engaged, adaptive, and mission-driven.

As higher education institutions around the world seek to operationalize the SDGs, the Valencia-UPV CUP offers a compelling model: one where place-based partnerships do not simply extend university influence into the city, but bring the urgency and complexity of urban challenges back into the university—reshaping its structures, knowledge production, and societal roles. In doing so, CUPs become more than collaborative mechanisms; they become instruments of mutual transformation, positioning HEIs as active agents in both local and global sustainability transitions.

6 Recommendations

The Valencia-UPV CUP's journey provides valuable insights that can strengthen CUPs as effective vehicles for advancing sustainable urban development. Drawing from this experience, several key recommendations emerge:

Build cross-party political support: securing political support across the spectrum is essential to protect CUPs from the volatility of changing administrations. Studies from cities like Copenhagen and Stockholm have shown that bipartisan backing enhances the resilience of partnerships, enabling sustainability initiatives to proceed uninterrupted, regardless of political changes (Evans et al., 2015). This insight is especially pertinent given the recent political shift in Valencia, where a more conservative administration has redirected focus toward technology-driven economic growth at the potential expense of comprehensive climate action. This trend reflects challenges observed in other CUPs, where political transitions have introduced new priorities that risk undermining established sustainability commitments (Evans et al., 2015).

Secure sustainable funding: long-term financial commitments from both university and city partners are vital for maintaining CUP activities and avoiding dependence on short-term grants. Successful examples, such as CUPs in Amsterdam and Freiburg, illustrate how stable funding supports ambitious, long-term projects that can weather political and economic fluctuations (Goddard & Vallance, 2013). Establishing multi-year funding streams ensures that partnerships are equipped to invest in enduring initiatives, fostering innovation and structural improvements rather than temporary, isolated efforts.

Institutionalize collaborative practices: embedding collaborative practices into the formal structures of CUPs is critical for long-term stability. Establishing joint governance boards, shared resources, and co-creation processes reduces reliance on specific individuals and supports continuity during personnel changes. The Vancouver CUP, for example, has shown that institutionalized practices enable partnerships to remain resilient and effective over time (Trencher et al., 2014). By formalizing collaboration, CUPs can maintain their momentum and prevent disruption due to turnover in leadership or key stakeholders.

Foster inclusive participation: ensuring diverse community involvement strengthens the relevance and democratic nature of CUP initiatives. Inclusive governance that incorporates voices from a broad range of stakeholders mitigates the risks of co-optation and reinforces community buy-in. Case studies from Birmingham and Manchester demonstrate that participatory governance models contribute to more equitable and effective decision-making, enhancing the overall impact of CUPs (Bulkeley et al., 2014). This participatory approach ensures that CUPs reflect the needs and priorities of all segments of the community, creating a more integrated and sustainable urban strategy.

Commit to internationally aligned sustainability goals: CUPs should anchor their commitments in internationally recognized and widely endorsed frameworks, such as the United Nations' SDGs or the EU Climate Missions. These global references provide legitimacy and strategic direction for CUPs, ensuring that their actions are aligned with broader transformational objectives that can contribute to systemic change and adapt to shifts in political and institutional priorities. This approach highlights the role of the broader "landscape" (Geels, 2019) in enabling local transformative initiatives, ensuring that CUPs are not becoming merely a consultancy arm for the current politics but contributing to a larger, cohesive movement toward sustainable development.

Adaptability and resilience in a changing landscape: the ability to adapt is essential for CUPs to remain relevant and effective amid the evolving challenges of urban development. The Valencia-UPV CUP, like many others, must be prepared to navigate changes in political, social, and economic contexts. Flexibility allows CUPs to adjust their strategies while maintaining their commitment to long-term goals. This adaptive capacity can be supported by mechanisms such as periodic reviews, flexible governance roles, and responsive funding models that allow partnerships to pivot as needed without losing their foundational focus on sustainability.

7 Future Research

In an era marked by rapid urbanization and pressing environmental challenges, the strategic evolution of City-University Partnerships (CUPs) is more critical than ever. These partnerships stand at the forefront of innovative solutions for sustainable urban development, but they must continuously adapt to the ever-changing global landscape. Future research is vital in guiding these adaptations, ensuring CUPs not only meet the current demands of urban sustainability but are also primed for future challenges and opportunities.

Future research should prioritize the evolution of CUPs' transformative goals, ensuring they are responsive to the shifting sands of political, economic, and environmental realities. This involves redefining focus areas to tackle emerging challenges such as just transitions and climate adaptation strategies. For this, developing agile governance frameworks is crucial in order for CUPs to pivot their strategies effectively in response to external changes, thus enhancing their responsiveness to political changes and allowing rapid and efficient adaption to unforeseen challenges.

Moreover, there is a pressing need to transform the traditional top-down experimental approach of CUPs by integrating more bottom-up, student-driven initiatives. The agency of students in these partnerships has been limited, underutilizing a valuable feature of universities—students' potential for innovation. Future research should focus on creating strategies to facilitate active engagement of students, which can be crucial both to the success of the climate initiatives and to the educational goals of the university. This engagement also contributes to democratizing the CUP and including new emerging values.

Finally, to provide a richer, more diversified understanding of CUPs, expanding the breadth of case studies to include various geographic and urban scales is essential. This broader perspective will uncover diverse implementation scenarios and impacts, offering invaluable insights into the functionality of CUPs. Additionally, examining the dynamics of different actors and the roles of non-traditional stakeholders like NGOs and community groups will help in forging robust, inclusive partnerships that address urban sustainability in a holistic manner. Research should also consider how CUPs can be replicated or scaled in different cultural and political contexts, identifying key factors that contribute to their success or failure.

Acknowledgements This research was partly funded by the European Union's Horizon Europe program under the Marie Skłodowska-Curie Post-Doctoral Fellowship Metamorphosis—101152245—DLV—101152245. The views and opinions expressed are those of the authors and do not necessarily reflect those of the European Union

References

Bulkeley, H., Castán Broto, V., & Maassen, A. (2014). Low-carbon transitions and the reconfiguration of urban infrastructure. *Urban Studies, 51*(7), 1471–1486.

Caughman, L., Withycombe Keeler, L., & Beaudoin, F. (2020). Real-time evaluation of city–university partnerships for sustainability and resilience. *Sustainability, 12*(21), 8796.

Evans, J., Jones, R., Karvonen, A., Millard, L., & Wendler, J. (2015). Living labs and co-production: University campuses as platforms for sustainability science. *Current Opinion in Environmental Sustainability, 16*, 1–6.

Flyvbjerg, B. (2006). Five misunderstandings about case-study research. *Qualitative Inquiry, 12*(2), 219–245.

Geels, F. W. (2019). Socio-technical transitions to sustainability: A review of criticisms and elaborations of the multi-level perspective. *Current Opinion in Environmental Sustainability, 39*, 187–201.

Goddard, J., & Vallance, P. (2013). *The university and the city*. Routledge.

Guenduez, A. A., Frischknecht, R., Frowein, S. C., & Schedler, K. (2024). Government-university collaboration on smart city and smart government projects: What are the success factors? *Cities, 144*, Article 104648.

Healey, P. (1998). Building institutional capacity through collaborative approaches to urban planning. *Environment and Planning A: Economy and Space, 30*(9), 1531–1546.

Keeler, L. W., Beaudoin, F., Cid, A., Cowley, R., Fahy, S., Lerner, A., & Torney, D. (2023). Building transformative city-university sustainability partnerships: The audacious partnerships process. *Urban Transformations, 5*(1), 1.

Keeler, L. W., Beaudoin, F., Wiek, A., John, B., Lerner, A. M., Beecroft, R., Tamm, K., Seebacher, A., Lang, D. J., Kay, B., & Forrest, N. (2019). Building actor-centric transformative capacity through city-university partnerships. *Ambio, 48*(5), 529–538.

Keeler, L. W., Beaudoin, F. D., Lerner, A. M., John, B., Beecroft, R., Tamm, K., & Lang, D. J. (2018). Transferring sustainability solutions across contexts through city–university partnerships. *Sustainability, 10*(9), 2966.

Loorbach, D. A., & Wittmayer, J. (2024). Transforming universities: Mobilizing research and education for sustainability transitions at Erasmus University Rotterdam, The Netherlands. *Sustainability Science, 19*(1), 19–33.

Moreno-Serna, J., Kordas, O., Lumbreras, J., Minoz, Å., Saniour, N., & Rohracher, H. (2024). National platforms to transform cities using collective experimentation and scale: The case of Sweden and Spain. In *Higher education and SDG11: Sustainable cities and communities* (pp. 145–158). Emerald Publishing Limited.

Peris-Blanes, J., Udovyk, O., Cerezo, F., Palau-Salvador, G., Cuesta-Fernandez, I., Ortiz-Miranda, D., Alapont, J. L., Domingo, D., Montagud, C., Escario-Chust, A., Segura-Calero, S., & Aránguiz-Mesías, P. (2024). Building a city-university partnership for accelerating urban climate neutrality: the case of València, Spain. In J. Lumbreras & J. Moreno-Serna (Eds.), *Higher education and SDG11. Sustainable cities and communities* (p. 184). Emerald Publishing Limited.

Torfing, J., Sivertsen, H., & Torvatn, H. (2024). Metagovernance of co-creation in city–university partnerships: How to avoid being stuck in the middle? *Public Administration, 102*(4), 1711–1727.

Trencher, G., Yarime, M., McCormick, K. B., Doll, C. N., & Kraines, S. B. (2014). Beyond the third mission: Exploring the emerging university function of co-creation for sustainability. *Science and Public Policy, 41*(2), 151–179.

Udovyk, O., Cerezo-Peco, F., Escario-Chust, A., Peris-Blanes, J. F., Palau-Salvador, G. C., & Segura-Calero, S. (2024). City-University Partnerships (CUPs) for climate urban transformation: The case of Valencia (Spain). *Journal of City Climate Policy and Economy* (Forthcoming).

Wolfram, M. (2016). Conceptualizing urban transformative capacity: A framework for research and policy. *Cities, 51*, 121–130.

Emerson, K., Nabatchi, T., & Balogh, S. (2012). An integrative framework for collaborative governance. *Journal of Public Administration Research and Theory, 22*(1), 1–29.

Empowering a Growth Mindset for Sustainable Development in Higher Education by Design Thinking

Iris Schmidberger and Sven Wippermann

Abstract In today's dynamic world, the focus on mindset is increasingly recognized as crucial for adapting to significant changes that affect various areas, especially sustainable economies and lifestyles (Rinne in Change management requires a change mindset. Havard Business Review, 2023; Uitto in Transformational change for people and the planet: Evaluating environment and development—introduction, Springer Nature, pp. 1–13, 2022). These changes, anchored in the concept of Sustainable Development, call for a change in our thinking patterns, known as mindset. The journey toward sustainability is greatly shaped by our individual and collective mindsets, which influence how we see the world, make decisions, and act. The concept of an innovation mindset, which combines attitudes, opinions, beliefs, and behaviors (Dosi et al. in Measuring design thinking mindset. University of Zagreb, Croatia; The Design Society, 2018), is crucial for developing the skills needed for this sustainable shift. This mindset is key to understanding and affecting human behavior in various fields, such as social and educational psychology, as well as innovation management (Krohn in Crucial Role of Mindsets in Innovation Efforts. Springer Fachmedien, 2023). The research by Dweck (2017) on transitioning from fixed to growth mindsets is crucial for an adaptable and resilient way of thinking to support sustainable change. Design Thinking is presented as a methodology, offering a framework that not only facilitates learning and skill-building for sustainable development but also encourages the growth mindset in higher education settings. The results of a pilot study with students at Ludwigsburg University of Education will provide initial insights into whether the Design Thinking process promotes the development of a Design Thinking mindset.

Keywords Education for sustainable development · Growth mindset · Design thinking · Design thinking mindset · Higher education

---

I. Schmidberger (✉) · S. Wippermann
Department for Educational Leadership, Ludwigsburg University of Education, Ludwigsburg, Germany
e-mail: iris.schmidberger@ph-ludwigsburg.de

S. Wippermann
e-mail: wippermann@ph-ludwigsburg.de

W. Leal Filho et al. (eds.), *Introducing the UN SDGs in Higher Education Institutions*, World Sustainability Series, https://doi.org/10.1007/978-3-032-23017-1_29

1 Introduction

Universities play a crucial role in promoting Education for Sustainable Development (ESD) (UNESCO, 2017). ESD refers to a holistic concept that addresses the global challenges of our interconnected world. It aims to inform and enable responsible decision-making in terms of ecological integrity, economic viability, and an equal-opportunity society (BNE-Portal, 2024).

It promotes participation, solidarity, future-oriented thinking, and action—these are key competencies for shaping the sustainable development of our society (BNE-Portal, 2024). Therefore, it is necessary to support the change in mindset towards sustainable development (Kanning & Meyer, 2022). This is where Design Thinking comes in as an innovative approach to support ESD in higher education.

This paper focuses on the impact of Design Thinking on the students' growth mindset in the context of sustainable development in higher education. The research question is 'How does the application of Design Thinking influence the students' growth mindset in a university seminar on sustainable development?'.

2 ESD and Design Thinking

ESD is a holistic concept that addresses the global—ecological, economic, and social—challenges of our interconnected world. The education campaign aims to inform and empower people to make responsible decisions regarding ecological integrity, economic viability, and equitable society (BNE-Portal, 2024). It enables people to recognize and evaluate problems of non-sustainable development and to apply knowledge about sustainable development. It aims to actively participate in the analysis and evaluation of non-sustainable development processes, orienting oneself towards sustainability criteria in one's own life and initiating sustainable development processes together with others, both locally and globally (de Haan, 2010). The concept of *Gestaltungskompetenz* (shaping competence), which goes back to de Haan and Harenberg (1999), is crucial in achieving these goals. *Gestaltungskompetenz* enables a change of perspective from reaction to action. This requires the use of prospective strategies since retrospective strategies prove to be less promising in the context of sustainable development. Retrospective strategies are based on proven strategies and search for facts that verify them once again. They prepare for successfully coping with the future by assuming that it is a linear continuation of the present. On the other hand, prospective strategies search for a wide range of information and use it to develop creative hypotheses that extend into the future. Due to the future-oriented nature of ESD, prospective strategies are preferable (de Haan, 2010). This is an important link to the innovation methodology of Design Thinking: "Everyone designs who devises courses of action aimed at changing existing situations into preferred ones" (Simon, 1996, p. 111). According to this interpretation, design can be characterized as a practice of transformation and shaping the future

(Mareis, 2016). Design Thinking takes up this view and aims to shape prospective development processes (Brown, 2019).

Today's understanding of Design Thinking as a methodology for initiating target group-oriented innovations for complex problems is largely based on the Hasso Plattner Institute of Design, called d.school, founded at Stanford University, USA, in 2005 (Meinel et al., 2015). Since 2007, Design Thinking has also been taught at the d.school of the Hasso Plattner Institute in Potsdam, Germany. Both institutions cooperate closely with each other and conduct joint research programs in the field of Design Thinking (Plattner et al., 2009). Meanwhile, Design Thinking has established itself internationally in teaching and research in the field of higher education and the methodology is already receiving increased attention in other fields of education (Schmidberger & Wippermann, 2022).

According to Meinel and Leifer (2011, p. xiv), Design Thinking can be characterized as follows: "Its human-centric methodology integrates expertise from design, social sciences, engineering, and business. It blends an end-user focus with multidisciplinary collaboration and iterative improvement to produce innovative products, systems, and services." Plattner et al. (2009) describe Design Thinking as inventive thinking in heterogeneous teams to develop creative solutions for complex problems. The essential core elements are collaboration in a multidisciplinary team, a flexible working environment, and an iterative process. Working in a multidisciplinary team opens different perspectives. A flexible working environment encourages creativity and out-of-the-box thinking, and the iterative process allows you to learn from feedback.

To create a solution for the design challenge, the emphasis is on exploring the problem space first. Then the focus shifts towards creating a solution space for the design challenge (see Fig. 1). This allows us to pay attention to the problem without rushing into finding solutions. At the beginning of the process, the problem is explored with an open mindset and with various perspectives from the observations of the target group. Following this, a shared point of view guided the transition into the solution space. Here, divergent thinking allows fresh solutions to develop prototypes. When it comes to refining and concretizing the solution ideas convergent thinking takes the lead. Throughout the entire process, a strong connection to the problem space is maintained. This iterative approach ensures that the solution aligns with the needs of the target group.

3 Design Thinking Mindset and Growth Mindset

To unleash the creative potential during the process, team members need a particular mindset (Schmidberger & Wippermann, 2022). A mindset can be defined as a set of attitudes, opinions, beliefs, and behaviors that collectively characterize an individual, a group, or an organization (Dosi et al., 2018). In her 2017 work, Dweck contrasts fixed and growth mindsets. She refers to the malleable perspective on mindsets as a growth mindset (Krohn, 2023). Dweck (2017) defines a fixed mindset as a belief

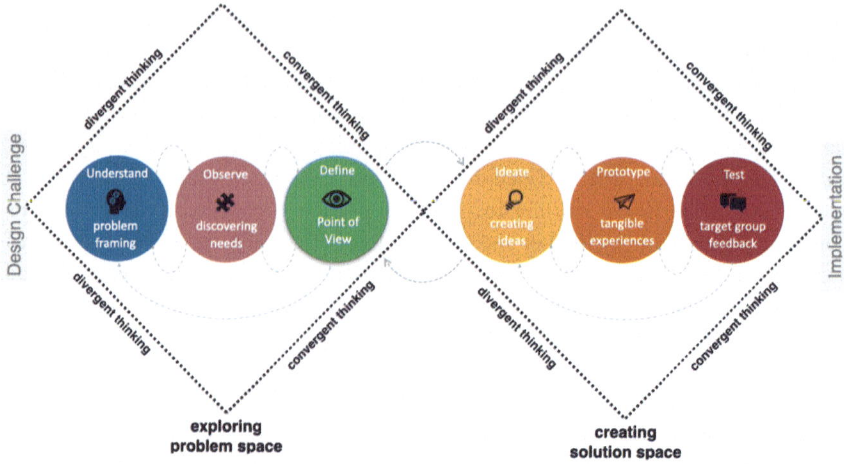

Fig. 1 Problem space and solution space (Schmidberger & Wippermann, 2022, p. 40)

in the unchangeable nature of one's inherent characteristics and abilities. A growth mindset is defined as the belief that one's inherent characteristics and abilities can be actively shaped through learning experiences.

In the context of Design Thinking, a growth mindset plays a pivotal role. It enables the team to alter their perspective and think as well as act in a human-centered manner. Furthermore, it fosters empathic collaboration in multidisciplinary teams and facilitates exploration of the problem space and learning through experimentation. Additionally, it encourages acceptance of uncertainties, ambiguities, and contradictions as inherent to the process. The following illustration encapsulates the core aspects of the design thinking mindset for innovation.

The Design Thinking mindset for innovation contains the following six aspects as shown in Fig. 2 (HPI Academy, 2024).

1. Think and act in a human-centered way: Design Thinking's strength lies in its people-centered focus, prioritizing a deep understanding of human needs, behaviors, and values to shape meaningful work. The methodology actively involves the target group in the design process, ensuring relevance and alignment with their needs.
2. Collaborate in Diverse Teams: Design Thinkers bring a genuine curiosity for user experiences into teamwork, valuing diversity for its power to clarify complex problems and inspire innovative solutions. Diverse teams leverage members' unique backgrounds and skills, creating a dynamic where shared insights and competencies intersect productively. The team members share knowledge, enhance collective understanding, and show openness to differing perspectives. They support team decisions even when these differ from their own and demonstrate empathy.

Fig. 2 Design Thinking mindset for innovation (HPI Academy, 2024)

3. Explore the problem space: Design Thinkers approach problems holistically to develop innovative solutions, aiming to understand the challenge's full context, including affected individuals' experiences, systemic factors, and regulatory constraints. By identifying and addressing their blind spots, the team members choose the most promising entry point to tackle the issue. If new insights emerge, they are prepared to reframe the challenge to better align with the needs of the target group. A problem-exploration mindset pursues discovery with openness and curiosity, unafraid to ask unconventional questions and explore diverse paths.
4. Learn through experimentation: Learning through experimentation emphasizes progressing a project through action rather than prolonged discussion. This approach values the quick testing of ideas, with iteration repeating steps and revisiting previous insights as essential. A positive attitude toward failure is also key, viewing mistakes as integral to the learning process. Experimenters make abstract concepts tangible, such as by prototyping ideas, and they extract insights from these experiments to better understand user needs. The team members employ spatial thinking and use tools like sketches, models, or roleplay to concrete ideas. Rapid iteration loops allow them to test, learn, and adapt continually, even if that means stopping a project when appropriate.
5. Embrace uncertainty: Design Thinking frequently addresses complex issues involving numerous variables and stakeholders, with no clear solutions. To navigate these challenges, the team members must approach them with openness and actively embrace uncertainty not only tolerating unpredictable outcomes but viewing them as opportunities. This mindset involves comfort with ambiguity, incomplete information, and potential contradictions, using these aspects constructively. Those who accept uncertainty exhibit courage and are skilled in abductive thinking, enabling them to make informed decisions based on

partial information. This adaptability allows them to progress constructively, remaining open to adjusting their opinions or direction as new insights emerge, understanding that results may be provisional.

6. Envision a radically new future: Radical envisioning describes the mindset essential for generating truly innovative ideas. It is marked by an optimistic belief in a better future and a firm conviction in achieving ambitious goals despite obstacles. Those who embrace radical envisioning demonstrate resilience, refusing to be limited by current realities and instead envisioning what could be. It extends their thinking beyond individual product or service redesigns to the conception of interconnected systems, enabling transformative possibilities. All elements of the Design Thinking mindset for Innovation described above relate to a growth mindset.

4 Self-Talk and Growth Mindset

A strategy to effectively access and understand one's mindset is reflective self-talk, as it involves consciously examining and verbalizing one's beliefs, attitudes, and responses to challenges (Dosi et al., 2018; Elliott-Moskwa, 2022; Fang et al., 2004; Gainsburg & Kross, 2020; Price, 2015). Self-talk can be defined as "[…] verbalizations directed at the self" (Latinjak et al., 2023, p. 2). Furthermore, other definitions add details on how the verbalizations are directed to the self and what functions they serve (Gainsburg & Kross, 2020; Morin, 2005; Oleś et al., 2020; Price, 2015). "Verbalizations directed to oneself, either aloud or silently" (Racy & Morin, 2024), and as a "dialogue [through which] the individual interprets feelings and perceptions, regulates and changes evaluations and convictions, and gives him/herself instructions and reinforcement" (Hackfort & Schwenkmezger, 1993, p. 355). Self-talk influences cognition and behavior and serves various functions, including self-regulation, problem-solving, and reinforcement (Brinthaupt & Morin, 2023; Gainsburg & Kross, 2020; Morin, 2005).

Consistent themes in self-talk can offer clues to one's prevailing mindset, in particular when facing challenges an awareness of one's self-talk can support both the analysis as well as the shifting of the mindset. A difficult situation "may jolt you into a fixed mindset, and your self-talk in this circumstance will give you a clue as to whether or not this has happened" (Elliott-Moskwa, 2022, p. 33). Typical situations leading to a fixed mindset include a valuable but difficult task ahead, experiencing it as challenging, with effort required, assessing one's progress, making mistakes, receiving praise or criticism from others (especially from those in authority), and becoming aware of others' successes and setbacks. Each of these situations states six fixed mindset thought patterns and growth mindset antidotes (Elliott-Moskwa, 2022).

1. All-or-none judgment of self when facing a challenging task vs. analysis of current skill or attribute. Operating from a fixed mindset one's self-talk engaging in challenging tasks is all black-and-white, either positive or negative, spinning

around a global judgment of oneself: "Am I or am I not good enough?" (Elliott-Moskwa, 2022, p. 35). With a growth mindset, one's focus is on how one can develop, based on an assessment of one's current level of skills and setting up plans to take realistic steps for growth: "What are my skills currently, and how do I improve?" (Elliott-Moskwa, 2022, p. 35).

2. Viewing efforts negatively versus viewing efforts positively when a task is difficult. A fixed mindset leads to thinking that success comes effortlessly, believing in a lack of abilities and skills when one must make an effort in dealing with a difficult task. When operating from a growth mindset in contrast one appreciates and embraces efforts, believing that effort is part of making progress and growing.
3. Perfectionist standards when assessing progress or performance versus any-percent is something standard. An additional indicator of a fixed mindset is the tendency to embrace a perfectionistic, all-or-nothing approach, whereby an outcome that falls below 100% is perceived as a lack of abilities or skills that are required. Additionally, a fixed mindset is marked by a constant sense of urgency regarding achieving one's goals. In contrast, a growth mindset entails a gradual and incremental standard for progress, whereby even minor improvements are acknowledged and accepted as part of the journey. Despite a desire to grow and progress, there is no overwhelming sense of urgency.
4. Magnification or minimization of mistakes versus mistakes are growth opportunities. This mindset focuses on how one deals with mistakes. In a fixed mindset, mistakes are perceived as indications of insufficient competence and are neither exaggerated as catastrophic occurrences nor dismissed as inconsequential. In either case, the underlying cause of the error is not subjected to analysis. Those with a growth mindset anticipate and accept mistakes as an inherent part of the learning process, analyzing them to gain insight and improve future performance. Mistakes are indicative of the process of developing one's abilities and skills.
5. Viewing others as judges versus viewing others as resources when you are praised or criticized. In a fixed mindset, one is preoccupied with appearing adequate, especially when it comes to praise and criticism from important others. The focus is on being perceived and judged by them as adequate or not. With a growth mindset, opinions and feedback from people you value are seen as a potential resource for growth, focusing on what you can learn from them and how to get specific feedback for your development, rather than feeling judged.
6. Competitive comparisons vs. constructive comparisons with peers when hearing of their successes or failures. A fixed mindset involves a competitive comparison of peers' achievements or setbacks, leading to thinking in terms of being inferior or respectively superior. In contrast, a growth mindset entails a focus on investigating how others improved or failed, intending to identify beneficial practices that can be applied for one's improvement. Such comparisons are constructive and do not involve competitive elements.

Applying self-talk as a tool for growth mindset development as described in this article is not explicitly explored in the context of developing a sustainable mindset in ESD. However, the sustainable mindset model for management education can be highlighted (Kassel & Rimanoczy, 2018; Kassel et al., 2018). It's a holistic framework designed to foster sustainable thinking, ethical decision-making, and responsible action in education, particularly in management. The model integrates three key dimensions—Thinking, Being, and Doing—with four major content areas: systems thinking, ecological worldview, emotional intelligence, and spiritual intelligence (Kassel et al., 2018). Design Thinking is closely connected with this framework as both approaches foster systems thinking, iterative problem-solving, and resilience. Its human-centered approach aligns with the "Doing" dimension of the sustainability mindset model, emphasizing action-oriented learning, stakeholder engagement, and adaptability (Kassel et al., 2018). Additionally, the reflection and self-awareness embedded in Design Thinking align with the "Being" dimension, fostering emotional intelligence, ethical considerations, and a purpose-driven perspective. Empowering a Growth Mindset for Sustainable Development in Higher Education by Design Thinking contributes to the existing body of research as it aligns with common aspects of developing a sustainable mindset on the one hand and hence opens up new perspectives for both, further research and practical application in HE contexts.

5 Research Project

In the summer term of 2024, we conducted a research project in our blended learning course on ESD for students from different master's programs, such as Early Childhood Education, Adult Education, Special Education, and Cultural Education, at the University of Education Ludwigsburg, Germany.

The course aimed at developing solutions to challenges in the field of Education for Sustainable Development through Design Thinking. It started with a kick-off event, which allowed the students to get to know each other personally, and introduced them to the basics of Design Thinking, emphasizing the Design Thinking mindset for Innovation and the importance of a growth mindset. Furthermore, the students were given access to various learning materials such as videos, study texts, and literature on Design Thinking and ESD via the university's learning management system. During those asynchronous self-learning phases, the students dealt with in-depth impulse questions and used the online learning platform as an exchange forum. Subsequently, the acquired learning content was discussed and critically reflected upon in synchronous classroom sessions.

An essential aspect was to gain practical experience in the application of Design Thinking in the context of ESD. The students worked on a real-life design challenge set by the Pro-Rector for Research, Innovation & Transfer of our university: What might the university curriculum look like to enable students to address global and societal challenges? To this end, the students formed four multidisciplinary teams (each consisting of six participants), each of which worked on the design challenge both synchronously in eight half-day face-to-face sessions in the classroom and asynchronously (e.g. in the interview phases) using the Learning Management System throughout the semester.

Based on the six-step Design Thinking process presented in this paper, the students first explored the problem space. This involved conducting interviews with the target group, using tools such as empathy maps to structure their insights, and creating personas to define a point of view. Creative techniques were then used to generate solution ideas and prototypes were created and tested with the target group. We supported each group as Design Thinking Coaches throughout the course.

To investigate our research question (How does the application of Design Thinking influence students' growth mindset in a university sustainable development seminar?), we created a digital workbook as an edible PDF document, which aimed to inform students about the different aspects of the mindset and also served as an individual journal for students to use for their self-talk and self-evaluation, providing guiding questions (e.g. When applying this aspect of the mindset, how do you rate the mistakes you make?) as well as impulses for reflections (What can I learn from my mistakes? What can I do differently?) for all aspects of the growth mindset (Elliott-Moskwa, 2022).

Within the journaling section, we used a mixed-methods research design. On the one hand, students recorded their self-talk regarding all elements of the growth mindset after each session to gain qualitative data. On the other hand, after each session, they self-rated all elements of the growth mindset on a 5-point Likert scale, as well as their overall rating of each element of the design thinking mindset to collect quantitative data (Fig. 3).

The timeline visualizes the students' self-reflections on self-talk and the growth mindset in general after each session using the individual workbooks. These were kept private and remained with the students.

To collect the data, we used two anonymous online questionnaires within our learning management system asking the students to refer to their journaling in their workbooks. The online questionnaire contains the following four sections:

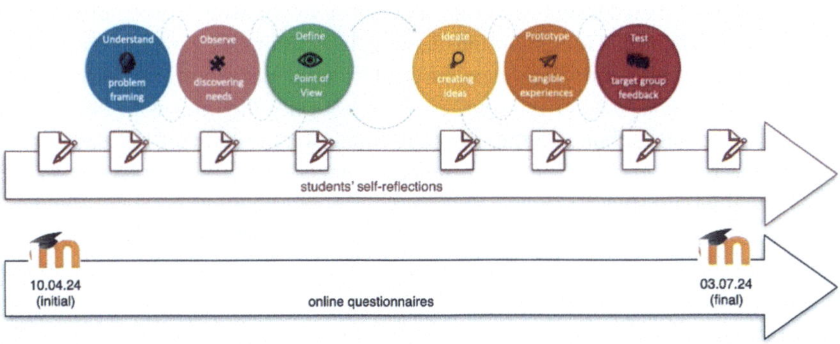

Fig. 3 Timeline of data collection (own illustration)

1. Self-assessment of all aspects of the Design Thinking mindset for innovation (where 1 is low and 5 is high): Please review the summary ratings for the different aspects of the mindset and provide your self-assessment for the initial and final session, each on a 5 point Likert scale, and enter your initial and final self-talk for each aspect of the mindset.
2. Development in each mindset aspect in terms of fixed mindset thought patterns and growth mindset antidotes (where 1 is the fixed mindset thought pattern and 5 is the growth mindset antidote): In which aspect of your mindset have you experienced the most growth? Please give your initial and final self-assessment, each on a 5-point Likert scale!
3. Promoting the mindset at different stages of the Design Thinking process: At which stages of the Design Thinking process were the different aspects of the Design Thinking mindset for innovation particularly promoted (multiple answers possible)?
4. Final evaluation: Do you think that the Design Thinking process has contributed to the development of your Design Thinking mindset for innovation (Yes/No? Why? Will you apply aspects of this mindset in other contexts? If yes, where and why? If no, why not?).

6 Results

All students reported that their growth mindset was empowered during the Design Thinking process. The following two examples illustrate the students' self-talk.

A2: "[…] by actively working with the method, you develop your mindset."

B6: "[…] the mindset adapts to new things and thus generates an open attitude that is helpful for Design Thinking (Fig. 4)."

This graph illustrates the frequency of changes in the six elements of the Design Thinking mindset for Innovation. All of these refer to changes between the initial and final evaluation, based on the students' self-evaluation. Students rated themselves as having improved their Design Thinking mindset in every aspect, especially in two elements of the Design Thinking mindset, namely 'explore the problem space' and 'embrace uncertainty', in which eleven out of 12 students reported improvements (92%). These two aspects of the mindset were not only the ones with the highest frequency of changes, but also the ones with the lowest initial self-evaluation (2.5 'explore the problem space' and 2.3 'embrace uncertainty', both referring to the mean), see diagram (Fig. 5).

In the following, the emphasis is on the element 'explore the problem space'. A comparison of the initial and final self-evaluation of a student's self-talk illustrates the improvement of this element.

A2 (initial self-talk): "I think I can quickly grasp problems and quickly define solutions. Nevertheless, I think that there are certainly still methods and ways that I can learn in this process to tackle problems in other ways".

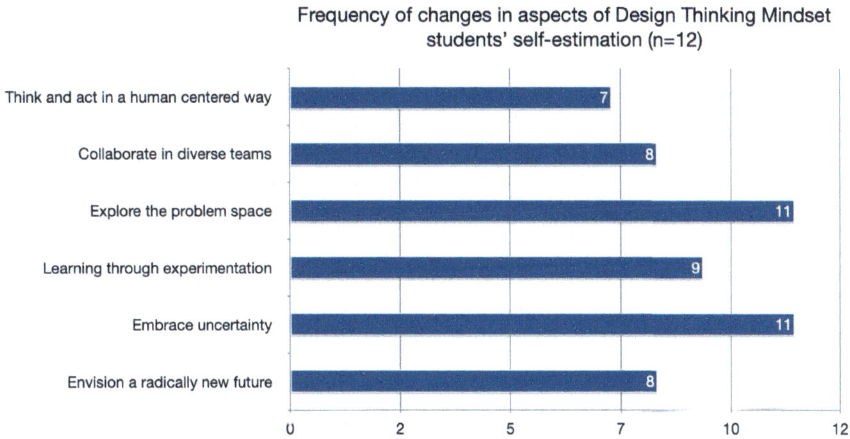

Fig. 4 Frequency of changes in aspects of Design Thinking mindset (own illustration)

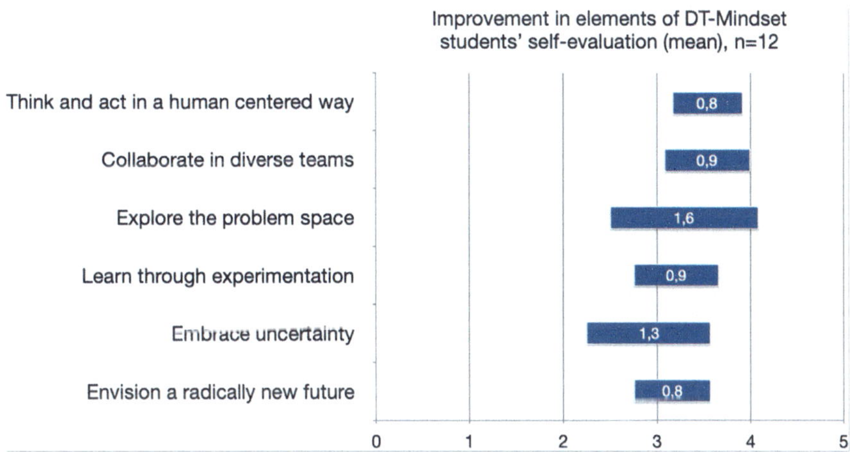

Fig. 5 Improvement in elements of Design Thinking mindset for innovation (own illustration)

Even this student's initial self-talk refers to a growth mindset ("I think there are certainly still methods and ways that I can learn") talking about enhancing his or her skills. Nevertheless, the student improved this element of the Design Thinking mindset for Innovation, reporting in the final self-talk.

A2 (final self-talk): "I learned that I often jumped too quickly to a solution. I was particularly impressed by the definition of needs, and their justification before we worked out a proposed solution".

Overall, the mean of all 12 students' initial self-evaluation was 2.5 with an improvement to 4.1 in the final self-evaluation reported in the online questionnaire, which was a change of 1.6 points. Regarding the element 'embrace uncertainty' the mean of the students's improvement increased by 1.3 points to a total of 3.6 on the 5-point Lickert scale. The two elements 'think and act in a human-centered way' and 'collaborate in diverse teams' were the ones with the highest initial self-evaluation (mean of 3.2 and 3.1). The changes amounted to 0.8 respectively 0.9 point (mean).

Furthermore, the students also reflected on each aspect of the Design Thinking mindset even more closely referring to the six fixed mindset thought patterns and growth mindset antidotes (Elliott-Moskwa, 2022) (Fig. 6).

The visualization illustrates that changes in the mindset thoughts and growth mindset antidotes occurred in each of the elements of the Design Thinking mindset for innovation. The students reported especially high frequencies of changes regarding the standard of perfectionism (7 changes in 'think and act in a human-centered way'), the comparisons with team members (7 changes in 'collaborate in diverse teams'), the evaluation of effort (7 changes in 'explore the problem space'), the dealing with mistakes (6 changes in 'learn through experimentation'), the attitude towards the task (6 changes in 'embrace uncertainty') and again regarding the standard of perfectionism (5 changes in 'envision a radically new future').

Focussing on the element 'explore the problem space' the following example of a student's change of self-talk illustrates the change regarding the effort it takes to improve towards a more growth mindset direction.

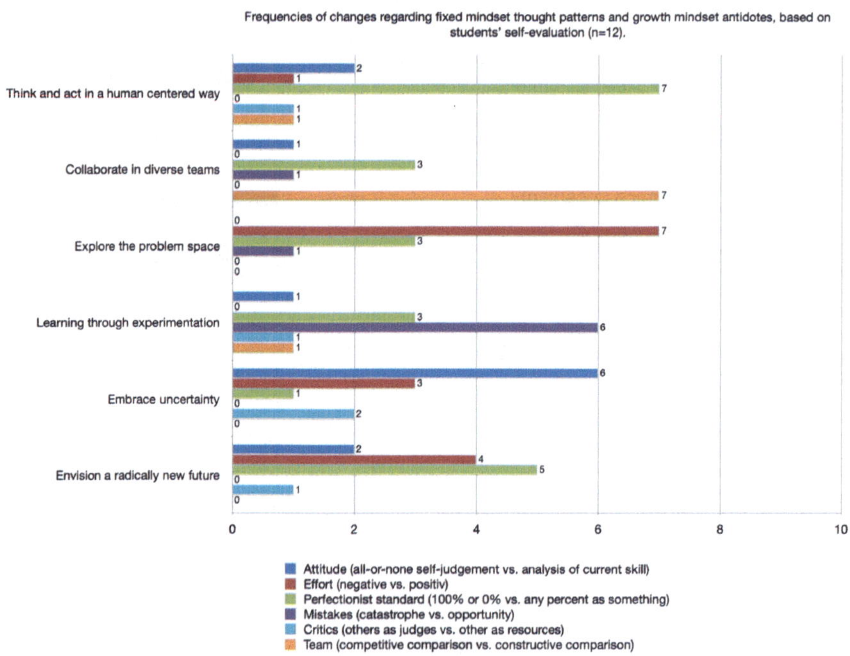

Fig. 6 Frequencies of changes regarding fixed mindset thought patterns and growth mindset antidotes (own illustration)

B5 (initial): "I can improve the aspect of asking questions that others might consider unusual and be more courageous."

B5 (final): "I am not afraid to initiate new ways of thinking. That makes me very proud."

Even the initial statement reveals a growth mindset, as it contains a clear commitment to develop ("I can improve"). In addition, the final self-talk contains a positive perspective regarding the effort it takes to grow ("I am not afraid to initiate new ways of thinking"). Interestingly, student B5 does not rate their development very highly, the change is only 1 point. This is less compared to the mean development of 2.2 points (see Fig. 7, where 1 is the fixed mindset thought pattern and 5 is the growth mindset antidote) points. However, across all elements of the Design Thinking mindset for innovation, this value is below the mean of 1.9 points (see Fig. 8).

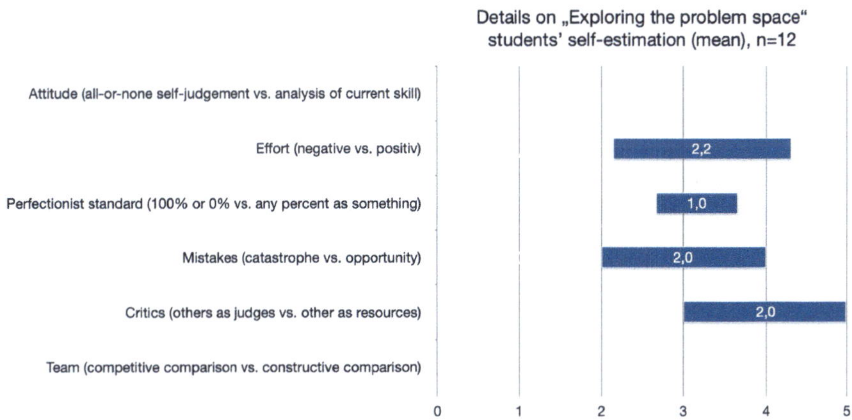

Fig. 7 Details on exploring the problem space (own illustration)

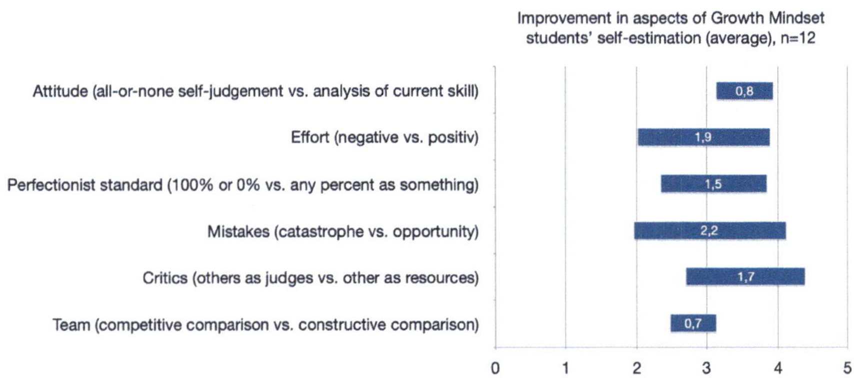

Fig. 8 Improvements in aspects of growth mindset (own illustration)

Broadening the perspective to changes of the mindset thoughts and growth mindset antidotes, the top three improvements of the students related to the way of coping with mistakes (2.2 points), the evaluation of the effort it takes (1.9 points) and the students' coping with critics (1.7 points) as shown in Fig. 8. Two elements of these had the lowest initial self-evaluations by the students (mistakes 1.9 points and mistakes 2.0 points). The aspect of coping with criticism was not only one of the top three aspects with the highest improvement but, was also the one with the highest final self-evaluation of the students (4.4 points).

7 Conclusion

The generalization of the results is limited by the small sample size and the fact, that the students were experienced in reflection and self-reflection as they were attending master's programs already. Additionally, self-assessment often reveals limitations, as individuals may lack self-awareness to accurately evaluate their mindset or may respond in ways they perceive as socially desirable rather than honest (Fang et al., 2004). Furthermore, research indicates, that the manifestation of self-talk, its purpose, and its effectiveness vary significantly across cultures shaped by their specific social norms, linguistic structures, and self-construals (e.g. Gainsburg & Kroos, 2020; Fernyhough, 2016). Taking these aspects into consideration is another limiting factor of our results as our sample represents German students only. In addition to cultural aspects the sample also only consisted of students from educational contexts, although their majors varied (Early Childhood Education, Adult Education, Special Education, and Cultural Education), which might limit the generalization of our findings.

However, the results indicate that the Design Thinking process has a positive impact on the students' growth mindset. The students' improvement in all six aspects of the Design Thinking Mindset for Innovation is consistent with their self-talk and even a detailed focus on the six fixed mindset thought patterns and growth mindset antidotes.

As awareness is key, the strategy of using self-talk seems to be very appropriate to access the mindset, become aware of it, and have a chance to shift towards a growth mindset. Applying Design Thinking in the context of sustainable development offers students a unique opportunity to experience an innovative methodology in action. Design Thinking fosters a human-centered and iterative approach to problem-solving, which is particularly suited to complex and uncertain challenges. Through Design Thinking, students can work on sustainable development projects that involve understanding user needs, iterating solutions, and embracing failures as learning opportunities. This practice cultivates a growth mindset, encouraging students to see challenges not as setbacks but as essential steps in the innovation process. In this way, Design Thinking fosters practical problem-solving skills and shapes a mindset conducive to continuous learning and adaptation. Design Thinking and its mindset is a way to evolve the necessary competencies of *Gestaltungskompetenz* on the one hand as well as developing innovative solutions to complex challenges on the other hand. Incorporating Design Thinking into educational systems can enhance key skills and mindsets, empowering individuals to contribute to and shape a sustainable future. This emphasizes the importance of mindset in moving towards sustainable development, illustrating its impact on our ability to innovate, adapt, and effect significant change. To empower students' growth mindset, they should be engaged in a real design challenge with real stakeholders facing real sustainability problems. It's crucial to introduce them to both the concept of the design thinking mindset for innovation and the growth mindset. In addition, self-talk should be introduced and practiced as a tool for analyzing mindsets (e.g. using a structured reflection workbook as described in the article). With these prerequisites in place, students will be well prepared to apply the Design Thinking process, supported by Design Thinking coaches.

However, further research utilizing our setting would be beneficial, employing a larger sample size to gain deeper insights. Additionally, it would be advantageous to apply this to students from diverse domains who may be less familiar with self-reflection, to ascertain whether their growth mindset is also enhanced through the application of Design Thinking. Qualitative data from student B5 indicates that further research should include emotions related to self-talk. This topic is also addressed by Elliott-Moskwa (2022). Finally, research should be conducted with an international and multi-major sample to investigate the generalization of the research results on empowering a growth mindset for sustainable development in Higher Education by Design Thinking.

References

BNE-Portal. (2024). *Das UNESCO Programm in Deutschland.* https://www.bne-portal.de/bne/de/bundesweit/das-unesco-programm-in-deutschland/das-unesco-programm-in-deutschland

Brown, T. (2019). *Change by design. How design thinking transforms organizations and inspires innovation.* Revised and Updated Edition. Harper.

Brinthaupt, T. M., & Morin, A. (2023). Self-talk: Research challenges and opportunities. *Frontiers in Psychology, 14*, 1210960. https://doi.org/10.3389/fpsyg.2023.1210960

de Haan, G. (2010). The development of ESD-related competencies in supportive institutional frameworks. *International Review of Education, 56*(2), 315–328. https://doi.org/10.1007/s11159-010-9157-9

de Haan, G., & Harenberg, D. (1999*). Bildung für eine nachhaltige Entwicklung, Gutachten zum Programm.* BLK. https://www.pedocs.de/volltexte/2008/218/pdf/heft72.pdf (Accessed on 04 Nov 2024)

Dosi, C., Rosati, F., & Vignoli, M. (2018). *Measuring design thinking mindset.* In Faculty of mechanical engineering and naval architecture, University of Zagreb, Croatia; The Design Society, 1991–2001.

Dweck, C. (2017). Mindset - updated edition: changing the way you think to fulfil your potential. Constable & Robinson Ltd.

Elliott-Moskwa, E. (2022). *The growth mindset workbook.* New Harbinger Publications, Inc.

Fang, F., Kang, S. P., & Liu, S. (2004). *Measuring mindset change in the systemic transformation of education* (pp. 298–304). Association for Educational Communications and Technology. https://eric.ed.gov/?id=ED484992

Fernyhough, C. (2016). *The voices within: The history and science of how we talk to ourselves*. Basic Books.

Gainsburg, I., & Kross, E. (2020). Distanced self-talk changes how people conceptualize the self. *Journal of Experimental Social Psychology, 88*, Article 103969. https://doi.org/10.1016/j.jesp.2020.103969

Hackfort, D., & Schwenkmezger, P. (1993). Anxiety. In R. N. Singer, M. Murphy, & L. K. Tennant (Eds.), *Handbook of research on sport psychology* (pp. 328–364). Macmillan.

HPI Academy. (2024). *Design thinking mindset for innovation*. https://hpi-academy.de/en/design-thinking/design-thinking-mindset-for-innovation/

Kanning, H., & Meyer, C. (2022). The understanding and meaning of knowledge transfer for research and education in the context of a great transformation. In: M. Abassiharofteh, J. Baier, A. Göb, I. Thimm, A. Eberth, F. Knaps, V. Larjosto, & F. Zebner (Eds.), *Spatial transformation: Processes, strategies, research design*, (pp. 9–26). ARL—Akademie für Raumentwicklung in der Leibniz-Gemeinschaft. https://www.ssoar.info/ssoar/handle/document/78748

Kassel, K., & Rimanoczy, I. (2018). *Developing a sustainability mindset in management education*. https://doi.org/10.4324/9781351063340

Kassel, K., Rimanoczy, I., & Mitchell, S. F. (2018). A sustainability mindset model for management education. In K. Kassel, & I. Rimanoczy (Eds.), *Developing a sustainability mindset in management education* (pp. 3–37). https://doi.org/10.4324/9781351063340

Krohn, M. (2023). *Crucial role of mindsets in innovation efforts*. Springer Fachmedien.

Latinjak, A. T., Morin, A., Brinthaupt, T. M., Hardy, J., Hatzigeorgiadis, A., Kendall, P. C., Neck, C., Oliver, E. J., Puchalska-Wasyl, M. M., Tovares, A. V., & Winsler, A. (2023). Self-talk: An interdisciplinary review and transdisciplinary model. *Review of General Psychology, 27*(4), 355–386. https://doi.org/10.1177/10892680231170263

Mareis, C. (2016). *Theorien des Designs zur Einführung. 2* (korrigierte). Junius.

Meinel, C., & Leifer, L. (2011). Design thinking research. In H. Plattner, C. Meinel, & L. Leifer (Eds.), *Design thinking: Understand–improve–apply* (pp. 13–21). Springer Science & Business Media. https://doi.org/10.1007/978-3-642-13757-0

Meinel; C., Weinberg, U., & Krohn, T. (2015). *Design thinking live*. Murmann Publisher.

Morin, A. (2005). Possible links between self-awareness and inner speech theoretical background, underlying mechanisms, and empirical evidence. *Journal of Consciousness Studies*. https://www.ingentaconnect.com/content/imp/jcs/2005/00000012/f0020004/art00007

Oleś, P. K., Brinthaupt, T. M., Dier, R., & Polak, D. (2020). Types of inner dialogues and functions of self-talk: Comparisons and implications. *Frontiers in Psychology*. https://www.frontiersin.org/articles/10.3389/fpsyg.2020.00227/full

Plattner, H., Meinel, C., & Weinberg, U. (2009). *Design thinking. Innovationen lernen—Ideenwelten öffnen*. FinanzBuch.

Price, E. L. (2015). *The development and application of a measure to assess the interpersonal qualities of self-talk*. Wilfrid Laurier University. https://scholars.wlu.ca/etd/1767

Racy, F., & Morin, A. (2024). Relationships between Self-Talk, Inner Speech, Mind Wandering, Mindfulness, Self-Concept Clarity, and Self-Regulation in University Students. Behav Sci (Basel), 14(1), 55. https://doi.org/10.3390/bs14010055

Rinne, A. (2023). *Change management requires a change mindset*. Havard Business Review. https://hbr.org/2023/05/change-management-requires-a-change-mindset

Schmidberger, I., Wippermann, S. (2022). Die innovationsmethodologie design thinking. In I. Schmidberger, S. Wippermann, T. Stricker, & U. Müller (Eds.), *Design thinking im Bildungsmanagement. Innovationen in Bildungskontexten erfolgreich entwickeln und umsetzen* (pp. 25–52). Springer Fachmedien Wiesbaden. https://doi.org/10.1007/978-3-658-36951-4

Simon, H. A. (1996). *The science of artificial* (3rd edn). MIT Press.

Uitto, J. I. (2022). Transformational change for people and the planet: Evaluating environment and development—introduction. In J. I. Uitto & G. Batra (Eds.), *Transformational change for people and the planet. Evaluating environment and development* (pp. 1–13). Springer Nature.

UNESCO. (2017). *Education for sustainable development goals. Learning objectives.* UNESCO. https://unesdoc.unesco.org/ark:/48223/pf0000247444

The Sustainable Reporting of a Higher Education Institution: A Pilot Project Where AI-Guided Document Analysis Facilitates the Environmental, Social, and Governance Reporting

Andries Paesen, Tjalling Rooks, Joos Kesters, Simon Van De Voorde, Jean-Pierre Segers, Marijke Maes, and Dirk V. Franco

Abstract Higher Education Institutions (HEIs) around the world are in the process of incorporating sustainability into their operations. New directives and legislations—such as the EU Green Deal, the AI-Act, the Corporate Social Responsibility Directive, and the attention for ESG policies—are examples of top-down approaches. However, there are also bottom-up initiatives emerging. A significant risk here is that these efforts may rely on a few individuals and that sustainability may be perceived as a trend, only gaining focus during the beginning. Therefore, after an enthusiastic start, sustainability should be embedded into the policies governing all activities of the HEI. Besides the inside-out approach, attention should also be given to the outside-in approach. Specifically, what does the external world believe the HEI should focus on regarding sustainability? Transparency around progress towards achieving the Sustainable Development Goals (SDGs) is crucial, and this is best done in accordance with European legislation on Environmental, Social, and Governance (ESG) reporting. Although HEIs in Belgium are not yet required to report on ESG, the University of Applied Science and Arts (PXL UAS) intends to start in an innovative way. Previous research on corporate social responsibility (CSR) investments has established its long-term financial benefits including fostering innovation. Similar

A. Paesen · T. Rooks · J. Kesters · S. Van De Voorde
RESULTX Bv, Oudsbergen, Belgium

J.-P. Segers
Faculty of Business Economics, UHasselt–Hasselt University, Agoralaan, Diepenbeek, Belgium

J.-P. Segers · M. Maes · D. V. Franco (✉)
PXL, Central Administration, Hasselt, Belgium
e-mail: dirk.franco@uhasselt.be

J.-P. Segers
HEC-ULg–Université de Liège, CEPE, Rue Louvrex, Liège, Belgium

D. V. Franco
Centre for Environmental Sciences (CMK), UHasselt–Hasselt University, Agoralaan, Diepenbeek, Belgium

W. Leal Filho et al. (eds.), *Introducing the UN SDGs in Higher Education Institutions*, World Sustainability Series, https://doi.org/10.1007/978-3-032-23017-1_30

trends are anticipated for ESG, reflecting the growing importance of sustainable investing. However, to stimulate this trend, one needs adequate solutions that help transform their operations efficiently. The main obstacle for businesses blocking them to follow up on their ESG reporting findings has been identified and indeed lies in the cost of the extensive data gathering process from various sources, which includes the business itself and its supply- and value chain. The developed tool in this paper by ResultX (in a co-creation with all the partners) aims to greatly improve the efficiency and accuracy of the reporting process by utilizing advanced AI techniques, including Natural Language Processing (NLP), Large Language Models (LLMs), Retrieval Augmented Generation (RAG) technology, and more. Implementing these techniques will enable to report findings more easily and save time and money that can be spent on incorporating impactful ESG strategies backed by data, acting upon them, resulting in increased resilience and long-term financial benefits. The PXL UAS deployed this newly developed tool in a pilot project on the one hand to report their achievements within the SDG ambassador track and on the other hand to also check them against the legislation in force.

Keywords CSR · ESG · Transition teams · AI-guided · Coalition of the willing

1 Introduction

The 17 Sustainable Development Goals (SDGs), adopted by the United Nations in 2015, (United Nations, 2015) represent an ambitious and essential roadmap for a future where no one is left behind. These goals aim to balance economic progress, social equality, and environmental protection—a truly holistic vision.

Higher education institutions (HEIs) play a pivotal role in advancing these goals (Leal Filho et al., 2023). Through research, teaching, and the implementation of sustainable practices, HEIs inspire students and staff to actively engage in this global movement. Moreover, HEIs are uniquely positioned to equip future leaders with the knowledge, skills, and mindset needed to address the challenges of sustainability.

In addition to these direct contributions, HEIs must focus on effective reporting to avoid common pitfalls such as:

1. Fragmentation: Lack of coordination and integration of internal initiatives.
2. Unclear Messaging: Inconsistent communication with external stakeholders.
3. Reduced Support: Declining engagement from both internal and external audiences.

To manage this internally and prevent issues, the principle of the coalition of the willing was applied within the framework of the multi-level perspective (which is further elaborated in Sect. 3.1). In this way, a core group was formed, functioning as a sensor to detect and consolidate various sustainability initiatives at the grassroots level. From this foundation, a coalition of change emerged—broadly represented within the institution's organizational structure and spanning multiple disciplines, including students.

At the same time, external stakeholders were involved (see also Sect. 3), exerting pressure on the HEI's policy (meso level) from both the internal micro level and the external macro level. As a result, the institution underwent a fundamental paradigm shift from a BUA HEI to an SDG-driven HEI.

As the GRI (2022) states, informing all stakeholders is good for business. Providing comparable, verifiable information on sustainability efforts is paramount. A well-structured approach to SDG reporting fosters alignment and impact both within the organization and externally. In essence, SDG reporting will help the HEIs align their mission with global sustainability priorities, showcasing leadership and responsibility in shaping a better future (Franco et al., 2019, 2025). It is important to have a diversity of input and perspectives and to consider an outside-in perspective. The ambitions of a higher education institution (HEI) should not be shaped solely by internal viewpoints coming from leadership bodies, faculty, employees and students. Multi-stakeholder credibility is supplemented by insights from external stakeholders, such as policymakers, industry representatives, social economy organizations and civil society.

What do these stakeholders believe the HEI should prioritize when it comes to sustainability? (Freire et al., 2021; Markovic et al., 2020). This external communication is critical for several reasons:

1. Demonstrating Accountability and Transparency: Builds trust and credibility by showcasing measurable impact.
2. Strengthening Stakeholder Engagement: Encourages collaboration with partners and communities.
3. Enhancing Reputation and Attractiveness: Positions the institution as a leader in sustainability, appealing to students, faculty, and funders.

This case study highlights a HEI effort to focus on sharing best practices and challenges, contributing to collective learning and accelerating SDG achievements. At the same time, it emphasizes the importance of continuous improvement and meeting student and societal expectations, which increasingly center on sustainability.

The internal approach adopted by PXL UAS aimed to align efforts across the institution. While bottom-up initiatives were abundant, the Multiple Level Perspective (MLP) approach streamlined these efforts through a "coalition of the willing." This coalition—comprising internal stakeholders from diverse backgrounds—ensured that sustainability efforts were embedded across the organization. Support from top management strengthened these initiatives, avoiding fragmentation, unclear messaging and reduced support by an internal buy-in (Avelino, 2017; Geels, 2011; Geels & Schot, 2007) analyses power dynamics in sustainability transitions.

Subsequently, PXL UAS engaged external stakeholders to determine priorities through a materiality index process (Calabrese et al., 2019). This collaborative step informed the integration of sustainability goals into the institution's operations and policies, encompassing education, research, and valorization services (Calabrese et al., 2019) (Franco et al., 2025).

Although Belgian HEIs are not yet required to report on Environmental, Social, and Governance (ESG) metrics, PXL UAS has proactively taken the lead. In a co-creation and through a pilot project, the HEI deployed a newly developed tool to achieve two key objectives:

1. Reporting Achievements: Documenting progress within the SDG Ambassador Track (CIFAL Flanders, n.d.)
2. Ensuring Legal Compliance: Verifying alignment with existing legislation.

Through innovative tools and forward-thinking practices, PXL UAS is setting an example for how HEIs can effectively embrace and communicate their sustainability commitments.

2 Frameworks

2.1 Regulatory Frameworks

In sustainable business, "the customer is the planet" (Daly et al., 2024). Sustainability is a long-term value driver. Empowering European companies to become change-makers is paramount. Companies—big and small—increasingly need to make more and better sustainable choices in their supply and value chains and in their operations. The European Union (EU) currently holds a special status in the world when it comes to enforcing new legislations and directives with respect to a net-zero society and industry and for addressing the UN Sustainable Development Goals 2030 Agenda (United Nations, 2015). Compliance with regulations is the most powerful ESG driver for most organizations. European policies include the European Sustainability Reporting Standards (ESRS) (European Commission, 2023), the Green Taxonomy (European Commission, n.d.–a), and the Corporate Sustainability Reporting Directive (CSRD): (European Commission, n.d.–b), which is a game changer. There are many more at a global, country, regional, and even sector level, such as the EU Green Deal (European Commission, n.d.–c) and the Green Deal Industrial Plan (European Commission, n.d.–d).

2.2 Environmental

The NFRD (environmental protection, employee treatment, human rights, board diversity, anti-corruption) (European Parliamentary Research Service, 2021) was applied to large public-interest entities, including banks, insurance firms and listed companies with over 500 employees. This directive covered approximately 11,700 companies and groups across the EU. However, it did not mandate a specific reporting framework, allowing companies flexibility in how relevant information was disclosed. This absence of detailed disclosure requirements led to inconsistencies in reporting practices across organizations. On the 21st of April 2021, the EU commission announced the adoption of the Corporate Sustainability Reporting Directive (CSRD) in line with the commitment made under the European Green Deal to respond to the NFRD's shortcomings. Its primary objective is to usher in a new era of sustainability reporting, characterized by enhanced uniformity, comparability, and reliability. Departing from the NFRD-approach, the CSRD not only introduces a standardized reporting framework but also carefully aligns it with globally recognized standards, such as the Global Reporting Initiative (GRI) and the Task Force on Climate-related Financial Disclosures (TCFD) (Bloomberg, 2022). This alignment lays the groundwork for more consistent reporting practices while emphasizing a commitment to international best practices in sustainable disclosure.

To address these limitations and to broaden the scope of sustainability reporting, the European Commission asked the European Financial Reporting Advisory Group (EFRAG, n.d.) to develop comprehensive standards. This initiative culminated in the creation of the European Sustainability Reporting Standards (ESRS) (European Parliamentary Research Service, 2021), aiming to standardize reporting practices, improve comparability of sustainability reports, and introduce requirements for external assurance and digitalization of reported data.

The CSRD reporting framework has become a key focus for large corporations. The directive mandates approximately 50,000 European companies, including 2,500 in Belgium, to maintain detailed "accounts" of their sustainability efforts across Environmental, Social, and Governance (ESG) dimensions. Its implementation is phased as follows:

1. Large public interest entities (with over 500 employees) already subject to the NFRD must begin compliance by January 2024, reporting on their 2024 financial year in 2025.
2. Large enterprises not yet covered by the NFRD (meeting two of the following three criteria for two consecutive financial years: over 250 FTEs, €50 million in net-turnover, or €25 million in total assets) will start reporting from January 2025, based on their 2025 financial year, with reporting due in 2026.
3. Listed small and medium-sized enterprises (SMEs) and other listed companies will be required to comply starting January 2026, reporting on their 2026 financial year in 2027. Listed SMEs can opt out during a transitional period, deferring reporting obligations until 2028.
4. Non-EU companies generating at least €150 million in EU net turnover and having at least one EU subsidiary or branch must report starting in 2029, based on their 2028 financial year.

The CSRD adopts a top-down approach to enforce a new sustainability reporting model, reflecting the EU legislator's intent to expand the scope of mandatory and higher-quality sustainability disclosures. This transition represents a significant shift from the NFRD, increasing the number of covered companies from 11,700 to approximately 50,000.

Important to note is that the CSRD mandates not only on report on the company's internal ESG performance but also across the entire value chain. This means that suppliers to CSRD-regulated companies must disclose their ESG data to maintain business relationships. Additionally, other stakeholders, such as banks, and financial institutions, are likely to require ESG reports from SMEs as part of credit and financing assessments.

2.3 Stakeholders Framework

The ongoing regulatory and technological transitions with regard to sustainability reporting and artificial intelligence are important both for higher education institutions as for small and medium-sized enterprises. Indeed, the CSRD reporting for SMEs is important for higher education institutions (HEIs) as the latter are strongly connected to the local network of SME's. This highlights best practices in sustainability reporting, fosters accountability, and encourages alignment with broader societal goals. By understanding and potentially adopting elements of SME reporting standards, HEIs can enhance their own sustainability strategies and multi-stakeholder credibility by effectively engaging with external stakeholders (policy-makers, industry representatives, social economy organizations and civil society), and prepare students, employees and staff for the realities of sustainable practices in professional environments. Moreover, as HEIs collaborate with SMEs, aligning their approaches can strengthen partnerships and mutual impact on sustainable development goals (SDGs). This approach also resonates with the outside in approach as mentioned above.

Retrieving information from both internal and external stakeholders can be done efficiently and without consuming excessive time using this AI tool. As a result, external stakeholders remain more engaged and are also more inclined to provide the necessary information. This, in turn, strengthens the network of change, enabling better monitoring and achievement of both internal and external sustainability goals.

A recurring concern is the fear of an administrative tsunami for small and medium-sized enterprises within the value chain. In addition, CSRD and ESG (Environmental, Social, and Governance) strategies are closely linked to one another, as the graphic below shows (see Fig. 1). It modernizes and strengthens the rules concerning the social and environmental information that companies must report, while also impacting suppliers and clients.

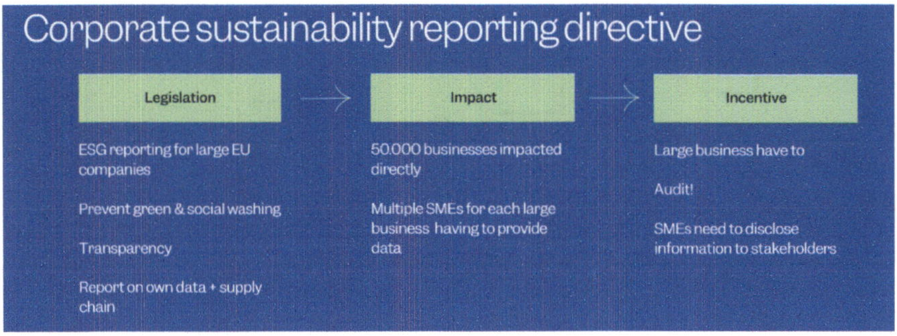

Fig. 1 Links between CSRD and ESG (Franco et al., 2024)

2.4 *Technology Framework*

Next to the regulatory frameworks, there is a wave of powerful, fast-proliferating new technologies. The regulatory environment for artificial intelligence in the European Union is characterized by a comprehensive and evolving framework aimed at fostering innovation while ensuring safety, ethical standards, and fundamental rights.

Suleyman and Bhaskar (2023) discuss the intersection of technology and power in the twenty-first century. Artificial Intelligence (AI) has indeed gained significant momentum over the past year(s). AI is expected to contribute significantly to improved productivity, economic growth and new job opportunities. Investments by major players and gatekeepers in the generative AI space have skyrocketed. The three types of artificial intelligence are: Narrow AI (ANI), General AI (AGI), and Artificial Superintelligence (ASI)/Artificial Capable Intelligence (ACI). ANI excels in specific tasks (Suleyman and Bhaskar 2023). AGI mirrors human cognition. On the other hand, ASI and ACI hold a futuristic promise (Suleyman and Bhaskar 2023). Currently, all AI is narrow. Overall, the EU's regulatory environment for AI is designed to balance the promotion of technological advancement with the protection of societal values, ensuring that AI development is both innovative and responsible. The Artificial Intelligence Act is the first-ever legal framework on AI, which addresses the risks of AI and positions Europe to play a leading role globally (Future of Life Institute, 2024).

3 Approach

3.1 *In the HEI*

In the beginning of the transition processing in the PXL UAS, the coalition of the change (Fig. 2) functions as a platform to bundle and strengthen innovation. However, it does not have to stop there. Powerful platforms can be built for inspiration and idea

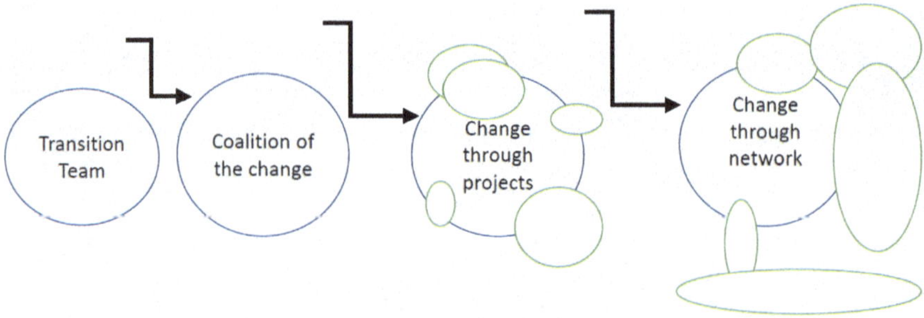

Fig. 2 Coalition for change—adapted by the authors from Van Yperen et al. (2017)

management, knowledge exchange and networking (Van Yperen et al., 2017). Those involved in the change projects can include regimes in the learning process, so that the latter can pick up the new way of working and the barriers that these projects must deal with can be addressed. In the next stage, to increase the pressure on this regime and to become more coherent, a transition team evolves via change projects towards a change network.

In May 2019, the PXL UAS installed a climate committee—SDG@PXL, which evolved into a transition team. Originally, the team consisted of five members (head of department Green & Tech, director of finance and facility management, research and education staff members and the policy advisor). The team quickly expanded to more than thirty internal stakeholders on a volunteer base, with divergent backgrounds and responsibilities, including students-junior colleagues. This became a *coalition of the willing* and later a *coalition for change* (Fig. 2). In this way, the transition team was strongly supported and anchored within the organization (both in terms of the different disciplines and the various responsibilities in the hierarchical structure). All members of the SDG@PXL (*coalition of change*) group took part in several digital (COVID-19 induced) lectures, workshops, webinars, and meetings in a close co-creation between CIFAL Flanders. The importance of facilitating continuous learning at all levels through empowering people, encouraging collaboration, and promoting open dialogue was a key feature in this process. Using this approach fragmentation, inconsistent communication with in-and external stakeholders and shrinking support was avoided.

In the next step the decision to apply for the UN certification of SDG Ambassador was taken (CIFAL Flanders, n.d.), in this step we moved towards a change trough project. At this moment the policy lines and the vision are based on the SDG's, consequently, also external stakeholders are involved, so we are now *a change through network* (see Fig. 2). By actively engaging with SMEs, we aim to bridge the gap between education, research, and entrepreneurship. These organizations are vital for regional economic growth and innovation, and their alignment with the SDGs creates shared value. In this project, the HEI not only aims to contribute to the sustainable development of our region but also to set an example for other institutions and organizations. By fostering meaningful partnerships and leveraging innovative technology, we strive to make a lasting impact on society and inspire others to do the same. To make these efforts even more visible, there is also a focus on the next step: reporting the progress in view SDG's. In this way the pitfalls mentioned earlier are avoided and moreover, it is also clear to the external stakeholders whether we are implementing the input and prioritization that they helped determine. Finally, this offers opportunities to identify trends and areas for improvement more quickly, helping us continuously evolve and enhance our environmental and societal contributions.

3.2 The Layer of AI

This sudden requirement for ESG reporting poses a significant challenge for many companies. The ESRS currently requires eligible companies to comply with around 84 disclosures (over 1200 data points) in a detailed and up-to-date manner. In addition, due diligence is of high importance as reports will be audited for correctness, completeness, and coherence. The complexity and volume of these tasks are often overwhelming, especially now, as ESG reporting is still a novelty. These challenges are addressed by automizing most of the reporting process, mitigating most of the administrative workload. Furthermore, due diligence and auditability are ensured through applying a thorough approach to completing disclosures, with clear references to all sources used.

This automation has mainly been achieved through the application of Artificial Intelligence (AI). To this end, a state-of-the-art framework has been developed in this case that handles data dynamically.

While generative AI holds great potential in automizing ESG reporting, there are certain limitations that need to be addressed. An important issue of generative AI, including current Large Language Models (LLMs) is its probabilistic nature (Vaswani, 2017). While useful for generating natural language, it comes with the risk of hallucinations. Generative AI systems, when unregulated, can produce inaccurate or entirely fabricated information. This is especially dangerous when generating ESG reports, which need to accurately reflect the company revenue, CO_2 emissions, among other things.

A second hurdle is its black-box property. Current models easily surpass 400 billion neurons in size (Grattafiori et al., 2023), making their inner workings practically impossible to understand. This is problematic particularly in context of ESG reporting because reports must be auditable, and companies must be able to provide reasons for their statements.

A third issue is naivety. Generic LLMs such as chatGPT (Dubey et al., 2024) lack contextual understanding of company data, which leads to naive and non-specific reports. A possible solution to this problem is fine-tuning domain- and company-specific data, but this is resource intensive and may not always be practical.

Beyond these primary technical challenges, additional critical limitations warrant consideration. Biases inherent in LLMs -arising from training on extensive datasets that may reflect social, cultural, or industry-specific prejudices- can skew the representation of company data and performance, thereby undermining the overall credibility and trustworthiness of ESG reports. Also, ensuring compliance with evolving regulatory frameworks such as ISO standards and generative AI guidelines remains a persistent challenge. Finally, the interpretability of machine-generated outputs poses a significant issue. Without clear insights into the rationale behind the outputs of LLMs, stakeholders and auditors may question the reliability and fairness of the reported information.

In this study a proprietary system has been developed that addresses these issues and ensures non-hallucinated, fully transparent reports. The system is deployable at minimal time, as soon as company data has been included into the platform, and the automized part of reporting will be finished well within 15 min. Furthermore, the system does not require human guidance which eliminates room for human error. Also, the generated reports are fully transparent. Each paragraph of the ESG report is enhanced with references to all sources in the company data. Consequently, reports are inherently audit friendly. Also of note, the system is built to be highly scalable and to handle numerous companies with substantial amounts of data. Furthermore, to address the issue of interpretability of machine-generated outputs, each AI-suggested paragraph needs to be human-reviewed before integration into the final report, and as previously mentioned, each suggestion is enhanced with the used source references.

Through this state-of-the-art architecture, the limitations of generative AI are addressed and a reliable solution to ESRS reporting compliance is provided.

ResultX distinguishes itself from existing AI-enhanced sustainability reporting platforms in multiple ways. A first one is by implementing a higher degree of automation. This is realized through two primary mechanisms: no requirement for prior data selection, and the utilization of live data connections instead of relying on traditional static file uploads.

First, companies can connect all company data they deem usable for ESG reporting without needing to manually select specific files and paragraphs. In contrast to conventional platforms—which often require users to manually link specific files to designated ESG topics and impose strict limits on file uploads—this approach offers greater flexibility. It simplifies data integration and reduces the risk of excluding crucial information due to overly restrictive limitations.

Second, the platform does not rely on static uploads, but instead on dynamic data connections. Any modifications made are to connected data sources such as OneDrive, SharePoint, and Dropbox are automatically detected by the platform, eliminating the need for manually re-uploading files whenever changes are made. This real-time updating system ensures that the reports reflect the most current information available.

A second innovation is the suggestion of impacts, risks, and opportunities (IROs). These IROs are a necessary part of the Double Materiality Assessment (DMA), the first step of ESG reporting under the ESRS legislation. IROs must be created by the company itself to scope out the impacts it has on the environment, and the risks and opportunities it experiences. This is often a cumbersome process, as it requires complete knowledge of the company information. AI is used to scan the company data for IROs, saving manual work.

Third, AI is used to acquire and refine sector-specific insights. As an increasing number of companies integrate their data with the platform, understanding of impacts, risks, and opportunities associated with certain industries. Is progressively developed. This iterative learning process allows the system to suggest tailored sector-specific recommendations. To ensure compliance with strict data protection protocols such as ISO, all Personally Identifiable Information (PII) is redacted before the data is utilized.

4 Findings from the PXL UAS Pilot

In early 2024, the PXL UAS launched a pilot project in collaboration with ResultX to assess the ESG reporting platform's capabilities and its influence on their sustainability strategy. A higher education institution (HEI) is not required to report under the CSRD, it wanted to conduct a CSRD gap assessment and produce a subsequent report to strengthen its strategic position, fitting in their vision on implementing the SDGs.

As a starting point, 16 different documents, procedures and policies, together with parts of the website, needed to be read and interpreted. Next, sustainability information had to be recognised and matched with the corresponding ESRS data point to obtain a CSRD gap assessment. To carry out the process manually, an estimated duration of 12–16 weeks can be considered.

The initial step of the pilot involved gathering the necessary documentation. It was determined that high-level management documents and policies would serve as an ideal starting point to gather a high-level overview of the PXL UAS ESG performance, since they covered policies overarching different departmentsd and data on environmental, social, and governance topics. In collaboration with the management and ESG experts, these documents were identified, collected, and uploaded to the platform for analysis. The starting point of the analysis is shown in Fig. 3.

Upon completion of the collection of the first set of documents, a first gap assessment was carried out. Although results were already promising, several issues were pointed out by the PXL UAS ESG experts and management team. These barriers hindered the ability to obtain a clean and accurate CSRD gap assessment, and included:

- Understanding context of retrieved data: contextual differences regarding full-time equivalent (FTEs),
- Consolidated reporting: collecting data from different departments,
- Security: restricting users from accessing documents they do not have permission to view.

The first barrier relates to the understanding of context. Several ESRS metrics assess FTE figures. The initial technology stack struggled to distinguish between

Fig. 3 Management and policies documents

students and employees, introducing noise into the results. Since this issue was identified and communicated immediately after reviewing the results of the first gap assessment run, ResultX was able to improve the data assessment process and test new technology implementation in close collaboration with the HEI in an iterative way. At the point in time when this paper was written, the platform could accurately differentiate between students and employee data. This is one example of how a technological challenge is identified and addressed. The result after implementing these improvements was an accurate gap assessment, shown in Fig. 4. An example consisting of one data point of the S1 ESRS standard, regarding 'own workforce' is given below.

Fig. 4 One data point of the S1 ESRS standard, regarding 'own workforce' is given below

Datapoint: Policies for Elimination of Discrimination and Promotion of Diversity.

Generated answer.

The specific policies aimed at the elimination of discrimination include:

1. Inclusivity in Hiring: Candidates are selected based on their qualities and skills, regardless of gender, origin, or disability. Job postings are formulated in a gender-neutral manner, and function titles are evaluated for gender neutrality.
2. Diversity and Inclusion Analysis: An analysis is planned to identify areas for improving inclusion and diversity within various departments and services.
3. Support for Non-Binary Employees: The organization accommodates employees who do not wish to disclose their gender.
4. Gender Equality Plan: The institution has a public gender equality plan that integrates gender dimensions into education and research, promoting gender equality across all activities.
5. Student Support Model: An inclusive student support model focuses on the well-being and development of students, ensuring that all students can thrive in a diverse environment.
6. Monitoring and Reporting: Regular monitoring of gender representation within the staff and leadership roles is conducted to ensure equitable representation and address any disparities.

All aspects of the generated answer were accurately marked with their source document, making this fully transparent. A potential auditor can see where data is stemming from, who uploaded the corresponding data, and if there were any conflicting sources of information.

Another barrier involved the challenge of consolidated reporting. The PXL UAS collects information at the departmental level (e.g., PXL Tech, PXL Business,) but aims to report at the level of the entire university college. This required the platform to distinguish between different departments and aggregate the data accurately. A clear example of the importance of correct data collection and aggregation is when greenhouse gas emissions from one department increase, while overall emissions for PXL decrease due to improvements in other departments. In such cases, the consolidated report must reflect this by addressing the emissions data at both the departmental level and in an aggregated manner. This ensures that the total greenhouse gas emissions for PXL UAS are accurately disclosed, while also providing transparency about variations across departments. However, this subject is quite relevant to social topics as well, since policies are often implemented on an overarching level.

Addressing this need required enhancements to the platform's capabilities, ensuring it could effectively differentiate departmental data and consolidate it for unified reporting. The importance of overcoming this challenge was evident, as many businesses, holdings, or groups construct consolidated reports containing all their subentities.

To tackle this barrier, ResultX introduced a department function, enabling data collection at the departmental (or subentity) level while allowing for automatic aggregation. This functionality streamlined the process of producing consolidated reports.

As a result, policies such as those related to diversity can be displayed and reported at the consolidated PXL UAS level, as they apply uniformly across all departments and do not need to be included in the documentation for each department separately.

A final example indicating the symbiosis of this pilot project related to security. In the first versions of the ResultX platform, any platform user was able to see documents that were used in the analyses. Unfortunately, ESG information contains often sensitive information, for example, summaries from management meetings or payroll information. Applied to the PXL UAS, some of the analyzed data was only accessible in their native data management system to management, while any user of the platform could read said documents once they were uploaded. This issue has been solved by adopting a new database structure that prioritizes privacy between company data. The new system ensures that data belonging to one company is securely isolated and cannot be accessed in report generation for another company. In addition to inter-company data security, intra-company data security has been implemented as well. ResultX recognizes that not all departments within one company should have access to all data. For example, sensitive data such as payroll data and unpublished intellectual property should not be accessible to everyone within the company. To this end, different member roles have been implemented. Managers can assign viewership rights to company members for specific files, enabling fine-grained control over who views what.

After considering the barriers mentioned above, and a constructive collaboration to resolve these challenges, the platform could generate in an hyperautomated way:

1. CSRD gap assessment,
2. Detailed description of disclosures,
3. Display information on a consolidated level,
4. Auto-generate summaries per ESRS sub(-sub) topic.

These summaries provided the HEI with a detailed and tailored overview of all ESRS subtopics and sub-subtopics where relevant data were identified for topical standard S1 (Own workforce)—subtopic Diversity.

Summary 'Diversity'

Elimination of Discrimination

The organization has established comprehensive policies aimed at eliminating discrimination in various forms. Key measures include:

1. Selection of candidates based on qualities and skills, irrespective of gender, origin, or disability.
2. Gender-neutral job titles and vacancy announcements.
3. Implementation of protocols to combat gender-related violence and inappropriate behavior.
4. Promotion of gender equality as a core aspect of sustainability goals.
5. Application of equal opportunities principles in recruitment, ensuring no discrimination based on minority status.

Addressing Harassment

To combat harassment, the organization has instituted specific policies that include:

1. Gender-neutral job postings and a widely communicated protocol against inappropriate behavior.
2. Special attention during onboarding conversations to ensure awareness.
3. Provision of psychosocial support for victims and a reporting point for students.
4. Training and awareness programs focused on unwanted behavior.
5. Implementation of gender equality in recruitment and career development policies.

Promoting Equal Opportunities and Diversity

The organization emphasizes equal opportunities and diversity through various initiatives:

1. Inclusivity in hiring practices and diversity analysis.
2. Support for non-binary employees and a comprehensive gender equality plan.
3. Student support models and ongoing monitoring and reporting of diversity initiatives.

Implementation Procedures

Policies are implemented through specific procedures designed to advance diversity and inclusion:

1. Equal opportunity recruitment and appointment of a diversity coordinator.
2. Annual monitoring of gender equality plans and data collection on gender representation.
3. Training programs aimed at fostering a positive work environment focused on gender equality.
4. Low-threshold student support for mental well-being, resilience, and career guidance.

Communication of Policies

The organization effectively communicates its diversity policies through various channels:

1. A diversity coordinator and the 'PXL Green Office' for stakeholder engagement.
2. Inclusion of sustainability principles in job vacancy texts.
3. Protocols available on the intranet and active communication about strategic goals related to sustainability and diversity.

Ensuring Understanding and Accessibility

To ensure understanding and accessibility of diversity policies, the organization undertakes the following measures:

1. Training and capacity-building activities, along with professional development policies.
2. Annual evaluations of training offerings and engagement in dialogue with employees.
3. Active communication about gender equality as part of sustainability goals.

Remarkably, the entire process following the upload of documentation was completed within minutes, delivering custom and tangible insights into PXL UAS ESG performance. This represents a significant advancement compared to the 12–16 weeks of manual effort previously required to achieve a gap assessment, which offered no immediate insights or analysis.

5 Critical Reflections and Next Steps

While AI brings many advantages, human interpretation remains essential. Connecting data to narratives ensures the context and stories behind the numbers are highlighted. When using AI, it is critical to safeguard privacy and address ethical concerns. Our approach must remain transparent and responsible to ensure trust among all stakeholders, including SMEs.

Beyond the internal "coalition of the willing," the PXL UAS aims to involve alumni, businesses, and communities more actively in the process. This not only broadens our support base but also strengthens our connection to society. For SMEs, this involvement creates opportunities for co-creating solutions, aligning their business goals with SDGs, and fostering innovation. The sustainably is now embed in our policies, curricula, and research practices. This vision is particularly relevant for SMEs as a stakeholder, as aligning their strategies with the SDGs policies of the HEI and helps in a co-creation to remain resilient and competitive in an evolving market.

By collaborating with other higher education institutions that are undertaking similar initiatives, the PXL UAS learn from each other's experiences and optimize their strategies. SMEs, too, can benefit from sharing insights and participating in knowledge networks that link academia and business.

6 Conclusion

The pilot project with ResultX demonstrated the potential of applying innovative technology to address the complexities of ESG reporting and lay the foundation towards compliance with CSRD requirements. Although PXL, as an HEI, is not mandated to report under CSRD, the collaboration proved how a strategic approach to sustainability reporting can strengthen institutional positioning and align with broader sustainability goals, such as implementing the SDGs, in an efficient and pragmatic manner. Throughout the pilot, significant barriers were identified and called out, including understanding contextual data differences, managing consolidated reporting, and ensuring robust data security. These challenges required iterative improvements and close collaboration between all partners, by addressing issues such as differentiating between students and employees for FTE metrics, enabling consolidated reporting across departments, and implementing advanced security features. As a result, the calculated 12–16 weeks to gain a first overview of PXL UAS ESG performance could be reduced to a few minutes, leaving the remaining time to analyze their ESG performance, investigate improvements and translate findings into strategy.

The pilot's outcomes underscored the platform's ability to provide a HEI with a comprehensive CSRD gap assessment, generate detailed ESG disclosures, and produce consolidated and department-specific reports, all in a fully auditable manner. This case highlights the value of addressing both technological and organizational barriers through innovation and collaboration, offering a scalable solution for other HEI and businesses.

Even in this initial phase, these pilot projects have already provided valuable insights into HEIs' own sustainability reporting. The tool has also been extensively tested in other sectors, and further expansion to more participants will undoubtedly enhance its accuracy and effectiveness. Moreover, this approach could help establish a meaningful benchmark, paving the way for greater standardization in sustainability reporting.

ESG and other reporting requirements are often perceived as a significant time investment (and therefore a cost) that leaves little room for strategic decision-making. However, ESG reporting should be seen as an opportunity rather than a burden. This approach helps to rebalance the traditional 80–20% resource allocation—ensuring that the majority of efforts are directed toward developing impactful ESG strategies that drive positive (financial) outcomes.

Furthermore, the system's ability to extract data from existing documents in a controlled manner eliminates the need for rigid formatting, allowing data to be retrieved seamlessly throughout the workflow. This ensures ongoing engagement both internally and externally, fosters clearer communication, and ultimately supports the broader mission of the coalition for change.

Acknowledgements The views expressed in this article are solely the responsibility of the authors. The authors are incredibly grateful to Christine Schoeters and Ann Knaepen (language), Cas Boyen, Isabelle Segers, and Inge Verheyen (inspiration).

References

Avelino, F. (2017). Power in sustainability transitions: Analysing power and (dis)empowerment in transformative change towards sustainability. *Environmental Policy and Governance, 27*(6), 505–520.

Bloomberg. (2022). *TCFD 2022 overview booklet.* https://assets.bbhub.io/company/sites/60/2022/12/tcfd-2022-overview-booklet.pdf

Calabrese, A., Costa, R., Levialdi Ghiron, N., & Menichini, T. (2019). Materiality analysis in sustainability reporting: A tool for directing corporate sustainability towards emerging economic, environmental, and social opportunities. *Technological and Economic Development of Economy, 25*(5), 1016–1038. https://doi.org/10.3846/tede.2019.10550

CIFAL Flanders. (n.d.). *SDG Wegwijzer/Start-to-SDG guide*. https://cifal-flanders.org/what-we-do/materials/sdg-publications/start-to-sdg-guide/

Daly, D., de Doncker, I., & Carlin, D. (2024). *The customer is the planet: A handbook for sustainable business*. Oak Tree Press. https://podcasts.apple.com/cm/podcast/the-customer-is-the-planet-a-handbook/id1620011535?i=1000674543309

Dubey, A., Jauhri, A., Pandey, A., Kadian, A., Al-Dahle, A., Letman, A., & Ganapathy, R., et al. (2024). *The llama 3 herd of models*. arXiv preprint arXiv:2407.21783

European Commission. (2023, July 31). *Commission adopts European sustainability reporting standards*. https://finance.ec.europa.eu/news/commission-adopts-european-sustainability-reporting-standards-2023-07-31

European Commission. (n.d.-a). *EU taxonomy for sustainable activities*. https://finance.ec.europa.eu/sustainable-finance/tools-and-standards/eu-taxonomy-sustainable-activities

European Commission. (n.d.-b). *Corporate sustainability reporting*. https://finance.ec.europa.eu/capital-markets-union-and-financial-markets/company-reporting-and-auditing/company-reporting/corporate-sustainability-reporting

European Commission. (n.d.-c). *European green deal*. https://commission.europa.eu/strategy-and-policy/priorities-2019-2024/european-green-deal

European Commission. (n.d.-d). *Green deal industrial plan*. https://commission.europa.eu/strategy-and-policy/priorities-2019-2024/european-green-deal/green-deal-industrial-plan

European Parliamentary Research Service. (2021). *Sustainability reporting in the EU: A comprehensive overview of current policy and challenges*. European Parliament. https://www.europarl.europa.eu/RegData/etudes/BRIE/2021/654213/EPRS_BRI(2021)654213_EN.pdf

EFRAG. (n.d.). *About EFRAG*. https://www.efrag.org/en

Franco, I., Saito, O., Vaughter, P., Whereat, J., Kanie, N., & Takemoto, K. (2019). Higher education for sustainable development: Actioning the global goals in policy, curriculum, and practice. *Sustainability Science, 14*(6), 1621–1642.

Franco, D. V., Segers, J. P., Wollaert, P., & De Deyn, M. (2025). Transition teams as an innovative tool in a higher education institution on the road to sustainability: A case study. In W. L. Filho, et al. (Eds.), *North American and European perspectives on sustainability in higher education* (pp. in press). Springer. ISBN 978-3-031-80433-5

Franco, D. V., Paesen, A., Rooks, T., Kesters, J., Van de Voorde, S., Segers, J. P., & Maes, M., (2024, October 10–11). The sustainable reporting of a Higher Education Institution: A pilot project where AI-guided document analysis facilitates the Environmental, Social, and Governance reporting. In *2nd transatlantic symposium on sustainable development in higher education, PXL, UHasselt, ResultX*.

Freire, P. D., Hawamdeh, S., & Dandolini, G. A. (Eds.). (2021). *Knowledge governance and learning for organizational creativity and transformation*. World Scientific. https://doi.org/10.1142/11933

Future of Life Institute. (2024). *EU artificial intelligence act*. https://artificialintelligenceact.eu/

Geels, F. W., & Schot, J. (2007). Typology of sociotechnical transition pathways. *Research Policy, 36*(3), 399–417.

Geels, F. W. (2011). The multi-level perspective on sustainability transitions: Responses to seven criticisms. *Environmental Innovation and Societal Transitions, 1*(1), 24–40.

Grattafiori, W. X., Défossez, A., Copet, J., Azhar, F., Touvron, H., Martin, L., & Synnaeve, G., et al. (2023). *Code Llama: Open foundation models for code*. ArXiv preprint arXiv:2308.12950.

GRI (2022). *The GRI perspective. A business case for environment & society.* (Issue 1). Amsterdam. https://www.globalreporting.org/

Leal Filho, W., Simaens, A., Paço, A., Hernandez-Diaz, P. M., Vasconcelos, C. R., Fritzen, B., & Mac-Lean, C. (2023). Integrating the sustainable development goals into the strategy of higher education institutions. *International Journal of Sustainable Development & World Ecology, 30*(5), 564–575.

Markovic, S., Bagherzadeh, M., Dubiel, A., Cheng, J., & Vanhaverbeke, W. (2020). Do not miss the boat to outside-in open innovation: Enable your employees. *Industrial Marketing Management, 91*, 152–161.

United Nations. (2015). *The 17 goals.* https://sdgs.un.org/goals

Suleyman, M., & Bhaskar, M. (2023). *The coming wave: Technology, power, and the 21st century's greatest dilemma.* Crown Publishing.

Van Yperen, M., Roorda, C., & Buchel, S. (2017). *Ondernemen voor transitie: Ruimte maken voor fundamentele vernieuwing.* Boom.

Vaswani, A. (2017). Attention is all you need. *Advances in Neural Information Processing Systems.*

The Sustainable Development Goals (SDGs) and Academic Activism: Organisational Culture and Teaching and Learning

Nicos Souleles

Abstract This chapter discusses universities' challenges in incorporating the Sustainable Development Goals (SDGs) into their organisational culture and educational practices. It specifically highlights the role of academic activism in supporting the SDGs. It posits that a discrepancy between academics' beliefs and problematic institutional cultures is a factor for sustainability activism. By employing participatory design principles, the paper critically examines why universities struggle to integrate the SDGs into their operations effectively. Academic activists often find themselves constrained within universities' existing structures and systems, limiting their ability to advocate for sustainability through curriculum design, teaching methodologies, and decision-making processes. The chapter also underscores the significance of transformative learning and sustainability-focused teaching approaches while exploring the challenges to their implementation and the restricted options available to academic activists. Key recommendations from the related literature stress the importance of fostering a culture of collaboration, breaking down disciplinary barriers, and promoting leadership that recognises and values grassroots initiatives and participation to facilitate positive transformations. Although the obstacles and hurdles to implementing the SDGs may differ among institutions and contexts, the prevalent bureaucratic and hierarchical structures in higher education institutions (HEIs) suggest an inherent mismatch between the effective adoption of SDGs and the prevailing organisational cultures, an ontological incompatibility. Academic activism is a response to this fundamental incompatibility.

Keywords SDGs · Sustainability · Higher education · Academic activism · Organisational culture · Transformative learning · Participatory culture

N. Souleles (✉)
Independent Researcher, Co-Editor in Chief DISCERN, the International Journal of Design for Social Change, Sustainable Innovation, and Entrepreneurship, Carlisle, United Kingdom
e-mail: nicos.souleles@cut.ac.cy

W. Leal Filho et al. (eds.), *Introducing the UN SDGs in Higher Education Institutions*, World Sustainability Series, https://doi.org/10.1007/978-3-032-23017-1_31

1 Introduction

The prevailing rhetoric is that HEIs can significantly contribute towards the SDGs through various policies and strategies, including the provision of education for sustainable development (ESD), fostering employment skills for implementing the SDGs, enhancing capacity building, mobilising learners, pursuing relevant research and technological innovations, as well as through contributions to civic, societal, and community-level initiatives (Leal Filho et al., 2023, p. 2; Seatter & Ceulemans, 2017, p. 49). According to the 2030 Agenda deadline stipulated by the United Nations (UN), higher education (HE) aims to guarantee that all learners will develop the necessary competencies to advocate for sustainability (Education, SDG 4, Target 4.7). This target includes ESD, human rights, gender equality, promotion of peace culture, global citizenship, and appreciation of cultural diversity (Rieckmann, 2017, p. 13). Thus, it is unsurprising that the SDGs feature prominently in the stated missions of many universities (Alm et al., 2021, p. 1451; Hoover & Harder, 2015, p. 2; Kestin et al., 2020, p. 5).

However, several studies have identified implementation obstacles vis-à-vis SDGs and HEIs at the micro (teaching and learning) and macro (organisational culture and management) levels. For example, there is the challenge of developing effective pedagogical approaches that embed SDGs into HE curricula (Alm et al., 2021, p. 1453). At the macro level, obstacles include the inability of HEIs to establish efficient organisational structures and policies to implement the SDGs in their functions (Gardner et al., 2021, p. 2; Rajabifard et al., 2021, p. 652). The implementation challenges present in each HEI are likely to vary in degree but not be dissimilar.

This paper focuses on organisational culture and teaching and learning and presents a critical understanding of several factors involved in academic activism and the implementation of SDGs. Teaching and learning are integral to an organisation's culture, particularly in educational institutions like universities. Academics committed to the SDGs face obstacles at both micro and macro levels. They assume the role of activists within their academic institutions, often unintentionally questioning bureaucratic obstacles, administrative approaches, and traditional but ineffective educational ideologies and methods. While the commitment and passion of academics can drive progress on the sustainability front, they are also susceptible to 'unsustainable' or 'unmanageable' work environments (Hoover & Harder, 2015, p. 10). Academics who strive to merge activism with their university duties may encounter various challenges, including threats, abuse, attempts to silence them, and peer pressure to discourage their involvement in activism (Flood et al., 2013). For this paper, the terms 'activist' and 'activism' entail a broad spectrum of activities whose common characteristic is seeking change that favours the 'worldview' of the activist (Fuad-Luke, 2013, pp. 5–6).

Spolander et al. (2022) reason that the influence of neoliberalism, commercialisation, and market forces on universities reduced opportunities for collaboration and caused competition between academics. This influence alienated academics from the traditional values of academic self-governance and collegiality. This alienation is described as a mismatch between the existing culture of an HEI and the realities of those who work within it, between the culture of competition and the individuality and worldviews of individual academics (Hoover & Harder, 2015, p. 11). In this disconcerting context, the authors describe activism in academia as involving a delicate balance—walking a tightrope—between the professional role of academics, their disciplines and the culture of their universities (Spolander et al., 2022, p. 68). If not confronted by the university hierarchy, academic activists tend to be 'invisible' in an organisational culture where alienation prevails (Spolander et al., 2022, p. 70). They can be isolated and work in fragmented ways, cut off from their other colleagues (Zamora-Polo & Sánchez-Martín, 2019, p. 5).

With 'overt' activism that potentially can interfere with or disrupt institutional policies, there is the apparent risk that disputes with university management can become disciplinary matters and undermine the career advancement of academic activists (Kjærgaard et al., 2024, p. 413). A heavy-handed management response can serve as a warning to early career academics. In contrast, 'subtle' activism refers to small-scale, limited, modest and personal initiatives that often occur unnoticed. This form of academic activism is described as "experimenting with quietly unsettling norms rather than overtly opposing or rejecting them" and as a "subversive compliance with institutional rules" (Kjærgaard et al., 2024, p. 416). There is debate among scholars as to whether 'subtle' activism can be classified as 'real' resistance or simply a form of compliance (Ybema & Horvers, 2017). Either overt or subtle, it is argued that the hidden causes behind academic activists seeking organisational change are the same barriers they strive to overcome in the broader society as citizens (Hoover & Harder, 2015, p. 11; Richter et al., 2020, p. 11; Rose, 2017, p. 72).

For activism to be relevant, it must offer an alternative to improve the current situation. A crucial aspect of implementing SDGs is engaging and involving all stakeholders, as these goals require broad participation (Kioupi & Voulvoulis, 2019, p. 5). Therefore, this paper draws on the principles of participatory design as the central reference point, and this includes giving a voice to marginalised individuals or those with weaker status within an organisation, promoting and valuing diverse perspectives and stakeholders, and fostering democratic practices and equality among all members of the university community (Luck, 2018, pp. 2–3). Van der Velden and Mörtberg (2015, p. 4) identified the 'guiding principles' of participatory design: equalising power and democratic practices, situation-based actions, mutual learning, tools and techniques, and alternative visions. This paper connects these principles and academic activists' actions in HEIs.

A further key area where the implementation of sustainability is important is research. Indeed, research on, about, and for sustainable development has grown significantly over recent years, encouraging various other sectors to adopt sustainable practices (Wichaisri & Sopadang, 2018). This is helpful, as research can bring together many disciplines and ensure the co-production of knowledge. The transdisciplinary action catalyzed by research enables researchers and students to develop practices that cater to the needs of multiple individuals and scenarios. Furthermore, research for sustainability has promoted innovation in many ways (Norström et al., 2020). Campus operations are also a promising area where many activities have been successfully undertaken (Leal Filho et al., 2015, 2018). The same applies to community engagement, where many higher education institutions are known to be active.

In all cases, the knowledge co-production is based on four main principles: context-based, pluralistic, goal-orientated, and interactive. These principles ensure that knowledge produced across all areas of action may reach a wide array of individuals while achieving specific goals. Furthermore, they cater to active practice rather than passive behavior. The former is especially beneficial towards ensuring the achievement of the SDGs (Norström et al., 2020).

2 Organisational Culture

In a comprehensive study, Holst (2023, p. 1015) articulated a whole institution approach (WIA) as the ideal model for implementing SDGs in HE. "[WIA is] an instrument for consistent organisational development in light of (un-)sustainability and as a keystone of integrated high-quality sustainability learning… [This is] the underlying organisational culture and critical conditions for successful implementation." The author elaborated on WIA as a multi-dimensional and comprehensive framework for all levels of organisational practice in HEIs to embed sustainability effectively. This organisational framework comprises the areas of coherence, continuous learning, participation, responsibility, and long-term commitment. In addition, governance, curriculum, campus, community, research, communication, and capacity building require elevated organisational collaboration and integration levels.

The WIA model for the effective implementation of sustainability is supported by a report compiled on behalf of UNESCO (Rieckmann, 2017, p. 53). In this report, the prerequisite for effective ESD is through a WIA that aims to mainstream sustainability into all institutional functions. Achieving this requires an organisation-wide participatory culture that enables all stakeholders—management, academics, learners, and administration—to jointly develop a vision and plan for the whole institution (Rieckmann, 2017, p. 54). Related to this last point, one can discern two guiding principles of participatory design in this case study: equalising power and democratic practices and alternative visions.

However, several studies on how HEIs deal with the complexities of embedding sustainability in organisational functions indicate that the WIA model is rare in practice. For example, Orlovic Lovren et al. (2020, p. 317), based on data comprising 300 university contributors from all continents, argues that the most significant barriers *vis-à-vis* the SDGs lie in the administration and management operation of HEIs. Considerable organisational shortcomings the author identified include a lack of institutional strategies and no management awareness of the importance of embedding sustainability. Similarly, in a systematic literature review compiled by Serafini et al. (2022, p. 652), the authors identified organisational barriers that impede the integration of sustainability across institutions, including lack of support from management, weak levels of awareness and concern, and no environmental committees. Further, in a systematic bibliometric review of SDGs and HEIs, Fauzi et al. (2023, pp. 30–32) identified institutional factors that inhibit and challenge institution-comprehensive approaches to embedding SDGs, and these encompass resistance to change and academic silos that hinder collaboration across disciplines and broad participation.

Regarding problematic organisational cultures in HEIs vis-à-vis the SDGs, another dimension involves the 'territories' and 'territoriality' concepts. These terms describe how academics defend and protect physical, intellectual, and symbolic disciplinary territories. Intellectual and disciplinary silos become distinct 'territories' within academia (Trowler, 2011). They also bring to the forefront the clash between the broad changes needed for sustainability initiatives across the university and the existing bureaucratic structures, assumptions, and practices attributed to territoriality, potentially resulting in differences among stakeholders, disputes and conflict or simply a lack of communication and apathy. A pertinent example is the conflict between the impact of neoliberalism on universities (emphasising competition and alienation) and the collaborative, collegial and participatory approach required for sustainability implementation (Hoover & Harder, 2015, p. 5).

A different aspect of a challenging organisational culture for implementing SDGs is 'top-down', mandated institutional change. The characteristic of this approach is that a presidential leadership style and executive decision-making drive it. According to Weiss et al. (2021, p. 1586), the top-down model generates a lower sense of stakeholder and participant ownership and less motivation for academic staff involvement in ESD. The same authors articulate common haphazard management responses due to autocratic approaches to implementing the SDGs. Firstly, a strategic plan is developed without comprehensive consultation, and a coordination team or an individual is appointed, often not representative of the wider academic community. Finally, some haphazard support mechanisms are implemented. Management is high on sustainability rhetoric (greenwashing). However, it only partially establishes ESD in strategic planning—if at all—and the emphasis is often superficially placed on the environmental aspects of sustainability and outcomes, not on gearing internal organisational processes towards the SDGs and ESD.

Combined with the insufficient involvement of academics, a lack of effective communication, and a lack of a unified vision, the top-down model can lead to cultural resistance manifested as apathy, passivity and indifference among academics. This model also hinders situated-based actions and mutual learning, i.e., sharing knowledge and competencies within the organisation as stakeholders act and share their experiences within their work environments. In some cases, professional development opportunities are offered to address any resistance and smooth the transition towards institute-wide implementation, but this is usually a bolt-on and ineffective, particularly regarding ESD. Overall, it is characteristic of problematic organisational cultures in universities that they lack a unified alternative vision, one that was developed through a process of participation, inclusion, equality, sharing and learning in common.

For institutions with a culture of collegiality, weak or no disciplinary silos or territories, and leadership that acknowledges and values bottom-up initiatives and participation, academic activists may consider the analysis of a case study by Cicmil et al. (2017, p. 302) based on the University of the West of England (UWE). At UWE, academics promoted a critical and activist discourse orientation through pedagogies that encouraged learners to question discursive power in the curricula and institutional decision-making vis-à-vis sustainability. The authors argue that this form of activism works better in classes that combine learners from different departments. These activist cells of learners coalesced around other academic groups through informal links and a shared vision of sustainability based on interdisciplinary collegiality, and eventually, collectively—and within a participatory management organisational framework—influenced positively curricula design and delivery for sustainability.

When publishing their insights about UWE, the authors (Cicmil et al., 2017) emphasised the need for 'devolved responsibility'. They stated that "the role of dialogue, partnerships, and transparency that support the requisite epistemic freedom and co-existence of different... viewpoints on the complex... concepts embedded in sustainable development would seem significant" (Cicmil et al., 2017, p. 304). At UWE, a team of academics facilitated a broad dialogue that gave voice to the invisible or weaker stakeholders and encouraged different perspectives, which are, according to the authors, the prerequisites of 'epistemic freedom'. The actions and strategies used to engage the academic community created an inclusive and democratic environment comprised of participatory design 'tools and techniques' (Velden & Mörtberg, 2015, p. 5). This case is also an example of combining situation-based actions and mutual learning.

Gardner et al. (2021) proposed recommendations to assist academic activists in overcoming the obstacles they encounter, and these comprise strategies universities can adopt to support academic staff in their advocacy efforts. One such strategy involves explicitly acknowledging advocacy as a legitimate aspect of academic work by modifying job descriptions and work allocation models. Additionally, the authors suggest facilitating engaged research sabbaticals and revising hiring and promotion policies to promote advocacy further. Furthermore, they emphasise the importance of providing effective training to enhance the effectiveness of engagement with sustainability issues.

Moreover, the authors argue that universities have a responsibility to safeguard the rights of academics to engage in protest and to resist any emerging threats to academic freedom. By taking these actions, the authors argue, universities can reinforce a long-standing tradition of academic protest and significantly enhance their contributions to the public good, extending beyond sustainability. However, universities' willingness to promote academic advocacy may be limited by increasing corporatisation and the need to comply with neoliberal agendas, argue Gardner et al. (2021).

For academic activists who experience various aspects of the problematic organisational cultures mentioned above, the principles underpinning participatory design can at least partially explain the failure to implement SDGs effectively. The literature on implementing SDGs emphasises that giving voice to all members of the academic community, encouraging and appreciating different perspectives, and facilitating the genuine expression of different views through meaningful, representative and participative practices are essential prerequisites for organisational progress with the SDGs

(Holst, 2023; Kioupi & Voulvoulis, 2019; Velden & Mörtberg, 2015, p. 4; Weiss et al., 2021).

Where the foundational elements for implementing SDGs are lacking, academic activists are limited to 'subversive compliance', adhering to the rules and regulations while indirectly challenging and reinterpreting them from within the system and overtly exposing its limitations and biases. This approach allows academic activists to question established norms subtly, implicitly propose innovative ideas, engage in sustainability-related research, and informally introduce SDGs in their teaching and learning while conforming to the organisation's regulations. They can undertake the task in a limited manner and within their subject teaching, research activities and networks. These initiatives will be confined within the boundaries of a school, department, programme of studies or network of colleagues (often external). They are not likely to be acknowledged by the management (and even many colleagues) as potential examples of best practices that could be supported and promoted. Subsequently, such initiatives tend not to progress beyond their limited application within a department or a faculty and can have a finite lifespan, particularly if the academics involved have moved on or retired.

3 Teaching and Learning

A critical dimension of the multifaceted implementation of SDGs in HE includes the development of appropriate graduate competencies through curricula that incorporate the complex facets of sustainability (Rieckmann, 2017, p. 7). It is essential to clarify that curricula development entails the content of subjects and the pedagogies used to deliver this content (Seatter & Ceulemans, 2017, p. 51). Regarding ESD, a familiar and unresolved debate revolves around the comparison between integrating sustainability issues across different disciplines (horizontal delivery) and creating separate and self-contained courses dedicated to incorporating discipline-specific sustainability principles (vertical delivery). While some academics believe in developing sustainability curricula based on their disciplinary frameworks to ensure a solid foundation and professional expertise in their area of specialisation (vertical delivery), others emphasise that sustainability should not be treated as a discipline-specific subject but as a holistic concept that demands a broader collaboration of knowledge beyond individual disciplines (horizontal delivery) (Orlovic Lovren et al., 2020, p. 316).

The former approach of a standalone course in sustainability offered exclusively within the framework of a discipline can indicate the existence of disciplinary silos or tribes, a detriment to the effective organisational implementation of SDGs. Vertical delivery of sustainability is contrary to the SDGs' inherently multi-disciplinary and multi-participatory dimensions. Gordon (2014, p. 86) describes this inward-looking tendency as 'disciplinary decadence'—the solidification of academic fields, causing them to become inflexible and isolated from other realms of broader scientific knowledge. Disciplinary decadence can result in an excessive fixation on methodology and an inability to acknowledge the constraints of a particular discipline, fostering for learners a narrow-minded approach to generating knowledge and understanding sustainability.

There is a broad consensus in the literature regarding the instructional strategies to cultivate the skills and competencies necessary for learners to actively contribute to developing a sustainable future. Transformative learning is acknowledged as the driving force behind change for sustainability at all educational levels (Kioupi & Voulvoulis, 2019, p. 3; Lotz-Sisitka et al., 2015; Sustainable Development Solutions Network [SDSN], 2020, pp. 12–13). In the same way, WIA is considered the ideal method to embed sustainability at the organisational (macro) level; 'transformative learning' is the model pedagogy that fosters SDG competencies among learners. For example, the Sustainable Development Solutions Network (SDSN) proposes pedagogies that entail "transformative learning approaches that are not currently standard practice within universities" and include interdisciplinarity, action-based learning and multi-actor involvement (SDSN, 2020, pp. 12–13).

Transformative learning entails instructional processes that empower learners to critically examine their preconceived notions and develop a more discerning, receptive, and introspective mindset. It facilitates profound transformations in cognition, emotions, behaviour, relationships, and self-identity while enabling the evaluation of values and assumptions about collective objectives. It encompasses the ability to comprehend diverse worldviews and strives to integrate action with one's newfound perspective, emphasising the crucial skills of self-reflection, embracing alternative viewpoints, and progressing towards a more inclusive and conscious frame of reference (Seatter & Ceulemans, 2017, p. 56).

Here, it is worth noting that there are overlaps between transformative learning and threshold concepts (TCs). TCs were initially articulated by Meyer and Land (2003) as akin to a gateway that unlocks a novel and previously inaccessible mode of thinking about a subject. Proficiency in TCs signifies a change in basic assumptions in how understanding is developed, explained, or recognised, without which progression for the learner is hindered. The assimilated perspective or newfound knowledge can influence how learners approach their field of study or interpret, comprehend, or engage with specific phenomena within their field of study. It is posited that this can result in a revised internalisation of the subject matter and a broader comprehension of it—this parallels transformative learning. This metamorphosis may unfold gradually or suddenly, with the journey towards comprehension often posing difficulties and presenting challenges for the learners.

Fang and O'Toole (2023, p. 3) conducted a study on integrating Sustainable Development Goals (SDGs) into HE curricula and concluded that experiential education through project-based learning (EEPBL) is a practical instructional approach vis-à-vis sustainability because it is not teacher-centred; it prioritises the learner and argues that learning is influenced by the context where learners explore real-life problems. Souleles (2017) argues that teacher-centred pedagogies are inappropriate for fostering learners' competencies to deal with real-life social challenges such as sustainability.

Despite the pervasiveness in the literature on the appropriate instructional strategies for ESD and the competencies learners need, there is often a contradiction in teaching and learning in HE. On the one hand, transformative learning is a tool that encourages critical thinking and initiative-taking learner engagement with real-life social challenges. On the other hand, the current pedagogies dealing with sustainability in HE do not foster sustainable thinking and actions due to their heavy reliance on lectures, one-way transmission of information and teacher-centred pedagogies (Alfirević et al., 2023, p. 17; Seatter & Ceulemans, 2017, p. 48). The need for the professional development of academics to develop teaching and learning abilities to support ESD is an often-repeated theme (Brundiers et al., 2021, p. 26; SDSN, 2020, p. 13), with one author arguing that professional development for ESD should become a prerequisite for applications and promotions of academics (Rieckmann, 2017, p. 52).

Where academic activists are confronted with the vertical delivery of sustainability within disciplinary silos—assuming this option even exists—and the prevalent pedagogies in the university are teacher-centred, subversive compliance can take place within the boundaries of a semester-long subject. The relative freedom of most academics to determine their pedagogies allows for the introduction of transformative learning opportunities to connect the curriculum to sustainability issues and, where possible, engage local community groups. It also allows for experimenting and using action research to incorporate sustainability-related TCs into curricula. Transformative learning can take the form of instructional activism cultivated by an academic for teaching and learning and requires practice. Any attempt to breach disciplinary silos by sharing instructional experiences across disciplines is likely to be met with stonewalling. In problematic organisational cultures, there are rare opportunities for mutual learning across disciplines regarding teaching and learning.

4 Closing Thoughts

By comparing the guiding principles of participatory design with the potential actions available to academic activists, it becomes evident that there is a hierarchy regarding the risk they face from work-related pressures (Table 1). These organisational pressures arise from the fundamental mismatch between the successful implementation of SDGs and ESD and the widespread hierarchical and bureaucratic structures in HEIs.

Inevitably, the challenge of implementing the SDGs exposes weaknesses in the hierarchical and bureaucratic structures of universities because the prerequisites for effective implementation include broad participation of all stakeholders and cross-disciplinary collaborations. These two prerequisites should be viewed as an ongoing process, an evolutionary synergy among the academic community members towards a common aim. The implementation of piecemeal and mandated approaches where there is a prevalence of academic silos may claim some success based on isolated outcomes, which may or may not have any measurable social impact both within

Table 1 Guiding principles of participatory design and level of risk for academic activism

Guiding principles of participatory design	Academic activism and level of risk
Equalising power and democratic practices	This principle refers to the core functions of the university and decision-making. Overt activism can potentially interfere with or disrupt institutional policies. This area implies high-risk activism
Situation-based actions	To observe and discern among colleagues' situation-based actions, the skills and knowledge that support positive change can occur through subversive compliance. This activity implies low-risk activism
Mutual learning	Mutual learning among a few academics who share similar concerns can occur within the context of subversive compliance
Tools and techniques	The development of tools and techniques to facilitate broader academic participation in the SDGs agenda is an action that can only occur unconcealed. These activities can range from low to elevated risk depending on the organisational culture
Alternative visions	Where one exists, the organisation has a monopoly on the stated vision. Depending on the organisational culture, any articulated alternative narrative implies overt activism that can potentially interfere with or disrupt institutional policies. This area infers high-risk activism

and outside the institution. Such problematic approaches disregard transformative learning at the micro and macro levels and indicate inherent implementation obstacles in the organisational culture. The challenge for universities is further compounded when considering the institutional environment in which ESD needs to thrive. Transformative learning can only take place in an organisation where SDG integration is evident across all functions of the institution, not merely limited to the classroom.

The 2023–2024 UN report on the 'Progress towards the Sustainable Development Goals' (2024) states that the world is severely off-track to achieve the 2030 Agenda. Specifically for SDG 4 (Education), there is a need to accelerate and prioritise progress because this goal can catalyse the 2030 Agenda. It is challenging to make optimistic predictions for this timetable. Even if we assume that universities will effectively implement an organisational culture that overcomes the challenges described in this paper vis-à-vis organisational culture and teaching and learning, a transitional period will be required before the organisation reaches the appropriate level of operation and our graduates have the sustainability competencies to apply in their professional careers.

Consequently, it is argued that an inherent mismatch—ontological incompatibility—exists between the organisational culture required to embed SDGs within the institution's functions effectively and the highly hierarchical structure and decision-making processes in many universities (Table 2). Academic activists operate in the area between these two contrasting realities, with at least some awareness on the one hand of what is required and, on the other, experiencing institutional factors that hinder progress in the implementation of SDGs. Within their limited scope for action,

Table 2 The ontological incompatibility between the effective adoption of SDGs and ESD and the hierarchical and bureaucratic systems prevalent in HEIs

Effective adoption of SDGs and ESD		Hierarchical and bureaucratic systems prevalent in HEIs
• Participatory culture, devolved responsibility • Transformative learning • Knowledge threshold concepts • Emphasis on process • Cross-disciplinary collaboration	Ontological incompatibility ↔	• Top-down mandated change • Teacher-centred pedagogies • Superficially plausible knowing • Emphasis on outcomes • Academic silos disciplinary territories

they can practice subversive compliance while taking all opportunities the curricula afford them to promote the competencies and knowledge the learners require. In addition, and to the extent possible, they can use university facilities and spaces for advocacy, mobilisation, and community-building activities.

In conclusion, it is crucial to contemplate the organisational cultures of universities, including the context of teaching and learning, to understand at least some reasons HEIs lag in implementing the 2030 Agenda. These cultures are deeply ingrained in many universities, and combined with organisational inflexibility and resistance to change, they play a significant role in the overall deficient performance of SDG implementation. Assuming HEIs are willing to transition to some level of WIA and thus decrease or even eliminate the ontological incompatibility that hinders effective adoption of the SDGs and feeds academic activism, a gradual process is more likely than a swift change. Foundational components to enable this gradual transformation vis-à-vis organisational culture and teaching and learning are listed in the left column in Table 2. In whatever form, academic activism arises and persists due to the lack of participatory culture, devolved responsibility, transformative learning, knowledge of threshold concepts related to SDGs, and weak or non-existent emphasis on process and cross-disciplinary collaboration.

References

Alfirević, N., Malešević Perović, L., & Mihaljević Kosor, M. (2023). Productivity and impact of sustainable development goals (SDGs)-related academic research: A bibliometric analysis. *Sustainability, 15*(9), Article 7434. https://doi.org/10.3390/su15097434

Alm, K., Melén, M., & Aggestam-Pontoppidan, C. (2021). Advancing SDG competencies in higher education: Exploring an interdisciplinary pedagogical approach. *International Journal of Sustainability in Higher Education, 22*(6), 1450–1466. https://doi.org/10.1108/IJSHE-10-2020-0417

Brundiers, K., Barth, M., Cebrián, G., Cohen, M., Diaz, L., Doucette-Remington, S., Dripps, W., Habron, G., Harré, N., Jarchow, M., Losch, K., Michel, J., Mochizuki, Y., Rieckmann, M., Parnell, R., Walker, P., & Zint, M. (2021). Key competencies in sustainability in higher education—Toward an agreed-upon reference framework. *Sustainability Science, 16*, 13–29. https://doi.org/10.1007/s11625-020-00838-2

Cicmil, S., Gough, G., & Hills, S. (2017). Insights into responsible education for sustainable development: The case of UWE, Bristol. *International Journal of Management Education, 15*(2), 293–305. https://doi.org/10.1016/j.ijme.2017.03.002

Fang, J., & O'Toole, J. (2023). Embedding sustainable development goals (SDGs) in an undergraduate business capstone subject using an experiential learning approach: A qualitative analysis. *International Journal of Management Education, 21*(1), Article 100749. https://doi.org/10.1016/j.ijme.2022.100749

Fauzi, M. A., Abdul Rahman, A. R., & Lee, C. K. (2023). A systematic bibliometric review of the United Nation's SDGS: Which are the most related to higher education institutions? *International Journal of Sustainability in Higher Education, 24*(3), 637–659. https://doi.org/10.1108/IJSHE-12-2021-0520

Flood, M., Martin, B., & Dreher, T. (2013). Combining academia and activism: Common obstacles and useful tools. *Australian Universities' Review, 55*(1), 17–26. https://search.informit.org/doi/10.3316/informit.143066509335335

Fuad-Luke, A. (2013). Design activism: Beautiful strangeness for a sustainable world. *Routledge*. https://doi.org/10.4324/9781849770941

Gardner, C. J., Thierry, A., Rowlandson, W., & Steinberger, J. K. (2021). From publications to public actions: The role of universities in facilitating academic advocacy and activism in the climate and ecological emergency. *Frontiers in Sustainability, 2*, Article 42. https://doi.org/10.3389/frsus.2021.679019

Gordon, L. R. (2014). Disciplinary decadence and the decolonisation of knowledge. *Africa Development, 39*(1), 81–92. https://www.ajol.info/index.php/ad/article/view/113311

Holst, J. (2023). Towards coherence on sustainability in education: A systematic review of Whole Institution Approaches. *Sustainability Science, 18*(2), 1015–1030. https://doi.org/10.1007/s11625-022-01226-8

Hoover, E., & Harder, M. K. (2015). What lies beneath the surface? The hidden complexities of organisational change for sustainability in Higher Education. *Journal of Cleaner Production, 106*, 175–188. https://doi.org/10.1016/j.jclepro.2014.01.081

Kestin, T., Lumbreras, J., & Puch M (2020). *Accelerating education for the SDGs in universities*. Sustainable Development Solutions Network. https://www.unsdsn.org/news/accelerating-education-for-the-sdgs-in-univerisities/

Kioupi, V., & Voulvoulis, N. (2019). Education for sustainable development: A systemic framework for connecting the SDGs to educational outcomes. *Sustainability, 11*(21), Article 6104. https://doi.org/10.3390/su11216104

Kjærgaard, A., Bergmann, R., Blasco, M., Padan, T., Elliott, C., Callahan, J., Robinson, S., & Wall, T. (2024). Subtle activism: Heterotopic principles for unsettling contemporary academia from within. *Organization, 31*(2), 412–424. https://doi.org/10.1177/13505084231116770

Leal Filho, W., Shiel, C., & Paço, A. do. (2015). Integrative approaches to environmental sustainability at universities: An overview of challenges and priorities. *Journal of Integrative Environmental Sciences, 12*(1), 1–14. https://doi.org/10.1080/1943815X.2014.988273

Leal Filho, W., Raath, S., Lazzarini, B., Vargas, V. R., de Souza, L., Anholon, R., ... & Orlovic, V. L. (2018). The role of transformation in learning and education for sustainability. *Journal of Cleaner Production, 199*, 286–295.

Leal Filho, W., Salvia, A. L., & Eustachio, J. H. P. P. (2023). An overview of the engagement of higher education institutions in the implementation of the UN Sustainable Development Goals. *Journal of Cleaner Production, 386*, Article 135694. https://doi.org/10.1016/j.jclepro.2022.135694

Lotz-Sisitka, H., Wals, A. E., Kronlid, D., & McGarry, D. (2015). Transformative, transgressive social learning: Rethinking higher education pedagogy in times of systemic global dysfunction. *Current Opinion in Environmental Sustainability, 16*, 73–80. https://doi.org/10.1016/j.cosust.2015.07.018

Luck, R. (2018). What is it that makes participation in design participatory? *Design Studies, 59*, 1–8. https://oro.open.ac.uk/57465/

Meyer, J. H. F., & Land, R. (2003). Threshold concepts and troublesome knowledge: Linkages to ways of thinking and practicing within disciplines. In C. Rust (Ed.), *Improving student learning: Ten years on* (pp. 412–414). Oxford Brookes University.

Orlovic Lovren, V., Maruna, M., & Stanarevic, S. (2020). Reflections on the learning objectives for sustainable development in the higher education curricula—three cases from the University of Belgrade. *International Journal of Sustainability in Higher Education, 21*(2), 315–335. https://doi.org/10.1108/IJSHE-09-2019-0260

Rajabifard, A., Kahalimoghadam, M., Lumantarna, E., Herath, N., Hui, F. K. P., & Assarkhaniki, Z. (2021). Applying SDGs as a systematic approach for incorporating sustainability in higher education. *International Journal of Sustainability in Higher Education, 22*(6), 1266–1284. https://doi.org/10.1108/IJSHE-10-2020-0418

Richter, J., Faragó, F., Swadener, B. B., Roca-Servat, D., & Eversman, K. A. (2020). Tempered radicalism and intersectionality: Scholar-activism in the neoliberal university. *Journal of Social Issues, 76*(4), 1014–1035. https://doi.org/10.1111/josi.12401

Rieckmann, M. (2017). *Education for sustainable development goals: Learning objectives.* UNESCO Publishing. https://doi.org/10.54675/CGBA9153

Rose, B. (2017). Moving from chasm to convergence: Benefits and barriers to academic activism for social justice and equity. *Brock Education, 27*(1), 67–78. https://files.eric.ed.gov/fulltext/EJ1165957.pdf

Sustainable Development Solutions Network. (2020). *Accelerating education for the SDGs in universities: A guide for universities, colleges, and tertiary and higher education institutions.* New Sustainable Development Solutions Network. https://ap-unsdsn.org/regional-initiatives/universities-sdgs/education-for-sdgs-guide/

Seatter, C. S., & Ceulemans, K. (2017). Teaching sustainability in higher education: Pedagogical styles that make a difference. *Canadian Journal of Higher Education, 47*(2), 47–70. https://doi.org/10.47678/cjhe.v47i2.186284

Serafini, P. G., de Moura, J. M., de Almeida, M. R., & de Rezende, J. F. D. (2022) Sustainable development goals in higher education institutions: A systematic literature review. *Journal of Cleaner Production, 370*, Article 133473. https://doi.org/10.1016/j.jclepro.2022.133473

Souleles, N. (2017). Design for social change and design education: Social challenges versus teacher-centred pedagogies. *Design Journal, 20*(Supp. 1), S927–S936. https://doi.org/10.1080/14606925.2017.1353037

Spolander, G., Garcia, M. L. T., Frizzera Delboni, T. M. Z. G., Teixeira, R. V., Khalil-Babatunde, M., & Adefila, A. (2022). Academic activism in higher education: A dialectic of resistance and surrender in a time of neoliberalism. *Philosophy and Theory in Higher Education, 4*(2), 61–86. https://doi.org/10.3726/PTIHE.022022.0006

Trowler, P. (2011, November 20). Rethinking academic tribes and territories. *University World News.* https://www.universityworldnews.com/post.php?story=20111118193403 77

Van der Velden, M., & Mörtberg, C. (2015). Participatory design and design for values. In *Handbook of ethics values, and technological design: Sources, theory, values and application domains* (pp. 41–66). Springer. https://doi.org/10.1007/978-94-007-6970-0_33

Weiss, M., Barth, M., & von Wehrden, H. (2021). The patterns of curriculum change processes that embed sustainability in higher education institutions. *Sustainability Science, 16*, 1579–1593. https://doi.org/10.1007/s11625-021-00984-1

Wichaisri, S., & Sopadang, A. (2018). Trends and future directions in sustainable development. *Sustainable Development, 26*(1), 1–17.

Ybema, S., & Horvers, M. (2017). Resistance through compliance: The strategic and subversive potential of frontstage and backstage resistance. *Organization Studies, 38*(9), 1233–1251. https://doi.org/10.1177/0170840617709305

Zamora-Polo, F., & Sánchez-Martín, J. (2019). Teaching for a better world. Sustainability and sustainable development goals in the construction of a change-maker university. *Sustainability, 11*(15), Article 4224. https://doi.org/10.3390/su11154224

Identifying and Developing Professional Competencies that Foster Sustainability Orientation

Ainhoa Saitua-Iribar, Javier Corral-Lage, Noemí Peña-Miguel, Sonia García-Delgado, and Izaskun Ipiñazar-Petralanda

Abstract The purpose of this work is to verify the relationship between cross-disciplinary professional competencies developed by university-level business management students and sustainability orientation. We identified key sustainability competencies and created an interrelation map between these competencies and those developed in a Serious Game (SG). We then measured the various competencies acquired by business management students through the SG "The Island". The results indicate that two competencies developed in the SG—analytical skills and people orientation—have a positive effect on the students' sustainability orientation. The novel contribution of this work consists of measuring the competences most related to sustainability through a serious game. Competencies should be promoted in Higher Education, particularly in the field of business management, as they have a positive influence on sustainability orientation. The development of these competencies will improve the sustainability orientation of the entities in which future graduates work.

Keywords Sustainability orientation · Professional competencies · Higher education · Serious games · Educational games

---

A. Saitua-Iribar · J. Corral-Lage · N. Peña-Miguel (✉) · S. García-Delgado · I. Ipiñazar-Petralanda
Financial Economy I Department, Faculty of Economic and Business, University of the Basque Country (UPV/EHU), Bilbao, Spain
e-mail: noemi.pena@ehu.eus

A. Saitua-Iribar
e-mail: ainhoa.saitua@ehu.eus

J. Corral-Lage
e-mail: javier.corral@ehu.eus

S. García-Delgado
e-mail: sonia.garcia@ehu.eus

I. Ipiñazar-Petralanda
e-mail: izaskun.ipinazar@ehu.eus

W. Leal Filho et al. (eds.), *Introducing the UN SDGs in Higher Education Institutions*, World Sustainability Series, https://doi.org/10.1007/978-3-032-23017-1_32

1 Introduction

Governments, businesses and individuals can no longer ignore the threat to the planet from climate change, biodiversity loss, and pollution (Carter, 2018). Sustainable development is "development that meets the needs of the present without compromising the ability of future generations to meet theirs" (Brundtland, 1987). The 2030 Agenda for Sustainable Development defined the Sustainable Development Goals (SDGs) which present the most important development challenges for humanity to ensure a sustainable, peaceful, prosperous and just life for all people, in the present but also in the future (UN, 2015).

A long-term strategy rooted in sustainability helps guarantee any organization's place in the future (IFAC, 2020). On the one hand, Global Citizenship Education (GCE) relates to finding answers as to how to live better together in a globalized world. However, it appears conceptually fragile and difficult to implement (Akkari & Maleq, 2019). On the other hand, individuals must become agents of change by acquiring knowledge, skills, values, and attitudes that empower them to contribute to sustainable development through GCE aimed at achieving the SDGs (Chung & Park, 2016; Goren & Yemini, 2017).

The University is a key agent in the process of transformation towards sustainability (Mataix et al., 2017). In this sense, universities are committed to fostering sustainable values among their students and as such, university curricula must cover sustainable development (Edwards & Fogelman, 2018; Kagawa, 2007). The aim is to train students in cross-disciplinary professional competencies to achieve the SDGs, and to ensure that all students acquire theoretical and practical knowledge to promote sustainable development. These competencies, as a cluster of knowledge, skills, abilities, and motivation (Buracas, 2020) become decisive sustainability factors for innovation in the society of the future. Systems thinking, collaboration, and critical thinking, among others, are key to the development of future graduates (Cebrián & Junyent, 2015; De Haan, 2010; Rieckmann, 2012).

This is notably true for Economics and Business students who are trained in business decision-making. Business schools have an opportunity to teach sustainability in a way that is positive and driven by self-interest, while also considering the benefits for multiple stakeholders (Larrán et al., 2018; Lourenço et al., 2013). With regards to professional accountants, a more holistic view of how people and the planet fit into this profession is essential (IFAC, 2020).

However, teaching GCE takes time, and there must be room to experiment, and the possibility to collaborate and exchange ideas (Willemse et al., 2015). Active learning processes are therefore crucial to achieving this type of development and a key component of Education for Sustainable Development (ESD) (Shulla et al., 2020).

The purpose of this work is to measure the level of competency acquired by students through a Serious Game (SG). We have reviewed previous literature on the development of sustainability competencies through an SG to provide background for this research. Following this, the aim is to analyse the influence of the professional competencies developed on students' Sustainability Orientation (SO) using linear regression models. We have done this with the objective of identifying which competencies should be promoted in Higher Education based on their positive relation with SO, with a particular focus on the field of business management.

2 Background

ESD has been increasingly recognized internationally over the past few decades, starting with the Rio and Johannesburg summits (Bunting et al., 2002; Cars & West, 2015; Weiss, 1992). From 2005–2014, the United Nations began to encourage changes in students' knowledge, values and attitudes to create a more sustainable society with their Decade of Education for Sustainable Development initiative (Bokova, 2015; Gutiérrez et al., 2006). Following this, Article 12 of the Paris Agreement in 2016 acknowledged the importance of education to achieve a sustainable planet (UNESCO, 2017). Today, there is a high demand for research to understand how ESD interacts with the SDGs in specific contexts (Shulla et al., 2020).

Within the specific field of action of Higher Education, the University has an important role in multi-stakeholder partnerships, in addition to leading by example and being a pioneer in the SDG proposals (Mataix et al., 2017). The University is an ecosystem that educates, transmits knowledge and skills, researches, innovates, and seeks solutions. All this is within a prism of values, such as education for peace and social justice. In Spain, the Conference of Rectors of Spanish Universities (CRSU) has the challenge of raising awareness among students (Bautista-Cerro & Díaz, 2017). How this is done is open to the creativity of each institution.

2.1 Identifying a Framework of Key Competencies for Sustainability

A conceptual framework to teach the SDGs at the Higher Education level should contribute to training undergraduate and postgraduate students from both a professional and a personal point of view (Zamora-Polo & Sánchez-Martín, 2019). Obtaining key professional competencies and putting them into practice will allow students to face the future challenges that the current globalized labour market and exponential technological growth will require. Weinert (2001) discusses the concept of competency and mentions specific economic, technological, technical, and methodological competencies; social competencies; creativity and innovation skills; and mobility and flexibility combined with persistence, reliability, and precision.

Within this work we will focus on the competencies identified by the main reference bodies for our case, business management. In this regard, the World Economic Forum (WEF) is committed to helping design education and training systems that

Table 1 Skills for the year 2025 jobs

N.	Skills
1	Analytical thinking and innovation
2	Active learning and learning strategies
3	Complex problem-solving
4	Critical thinking and analysis
5	Creativity, originality and initiative
6	Leadership and social influence
7	Technology use, monitoring and control
8	Technology design and programming
9	Resilience, stress tolerance and flexibility
10	Reasoning, problem-solving and ideation

Source Adapted from WEF (2020)

enable people to keep pace with rapidly changing labour markets. The top 10 skills for the year 2025 jobs identified by WEF are therefore focused on preparing workers for technological disruption (Table 1).

In our case, the list of cross-disciplinary competencies (Table 2) that our university has established (UPV/EHU, 2019) is the result of a collaborative process. This process involved the participation of undergraduate coordinators from Bachelor degree courses in the different knowledge areas, members of the Master's degree academia, Academic Committees of Master's degrees, and representatives of Doctoral Programs.

On the other hand, there is a need to identify key competencies for the progress of Sustainability (De Haan, 2010; Wiek et al., 2011). By adopting sustainable competencies, students will be able to participate in creating an economically, socially, politically, and environmentally sustainable future (Barr, 2016; Barth et al., 2007; Carter, 2018; Knox et al., 2016).

The results of a Delphi study identified twelve key competencies that are crucial for sustainable development; the most relevant competencies are systemic thinking, anticipatory thinking, and critical thinking (Rieckmann, 2012). Semerikov et al. (2020) identify fifteen competencies for sustainability. In any case, the key competencies for sustainability put forward by UNESCO (2017) in the Agenda 2030 are taken as one of the main references (Table 3). These competencies describe specific attributes that students require for action and autonomy in different complex situations and contexts, including not only motivational, but also cognitive, affective, and volitional elements (UNESCO, 2017).

As discussed in previous literature, students acquire key professional competencies through action, sustained by the basis of experience and reflection, i.e., competencies are not taught, but developed through studies (UNESCO, 2015). In this sense, Lozano et al. (2017) use hermeneutics and grounded theory to create a framework that connects sustainability competencies and pedagogical approaches. Wu and Shen (2016) use a mixed methods approach. Similarly, Figueiró and Raufflet (2015)

Table 2 Cross-disciplinary competencies to be developed at the University of the Basque Country (UPV/EHU)

Competency	Description
Critical thinking	Intellectual and systematic process that requires the mobilization of diverse mental actions such as questioning, analysis, interpretation, synthesis, evaluation, and issuing reasoned judgments
Social commitment	Empathetic and responsible action in the face of social, environmental, and economic challenges at the local, national, and global levels, as well as the fundamental democratic values of today's society
Information management and digital citizenship	The handling of information sources with a critical and responsible attitude that guarantees the reliability of the scientific information
Innovation and entrepreneurship	Generation, transformation, and implementation of an idea, a procedure, or an attitude to provide a satisfactory response to different needs
Ethics and professional responsibility	Ethical and deontological, and attitudes related to intellectual integrity
Autonomy and self-regulation	The metacognition processes that people carry out to understand the actions they develop in relation to their way of learning, to make effective decisions in different contexts, and to transfer knowledge to new situations
Teamwork	The relationship and integration in a group, actively collaborating to achieve common objectives, exchanging information, assuming responsibilities and leadership roles, solving difficulties, and contributing to the achievement of common goals
Communication and multilingualism	The use of verbal and non-verbal language in contexts of interaction to interpret and exchange meanings with a given mode of action

Source Adapted from UPV/EHU (2019)

develop a systematic review to map and assess the state of sustainability in relation to competencies. Barth and Rieckmann (2016) outline a bibliometric overview, combining quantitative analysis with a qualitative analysis of content areas and research methodologies in the field of sustainability. A systematic literature review by Mindt and Rieckmann (2017) outlined the state of the art regarding teaching–learning approaches and methods for sustainability-driven entrepreneurship.

Table 3 Key competencies for sustainability from UNESCO

Competency	Description
Critical thinking skills	Every student must question the norms, practices and opinions that surround them; reflect on values, perceptions and above all on their own actions; adopt a position based on a sustainability discourse
Systems thinking competency	Students should be able to recognize and understand relationships that originate in society; analyse complex and disparate systems; understand how such systems integrate within various domains and scales; grapple with risk and uncertainty
Anticipatory skills	Students must have the ability to understand, analyse and evaluate different future scenarios; differentiate between the possible, the probable and the desirable; create their own visions of what will happen in the future; be cautious; evaluate the consequences of the actions they take and deal with the risks and changes that arise
Regulatory competency	Future graduates must acquire the necessary skills to understand and reflect on the norms and values underlying their own actions
Strategic competency	Universities should encourage their students to collectively develop and implement innovative actions that promote sustainability
Integrated problem-solving skills	Students should learn to apply different problem-solving frameworks to complex sustainability problems and devise equitable solutions that promote sustainable development by integrating the other key competencies described above
Self-awareness competency	Future graduates should reflect on their role in the community and society; constantly evaluate and encourage their own actions; deal with personal feelings and desires
Collaborative competency	Students must learn from each other, understand and respect each other's needs, perspectives and actions through empathy and empathetic leadership, addressing group conflicts and facilitating problem-solving in a collaborative and participatory manner

Source Adapted from UNESCO (2017)

2.2 *Developing Competencies for Sustainability Through an SG*

According to the UNESCO (2017), current and future challenges include a higher degree of complexity and uncertainty than in past times. Thus, new generations must face greater individualization, social diversity, social and environmental degradation, and vulnerability, among others. Facing these decisions requires a creative and autonomous attitude, as well as accurate, clear, and concise information on all the variables that can affect how the problems and challenges they face are resolved. University students must therefore have the necessary and sufficient capacity to collaborate, manifest themselves, and act with the objective of obtaining a positive change in society and the environment that surrounds them (UNESCO, 2015). That is, to become true sustainable citizens (Wals, 2015; Wals & Lenglet, 2016). However, there has been much criticism that education has failed to bring real-world experience into the classroom (Chiang et al., 2021).

The use of Information and Communication Technologies (ICTs) to promote active learning processes in higher education improves education and allows teachers to develop new competencies in students (Goldin & Katz, 2018). Specifically, SG are virtual simulations of real-world activities that can both educate users and prompt behavioural change (Orland et al., 2014). Gamification is therefore one of the methodologies with the greatest impact on the learning process (López-Belmonte et al., 2020).

In the case of ESD, using an SG as an e-learning tool can allow for a more positive implementation of the key cross-disciplinary competencies for sustainability (Breuer & Bente, 2010; Corrigan et al., 2015; Katsaliaki, 2013). In the Netherlands, the effects of an SG that addresses the urgency and complexities of providing affordable energy from renewable sources were analysed through a pre-test and post-test completed by a hundred university students. The results revealed positive outcomes on players' awareness, understanding, and perception of efficacy (Ouariachi & Elving, 2020).

Acquiring competencies does not occur solely on an individual level, but also in social contexts and, at least in part, collaboratively (Barth et al., 2007). However, we do not yet know enough about the structure of naturally occurring connected learning ecologies that successfully connect youth learning across contexts and help them follow pathways toward careers and other desired outcomes (Bender & Peppler, 2019).

In Spain, the use of gamification as a complement in flipped learning has led to improvements in various academic indicators; flipped learning has proven to improve motivation, interaction with teachers, and student exchanges in face-to-face scenarios (Pozo Sánchez et al., 2020). At the university level, the implementation of active learning methodologies has proven to be effective in improving student engagement and performance. Among these methodologies, SGs, such as 'The Island', stand out for their ability to simulate complex scenarios related to sustainability, allowing students to make decisions and observe their consequences in a controlled environment. This form of experiential learning facilitates a deeper understanding of theoretical concepts and their practical application (Martínez Sánchez, 2025).

Comparatively, case studies present students with real or hypothetical situations to analyse and solve, promoting critical thinking and synthesis skills. However, their effectiveness should depend on the tutor's ability to guide the learning process and the active participation of students (Armellini et al., 2021).

On the other hand, flipped classes require students to study theoretical content outside the classroom and use class time for practical and collaborative activities. This approach encourages autonomous learning, but requires better student preparation before class and positive student perceptions which is a challenge how to undertake it (Akçayır & Akçayır, 2018).

Gamification in higher education for sustainability offers unique benefits, such as increased student motivation and engagement (Langendahl et al., 2016). By incorporating gamification elements, the immediate feedback provided, allows university students to adjust their learning strategies in real time. Moreover, a more engaging learning environment is created that facilitates the assimilation of complex concepts and the development of skills such as collaboration and critical thinking (Mposula, 2019).

Thus, the application of the SG, specifically The Island, as an educational tool, will allow us to achieve several educational benefits: on the one hand, students have detailed information on the current and future needs (number of people that will increase the population of the island), the costs and characteristics related to the social impact (number of jobs created) and environmental impact, etc. The SG will provide them with information on the results of their decisions in a safe environment (i.e. they are not exposed to the real risk of polluting rivers, for example) but they will be confronted with the written report showing these results. In addition, the whole learning process is done in a collaborative environment, although each person also has to develop their competences individually.

2.3 Sustainability Competency Map for the SG: The Island

In our study, "The Island" is an SG in which business management students, in a set time period, make economic, social, and environmental decisions to effectively and efficiently govern a defined territory. The students as a group must decide what, how, and when to implement actions in "The Island" that satisfy the citizens, do not have a negative ecological impact, and that are limited to the available budget. They must do this with the aim of achieving as many SDGs as possible. Thus, this SG helps to develop sustainable competencies, since the students can observe and evaluate the various future scenarios presented in the game. They perceive and learn about the relationship between the decisions made and citizenship satisfaction, using different problem-solving frameworks in the various situations outlined by the SG. Students can also question practices and opinions taken by themselves or by others, reaching conscious decisions and acting responsibly through an economic, social, and environmental vision.

The following competencies are measured by this SG (Table 4).

We have designed a competency map based on the result of this SG (Fig. 1). In this way, we understand that if our students develop the competencies through the SG (Table 4), they will also have developed the corresponding competencies required by the UPV/EHU (2019) (Table 2), and in turn, those corresponding to the Agenda 2030 (UNESCO, 2017) (Table 3), and the WEF (2020) (Table 1).

Table 4 Competencies measured by the SG The Island

Competency name and code	Description	Relative value (%)
GV_Global Vision	The SG mainly assesses that all indicators or needs are met. It is of little use to have some good outcomes while others are bad, even if the average result is good. The SG will assess that all indicators reach a minimum level, taking economic variables into account	15
AS_Analytical Skills	This mainly evaluates whether the user manages to ensure that nothing is lacking or is surplus, i.e., that the deviation, whatever it may be, is minimal In other words, resources should be optimized	15
PO_People Orientation	This competency checks that the user manages to satisfy the needs of the citizens in terms of the needs of the cities, the environment (that they live in a dignified environment), and the inhabitants' income (that it is as high as possible)	15
PL_Planning	It measures that users anticipate needs and act before problems arise It mainly evaluates whether users have managed to ensure that nothing is missing (housing, employment, etc.). If they can make sure that everyone is on top of everything, they have valuable planning capacity	15
QO_Quality Orientation	It checks the user's choice of the different existing types of industry, housing, telecommunications, etc. Users who choose type 3 (highest quality) more often than others should be rewarded	15
IO_Innovation Orientation	It checks that the user considers investment in research and development to improve living conditions, labour, and supply citizen needs	10
SE_Speed of Execution	It evaluates the time taken. It penalizes more for taking longer than it rewards for taking less time	15

Source Own work based on information provided by the SG developer

3 Research Questions and Methodology

The University can implement active methodologies in the classroom and use new ICTs. There are tools that can help develop sustainable professional competencies (see Table 5). However, according to Redman et al. (2021), these tools are not yet sufficiently developed to assess sustainability competencies. These tools are rarely selected with a clear and informed intention, largely due to a lack of guidance in the literature (Besong & Holland, 2015). In addition, there is little empirical evidence on whether and how these pedagogies are successful in developing students' sustainability competencies (Garrecht et al., 2018; Mindt & Rieckmann, 2017; Osagie et al., 2016).

Several sources point out that the use of an SG as an e-learning tool can allow for a more positive implementation of the key cross-disciplinary competencies for

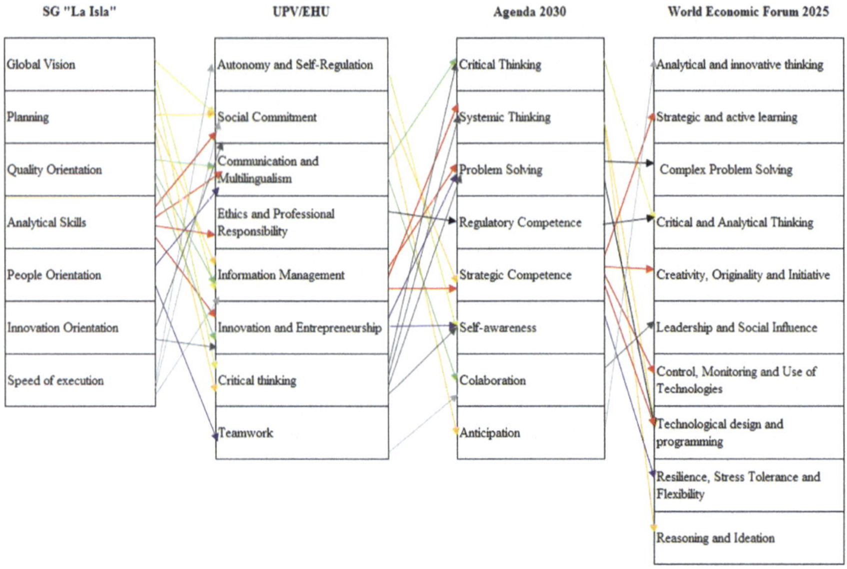

Fig. 1 Map of sustainability competencies to be developed through the SG "The island" and their interrelation. *Source* Own elaboration

Table 5 Tools currently used to assess sustainable competencies in learners

Tool	Brief description
Scaled self-assessment	Learners are asked to rate their own competency development on a pre-determined scale
Reflective writing	Students respond in writing to prompts that reflect their competency development
Scenario/case test	Students are presented with a case and asked to respond to specific prompts that require competencies
Focus group/ interview	Students respond to prompts verbally reflecting on their competency development
Performance observation	Students are assessed for competency while performing course activities inside or outside the classroom (e.g., professional setting)
Concept mapping	Students are given a prompt and asked to create a two-dimensional picture with nodes and connections (specific to systems thinking competency)
Conventional test	Students take a test that may include multiple choice or short answers that are linked to the competencies
Regular coursework	Students complete regular coursework that is analysed for evidence of competencies

Source Redman et al. (2021)

sustainability (Breuer & Bente, 2010; Corrigan et al., 2015; Katsaliaki, 2013). Thus, business simulators or SG are one of the most-used tools in the development of competencies (Calabor et al., 2018; Katsialaki & Mustafee, 2015; Liarakou et al., 2012; Ouariachi & Elving, 2020; Thirouard et al., 2015). In any case, there is a need for studies to support proposed competencies with empirical evidence showing they enable successful real-world sustainability research and problem-solving (Wiek et al., 2011).

This study focuses on measuring the level of competency achieved by students through an SG with the aim of contributing to research on this topic. Furthermore, we aim to analyse the competencies most related to sustainability to improve ESD through an SG. Accordingly, the research question that this study aims to investigate is:

Which Cross-disciplinary Professional Competencies measured by the SG "The Island" affect Sustainability Orientation (SO)?

3.1 Sample Design

In order to predict the students' levels of SO and to identify the factors that lead to an increase in SO based on the competencies mentioned above, we have based our study on the levels of competencies acquired through the SG "The Island".

The characteristics of the sample are as follows: the population is 65 groups of students, forming a total of almost 300 students from the Faculty of Economics and Business. The students carried out a test through the SG "The Island" in spring 2021. Stratified random sampling was used in the study. The sampling error was ±5%, for a confidence level of 95% ($K = 1.96$) in the most unfavourable condition ($p = q = 0.5$).

3.2 Model for Hypothesis Testing

This study analyses the influence of cross-disciplinary competencies on SO. That is, the degree of SO acquired by each group of students based on the achievement of different competencies that are developed and measured through the SG.

Therefore, the dependent variable is SO, while the independent variables (Vx) are the competencies: Global vision (GV); People orientation (PO); Planning (PL); Speed of execution (SE); Analytical skills (AS); Innovation orientation (IO); and Quality orientation (QO).

Thus, the regression model is as follows:

$$SO = \alpha + (\beta_0 + \beta_1 * GV + \beta_2 * PO + \beta_3 * PL + \beta_4 * SE + \beta_5 * AS + \beta_6 * IO + \beta_7 * QO) + \varepsilon$$

That is,

$$SO = \alpha + \sum_{x=1}^{7} \beta_h \cdot V_x + \varepsilon$$

where,

β_h: Partial relationship of each h-th explanatory variable with the explained variable. This is the average effect of a one-unit increase in the predictor variable Vx on the dependent variable "SO", with all other variables remaining constant.

ε: The residual or error, the difference between the observed value and the value estimated by the model.

3.3 Data Analysis

The variables have been selected in steps with the support of the SPSS Statistics 26 software; this is done to find the model that generates the most accurate and valid prediction, with the smallest number of possible variables (independent and control) to approach a multiple linear regression analysis. Thus, the null hypothesis to be tested is formulated:

H0: There is no relationship between Sustainability Orientation (SO) and Vx,

$$H0 = ßh = 0$$

4 Findings

4.1 Predictor Model. Model Estimation and Overall Assessment of Fit

(a) **Selection of independent variables**

We obtained the proposed model using the stepwise regression procedure. This model is the one with the highest multiple correlation coefficient (R), with the best fit to the multiple linear regression model. We found that the variables People Orientation, Planning, and Analytical skills are statistically significant as they all have significance values (Sig.) of 0.000 (Table 6). Based on this, we constructed the regression in stages by adding explanatory variables (predictors) to each model.

Table 6 Initial coefficients

Model		Unstandardized coefficients		Standardized coefficients	t	Sig.	95.0% confidence interval for B	
		B	Std. error	Beta			Lower bound	Upper bound
1	(Constant)	4.332	1 128		3.840	0.000	2.073	6.590
	GV_Global vision	−0.148	0 103	−0.396	−1.433	0.157	−0.354	0.059
	PO_People orientation	1.472	0 225	3.563	6.540	0.000	1.021	1.922
	PL_Planning	−1.433	0 187	−3.860	−7.668	0.000	−1.807	−1.059
	SE_Speed of execution	0.024	0 126	0.016	0.192	0.848	−0.228	0.277
	AS_Analytical skills	0.353	0.064	0.599	5.556	0.000	0.226	0.480
	IO_Innovation orientation	0.044	0.050	0.072	0.893	0.375	−0.055	0.144
	QO_Quality orientation	−0.013	0.040	−0.027	−0.326	0.746	−0.093	0.067

Source Own elaboration

[a] Dependent variable: Sustainability Orientation (SO)

The magnitude of each partial regression coefficient depends on the units in which the predictor variable to which "V_X" corresponds is measured. Its magnitude is therefore not associated with the importance of each predictor. Standardized partial coefficients are used to determine the impact of each of the variables on the model, which are obtained by standardizing (subtracting the mean and dividing by the standard deviation) the predictor variables after adjusting the model (Table 7).

In this case, the final adjusted model, after three steps (Table 8), explains approximately 65.3% of the variability of SO.

As previously seen, in the final adjusted model, the variables (predictors) are linearly related to the dependent variable in all three models, as the significance value (Sig.) is less than 0.05 (see Table 8).

Table 7 Summary of the model for dependent variable: Sustainability Orientation

Model	R	R^2	Adjusted R^2	Standard error of the estimate	Statistics of change				
					Change in R-squared	Change in F	df1	df2	Sig. change in F
1	0.329[a]	0.108	0.094	0.82108	0.108	7.752	1	64	0.007
2	0.719[b]	0.517	0.502	0.60885	0.409	53.396	1	63	0.000
3	0.818[c]	0.669	0.653	0.50826	0.152	28.405	1	62	0.000

Source Own elaboration
[a] Predictors: (Constant), Planning (PL)
[b] Predictors: (Constant), Planning (PL), People Orientation (PO)
[c] Predictors: (Constant), Planning (PL), People Orientation (PO), Analytical skills (AS)

Table 8 ANOVA test with one dependent variable: Sustainability Orientation (SO)

Model		Sum of squares	Degree of freedom	Root mean square	F-ratio	Sig.
1	Regression	5.226	1	5.226	7.752	0.007[a]
	Residue	43.148	64	0.674		
	Total	48.374	65			
2	Regression	25.020	2	12.510	33.748	0.000[b]
	Residue	23.354	63	0.371		
	Total	48.374	65			
3	Regression	32.358	3	10.786	41.754	0.000[c]
	Residue	16.016	62	0.258		
	Total	48.374	65			

Source Own elaboration
[a] Predictors: (Constant), Planning (PL)
[b] Predictors: (Constant), Planning (PL), People Orientation (PO)
[c] Predictors: (Constant), Planning (PL), People Orientation (PO), Analytical skills (AS)

The stepwise study also shows that all the variables introduced are significant (Sig. <0.05), including the constant (see Table 9).

(b) **Interpretation of the final adjusted model**

The final adjusted model, with a goodness-of-fit of 65.3% (see Table 9) is:

$$SO = 4.817 - 1.467PL + 1.367\ PO + 0.319\ AS$$

On the one hand, the following competencies are excluded from the model (Table 8): GV_Global Vision ($t = -1.510$; $p > 0.05$); SE_Speed of Execution ($t = 0.153$; $p > 0.05$); IO_Innovation Orientation ($t = 0.869$; $p > 0.05$); and QO_Quality Orientation (t0.315; $p > 0.05$). On the other hand, it is corroborated that the competencies PO_People-Orientation ($t = 6.821$, $p < 0.05$) and AS_Analytical Skills ($t = 5.330$, $p < 0.05$) explain the independent variable SO_Sustainability-Orientation more strongly, by means of a positive linear association. In contrast, the competency PL_Planning ($t = -8{,}415$, $p < 0.05$) explains the independent variable "SO Sustainability Orientation" more strongly but through a negative linear association (see Table 9).

Likewise, the coefficient of determination (R^2) shows that the variability explained by the model is 65.3%, which is very close to the adjusted coefficient of determination (Table 8). Based on the analysis of variance, the ANOVA test was performed (Table 8), where it is noted that for model 3 (the model that integrates these three characteristics) the p-value associated with the F statistic is lower than the significance level (Sig.) (Table 9).

Therefore, the null hypothesis "H0 There is no relationship between SO and Vx" is rejected, thus showing that there is a relationship.

Table 9 Coefficients for dependent variable Sustainability Orientation (SO)

Model		Unstandardized coefficients		Standardized coefficients	t	Sig.	95.0% confidence interval for B	
		B	Std. error	Beta			Lower bound	Upper bound
3	(Constant)	4.817	0.340		14.157	0.000	4.137	5.497
	PL_ Planning	−1.467	0.174	−3.954	−8.415	0.000	−1.816	−1.119
	PO_People orientation	1.367	0.200	3.310	6.821	0.000	0.966	1.768
	AS_ Analytical skills	0.319	0.060	0.541	5.330	0.000	0.199	0.439

Source Own elaboration

4.2 Examination of Correlations

To detect possible multicollinearity and which variables are most correlated with the dependent variable "SO Sustainability Orientation", we must take Pearson's correlation into account. Thus, we can see that the variables most correlated with the dependent variable "SO Sustainability Orientation" are "GV Global Vision" and "PL Planning". On the other hand, the highest correlation is observed between "GV Global Vision" and " PO People Orientation", "PL Planning" and "AS Analytical Skills"; between "PO People Orientation" and "PL Planning" and "AS Analytical Skills"; between "PL Planning" and "AS Analytical Skills"; and finally, between "SE Speed of Execution" and "IO Innovation Orientation" (see Table 10).

4.3 Discussion of the Results

Table 8 shows the empirical results of the regression model. We can point out that some of the competencies that have been measured in this study are important in business management but that they do not necessarily promote sustainability, such as Quality Orientation (QO) and Innovation Orientation (IO). One can guarantee a certain quality, or innovate in products and processes, regardless of whether it pollutes, consumes excessive resources, and so forth. Similarly, Speed of Execution (SE) is important in production and business management processes, but does not necessarily consider sustainability, which requires a more long-term vision. As such, the variables we have identified are those that will best ensure a long-term vision, such as People Orientation (PO), and Analytical Skills (AS).

In this case, analytical skills (AS) are important to foster sustainability orientation (SO), because people making economic investment and resource allocation decisions need to analyse in depth the information (costs and benefits, economic as well as social and environmental) related to each investment and management alternative. In addition, People Orientation (PO) is important in order to foster sustainability orientation (SO), because people making economic investment and resource allocation decisions must take into consideration the social and environmental impacts (both positive and negative) that the company can have on its stakeholders, and not only the economic impact that benefits the company itself. If curricula do not explicitly incorporate sustainability into planning competences, students may not recognise the importance of integrating the two concepts. The absence of an education that links planning and sustainability may result in a limited understanding of how these elements can coexist (Kioupi & Voulvoulis, 2019).

On the other hand, the Planning (PL) competency explains the independent variable Sustainability Orientation (SO) more strongly, but through a negative linear association. It is possible that students associate planning (PL) with conventional methods that do not integrate sustainability principles. Traditional planning often prioritises economic and operational efficiency without considering environmental

Table 10 Correlations

		SO	GV	PO	PL	SE	AS	IO	QO
SO_Sustainability Orientation	Pearson correlation	1							
	Sig. (bilateral)								
	N	66							
GV_Global vision	Pearson correlation	−0.245*	1						
	Sig. (bilateral)	0.048							
	N	66	66						
PO_People orientation	Pearson correlation	−0.224	0.949**	1					
	Sig. (bilateral)	0.071	0.000						
	N	66	66	66					
PL_Planning	Pearson correlation	−0.329**	0.945**	0.988**	1				
	Sig. (bilateral)	0.007	0.000	0.000					
	N	66	66	66	66				
SE_Speed of execution	Pearson correlation	0.202	−0.052	−0.002	−0.045	1			
	Sig. (bilateral)	0.103	0.678	0.986	0.718				
	N	66	66	66	66	66			
AS_Analytical skills	Pearson correlation	0.204	0.711**	0.685**	0.659**	0.037	1		
	Sig. (bilateral)	0.100	0.000	0.000	0.000	0.771			
	N	66	66	66	66	66	66		
IO_Innovation orientation	Pearson correlation	−0.096	−0.099	−0.103	−0.070	−0.353**	−0.175	1	
	Sig. (bilateral)	0.444	0.427	0.412	0.576	0.004	0.159		
	N	66	66	66	66	66	66	66	
QO_Quality orientation	Pearson correlation	0.023	−0.039	0.099	0.083	−0.086	−0.001	0.043	1

(continued)

Table 10 (continued)

		SO	GV	PO	PL	SE	AS	IO	QO
	Sig. (bilateral)	0.856	0.756	0.429	0.509	0.491	0.993	0.732	
	N	66	66	66	66	66	66	66	66

Source Own elaboration
* The correlation is significant at the 0.05 level (bilateral)
** The correlation is significant at the 0.01 level (bilateral)

or social impacts. This fact could lead to a negative perception of planning in relation to sustainability. Therefore, there is a need to support integrative approaches to sustainability in a higher education context (Leal et al., 2016).

Among the possible strategies to be implemented to avoid this negative correlation, some researchers (Kioupi & Voulvoulis, 2019; Tejedor et al., 2019) point out the following:

(a) Incorporate sustainability principles in a cross-cutting manner in all subjects, especially in those related to planning. This can be achieved by reviewing and adapting curricula to include the Sustainable Development Goals (SDGs) and case studies that reflect their application in real contexts.
(b) Project-based learning allows students to apply knowledge and plan sustainable solutions to real problems. This approach encourages active participation and critical thinking, helping to understand the relevance of sustainability in planning.
(c) Training teachers in teaching methodologies that integrate sustainability in order to ensure that concepts are transmitted in an effective and up-to-date manner. Teachers have a key role in promoting education that prepares students to face environmental and social challenges.
(d) Teaching students to consider the interconnections between social, economic and environmental systems in their planning processes is essential. Systems thinking enables the development of more holistic and sustainable solutions by understanding the complexity of current problems.
(e) Incorporating activities that foster values such as environmental responsibility and social equity in university education can help students internalise the importance of sustainability in their professional and personal lives.

It can be explained because managing all the resources is complex. Decision-making is not easy, especially when the aim is to manage economic profit, society, and the environment. Business management students tend to focus on economic maximization, so they prioritize covering needs and planning to maximize profit and reduce costs, and tend not to make optimal decisions from a social and environmental point of view.

For this reason, activities such as the use of this SG may help students to make decisions not only from an economic, but also from a social and environmental point of view.

5 Conclusions

This study demonstrates that the use of an SG contributes to the development of key competencies for sustainability that have been defined by some of the main reference bodies for ESD in Higher Education. We have also identified the competencies that need to be promoted and must be trained over a long period of time, rather than in the short-term. The competencies that should be promoted are identified specifically because they are more associated with Sustainability Orientation (SO) in the field of business management.

We have demonstrated that the methodology used through an SG counteracts the weaknesses of other active teaching methodologies. Therefore, the use of an SG confirms the need to promote the use of such active methodologies to complement more traditional theoretical training.

This work advances existing sustainability frameworks, such as those of UNESCO or the World Economic Forum, in that it presents the SG the Island as a competency development tool for university students in business management degrees.

Thus, we also believe that the competences developed through the SG will be able to influence real-world behaviour and drive systemic change beyond the classroom, as it will produce real changes in the behaviour of future professionals when they have to make economic decisions on environmental and social resource management and investment in their future working lives.

Our study contributes significantly to the advancement of sustainability frameworks established by international organisations such as UNESCO and the World Economic Forum. On the one hand, UNESCO, through its Education for Sustainable Development (ESD) programme, emphasises the importance of integrating sustainability into all levels of education, promoting competencies that enable individuals to actively contribute to a sustainable future (UNESCO, 2024). This study aligns with these objectives by identifying and developing professional competencies that foster sustainability orientation in university students, providing an empirical basis that supports and enriches UNESCO's theoretical frameworks. On the other hand, the World Economic Forum has highlighted the need to close the sustainability skills gap to transform supply chains and make them more inclusive and sustainable (WEF, 2024).

The skills developed through this study, such as analytical skills (AS) and people orientation (PO), prepare future professionals to address environmental, governance and social challenges in their respective labour markets, thereby driving more responsible and sustainable practices in the real world.

The influence of these competences transcends academia, as students who acquire them are better equipped to implement sustainable practices in their future professions. This can lead to systemic changes in various sectors, promoting a culture of sustainability that extends beyond the classroom and contributes to the achievement of the Sustainable Development Goals (SDGs) set by the international community.

However, this study has some limitations. Firstly, the chosen SG does not allow work to be done on SDG 5 (gender equality) and SDG 16 (peace, justice and strong institutions), which may reduce the competencies achieved. In the future, consideration should be given to expanding 'The Island' to include SDGs that have been overlooked, such as gender equality (SDG 5) and strong institutions (SDG 16). Such an expansion would increase the comprehensiveness and applicability of the tool. Secondly, cultural, institutional or regional differences may influence the results of the study and they should be undertaken. Each context has unique characteristics that affect the implementation and effectiveness of learning methodologies such as SG.

Therefore, higher education institutions can promote cultural diversity and interculturality adapting their programmes to meet the specific needs of their communities. Thus, future studies may analyse the impact of different factors such as country, or the subject area of the relevant university degree.

In the future, the aim would be to apply the SG "The Island" to other educational fields such as Engineering, Communications, Arts, and so forth. The simulation could even be addressed with multidisciplinary teams, i.e. ensuring there is a student from a different faculty in each group.

References

Akçayır, G., & Akçayır, M. (2018). The flipped classroom: A review of its advantages and challenges. *Computers & Education, 126*, 334–345. https://doi.org/10.1016/j.compedu.2018.07.021

Akkari, A., & Maleq, K. (2019). Global citizenship: Buzzword or new instrument for educational change? *Europe's Journal of Psychology, 15*(2), 176–182. https://doi.org/10.5964/ejop.v15i2.1999

Armellini, A., Teixeira Antunes, V., & Howe, R. (2021). Student perspectives on learning experiences in a higher education active blended learning context. *TechTrends, 65*(4), 433–443. https://link.springer.com/article/10.1007/s11528-021-00593-w

Barr, S. (2016). *Environment and society: Sustainability, policy and the citizen*. Routledge.

Barth, M., Godemann, J., Rieckmann, M., & Stoltenberg, U. (2007). Developing key competencies for sustainable development in higher education. *International Journal of Sustainability in Higher Education, 8*(4), 416–430. https://doi.org/10.1108/14676370710823582

Barth, M., & Rieckmann, M. (2016). State of the art in research on higher education for sustainable development. In *Routledge handbook of higher education for sustainable development* (pp. 100–113).

Bautista-Cerro Ruiz, M. J., & Díaz González, M. J. (2017). La sostenibilidad en los grados universitarios: presencia y coherencia. Teoría de la Educación. *Revista Interuniversitaria, 29*(1), 161–187. https://doi.org/10.14201/teoredu291161187

Bender, S., & Peppler, K. (2019). Connected learning ecologies as an emerging opportunity through Cosplay. *Comunicar, 58*, 31–40. https://doi.org/10.3916/C58-2019-03

Besong, F., & Holland, C. (2015). The dispositions, abilities and behaviours (DAB) framework for profiling learners' sustainability competencies in higher education. *Journal of Teacher Education for Sustainability, 17*(1), 5–22.

Bokova, I. (2015). Foreword. In: UNESCO (Ed.), *Rethinking education—Towards a global common good?* Paris, France. https://unevoc.unesco.org/e-forum/RethinkingEducation.pdf

Breuer, J., & Bente, G. (2010). Why so serious? On the relation of serious games and learning. *Journal for Computer Game Culture, 4*, 7–24. https://eludamos.org/index.php/eludamos/article/view/78

Bunting, C., Hunt, J., Walker, L., & de Roode, J. (2002). World summit on sustainable development. *Times Higher Education Supplement, 1552*, 16–19.

Buracas, A. (2020). The influence of professional competencies on social sustainability. *European Journal of Workplace Innovation, 5*(2), 67–95. https://journal.uia.no/index.php/EJWI/article/view/757

Brundtland, G. H. (1987). What is sustainable development. *Our common future, 8*(9).

Calabor, M. S., Mora, A., & Moya, S. (2018). Adquisición de competencias a través de juegos serios en el área contable: un análisis empírico. *Revista de Contabilidad, 21*(1), 38–47. https://doi.org/10.1016/j.rcsar.2016.11.001

Cars, M., & West, E. E. (2015). Education for sustainable society: Attainments and good practices in Sweden during the United Nations Decade for Education for Sustainable Development (UNDESD). *Environment, Development and Sustainability, 17*(1), 1–21. https://doi.org/10.1007/s10668-014-9537-6

Carter, N. (2018). The politics of the environment: Ideas, activism, policy. *Cambridge University Press*. https://doi.org/10.1017/9781108642163

Cebrián, G., & Junyent, M. (2015). Competencies in education for sustainable development: Exploring the student teachers' views. *Sustainability, 7*(3), 2768–2786. https://doi.org/10.3390/su7032768

Chiang, C., Wells, P. K., & Xu, G. (2021). How does experiential learning encourage active learning in auditing education? *Journal of Accounting Education, 54*, Article 100713. https://doi.org/10.1016/j.jaccedu.2020.100713

Chung, B. G., & Park, I. (2016). A review of the differences between ESD and GCED in SDGs: Focusing on the concepts of global citizenship education. *Journal of International Cooperation in Education, 18*(2), 17–35. https://cice.hiroshima-u.ac.jp/wp-content/uploads/2016/12/18-2-2.pdf

Corrigan, S., Zon, G. D. R., Maij, A., McDonald, N., & Mårtensson, L. (2015). An approach to collaborative learning and the serious game development. *Cognition, Technology and Work, 17*(2), 269–278. https://doi.org/10.1007/s10111-014-0289-8

De Haan, G. (2010). The development of ESD-related competencies in supportive institutional frameworks. *International Review of Education, 56*(2–3), 315–328. https://doi.org/10.1007/s11159-010-9157-9

Edwards, J., & Fogelman, K. (Eds.). (2018). *Developing citizenship in the curriculum*. Routledge. https://doi.org/10.4324/9780429454653

Figueiró, P. S., & Raufflet, E. (2015). Sustainability in higher education: A systematic review with focus on management education. *Journal of Cleaner Production, 106*, 22–33.

Garrecht, C., Bruckermann, T., & Harms, U. (2018). Students' decision-making in education for sustainability-related extracurricular activities—A systematic review of empirical studies. *Sustainability, 10*(11), 3876.

Goldin, C. D., & Katz, L. F. (2018). The race between education and technology. In *Inequality in the 21st century* (pp. 49–54). Routledge. https://doi.org/10.4324/9780429499821-10

Goren, H., & Yemini, M. (2017). Global citizenship education redefined—A systematic review of empirical studies on global citizenship education. *International Journal of Educational Research, 82*, 170–183. https://doi.org/10.1016/j.ijer.2017.02.004

Gutiérrez, J., Benayas, J., & Calvo, S. (2006). Educación para el desarrollo sostenible: evaluación de retos y oportunidades del decenio 2005–2014. *Revista Iberoamericana de educación, 40*(1), 25–60. https://dialnet.unirioja.es/servlet/articulo?codigo=2162363

IFAC (International Federation of Accountants). (2020). *Accountancy skills evolution: Impact of Covid-19 and the path forward*. November, 4. https://www.ifac.org/system/files/publications/files/IFAC-Accountancy-Skills-Evolution.pdf

Kagawa, F. (2007). Dissonance in students' perceptions of sustainable development and sustainability: Implications for curriculum change. *International Journal of Sustainability in Higher Education, 8*(3), 317–338. https://doi.org/10.1108/14676370710817174

Katsaliaki, K. (2013). Serious games for sustainable development. *Journal of Management Education, 37*(6), 889–894. https://doi.org/10.1177/1052562913509219

Katsaliaki, K., & Mustafee, N. (2015). Edutainment for sustainable development: A survey of games in the field. *Simulation & Gaming, 46*(6), 647–672. https://doi.org/10.1177/1046878114552166

Kioupi, V., & Voulvoulis, N. (2019). Education for sustainable development: A review of the field and the role of higher education institutions. *Sustainability, 11*(1), 1–19. https://www.mdpi.com/2071-1050/11/21/6104

Knox, P. L., Marston, S. A., & Imort, M. (2016). *Human geography: Places and regions in global context*. Pearson.

Langendahl, P. A., Cook, M., & Mark-Herbert, C. (2016). *Gamification in higher education*. Working Paper Series/Swedish University of Agricultural Sciences, Department of Economics (2016: 6). https://publications.slu.se/?file=publ/show&id=76235

Larrán, M., Andrades, J., & Herrera, J. (2018). An examination of attitudes and perceptions of Spanish business and accounting students toward corporate social responsibility and sustainability themes. *Revista de Contabilidad, 21*(2), 196–205. https://doi.org/10.1016/j.rcsar.2018.02.001

Leal Filho, W., Shiel, C., & Paço, A. (2016). Implementing and operationalising integrative approaches to sustainability in higher education: The role of project-oriented learning. *Journal of Cleaner Production, 133*, 126–135. https://doi.org/10.1016/j.jclepro.2016.05.079

Liarakou, G., Sakka, E., Gavrilakis, C., & Tsolakidis, C. (2012). Evaluation of serious games, as a tool for education for sustainable development. *European Journal of Open, Distance and E-learning, 15*(2).

López-Belmonte, J., Parra-González, M. E., Segura-Robles, A., & Pozo-Sánchez, S. (2020). Scientific mapping of gamification in web of science. *European Journal of Investigation in Health, Psychology and Education, 10*(3), 832–847. https://doi.org/10.3390/ejihpe10030060

Lourenço, F., Jones, O., & Jayawarna, D. (2013). Promoting sustainable development: The role of entrepreneurship education. *International Small Business Journal, 31*(8), 841–865. https://doi.org/10.1177/0266242611435825

Lozano, R., Merrill, M. Y., Sammalisto, K., Ceulemans, K., & Lozano, F. J. (2017). Connecting competences and pedagogical approaches for sustainable development in higher education: A literature review and framework proposal. *Sustainability, 9*(10), 1889.

Martínez Sánchez, R. (2025). *Classcraft: The impact of gamification in higher education. Gamification and Augmented Reality. 3, 2.* https://doi.org/10.56294/gr2025100

Mataix, C., Rojo, A., Pastor, M., Gurmu, D., Stott, L., & Mazorra, J. (2017). Alianzas multiactor para la innovación social en la consecución de los ODS: la universidad y su gran potencial como facilitador y promotor. *Revista española de desarrollo y cooperación, 41*, 27–38.

Mindt, L., & Rieckmann, M. (2017). Developing competencies for sustainability-driven entrepreneurship in higher education: A literature review of teaching and learning methods. *Teoría de la Educación; Revista Interuniversitaria, 29*(1), 129.

Mposula, N. F. (2019). *Gamification as a tool for developing critical thinking among ICT students at a tertiary institution in South Africa* (Doctoral dissertation, Cape Peninsula University of Technology). https://etd.cput.ac.za/handle/20.500.11838/3020

Orland, B., Ram, N., Lang, D., Houser, K., Kling, N., & Coccia, M. (2014). Saving energy in an office environment: A serious game intervention. *Energy and Buildings, 74*, 43–52. https://doi.org/10.1016/j.enbuild.2014.01.036

Osagie, E. R., Wesselink, R., Blok, V., Lans, T., & Mulder, M. (2016). Individual competencies for corporate social responsibility: A literature and practice perspective. *Journal of Business Ethics, 135*(2), 233–252.

Ouariachi, T., & Elving, W. (2020). Accelerating the energy transition through serious gaming: testing effects on awareness, knowledge and efficacy beliefs. *Electronic Journal of e-Learning, 18*(5), 410–420. https://doi.org/10.34190/JEL.18.5.004

Pozo Sánchez, S., López Belmonte, J., Fuentes Cabrera, A., & López Núñez, J. A. (2020). Gamification as a methodological complement to flipped learning—An incident factor in learning improvement. *Multimodal Technologies and Interaction, 4*(2), 12. https://doi.org/10.3390/mti4020012

Redman, A., Wiek, A., & Barth, M. (2021). Current practice of assessing students' sustainability competencies: A review of tools. *Sustainability Science, 16*(1), 117–135.

Rieckmann, M. (2012). Future-oriented higher education: Which key competencies should be fostered through university teaching and learning? *Futures, 44*(2), 127–135. https://doi.org/10.1016/j.futures.2011.09.005

Semerikov, S., Striuk, A., Striuk, L., Striuk, M., & Shalatska, H. (2020). Sustainability in Software Engineering Education: a case of general professional competencies. In *E3S Web of Conferences*

(Vol. 166, no. 2, p. 10036). https://www.e3s-conferences.org/articles/e3sconf/abs/2020/26/e3sconf_icsf2020_10036/e3sconf_icsf2020_10036.html

Shulla, K., Leal Filho, W., Lardjane, S., Sommer, J. H., & Borgemeister, C. (2020). Sustainable development education in the context of the 2030 Agenda for sustainable development. *International Journal of Sustainable Development and World Ecology, 27*(5), 458–468. https://doi.org/10.1080/13504509.2020.1721378

Tejedor, G., Segalàs, J., & Rosas-Casals, M. (2019). Transdisciplinarity in higher education for sustainability: How discourses are approached in engineering education. *Journal of Cleaner Production, 175*, 29–37. https://doi.org/10.1016/j.jclepro.2017.11.085

Thirouard, M., Bernaert, O., Dhorne, L., Bianchi, S., Pidol, L., & Petit, Y. (2015). Learning by doing: Integrating a serious game in a MOOC to promote new skills. In *Proceedings Papers* (p. 92).

UN (United Nations) (2015). *UN adopts 2030 Agenda for Sustainable Development*. https://sustainabledevelopment.un.org/sdinaction/newsletter/september2015#1

UNESCO (United Nations Educational, Scientific and Cultural Organization). (2015). *Indicadores Temáticos para monitorear la Agenda de Educación 2030: Propuesta del Grupo Consultivo Técnico (TAG)*, Madrid, España. https://repositorio.minedu.gob.pe/bitstream/handle/20.500.12799/3665/Indicadores%20tem%C3%A1ticos%20para%20el%20monitoreo%20de%20la%20Agenda%20Educativa%202030%20propuesta%20del%20Grupo%20Consultivo%20T%C3%A9cnico%20%28TAG%29.pdf?sequence=1&isAllowed=y

UNESCO (United Nations Educational, Scientific and Cultural Organization). (2017). *Education for Sustainable Developments Goals—Learning Objectives*. Paris, France. https://unesdoc.unesco.org/ark:/48223/pf0000247444

UNESCO (United Nations Educational, Scientific and Cultural Organization). (2024). *What you need to know about education for sustainable development*. https://www.unesco.org/en/sustainable-development/education/need-know

UPV/EHU (University of the Basque Country—Universidad del País Vasco—Euskal Herriko Unibertsitatea) (2019). *Catálogo de Competencias Transversales de la UPV/EHU*. https://www.ehu.eus/documents/1432750/12757375/Cat%C3%A1logo+de+Competencias+trasnversales_cas.pdf/5dd00732-9e32-5e52-0ee2-2128d9a6867b

Wals, A. E. J. (2015). *Beyond unreasonable doubt. Education and learning for socio-ecological sustainability in the Anthropocene*. Wageningen University. https://edepot.wur.nl/365312

Wals, A. E., & Lenglet, F. (2016). Sustainability citizens: collaborative and disruptive social learning. In *Sustainability citizenship in cities* (pp. 72–86). Routledge.

WEF (World Economic Forum). (2020). *The future of jobs report 2020, October*. https://www.weforum.org/agenda/2020/10/top-10-work-skills-of-tomorrow-how-long-it-takes-to-learn-them/

WEF (World Economic Forum). (2024). *How to bridge the sustainability skills gap across global value chains*. https://www.weforum.org/stories/2024/09/bridge-sustainability-skills-gap-across-global-value-chains/

Weinert, F. E. (2001). Concept of competence: A conceptual clarification. In D. S. Rychen & L. H. Salganik (Eds.), *Defining and selecting key competencies* (pp. 45–65). Hogrefe & Huber Publishers. https://psycnet.apa.org/record/2001-05275-003

Weiss, E. B. (1992). United Nations conference on environment and development. *International Legal Materials, 31*(4), 814–817. https://www.jstor.org/stable/20693712

Wiek, A., Withycombe, L., & Redman, C. L. (2011). Key competencies in sustainability: A reference framework for academic program development. *Sustainability Science, 6*(2), 203–218. https://doi.org/10.1007/s11625-011-0132-6

Willemse, T. M., Ten Dam, G., Geijsel, F., Van Wessum, L., & Volman, M. (2015). Fostering teachers' professional development for citizenship education. *Teaching and Teacher Education, 49*, 118–127. https://doi.org/10.1016/j.tate.2015.03.008

Wu, Y. C. J., & Shen, J. P. (2016). Higher education for sustainable development: a systematic review. *International Journal of Sustainability in Higher Education*.

Zamora-Polo, F., & Sánchez-Martín, J. (2019). Teaching for a better world. Sustainability and sustainable development goals in the construction of a change-maker university. *Sustainability, 11*(15), 4224. https://doi.org/10.3390/su11154224

Transdisciplinary Approaches to Water Security and Food Security: Insights from ASEAN Universities

Mohd Yusoff Ishak

Abstract This chapter explores the critical role of ASEAN universities in addressing water and food security challenges through transdisciplinary approaches that align with the United Nations Sustainable Development Goals (SDGs). With Asia's unique socio-cultural and environmental landscapes, higher education institutions serve as hubs for fostering innovative solutions and bridging academic knowledge with real-world applications. The chapter highlights how universities integrate diverse fields to tackle pressing issues such as water scarcity, agricultural sustainability, and climate resilience. Through case studies focusing on Universiti Putra Malaysia (UPM), the chapter illustrates models of transdisciplinary research and education, emphasizing partnerships between academia, government, and communities. Examples include citizen science (CS) initiatives for river monitoring, sustainable community farming practices, and technological innovations like precision agriculture and water-efficient technologies. Moreover, the chapter underscores the importance of culturally contextualized strategies, leveraging local knowledge alongside modern scientific methods. By showcasing UPM distinctive contributions to SDG 6 (Clean Water and Sanitation) and SDG 2 (Zero Hunger), the chapter aims to inspire global academia to adopt collaborative approaches and positions higher education institutions as catalysts for sustainable development, advancing water and food security within the broader SDG framework.

Keywords Sustainable development · Water security · Food security · SDG2030

M. Y. Ishak (✉)
Department of Environmental Management, Faculty of Forestry and Environment, Universiti Putra Malaysia, Serdang, Malaysia
e-mail: m_yusoff@upm.edu.my

W. Leal Filho et al. (eds.), *Introducing the UN SDGs in Higher Education Institutions*, World Sustainability Series, https://doi.org/10.1007/978-3-032-23017-1_33

1 Introduction

1.1 Global Challenges in Water and Food Security

Water and food insecurity are complex, interconnected challenges requiring transdisciplinary research that integrates socio-hydro-ecological perspectives (Agramont et al., 2019). The Asia region is experiencing rapid economic growth alongside increasing environmental challenges that have global implications (Miao et al., 2025). This chapter focuses on ASEAN universities, particularly in Malaysia, as a case study within the broader Asian context—a region housing over 60% of the world's population and facing acute water and food security pressures due to rapid urbanization and climate change (UN, 2024). While rooted in Southeast Asia, the approach here offers insights for higher education institutions, particularly in the Global South, where similar socio-economic and environmental dynamics prevail.

Malaysia is part of The Association of Southeast Asian Nations (ASEAN) which comprises 10 member countries: Brunei, Darussalam, Cambodia, Indonesia, Laos, Malaysia, Myanmar, the Philippines, Singapore, Thailand, and Vietnam. ASEAN's combined population reached 698 million in 2024, making it one of the most populous regions globally (The ASEAN Statistical Yearbook, 2023). This region's population is projected to reach 723 million people by 2030 adding pressure on food systems to ensure sufficient and nutritious food for all (World Economic Forum, 2020).

There is abundant scientific data to suggest that without substantial reforms in water and food resources management, this region will face serious challenges arising from limited supply. As regional specificities of ASEAN require careful consideration when adapting these approaches elsewhere, readers from different regions, such as temperate zones or large-scale farming systems, should note that Malaysia's humid climate shape these solutions, necessitating adjustments for local conditions elsewhere.

By 2050, rapid urbanization will add 1.2 billion people to cities in Asia and the Pacific, intensifying pressure on already strained infrastructure and services, compounded by the impacts of climate change (WEF, 2024). This complexity is particularly evident in the Global South, where socio-economic inequalities and cultural differences create additional challenges, while simultaneously offering unique opportunities for innovative research approaches that bridge the gap between science and society (Agramont et al., 2019).

The need for integrating multiple sustainable development sectors has expanded the scope of interdependent dynamics within nature and society (Karpouzoglou et al., 2017) and this requires implementing the Sustainable Development Goals (SDGs) in an integrated way. For example, water and food production both require significant energy inputs where energy production often relies on water resources for cooling and steam generation (Verma et al., 2024). If current consumption and demand patterns continue in Southeast Asia, food production will need to increase by up to 70% from 2000 levels by 2050 (Msangi & Rosegrant, 2009). To address these challenges, an increasing number of stakeholders are adopting innovative nexus approach as the key for water, energy and food security.

1.2 The Water Security-Food Security Nexus

Extensive research has examined the effects of climate change-induced variations in food production (Leal et al., 2022). However, the combined impact of water security challenges arising from both climatic and economic changes on Asia's future food security remains insufficiently understood. The nexus framework provides a holistic approach to understanding the interconnectedness and interdependencies of complex natural systems and their sustainable management. By embedding nexus into research and teaching, universities can bridge knowledge gaps and drive long-term food and water sustainability (Albrecht et al., 2018).

Universities have been amongst the earliest signatories to sustainability declarations, beginning in 1972 with the Stockholm Declaration on the Human Environment (Adams et al., 2018) and later the Talloires Declaration (Adlong, 2013). The current practice is to incorporate the element of Education for Sustainable Development (ESD) into university curriculum in response to the changing demands of society and key stakeholders. This is especially pertinent for tackling the SDG goals namely SDG2 Zero Hunger which aims to end hunger, achieve food security and improved nutrition where universities contribute by developing innovative water-efficient agriculture (Dai et al., 2024) and promoting renewable energy solutions in farming (Amusa et al., 2024).

The transformative role of universities toward a more sustainable society is attested in both the scholarly and practitioner literature (Ramos et al., 2015) where universities hold the key to addressing planetary challenges. The Times Higher Education (THE) University Impact Rankings which was introduced in 2019 measures the success of universities across the world in delivering the SDGs and in tackling these global challenges.

1.3 The Role of ASEAN Higher Education in SDGs

While universities worldwide have increasingly aligned their activities with the SDGs and implemented various sustainability initiatives, our academic contributions are unevenly represented (Cottafava et al., 2022) where the analysis of the impact will surely be influenced by the global north viewpoint. In this paper, the author prioritises references from Malaysian scholars, as a step toward correcting this imbalance, although in some cases this was not possible due to research gaps. Indeed, the way forward is that we need more North–South partnerships in research (Delina, 2021).

Even though universities are seen as appropriate leaders in championing ESD, with success dependent on linkages between government, communities, NGOs, and the private sector (Khelghat-Doost et al., 2011), the sustainability initiatives being carried out by global south still needs to be examined in terms of teaching, research, community involvement, and campus operations. This could be analysed through transdisciplinary approaches in tackling water and food security related SDGs. Malaysian universities have demonstrated sustainable projects engaging students and communities in SDG implementation (Jasmin Irisha Jim Ilham et al., 2019). Thus, expanding on this would assist our understanding on how Asia's higher education may fulfil this through pedagogical, research and engagement roles.

1.4 Malaysian University Transdisciplinary Approaches

Transdisciplinary approaches in higher education involve collaboration across disciplines to address complex problems and create innovative solutions (Bain et al., 2019; Baumber et al., 2020). This solution is particularly needed in the Global South, where socio-economic inequalities and cultural differences create additional challenges for researchers, while simultaneously offering unique opportunities for innovative research approaches that bridge the gap between science and society (Agramont et al., 2019). The transdisciplinary challenges traditional fragmented academic structures, promoting a more holistic and interconnected view of knowledge.

In view of that, universities are increasingly combining scientific methods with local ecological knowledge. For example, to tackle complex water industry challenges, university-industry partnerships are essential, requiring a coordinated and integrated approach (Padfield et al., 2014). The Jeffrey Sachs Center at Sunway University, Malaysia collaborates with government and international partners on SDGs 3, 6, 13, and 17 (Woo et al., 2019) while the Malaysian Youth SDG Summit leverages partnerships to expose youth to sustainable development through capacity-building initiative (Ng et al., 2023). These approaches emphasize integrating diverse perspectives, knowledge types, and industry insights into universities curriculum, research and community engagement.

Agricultural universities in developing countries like Nigeria contribute to food security through research, training, and extension services, despite facing constraints in resources and facilities (Okolo, 2010) while urban universities, such as Ohio State University, contributes to food security by fostering intentional linkages and addressing complex food system challenges (Fox, 2017). However, in meeting present challenges, the underlying problem lies in the large gap between scientific findings and the user requirements of government authorities and receiving organizations (Enenkel et al., 2015) where approach by universities is primarily and intrinsically focused on generating new knowledge, which does not necessarily lead to applicable information in water and food security context.

This is most apparent in Asia where most of farmers are operating on small scale farms which are defined as farms with an area of less than 2 ha, make up 84% of the world's farms (Lowder et al, 2021). These small-scale farmers are hampered by a lack of educational resources, limited financial resources, and costly technological advancements, which causes them to be hesitant to experiment the new agricultural methods that they believe to be risky for their ability to provide for themselves financially (OECD, 2001). This is where universities could play a crucial role in addressing food security challenges through research, education, and collaborative initiatives as they serve as learning laboratories for sustainable resource use, particularly in water management and agriculture (Rushforth & Phillips, 2012). Magno (2024) highlights the need for integrated approaches to address the interdependencies of food, energy, and water systems. Such work reveals that integration enhances resilience to global environmental change and helps tackle complex problems like water scarcity and climate resilience (Gómez-Baggethun et al., 2013) while interdisciplinary science efforts are crucial for understanding food-energy-water nexus interactions and informing coordinated management strategies (Scanlon et al., 2017).

Numerous research highlights the critical role of water management in ensuring food security across Asia (Montesclaros & Teng, 2021) with water security is a growing concern in Malaysia, despite uniform temperature, high humidity and annual rainfall distribution between 1800 to 3500 mm that characterises its climate. A growing population and changing climate and water scarcity poses significant challenges to agricultural productivity (Khanal, 2021).

Agriculture contributes significantly to ASEAN economies where agriculture constituted approximately 11% of the ASEAN region's GDP in 2022, with Cambodia and Myanmar, relying on it for over 20% of their national GDP (Sundram, 2023). However, Southeast Asia's agricultural landscape has also faced challenges posed by climate change, resulting in significant production losses of about USD 21 billion due to extreme weather events between 2008 and 2018 (Asian Development Bank, 2021). Compounding the problem, the agriculture sector showed a declining of the working-age population, suffers labour shortages and becomes more dependent on foreign workers with the share of agricultural employment to the total workforce in Malaysia fell from 26 to 10.5% (Penang Institute, 2022).

Enhancing food security has become a critical agenda for ASEAN to mitigate the impacts of future crises and build resilience that ASEAN has prioritized "food security in its integration agenda as the association strives to build resilience against possible crises" (Economic Research Institute of ASEAN & East Asia, 2022). Nonetheless, while ASEAN's diversity is a strength, differing national priorities such as Singapore's technology focus versus Myanmar's rural needs may complicate regional solutions.

Pioneering research and technological development in food and water security have been major focuses of UPM. Digital technologies have the potential to significantly enhance agricultural productivity, efficiency, and resilience by enabling real-time monitoring of various agricultural parameters, facilitating better decision-making and improved yields (Abiri et al., 2023) where innovations contribute to reduced environmental footprints and higher farm resilience (Finger, 2023). Precision agriculture technologies, for example, enable targeted application of inputs, reducing overuse and environmental contamination (Finger et al., 2019). Automation in agriculture has been shown to increase food production levels and enable precision farming practices, leading to more efficient use of resources like water, fertilizers, and pesticides (Mentsiev et al., 2019). However, the assumption that technology adoption is straightforward overlooks cultural resistance and economic barriers.

2 Methodology

This chapter adopts a qualitative case study approach, focusing on transdisciplinary initiatives at UPM. Data were gathered through a synthesis of peer-reviewed literature, reports, and publicly available policy documents (e.g., ASEAN Statistical Yearbook, 2023), supplemented by examples of UPM's research outputs (e.g., Pertanika Journal). Case studies were selected based on their alignment with SDG2030, and their demonstration of collaboration across disciplines where rigor was ensured by triangulating findings across multiple sources, while robustness stems from grounding the analysis in the water-food nexus framework (Albrecht et al., 2018), which provides a holistic lens for interdependent sustainability challenges. A case study approach was chosen to provide in-depth and context-specific insights into how universities operationalize transdisciplinary principles while offering practical examples. The nexus framework was adopted to reflect the interconnected nature of water and food security, aligning with the SDGs' integrated approach.

2.1 Innovative Research and Development by Universities for Advancements in Agriculture

In Malaysia, UPM plays an essential role in addressing food security issues by researching, developing new technologies, and providing education and training to farmers and other stakeholders (BERNAMA, 2024). In addition, UPM offer integrated farming creating new technologies for improving crop varieties, and student entrepreneur programs to improve agricultural productivity and promote sustainable practices (Sultan et al., 2024). For research output, the most productive journal in agriculture of this region is Pertanika Journal of Science and Technology which is published by UPM.

On one hand, the agricultural research and development at UPM received considerable funding for maintaining a steady supply of food and animal products to match the increasing pace of human population (Mohd-Azmi et al., 2016) while on the other, universities in Malaysia have established research platforms with highly qualified researchers in collaboration with local and international partners to alleviate persistent food insecurity through integrated farming (Sultan et al., 2024).

UPM is also pioneering sustainable water management, contributing to SDG6 and SDG2 in addition to collaboration with UK institutions on financing from the British Council. While Universiti Sains Malaysia, as a UNESCO Regional Centre of Expertise, has developed programs promoting SDGs, including research on stormwater management and climate change impacts many local universities offer CS projects involving diverse participants as part of global initiatives. For instance, the DRYRivers app, used to monitor intermittent rivers and ephemeral streams, that has been applied in 19 countries, including Malaysia, with over 3600 observations recorded across 1900 river reaches is a strategy transferable for other watershed management offering CS as a low-cost yet scalable approach for agro-ecological monitoring in resource-limited settings globally fostering inclusivity (SDG 17). While CS empowers communities, its reliance on volunteerism may limit sustainability without institutional support.

Hence, in recognition of Sunway University's (Malaysia) commitment to advancing the sustainability agenda in Asia, Sunway City Kuala Lumpur has been selected as one of the three global United Nations Sustainable Development Solutions Network (SDSN) centres alongside New York City and Paris to coordinate continent-wide sustainability initiatives for Asia, the Americas and Europe/Africa respectively. SDSN for Asia is based at Jeffrey Sachs Center on Sustainable Development at Sunway University, to host the KL office of the SDG Academy, and to lead flagship programmes of SDSN in the Asia region. The SDSN Asia work with Sunway University and Jeffrey Cheah Foundation on transdisciplinary educational collaborations.

The selection for SDSN is an indication of how SDGs can be achieved through higher education with universities designated as SDG Hubs for addressing SDGs and educating future generations about global challenges. As these initiatives excel in collaboration, continuous evaluation of long-term impacts is needed to ensure applicability to many other SDG projects.

3 Key Outcomes

3.1 Development of High-Yielding Varieties

Sustainable food security requires balancing agricultural productivity, environmental conservation, and economic viability to meet the growing demands. In this pursuit, researchers have advanced the development of high-yielding, disease-resistant rice varieties namely MR263, MR220CL1, and MR220CL2 with 1.5 times higher yield compared to local varieties (Adedoyin et al., 2016). Also in rice, the bacterial leaf blight resistant gene Xa7 has been successfully introgressed into the high-yielding variety MR219, resulting in resistant lines with similar yield performance (Yazid et al., 2021). Through a partnership with industry, UPM has developed precision irrigation systems incorporating IoT soil moisture sensors and automated drip irrigation that reduced water consumption by 30% which is equivalent to 3,000 L saved per hectare per cropping cycle while maintaining crop yields. This directly enhances water security for smallholder farmers (Sultan et al., 2024). A 35% reduction in water usage was also observed through a smart irrigation system that integrates IoT sensors

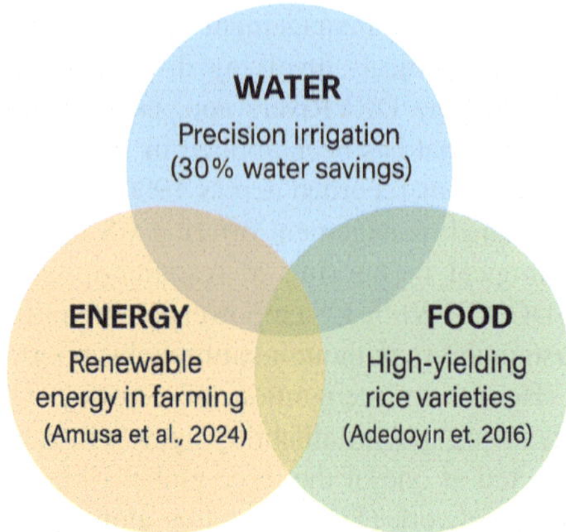

Fig. 1 The Water–Energy–Food Nexus at Universiti Putra Malaysia (UPM) diagram illustrating the interconnections among water, energy and food systems

and machine learning, optimizing water management in small-scale farms (Monchusi et al., 2024) which is particularly beneficial for arid regions facing water scarcity while a review of smart irrigation monitoring and control strategies in precision agriculture by Bwambale et al (2022) showed strong potential for model transferability across agro-ecological zones.

With approximately 195,000 farmers in Malaysia actively involved in rice cultivation (KRI, 2019), the adoption of high-yielding crop varieties holds significant potential to boost domestic productivity. This shift is crucial not only for strengthening national food security but also for reducing dependence on imported rice—a priority similarly pursued by countries such as India and Vietnam, which have long emphasized self-sufficiency in staple crop production. These in combination with UPM case demonstrates a scalable solution replicable for diverse agroecological contexts, aligning with the SDG2 and SDG6 as highlighted in Fig. 1.

3.2 *Crop Diversification and Improvement*

Diversification of crop has been a long-standing objective in Malaysian agriculture, aimed at improving production stability and nutrition security (Pin Keong et al., 2022). Efforts to enhance crop diversity and improvement in Malaysia have included the use of gamma irradiation techniques, in inducing mutations for trait improvements in various plant species (Ahmad et al., 2018). Additionally, the importance of neglected and underutilized species in agricultural diversification has been recognized for their potential to combat food and nutritional insecurity, particularly among rural communities (Mal, 2007).

3.3 Sustainable Farming Practices

Policies in support of sustainable agricultural practices includes the National Agrofood Policy 2021–2030 which reflects Malaysia's commitment to ensuring a resilient and sustainable agrofood industry. In addition, agriculture is considered a National Key Economic Area (NKEA) in Malaysia while universities have made notable contributions to conservation agriculture, water management, climate-resilient varieties, agricultural diversification, integrated pest management, nutrient management, precision farming, agrivoltaics, and livestock management (Nguyen et al., 2025). By coupling policy support, improving technical assistance, and fostering collaboration among stakeholders with these methods, Malaysia can overcome barriers to low-carbon farming and facilitate the transition towards a more resilient and sustainable agricultural future.

3.4 Comprehensive Agricultural Education and Training

Agricultural education and training form a crucial part in developing competent human resource for the agricultural sector which indirectly addresses water and food security challenges. Numerous undergraduate, graduate, and doctorate programs in agronomy, horticulture, soil science, agricultural engineering, and plant biotechnology are available at UPM and other agricultural universities throughout ASEAN to provide specialized education like precision farming and artificial intelligence that equips not only students but also community at large and stakeholders with the knowledge and skills needed.

There is an interest amongst the stakeholders, namely, administrators and academicians in education for sustainable development (Kanapathy et al., 2021) with university's social programmes on SDGs highlights the 'Green Campus' philosophy and Local Agenda 21 integration into its mission. The UPM-Yunus Social Business Centre for example, addresses SDG1 on eradicating poverty, SDG8 on reducing unemployment and SDG11 on carbon reduction (Khalid & Maidin, 2022).

The research on enhancement of Malaysian Good Agricultural Practices (MyGAP) were mainly conducted by researchers from the UPM to evolving agricultural practices and challenges. The number of farms who maintain MyGAP certified status is the second largest among the public GAP standards in Southeast Asia (Islam et al., 2012). UPM actively integrates MyGAP principles into agricultural and related degree programs by acting as centers of knowledge creation, dissemination and capacity building to equip future farmers and agricultural professionals with the necessary knowledge and skills for water and food security in the region.

In addition to achieving myGAP certification in the livestock sector, UPM has supported in assisting a total of 11,832 farms under the plantation subsector to achieve myGAP certification, and 326 farms for myOrganic certification from 2003 until

2022. This certification not only ensures the production of safe and high-quality products but could inspire agricultural certification programs in regions like Latin America adopting model applicable for global south countries aiming to meet export standards, enhance market access and gaining consumer trust while enhancing UPM's competitiveness based on international standards such as the ASEAN GAP and the Global GAP. Similarly, Universiti Teknologi Malaysia (UTM) has received myGAP certification for its research farm at the Innovation Centre in Agritechnology for Advanced Bioprocessing (ICA). By incorporating myGAP principles into their curricula and obtaining relevant certifications, universities are preparing graduates for sustainable agricultural practices. However, knowledge, cost constraints and the role of extension agents, the certification process emerges as the main barrier for adoption (Mohd Ariffin et al., 2024) (Table 1).

As the existing agriculture practices faced many challenges and fail to address some of the most critical needs of the growing population, universities have been tasked to develop and test innovative technologies and practices that can improve agricultural productivity, efficiency, and sustainability, aligning with MyGAP principles. Examples include precision agriculture, water-efficient irrigation systems, and integrated pest management strategies. For this, UPM works with communities to create novel drought-tolerant paddy as climate solutions, boosting agricultural productivity while conserving water in a nation where food security is threatened by water scarcity. This pragmatic research methodology demonstrates how academic institutions can effectively tackle regional and worldwide issues.

Universities offer training for farmers, extension officers, and other stakeholders on myGAP implementation, certification processes, and best practices. Such training includes evaluation of the effectiveness of the Basic Student Entrepreneurial Programme (BSEP) among local university graduates who have undergone the training programme in entrepreneurship development.

Beyond formal education, UPM provide practical experience with the newest agricultural technologies and techniques through its PUTRA branding where outreach programs were conducted to help farmers adopt improved cultivation techniques, leading to higher productivity and better income. In addition, Universiti Malaysia Sabah conducted programmes modernising sustainable seaweed cultivation to

Table 1 Growth of MyGAP-certified farms in Malaysia (2003–2022)

Year	Number of MyGAP-certified farms	Notes
2003	Data not specified	MyGAP certification was introduced
2012	Increasing trend begins	Marked rise in adoption rates from this year onward
2015	4,024	Significant growth since 2012
2020	6,792	Reported total by end of 2020
2022	11,832 (cumulative since 2003)	Total number of farms certified under MyGAP since inception

enhance the socio-economic status of the community through seaweed cultivation (Hussin et al., 2015).

Agricultural universities such as UPM also play a pivotal role in nurturing the next generation of agripreneurs by integrating agribusiness management courses into agriculture curriculum and offering incubation support for start-up. Among the key motivators for the respondents to start their agro-based businesses were pursuing their dream, elevating their socio-economic status, and challenging themselves (Rahim et al., 2020). These agripreneurs are developing innovative solutions to address pressing agricultural challenges, thereby driving efficiency and sustainability.

Extension services provided by universities is crucial for translating research findings into practical applications. Both extension services and climate change are vital factors that can affect the farm level outcomes. In Malaysia, the commercialization and farm literacy significantly influence the adoption of extension services while the social influence and information quality were found to mediate the associations positively as well (Shibli et al., 2021). Universities can integrate agriculture and agronomic models with social and demographic modeling approaches to increase agricultural sustainability (Sultan et al., 2024). While farmers rely on these services to assist them embrace new technology and methods that can increase their output and profitability, they also highlighted that extension staff need training to improve their capacity (Ali et al., 2018). As technology continue to evolve, farmers could ask for support from universities to integrate big data analytic for agricultural risk management.

3.5 Transdisciplinary Frameworks

As university houses diverse expertise, interdisciplinary research culture aligns with transdisciplinary approaches such as in Universiti Sains Malaysia had been successful in promoting the sustainable development agenda through leadership role in the Regional Centre of Expertise (RCE) Penang using an "inside-out approach" (Abidin Sanusi & Khelghat-Doost, 2008). RCE network provides an impetus for transformative role of the university, serves as platform for the university-community partnership programme. Transdisciplinary learning has seen river as co-learning arenas (RCAs) providing spaces for dialogue and action, facilitating transdisciplinary knowledge co-creation and promoting socio-ecological justice (Souza et al., 2024).

4 Challenges and Opportunities

Transdisciplinary research can be time-consuming and challenging, requiring reflexivity and careful attention to research design (Wolff et al., 2019). While universities in developed countries have achieved much, their ASEAN counterpart face significant challenges to establish international collaborations to exchange knowledge and compete for dwindling grant opportunities. These limitations can be addressed through cooperative resource sharing, such as cooperative research projects. The Southeast Asian Regional Centre for Graduate Study and Research in Agriculture (SEARCA) is an example of an agency that reflects how transdisciplinary approaches has evolved over the last fifty years in the region.

Barriers such as funding limitations, institutional silos, and policy gaps significantly hinder the effective implementation of transdisciplinary approaches in addressing water and food security challenges in Malaysia. Funding constraints limit long-term research sustainability, particularly for collaborative projects that require continuous investment in infrastructure, technology, and capacity building (Sultan et al., 2024). The constraints faced by universities in the Global South underscore a persistent structural imbalance in global academic ecosystems, where institutions in the Global North often dominate in terms of resources, visibility, and influence. To move forward, it is essential for universities in the Global South to harness regional strengths and cultivate North–South collaborations to leverage their unique socio-environmental contexts.

Institutional silos create fragmentation between academic disciplines and external stakeholders, reducing the efficiency of knowledge exchange and problem-solving efforts (Padfield et al., 2014). Additionally, policy gaps including the lack of clear governance frameworks and coordination between government agencies further impede the translation of research into actionable solutions for water and food security (Anuar et al., 2022).

For example, the existing governance of river basins fails to ensure that rivers are clean and of good ecological quality where water agencies in Malaysia are not integrated and are often locked in constitutional conflict (Anuar et al., 2022). The author pointed to lessons learned from other jurisdictions such as in the Philippines where court had strengthened river basin governance by integration of transdisciplinary collaboration to improve water quality by reducing pollution (SDG 6).

Universities play a crucial role in implementing SDGs through research, community outreach, and youth engagement (Woo et al., 2019; Jasmin Irisha Jim Ilham et al., 2019). UPM has played a pivotal role in advancing sustainable farming practices that align SDG 2 (Zero Hunger) and SDG 6 (Clean Water and Sanitation) where research on integrated farming systems and water-efficient agricultural techniques has contributed to improving food security while minimizing environmental impact (Sultan et al., 2024). However, scaling up these efforts requires stronger academia-industry-government partnerships, where universities collaborate with policymakers and private sectors to translate research into large-scale implementation (Padfield et al., 2014).

Successful models in ASEAN demonstrate the value of stakeholder engagement, as seen in Philippines's cooperative agricultural research programs and Singapore's precision farming initiatives, which integrate technology, policy support, and academic expertise to enhance water and food security. By fostering transdisciplinary collaborations, ASEAN and Malaysian universities can contribute to global solutions for sustainable water and food systems, ensuring resilient and climate-adaptive agricultural practices across diverse socio-environmental landscapes (Woo et al., 2019).

5 Conclusion

This chapter illustrates how ASEAN universities, exemplified by UPM, employ transdisciplinary approaches to advance water and food security within the SDG framework. Concrete outcomes, such as a 30% reduction in water use through precision irrigation, increased yields from improved rice varieties, and enhanced sustainability via MyGAP certification, highlight the transformative role of academia. Rooted in Malaysia's smallholder context, these initiatives offer adaptable models for the Global South, where similar resource constraints and climate challenges persist. However, broader applicability requires context-specific adaptation such as modifying irrigation technologies for local environments or tailoring certification systems for large-scale farms. Crucially, success depends on addressing persistent barriers, including fragmented funding, policy disconnects, and uneven community engagement—challenges that extend beyond Malaysia and resonate globally.

References

Abidin Sanusi, Z., & Khelghat-Doost, H. (2008). Regional Centre of Expertise as transformational platform for sustainability: A case study of Universiti Sains Malaysia, Penang. *International Journal of Sustainability in Higher Education, 9*(4), 487–497. https://doi.org/10.1108/14676370810905580

Abiri, R., Rizan, N., Balasundram, S. K., Shahbazi, A. B., & Abdul-Hamid, H. (2023). Application of digital technologies for ensuring agricultural productivity. *Heliyon, 9*. https://doi.org/10.1016/j.heliyon.2023.e22601

Adams, R., Martin, S., & Boom, K. (2018). University culture and sustainability: Designing and implementing an enabling framework. *Journal of Cleaner Production, 171*, 434–445. https://doi.org/10.1016/j.jclepro.2017.10.032

Adedoyin, A. O., Shamsudin, M. N., Radam, A., & Latif, I. A. (2016). Effect of improved high yielding rice variety on farmers productivity in MADA, Malaysia. Corpus ID: 58892284.

Adlong, W. (2013). Rethinking the Talloires declaration. *International Journal of Sustainability in Higher Education, 14*(1), 56–70. https://doi.org/10.1108/14676371311288958

Agramont, A., Craps, M., Balderrama, M., & Huysmans, M. (2019). Transdisciplinary learning communities to involve vulnerable social groups in solving complex water-related problems in Bolivia. *Water, 11*(2), 385.

Ahmad, F., Ahmad, Z., Hassan, A. A., Ariffin, S., Noordin, N., Salleh, S., Hussein, S. B., Akil, M., Sani, M. Z., Harun, A. R., & Rahim, K. A. (2018). A review on gamma greenhouse as a chronic gamma irradiation facility for plant breeding and improvement program. *Jurnal Sains Nuklear Malaysia, 30*(1), 8–18, ISSN: 2232-0946.

Albrecht, T. R., Crootof, A., & Scott, C. A. (2018). The Water-Energy-Food Nexus: A systematic review of methods for nexus assessment. *Environmental Research Letters, 13*(4), Article 043002.

Ali, M., Man, N., Abd Latif, I., Muharam, F. M., et al. (2018). The use of information and communication technologies in agricultural risk management by the agricultural extension services in

Malaysia. *International Journal of Agriculture Environment and Food Sciences, 2*(1), 29–35. https://doi.org/10.31015/jaefs.18005

Amusa, A. A., Johari, A., Jalil, A. A., Abdullah, T. A. T., Adeleke, A. O., Katibi, K. K., et al. (2024). Sustainable electricity generation and farm-grid utilization from photovoltaic aquaculture: A bibliometric analysis. *International Journal of Environmental Science and Technology, 21*(11), 7797–7818.

Anuar, H. M., Yaacob, N., Wahab, H. A., & Khalid, R. M. (2022). Strengthening river basin governance through Green Courts towards ensuring water security (Goal 6). In *Good governance and the sustainable development goals in Southeast Asia* (pp. 69–83). Routledge. https://doi.org/10.4324/9781003230724

Asian Development Bank. (2021). *Asian Development Outlook 2021 update transforming agriculture in Asia.* Retrieved January 2, 2025, from https://www.adb.org/sites/default/files/publication/726556/ado2021-update.pdf

Baumber, A., Kligyte, G., van der Bijl-Brouwer, M., & Pratt, S. (2020). Learning together: A transdisciplinary approach to student–staff partnerships in higher education. *Higher Education Research & Development, 39*(3), 395–410. https://doi.org/10.1080/07294360.2019.1684454

Bain, B., Griffith, K., & Varney, J. (2019). Transdisciplinarity practice in higher education. In *Handbook of research on transdisciplinary knowledge generation* (pp. 115–131). IGI Global. https://doi.org/10.4018/978-1-5225-9531-1.CH009

BERNAMA. (2024). UPM food security blueprint to drive tech-based agriculture strategy. Retrieved December 12, 2024, from https://bernama.com/en/news.php?id=2305528

Bwambale, E., Abagale, F. K., & Anornu, G. K. (2022). Smart irrigation monitoring and control strategies for improving water use efficiency in precision agriculture: A review. *Agricultural Water Management, 260*, Article 107324.

Dai, Z., Wang, Q., Jiang, J., & Lu, Y. (2024). Influence of university agricultural technology extension on efficient and sustainable agriculture. *Scientific Reports, 14*(1), 4874.

Cottafava, D., Ascione, G. S., Corazza, L., & Dhir, A. (2022). Sustainable development goals research in higher education institutions: An interdisciplinarity assessment through an entropy-based indicator. *Journal of Business Research, 151*, 138–155.

Delina, L. L. (2021). Southeast Asian expert perceptions of solar radiation management techniques and carbon dioxide removal approaches: Caution, ambivalence, risk precaution, and research directions. *Environmental Research Communications, 3*(12), Article 125005. https://doi.org/10.1088/2515-7620/ac3dc1

Economic Research Institute of ASEAN and East Asia. (2022). *Ensuring food security in post pandemic ASEAN*. Retrieved January 2, 2025, from https://www.eria.org/uploads/media/ASEAN-on-Point/2022_Sep_AoP_Report.pdf

Enenkel, M., See, L., Bonifacio, R., Boken, V., Chaney, N., Vinck, P., et al. (2015). Drought and food security—Improving decision-support via new technologies and innovative collaboration. *Global Food Security, 4*, 51–55.

Pin Keong, E., Ehsan Rana, M., & Abdul Hameed, V. (2022). Crop suitability prediction model for Malaysian crop diversification. In *2022 International Conference on Decision Aid Sciences and Applications (DASA)* (pp. 1004–1012). https://doi.org/10.1109/DASA54658.2022.9765283

Finger, R. (2023). Digital innovations for sustainable and resilient agricultural systems. *European Review of Agricultural Economics*. https://doi.org/10.1093/erae/jbad021

Finger, R., Swinton, S. M., El Benni, N., & Walter, A. (2019). Precision farming at the nexus of agricultural production and the environment. *Annual Review of Resource Economics, 11*(1), 313–335.

Fox, J. (2017). Collective approach to complex food system issues, the Case of the Ohio State University. *Metropolitan Universities, 28*(1), 94–112. https://doi.org/10.18060/21470

Gómez-Baggethun, E., Corbera, E., & Reyes-García, V. (2013). Traditional ecological knowledge and global environmental change: Research findings and policy implications. *Ecology and Society: A Journal of Integrative Science for Resilience and Sustainability, 18*(4). https://doi.org/10.5751/ES-06288-180472

Hussin, R., Yasir, S. M., Kunjuraman, V., & Hossin, A. (2015). Enhancing capacity building in seaweed cultivation system among the poor fishermen: A case study in Sabah, East Malaysia. *Asian Social Science, 11*(18), 1.

Ilham, J. I., Zaihan, M. H., Hakimi, S. M., Ibrahim, M. H., & Shahrul, S. (2019). Mobilising the sustainable development goals through universities: Case studies of sustainable campuses in Malaysia. *World Sustainability Series*. https://doi.org/10.1007/978-3-030-15604-6_8

Islam, G. M. N., Arshad, F. M., Radam, A., & Alias, E. F. (2012). Good agricultural practices (GAP) of tomatoes in Malaysia: Evidences from Cameron Highlands. *African Journal of Business Management, 6*(27), 7969–7976. https://doi.org/10.5897/AJBM10.1304

Kanapathy, S., Lee, K. E., Mokhtar, M., Syed Zakaria, S. Z., & Sivapalan, S. (2021). A framework for integrating sustainable development concepts into the chemistry curriculum towards achieving education for sustainable development in Malaysia. *International Journal of Sustainability in Higher Education, 22*(6), 1421–1449. https://doi.org/10.1108/IJSHE-07-2020-0241

Karpouzoglou, T., Pereira, L. M., & Doshi, S. (2017). Bridging ICTs with governance capabilities for food-energy-water sustainability. In *Food, energy and water sustainability: emergent governance strategies* (pp. 222–238). Earthscan. https://doi.org/10.9774/GLEAF.9781315696522_13

Khalid, R. M., & Maidin, A. J. (2022). Introduction: Governance issues towards achieving SDGs in Southeast Asia. In *Good governance and the sustainable development goals in Southeast Asia* (pp. 1–9). Routledge. https://doi.org/10.4324/9781003230724

Khanal, P. R. (2021). Water, food security and Asian transition: A new perspective within the face of climate change. https://doi.org/10.1007/978-3-319-54612-4_1

Khazanah Research Institute. (2019). *The status of the paddy and rice industry in Malaysia*. Khazanah Research Institute.

Khelghat-Doost, H., Sanusi, Z. A., Faridd, T. F., & Jegatesen, G. (2011). Institutions of higher education and partnerships in education for sustainable development: Case study of the Regional Center of Expertise (RCE) Penang, Malaysia. *Journal of Sustainable Development, 4*, 108. https://doi.org/10.5539/JSD.V4N3P108

Leal Filho, W., Setti, A. F. F., Azeiteiro, U. M., Lokupitiya, E., Donkor, F. K., Etim, N. N., et al. (2022). An overview of the interactions between food production and climate change. *Science of the Total Environment, 838*, Article 156438.

Lowder, S. K., Sánchez, M. V., & Bertini, R. (2021). Which farms feed the world and has farmland become more concentrated? *World Development, 142*, Article 105455.

Magno, R. M. (2024). Sustainability as a transdisciplinary. In *Transdisciplinary teaching and technological integration for improved learning: Case studies and practical approaches*, 84.

Mal, B. C. (2007). Neglected and underutilized crop genetic resources for sustainable agriculture. *Indian Journal of Plant Genetic Resources, 20*, 1–14. Corpus ID: 110119157.

Mentsiev, A. U., Isaev, A. R., Supaeva, K., Yunaeva, S., & Khatuev, U. (2019). Advancement of mechanical automation in the agriculture sector and overview of IoT. *Journal of Physics:* Conference Series, 1399. https://doi.org/10.1088/1742-6596/1399/4/044042

Miao, Y., Bukhari, A. A. A., Bukhari, W. A. A., Ahmad, S., & Hayat, N. (2025). Why fossil fuels stifle green economic growth? An environmental management perspective in assessing the spatial spillover impact of energy consumption in South Asia. *Journal of Environmental Management, 373*, Article 123471.

Mohd Arifin, N. A., Ismail, N. A., Bakar, M. A., Syahlan, S., & Zulkifli, F. (2024). Barriers to myGAP certification: A study of farmer knowledge, intentions, and challenges in Kelantan, Malaysia. *Borneo Akademika, 8*(1), 53–62.

Mohd-Azmi, M. L., Jesse, F. F. A., Sarah, S. A., Roslan, S., Zuraidah, A., & Hambali, I. U. (2016). Transforming agriculture research into commercialisation: Experience of Universiti Putra Malaysia. *Pertanika Journal of Social Sciences and Humanities, 24*(3), 907–921.

Monchusi, B. B., Kgopa, A. T., & Mokwana, T. I. (2024). Integrating IoT and AI for precision agriculture: Enhancing water management and crop monitoring in small-scale farms. In *Proceedings*

of the International Conference on Intelligent and Innovative Computing Applications, 2024 (pp. 151–158). https://doi.org/10.59200/iconic.2024.017
Montesclaros, J. M. L., & Teng, P. S. (2021). Agriculture and food security in Asia. In *Climate change, disaster risks, and human security: Asian experience and perspectives* (pp. 137–168).
Msangi, S., & Rosegrant, M. (2009). *World agriculture in a dynamically-changing environment: IFRPI long-term outlook for food and agriculture under additional demand and constraints.* Paper presented at the FAO expert meeting on how to feed the world in 2050.http://ftp.fao.org/docrep/fao/012/ak542e/ak542e05.pdf
Ng, Z., George, P., & Thomas (2023). Leveraging on partnership to expose youth to sustainable developments: A case study. *Journal of Business Management and Accounting*. https://doi.org/10.32890/jbma2023.13.2.5
Nguyen, H. T. T., Le, T. Q. A., Tuyen, M. C., & Hung, P. X. (2025). A review of climate-smart agriculture in Asia: Critical achievements, key challenges, and potential prospects. *Journal of Agriculture and Rural Development in the Tropics and Subtropics (JARTS), 1*(126), 25–42.
OECD (Organization of Economic Cooperation and Development). (2001). Workshop for adoption of technologies for sustainable farming systems, hosted by the Netherland's ministry of agriculture. In *Nature Management and Fisheries on 4–7 July 2000* (pp. 15–25). Retrieved January 5, 2025, from https://www.researchgate.net/profile/Anoop-Srivastava/post/What-are-the-appropriate-recent-models-for-analyzing-adoption-of-a-given-technology/attachment/5b4e0c88b53d2f89289a8b4b/AS%3A649405195644931%401531841672874/download/2739771.pdf
Okolo, B. N. (2010). Nigerian Universities and agriculture: Their role in the development of agriculture for sustainable food security. *Journal of Developments in Sustainable Agriculture, 5*(1), 23–30. https://doi.org/10.11178/jdsa.5.23
Padfield, R., Richards, C., Yusop, Z., Parkey, J. R., Harrison, O., & Preece, C. (2014). Tackling wicked problems in the Malaysian water industry: A framework for university-industry research partnerships. *Journal of Governance and Development, 10*, 99–119.
Penang Institute. (2022). Automation and the future of work in agriculture. Retrieved January 2, 2025, from https://penanginstitute.org/publications/issues/automation-and-the-future-of-work-in-agriculture
Rahim, M. F. A., Chew, K. W., Zainuddin, M. N., & Bujang, A. S. (2020). Motivation to become agropreneurs among youths in Malaysia. In *Understanding digital industry* (pp. 350–353). Routledge.
Ramos, T. B., Caeiro, S., Van Hoof, B., Lozano, R., Huisingh, D., & Ceulemans, K. (2015). Experiences from the implementation of sustainable development in higher education institutions: Environmental management for sustainable universities. *Journal of Cleaner Production, 106*, 3–10. https://doi.org/10.1016/j.jclepro.2015.05.110
Rushforth, R., & Phillips, C. F. (2012). Gathering under a green umbrella: collaborative rainwater harvesting at the University of Arizona. In *Sustainable policy applications for social ecology and development* (pp. 139–149). IGI Global. https://doi.org/10.4018/978-1-4666-1586-1.ch011
Scanlon, B. R., Ruddell, B. L., Reed, P. M., Hook, R. I., Zheng, C., Tidwell, V. C., & Siebert, S. (2017). The food-energy-water nexus: Transforming science for society. *Water Resources Research, 53*(5), 3550–3556. https://doi.org/10.1002/2017WR020889
Shibli, R., Saifan, S., Ab Yajid, M. S., Khatibi, A., & Azam, S. F. (2021). Commercialization and farmer literacy's role on agriculture extension services in Malaysia. *AgBioForum, 23*(2), 82–91. https://agbioforum.org/menuscript/index.php/agb/article/download/64/43/122
Souzsm Pante de, D. T., Hommes, L., Wals, A. E. J., Hoogesteger, J., Boelens, R., Duarte-Abadía, B., Hidalgo-Bastidas, J. P., Huijbens, E. H., Harris, L. M., Suhardiman, D., Melsen, L., Buijse, A. D., de Castro, F., Cremers, L., Hogenboom, B., García Vargas, M., García Vargas, M., Roca-Servat, D., Veldwisch, G. J., & Joy, K. J. (2024). River co-learning arenas: principles and practices for transdisciplinary knowledge co-creation and multi-scalar (inter)action. *Local Environment*, 1–23. https://doi.org/10.1080/13549839.2024.2428215

Sultan MTH, Shahar FS, Zain MIM, Komoo I. (2024 Mar 29). A systematic review of the role of integrated farming and the participation of universities in ensuring food security: Malaysia's effort. *Italian Journal of Food Safety, 13*(2), 11854. https://doi.org/10.4081/ijfs.2024.11854

Sundram, P. (2023). Food security in ASEAN: Progress, challenges and future. *Frontiers in Sustainable Food Systems, 7*, 1–14. https://doi.org/10.3389/fsufs.2023.1260619

The ASEAN Statistical Yearbook. (2023). Retrieved January 2, 2025, from https://asean.org/wp-content/uploads/2023/12/ASEAN-Statistical-Yearbook-2023.pdf,%20

United Nations. (2024). *World population prospects 2024: Summary of results*. Department of Economic and Social Affairs, Population Division.

Verma, N., Talwar, P., Upadhyay, A., Singh, R., Lindenberger, C., Pareek, N., et al. (2024). Food-Energy-Water Nexus in compliance with Sustainable Development Goals for integrating and managing the core environmental verticals for sustainable energy and circular economy. *Science of the Total Environment, 930*, Article 172649. https://doi.org/10.1016/j.scitotenv.2024.172649

WEF. (2024). Retrieved December 31, 2024, from https://www.weforum.org/stories/2024/11/shaping-cities-future-inclusive-sustainable-and-resilient/

Woo, W. T., Koh, H. L., & Teh, S. Y. (2019). Achieving excellence in sustainable development goals in Sunway University Malaysia. *World Sustainability Series*. https://doi.org/10.1007/978-3-030-15604-6_17

Wolff, M. G., Finca, A., Cockburn, J. J., De Wet, C., Carlos Bezerra, J., Weaver, M. J., et al. (2019). Exploring and expanding transdisciplinary research for sustainable and just natural resource management. *Ecology and Society, 24*(4).

World Economic Forum. (2020). *Future of consumption in fast-growth consumer markets: ASEAN. Insights report*. Retrieved January 2, 2025, from https://www3.weforum.org/docs/WEF_Future_of_Consumption_in_Fast_Growth_Consumer_Markets_ASEAN_2020.pdf

Yazid, S. N., Ahmad, K., Razak, M. S., Rahman, Z. A., Ramachandran, K., Mohamad, S., & Ghaffar, M. B. (2021). Introgression of bacterial leaf blight (BLB) resistant gene, Xa7 into MARDI elite variety, MR219 by marker assisted backcrossing (MABC) approach. *Brazilian Journal of Biology = Revista brasleira de biologia, 84*, e248359. https://doi.org/10.1590/1519-6984.248359

Assessing the Carbon Footprint of Thai Nguyen University of Agriculture and Forestry: A Step Toward Sustainability in Vietnam's Higher Education

Tobias Matusch and Ho Ngoc Son

Abstract Undoubtful Vietnamese higher education institutions (HEIs) have undoubtful made substantial contributions to sustainable development and climate protection. However, the growth in the number of universities and enrolled students aligns with a nationwide surge in energy consumption. While recent efforts by HEIs have focused on the expansion and improvement of facilities, future priorities must emphasize greater energy efficiency and transformative actions. The Vietnamese government's commitment to achieving net-zero emissions by 2050 underscores the necessity for HEIs to adopt clear sustainability targets and implement climate protection measures. Currently, many universities lack the human resources and expertise to conduct comprehensive carbon footprint assessments, due in part to the absence of international standards tailored to universities and a lack of region-specific emission factors. As a result, HEIs frequently lack individual carbon footprints and concrete emissions reduction targets. This study presents a practical approach for calculating a university's carbon footprint in Vietnam, encompassing Scope 1, 2, and 3 emissions. Moreover, the methodology is transferable to other HEIs in the region, enabling sector-wide emission assessments that provide the necessary foundation for identifying key steps towards transforming institutional facilities and management for greater sustainability.

Keywords Higher education institution · Sustainability · Corporate carbon footprint · GHG accounting and reporting · Net-zero · Vietnam

T. Matusch (✉)
Hochschule Ruhr West, Mülheim an der Ruhr, Germany
e-mail: tobias.matusch@hs-ruhrwest.de

H. N. Son
Thai Nguyen University of Agriculture and Forestry, Thai Nguyen City, Vietnam
e-mail: hongocson@tuaf.edu.vn

W. Leal Filho et al. (eds.), *Introducing the UN SDGs in Higher Education Institutions*, World Sustainability Series, https://doi.org/10.1007/978-3-032-23017-1_34

1 Introduction

Severe impacts of climate change are forecasted for Vietnam, with significant social and economic consequences (The World Bank Group and the Asian Development Bank, 2021). It is in Vietnam's best interest to limit global warming to 1.5 °C by the end of this century as outlined in the 2015 Paris Agreement (United Nations, 2015), which the Vietnamese government ratified in November 2016. Five years later, during the Conference of the Parties (COP26) summit, Vietnam announced its target to achieve net-zero emissions by 2050, followed by the National Strategy for Climate Change (Decision No. 896/QD-TTg on approving the National Strategy for Climate Change until 2050, 2022/July 26, 2022). The strategy's primary objectives include effective adaptation to climate change, reducing vulnerability, minimizing climate-related damages, lowering greenhouse gas (GHG) emissions, and contributing to global climate protection efforts. However, both globally and in Vietnam, annual carbon dioxide (CO_2) emissions from fossil fuels and industry saw a significant increase in 2023 after a temporary decline due to the COVID-19 pandemic (Global Carbon Budget, 2024). With total emissions of 335 $MtCO_2$, Vietnam ranks 17th globally and 9th in Asia. At the same time, primary energy consumption in Vietnam reached an all-time high of 1,359 TWh in 2023, with over 77% derived from fossil fuels (Energy Institute, 2024).

The long-term period of economic growth in Vietnam is also reflected in the increasing number of higher education institutions (HEIs) and enrolled students. After peaking in 2014, student enrollment initially declined but has since resumed growth (Ministry of Education and Training [Vietnam], 2024). To remain competitive, Vietnamese HEIs have prioritized international research projects, external funding, and partnerships with prestigious research institutions. Consequently, similar to other countries, emissions from HEIs contribute significantly to national emissions (Robinson et al., 2015). However, unlike the rapid digital transformation that surged during the COVID-19 pandemic, the transformation of university infrastructure and management toward sustainability remains a fundamental challenge.

1.1 Vietnamese Universities in International Comparison

Globally, only a small percentage of universities worldwide have made significant progress toward sustainability, despite their leadership role in promoting a carbon–neutral society (Helmers et al., 2021). The absence of a transparent legal framework, clear accountability mechanisms, and targeted implementation often hampers progress. According to the Times Higher Education global impact rankings (THE, 2024), which assess universities based on their commitment to the UN Sustainable Development Goals (SDGs), 2,152 universities from 125 countries participated in the 2025 ranking—about 10% of all universities worldwide (International Association of Universities [IAU], 2025). Leading institutions, such as Western Sydney University or the University of Manchester, retained top positions. Among Vietnamese universities, 13 were evaluated, with Nguyen Tat Thanh University in Ho Chi Minh City ranked as the most sustainable. Six other Vietnamese universities were placed between 401 and 600 globally (THE, 2024). Other international alliances, such as the "International Universities Climate Alliance" (Secretariat of the International Universities Climate Alliance, 2025) and the "Race to Zero" initiative (EAUC—The Alliance for Sustainability Leadership in Education, 2025), offer additional networks with specific sustainability commitments.

When considering the SDGs as the superior framework, with SDG 13 on climate action as a specific category, the target for net-zero emissions appears to be considerably easier to monitor. Based on the GHG Protocol (WRI), corporate carbon footprints (CCFs) are created globally as baselines for climate protection plans and reduction targets (Valls-Val & Bovea, 2022). While companies regularly prepare CCFs, examples from HEIs, especially on a regional basis, are still rarely published, with highly diverse results (Helmers et al., 2021; Kiehle et al., 2023; Valls-Val & Bovea, 2021). Similarly, there is only limited coverage of universities with regular sustainability reports (Udas et al., 2018).

1.2 Vietnam's Emission Targets and Its Universities

The most ambitious HEIs have already implemented sustainability reporting and achieved carbon neutrality, with many others aiming for 2025 or 2030 (EAUC—The Alliance for Sustainability Leadership in Education, 2025). In East Asia and Southeast Asia, most committed HEIs target net-zero emissions by 2040 or 2050. In alignment with Vietnam's national climate strategy and Decision 896/QD-TTg (Decision No. 896/QD-TTg on approving the National Strategy for Climate Change until 2050, 2022/July 26, 2022), Vietnamese HEIs must establish specific climate protection targets in the coming years.

At present, the emissions of universities are neither known at the sectoral level nor at the individual level. No case study of a Vietnamese university has been published regarding its carbon footprint. Due to the lack of data, it remains unclear how the emissions of Vietnamese universities compare to those of other institutions, especially considering the current lack of a standardized methodological framework or appropriate emission factors. As a result, international comparisons, such as those by Times Higher Education, can only provide vague information in this regard. Consequently, the inadequate data makes it impossible for Vietnamese universities to identify priority areas for action or implement appropriate climate protection measures to reduce their emissions. This study offers an initial framework for determining the CCF of a Vietnamese university, providing insights into its emissions and identifying steps toward climate neutrality, with Thai Nguyen University of Agriculture and Forestry as a case study. The approach applied here can be adapted and modified by other HEIs in the region to assess their individual carbon footprints. This could also increase the pressure to develop more accurate and up-to-date emission factors for regions outside the US and Europe. Many other important areas of action, such as education for sustainable development, curricula development, or knowledge transfer, have been intentionally excluded from the focus of this paper, through their immense importance is not overlooked.

2 Case Study—Thai Nguyen University of Agriculture and Forestry

Located in northern Vietnam, the Thai Nguyen University of Agriculture and Forestry (TUAF) was established in 1969. Currently, TUAF enrolls approximately 5,000 students and employs 454 staff members. The university comprises eight specialized faculties, an office for administering advanced educational programs, seven functional departments, two research institutes, and 13 research centers. In 2024, TUAF offers 24 undergraduate programs, ten master's programs, and eight doctoral programs. In recent years, TUAF has been recognized as a leading center for education, scientific research, and technology transfer in agriculture, forestry, natural resources, and environmental management. Notably, approximately 40% of TUAF's students belong to ethnic minority groups, predominantly from economically disadvantaged mountainous regions in northern Vietnam.

TUAF has extensive infrastructure and diverse research, teaching and technology transfer activities, which makes the assessment of Scope 3 emissions particularly challenging. The campus includes multi-story academic buildings, 60 classrooms, and four main lecture halls covering a total area of 10,089 m^2. The university operates 34 laboratories (4,811 m^2) across its specialized faculties and research institutes, along with five computer labs and an 840 m^2 animal husbandry facility. Practical training sites span approximately 575,000 m^2, supporting both education and research. TUAF's library covers 1,900 m^2 and offers 500 study spaces. Student dormitories consists of 363 rooms in multi-story buildings. In addition, the campus includes a sports stadium, a 3,700 m^2 multipurpose court, four artificial turf fields, two tennis courts, and various outdoor training areas—all of which are used by both students and faculty, as physical education is integrated into the curriculum.

For a long time, the contribution of the agriculture, forestry and fishing sector to Vietnam's gross domestic product (GDP) has been steadily declining (General Statistics Office of Vietnam, 2024). Exacerbated by harsh working conditions in the sector and improved job prospects in urban areas, interest in studying agricultural or forestry degree programs at TUAF has declined in recent years, along with student enrollment numbers. Consequently, the university is now confronted with a triple challenge: maintaining its appeal to students by offering innovative programs, modernizing its facilities and infrastructure, and preparing for greater financial independence, as public universities in Vietnam are expected to become more autonomous and receive reduced state funding in the future (Trinh et al., 2025).

3 Methodology

The presented approach primarily follows the GHG Protocol for calculating emissions across Scope 1, 2, and 3 (WRI, 2013). Therefore, the main emission types considered are carbon dioxide (CO_2), methane (CH_4), nitrous oxide (N_2O), hydrofluorocarbons (HFCs), perfluorinated carbons (PFCs), and sulphur hexafluoride (SF_6). The GHG Protocol is increasingly becoming the accounting standard and offers a clear division of emissions into three scopes (Paredes-Canencio et al., 2024; Valls-Val & Bovea, 2021). The ISO 14064-1:2018 standard (International Organization for Standardization, 2024) was not used due to its rigid reporting guidelines, as the TUAF management board also decided against certification or auditing. Other standards, such as PAS 2060, SIMAP, or COPE, were avoided due to their lack of flexibility or focus on carbon assessment, limitations in covering specific emission types, or limited use, which reduces comparability with other HEIs (Helmers et al., 2021; Valls-Val & Bovea, 2021).

The objective of the applied approach was to comprehensively account for all existing emissions without imposing restrictions due to data limitations or lack of direct university affiliation. Given that Scope 3 emissions often constitute the predominant source of university-related emissions (Robinson et al., 2018), a complete CCF was developed despite known limitations associated with data availability and data accuracy. Despite the inherent complexity of applying the GHG Protocol to HEIs—particularly regarding Scope 3 emissions—we ensured rigor and robustness by systematically assessing all three scopes in detail across all relevant sub-categories. Due to the absence of a standardized international methodology for educational institutions (Valls-Val & Bovea, 2021), an independent spreadsheet model incorporating established emission factors was used, as detailed in Table 1. However, due to varying accounting methods, direct comparisons between HEIs remain of limited value. While these limitations persist, our methodology provides a comprehensive carbon footprint assessment, offering a solid foundation for future refinements and institutional climate strategies.

The data presented in Table 1 are derived from measured, sampled, or, in some cases, estimated values. The accuracy of the data is categorized on a scale from very low (5) to very high (1), as indicated in the final column of Table 1. Data with lower accuracy should be interpreted with caution. The relevance of each emission source is assessed based on total emission volumes. The selected emissions factors, sourced from various databases, are provided in Table 1 for transparency. In addition to Table 1, multiple materiality matrices were developed. The first matrix prioritizes data availability and quality by linking emission relevance with data accuracy. The second matrix associates key emission sources with a feasibility factor, evaluating the potential for mitigation actions. The third matrix focuses on prospective trends and necessary interventions for the university.

Assessments of relevance, data accuracy, feasibility, and future trends are based on an average of personal evaluations by the authors and key stakeholders of TUAF. These assessments do not claim absolute validity, but rather reflect the current

Table 1 Overview of calculated emission sources, relevance, emissions factors and resulting emissions, and accuracy level

Scope	Category	Source of emission	Relevance	Emission factor	Source of the emission factor	Amount	Accuracy
Scope 1	Laboratory gases	Laboratory gases (estimation)	Low (2)	Varying	Varying	5 tCO_{2e}	Very low (5)
	Refrigerant	Fridges (25 with 0.05 kg, with leakage of 10% per year)	Very low (1)	Assumption R134a GWP 1430 $kgCO_{2e}/kg$	www.infraserv.com	0.18 tCO_{2e}	Medium (3)
		Air conditions (129 with 0.5 kg, with leakage of 10% per year)	Low (2)	Assumption R32 GWP 675 $kgCO_{2e}/kg$	www.infraserv.com	4.35 tCO_{2e}	Medium (3)
	Fossil fuel	Diesel (5,000 l of diesel)	Medium (3)	2.626 $kgCO_{2e}/l$	GHG Protocol	13.13 tCO_{2e}	Low (4)
		Tractor (5,000 l of diesel)	Medium (3)	2.626 $kgCO_{2e}/l$	GHG Protocol	13.13 tCO_{2e}	Low (4)
Scope 2	Electricity	Location-based (1,338,489 kWh)	Very high (5)	0.43714 $kgCO_{2e}/kWh$	CaDI	585.11 tCO_{2e}	Very high (1)
Scope 3	Category 1—purchased goods and services	Purchased goods (200,000 USD; exchange rate 1.05 USD/€)	High (4)	0.161 $kgCO_{2e}$/€	EXIOBASE 3	30.67 tCO_{2e}	Low (4)
		Fertilizer (1,200 kg/ha/a; 15 ha)	Low (2)	0.39 $kgCO_{2e}/kg$ (NPK (15:15:15))	(Wood & Cowie, 2004)	7.02 tCO_{2e}	Low (4)
		Water (29,580 m^3)	Low (2)	0.1531 $kgCO_{2e}/m^3$	BEIS	4.53 tCO_{2e}	Very high (1)
	Category 1.1—dining/canteen	Meals (all kinds) (1,000 meals * 300 days) (1.5 USD/meal)	High (4)	0.132 $kgCO_{2e}$/USD	EPA 2024/Climatiq	59.40 tCO_{2e}	Low (4)
	Category 2—capital goods	–	–	–	–	–	–

(continued)

Table 1 (continued)

Scope	Category	Source of emission	Relevance	Emission factor	Source of the emission factor	Amount	Accuracy
	Category 3—fuel and energy-related emissions		–	–	-	–	–
	Category 4—transport and distribution (upstream)		–	–	-	–	–
	Category 5—disposal and recycling of (production) waste	Solid waste (2 t)	Low (2)	580 $kgCO_{2e}$/short ton (mixed MSW—landfill)	EPA 2024/ Climatiq	1.16 tCO_{2e}	Medium (3)
		Organic/food waste (15 t)	Low (2)	110 $kgCO_{2e}$/short ton (food waste composted)	EPA 2024/ Climatiq	1.82 tCO_{2e}	Medium (3)
		Organic/food waste (15 t)	Medium (3)	680 $kgCO_{2e}$/short ton (food waste landfill)	EPA 2024/ Climatiq	11.24 tCO_{2e}	Medium (3)
		Paper (office) (2 t)	Very low (1)	50 $kgCO_{2e}$/short ton (mixed paper (primarily from offices) combusted)	EPA 2024/ Climatiq	0.1 tCO_{2e}	Medium (3)
		Organic garden waste (combusting) 250 kg/ 230 days	Low (2)	21.28 $kgCO_{2e}$/t (combusting)	BEIS	1.22 tCO_{2e}	Low (4)
	Category 6—business trips	Domestic flights (70) (total 121,627 km)	High (4)	0.2726 $kgCO_{2e}$/km	BEIS/Defra	33.16 tCO_{2e}	Medium (3)
		International flights (122) (total 1,112,857 km)	Very high (5)	0.2001 $kgCO_{2e}$/km	BEIS/Defra	222.68 tCO_{2e}	High (2)
		Hotel stays (886)	High (4)	Diverse	BEIS/ Greenview/ Climatiq	31.62 tCO_{2e}	High (2)

(continued)

Table 1 (continued)

Scope	Category	Source of emission	Relevance	Emission factor	Source of the emission factor	Amount	Accuracy
	Category 7—commuting of employees	Motorbike (small) (228*184 days*16 km; total 671,232 km)	High (4)	0.0832 $kgCO_{2e}$/km	BEIS	55.85 tCO_{2e}	Low (4)
		Car (diesel, 150) (total 441,600 km)	High (4)	0.1838 $kgCO_{2e}$/km	GHG protocol	81.17 tCO_{2e}	Low (4)
		Car (petrol, 150) (total 441,600 km)	High (4)	0,1631 $kgCO_{2e}$/km	GHG protocol	72.03 tCO_{2e}	Low (4)
	Category 8—rented or leased property, plant, and equipment	Rented cars (diesel) (90,000 km)	Medium (3)	0.1838 $kgCO_{2e}$/km	GHG protocol	16.54 tCO_{2e}	Low (4)
	Category 9—transport and distribution	Students (motorbike) (1,000*184 days*10 km)	Very high (5)	0.0832 $kgCO_{2e}$/km	BEIS	153.09 tCO_{2e}	Low (4)
	Category 10—processing of products sold by the university	Production (1) Cultivation of Cordyceps (2,400 kg fresh fruit; 400 kg dried fruit)	Very low (1)	–	–	Included in other categories	Low (4)
		(1) Packaging (20,000 packages a 50 g)	Very low (1)	1,283 kgCO2e/t	BEIS	1.28 tCO_{2e}	Medium (3)
		(1) Transport (20 g + 50 g*20,000*1,000 km)	Very low (1)	0.6226 $kgCO_{2e}$/t/km	BEIS	0.87 tCO_{2e}	Medium (3)
	Category 11—use of the products sold	–	–	–	–	–	–

(continued)

Table 1 (continued)

Scope	Category	Source of emission	Relevance	Emission factor	Source of the emission factor	Amount	Accuracy
	Category 12—handling of sold products at the end of the life-cycle	(1) Waste (50 g*20,000)	Very low (1)	6.411 $kgCO_{2e}/t$	BEIS	0.01 tCO_{2e}	Medium (3)
	Category 13—rented or leased property, plant, and equipment (downstream)	–	–	–	–	–	–
	Category 14—franchise[1]	Satellite sites	(1) Aquaculture farm	–	–	–	–
			(2) Plantation forest	–	–	–	–
	Category 15—investments	–	–	–	–	–	–

[1] As potential franchise or satellite sites the aqua farm and plantation forest can be defined differently (either inside or outside the system boundaries).

consensus among involved stakeholders. Consequently, the evaluation of the actors involved with regard to the relevance of the emission sources, the data quality, and the feasibility by the university may be subject to change over time. It is therefore logical for HEIs to utilize a double materiality analysis in the implementation process, as stipulated by the CSRD, among other directives, to ensure the reliability of the information and clarity among the stakeholders.

4 Results

Considering the current size and operational emissions of TUAF, the university emits a total of 1,406.4 tCO_{2e} annually. Using the tCO_{2e} per student indicator (Valls-Val & Bovea, 2021), TUAF records emissions of 0.28 tCO_{2e} per student and 3.1 tCO_{2e} per staff member (0.26 tCO_{2e} per capita). Alternative metrics, such as CO_{2e} emissions per monetary unit or per square meter (Helmers et al., 2021), could not be utilized due to imprecise data. As shown in Fig. 1, TUAF exhibits a particularly low CCF in comparison to other international universities (Helmers et al., 2021; Valls-Val & Bovea, 2021; Wilk et al., 2020), notably given that the presented footprint comprehensively includes all scopes and emission types. In addition, the figure reveals relatively low emissions from European and Latin American universities and considerably higher emissions from universities in the United States, Australia, and specific countries such as Saudi Arabia and Singapore.

As is typical for most universities, Scope 1 emissions at TUAF are of minor significance. The university maintains a limited number of vehicles with low mileage, and laboratory gases, though present, are used minimally for research purposes only.

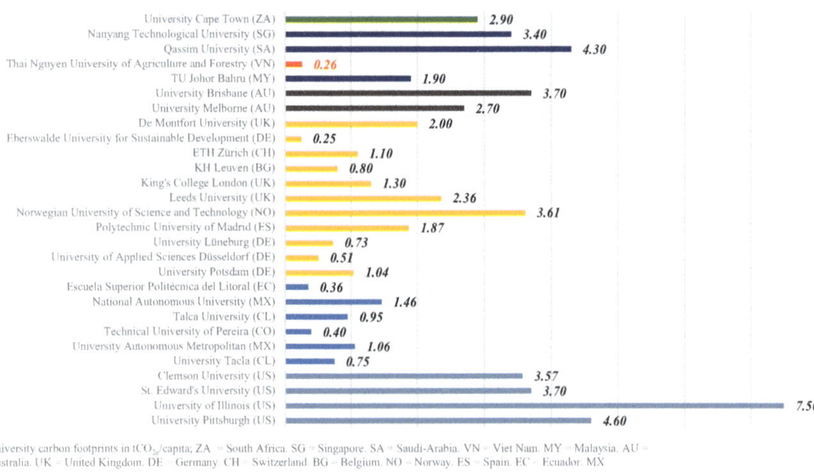

Fig. 1 Emissions of selected universities in tCO_{2e} per Capita (partly derived from: Helmers et al., 2021; Valls-Val & Bovea, 2021; Wilk et al., 2020)

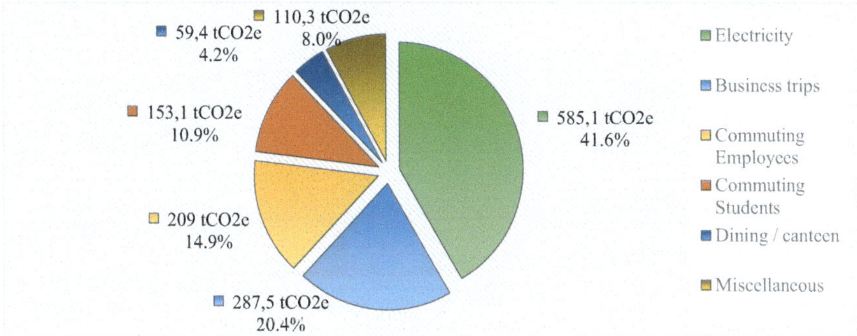

Fig. 2 Main sources of emissions at TUAF in 2024

Refrigerant leakage, estimated at 10% per year, contributes approximately 4.5 tCO_{2e}, posing a considerable environmental risk due to its high global warming potential (GWP). However, in comparison to other Scope 1 emissions, as all other energy consumption is sourced from externally supplied electricity.

Scope 2 emissions, which stem exclusively from purchased electricity, play a substantial role in TUAF's carbon footprint. Unlike many universities, TUAF does not require heating infrastructure due to its warm climate. Scope 2 emissions for electricity (585 tCO_{2e}) account for about 42% of total emissions in 2024 (Fig. 2).

Scope 3 emissions present a more complex challenge, contributing approximately 56% of TUAF's total emissions. A spend-based approach was used to estimate emissions from purchased goods and services, amounting to 42.2 tCO_{2e}. This includes fertilizer (7 tCO_{2e}) and water supply (4.5 tCO_{2e}). Dining and catering activities, primarily from the university canteen, contribute 59.4 tCO_{2e}. Waste disposal and the recycling of (production) waste generate 15.5 tCO_{2e}. A significant contributor to Scope 3 emissions is business travel and international flights, which—essential for academic collaboration and exchange—result in 287.5 tCO_{2e}. Similarly, staff and student commuting represents a major source of emissions, totaling 362.1 tCO_{2e}, based on stakeholder estimates. Additionally, TUAF's role in knowledge transfer and product development—such as greenhouse farming and agricultural production—contributes further emissions. For example, the production and processing of Cordyceps fungi results in 2.2 tCO_{2e}. Franchise facilities, including aquaculture farms and forest plantations, were assessed but excluded from the CCF, as they fall outside TUAF's defined system boundaries (Fig. 3).

As illustrated in Table 1, several data points exhibit low accuracy and requiring further refinement. Consequently, a materiality matrix (Table 2) was developed to identify high-priority emission sources that necessitate improved data collection and measurement processes. In particular, emissions from purchased goods and services, commuting, waste management, and business travel demand greater precision to enable effective mitigation strategies. Scope 1 emissions—specifically from refrigerants and fossil fuel usage—are also relevant but are considered of lower materiality compared to Scope 3 emissions.

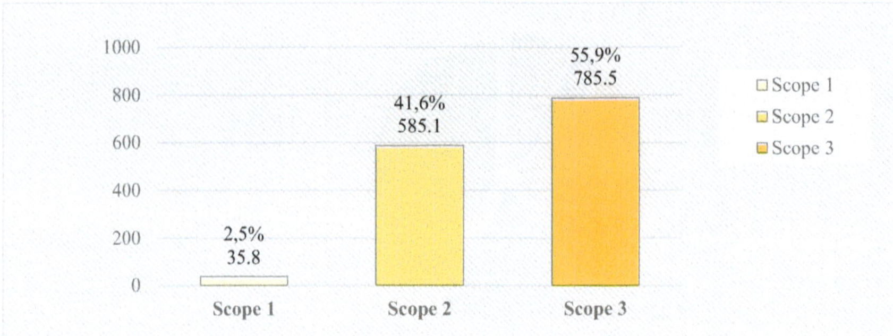

Fig. 3 CO_{2e} emissions by scope at TUAF in 2024 (in tCO_{2e})

Table 2 Materiality matrix: relevance and data accuracy of emission sources

Scope	Category	(A) Relevance from 1 (very low) to 5 (very high)	(B) Accuracy from 1 (very high) to 5 (very low)	Sum (A*B)
Scope 1	Laboratory gases	Low (2)	Very low (5)	10
	Refrigerant (fridges)	Low (2)	Medium (3)	6
	Fossil fuel	Medium (3)	Low (4)	12
Scope 2	Electricity	Very high (5)	Very high (1)	5
Scope 3	(1) Purchased goods and services	High (4)	Low (4)	16
	(1.1) Dining/canteen	High (4)	Medium (3)	12
	(5) Disposal and recycling of (production) waste	Medium (3)	Medium (3)	9
	(6) Business trips	Very high (5)	Medium (3)	15
	(7) Commuting of employees	Very high (5)	Low (4)	20
	(8) Rented or leased property, plant, and equipment	Medium (3)	Low (4)	12
	(9) Transport and distribution	Very high (5)	Low (4)	20
	(10) Processing of products sold by the university	Low (2)	Low (4)	8
	(12) Handling of sold products at the end of the life-cycle	Very low (1)	Medium (3)	3

Given that universities operate within legal and institutional frameworks, they often have limited autonomy in addressing certain sustainability challenges. Table 3 contextualizes the relevance of each emission source in relation to TUAF's capacity to implement mitigation measures. Purchased electricity emerges as a critical factor, alongside emissions from purchased goods and services, business travel, and student commuting. Waste management policies and procurement strategies for rentals and leases also present opportunities in terms of CO_{2e} saving potential. While emissions from dining and staff commuting are substantial, the university has limited influence over individual mobility and dietary choices.

Considering future trends and stakeholder predictions, emissions from commuting, electricity consumption, and the procurements of goods and services are expected to increase significantly. Consequently, Table 4 outlines the key areas requiring urgent intervention at TUAF. A slight increase is also anticipated in emissions from refrigerants, fossil fuel consumption (both Scope 1), as well as from dining

Table 3 Materiality matrix: relevance and feasibility of emission reduction measures

Scope	Category	(A) Relevance from 1 (very low) to 5 (very high)	(C) Feasibility from 1 (very low) to 5 (very high)	Materiality (A*C)
Scope 1	Laboratory gases	Low (2)	Medium (3)	6
	Refrigerant (fridges)	Low (2)	Medium (3)	6
	Fossil fuel	Medium (3)	High (4)	12
Scope 2	Electricity	Very high (5)	High (4)	20
Scope 3	(1) Purchased goods and services	High (4)	High (4)	16
	(1.1) Dining/canteen	High (4)	Low (2)	8
	(5) Disposal and recycling of (production) waste	Medium (3)	Medium (3)	9
	(6) Business trips	Very high (5)	Medium (3)	15
	(7) Commuting of employees	Very high (5)	Low (2)	10
	(8) Rented or leased property, plant, and equipment	Medium (3)	High (4)	12
	(9) Transport and distribution	Very high (5)	Medium (3)	15
	(10) Processing of products sold by the university	Low (2)	Medium (3)	6
	(12) Handling of sold products at the end of the life-cycle	Very low (1)	Low (2)	2

services, waste disposal, and business travel (all Scope 3). No emission sources are expected to decline without targeted action.

Overall, the results indicate that TUAF currently maintains a comparatively low carbon footprint, though emissions are projected to rise across various categories. At the same time, the findings underscore the need for TUAF to refine its emission data, prioritize high-impact mitigation strategies, and align its sustainability goals with Vietnam's broader climate objectives. Future research at TUAF should also focus on enhancing data accuracy, strengthening institutional policies, and ultimately implementing systematic carbon reduction measures.

Table 4 Materiality and predicted trends of emission sources

Scope	Category	Materiality (A*C)	Predicted trend from 1 (strongly declining) to 5 (strongly increasing)	Need for action for TUAF
Scope 1	Laboratory gases	6	Constant (3)	18
	Refrigerant (fridges)	6	Increasing (4)	24
	Fossil fuel	12	Increasing (5)	60
Scope 2	Electricity	20	Strongly increasing (5)	100
Scope 3	(1) Purchased goods and services	16	Strongly increasing (5)	80
	(1.1) Dining/canteen	8	Increasing (4)	32
	(5) Disposal and recycling of (production) waste	9	Increasing (4)	36
	(6) Business trips	15	Increasing (4)	60
	(7) Commuting of employees	10	Strongly increasing (5)	50
	(8) Rented or leased property, plant, and equipment	12	Constant (3)	36
	(9) Transport and distribution	15	Strongly increasing (5)	75
	(10) Processing of products sold by the university	6	Increasing (4)	24
	(12) Handling of sold products at the end of the life-cycle	2	Constant (3)	6

5 Discussion

The analysis of TUAF's carbon footprint reveals a complex framework characterized by both challenges and opportunities. Key considerations include the materiality and completeness of emission types analyzed, the accuracy of provided data, and the evolving nature of emission sources. Importantly, TUAF is not only a source of emissions but may also serve as a potential carbon sink due to its extensive land holdings—an aspect not yet incorporated into the current footprint assessment due to the current state of land use planning.

5.1 *Results of the CCF*

The findings reveal that TUAF's activities contribute to significant CO_{2e} emissions, although considerably below international average, as illustrated by Fig. 1 (Helmers et al., 2021; Wilk et al., 2020). The largest share of emissions results from purchased electricity (approximately 42%), primarily due to the absence of on-site renewable energy generation. Without the implementation of photovoltaic systems or alternative renewable energy sources, these emissions are likely to increase. Business travel (20%), staff commuting (15%), and student commuting (11%) also represent substantial contributors. However, viable mitigation measures for travel-related emissions often involve considerable financial and logistical challenges.

Improving the accuracy of emissions calculations—particularly for the procurement of goods and services (Scope 3) and fossil fuel consumption (Scope 1)—is essential for more effective emissions tracking and the development of targeted reduction strategies. Prompt action is required to address these rising emissions and avoid future emission peaks. Emissions from laboratory gases (Scope 1) and the processing of products sold by the university are currently minimal. However, defining system boundaries that account for potential future emission sources remains a critical task. Given TUAF's extensive land holdings, which could serve as both emissions sources and carbon sinks, integrated and forward-looking land-use planning is necessary.

5.2 *Current Challenges at TUAF*

Initial CCF assessments frequently lack precision, mainly due to incomplete datasets and unclear frameworks. These shortcomings must be systematically addressed over time through annual assessments. A key challenge lies in the absence of a universal international standard for universities, unlike frameworks such as the Corporate Social Responsibility Directive (CSRD) (Directive (EU) 2022/2464, 2022) and the German Climate Protection Act (Federal Climate Action Act, 2019/15 July 2024). While the GHG Protocol (WRI) is widely recognized as a standard, it is not legally binding and is often individually tailored to meet the specific requirements of each institution. The lack of an appropriate framework poses the risk to undermine the achievement of reliable and comprehensive CCFs, which are crucial for long-term emission reduction.

A major issue for HEIs in the Global South is the selection of feasible and accurate emission factors. Many databases provide exclusive, fee-based access to emission factors, which most universities cannot afford at the initial stages of assessment. Furthermore, as shown in Table 1, most emission factors are derived from Global North countries, with limited data available for Asia, despite significant regional disparities.

Beyond teaching and research, knowledge transfer has become a third mission for HEIs, adding further complexity (Compagnucci & Spigarelli, 2020; Davey, 2017). Even before the third mission itself began, TUAF's extensive third-party funding activities had led to complex organizational structures, including semi-autonomous facilities and independent research centers. Establishing appropriate system boundaries for a comprehensive CCF is crucial, particularly given TUAF's franchise-like structures. Some affiliated entities, such as plantation forestry and aquaculture companies, offer benefits but also complicate carbon accounting. Based on stakeholder feedback, these entities were excluded from TUAF's system boundaries. However, data availability and quality remain problematic also due to insufficiently trained stakeholders. This highlights the need for clear guidelines within Scope 3 accounting to evaluate emissions from affiliated enterprises.

Another challenge is, among other issues, the lack of data on maintenance and construction processes at the university. This includes not only missing maintenance logs and incomplete records of construction materials but also untracked refrigerant use in the air conditioning system, which hinder accurate emissions accounting.

5.3 Carbon Offsetting

Carbon offsetting provides an option for institutions to achieve net-zero status despite ongoing operational emissions (Helmers et al., 2021). Strategies include aquifer thermal energy storage (Opel et al., 2017), purchasing carbon credits (University of Maryland, 2024), and reforestation or alternative land use practices (Udas et al., 2018). As illustrated in Table 5, TUAF has several options, including 57.5 ha of land for potential plantation forestry or sustainable agriculture. However, land-use choices are critical, as some practices, such as rice cultivation, may result in considerable methane emissions (Qian et al., 2023). Other viable options include forest plantations (Chen et al., 2024), fruit tree cultivation (Gelaye & Getahun, 2024; Wu et al., 2012), and pond management (Holgerson et al., 2023). Utilizing approximately 40 ha for forest plantations could sequester about 55.6 tCO_{2e} annually, but this would represent only around 4% of TUAF's total annual emissions. Other climate protection measures are therefore currently more important for TUAF, as carbon offsetting on its own premises can only be a niche solution for achieving carbon neutrality.

Table 5 Opportunities of land use types for TUAF and the impact on carbon emissions and sequestration

Type of land use	Relevance	Emission factor	Source of emission factor	Estimation	Accuracy
Farmland (crops, rice)	High-emission source (on max. 15 ha)	7,870 $kgCO_{2e}$/ha/a	(Qian et al., 2023)	Emission of about 118.1 tCO_{2e}/a	High
Forest plantation	High-CO_2 sink (on max. 40 ha)	1,390 $kgCO_{2e}$/ha/a	(Chen et al., 2024)	Sequestration of about 55.6 tCO_{2e}/a	Medium
Fruit trees	Medium-CO_2 sink (on max. 15 ha)	51,380 $kgCO_{2e}$/ha (26.7 tCO_2/ha to 206.6 tCO_2/ha)	(Gelaye & Getahun, 2024; Wu et al., 2012)	Sequestration of about 770.7 tCO_{2e} or 25.7 tCO_{2e}/a (30a)	Low
Pond	Low-CO_2 sink (about 0.5 ha)	79–247 gC/m^2/a = 5.21 tCO_2/ha/a (67 gC/m^2/a)	(Holgerson et al., 2023)	Sequestration of about 2.6 tCO_{2e}	Low

However, comparative studies show that this form of compensation has great potential (Udas et al., 2018). While TUAF itself cannot offset all its emissions, universities with larger land holdings—particularly those recently relocated to suburban areas in Vietnam—may have greater offsetting potentials. Accordingly, other universities could significantly benefit from case studies and shared methodologies on GHG emissions and sequestration potential carried out on TUAF's land holdings.

5.4 *Perspectives and Solutions*

To ensure comparability of CCFs, establishing a standardized framework for the more than 200 Vietnamese HEIs is essential. A national monitoring system would facilitate practical implementation and improve data collection. Initiatives such as the "Climate-neutral state administration of North Rhine-Westphalia" ("Klimaneutrale Landesverwaltung Nordrhein-Westfalen", Gesetz zur Neufassung des Klimaschutzgesetzes Nordrhein-Westfalen/8. Juli 2021) provide useful templates for this purpose. Additionally, university must form alliances and train staff in CCF accounting and emissions documentation, such as maintenance processes. Government support will be necessary to achieve net-zero emissions by 2050, considering variations between HEIs and the required staff capacity.

At the governmental level, incentive mechanisms can promote cost-effective climate protection measures. The implementation of the CSRD in Europe demonstrates that legal requirements often drive participation beyond flagship institutions. However, as mentioned, universities possess a vital role on education and knowledge transfer to other stakeholders. Accordingly, to foster systematic enhancement both within the higher education sector and across society, climate education and sustainability must be integrated into most curricula (Molthan-Hill et al., 2019). Universities are uniquely positioned to educate future decision-makers, entrepreneurs, and corporate leaders on climate protection, sustainability, and education for sustainable development. To capitalize on this position, universities must also implement climate change policies and sustainability initiatives within their operations.

6 Conclusion

The Vietnamese higher education sector is in the early stages of monitoring CO_{2e} emissions and striving for carbon neutrality. Various processes remain ambiguous due to a lack of information, limited staff capacity, and insufficient awareness about existing frameworks, tools, and networks. Establishing a comprehensive CCF for a university such as TUAF is particularly complex due to fragmented institutional structures, missing data, and the absence of appropriate emission factors. However, as key actors in research and education, universities are well-positioned to drive sustainability and carbon neutrality initiatives.

TUAF exemplifies an ambitious university in the Global South, characterized by institutional expansion and modernization while also facing the triple challenges of declining student enrollment, limited financial resources, and increasing emissions. The findings indicate that emissions across all scopes are currently increasing rather than declining. Of particular concern are high emissions from electricity consumption (Scope 2), procurement of goods and services, and travel (both Scope 3). Achieving carbon neutrality will require prioritizing these emissions through measures such as photovoltaic installations, e-mobility expansion, and awareness campaigns. However, reducing the carbon footprint remains a complex process requiring substantial financial and human resources.

In the medium term, universities in Vietnam are likely to compete for recognition as the most sustainable institutions. Sustainability credentials will not only enhance their attractiveness to students and staff but also support the acquisition of third-party funding. However, without a standardized accounting framework, benchmarking between universities will remain unclear. System boundaries and emission factor selection significantly impact reported emissions, highlighting the need for improved data quality and verification in future assessments. Despite current limitations, TUAF's CCF provides an initial baseline, enabling year-over-year comparison and serving as a foundation for critical juxtapositions. Additionally, the findings offer valuable insights for other universities in Vietnam and internationally.

Universities play a pivotal role in the global conversation about sustainability and carbon neutrality, with contributions to research and education spanning decades. At TUAF, these efforts extent to sustainable agricultural, forest management, and support for ethnic minorities. However, integrating sustainability into university operations remain a work in progress. As Vietnamese universities navigate the transition between institutional growth and operational efficiency improvements, developing a long-term carbon neutrality strategy—supported by high quality CCF data—will be essential.

References

Chen, Z., Dayananda, B., Du, H., Zhou, G., & Wang, G. (2024). Assessing carbon sequestration potential in state-owned plantation forests in China and exploring feasibility for carbon offset projects. *Forests, 15*(5), 886. https://doi.org/10.3390/f15050886

Compagnucci, L., & Spigarelli, F. (2020). The Third Mission of the university: A systematic literature review on potentials and constraints. *Technological Forecasting and Social Change, 161*, Article 120284. https://doi.org/10.1016/j.techfore.2020.120284

Davey, T. (2017). Converting university knowledge into value: How conceptual frameworks contribute to the understanding of the third mission role of European universities. *International Journal of Technology Transfer and Commercialisation, 15*(1), Article 84637, 65. https://doi.org/10.1504/IJTTC.2017.084637

Decision No. 896/QD-TTg on approving the National Strategy for Climate Change until 2050, 2022.

Directive (EU) 2022/2464. (2022). *Official Journal of the European Union.*

EAUC—The Alliance for Sustainability Leadership in Education. (2025). *Race to zero: Current signatories.* EAUC—The Alliance for Sustainability Leadership in Education. https://www.educationracetozero.org/current-signatories

Energy Institute. (2024). *Statistical Review of World Energy—With major processing by Our World in Data: Share of primary energy consumption that comes from fossil fuels—Using the substitution method.*

Federal Climate Action Act, Federal Law Gazette. (2019).

Gelaye, Y., & Getahun, S. (2024). A review of the carbon sequestration potential of fruit trees and their implications for climate change mitigation: The case of Ethiopia. *Cogent Food & Agriculture, 10*(1), Article 2294544. https://doi.org/10.1080/23311932.2023.2294544

General Statistics Office of Vietnam. (2024). *GDP contribution of the agriculture, forestry and fishing sector in Vietnam from 2011 to 2023.* https://www.statista.com/statistics/1027971/vietnam-gdp-contribution-of-agriculture-forestry-and-fishing-sector/

Gesetz zur Neufassung des Klimaschutzgesetzes Nordrhein-Westfalen.

Global Carbon Budget. (2024). *Annual CO_2 emissions—GCB* [- with major processing by Our World in Data.].

Helmers, E., Chang, C. C., & Dauwels, J. (2021). Carbon footprinting of universities worldwide: Part I—Objective comparison by standardized metrics. *Environmental Sciences Europe, 33*(1). https://doi.org/10.1186/s12302-021-00454-6

Holgerson, M., Ray, N., & Russ, C. (2023). *Data from: Carbon burial in Cornell's experimental ponds.* https://doi.org/10.5061/dryad.wstqjq2rc

International Association of Universities. (2025). *IAU WHED: The world of higher education at your fingertips.* International Association of Universities. https://whed.net/home.php

International Organization for Standardization (2024). *ISO 14064-1:2018, Greenhouse gases Part 1: Specification with guidance at the organization level for quantification and reporting of greenhouse gas emissions and removals* (2nd ed.). https://www.iso.org/standard/66453.html

Kiehle, J., Kopsakangas-Savolainen, M., Hilli, M., & Pongrácz, E. (2023). Carbon footprint at institutions of higher education: The case of the University of Oulu. *Journal of Environmental Management, 329*, Article 117056. https://doi.org/10.1016/j.jenvman.2022.117056

Ministry of Education and Training. (2024). *Number of university students in Vietnam from 2013 to 2021.* https://www.statista.com/statistics/815091/number-of-university-students-in-vietnam/

Molthan-Hill, P., Worsfold, N., Nagy, G. J., Leal Filho, W., & Mifsud, M. (2019). Climate change education for universities: A conceptual framework from an international study. *Journal of Cleaner Production, 226*, 1092–1101. https://doi.org/10.1016/j.jclepro.2019.04.053

Opel, O., Strodel, N., Werner, K. F., Geffken, J., Tribel, A., & Ruck, W. (2017). Climate-neutral and sustainable campus Leuphana University of Lueneburg. *Energy, 141*, 2628–2639. https://doi.org/10.1016/j.energy.2017.08.039

Paredes-Canencio, K. N., Lasso, A., Castrillon, R., et al. (2024). Carbon footprint of higher education institutions. *Environment, Development and Sustainability, 26*, 30239–30272. https://doi.org/10.1007/s10668-024-04596-4

Qian, H., Zhu, X., Huang, S., Linquist, B., Kuzyakov, Y., Wassmann, R., Minamikawa, K., Martinez-Eixarch, M., Yan, X., Zhou, F., Sander, B. O., Zhang, W., Shang, Z., Zou, J., Zheng, X., Li, G., Liu, Z., Wang, S., Ding, Y., et al. (2023). Greenhouse gas emissions and mitigation in rice agriculture. *Nature Reviews Earth & Environment, 4*(10), 716–732. https://doi.org/10.1038/s43017-023-00482-1

Robinson, O., Kemp, S., & Williams, I. (2015). Carbon management at universities: A reality check. *Journal of Cleaner Production, 106*, 109–118. https://doi.org/10.1016/j.jclepro.2014.06.095

Robinson, O. J., Tewkesbury, A., Kemp, S., & Williams, I. D. (2018). Towards a universal carbon footprint standard: A case study of carbon management at universities. *Journal of Cleaner Production, 172*, 4435–4455. https://doi.org/10.1016/j.jclepro.2017.02.147

Secretariat of the International Universities Climate Alliance. (2025). *Members.* International Universities Climate Alliance. https://universitiesforclimate.org/member

The World Bank Group and the Asian Development Bank. (2021). *Climate risk country profile Vietnam.*

THE. (2024). *Times higher education—Top universities pursuing sustainable development goals: University impact rankings.* Times Higher Education. https://www.timeshighereducation.com/impactrankings

Trinh, T., Nguyen, H., Pham, O., & Pham, H.-H. (2025). The financial sustainability of Vietnamese higher education institutions: Unveiling the hidden factors. *Asian Education and Development Studies, 14.* https://doi.org/10.1108/AEDS-07-2024-0136

Udas, E., Wölk, M., & Wilmking, M. (2018). The "carbon-neutral university"—A study from Germany. *International Journal of Sustainability in Higher Education, 19*(1), 130–145. https://doi.org/10.1108/IJSHE-05-2016-0089

United Nations. (2015). *Paris agreement.*

University of Maryland. (2024). *Climate Action Plan 2.0: University-wide carbon neutrality by 2025.* https://sustainingprogress.umd.edu/progress-commitments/climate-action-plan

Valls-Val, K., & Bovea, M. D. (2021). Carbon footprint in Higher Education Institutions: A literature review and prospects for future research. *Clean Technologies and Environmental Policy, 23*(9), 2523–2542. https://doi.org/10.1007/s10098-021-02180-2

Valls-Val, K., & Bovea, M. D. (2022). Carbon footprint assessment tool for universities: CO2UNV. *Sustainable Production and Consumption, 29,* 791–804. https://doi.org/10.1016/j.spc.2021.11.020

Wilk, A., Pantiou, M., Göttlicher, S., & Küster, S. (2020). *Kurzstudie zu Klimaschutzkonzepten an Hochschulen.*

Wood, S., & Cowie, A. (June 2004). *A review of greenhouse gas emission factors for fertiliser production* (IEA Bioenergy Task No. 38). Research and Development Division, State Forests of New South Wales; Cooperative Research Centre for Greenhouse Accounting.

World Resource Institute, W. *The greenhouse gas protocol: A corporate accounting and reporting standard.* Revised Edition.

World Resource Institute, W. (2013). *Technical guidance for calculating scope 3 emissions (version 1.0): Supplement to the corporate value chain (scope 3) accounting & reporting standard.* World Resources Institute.

Wu, T., Wang, Y., Yu, C., Chiarawipa, R., Zhang, X., Han, Z., & Wu, L. (2012). Carbon sequestration by fruit trees—Chinese apple orchards as an example. *PLoS One, 7*(6), Article e38883. https://doi.org/10.1371/journal.pone.0038883

Sustainable Synergy: Virtual Exchanges for Technical Students

Oksana Polyakova and Ruzana Galstyan-Sargsyan

Abstract This study explores how sustainability can be embedded into Virtual Exchange (VE) programmes in higher education, focusing on technical university students' development of transversal competences, intercultural awareness, and alignment with the United Nations Sustainable Development Goals (SDGs). Framed within the Sustainable Active Virtual Learning (SAVL) model, the research draws on educational theories related to active learning, intercultural communication, and online collaborative pedagogy. SAVL integrates four interconnected pillars—sustainability principles (S), active learning strategies (A), virtual collaboration tools (V), and language and competence development (L)—within an inclusive and interdisciplinary instructional design. The empirical research was conducted at the Universitat Politècnica de València (Spain), involving two distinct VE case studies implemented as part of curricular innovation. Case Study 1 brought together engineering students from Spain and English studies students from the University of Zadar (Croatia) to collaboratively examine and compare educational systems through project-based tasks. Case Study 2 facilitated a virtual laboratory experience between engineering students from Spain and Ukraine (Dnipro University of Technology), where synchronous interaction supported remote observation and discussion of scientific experiments. Both cases explicitly embedded SDG 4 (Quality Education) and SDG 17 (Partnerships for the Goals), while the second also addressed SDG 16 (Peace, Justice and Strong Institutions). The research design followed a four-phase methodology: (1) alignment of SDGs and institutional coordination; (2) joint planning and material development; (3) implementation through virtual sessions using Microsoft Teams, Canvas, and PowerPoint; and (4) evaluation using pre- and post-intervention Likert-scale and open-response questionnaires, supplemented with sentiment analysis (Lingmotif tool) and qualitative reflections. The results indicate significant positive shifts in students' self-perceived communication skills, digital collaboration, sustainability awareness, and intercultural sensitivity. In Case Study 1, participants

O. Polyakova (✉) · R. Galstyan-Sargsyan
Applied Linguistics Department, Universitat Politècnica de València, Valencia, Spain
e-mail: okpolnes@upv.es

R. Galstyan-Sargsyan
e-mail: rugalsar@upvnet.upv.es

W. Leal Filho et al. (eds.), *Introducing the UN SDGs in Higher Education Institutions*, World Sustainability Series, https://doi.org/10.1007/978-3-032-23017-1_35

showed improved understanding of SDGs and reported enjoyment and motivation tied to international teamwork. In Case Study 2, the technical students emphasized the usefulness of the VE for language acquisition, technical terminology, and collaborative problem-solving across borders. Sentiment analysis showed a consistently positive tone before and after the exchanges, reinforcing the learners' appreciation of the experience. Keywords such as "motivation," "sustainability," "collaboration," and "interactive" emerged repeatedly in qualitative feedback and word cloud visualisations. The study concludes that embedding sustainability into VE for technical education fosters not only academic and professional growth but also global citizenship and intercultural understanding. It calls for the adoption of clear methodological frameworks, critical self-reflection, and broader contextual explanations to ensure the transferability of results across educational regions. These insights can support educators and institutions in designing robust, inclusive, and future-ready VE programmes aligned with global sustainability objectives.

Keywords Sustainable development · Sustainable development goals · Sustainable education · Technical students · Virtual exchange

1 Introduction

Over the past decades, higher education has undergone profound transformations, driven by globalisation, digitalisation, and the increasing emphasis on sustainability. Since the inception of the Bologna Process in 1998, the European Higher Education Area (EHEA) has advanced toward a more harmonised, inclusive, and quality-driven academic system. Core initiatives—such as the adoption of the three-cycle degree structure, improved mobility, and qualification recognition—have significantly reshaped institutional practices (Grek & Russell, 2024). In line with these objectives, the European Commission's, 2024 education agenda promotes sustainability, inclusivity, and innovation in academic structures, underscoring the need to integrate global values into pedagogical design (European Commission, 2024).

Within this evolving landscape, sustainability has emerged as a foundational principle for reimagining educational content, teaching practices, and student engagement (Leal Filho et al., 2018). The integration of the United Nations' Agenda 2030 (UN General Assembly, 2015) and its 17 Sustainable Development Goals (SDGs) into university curricula signals a shift toward Higher Education for Sustainable Development, which seeks to transform learning environments and foster global citizenship (Tilbury, 2011; Žalėnienė & Pereira, 2021). Universities are no longer merely centres of knowledge transfer but agents of societal change, entrusted with equipping students to address complex global challenges such as climate change, inequality, and peacebuilding. Moreover, a study relating sustainable development indicators and higher education institutions proved the existence of 61 indicators (Griebeler et al., 2022).

In this context, Virtual Exchange (VE) has gained prominence as an innovative pedagogical approach that promotes intercultural communication, collaborative learning, and digital literacy through structured online interaction between students from different cultural and academic backgrounds (Helm, 2020; O'Dowd, 2023). VE is particularly relevant for technical education, where cross-border cooperation, interdisciplinary problem-solving, and sustainability awareness are increasingly viewed as essential graduate attributes. Furthermore, VE offers a more accessible and environmentally responsible alternative to physical mobility programmes, aligning closely with institutional sustainability goals (Helm & Guth, 2020).

The present study contributes to this academic and pedagogical discourse by examining the integration of sustainability into virtual exchanges for technical university students. It introduces the Sustainable Active Virtual Learning (SAVL) model—an instructional framework combining sustainability values (S), active learning (A), digital collaboration (V), and language/competence development (L)—and evaluates its implementation in two international VE initiatives involving Spanish, Croatian, and Ukrainian higher education institutions. Grounded in the theoretical foundations of transformative pedagogy and competence-based learning, this research investigates how VE can serve as a platform for embedding SDGs into technical education and fostering transversal skills such as teamwork, intercultural awareness, and digital communication.

The study is guided by the following research questions:

- **RQ1**. How can we bring sustainability to the heart of virtual exchanges?
- **RQ2**. Will it make a difference in the competence-based preparation of future engineers?
- **RQ3**. Does common interest in Sustainable Development Goals encourage better intercultural communication?

By addressing these questions, this chapter aims to demonstrate the value of sustainability-oriented VE models for technical education and provide a structured, evidence-based perspective on how such programmes can be designed, implemented, and assessed across institutions and regions. The following sections present the theoretical background on VE, describe the SAVL model and its application in two case studies, and analyse the learning outcomes based on student feedback and perception data.

2 Virtual Exchange in the Context of Sustainability and Critical Pedagogy

From classical models of teacher-centred instruction to the current landscape of digitally mediated education, knowledge transmission has remained a defining pillar of academia. Yet the evolving modalities through which learning occurs—particularly in response to global challenges—demand not only innovation but critical scrutiny.

As Sandri (2022) suggests, educators' values are not merely embedded in pedagogy; they serve as a linchpin for fostering sustainable and equitable education. In this context, Virtual Exchange (VE) has emerged as more than a mechanism for intercultural dialogue—it is increasingly positioned as a transformative pedagogical strategy with global relevance.

Virtual Exchange (VE), defined by O'Dowd (2023, p. 11) as "sustained online collaborative learning and interaction with partners from different cultural backgrounds," is often promoted as a vehicle for internationalisation at home. However, this framing tends to obscure its embeddedness in broader structural conditions, including digital inequities, institutional asymmetries, and varying pedagogical traditions. As De Wit and Altbach (2020) note, VE offers an alternative to physical mobility, yet this potential is tempered by unequal access to infrastructure, faculty training, and institutional commitment.

While the literature celebrates VE's capacity to enhance digital literacy, intercultural awareness, and language proficiency (Helm, 2020), there is insufficient engagement with its limitations. For instance, claims about accessibility often rely on normative assumptions about technology use, disregarding global disparities in connectivity, device access, and digital competence. A critical lens is therefore necessary to interrogate whether VE disrupts or inadvertently perpetuates existing educational hierarchies.

Multiple VE models have been developed in response to pedagogical and contextual demands. Telecollaboration, initially grounded in language education, fosters structured intercultural interactions. However, its efficacy is contingent upon reflexive and culturally responsive pedagogy (Belz & Thorne, 2021). Without deliberate attention to power dynamics, these exchanges risk reinforcing essentialist views of culture or devolving into superficial engagements.

COIL (Collaborative Online International Learning) represents a curricular integration of VE, emphasising co-creation and interdisciplinary collaboration (Rubin & Guth, 2022). While this model enables deep learning, it requires institutional alignment in terms of calendars, workload, and assessment criteria—factors often overlooked in implementation. Moreover, its success is predicated on flexible curricula and cross-cultural pedagogical literacy, both of which vary widely across institutions.

Global Virtual Teams replicate professional scenarios and offer students opportunities to develop transversal skills, including project management and digital leadership (Häkkinen & Rienties, 2021). Yet these scenarios assume a baseline of digital fluency and self-regulation that may not be present in all student cohorts. Without intentional scaffolding, VE may exacerbate inequalities among participants, marginalising those with less experience or access.

Hybrid models combining online and physical mobility aim to bridge gaps in engagement and immersion (De Wit & Leask, 2021). However, such models require scrutiny regarding their ecological and financial costs. While they promise inclusivity, they may reproduce the exclusivity of traditional mobility pathways, particularly for students from underrepresented or economically disadvantaged backgrounds (De Wit & Beelen, 2021).

The sustainability claims of VE must also be critically assessed. Although reduced travel suggests environmental benefits, the environmental impact of digital infrastructure—servers, data centres, and high bandwidth usage—is rarely acknowledged. As Leal Filho (2022) argues, digitalisation is not inherently sustainable. It must be supported by governance structures and critical engagement that address the material and ethical dimensions of technological expansion.

In addition, VE's alignment with the United Nations Sustainable Development Goals (SDGs) is often taken for granted. A more robust integration of Education for Sustainable Development (ESD) would require not only thematic content on sustainability but also critical engagement with global challenges such as climate ethics, social justice, and postcolonial power asymmetries. Leal Filho (2022) reminds us that interdisciplinary and transdisciplinary engagement, grounded in knowledge co-production, is essential for meaningful sustainability education. VE, when designed with this ethos, can act as a participatory platform for negotiating diverse epistemologies and co-creating more just futures.

While Virtual Exchange holds significant promise as a strategy for advancing equity, sustainability, and internationalisation in higher education, its current conceptualisation often lacks critical depth. Much of the literature and practice surrounding VE is celebratory, focusing on positive student outcomes, but failing to interrogate the underlying structural, institutional, and geopolitical forces that shape its implementation. This uncritical optimism risks masking persistent inequities in digital access, institutional readiness, and curricular flexibility.

The tendency to equate digital modes of delivery with sustainability is particularly problematic. Drawing on Leal Filho (2022), it becomes evident that digitalisation—though a powerful enabler—carries environmental and ethical implications that must be scrutinised. The carbon footprint of digital infrastructures, issues of data privacy, and the unequal global distribution of digital resources are often overlooked in VE design and policy discourse.

Moreover, VE is frequently framed as a neutral or universally accessible pedagogical tool. This assumption disregards the power asymmetries embedded in intercultural communication and the epistemic hierarchies that shape knowledge production. Without embedding critical pedagogy and reflective practice into VE models, these exchanges risk becoming performative or tokenistic, reproducing rather than disrupting dominant worldviews.

To move VE from a technocratic tool to a transformative practice, a deeper commitment to contextual relevance, ethical intentionality, and interdisciplinary integration is required. This includes engaging students with urgent global issues—climate justice, digital inequality, postcolonial perspectives—not as abstract learning themes, but as contested terrains that require agency, empathy, and critique.

Ultimately, sustainability in VE must be reconceptualised not only as an environmental or operational concern, but as a pedagogical stance. This means fostering learning environments where complexity is embraced, diverse knowledges are valued, and education becomes a space for co-creating futures that are not only connected—but just and inclusive.

3 Methodological Framework. Applying the Sustainable Active Virtual Learning Model

The methodology adopted in this study was purposefully grounded in the Sustainable Active Virtual Learning (SAVL) model, which functions as both a pedagogical framework and a methodological structure. This model was selected to align the research design with our core objective: to embed sustainability principles into Virtual Exchange (VE) programmes for technical students, while fostering transversal competences and intercultural collaboration.

The SAVL framework is composed of four interrelated pillars—Sustainability (S), Active learning (A), Virtual collaboration (V), and Language and competence development (L)—which map directly onto current theories in transformative pedagogy (Tilbury, 2011), experiential learning, and digital education (Helm, 2020). By adopting this model, we ensured internal coherence between theoretical foundations, instructional practices, and research methods.

The choice of a mixed-methods design, including pre- and post-intervention questionnaires, sentiment analysis, and qualitative reflections, was justified by the need to capture both quantifiable changes in students' competence development and nuanced perceptions of sustainability and intercultural exchange. This triangulation of methods enhances the validity and transferability of findings, responding to recent calls in VE research for more evidence-based, critically grounded, and context-sensitive approaches (Helm & Acconcia, 2021).

Moreover, by implementing the SAVL model across two distinct case studies—involving interdisciplinary and international VE partnerships—we demonstrated the model's scalability and flexibility across different learning contexts, making it a suitable methodological vehicle for this research. The SAVL model, therefore, not only structures the intervention but serves as the overarching methodological framework guiding the study's rationale, execution, and evaluation.

Despite the promising trends of the virtual exchange options listed above, there is a clear need for effective communication around sustainable goals in technical training. Likewise, an open, simple and flexible methodological framework for engaging instructors and universities was needed. While critically rethinking technical education and language learning, we envisioned the following system.

The Sustainable Active Virtual Learning (SAVL) system represents an innovative educational framework that integrates sustainability principles, active learning methodologies, and virtual technologies to foster practical and environmentally conscious learning experiences (Polyakova & Galstyan-Sargsyan, 2022).

Specifically, this training system was designed to enhance inter-university collaboration cohesion; it comprises four key components: (S) sustainability, (A) active learning, (V) virtual exchange, and (L) foreign language and competence development. We aligned this general approach with the course content and professional communication, as Fig. 1 illustrates.

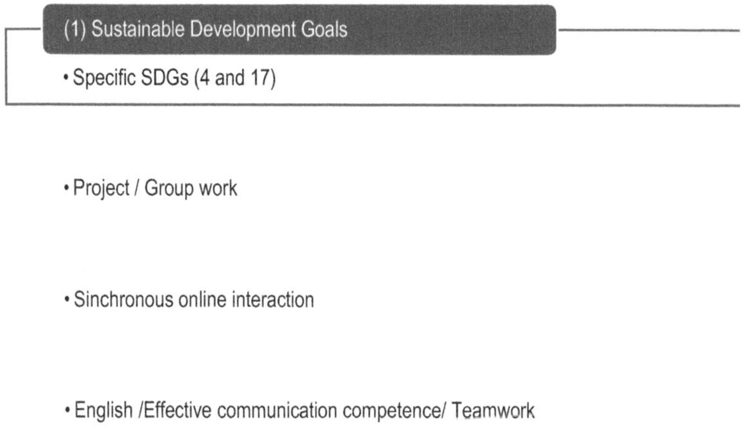

Fig. 1 SAVL (Sustainable Active Virtual Learning) model adaptation. *Source* Designed by authors

Furthermore, the general instructional scheme was created to bring sustainability and competences into line with the current educational context in the following or SAVL-oriented way:

(S) Sustainability: values, goals and philosophy

Sustainable Development Goals 4, "Quality education", and 17, "Global partnerships", are the core objectives for the present studies, although the additional SDGs will also be addressed in the second case. The expected training is implicit as we consider these values to be the essential components of the learning scenario while undergraduates interact through digital environments.

(A) Active learning methods

On a bigger scale, the SAVL model is suitable for any teaching method, scaffolding the training process and progressing from LOTS (Lower Order Thinking Skills: understanding and applying) to HOTS (Higher Order Thinking Skills: analysing and creating).

Here, we provide students with learning-by-doing scenarios where they implement project-based and experiment-based approaches. The dynamic process of interactions is supported by formative assessment.

(V) Virtual exchange and digital tools

The digital process utilises the Microsoft Teams environment for discussing exchange topics to the Canvas and PowerPoint tools for designing final presentations.

(L) Learning of second languages and competences

Current research aligns global language acquisition with capacity building for university students. Further, we incorporate the core values of two transversal competences of Effective communication in digital environments as well as teamwork (Universitat Politècnica de València, 2022).

We put the above training model into practice to better understand the connection between transformative didactics, virtual exchanges and sustainability. Hence, we hypothesise that the investigation of three different cases of learning scenarios at a polytechnic university can help understand the perception of SDGs by students.

4 Implementing Virtual Learning Scenarios for Sustainable Education

Motivated by the need for an empirical approach towards such learning systems, we now proceed with a brief description of the case studies representing a diversity of the SAVL application embedded in the technical curriculum. Further, the core aspects selected for the current overview include specific Sustainable Development Goals, competences and virtual exchange interaction in English.

4.1 Case Study 1: Technology-Humanities Learners' Cooperation

From November to December of 2022, a group of 11 engineering students from Universitat Politècnica de València—Polytechnic University of Valencia (Spain) and 9 English studies undergraduates from the University of Zadar (Croatia) took part in the experimental project "Education systems in different countries".

Grounded on Sustainable Active Virtual Learning, the training was rooted in two specific sustainability goals: SDG 4, "Quality education", and SDG 17, "Partnerships for the Goals". 20 participants reported on all five dimensions of transversal competences before and after the collaboration (Polyakova & Lovrović, 2024).

The overall exchange dynamics included three synchronous online sessions where four interuniversity teams explored specific areas of the main topic and presented the visual project work findings during the final meeting. Among the issues discussed by each team were the comparisons of Spanish and Croatian education systems and their resemblance to the UK and US frameworks.

Moreover, the collaboration structure evolved around four main stages, as previously detailed in methodological sections on the SAVL exchange model:

(S) Phase 1, Sustainability values and SDGs. Establishing a common background within Goals 4 and 17, exchange objectives, didactics, and research. This stage helped align both universities' curricular settings, schedules and availability of resources.

(A) Phase 2, Active learning methods. Planning and designing the telecollaboration process included foundational principles of the SAVL framework. Additionally, partnering institutions created and validated training materials and questionnaires here.

(V) Phase 3, Virtual exchange and digital tools. Implementing the virtual interaction required hosting synchronous online meetings, organising inter-university teamwork and monitoring the learning process. We emphasise information reliability alongside creativity and open communication in English. Besides, pre- and post-experiment enquiries gave us a vivid understanding of capacity-building.

(L) Phase 4, Language learning and competence development. Evaluating the final VE results was based on exploring outcomes through presentations, visual reports, and critical reflections. As mentioned earlier, data analysis and dissemination of findings were key components of the questionnaires.

Hence, the study used educational research elements such as pre-and post-experimental questionnaires containing close (Likert scale) and open-ended questions. In this chapter, special attention has been paid to the sentiment analysis of the learners' opinions to better understand better perceptions (Moreno Ortiz, 2019; Moreno-Ortiz, 2017).

This collaboration supported meaningful and capacity-based learning by shifting the focus from language proficiency to sustainability emphasis. We also identified some practical limitations.

- Curricular alignment across the institutions: in terms of official curricula, grading and scheduling, both institutions had to make adjustments while adding this setting to the learning process advancements and assessment. Croatian participants received speaking progress marks, whereas their Spanish counterparts were monitored and evaluated for Computer practicum and interaction abilities.
- Topic choice and training materials: due to the collaboration of technical and humanities undergraduates, the interaction topics were not related to technical areas but rather to the joint ones connected to SDGs. Additionally, the educators hand-picked or prepared training materials because of the lack of proper open-source resources for higher education institutions.
- Equipment and troubleshooting: even though the same Microsoft Team environment was used for all sessions, some learners had connection, microphone or webcam use issues. These matters were addressed by providing university equipment for the project work.

4.2 Case Study 2: Technology—Technology Students' Collaboration

As we will soon uncover, implementing the SAVL model may be manifold. In the upcoming setting, this framework was also linked to the virtual laboratory interaction between 25 Spanish engineering undergraduates from Universitat Politècnica de València (UPV) and 5 Ukrainian counterparts from Dnipro University of Technology (DUT) (Klyatskina et al., 2024).

Building a feasible and relevant training environment took time, but it was worthwhile. The didactic innovation process based on the above-mentioned SAVL framework followed the stages of the prior case. However, some learning and laboratory experiments required attention to detail during all four phases depicted earlier in the study methodology:

(S) Phase 1, Sustainability values and SDGs. The general implementation scheme embedded three core sustainability goals. Goal 4, "Quality education", provided the strive for excellence component; Goal 16 ", Peace", promoted peaceful educational institutions; and Goal 17 ", Global partnership", strengthened inter-university cooperation.

(B) Phase 2, Active learning methods. Planning and design. Initially, the educators from both institutions designed the Materials Science course structure so that the theory was taught separately in the local (Spanish and Ukrainian) languages. Then, they aligned schedules for joining online Team meetings at UPV performed. Specifically, UPV teammates performed the experimental part, transmitting the process and the results to the DUT counterparts. Finally, the participants offered hybrid presentations of experimental results and answered some questions.

(V) Phase 3, Virtual exchange and digital tools. The "ELVI—Experimental Laboratory of Virtual Interaction" was performed as an experimental virtual interaction laboratory conducted as a dynamic and proactive collaboration between two higher education institutions: Universitat Politècnica de València (UPV) in Spain and Dnipro University of Technology (DUT) in Ukraine. Spanish learners conducted and retransmitted the lab experiments while commenting on the specific measurements with Ukrainian students. During the final session, each inter-university team offered a short presentation of one of the experiments.

(L) Phase 4, Language learning and competence development. Global language learning and teamwork provided a strong background for assessing the final results of the training process. The questionnaire delivered at the end of the training process highlighted the primary outcomes of the initiative.

Regarding Case Study 2 limitations, besides those listed above (curricular alignment, topic and materials, equipment), we educators had to arrange interaction during the laboratory practicum setting.

5 Analysing Learning Outcomes and Perceptions of SDGs

5.1 Case Study 1 Results: Pre- and Post-Study Questionnaires

The tech-humanities cooperation (Spain-Croatia) assessment based on a short questionnaire included two key perceptions: the SDG achievement of Goals 4 and 17 (2 Likert scale ranging from 1 to 5 questions). As the charts below illustrate (see Figs. 2 and 3), the data indicates a considerable advancement in learners' comprehension and application of Sustainable Development Goals.

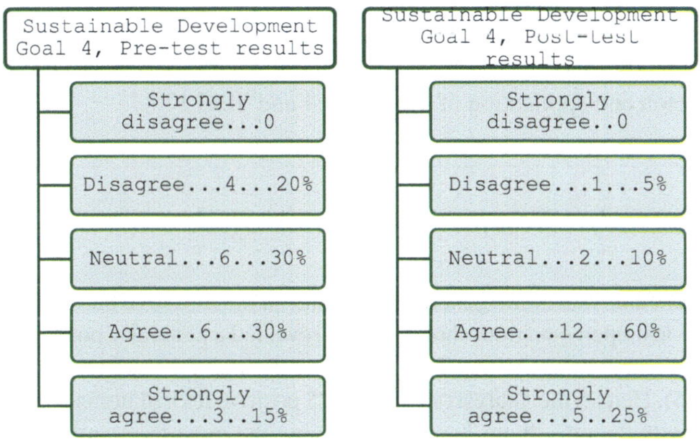

Fig. 2 Study 1, SDG 4 achievement perceptions, pre-/post-experimental results (participants' responses). *Source* Designed by authors

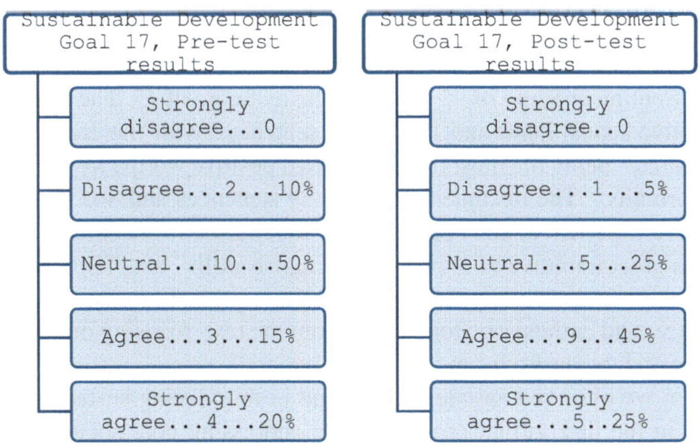

Fig. 3 Study 2, SDG 17 achievement perceptions, pre-/post-experimental results (participants' responses). *Source* Designed by authors

After the intervention, the data reveals a significant enhancement in participants' understanding and implementation of Sustainable Development Goal (SDG) 4, Quality Education. In the post-test, 25% of participants strongly agreed, and 60% agreed with statements concerning SDG 4, an increase from 15 and 30%, respectively, in the pre-test. Neutral responses declined from 30% before the intervention to 10% afterwards. Moreover, disagreement levels decreased, with 20% disagreeing and 5% strongly disagreeing in the pre-test, reducing to 5% disagreeing and none strongly disagreeing in the post-test.

Similarly, there was a positive trend for SDG 17, Partnerships for the Goals. Post-test results showed that 25% of participants strongly agreed and 45% agreed, up from 20 and 45% in the pre-test. Neutral responses dropped from 50 to 25%, and disagreement levels fell from 10% disagreeing and 5% strongly disagreeing in the pre-test to 5% disagreeing and none strongly disagreeing in the post-test.

These outcomes indicate that the intervention effectively improved participants' comprehension and application of both SDG 4 and SDG 17.

5.2 *Case Study 1 Results: Sentiment Analysis*

The sentiment analysis of the general comments on the initiative based on Lingmotif reports (pre-test open question, short answer) revealed a generally positive attitude at the initiative's beginning, with a Text Sentiment Score of 92 and an Intensity Score of 91 (Fig. 4). Hence, the analysis identified 25 positive lexical items and 19 positive sentences, with "good," "help," and "nice" being the most frequently used positive words. Key topics include "good initiative" and "great opportunity." The document contains 23 sentences and 319 words, emphasising positive experiences and opportunities, as reflected in statements like "It is nice to be a part of this initiative" and "Great opportunity." Overall, the report highlights a highly positive and enthusiastic tone throughout the text.

After the virtual exchange, the data indicated an extremely positive sentiment with a Text Sentiment Score of 96 and an Intensity Score of 75. The analysis identified 30 positive lexical items and 19 positive sentences, with "be great," "improve," and "collaborate" being the most frequently used positive words. Key topics include "great opportunity." The document contains 29 sentences and 462 words, emphasising positive experiences and opportunities, as reflected in statements like "I think that this is a great initiative to improve our English skills" and "This initiative was great and useful for practising English language." Overall, the report highlights a highly positive and enthusiastic tone throughout the text, focusing on the benefits of the initiative and the desire for more such activities.

Moreover, we observed an interesting trend in introducing sustainable development goals in the first virtual project. While addressing core societal challenges, higher education students join forces in enhancing their specific communication and intercultural competences.

5.3 *Case Study 2 Results: Post-experimental Questionnaire*

While tackling tech-tech virtual laboratory (Spain-Ukraine), the scholars collected learners' self-reported capabilities in sustainability, education, teamwork, professional vocabulary, communication, and report writing. The categories varied from Strongly disagree, Disagree, Neither agree nor disagree, Agree, to Strongly agree according to the Likert scale (see Figs. 4 and 5)

Most students reported high levels of agreement (Agree or Strongly agree) in all assessed competences, particularly in communicating effectively in English (83%) and working in online teams (83%). Similar levels of positive perceptions agreement (Agree or Strongly agree) are correlated with SDG 17 (83%) and SDG 4 (90%).

Fig. 4 Study 1, virtual initiative perceptions, pre-/post-experimental results (participants' responses). *Source* Designed by authors

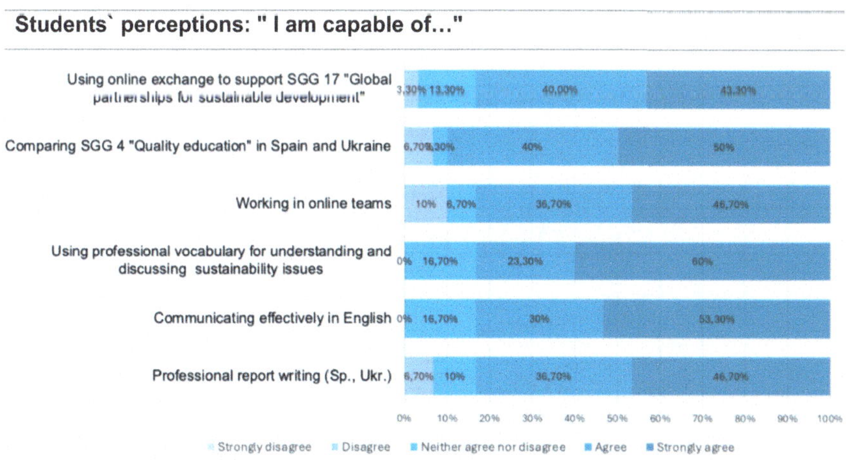

Fig. 5 Study 2, virtual initiative perceptions, final experimental results (participants' responses). *Source* (Klyatskina et al., 2024)

Using professional vocabulary for discussing sustainability proved to be a valuable outcome, too (83%).

5.4 Case Study 2 Results: Word Cloud Analysis

Additionally, the word cloud (five keywords related to the virtual exchange) visually represents the central themes and concepts identified in the research findings, highlighting the most frequently mentioned points in students' feedback on virtual collaboration for technical education. More prominent words indicate higher frequency and significance, suggesting dominant aspects of the learning experience (Fig. 6).

Notably, terms such as "sustainability," "communication," "collaboration," and "teamwork" appear prominently, reflecting students' emphasis on the importance of cooperative learning and global engagement. The presence of words like "motivation," "interactive," and "engaging" suggests that participants found the virtual exchange stimulating and beneficial for their academic growth. Additionally, technical and educational terms such as "engineering," "problem-solving," "technical skills," and "digital tools" indicate that the exchange effectively supported both professional and transversal skill development.

While the word cloud demonstrates positive outcomes, words like "challenges" and "intercultural" suggest that students encountered difficulties related to cross-cultural communication and adaptation. These insights align with the study's findings

Fig. 6 Study 2, virtual initiative perceptions, five keywords (participants' responses). *Source* Own elaboration based

on the need for structured support to overcome technical and curricular barriers in international virtual collaboration.

Overall, the word cloud reinforces the research conclusion that sustainability-driven virtual exchanges are valuable in fostering interdisciplinary learning, intercultural competence, and professional ability-building among technical students.

Generally, the results effectively illustrate students' self-perceptions of their academic and professional competences, highlighting areas of strength and potential gaps for further development.

6 Discussion

As demonstrated above, our primary source of inspiration is the active social role of university students, professors, and higher education institutions linked to SDGs. Hence, by geographically broadening the scope of the study, we can work on embedding sustainability in virtual exchanges worldwide.

Furthermore, the present study focuses on specific regions (Spain, Croatia and Ukraine) and synchronous interactions in English, whereas readers might combine other territories, degree levels, educational environments and technological scenarios.

We drew on the virtual exchange model for inter-university cooperation in designing our experimental setting. Continuing with this rationale, the setting can be expanded to:

[Sustainability]: dimensions, five pillars, Education for Sustainability, among others,

[Active learning & cooperation]: experiential, enquiry-based, problem-based, case-based and service learning, gamification, and simulation, combined with joint research projects, cooperative groups and learning communities, overseas universities,

[Virtual Exchange and Digital tools]: Erasmus Virtual Learning, COIL, UNIcollaboration embedded in the Learning Management Systems, Video conferencing platforms, Virtual Labs and project management, collaboration tools, or online discussion forums,

[Language Learning and competences]: bilingual and plurilingual focuses, translanguaging, general and specific purposes language education, language tandems, virtual reality interaction, debate clubs, cultural exchange programmes, elevators' pitches, podcasting, vlogging together with mentorship and electronic portfolios.

On the whole, there should be a straightforward methodological approach, detailed didactic planning, and critical self-reflection to transfer study results better. Even though other practitioners might choose other analytical tools for tracking the study results, we highlight the importance of measuring outcomes. These insights demonstrate the ability and possibility of enriching technical education from the SDG's perspective.

7 Conclusions

Virtual collaboration has become a vital technical education component in an increasingly interconnected world, fostering interdisciplinary learning and global engagement. Sustainable synergy in this context refers to integrating sustainability principles into virtual exchanges, enabling students to develop technical expertise and cross-cultural competencies. By leveraging digital platforms, institutions can facilitate meaningful cooperation among students from diverse academic backgrounds, enhancing their problem-solving abilities and preparing them for the challenges of a globalised workforce.

This study highlights the potential of sustainability-focused virtual exchanges in enhancing the competence-based preparation of future engineers. By integrating the Sustainable Active Virtual Learning (SAVL) system into two distinct virtual exchange programmes—one focused on technology-humanities students' cooperation and the other on a technology-technology undergraduate's virtual laboratory—students were able to develop both technical and transversal capacities while engaging with the principles of Agenda 2030.

Despite challenges such as intercultural differences, technical limitations, and curriculum alignment, the results demonstrate that virtual exchanges foster motivation, effective communication, and collaborative problem-solving. Furthermore, these initiatives strengthen inter-university cooperation, providing a valuable framework for globalised technical education.

The findings suggest that sustainability-oriented virtual exchanges can promote global communication, academic growth, cross-cultural understanding, and interdisciplinary collaboration. Future research should explore strategies to optimise these programmes and address existing challenges, ensuring their long-term effectiveness in engineering education.

Acknowledgements We wish to convey our sincere appreciation to the technical learners and instructors from Universitat Politècnica de València—Polytechnic University of Valencia (UPV), University of Zadar (UniZd) and Dnipro University of Technology (DUT) for their valuable participation in these experiments. We would also like to acknowledge the Institute of Education Sciences (UPV) for their generous support and financial backing. Besides, we are grateful to the Didactic Innovation Programme PIME/23-24/380 "SostenibleEdu", PIME/25-26/580, "IncludED: Virtual Collaboration for Global and Diverse STEM Learners" and the Innovation and Quality Education Team ""Innovation, Plurilingualism, Universities, and Sustainability" (i-PLUS) at UPV for their significant contributions and assistance. The collective efforts have been instrumental in advancing our research.

References

Beelen, J., & Jones, E. (2021). *Redefining internationalization at home: Future perspectives.* Springer.

Belz, J. A., & Thorne, S. L. (2021). *Internet-mediated intercultural foreign language education*. Routledge.

De Wit, H., & Altbach, P. G. (2020). Internationalization in higher education: Global trends and recommendations. *International Higher Education, 102*, 2–4.

De Wit, H., & Beelen, J. (2021). Virtual exchange and the future of internationalisation in higher education. *Journal of International Education Research, 17*(3), 212–234.

De Wit, H., & Leask, B. (2021). Hybrid models of internationalization in higher education. *Journal of Studies in International Education, 25*(2), 123–140.

European Commission. (2024). *The European higher education area in 2024: Bologna process implementation report*. Publications Office. https://doi.org/10.2797/351309

Grek, S., & Russell, I. (2024). Beyond Bologna? Infrastructuring quality in European higher education. *European Educational Research Journal, 23*(2), 215–236. https://doi.org/10.1177/14749041231170518

Griebeler, J. S., Brandli, L. L., Salvia, A. L., Leal Filho, W., & Reginatto, G. (2022). Sustainable development goals: A framework for deploying indicators for higher education institutions. *International Journal of Sustainability in Higher Education, 23*(4), 887–914. https://doi.org/10.1108/IJSHE-03-2021-0088

Häkkinen, P., & Rienties, B. (2021). *Global virtual teams and employability skills development*. Routledge.

Helm, F., & Guth, S. (2020). The multifarious goals of Telecollaboration 2.0: Education, training, and employability. *ReCALL, 32*(2), 181–198.

Helm, F. (2020). The evolving role of virtual exchange in education. *Language Learning & Technology, 24*(3), 205–229.

Helm, F., & Acconcia, G. (2021). Critical virtual exchange: Exploring equity and social justice through digital collaboration. *Journal of Virtual Exchange, 5*(1), 45–67.

Klyatskina, E., Polyakova, O., & Pismenkova, T. (2024, July 11). *Fomentando la diversidad, sostenibilidad y digitalización: Experiencias virtuales interactivas en Ciencia de Materiales*. In-Red 2024-X Congreso Nacional de Innovación Educativa y Docencia En Red. IN-RED 2024: X Congreso de Innovación Educativa y Docencia en Red. https://doi.org/10.4995/INRED2024.2024.18510

Leal Filho, W., Raath, S., Lazzarini, B., Vargas, V. R., de Souza, L., Anholon, R., Quelhas, O. L. G., Haddad, R., Klavins, M., & Orlovic, V. L. (2018). The role of transformation in learning and education for sustainability. *Journal of Cleaner Production, 199*, 286–295. https://www.sciencedirect.com/science/article/pii/S095965261831984X

Leal Filho, W. (2022). Pursuing sustainable development across disciplines. In *Handbook of best practices in sustainable development at university level* (pp. 495–504). Springer. https://doi.org/10.1007/978-3-031-04764-0_27

Moreno Ortiz, A. (2019). Mi opinión cuenta: La expresión del sentimiento en la red. In *Comunicación mediada por ordenador: La lengua, el discurso y la imagen* (S. Robles Avila (ed. lit.), A. Moreno Ortiz (ed. lit.), pp. 38–74). Cátedra.

Moreno-Ortiz, A. (2017). Lingmotif: A user-focused sentiment analysis tool. *Procesamiento Del Lenguaje Natural, 58*, 133–140. https://rua.ua.es/dspace/handle/10045/64038

O'Dowd, R. (2023). *Internationalising higher education and the role of virtual exchange*. Routledge.

Polyakova, O., & Galstyan-Sargsyan, R. (2022). Towards a sustainable active virtual learning model. In J. M. Esteve Flaubel, A. Fernández Sogorb, R. Martínez Roig, & J. F. Álvarez Herrero (Eds.), *Transformando la educación a través del conocimiento* (pp. 1020–1035). Octaedro. http://hdl.handle.net/10045/129549

Polyakova, O., & Lovrović, L. (2024). Sustainable development goals and transversal competences through L2/3 virtual exchange. *Language Value, 17*(1). https://doi.org/10.6035/languagev.7928

Rubin, J., & Guth, S. (2022). *Collaborative online international learning (COIL): Strategies for 21st-century education*. Peter Lang.

Sandri, O. J. (2022). Sustainability education and the role of virtual exchange. *Sustainability Journal, 14*(7), 1783–1799.

Tilbury, D. (2011). Higher education for sustainability: A global overview of commitment and progress. *Higher Education in the World*, *4*(1), 18–28. https://www.academia.edu/download/9551840/guni%20he%20in%20the%20world%204%20hes%20committment%20to%20sus.pdf

UN General Assembly. (2015). *Transforming our world: The 2030 agenda for sustainable development, A/RES/70/1*. https://www.refworld.org/docid/57b6e3e44.html

Universitat Politècnica de València. (2022). *Acuerdo relativo a la ordenación e integración de las competencias transversales en las titulaciones oficiales*. Universitat Politècnica de València. https://www.upv.es/entidades/vecal/wp-content/uploads/2022/10/Ordenacion-e-integracion-de-las-CT-en-las-titulaciones-oficiales-Consejo-Gobierno-21-07-22.pdf

Žalėnienė, I., & Pereira, P. (2021). Higher education for sustainability: A global perspective. *Geography and Sustainability*, *2*(2), 99–106. https://doi.org/10.1016/j.geosus.2021.05.001

Bridging Cultures in Higher Education: The Role of Cross-Cultural Communication in Achieving the UN Sustainable Development Goals

Jintao Yang, Nikolai Gorbatchev, Bilal Ahmed Memon, and Pengfei Zhang

Abstract This research explores the role of cross-cultural communication in higher education as a means to advance the United Nations Sustainable Development Goals (SDGs). The international authors focus on the perspectives and experiences of students from Belarus, China, and Uzbekistan, who represent diverse cultural backgrounds and educational contexts. Through a mixed-methods approach, including surveys and interviews, the chapter investigates how cross-cultural interactions within universities contribute to a deeper understanding of sustainability issues and foster collaborative efforts toward achieving the SDGs. By highlighting the challenges and opportunities faced by these institutions in promoting intercultural dialogue, the chapter underscores the importance of developing intercultural competencies among students and educators. It discusses strategies for integrating SDG-related content into curricula and creating inclusive learning environments that encourage mutual respect and understanding. The findings reveal that effective cross-cultural communication not only enhances academic collaboration but also empowers students to become active global citizens committed to sustainable development.

Keywords Cross-cultural communication · Higher education · SDGs · Intercultural competencies · Global citizenship

---

J. Yang (✉)
University of Latvia, Riga, Latvia
e-mail: jy22003@edu.lu.lv

N. Gorbatchev
Minsk Branch, Plekhanov Russian University of Economics, Minsk, Belarus

N. Gorbatchev · P. Zhang
International Institute of Management and Business, Minsk, Belarus

B. A. Memon
Westminster International University in Tashkent, Tashkent, Uzbekistan

W. Leal Filho et al. (eds.), *Introducing the UN SDGs in Higher Education Institutions*, World Sustainability Series, https://doi.org/10.1007/978-3-032-23017-1_36

1 Introduction

In today's world, where globalization and cultural integration have become an integral part of our lives, intercultural communication is of particular importance in the context of higher education. It is becoming a key tool for strengthening mutual understanding and collaboration between students and teachers (Adhikari & Shrestha, 2023).

As the global community strives to achieve the Sustainable Development Goals (SDGs) set by the United Nations by 2030, higher education institutions play an important role in developing the knowledge, skills, and thinking needed to address global challenges such as climate change, inequality, and sustainable development (Heleta & Bagus, 2021).

This article presents perspectives from Belarus, China, and Uzbekistan, countries with unique cultural, historical, and educational experiences, to better understand how intercultural communication can help overcome differences and advance shared goals. These states are shaped by their historical contexts, particularly the Soviet legacy that emphasizes STEM fields. In Belarus, collaborations between universities and Confucius Institutes facilitate Chinese language learning and cultural exchange, fostering closer ties with Chinese students. Uzbekistan, as part of the Belt and Road Initiative, offers programs focused on the historical Silk Road, inviting international students to explore its rich cultural heritage. Specific initiatives like joint research projects between these countries promote academic collaboration and intercultural understanding.

It is important to note that intercultural communication is not limited to exchanging information about culture and traditions. It also includes understanding and respect for different values, norms and customs (Aririguzoh, 2022). This helps to create more harmonious and effective interactions between representatives of different cultures. Intercultural communication is therefore an important tool for achieving the Sustainable Development Goals and creating a fairer and more harmonious world. It helps to strengthen mutual understanding and cooperation between countries and regions, as well as to develop a more tolerant and respectful attitude towards other cultures.

The purpose of this paper is to analyze the role of intercultural communication in higher education as a tool for promoting the SDGs, improving the quality of education and increasing the export of educational services. The experience of students from Belarus, China and Uzbekistan is analyzed to understand the mechanisms of intercultural interaction, ways of integrating the SDGs into educational programs, and the formation of a responsible attitude to achieving the SDGs. The problems and opportunities that universities face in promoting intercultural dialogue and creating an inclusive learning environment are also considered.

To achieve the aim of the study, we attempted to complete the following tasks:

1. To study the current state of intercultural communication among students and teachers in Belarus, China and Uzbekistan.

2. To analyze the strategies of higher education institutions to promote intercultural dialogue and evaluate their effectiveness.
3. To develop practical recommendations for improving intercultural competencies in university programs.

The discussion addresses how improved cross-cultural communication can contribute to the following SDGs (UN, 2025):

SDG 4 (Quality Education): Ensuring inclusive and equitable access to quality education while supporting lifelong learning for all.

SDG 10 (Reduced Inequalities): Promoting inclusivity and understanding among diverse student groups, thereby reducing disparities and advancing equality.

SDG 13 (Climate Action): Encouraging collaborative efforts to address climate change through various cultural perspectives and shared solutions.

SDG 17 (Partnerships for the Goals): Strengthening global partnerships through enhanced intercultural communication and collaboration among universities and their surrounding communities.

2 Literature Review

In today's world, where globalization and cultural integration have become an integral part of our lives, higher education is undergoing significant changes (Crawford & Cifuentes-Faura, 2022). The increasing interconnectedness between different countries and cultures requires a deep understanding of diverse cultural contexts. As institutions actively attract international students, the SDGs become a guiding principle for initiatives aimed at promoting sustainability, equity, and cultural awareness. Integrating the SDGs into higher education goes beyond simple compliance; it offers a unique opportunity to improve educational outcomes by promoting inclusion and building a sense of global responsibility among students (Monzó-Martínez et al., 2024).

In the context of globalization and China's growing role on the world stage, cross-cultural competence is becoming an important factor for success in international education, diplomacy and cooperation. The Belt and Road Initiative (BRI) is promoting China's influence, which requires that participants in international cooperation have deep knowledge of cultural differences (Woo, 2023). Educational institutions should pay special attention to developing cross-cultural competence in their curricula and plans to prepare professionals to meet the challenges of international cooperation. The growing presence of Chinese enterprises along the China-EU transport corridor highlights the importance of good communication and cultural awareness.

According to Deardorff, intercultural competence is "the ability to communicate and interact effectively and appropriately with people from different cultures"

(Deardorff, 2006). Understanding the cultural differences that influence negotiation styles, business practices, and relationship building is critical for international partners not only to ensure compliance but also to achieve success in their interactions with Chinese companies. Integrating intercultural competences into educational programs promote inclusion, equality, and awareness of cultural diversity. As higher education institutions around the world attract more international students, including those from China, there is a valuable opportunity to promote global citizenship and social responsibility.

In recent years, the number of students going abroad for higher education has increased significantly. The increase in the number of students from China choosing to study in post-Soviet countries such as Belarus and Uzbekistan are particularly noticeable. The main factors contributing to this phenomenon are the globalization of education, high competition in Chinese universities, the search for quality education, and the desire to strengthen global understanding and cooperation (Altbach & Knight, 2007). Intercultural communication plays a key role. It allows students from different cultural and educational backgrounds to exchange ideas, values, and information (Deardorff, 2006). Chinese students studying at universities in Belarus and Uzbekistan face both advantages and challenges, especially when it comes to reconciling academic goals with international sustainable development projects.

"Higher education institutions play a key role in achieving these goals as they prepare future leaders, foster innovation, and promote intercultural understanding" (Leal Filho et al., 2019a, 2019b). Creating inclusive learning environments and incorporating SDG-related content into curricula are important ways to equip students with the knowledge and skills needed to contribute to sustainable development (Tilbury, 2011). The success of these initiatives depends on the ability of institutions to manage cultural differences and facilitate intercultural communication between faculty and students. Intercultural communication is a key element of intercultural competence, i.e. the ability to interact effectively and appropriately with people from other cultures (Bayram, 1997). The experiences of students from China, Belarus, and Uzbekistan show how important it is to develop intercultural competence to overcome problems of cultural adaptation, language barriers, and differences in educational expectations (Guo & Chase, 2011). Effective intercultural communication not only enhances academic cooperation, but also enables students to become active global citizens who are committed to the ideas of sustainable development (Shonfeld et al., 2021).

In today's world, where globalization and cultural integration have become an integral part of our lives, the development of intercultural dialogue in higher education institutions is becoming especially relevant. Students from different countries face certain difficulties in adapting to a new education system, social norms and communication styles (Spencer-Oatey & Franklin, 2009).

Universities play an important role in developing intercultural competences. This includes the ability to effectively interact with representatives of other cultures (Leal Filho et al., 2019a, 2019b). They must ensure that competencies related to the SDGs are integrated into curricula, making them relevant and accessible to students from different cultural contexts (Ryan & Viete, 2009). These tasks also open up new opportunities for innovation and development.

By promoting an inclusive educational environment and encouraging intercultural interaction, universities can improve the quality of education and contribute to achieving the SDGs (Hernández-Torrano et al., 2021). The increase in the export of educational services in post-Soviet countries indicates a growing recognition of the importance of internationalization in higher education. By attracting students from China and other countries, these universities not only provide quality education, but also promote intercultural understanding (Marginson, 2014).

Joint educational projects are an effective tool for teaching and integrating the SDGs into educational programs. They help to form a new generation of leaders who are ready to solve complex problems related to achieving the SDGs (Altbach & de Wit, 2015).

Leading universities around the world integrate international students into educational programs, which contributes to the formation of a multicultural and inclusive environment. Universities have strategies aimed at supporting international students, including language courses, orientation programs and cultural events (Altbach, 2002). This helps students adapt to the educational system and culture, and facilitates their integration into the academic community. Joint educational projects, which involve students from different countries, play an important role in teaching and integrating the SDGs into educational programs (Leal Filho et al., 2021). They facilitate the exchange of knowledge, experience and cultural values between participants, which helps to form a deeper understanding of the SDGs and their significance for society.

Such projects provide students with the opportunity to work together to solve real-life problems related to sustainable development, which helps to develop critical thinking skills, collaboration and intercultural communication (Salvia et al., 2019).

Intercultural competence is becoming a prerequisite for success in higher education and international cooperation. With the expansion of global initiatives such as China's Belt and Road Initiative, universities play a key role in achieving the SDGs. Such initiatives also contribute to the formation of an inclusive learning environment by integrating SDG-related competencies into their curricula. Inclusive environment and supporting international students through language courses, orientation programs, and cultural events improve learning outcomes. In addition, these factors prepare students to become leaders who can solve complex global problems and take responsibility for the sustainable development of our planet.

Based on the results of previous studies, we formulated the scientific hypothesis of the study (H1) and conducted a survey among students: Integrating intercultural communication and SDG-related content into higher education curricula improves students' intercultural competence and facilitates joint efforts to achieve the SDGs.

3 Methodology

This study investigates the role of intercultural communication in higher education and its impact on the promotion of the SDGs, guided by the research hypothesis. A mixed-method approach was chosen to achieve this objective.

Data collection involved students from Belarus, China, and Uzbekistan, which allowed us to capture their perspectives and experiences within their respective educational contexts. Qualitative data were obtained through interviews with students from various educational levels and disciplines. These interviews provided in-depth information about the experiences of intercultural interaction and its relationship to sustainability issues and the SDGs. The interview questions emphasized the significance of intercultural dialogue in fostering collaborative initiatives for sustainable development.

A multi-method approach was implemented in this research. An online survey was administered to students, collecting preliminary data. The survey included closed-ended questions to gather demographic information and scaled questions to assess perceptions of the effectiveness of intercultural communication in relation to the SDGs. Open-ended questions allowed participants to offer detailed feedback and recommendations. This mixed-methods approach facilitated data triangulation, enhancing the validity and reliability of the findings by integrating qualitative and quantitative information. The domains and subthemes covered in the survey questionnaire are outlined in Table 1.

Several limitations of the survey should be acknowledged. Participants were selected through convenience sampling, targeting higher education students in the three countries. To avoid misunderstandings, the survey questions were translated by native speakers. The average response rate was 86.2% (initially, we aimed to obtain responses from 400 students from Belarus, 300 from China, and 200 from Uzbekistan), which is considered adequate. However, the distribution of participants was unbalanced, with 368 individuals from Belarus, 252 from China, and 156 from Uzbekistan. This variance may limit the applicability of the findings to other regions or cultural contexts.

A high percentage of respondents (74.6%) were bachelor's degree students, while fewer were master's (22.7%) or PhD candidates. This focus may result in limited insights regarding advanced research or professional practices. The distribution of respondents by specialty is detailed in Table 2.

The extent to which SDG-related topics are addressed in the survey is influenced by how educators incorporate these themes into their teaching as well as the students' individual interests in sustainability. This could lead to varying responses and affect the reliability of the data concerning SDG participation. The survey-maintained anonymity, and no personal data were collected. The questions were translated into the national languages of the participants by native speakers to prevent misunderstandings.

Table 1 The domains and subthemes in the survey questionnaire

General section: age, location, specialty, gender etc.	
Domain: Curriculum design	Domain: Outcomes and impact assessment
Subtheme: Courses' content	*Subtheme: Intercultural competency*
– Integration of SDG-related topics	– Cross-cultural communication skills
– Inclusion of global perspectives	– Growth in cultural awareness and sensitivity
– Alignment with sustainability goals	– Development of global citizenship values
Subtheme: Delivery methods	*Subtheme: Collaboration and outcomes*
– Online, in-person, or hybrid modes	– Improved teamwork across cultures
– Flexibility for diverse learners	– Academic collaboration on SDG-related projects
– Language accessibility	– Enhanced problem-solving
Subtheme: Inclusivity	*Subtheme: SDG awareness and engagement*
– Accessibility	– Understanding of sustainability challenges
– Geographic and cultural inclusivity	– Active participation in SDG initiatives
– Availability of advising, tutoring	– Sustainable development practices
Subtheme: Stakeholder engagement	*Subtheme: Policy and institutional alignment*
– Student and educator perceptions	– Alignment of curricula with SDGs
– Employer and industry acceptance	– Institutional support for intercultural initiatives
– Feedback from alumni and stakeholders	– Compliance with global educational standards
Subtheme: Challenges and opportunities	*Subtheme: Long-term impact*
– Barriers to cross-cultural communication	– Career progression with a focus SDGs
– Opportunities for intercultural dialogue	– Promotion of environmental consciousness

Source Developed by authors

Table 2 Distribution of respondents by specialties

Major	Belarus (N = 368) (%)	China (N = 252) (%)	Uzbekistan (N = 156) (%)
Public administration	15	28	28
Economics	25	14	38
Management	32	38	34
Business informatics	28	20	–

Source Developed by authors

4 Results

The first sections focus on the perceived importance of cultural understanding and cross-cultural interactions, providing insights into how students view these factors as integral to their academic success and learning experiences. The survey results reveal significant differences in perceptions of the role of cross-cultural communication in higher education among participants from Belarus, China, and Uzbekistan (see Table 3).

Table 4 presents the survey responses regarding participants' perceptions of the importance of understanding cultural differences for success in their academic careers. Notably, the findings indicate that a substantial proportion of participants across all countries value cultural understanding as essential, with the highest endorsement observed among Chinese respondents.

Table 5 presents the distribution of responses regarding participation in cultural events or activities related to cultural backgrounds among university students from Belarus, China, and Uzbekistan. The findings reveal notable differences in engagement across these countries.

In Belarus, a significant 42% of respondents reported low participation, indicating that many students seldom engage in cultural events, with only 18% stating they participate frequently (levels 4 and 5 combined). In contrast, Chinese respondents show a higher level of engagement, with 44% participating at a moderate level (3), underscoring a greater frequency of involvement in cultural activities compared to

Table 3 "How strongly do you feel that your culture influences your behavior and interactions in academic settings?"

Country	5 (Max) (%)	4 (%)	3 (%)	2 (%)	1 (Min) (%)
Belarus	12.5	56.2	31.25	0	0
China	30	35	30	5	0
Uzbekistan	20	45	30	5	0

Source Developed by authors

Table 4 To what extent do you believe that understanding cultural differences is important for success in your academic career?

Country	5 (Max) (%)	4 (%)	3 (%)	2 (%)	1 (Min) (%)
Belarus	44	29	24	3	0
China	57	18	23	2	0
Uzbekistan	50	25	20	5	0

Source Developed by authors

Table 5 How often do you participate in cultural events or activities (e.g., festivals, workshops, seminars) related to your cultural background while at university?

Country	1 (Never) (%)	2 (%)	3 (%)	4 (%)	5 (Very often) (%)
Belarus	18	42	25	18	15
China	5	22	44	20	14
Uzbekistan	10	30	35	15	10

Source Developed by authors

Belarus. The Chinese data also highlights that only 5% of respondents never participate, suggesting a more active cultural scene within Chinese universities. Uzbekistan's results reflect a middle ground, with 10% of participants indicating they never engage in cultural events, 30% at level 2, and 35% at level 3, demonstrating moderate participation.

However, 100% of survey respondents indicated that their universities often organize cultural events, which reflects a strong institutional commitment to promoting cultural awareness and engagement among students. This finding suggests that universities recognize the importance of fostering an inclusive environment that values diverse cultural backgrounds, which is essential for enhancing the educational experience and personal development of students.

Table 6 illustrates the survey responses regarding the perceived importance of effective communication for successful collaboration in a multicultural academic environment among students from Belarus, China, and Uzbekistan. The data reveals that a significant majority of respondents across all three countries recognize effective communication as a vital element of collaborative success in diverse educational settings.

In Belarus, no respondents rated effective communication as "not important at all," with a combined total of 64% rating its importance at levels 4 and 5, suggesting a strong consensus on its significance. Similarly, Chinese respondents show a robust appreciation for effective communication, with 42% acknowledging its high importance (levels 4 and 5). Although Uzbekistan has slightly lower percentages in the top two categories, with 54% of respondents rating it as important (levels 4 and 5), it still demonstrates a comparable recognition of the necessity for effective communication in multicultural contexts.

Table 7 presents survey responses regarding the effectiveness of group projects in promoting understanding and appreciation of cultural differences among students. The findings indicate a generally positive perception of the role of group projects in enhancing cultural understanding, though some notable differences emerge based on the context of international students, particularly in Belarus. In Belarus, an encouraging 56% of respondents rated group projects at levels 4 and 5, indicating that they believe these collaborative efforts significantly contribute to understanding cultural differences. However, it's important to consider that Belarus hosts a substantial number of international students from China, many of whom face language barriers as they mentioned in this survey.

Table 6 How important do you think effective communication is for successful collaboration in a multicultural academic environment?

Country	1 (Not important at all) (%)	2 (%)	3 (%)	4 (%)	5 (Very important) (%)
Belarus	0	0	36	41	23
China	2	11	47	28	14
Uzbekistan	1	5	40	35	19

Source Developed by authors

Table 7 In your opinion, how well do group projects in your courses promote understanding and appreciation of cultural differences?

Country	1 (No at all) (%)	2 (%)	3 (%)	4 (%)	5 (High extend) (%)
Belarus	0	5	39	46	10
China	0	2	44	27	27
Uzbekistan	5	15	45	25	10

Source Developed by authors

Table 7 reveals that the majority of Belarusian and Chinese students evaluate group projects as moderately to extremely helpful in increasing awareness and appreciation of cultural differences, with 56% and 54% rating them as levels 4 or 5, respectively. In contrast, responses from Uzbekistan show a poorer view, with just 35% rating their efficacy at these higher levels and a substantial 15% rating them as a 2, implying that some students do not believe these projects greatly improve their intercultural learning opportunities. Overall, while the data show a generally positive attitude towards group projects, the middling scores and presence of critical replies underscore the need for improvements in their design and implementation to better correspond with the goals of developing intercultural understanding.

Table 8 presents survey results regarding the familiarity of students in Belarus, China, and Uzbekistan with the SDGs. Understanding the level of awareness of the SDGs among students is crucial, as these goals serve as a global framework aimed at addressing pressing social, economic, and environmental challenges. Familiarity with the SDGs can influence students' engagement with sustainability initiatives and inform their perspectives on global citizenship.

Table 8 presents the survey results regarding students' familiarity with the SDGs in Belarus, China, and Uzbekistan. The data indicate varying levels of awareness, with the majority of students in Belarus (72%) reporting that they are "somewhat familiar" with the SDGs, while a significant portion of students in Uzbekistan (40%) fall into the "not very familiar" category. Understanding these familiarity levels is essential, as they can directly impact students' engagement with sustainability initiatives and shape their perspectives on global citizenship, highlighting areas where educational institutions may need to enhance outreach and education regarding the SDGs.

The results of the question "Which of the following SDGs do you think is the most relevant to your academic studies?" indicate that among students in Belarus,

Table 8 How familiar are you with the UN SDGs?

Country	Very familiar (%)	Somewhat familiar (%)	Not very familiar (%)	Not at all familiar (%)
Belarus	18	72	25	3
China	16	41	11	7
Uzbekistan	10	35	40	15

Source Developed by authors

China, and Uzbekistan, Quality Education (Goal 4) stands out as the SDG deemed most relevant, particularly among Belarusian respondents, who show the highest level of support at 55%. Respondents from Belarus and Uzbekistan mentioned the significance of Partnerships for the Goals (Goal 17), with 40% of Belarusian students recognizing its relevance to their education, though the figure decreases to 15% in Uzbekistan.

Reduced Inequalities (Goal 10) holds more significance for Uzbekistan students, who rate its importance at 25%, in contrast to only 5% from Belarus, suggesting a greater focus on tackling inequalities in Uzbekistan's academic environment. Climate Action (Goal 13) garnered little interest, with just 12% of Chinese students viewing it as relevant, indicating a potential gap in climate-related education that could be addressed in the curricula of all three nations. In summary, the responses demonstrate differing perceptions of the SDGs, emphasizing Quality Education as a shared priority while also showcasing unique regional concerns about partnerships and inequality issues.

Table 9 presents the findings related to the most effective strategies for enhancing cross-cultural communication in connection with the SDGs. Participants had the option to select multiple responses, revealing a range of views on how to best encourage understanding and collaboration across cultures. The results indicate the level of endorsement for different strategies. They the importance of interactive and collaborative approaches in fostering engagement with the SDGs.

The low number of responses for the "Other" category (just 6 in total) indicates a general agreement on the main strategies proposed. This underlines the necessity for universities to emphasize the development of collaborative and experiential learning opportunities to improve cross-cultural communication regarding the SDGs.

Table 9 Which of the following strategies do you believe would be most effective in promoting cross-cultural communication in relation to the SDGs?

Answer	Belarus (N = 368) (%)	China (N = 252) (%)	Uzbekistan (N = 156) (%)
Workshops and training sessions focused on SDGs and cultural awareness	10.9	34.9	33.3
Collaborative projects combining multiple cultural perspectives	32.6	44.4	62.8
Guest lectures from experts on SDGs and cultural issues	21.7	20.6	30.8
Cultural exchange programs	32.6	17.5	21.2
Other (please specify)	0	1.6	1.3

Source Developed by authors

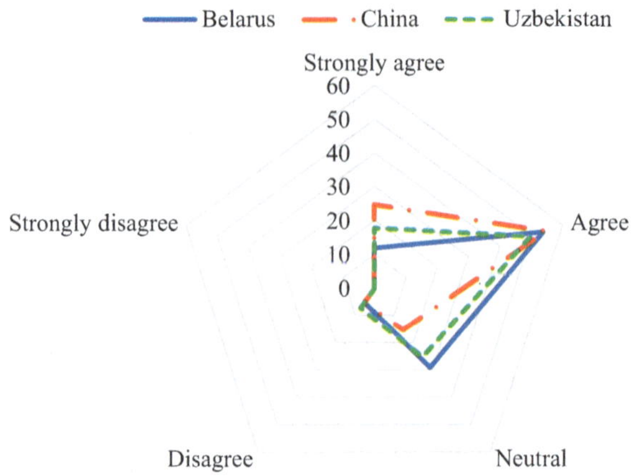

Fig. 1 Do you think that greater emphasis on SDGs can help improve the cross-cultural communication skills of students? (in %) *Source* Developed by authors

The survey findings (see Fig. 1) reveal a clear agreement among students from Belarus, China, and Uzbekistan that placing a greater focus on SDGs improves their skills in cross-cultural communication. A large majority from each country expressed their support, with 54% of students from Belarus, 55% from China, and 50% from Uzbekistan indicating their agreement. 12% of respondents from Belarus and 25% from China strongly supported this notion, indicating their belief in the potential benefits of the SDGs for improving communication among various cultures. The absence of strong opposition—demonstrated by the lack of participants who strongly disagreed—bolsters the idea that students generally view the promotion of SDGs favorably in terms of their interactions with different cultures. This prevailing positive attitude toward the SDGs reflects a shared acknowledgment of their importance in enhancing cross-cultural understanding among students.

The data reveals that a major challenge for universities in Belarus, China, and Uzbekistan is the limited awareness of cultural diversity among students, indicating the need for enhanced education on this topic (see Fig. 2). Additionally, organizational constraints hinder institutions' ability to implement effective cross-cultural initiatives, emphasizing the requirement for better support and resources. These insights indicate that universities should focus on initiatives aimed at enhancing cultural awareness, obtaining adequate funding, and encouraging student participation in cross-cultural experiences to cultivate a more inclusive academic atmosphere.

What do you consider to be the biggest challenge universities face in promoting cross-cultural communication among students?

The survey results provide some insights into the barriers faced by students in Belarus, China, and Uzbekistan regarding their engagement in initiatives centered on the SDGs and cross-cultural issues (see Table 10). The most significant barrier identified across all three countries is the lack of time due to academic workloads, with the highest percentage reported in China (42%), closely followed by Uzbekistan (40%) and Belarus (35%). This trend underscores the pressure that students in these

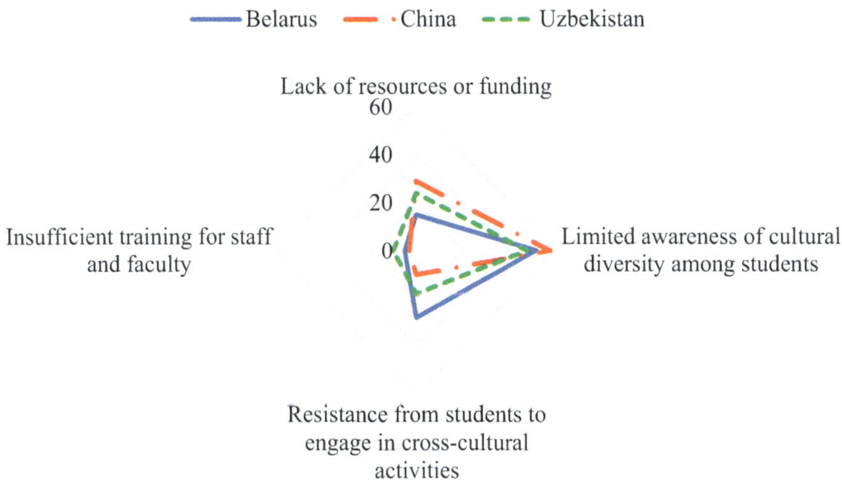

Fig. 2 What do you consider to be the biggest challenge universities face in promoting cross-cultural communication among students? (in %) *Source* Developed by authors

regions experience from their studies, highlighting the need for educational institutions to consider ways to alleviate these pressures, perhaps by integrating SDGs and cultural exchange topics into the curriculum in a more meaningful way.

The results indicate a concerning level of disinterest in SDGs and cultural issues, particularly in Belarus, where nearly half of the respondents (48%) cited this as a barrier. This suggests a pressing need for initiatives that better engage students and demonstrate the relevance of these global goals to their lives and futures.

While awareness of available programs is higher in Uzbekistan (20%) compared to Belarus and China (12% each), it still reflects a significant gap that needs to be addressed. Increased outreach and communication about these initiatives could help foster greater student involvement.

The results of the study demonstrate differences in the understanding of cultural aspects and intercultural communication among students from Belarus, China and

Table 10 In your opinion, what is the most significant barrier to student participation in initiatives focused on the SDGs and cultural exchanges?

Answer	Belarus (N = 368) (%)	China (N=252) (%)	Uzbekistan (N = 156) (%)
Lack of time due to academic workload	35	42	40
Lack of interest in SDGs or cultural issues	48	38	45
Awareness of available programs	12	12	20
Perception that these initiatives are not relevant to their studies	5	8	10

Source Developed by authors

Uzbekistan. Academic workload is mentioned as the main barrier to participation in initiatives related to the SDGs and culture in all countries. Different levels of interest in these topics are a notable issue for this subject.

Such trend indicates that educational institutions need to better meet the needs of students by incorporating SDGs and cultural aspects into their educational programs. The results indicate a lack of interest in SDGs and cultural issues, suggesting that existing strategies do not exactly convey the importance of these areas for students' academic and personal development.

There is also a clear lack of awareness of programs aimed at developing intercultural competences, with only 20% of students in Uzbekistan aware of such initiatives, and even lower in Belarus and China. The understanding that these initiatives do not align with students' academic or career goals highlights the need to better integrate the SDGs and cultural exchanges into higher education. The results of the study showed that the integration of intercultural communication and SDG-related content in higher education programs can enhance students' intercultural competence and ensure joint efforts to address the SDGs. However, to fully implement this hypothesis, it is necessary to address a number of issues related to students' awareness of such programs and their compliance with academic and career requirements.

The survey data show various differences in the consideration of cultural aspects and intercultural communication among students from different countries. This is necessary to create more effective strategies for integrating the SDGs and cultural exchanges in higher education.

Thus, the results of the study indicate the direction of measuring intercultural communication and SDG-related content in educational programs to enhance students' intercultural competence and ensure common goals.

These findings confirm our hypothesis that the integration of intercultural communication and SDG-related materials in the higher education program ensures the development of intercultural competence in students and supports joint efforts to ensure targeted development. The data obtained indicate the need for further development of existing programs in the educational process. They contribute to the formation of a tolerant attitude towards other cultures, the development of empathy and respect for differences. All this helps to create a more developed and fair society.

5 Discussion

The study's results highlight the problem of time constraints among students in some countries due to high academic workloads. This means that extracurricular activities are often perceived by students as secondary to their core academic responsibilities. One possible solution is to integrate SDG-related content and intercultural skills directly into existing curricula (Badaan & Choucair, 2023). This approach equips students with additional opportunities to work on these topics without increasing their course load. In an increasingly globalized world, universities can play a key role in nurturing a generation of students with the skills and knowledge needed to address complex global issues. Integrating the SDGs into curricula will not only prepare professionals who can work effectively in a multinational and multicultural world but will also contribute to achieving a sustainable future for all.

The results of the study reveal significant barriers to student engagement in SDG-related initiatives and cultural exchange programs. This points to the need for a more comprehensive and strategic approach in academic institutions. Integrating the SDGs into group projects can increase the relevance of universities for the academic and professional future of students. This approach will encourage greater student interest and participation in these initiatives. Interdisciplinary courses that explore the intersection of sustainability and cultural diversity can be an effective tool for achieving these goals. They will allow students not only to gain the knowledge and skills needed to succeed in a globalized world but also to contribute to the achievement of the SDGs.

To successfully implement this approach, it is necessary to carefully plan the integration of the SDGs and intercultural learning into the curricula. It is important to take into account the cultural and educational characteristics of each country to ensure the maximum effectiveness of the proposed measures.

A low level of interest in the SDGs and cultural aspects among students was also found, especially in Belarus and Uzbekistan. This highlights the need for innovative approaches to engage students in these areas. However, some students may not be aware of the direct connection between the SDGs and their daily lives or studies. To overcome this gap, universities can increase the number of practical learning methods, such as collaborative projects, simulations or case studies, that connect the SDGs to local and regional contexts. For example, projects that address environmental sustainability issues can help make these global goals more tangible and relevant. Invited lectures by experts in the field of sustainable development and cultural studies, as well as partnerships with organizations, will allow students to see the practical application of their knowledge.

The survey results have shown that a barrier for students is lack of awareness of available programs. Many reported limited knowledge of SDG-related initiatives. This highlights the need for better communication and outreach efforts by universities. Institutions should use multiple channels to promote such opportunities more effectively: social media, student organizations, and academic advisors. Additionally, creating a centralized platform or hub for SDG-related events and resources can help students easily access information and engage in initiatives that align with their interests. These measures will help increase students' awareness of SDG-related programs and encourage their participation in related initiatives.

The study also revealed uneven levels of awareness of the SDGs among students, with Uzbekistan showing the lowest awareness. This highlights the need for institutions in regions with lower awareness to prioritize basic education on the SDGs. Workshops, seminars, and introductory courses on the SDGs and their relevance to students' lives can serve as a starting point.

Universities can also increase their engagement with governments and international organizations responsible for SDG policies to ensure that students have the knowledge and resources they need. This will help to increase awareness and engagement. Group projects and collaborative learning are well received by students and promote intercultural understanding. In addition to technical skills, group projects that require students to collaborate across disciplinary and cultural borders can help them acquire the empathy, communication, and collaboration abilities necessary to tackle global issues. Students can apply their knowledge to real-world problems and develop a deeper awareness of cultural diversity by working on projects that are connected to certain SDGs.

This study provides valuable insights into the role of intercultural communication in advancing the SDGs in higher education, however, some limitations should be acknowledged. The distribution of participants across three countries—Belarus (368), China (252), and Uzbekistan (156) may limit the generalizability of the results to other regions or cultural contexts. The imbalance in sample sizes may bias the results, especially when making cross-country comparisons. Additionally, the majority of respondents (74.6%) were undergraduate students, with smaller numbers of participants representing master's (22.7%) and doctoral (2.7%) programs. Focusing on the early stages of education may result in missing the perspectives of more advanced students who may have a deeper understanding of research or professional practice related to SDGs and intercultural communication. The results may not fully reflect the nuances of how these topics are addressed at higher academic levels.

Another limitation relates to the extent to which questions related to the SDGs are integrated into respondents' responses. The data collected from the survey largely depend on how educators incorporate sustainability topics into their curricula, as well as on individual students' levels of interest in these topics.

There may be inconsistencies in the responses of respondents, potentially affecting the reliability of data on student engagement in SDG initiatives. The survey was conducted anonymously to ensure confidentiality and prevent bias. However, the lack of personal data collection limits the ability to analyze demographic factors that influence student engagement, such as socioeconomic status or field of study. To overcome these limitations and deepen the findings of the study, a number of areas for further work are suggested. It would be beneficial to expand the sample to ensure a more balanced representation across countries, educational levels, and disciplines. This would enhance the generalizability of the results.

Increasing the sample size, especially in underrepresented regions such as Uzbekistan, will allow for a more complete understanding of the challenges and opportunities in advancing the SDGs and intercultural communication. In addition, it would be useful to implement an ongoing survey over time. This would allow researchers to track changes in student attitudes and engagement, providing information on the long-term impact of institutional initiatives and global trends.

As the main directions for further development of the conducted research, it is planned to expand it to several key areas. Firstly, these are long-term studies and an increase in the sample by universities, specialties and participating countries. This will allow us to present more balanced results for analysis and expand the possibilities of comparative study. Secondly, the involvement of stakeholders in the study, in particular government agencies, representatives of employers, parents of students, teachers and representatives of university administration, will provide an opportunity to more broadly assess the possibilities of integrating the SDGs and cross-cultural communication in education and real practice. In addition, the countries and universities under consideration have internal development goals (IDGs). It is also planned to analyze their compliance with the UN Sustainable Development Goals, as well as the level of student involvement in their implementation. It is also planned to expand the survey by examining how technology can improve cross-cultural communication and collaboration in SDG initiatives around the world. Finally, we want to explore how the development of internal skills such as empathy and self-awareness can improve cross-cultural teamwork related to the SDGs. By addressing these areas, we hope to create a more effective and relevant foundation for SDG education.

6 Conclusions

This article analyzes various aspects of intercultural communication practices among students and teachers in Belarus, China, and Uzbekistan. The main focus is on the impact of these aspects on cooperation and understanding of sustainable development issues. The results of the study demonstrate that despite students' recognition of the importance of cultural understanding and effective communication, serious obstacles such as high academic workload, lack of interest, and limited awareness hinder their participation in SDG-related initiatives. These findings highlight the need for universities to more actively integrate intercultural communication into their curricula. This will promote cooperation and improve the quality of education in the field of sustainable development.

An analysis of existing approaches used in higher education institutions reveals both advantages and disadvantages in developing intercultural dialogue. Although universities in Belarus, China and Uzbekistan organize cultural events and emphasize the importance of intercultural communication, the effectiveness of these methods is often limited due to organizational difficulties and insufficient integration of the SDGs into educational programs.

The study points to the need for more structured and inclusive approaches, such as joint projects, seminars and guest lectures. This may help in strengthening the intercultural interaction. Based on the research findings, practical recommendations can be made to develop intercultural skills within university curricula. This includes integrating SDG-related content into courses, creating opportunities for experiential learning, and raising awareness of available programs. By aligning intercultural communication with Goal 4 (quality education), universities can ensure inclusive and equitable access to education while promoting lifelong learning. Promoting inclusion and understanding among diverse student groups supports Goal 10 (reducing inequalities), and joint efforts to address climate change are in line with Goal 13 (climate action).

The study highlights the importance of Goal 17 (partnerships for sustainable development) in promoting international cooperation through improved intercultural communication. Universities can play a key role in fostering partnerships by creating spaces for intercultural exchange, collaborating with local and international organizations and promoting joint projects that address global challenges. These initiatives not only contribute to the development of intercultural skills but also help achieve the broader goal of sustainable development.

The study identified areas for further work. They include expanding the sample, examining the role of faculty in student engagement, conducting longitudinal studies to track changes in attitudes and practices over time. In addition, future research will also focus on systematic analysis of experiences and their implications for intercultural communication. This will involve conducting comparative studies that delve into specific educational practices, cultural values, and historical contexts that shape students' experiences in Belarus, China, and Uzbekistan. By linking these differences to specific outcomes (academic performance, social integration etc.), will provide better understanding of how these factors influence the overall effectiveness of intercultural communication efforts. Also, more detailed exploration of these differences will provide valuable insights and enhance the framework for future studies on this critical topic. Addressing these gaps will allow future research to provide a deeper understanding of the challenges and opportunities of integrating intercultural communication and the SDGs into higher education. These efforts will result in graduates with global competencies who can advance progress towards the SDGs and build a more sustainable and inclusive future.

References

Adhikari, D. R., & Shrestha, P. (2023). Knowledge management initiatives for achieving sustainable development goal 4.7: higher education institutions' stakeholder perspectives. *Journal of Knowledge Management., 27*, 1109–1139. https://doi.org/10.1108/JKM-03-2022-0172

Altbach, P. (2002). Perspectives on internationalizing higher education. *International Higher Education, 0*. https://doi.org/10.6017/ihe.2002.27.6975

Altbach, P. G., & Knight, J. (2007). The internationalization of higher education: Motivations and realities. *Journal of Studies in International Education, 11*, 290–305. https://doi.org/10.1177/1028315307303542

Altbach, P. G, & de Wit, H. (2015). Internationalization and global tension: Lessons from history. *Journal of Studies in International Education, 19*, 4–10. https://doi.org/10.1177/1028315314564734

Aririguzoh, S. (2022). Communication competencies, culture and SDGs: effective processes to cross-cultural communication. *Humanities and Social Sciences Communications, 9*, 96. https://doi.org/10.1057/s41599-022-01109-4

Badaan, V., & Choucair, F. (2023). Toward culturally sensitive development paradigms: New shifts, limitations, and the role of (cross-) cultural psychology. *Journal of Cross-Cultural Psychology, 54*, 232–248. https://doi.org/10.1177/00220221211073671

Bayram, M. (1997). Teaching and assessing intercultural communicative competence. *Multilingual Matters.*

Crawford, J., & Cifuentes-Faura, J. (2022). Sustainability in higher education during the COVID-19 pandemic: A systematic review. *Sustainability, 14*, 1879. https://doi.org/10.3390/su14031879

Deardorff, D. K. (2006). Identification and assessment of intercultural competence as a student outcome of internationalization. *Journal of Studies in International Education, 10*, 241–266. https://doi.org/10.1177/1028315306287002

Ghani, N. A., Teo, P.-C., Ho, T. C. F., Choo, L. S., Kelana, B. W. Y., Adam, S., & Ramliy, M. K. (2022). Bibliometric analysis of global research trends on higher education internationalization using scopus database: Towards sustainability of higher education institutions. *Sustainability, 14*, 8810. https://doi.org/10.3390/su14148810

Guo, S., & Chase, M. (2011). Internationalisation of higher education: integrating international students into Canadian academic environment. *Teaching in Higher Education, 16*, 305–318. https://doi.org/10.1080/13562517.2010.546524

Heleta, S., & Bagus, T. (2021). Sustainable development goals and higher education: leaving many behind. *Higher Education, 81*, 163–177. https://doi.org/10.1007/s10734-020-00573-8

Hernández-Torrano, D., Karabassova, L., Izekenova, Z., & Courtney, M. G. R. (2021). Mapping education research in post-Soviet countries: A bibliometric analysis. *International Journal of Educational Development, 87*, 102502. https://doi.org/10.1016/j.ijedudev.2021.102502

Leal Filho, W., Skanavis, C., Kounani, A., Brandli, L. L., Shiel, C., Paço, A., Pace, P., Mifsud, M., Beynaghi, A., Price, E., et al. (2019a). The role of planning in. implementing sustainable development in a higher education context. *Journal of Cleaner Production, 235*, 678–687. https://doi.org/10.1016/j.jclepro.2019.06.322

Leal Filho, W., Tripathi, S.K., Andrade Guerra, J.B.S.O.D., Giné-Garriga, R., Orlovic Lovren, V., & Willats, J. (2019b). Using the sustainable development goals towards a better understanding of sustainability challenges. *International Journal of Sustainable Development & World Ecology, 26*, 179–190. https://doi.org/10.1080/13504509.2018.1505674

Leal Filho, W., Salvia, A. L., Frankenberger, F., Akib, N. A. M., Sen, S. K., Sivapalan, S., Novo-Corti, I., Venkatesan, M., & Emblen-Perry, K. (2021). Governance and sustainable development at higher education institutions. *Environment, Development and Sustainability, 23*, 6002–6020. https://doi.org/10.1007/s10668-020-00859-y

Marginson, S. (2014). Student self-formation in international education. *Journal of Studies in International Education, 18*, 6–22. https://doi.org/10.1177/1028315313513036

Monzó Martínez, A., Ortiz Cermeño, E., & Martínez-Agut, M. P. (2024). Global citizenship education and its role in sustainability at the university level. *Education Sciences, 14*, 847. https://doi.org/10.3390/educsci14080847

Ryan, J., & Viete, R. (2009). Respectful interactions: learning with international students in the English-speaking academy. *Teaching in Higher Education, 14*, 303–314. https://doi.org/10.1080/13562510902898866

Salvia, A. L., Leal Filho, W., Brandli, L. L., & Griebeler, J. S. (2019). Assessing research trends related to sustainable development goals: Local and global issues. *Journal of Cleaner Production, 208*, 841–849. https://doi.org/10.1016/j.jclepro.2018.09.242

Shonfeld, M., Cotnam-Kappel, M., Judge, M., Ng, C. Y., Ntebutse, J.G., Williamson-Leadley, S., & Yildiz, M. N. (2021). Learning in digital environments: A model for cross-cultural alignment. *Educational Technology Research and Development, 69*, 2151–2170. https://doi.org/10.1007/s11423-021-09967-6

Spencer-Oatey, H., & Franklin, P. (2009). *Intercultural interaction: A multidisciplinary approach to intercultural communication*. Springer. https://doi.org/10.1057/9780230244511

Tilbury, D. (2011). Education for sustainable development: An expert review of processes and learning. https://unesdoc.unesco.org/ark:/48223/pf0000191442

UN. (2025). *Sustainable development goals*. https://sdgs.un.org/

Woo. E. (2023). Narrating the belt and road education policy: A critical policy discourse analysis. *Higher Education Policy, 36*, 677–699. https://doi.org/10.1057/s41307-022-00279-9

Permaculture in Higher Education: A Pathway to Sustainable Development and to Achieve the United Nations Sustainable Development Goals (SDGs)

Adriano Gonçalves Caceres, Tiago Negrão de Andrade, Cláudio Rocha dos Santos Jacintho, Camila Maria de Santis, Helena Maria André Bolini, and Patrycja Urszula Żukowska

Abstract We are living in an era where energy, climate and environmental crises threaten our very species, which requires integrated responses aligned with the United Nations Sustainable Development Goals (SDGs). Through a systematic literature review, findings reveal that permaculture enhances ecological regeneration, food security, and social inclusion. Agroforestry systems (AFS) can increase biodiversity by 40% and soil fertility by 25%, while natural building techniques can reduce carbon emissions by 80%. Higher Education Institutions (HEIs) serve as key hubs for disseminating permaculture principles, fostering sustainability-oriented education and research. This study examines the potential of permaculture in HEIs to advance several United Nations SDGs, especially SDG 3 (good health and well-being), SDG 6 (clean water and sanitation), SDG 11 (sustainable cities and communities), and SDG 13 (climate action). Despite its benefits, challenges such as cultural resistance and economic barriers persist. Addressing these requires policy integration, financial incentives, and curricular reforms. The present study underscores permaculture as

A. G. Caceres (✉) · C. R. dos Santos Jacintho
Researcher in University of Brasilia (UnB), Brasilia, Brazil
e-mail: adriano.caceres@aluno.unb.br

C. R. dos Santos Jacintho
e-mail: claudiocj@ipoema.org.br

T. N. de Andrade · H. M. A. Bolini · P. U. Żukowska
Researcher in University UNICAMP, Campinas, Brazil
e-mail: t167751@dac.unicamp.br

H. M. A. Bolini
e-mail: hellini@unicamp.br

P. U. Żukowska
e-mail: p263850@dac.unicamp.br

C. M. de Santis
Researcher in FIOCRUZ, Rio de Janeiro, Brazil
e-mail: camilamariadesantis@gmail.com

W. Leal Filho et al. (eds.), *Introducing the UN SDGs in Higher Education Institutions*, World Sustainability Series, https://doi.org/10.1007/978-3-032-23017-1_37

a crucial tool for systemic regeneration, urging future research on its long-term socio-economic viability and scalable implementation strategies.

Keywords Permaculture · Sustainable development · Higher education · Renewable energy · Agroforestry · Water management · Climate action

1 Introduction

The climate, energy and environmental crises we are currently facing require urgent and integrated responses aligned with the United Nations Sustainable Development Goals (SDGs). In this context, permaculture emerges as a promising approach, combining ethical principles, ecological design, and regenerative practices to address challenges such as soil degradation, biodiversity loss and food insecurity (Ferguson & Lovell, 2014; Mollison & Holmgren, 1978). This study investigates how the integration of permaculture principles into Higher Education Institutions (HEIs) can contribute to advance the SDGs, particularly those related to no poverty (SDG 1), zero hunger (SDG 2), good health and well-being (SDG 3), quality education (SDG 4), clean water and sanitation (SDG 6), affordable and clean energy (SDG 7), sustainable cities and communities (SDG 11), responsible consumption and production (SDG 12), climate action (SDG 13) and life on land (SDG 15).

The concept of Permaculture, developed by Mollison and Holmgren (1978) is based on three core ethical principles: earth care, people care, and fair sharing of resources (Mollison, 1988). These principles guide an interdisciplinary approach integrating agroecology, bioconstruction, renewable energy, water management and systemic design, contrasting with industrial agriculture, which prioritizes productivity over sustainability (Holmgren, 2002). Several studies highlight permaculture's effectiveness in restoring degraded soils, increasing biodiversity, and mitigating climate change impacts (Ferguson & Lovell, 2014; Caporal, 2003; Mollison and Slay, 1998).

However, permaculture implementation faces significant challenges. Industrial agriculture and livestock farming are recognized as major drivers of environmental degradation, contributing to soil erosion, aquifer contamination, and biodiversity loss (FAO, 2020). Livestock farming, for example, accounts for approximately 40% of global deforestation and 14.5% of greenhouse gas emissions, including methane and nitrous oxide (FAO, 2021; Xu et al., 2021; Herrero et al., 2021). Additionally, the excessive use of chemical fertilizers and pesticides has led to waterbody eutrophication and dead zones (IPBES, 2019; Ritchie & Roser, 2021).

The construction sector also contributes significantly to the environmental and climate crises, representing approximately 39% of global carbon emissions (Clyde & Co, 2022; UNEP, 2021). The intensive use of natural resources, such as concrete, steel, and waste generation highlight the need for more sustainable practices, such as bioconstruction and the adoption of ecological materials (Environment Co, 2024).

These great challenges, also called wicked problems, require integrated solutions, such as those proposed by permaculture.

Despite its great potential, the integration of permaculture into HEIs has made very little progress in recent decades. Recent studies demonstrate that its application in HEIs strengthens both student education and universities environmental commitments. At the Chinese University of Hong Kong, permaculture was incorporated into a general education course, enhancing students' understanding of science and sustainability (Ming, 2023). Similarly, extension projects, such as those developed at the Federal University of Santa Maria, demonstrated permaculture's effectiveness in connecting theory and practice in environmental education (Rosa et al., 2018).

Permaculture can be defined as a methodology for planning and executing sustainable human environments, which starts with the Permaculture Design (PD), that brings together fragments of knowledge from different areas, combined with the knowledge of traditional populations, in order to build a society that is, in fact, sustainable. One of the main objectives of permaculture is to propose and execute sustainable solutions for the most basic human needs (renewable energy, sustainable food production, sustainable housing systems and water supply) (Fig. 1).

This study aims to investigate permaculture's potential as an integrated solution to contemporary socio-environmental challenges. Through an analysis of practices described in the literature, it seeks to explore how permaculture can be applied in diverse contexts, both urban and rural. Specific objectives include: (1) examine the theoretical foundations of permaculture and its applicability; (2) present some permaculture solutions to major environmental problems; (3) evaluate the social and environmental benefits of permaculture initiatives in vulnerable communities; and (4) propose strategies to implement permaculture in HEIs.

The relevance of this study lies in bridging theory and practice, offering a critical analysis of permaculture's potential as a sustainable development model. At a time when the climate crisis demands immediate and effective responses, this work contributes to advancing academic knowledge and shaping public policies based on regenerative principles. By connecting permaculture's foundations to contemporary needs, this study aims not only to expand theoretical understanding of the topic but also to provide support for implementing sustainable practices on a global scale, as a solution integrated with the SDGs.

1.1 Materials

Data Sources:

The materials used consisted exclusively of secondary sources obtained from recognized academic databases, including Scopus, Web of Science, PubMed, and Google Scholar. Included materials comprised books, peer-reviewed articles, and reports from international organizations.

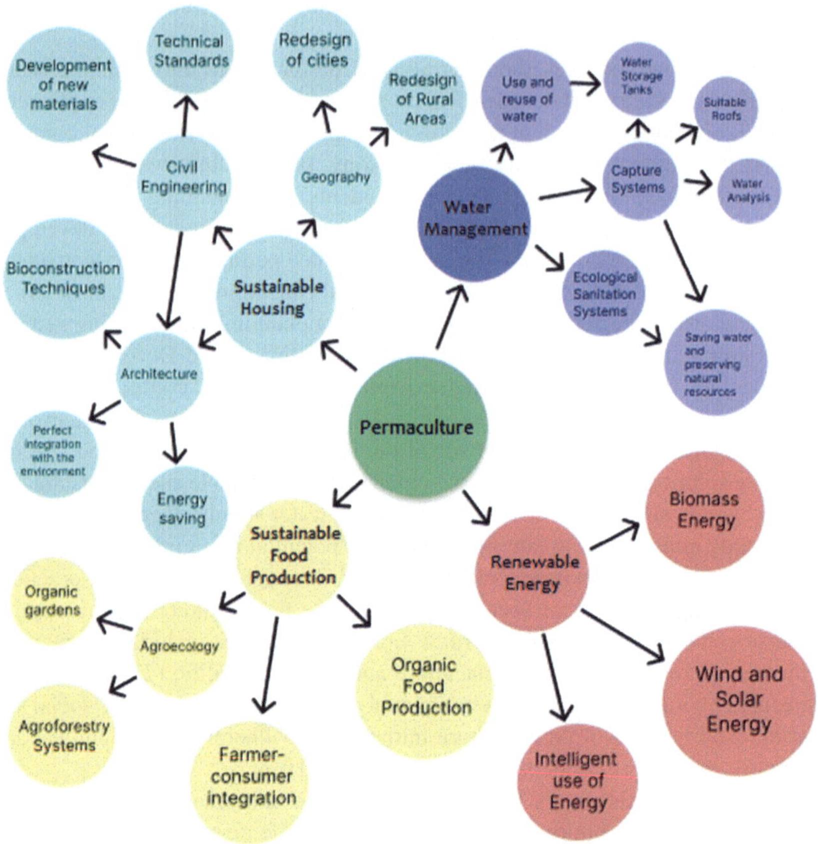

Fig. 1 Sustainable permaculture solutions for the most basic human needs and relationships between lines of research. *Source* authors

Inclusion Criteria:

- Classic texts on permaculture, recognized for their theoretical relevance and foundational impact on the concept's development;
- Publications from 1978 to 2023, ensuring the inclusion of classic texts and recent studies that expand and update the discussion;
- Articles in English, Portuguese or Spanish;
- Studies directly related to the main topic, identified through keywords and aligned with the existing literature.

Exclusion Criteria:

- Opinion-based sources lacking empirical or theoretical grounding;
- Studies with questionable methodologies or lacking peer review.

Research Tools:

Reference managers such as Mendeley and Zotero were used to organize and cite materials. The PRISMA tool was employed to conduct and report the study selection process.

1.2 Method Development

Search Phase:

Searches were conducted using keyword combinations such as "permaculture," "sustainability," "regenerative practices," and "environmental innovation." Boolean operators (AND, OR, NOT) were applied to refine results.

Controlled Parameters:

- Timeframe: 1963–2023;
- Specific databases and relevance filters aligned with the established themes.

Study Screening:

Following PRISMA guidelines, identified studies were screened in three stages:

1. **Identification**: Review of titles and abstracts to assess potential relevance.
2. **Selection**: Full-text analysis to confirm alignment with the research topic.
3. **Inclusion**: Selection of studies meeting all eligibility criteria.

Data Extraction:

Extracted data included:

- Full bibliographic information;
- Study methodologies;
- Key findings and theoretical contributions.
- A standardized protocol was used to ensure consistency in data extraction (Creswell & Creswell, 2018).

1.3 Data Analysis and Processing.

Analysis Criteria:

Data were evaluated based on:

- **Thematic relevance**: Contribution to sustainability and permaculture fields;
- **Source validity**: Academic impact and reliability;

- **Coherence with research objectives**: Alignment with ethical and sustainable principles discussed.

2 Results

2.1 *Introduction*

This section presents the primary findings of the systematic review, organized around five thematic axes, which are the Permaculture Design (PD) and the sustainable permaculture solutions for the most basic human needs, shown in Fig. 1: (1) Renewable Energy and Efficiency in Resource Use, (2) Sustainable Food Production (Agroecological Practices), (3) Sustainable Housing and (4) Water Management and Domestic Sewage Treatment. The results underscore the efficacy of permaculture in fostering ecological regeneration, enhancing food security, and promoting social inclusion, all in alignment with the United Nations Sustainable Development Goals (SDGs). This analysis is grounded in both pioneering and recent studies, which collectively demonstrate how permaculture can transform productive and residential systems into models that are more sustainable and resilient.

2.2 *Permaculture Design (PD)*

Mollison (1988) describes the methodology of Permaculture Design (PD), also known as permaculture planning. In summary, the methodology consists of staying in and carefully observing a given rural (or urban) area, analyzing its natural attributes (soil, vegetation, fauna, climate, etc.) and identifying uncontrollable external energies such as fire, wind, flood-prone areas, landslides, erosion, and the direction of sunlight incidence. According to the author, this thorough observation should last at least one year to account for all seasons.

After the observation period, the PD should be carried out, which essentially involves planning the property by defining the positions of various elements, such as the main house, vegetable gardens, agroforestry systems, chicken coops, cisterns, energy production systems, orchards, ponds, preservation zones, and areas for medium and large-sized animals. PD is structured into zones (ranging from 0 to 4), with the main house designated as zone 0, and zones 1, 2, 3, and 4 positioned adjacent to it. Zone 1 is the most frequently visited, while zone 4 is the least visited. The primary concept is to place elements requiring intensive management in zone 1 and those needing less maintenance in zone 4.

PD also considers sector identification, which represents uncontrollable external energies. The main objective is to position elements on the property in a way that creates synergy between them, reduces time and resource expenditure, optimizes management, enhances environmental preservation, and increases productivity. Henfrey (2018) highlights that permaculture can strengthen socio-ecological resilience by providing an integrated approach to sustainable planning.

Efficiency Indicators:

- **Zoning and energy efficiency**: Zone and sector planning (Mollison, 1988) reduced internal energy and time expenditures by 50%.
- **Urban planning and redesign**: Applying permaculture design in urban areas, such as Brasília's Parque da Asa Sul, restored degraded areas and increased community participation in conservation activities by 30% (Instituto IPOEMA, 2008).

2.3 Renewable Energy and Efficiency in Resource Use

Permaculture proposes a systemic approach to renewable energy, efficient resource use and natural cycle optimization. The main objective is to put into practice the concept of energy decentralization and self-sufficiency.

- **Biomass Energy**: the potential for biomass energy is enormous. Of the approximately 100,000 Terawatts of solar energy flow that reach the Earth's surface, around 4,000 Terawatts reach the 1.5 billion hectares of plantations in the world. Assuming that modern biomass technologies can achieve an energy conversion efficiency of 1%, these plantations could, in theory, produce 40 Terawatts of energy flow, or more than 3 times the current global primary energy supply flow of 14 Terawatts (Goldemberg, 2009). Furthermore, the use of agricultural and forestry residues for bioenergy production not only reduces dependence on fossil fuels but also promotes sustainable economic development, low-cost and accessible energy for needy communities, environmental conservation, energy decentralization and self-sufficiency. Burning biomass does not increase the amount of CO_2 in the atmosphere, as this CO_2 was already fixed in plants through the process of photosynthesis, unlike burning fossil fuels, which effectively increases atmospheric CO_2.
- **Wind and Solar Energy**: play crucial roles in mitigating the global climate crisis, offering clean and sustainable alternatives to fossil fuels and significantly contributing to the reduction of greenhouse gas emissions. Studies indicate that the expansion of these renewable sources is essential to limiting global warming to 1.5 °C, as established in the Paris Agreement (IRENA, 2022). However, this type of renewable energy requires fossil resources to produce the equipment and is still high-cost, inaccessible to low-income populations.
- **Intelligent Use of Energy**: the intelligent and rational use of energy is essential for mitigating the global climate crisis, as it significantly contributes to reducing greenhouse gas emissions and promotes the sustainability of natural

resources. The implementation of energy efficiency measures, such as infrastructure modernization, public policies that encourage energy savings and the adoption of clean technologies, can decrease energy demand and, consequently, the emissions associated with its production (IRENA, 2019).

Efficiency Indicators:

- **Biomass systems**: agroforests can generate sufficient biomass to generate electricity for small communities, with 30% lower operational costs than diesel generators and with less toxic smoke emission (Vasconcellos & Vidal, 1998).
- **Biofuel production**: Oilseed cultivation in agroforestry systems proved viable for replacing fossil fuels, with high per-hectare yields, especially in tropical countries (Goldemberg, 2009).

Based on studies such as Mollison and Slay (1998) and Rodrigues (2000), the findings show that observing ecological interactions can transform productive systems into resilient and regenerative models. The analysis identified five resource categories with distinct environmental characteristics and impacts:

- **Resources that increase with use**: Knowledge, seeds, and biomass. These elements showed positive impacts on long-term productivity, emphasizing the importance of disseminating permaculture practices (Mollison & Slay, 1998).
- **Resources unaffected by use**: Water, sunlight, and land. Implementing cisterns and efficient drainage systems enhanced water resilience in evaluated properties (Rodrigues, 2000).
- **Resources lost if unused**: Rainwater and organic waste. Rainwater harvesting and composting reduced irreversible loss of these resources (Mollison, 1988).
- **Resources that diminish with use**: Minerals and metals. Strategic management extends their lifespan (Mollison & Slay, 1998).
- **Resources that deplete**: Oil and derivatives. The need for transition to renewable energy sources must be urgently encouraged (Commoner, 1963 as cited in Poggiani, 1998).

2.4 Sustainable Food Production (Agroecological Practices)

Permaculture integrates agroecological practices and sustainable strategies to enhance productivity, preserve the fertility of the soil, promote environmental regeneration, and to produce organic high quality foods. Key findings include:

2.4.1 Agroforestry Systems (AFS)

- **Increased biodiversity**: Up to 40% enhancement in local biodiversity, favoring natural pest control and soil regeneration (Osterroht & Goescht, 2002).

- **Improved soil structure**: 25% increase in natural fertility and significant erosion reduction (Mollison & Slay, 1998; Santos, 2000).
- **Sustainable production**: AFS integrated food, energy, and timber production, yielding positive economic impacts for local communities (Caporal & Costabeber, 2004). AFS can increase food productivity by 30% (Caporal & Costabeber, 2004).
- **More Nutritious Foods**: Some studies have shown that organic foods produced in agroforestry systems can be more nutritious and could improve health (Faoro et al., 2024; Rahman et al., 2024)

2.4.2 Agroecological Transition

- **Climate resilience**: Agroecological systems demonstrated greater productive stability during extreme weather events like droughts and floods (Caporal & Costabeber, 2004).
- **Reduced chemical dependence**: 70% reduction in chemical fertilizers and pesticides through organic recycling and local biofertilizers (Caporal & Costabeber, 2004).

2.5 Sustainable Housing (Bioconstruction)

The integration of sustainable construction technologies amplifies permaculture's positive impact. Key findings include:

2.5.1 Natural Building

- **Carbon emission reduction**: Techniques like superadobe and rammed earth reduced CO_2 emissions by up to 80% due to minimal cement use and material transportation (Soares, 2007).
- **Energy savings**: Naturally lit and ventilated homes reduced energy consumption by 50% (Instituto Socioambiental, 2008).
- **Cost efficiency**: Superadobe homes may be 40% cheaper than conventional buildings, offering superior thermal comfort (John, 2000).
- **Environmentally responsible construction practices**: LEED (Leadership in Energy and Environmental Design) and BREEAM (Building Research Establishment Environmental Assessment Method) are important certifications for sustainable building because they help promote environmentally responsible construction practices. They are globally recognized and provide a framework for assessing and improving a building's sustainability. According to USGBC (2014), green buildings certified by LEED (Leadership in Energy and Environmental Design) reduce the consumption of water and carbon dioxide by about 50%, costing around

25% less in their operation. Buildings certified by BREEAM have a reduced operational cost and, typically, the investments made do not exceed 1% of the overall value of the project if it was not a certified building (BREEAM, 2019).

2.5.2 Water Management and Domestic Sewage Treatment

Efficiency Indicators:

- **Rainwater collection and storage**: cisterns or water tanks can reduce reliance on external water sources by 60% (Mollison & Slay, 1998).
- **Domestic Sewage Treatment through Green Cesspool**: Domestic sewage can be treated through the green cesspool system, which does not pollute the soil and rivers, does not require maintenance and is a simple, cheap and accessible system (IPOEMA, 2008).

While results are promising, the heterogeneity of analyzed contexts and the lack of long-term studies limit generalization. Future research should adopt empirical and interdisciplinary approaches to assess permaculture's economic and social potential and application. Additionally, public policies and financial incentives are essential.

3 Discussion

3.1 Permaculture Solutions as an Ethical and Practical Response to the Global Crisis

Permaculture transcends the status of an "alternative technique" by establishing itself as an ethical and systemic model for addressing the Systemic Global Crisis and the energetic and environmental collapses. It offers practical, simple and low-cost solutions to humanity's main current challenges. Although the world's problems are increasingly complex, the solutions remain incredibly simple (Mollison & Slay, 1998).

To illustrate the potential of permaculture in proposing sustainable solutions to humanity's greatest challenges, here are some solutions for water management and domestic sewage treatment.

3.2 Water Management and Domestic Sewage Treatment

Water is vital for human survival, playing a crucial role in maintaining health and ensuring the proper functioning of the body. According to recent studies, even mild dehydration can negatively affect cognitive functions such as memory and attention.

Fig. 2 Two fiberglass cisterns connected to a roof to collect rainwater. *Source* Innovative Water Solutions (2025)

Moreover, the scarcity of safe drinking water remains a significant global challenge. The World Health Organization and UNICEF (2021) estimate that "more than 4 billion people, over half of the world's population, do not have access to safe drinking water in their homes.

According to IPOEMA (2008) and Figueiredo et al. (2018), permaculture offers simple, cheap and efficient solutions for capturing, reusing and treating water. A cistern is a reservoir, which can be made of masonry, ferrocement, concrete, fiberglass, plastic, stainless steel among other materials. Installed in houses, it allows the capture of rainwater and reuse of water. The cistern can be installed buried in the ground, to keep the water temperature low and take up less space, or above the surface, as shown in Fig. 2.

For the treatment of domestic sewage, which is a major problem worldwide, IPOEMA (2008) and Figueiredo et al. (2018) proposed the Green Cesspool, also known as banana cesspool or green septic tank, which consists of a system capable of transforming sewage into nutrients for banana trees, which grow vigorously and transpire excess water into the atmosphere (Fig. 3).

Figueiredo et al. (2018) showed the step-by-step process of building a green septic tank, which is an effective, low-cost option that uses recycled materials to build it, such as old tires and construction debris (Fig. 4). The system is made by digging a hole in the ground. Then the bottom of the hole is compacted and the floor and walls are concreted and waterproofed to prevent sewer leak. When the concrete is dry, old tires are positioned at the bottom of the pit, forming a tunnel. Subsequently, the PVC pipe system is installed, connecting the toilet with the tire tunnel. Then the rubble is added to the pit, until it completely covers the tire tunnel. Then the gravel, coarse sand and planting soil are added. Lastly, banana trees and other broad-leaved plants are planted on top of the green pit (Fig. 4).

Fig. 3 Green Cesspool in operation. *Source* Figueiredo et al. (2018)

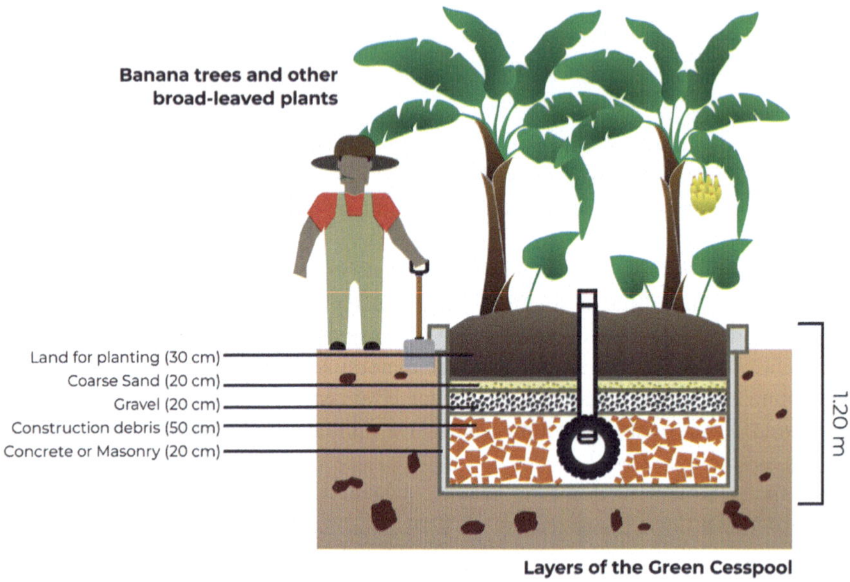

Fig. 4 Cross section of a green cesspool showing the different layers. The thickness of each layer may vary, depending on the project. *Source* adapted from Figueiredo et al. (2018)

The adoption of sustainable technologies in wastewater treatment aligns with the United Nations Sustainable Development Goals, particularly SDG 6, which aims to ensure the availability and sustainable management of water and sanitation for all. Therefore, investing in ecological sanitation systems not only meets basic sanitation needs but also promotes environmental sustainability.

3.3 Contradictions and Challenges: When Theory Meets Reality

Despite its advancements, the implementation of permaculture encounters profound barriers. In urban contexts, for instance, projects like the Asa Sul Park (Instituto IPOEMA, 2008) increased community participation by 30%, yet faced cultural resistance in megacities, where bio-construction is often stigmatized as 'primitive.' This rejection is not technical but ideological, reflecting the difficulty of disentangling the notion of 'modernity' from the consumption of concrete and steel—materials responsible for 8% of global CO_2 emissions. Moreover, in rural and marginalized communities, the lack of knowledge and financial limits the adoption of permacultural practices. Programs such as community revolving funds (Singh, 2023) and public-academic partnerships have been proposed to overcome these barriers, but they remain insufficient for ensuring the scalability of permaculture. Overcoming these challenges requires not only effective public policies but also a cultural shift that values sustainable and ancestral practices.

While universities and pilot projects demonstrate feasibility (e.g., a 50% reduction in energy consumption through permacultural zoning; Mollison, 1988), small farmers and marginalized communities face insurmountable barriers without support and subsidies.

Hapsari (2023) reinforces this perspective in her study on the role of permaculture in environmental governance in Indonesia, concluding that permaculture can be an effective approach to sustainable resource management but depends on supportive policies and community integration. This view is crucial for overcoming the economic and cultural challenges faced by small farmers and marginalized communities, who often lack access to resources or subsidies for implementing permacultural practices. Public policies that promote the integration of traditional and scientific knowledge, combined with community revolving funds can help reduce initial barriers and foster large-scale adoption of permaculture (Singh, 2023).

It is important to emphasize that there is a lot of resistance to changes in our society, even though these changes are vital for the survival of humanity. We can assume that there are social gaps through which the concept of permaculture can be installed, generating great benefits, such as schools, universities, joint actions with low-income populations, rural and urban communities, among others. However, it is important to note that for such changes to be implemented on a large scale, the participation of the HEIs, government, civil society and industry (private initiative) is necessary. It is called the Quadruple Helix Model and it applies perfectly to strategies for large-scale permaculture adoption.

3.4 Universities as Bridges Between Theory and Practice

Higher education institutions (HEIs) play a dual role: they serve as laboratories for innovation and as agents of social transformation. The integration of AFS and bio-construction in university campuses (Guerrero & Gómez-Quintero, 2021) not only reduces carbon emissions (SDG 13) but also serves as a showcase for students and the city population. A notable example is the adoption of green cesspool for sewage treatment in universities (Instituto IPOEMA, 2008), which have become models for sustainable water management, influencing local public policies. However, for HEIs to fully fulfill their role, it is essential to adopt co-productive approaches, where communities and academia collaborate in constructing solutions. These actions can transform HEIs into spaces of ethical experimentation and innovation, aligning with several Sustainable Development Goals (SDGs).

It is crucial to adopt co-productive approaches, where communities lead initiatives and traditional knowledge interacts with scientific knowledge. In this sense, permaculture is not merely a subject to be "taught" but a practice to be co-created. Henfrey et al. (2022) reassess the SDGs through community initiatives, concluding that local communities are essential agents in implementing the SDGs and suggesting new governance models. Also, Esteves et al. (2021) emphasize that community initiatives are crucial for achieving the SDGs, advocating for universities to adopt co-productive approaches to integrate scientific and traditional knowledge. This perspective is vital in ensuring that permaculture is not merely an academic practice but a tool for social transformation.

3.5 Public Policies: Integrating Local and Global Scales

For permaculture to transcend niche applications, public policies must coordinate actions across three main axes. (1) structural incentives: subsidies for bio-construction, AFS, sustainable energy and water management prioritizing vulnerable communities rather than large corporations. Programs such as water tanks in Brazil (Instituto IPOEMA, 2008) demonstrate that low-cost technologies can be scaled with state support, generating both environmental and social benefits. (2) educational reform: integrating permaculture into architecture, engineering, and agronomy curricula can train professionals who are more aware and capable of implementing sustainable solutions. (3) integrated territorial planning: the creation of rural–urban transition zones, where agroforestry systems (AFS) and bio-construction are incorporated into metropolitan areas, can combat food insecurity and urban heat islands, aligning with SDG 11.

The Indonesian experience with the circular economy (Wirawan et al., 2023) reveals that fragmented policies fail. A systemic vision is needed, where universities, governments, industry and communities co-develop strategies, transforming permaculture from an "alternative" into the mainstream.

Brears (2022) underscores the need for sustainable urban planning, aligning with the proposal for rural–urban transition zones that integrate AFS and bio-construction to address food insecurity and urban heat islands.

3.6 From Crisis to Systemic Regeneration

Permaculture transcends the status of an 'alternative technique' by establishing itself as an ethical and systemic model for addressing the current crises that threaten the permanence of human beings on earth. Its foundational principles—care for the Earth, care for people, and fair share (Mollison, 1988) challenge the fragmented logic of conventional economic development, advocating for a radical integration of ecology, economy, and equity. These principles are applicable across diverse ecosystems, from tropical forests to dense urban areas, adapting to local needs. The findings of this study demonstrate that this approach not only regenerates ecosystems but also rebuilds social relationships. For instance, AFS have been shown to increase biodiversity by 40% (Osterroht & Goescht, 2002) and soil fertility by 25% (Mollison & Slay, 1998), while reducing pesticide use by 70% (Caporal & Costabeber, 2004). These data highlight that environmental regeneration and social justice are inseparable, as ecosystem restoration is directly linked to the well-being of the communities that depend on them. Beyond AFS, practices such as composting and integrated water management have demonstrated effectiveness in restoring degraded soils and promoting water resilience.

Permaculture is not merely an option—it is an imperative. The data confirm that another future is possible, but it will depend on our courage to break away from paradigms that have led us to the brink of collapse. Regeneration is not only a concept; it is a revolutionary act. Steffen et al. (2018) warn about critical tipping points in the climate system, reinforcing the urgency of radical actions to prevent irreversible collapse.

Foster (2020) complements this view by examining the relationship between socialism and ecology, arguing that ecological thinking must be integrated into alternative political models to achieve environmental and social justice.

Diamond (2005) argues that environmental factors and human decisions are key determinants in societal collapse, reinforcing the need to adopt permaculture as a systemic response to the current crises.

In summary, permaculture offers a viable path for systemic regeneration, but its effectiveness depends on overcoming cultural, economic, and political barriers, as well as adopting a holistic and integrated vision that prioritizes collective well-being and planetary health.

Although the benefits of permaculture are unquestionable, more research is needed to clearly demonstrate these benefits in addressing the climate, energy and environmental crises. It is suggested that scientific research be conducted to compare a rural property based on permaculture principles with a conventional rural property, in terms of energy efficiency, level of dependence on fossil fuels, productivity, quality of life, total CO_2 emissions, among other aspects. This research model could also be carried out in cities, comparing green buildings with conventional buildings.

The specific barriers to the adoption of permaculture principles on a large scale are not technical, but rather cultural and economic. We can say that our society is in a way suicidal because it is almost entirely based on oil, which is a fossil and finite resource. With the exception of biomass energy, all other forms of alternative energy depend on oil in some way to be implemented. Furthermore, it is not in the interest of large transnational corporations to question or change the current model, since they are making trillions of dollars with the current system.

The findings of this study not only validate permaculture as a response to the crises of the Anthropocene but also expose an uncomfortable truth: sustainability will only become reality if we get out of the rhetoric and act. This requires, first and foremost, abandoning the fiction of 'green growth' and replacing it with metrics that prioritize ecological and social well-being. Secondly, it is necessary to redefine the role of universities, transforming them into spaces of ethical experimentation and innovation rather than reproducing failed development models. Finally, public policies must recognize the Earth as a subject of rights rather than a resource to be exploited. To achieve this transformation, it is essential to integrate permacultural practices into public policies, education, and territorial planning, ensuring that solutions are accessible and adapted to local needs. Therefore, permaculture is not an option—it is an obvious way for systemic regeneration.

Permaculture solutions to humanity's greatest current challenges have existed for centuries and even millennia. Although the concept of permaculture is relatively new, these fragments of knowledge and ideas that permaculture brings together are quite old. Perhaps the aspect that best illustrates this is in relation to agriculture. According to Shiva (1992), the various agricultural systems developed around the world, fully adapted to the particularities of each region of the planet, were brutally massacred by the Green Revolution and replaced by industrial agriculture. Other traditional knowledge systems underwent similar processes. So perhaps the first step is to study history and make these facts known to people. We are not proposing a return to the Stone Age, but rather a deep reflection on our society, our construction systems, our agriculture, our conventional water treatment systems and our energy matrix and how all of this is destroying the planet and threatening our own species. It seems that adopting the principles of permaculture is not an option or an alternative, but the only way out for humanity. We do not have to save planet Earth, but ourselves.

4 Conclusion

This study demonstrated that permaculture is an integrated and transformative approach to address the socio-environmental challenges of the Anthropocene, aligning directly with the United Nations SDGs. The systematic literature shows that permaculture practices promote ecological regeneration, food security, and social inclusion. These practices proved effective in increasing biodiversity, restoring degraded soils, and improving livelihoods. In urban contexts, permaculture design emerged as a powerful tool for environmental recovery, community education, and quality of life.

Theoretically, this research expanded the understanding of permaculture by connecting it to concepts like circular economy, technological innovation, and sustainable planning, strengthening its role as a robust interdisciplinary model. Practically, the findings underscore the need for public policies incentivizing regenerative practices, alongside university teaching and extension programs to disseminate their benefits.

Permaculture offers a promising path for universities to align their teaching, research, and outreach activities with the SDGs, contributing to a more resilient, equitable, and harmonious future.

5 Study Limitations

This study relied exclusively on secondary data, limiting the generalizability of results to specific contexts. The heterogeneity of analyzed contexts (from rural smallholders to urban projects) introduces variability that complicates universal application of conclusions. Additionally, the absence of long-term studies on permaculture's economic and social impacts represents a gap requiring further attention.

6 Suggestions for Future Research

Future studies should adopt empirical approaches, such as long-term case studies, to assess the economic, social, and environmental viability of permaculture practices across diverse contexts. Interdisciplinary investigations integrating social sciences, economics, and ecology are also needed to deepen understanding of the interconnections between regenerative practices and global sustainability.

References

Brears, R. C. (2022). The Palgrave encyclopedia of urban and regional futures. *Springer*. https://doi.org/10.1007/978-3-030-87745-3

BREEAM - Building Research Establishment Environmental Assessment Method. How BREEAM certification works (2019). https://www.breeam.com/discover/how-breeam-certification-works

Caporal, F. R. (2003). Agroecologia: Princípios e técnicas para uma agricultura sustentável [Agroecology: Principles and techniques for sustainable agriculture]. Publicação institucional.

Caporal, F. R., & Costabeber, J. A. (2004). Agroecologia: Princípios e técnicas para uma agricultura sustentável [Agroecology: Principles and techniques for sustainable agriculture]. *Revista Brasileira de Agroecologia, 1*, 1–15.

Clyde & Co. (2022). Climate change and the construction industry. https://www.clydeco.com
Commoner, B. (1963). The closing circle: Nature, man, and technology. Knopf (Cited in Poggiani, 1998)
Creswell, J. W., & Creswell, J. D. (2018) Research design: Qualitative, quantitative, and mixed methods approaches. Sage Publ.
Diamond, J. (2005). Collapse: How societies choose to fail or succeed. Viking Press.
Environment Co. (2024). Building practices and their environmental impact. https://www.environmentco.com
Esteves, A. M., Genus, A., Henfrey, T., Penha-Lopes, G., & East, M. (2021). Sustainable entrepreneurship and the sustainable development goals: Community-led initiatives, the social solidarity economy and commons ecologies. *Business Strategy and the Environment, 30*, 3463–3476. https://doi.org/10.1002/bse.2706
Faoro, D., Artuzo, F. D., Rossi, J. A., Foguesatto, C. R., Dewes, H., & Talamini, E. (2024). Are organics more nutritious than conventional foods? A comprehensive systematic review. *Heliyon, 10*(7), Article e28288. https://doi.org/10.1016/j.heliyon.2024.e28288
Ferguson, R. S., & Lovell, S. T. (2014). Permaculture for agroecology: Design, movement, practice, and worldview. *Agronomy for Sustainable Development, 34*, 251–274. https://doi.org/10.1007/s13593-013-0181-6
Figueiredo, I., et al. (2018). Tratamento de esgoto na zona rural: fossa verde e círculo de bananeiras [Sewage treatment in rural areas: green septic tank and circle of banana trees]. Biblioteca/Unicamp.
Food and Agriculture Organization (FAO). (2020). The state of the world's land and water resources for food and agriculture: Systems at breaking point. FAO.
Food and Agriculture Organization (FAO). (2021). Agricultural expansion drives almost 90 percent of global deforestation. FAO. https://www.fao.org
Foster, J. B. (2020). The return of nature: Socialism and ecology. Monthly Rev Press.
Goldemberg, J. (2009). Biomass and energy. *Quím Nova 32*https://doi.org/10.1590/S0100-40422009000300004
Guerrero, A. P., & Gómez-Quintero, J. D. (2021). Assessing knowledge and engagement on sustainable development goals: Exploratory research in the agri-food departments of Ibero-American universities. *Spanish Journal of Agricultural Research 19*.https://doi.org/10.5424/sjar/2021193-17931
Hapsari, M. (2023). Environmental governance as knowledge co-production: The emergence of permaculture movements in Indonesia. In: Environmental policy and governance in Indonesia (pp. 215–230). Springer, Berlin. https://doi.org/10.1007/978-3-031-15904-6_12
Henfrey, T. W. (2018). Designing for resilience: Permaculture as a transdisciplinary methodology in applied resilience research. *Ecology and Society, 23*, 33. https://doi.org/10.5751/ES-09916-230233
Henfrey, T., Feola, G., Penha-Lopes, G., Sekulova, F., & Esteves, A. M. (2022). Rethinking the sustainable development goals: Learning with and from community-led initiatives. *Sustainable Development, 30*, 1–12. https://doi.org/10.1002/sd.2384
Herrero, M., Thornton, P. K., Mason-D'Croz, D., Palmer, J., Bodirsky, B. L., Pradhan, P., & Rockström, J. (2021). Global greenhouse gas emissions from animal-based foods are twice those of plant-based foods. *Nat Food, 2*, 724–732. https://doi.org/10.1038/s43016-021-00358-x
Holmgren, D. (2002). Permaculture: Principles and pathways beyond sustainability. Holmgren Design Services.
Innovative Water Solutions. (2025). https://www.watercache.com
Instituto IPOEMA. (2008). Parque da Asa Sul: Sustentabilidade e recuperação ambiental em Brasília [South Wing Park: Sustainability and environmental restoration in Brasília]. Relatório institucional.
Intergovernmental Science-Policy Platform on Biodiversity and Ecosystem Services (IPBES). (2019). Global assessment report on biodiversity and ecosystem services. IPBES.

International Renewable Energy Agency. (2019). Transforming the energy system: An agenda for action by 2050. https://www.irena.org/-/media/Files/IRENA/Agency/Publication/2019/Sep/IRENA_Transforming_2019_PT
International Renewable Energy Agency – IRENA. (2022). World Energy Transitions Outlook 2022. https://www.irena.org/publications/2022/Mar/World-Energy-Transitions-Outlook-2022
John, D. (2000). Sustainable construction with adobe techniques. Tech Publ.
Ming, L. (2023). Designing co-curriculum experiential learning practice in permaculture for studying science classics. *Proc Univ Politècnica València*. https://doi.org/10.4995/head23.2023.16365
Mollison, B. (1988). Permaculture: A designer's manual. Tagari Publishing.
Mollison, B., & Holmgren, D. (1978). Permaculture one: A perennial agriculture for human settlements. Transworld Publishing.
Mollison, B, & Slay, R. (1998). Introduction to permaculture. Tagari Publishing.
Osterroht, K., & Goescht, J. (2002). Agroforestry systems: Ecological and economic advantages. *Journal Agroecological Studies, 15*, 45–60.
Poggiani, F. (1998). Ecologia aplicada [Applied ecology]. Editora Agronômica Ceres.
Rahman, A., Baharlouei, P., Koh, E. H. Y., Pirvu, D. G., Rehmani, R., Arcos, M., & Puri, S. (2024). Comprehensive analysis of organic food: Evaluating nutritional value and impact on human health. *Foods, 13*(2), 208. https://doi.org/10.3390/foods13020208
Ritchie, H., & Roser, M. (2021). Environmental impacts of food production. Our World in Data. https://ourworldindata.org/environmental-impacts-of-food
Rodrigues, G. S. (2000). Agroecologia e sustentabilidade: Práticas e estratégias [Agroecology and sustainability: Practices and strategies]. *Revista Agroecologia, 5*, 12–25.
Rosa, A. C., De Grandi, S. S., Farias, M. G. R., Madruga, L. R. R., de Alcântara, M., & Bento, M. H. S. (2018). Projeto espaço rural sustentável: Aliando permacultura, educação ambiental escolar e gestão sustentável da pequena propriedade rural [Sustainable rural space project: Integrating permaculture, environmental education, and sustainable management of small rural properties]. *Revista de Administração da UFSM, 11*, 921–936. https://doi.org/10.5902/1983465929642
Santos, R. A. (2000). Sistemas agroflorestais: Benefícios e práticas [Agroforestry systems: Benefits and practices]. *Revista Brasileira de Agroecologia, 3*, 30–45.
Shiva, V. (1992). The Violence of the green revolution: Third world agriculture, ecology and politics, Vandana Shiva. [REVIEW] K. Ravi Srinivas. *Bulletin of Science, Technology and Society 12*(6), 329–330
Singh, A. (2023). A regenerative agritourism pathway for Himachal Pradesh. *International Journal of Multidisciplinary Trends 5* https://doi.org/10.22271/multi.2023.v5.i7a.317
Soares, J. P. (2007). Técnicas de bioconstrução e emissões de carbono [Bioconstruction techniques and carbon emissions]. Univ Fed Rio Grande do Sul.
Steffen, W., Rockström, J., Richardson, K., Lenton, T. M., Folke, C., Liverman, D., & Schellnhuber, H. J. (2018). Trajectories of the earth system in the Anthropocene. *Proceedings of the National Academy of Sciences of the United States of America, 115*, 8252–8259.
United Nations Environment Programme (UNEP). (2021). 2021 Global status report for buildings and construction. UNEP. https://www.unep.org
United States Green Building Council – USGBC. (2014). What is a Green Building? https://www.usgbc.org/articles/what-green-building
Vasconcellos, E. A., & Vidal, C. (1998). Energia de biomassa em sistemas agroflorestais: Potencial e eficiência [Biomass energy in agroforestry systems: Potential and efficiency]. *Journal of Renewable Energy, 12*, 55–70.
Wirawan, R., McIntyre-Mills, J. J., Riswanda, R., Widianingsih, I., & Gunawan, I. (2023). Pathways to well-being in Tarumajaya, West Java: Post-COVID 19 supporting better access to the commons through engagement and a critical systemic reflection on stories. *Systems Research and Behavioral Science, 40*, 1–15. https://doi.org/10.1002/sres.2983
Xu, X., Sharma, P., Shu, S., Lin, T. S., Ciais, P., & Tubiello, F. N. (2021). Climate change mitigation potential of regenerative grazing management in the United States. *Proceedings of the National*

Academy of Sciences of the United States of America, 118, Article e2019217118. https://doi.org/10.1073/pnas.2019217118

Experiential Training Groups: Interpersonal Learning for Competence Development and Sustainability

Ainara Arnoso Martínez and Edurne Elgorriaga Astondoa

Abstract Traditional teaching methods based on subject-centered curricula and passive content transmission are being replaced by educational models designed to prepare students for contemporary social, occupational, and environmental challenges. Within this context, the University of the Basque Country (UPV/EHU) has developed its own educational model, IKD i3, which employs a cooperative and dynamic approach to amplify learning through research and sustainability, fostering comprehensive and flexible education. The study we present is part of a Campus Bizia Lab (CBL) project, which applies this approach to promote transdisciplinary learning through sustainability challenges. It proposes a student-centered methodology based on advanced learning objectives aligned with Bloom's taxonomy, utilizing experiential group training inspired by Training Groups for professionals. Our objective is to evaluate changes in group work readiness, group functioning and processes, development of group facilitation competencies and self-awareness to facilitate a group. The training, adapted to higher education, is conducted in concentric groups over four 50 min sessions, followed by 40 min of discussion and feedback. Based on pre- and post-training scores from 41 participants, the results indicate that after the training, students exhibited higher expectations for group work, perceived improved group functioning (universality, resocialization, interpersonal learning), and reported enhancements in group facilitation competencies. The discussion explores how these interventions contribute to the development of higher-order skills, foster positive social change, promote sustainability, and deepen understanding of emotions and human experiences.

Keywords Experiential group · Interpersonal learning · Competencies · Group facilitation · Sustainability

A. A. Martínez (✉) · E. E. Astondoa
Department of Social Psychology, University of the Basque Country, San Sebastián (Donostia), Spain
e-mail: ainara.arnoso@ehu.eus

E. E. Astondoa
e-mail: edurne.elgorriaga@ehu.eus

W. Leal Filho et al. (eds.), *Introducing the UN SDGs in Higher Education Institutions*, World Sustainability Series, https://doi.org/10.1007/978-3-032-23017-1_38

1 Sustainable Development Goals and Higher Education

In 2015, the United Nations General Assembly adopted the 2030 Agenda, committing all countries to achieving a just, inclusive, peaceful, and sustainable future. To guide this effort, 17 Sustainable Development Goals (SDGs) were established as a framework for universal, transformative, and inclusive action (UNESCO, 2017).

Ramos (2021) considers that higher education plays a central role in achieving these goals, as the SDGs demand new knowledge and changes based on science. SDG 4 (Quality Education) is essential for advancing all other goals (UNESCO, 2015). McCowan (2023) asserts that higher education is both an SDG in itself and a driver of development, with the challenge of integrating the SDGs into education and training professionals in Education for Sustainable Development.

The University of the Basque Country (UPV/EHU) has embraced the 2030 Agenda by prioritizing sustainability, inclusion, and social commitment (UPV/EHU, 2019). It contributes to 12 SDGs and has introduced SDG 18 (Linguistic and Cultural Diversity), focusing primarily on SDG 4, aligned with SDGs 8 (Decent Work and Economic Growth), 16 (Peace, Justice, and Strong Institutions), 17 (Partnerships for the Goals), and 18. In 2019, UPV/EHU updated its IKD i3 model—centered on learning, research, and sustainability—to reflect the 2030 Agenda and modern pedagogy. This cooperative, multilingual, and inclusive model fosters active student engagement and innovative teaching approaches, requiring coordination in faculty training, institutional policies, and student roles.

Bloom et al. (1956) developed a taxonomy outlining six hierarchical levels of cognitive objectives, from basic to advanced: remembering, understanding, applying, analyzing, evaluating, and creating. Additionally, it is understood that knowledge progresses from concrete to abstract across four levels: factual, conceptual, procedural, and metacognitive (Tourón & Martín, 2019). Finally, skills can also be categorized as hard (technical expertise) or soft (interaction and adaptability) (McClelland, 1973).

UNESCO (2017) identifies key competencies such as critical thinking, self-awareness, systemic thinking, and collaboration. Developing these requires complex cognitive engagement and soft skills, reinforcing the need for active, student-centered pedagogies aligned with the SDGs and UPV/EHU's educational model (UPV/EHU, 2019).

This study was developed within the framework of the,Campus BiziaLab Project, titled *"Experiential Training Groups as a Tool for Interpersonal Learning and Social Transformation"*, a program that promotes high-impact practices involving transdisciplinary learning based on sustainability challenges.

2 T-Groups as an Active, Student-Centered Methodology

In recent years, traditional subject-based curricula and passive teaching methods have been increasingly replaced by active, student-centered approaches. In this context, experiential training has emerged as an effective tool for developing interpersonal and professional competencies (Kolb, 1984; Kolb & Kolb, 2005, 2009). Kolb (1984) defined experiential learning as the process through which knowledge is created via the transformation of experience. His experiential learning cycle provides a structured framework that guides participants through four stages: concrete experience, reflective observation, abstract conceptualization, and active experimentation.

Within this framework, T-Groups, or Training Groups, have gained recognition as a methodology that facilitates self-awareness, enhances interpersonal sensitivity, and strengthens group dynamics by immersing participants in interactive and reflective learning processes. Conceived by Kurt Lewin (1946), the T-Group was based on the idea that when individuals participate in a group, certain phenomena emerge. When these phenomena are managed by an expert in group dynamics, they foster both reflection and learning—about group processes and self-awareness. The T-Group is thus seen as a tool for raising awareness about group work and a technique that promotes self-knowledge. Additionally, it facilitates the development of critical thinking and self-awareness, making it a valuable method for personal and interpersonal growth. Experiential groups are small groups, typically consisting of 8 to 10 members, that usually last several weeks. Through interpersonal dialogue, they foster greater interaction and reflection among members, promoting a deeper understanding of group dynamics, personal growth, and the skills necessary for group facilitation (Council for Accreditation of Counseling & Related Programs, 2016; Lewin, 1946; Merta et al., 1995).

Unlike traditional teaching methods, experiential approaches emphasize learning through direct participation, leveraging group dynamics, feedback processes, and reflective integration to prepare learners.

Various studies highlight the positive impact of experiential training groups on the education of counselors and group therapists, as they facilitate both personal and professional growth among participants. It has been shown that these programs improve self-awareness, group leadership, empathy, and professional identity (Chang et al., 2017; Ieva et al., 2009; Pamukçu, 2022; Urkmez & Singhani, 2023). These groups also contribute to the development of key competencies, such as group facilitation and the management of group dynamics, providing a practical understanding of group processes that complements the theoretical knowledge gained in regular courses (Corey, 2016; Nathan & Poulsen, 2004). These experiences allow group members to become actively involved in interpersonal interactions, fostering critical reflection on their professional practice and strengthening essential skills such as group cohesion and the exchange of feedback, which are fundamental for change and learning in educational and therapeutic contexts (Corey & Corey, 2016; Falgares et al., 2017; Yalom & Leszcz, 2020).

Most group counselor training programs integrate experiential training groups into their curricula (Anderson & Price, 2001; Merta et al., 1993; Shumaker et al., 2011). However, as Fuhriman and Burlingame (2001) point out, the major disciplines responsible for training group therapists—such as psychiatry, psychology, and social work—adopt varied approaches regarding the role, significance, and intensity of group training needed to prepare effective professionals.

There is no universally accepted framework for structuring experiential group training. Programs vary in design, therapeutic approach (e.g., group analysis, person-centered, hybrid), and duration (Feiner, 1998; Ieva et al., 2009; Pamukçu, 2022). Counseling training programs emphasize the importance of experiential learning in preparing future group facilitators, although the implementation of the experiential component is often left to the discretion of the instructor or institution. Previous studies have shown better outcomes when facilitators actively engage with group processes during experiential training sessions, compared to approaches where the leader remains silent. Additionally, clear group norms, such as cohesion, present-focused self-disclosure, and interpersonal feedback exchange, are considered essential group change processes in the group framework (Yalom & Leszcz, 2020).

The training of group facilitators plays a critical role in ensuring a safe environment where members can share and process their experiences. A facilitator with a strong theoretical foundation and specialized training can effectively guide group dynamics, fostering feedback, interpersonal exploration, and the development of new behaviors that support both personal and collective growth (Corey & Corey, 2016). Furthermore, experienced facilitators serve as role models within the group, and observing their leadership style provides a valuable learning experience, allowing members to absorb skills through observation and reflection (Chang et al., 2017).

The Association for Specialists in Group Work (ASGW) and the American Group Psychotherapy Association (AGPA) are currently updating their guidelines to improve supervision in group psychotherapy training. Recently, Hahn et al. (2022) examined the responses of expert therapists about how one develops into a group therapist. Participants agreed that, to be effective, a therapist must possess personal characteristics such as self-reflection and vulnerability, have a solid understanding of group theory, and be able to apply it in practice. Experiential training was identified as a key element in developing these competencies (Hahn et al., 2022).

The experience of T-Groups can be considered a specialized form of experiential training for group facilitators which has been widely studied as a tool for developing interpersonal and professional competencies. However, it shares a common philosophy with other active learning methodologies, such as peer learning, and cooperative learning. These methodologies, like T-Groups, are based on collaboration and horizontal interaction among participants, promoting the development of social, communication, and emotional competencies essential for the training of future professionals. In this context, it is worth noting the study by Chang et al. (2017), which compares peer-led and instructor-led experiential groups, finding similar improvements in empathy and group climate, although with a significant increase in confidence in using group work skills in externally led groups.

3 Context for the Current Study on Experiential Groups

The potential of experiential groups to enhance group training is widely recognized in graduate-level training programs. However, there has been little empirical research on the impact of implementing such training at the undergraduate level.

Recent studies with undergraduate students use pre-and-post intervention designs and mixed methods approaches to assess personal and professional growth (Arnoso & Gonzalo, 2022; Chang et al., 2017; Pamukçu, 2022). In Pamukçu's study (2022), conducted with 62 undergraduate counseling students, an increase in empathy levels was observed after participating in an experiential training group. The qualitative findings indicated that participants engaged in various behaviors associated with group roles, such as self-disclosure, active listening, and providing feedback. They also highlighted the personal and professional benefits gained during the process.

The study by Arnoso and Gonzalo (2022) focused on a group of 11 undergraduate psychology students from the University of the Basque Country who participated in an experiential training group. The qualitative analysis revealed an expected progression in the group process, from hesitation and superficial interactions at the beginning to deep, emotionally charged interventions, with themes related to closure and integration. Learning mechanisms such as practicing the observation role and using silence within the group were identified. The experiential approach allowed students to learn about the group process, develop interpersonal skills, experience personal growth, and acquire facilitation and social transformation skills. Quantitative results showed a significant increase in levels of altruism, universality, and cohesion after the training, emphasizing the value of this type of training in both the personal and professional development of the students.

We initiated the present study to contribute to the broader field of experiential training with undergraduate students, adding to the existing literature on its effectiveness. One of the challenges of incorporating T-groups into higher education is that they require several weeks to complete. With this in mind, we designed a short four-session experience to ensure it can be easily adapted to university schedules and ECTS credits. Quantitative data were collected using pre- and post-training scores (after 4 weeks) to assess the impact of these groups on readiness for group work, group change processes and group facilitation competencies and to verify that short group training experiences facilitate significant changes.

In summary, the main objectives of the present study are as follows:

1. To examine the level of participation in the program and changes in readiness for group work.
2. To evaluate the changes in group processes reported by participants, analyzing the data from pre- to post-experiential training.
3. To evaluate how participation in an experiential group impacts the development of group facilitation competencies in participants.

4 Development and Implementation of the Activity

The activity is part of a core second year course, Group Psychology, and is conducted within a teaching modality consisting of 20–25 students per group, with four 2 h sessions. Considering the ideal number for an experiential training group (8–10 participants), each group is divided into two subgroups (randomly), and the dynamics are carried out in a 'fishbowl' format. In the fishbowl format, two concentric circles are formed, with participants in the inner circle and observers in the outer circle.

Once the chairs are arranged, the facilitators wait seated for the participants in the inner circle, and the observers sit in the outer circle at a sufficient distance to observe but not interact with the participants. Once the session begins, the facilitators set the group framework, communicating the established rules, guidelines, and emphasizing the importance of confidentiality, respecting the turn to speak, and punctuality. The external group remains silent throughout the session, making observations. At the end of the session, all members come together to discuss what they observed during the session. The facilitators (course instructors) lead the discussion while the observers share their impressions in an open dialogue with the participants.

The groups run over four weeks, with a total of four 50 min experiential group sessions, followed by 40 min of discussion and feedback with the observers in the external group. During the group training, the two subgroups exchange roles, so the observers in the first session participate in the inner circle in the second session, and the roles are exchanged again in the third and fourth sessions.

The stated goal of the experiential training groups is to provide opportunities for reflection on the emotional aspects activated in groups, interactions, and other important learnings for future group facilitators. Participation is voluntary, and students can opt to complete a theoretical assignment instead of participating in the experiential group. The facilitators explicitly communicate the educational purposes of the groups, providing students with instruction and training on the types of self-disclosure considered appropriate for the experiential group setting.

Informed consent is developed following the recommendations of Pierre (2014). This includes a statement of purpose and objectives of the experiential groups; a statement of potential risks and benefits associated with participation; a statement of grading criteria (meaning the content of personal reflection journals would not affect course grades); and a statement about confidentiality.

There is no predetermined theme or content for the group, thus ensuring that it evolves naturally and allowing the process to be the dominant facilitating aspect. The group is led by the course instructors, who are trained in group analytic psychotherapy, specifically the group analysis approach proposed by Foulkes (1948), Foulkes and Anthony, (1965), a group therapy model widely practiced today in the European context (Leszcz, 2018).

5 Method

5.1 Participants

The participants were second-year psychology students at UPV/EHU, enrolled in the "Psychology of Groups" course during the 2023–2024 academic year. A total of 96 students, distributed across six groups (with 14–18 students per group), participated in the experiential training group as part of a foundational course on group facilitation.

As a selection criterion for participants, participation in the study was voluntary. All students enrolled in the course were eligible to participate, provided they were willing to complete the informed consent process and the study surveys. To be included in the final analysis of the intervention effects, participants had to have completed all four sessions and provided responses to both the pre- and post-training questionnaires. Participants who did not meet these requirements were excluded from the final analysis.

Of the 71 participants who agreed to take part in the study, 82.6% identified as women, 14.5% as men, and 2.9% as non-binary. The majority of participants were between the ages of 19 and 21 years ($M = 20.46$; $SD = 5.32$). All were born in Spain, except for two individuals from France. Most of the participants resided in the Basque Country or neighboring northern regions of Spain. Regarding prior experience, 88.1% had no previous training in group dynamics, and 68.7% had not participated in any therapeutic or personal development experiences.

Of the participants, 71 completed the pre-training evaluation, while 56 completed the post-training evaluation. In total, 41 participants completed both the pre- and post-training questionnaires and completed all four sessions.

5.2 Variables and Instruments

To standardize the response format, a five-point Likert scale (1 = Never; 5 = Very often) was used for all instruments. The same items were used in both the pre-test and post-test.

Pre- and Post-training

- Socio-Demographic Data. A customized questionnaire was used to collect participants' identification data (code) and socio-demographic information, including gender, age, country of origin, place of residence, and prior experience in group training and/or therapeutic settings. The questionnaire also included a question about the number of experiential training sessions each participant had attended.

- Group Readiness Questionnaire (Baker et al., 2013). The Group Readiness Questionnaire is a 19-item self-report tool that assesses three domains: 13 items evaluate *positive interpersonal skills* (e.g., "I am an open person"), 3 items assess *negative* *interpersonal skills* (e.g., "I avoid speaking in public"), and 3 items gauge *expectations about the benefits of group therapy* (e.g., "I believe working in a group will help me a lot"). The questionnaire has shown preliminary success in predicting process, outcome, and retention in group therapy settings (Baker et al., 2013). The total score of the scale reflects the participant's overall readiness for group participation.

- **Group Functioning Assessment Scale (EFG-25, Zukerfeld,** 1999**, based on Yalom,** 1995). The questionnaire includes factors that have been shown to be effective as change processes, regardless of the type of group approach. It consists of 25 Likert-type items that measure the presence of Cohesion, Universality, Hope, Altruism, Imitation, Information, Resocialization, Existential Factor, Catharsis, Corrective Emotional Experience, and Interpersonal Learning. The items are presented as statements that describe reasons why many people believe a group can help them (e.g., "The group offers the opportunity to learn how to communicate"). These factors are grouped into three axes: help through identification (e.g., Universality, Altruism), help through information (e.g., Interpersonal Learning, Resocialization), and help through emotional expression (e.g., Catharsis, Corrective Emotional Experience). The EFG provides scores ranging from 1 to 125 (maximum perceived help), with a score of 75 or higher considered indicative of a good perception of help for adult participants.

- **Group Psychology Competencies (ANECA,** 2005). The instrument used was a self-administered questionnaire based on the Group Psychology competencies outlined in the White Book for the first degree in Psychology, sponsored by the National Agency for Quality Assessment and Accreditation of Spain (ANECA, 2005). This framework describes the competencies included in the Psychology degree in Spain to align with the European Higher Education Area. The questionnaire included both the general competencies for the Group Psychology subject (ANECA, 2005) and specific competencies related to group facilitation. It consisted of 16 Likert-type items. The items were divided into two categories: 8 items addressing general competencies (e.g., "*Being able to identify conflicts, needs, and group objectives, and choosing appropriate intervention strategies*") and 8 items focused on specific group facilitation competencies (e.g., "*Facilitating effective interpersonal exchanges within the group*"). The items were presented in the form of statements reflecting the competencies that students believe they possess.

- **Self-Awareness to Guide a Group**. The importance of group training was assessed through three Likert-type items, evaluating participants' awareness of its significance. These items considered the level of ability for group facilitation ("*I feel that I have the skills to lead a group*"), perceived difficulty ("*I consider leading a group to be difficult*"), and the need for training ("*I believe specific training is necessary to lead a group*").

5.3 Procedure

At the beginning of the course, students were informed about the study associated with the experiential training groups. They were told that participation was voluntary, that the collected data would remain anonymous, and that their participation would not affect their course grades. The course coordinator sent a request to complete the survey via educational platforms and email to all enrolled students. A brief cover letter was attached, explaining the purpose of the study and the procedures for completing the survey. Potential participants were asked to sign an informed consent form to confirm their voluntary participation, and were then provided with a link to the survey via SurveyMonkey, a widely used online data management platform for social science research. The survey took approximately 25 min to complete.

5.4 Data Analysis

The analyses of participant profiles were conducted using the total sample of 96 individuals. However, to assess the training outcomes, a subset of 41 participants who completed both the pre- and post-training evaluations was considered. All statistical analyses were performed using IBM SPSS Statistics (version 28).

Although this study did not employ a control group, several steps were taken to strengthen the reliability of the findings. Participants were drawn from the same academic cohort to minimize variability in educational background and prior exposure to group dynamics. Data collection was standardized by administering all surveys online under uniform conditions. Additionally, gender and prior exposure to group dynamics were considered as potential covariates in preliminary analyses to account for their possible influence on the outcomes. Before conducting parametric tests, data normality was verified to ensure the assumptions were met. Only participants who attended all four experiential sessions and completed both the pre- and post-training evaluations were included in the final analysis. Finally, effect sizes (Cohen's d) were reported alongside statistical significance to provide a measure of practical impact.

To assess the effects of the experiential group training, descriptive analyses were conducted using frequencies, percentages, means, and standard deviations. Repeated-measures Student's t-tests were then performed, with Cohen's d used to calculate effect sizes. Following Cohen's (1977) recommendations, effect size values were interpreted as small ($\eta^2 = 0.20$), medium ($\eta^2 = 0.50$), or large ($\eta^2 = 0.80$) when assessing the impact of an intervention.

6 Results

This section presents information on the development of the training program and the changes at both individual and group levels, including attendance and dropout rates, as well as the impact of these groups on readiness for group work, group change processes, competencies related to group facilitation and self-awareness to facilitate a group.

6.1 Development of the Training Program, Attendance, and Dropout Rate

Out of the total enrolled students, 96 (81.36%) participated in the experience. Among them, 64 participants (66.67%) completed the four-session training program; 18 participants (18.75%) attended three sessions; 6 participants (6.25%) attended two sessions; and 8 participants (8.33%) attended only the first session (Fig. 1).

The overall dropout rate for the group training was 14.58%. This rate was calculated based on participants who attended fewer than 50% of the sessions (2 sessions or less). The reasons provided for non-participation or dropout after the first session included schedule incompatibility with work, membership in a high-performance sports group, overlap with other courses, or having previously participated in the experience in earlier years. Specifically, only 3 individuals cited their shyness in meeting new people and/or difficulty participating in group spaces like this as the reason for their dropout.

Fig. 1 Participation rate

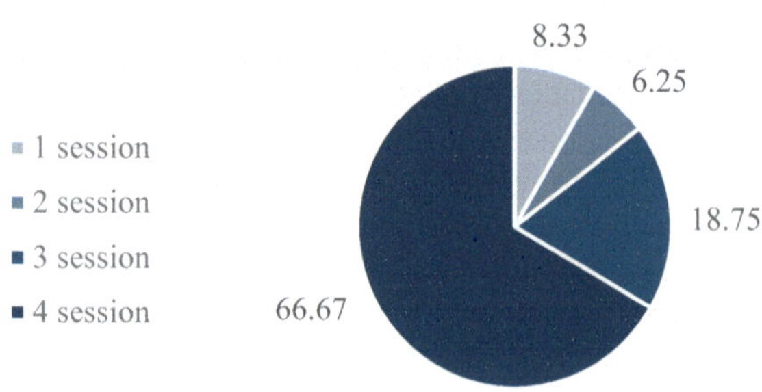

Table 1 Comparison of means between the pre-training and post-training conditions for readiness to work in a group, with the standard deviation in parentheses

Variables	Pre	Post	*t*	*p*	*d*
Readiness to work in a group					
Positive skills	3.27 (0.54)	3.28 (0.43)	−0.11	0.916	0.020
Critical skills	2.00 (0.66)	2.07 (0.79)	−0.56	0.585	0.09
Expectations for group work	3.35 (0.70)	3.67 (0.85)	−3.23	0.003	0.143

6.2 *Effects of the Group Training Program*

6.2.1 Readiness to Work in a Group

Based on the results in Table 1, participants in the post-training condition had higher *expectations for group work* compared to the pre-training condition. However, the changes in *positive* and *critical skills* were not statistically significant.

6.2.2 Group Change Processes

Considering the total scores, it was found that in the post-training condition, participants perceived better group functioning compared to the pre-training condition. Specifically, they perceived greater *universality*, *resocialization*, as well as *interpersonal learning*. All the comparisons of means regarding perceived help were significantly higher in the post-training condition than in the pre-training condition: *help through identification*, *help through information*, and *help through expression*.

The other factors reflecting good group functioning also improved, but the differences were trend-level for *altruism*, *imitation*, *information*, and *corrective emotional experience*. The differences were not statistically significant for *cohesion*, *hope* and *catharsis* (Table 2).

Figures 2 and 3 clearly show that after the training, the factors of universality, resocialization, and interpersonal learning improved (Fig. 2), as is the case with help through identification, information, and expression (Fig. 3).

6.2.3 Group Psychology Competencies and Group Facilitation Competencies

According to the results in Table 3, participants in the post-training condition perceived greater competencies in the field of Group Psychology. The competencies of *Being able to identify group conflicts, needs, and objectives, choosing appropriate intervention systems*, and *Developing skills to coordinate group work, manage conflicts, and improve performance* also improved, but the differences were not statistically significant.

Table 2 Comparison of means between pre-training and post-training situations for group change processes and help axes, with standard deviation in parentheses

Variables	Pre	Post	*t*	*p*	*d*
Group change processes	90.38 (14.80)	96.73 (16.26)	−2.74	0.009	−0.43
Altruism	3.37 (0.89)	3.60 (0.80)	−1.79	0.081	−0.28
Cohesion	3.77 (0.75)	4.00 (0.77)	−1.64	0.108	−0.26
Universality	3.93 (0.86)	4.33 (0.81)	−3.22	0.003	−0.51
Imitation	3.29 (0.61)	3.50 (0.80)	−1.73	0.091	−0.27
Hope	3.82 (0.83)	4.04 (0.92)	−1.55	0.130	−0.24
Resocialization	3.74 (0.77)	4.10 (0.75)	−2.78	0.008	−0.43
Information	3.56 (0.72)	3.80 (0.82)	−1.75	0.088	−0.27
Catharsis	3.82 (0.87)	3.88 (0.95)	−0.383	0.704	−0.06
Interpersonal learning	3.31 (0.86)	3.71 (0.92)	−2.84	0.007	−0.45
Corrective emotional experience	3.76 (0.91)	4.07 (0.86)	−1.82	0.076	−0.28
Help axes					
Help through identification	3.62 (0.56)	3.85 (0.61)	−2.63	0.012	−0.41
Help through information	3.63 (0.70)	3.91 (0.74)	−2.29	0.028	−0.35
Help through emotional expression	3.60 (0.74)	3.88 (0.80)	−2.18	0.035	−0.34

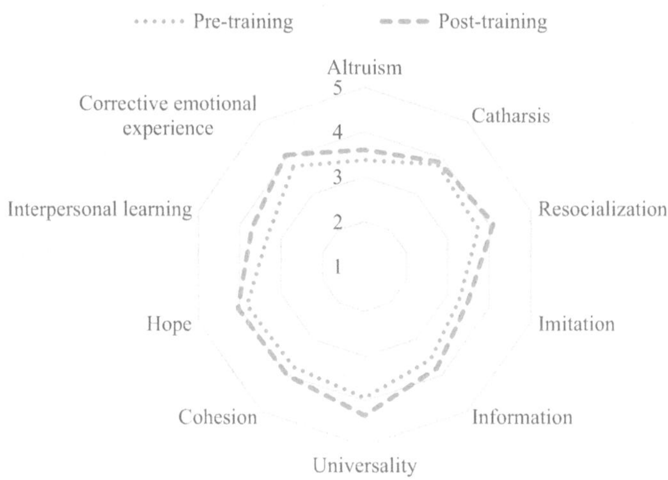

Fig. 2 Development in group change processes

Regarding the specific group facilitation competencies, in the post-training condition, participants perceived higher scores than in the pre-training condition in the following competencies: *Promoting self-development and personal growth, as well as the ability to perceive and understand interpersonal interactions more deeply;*

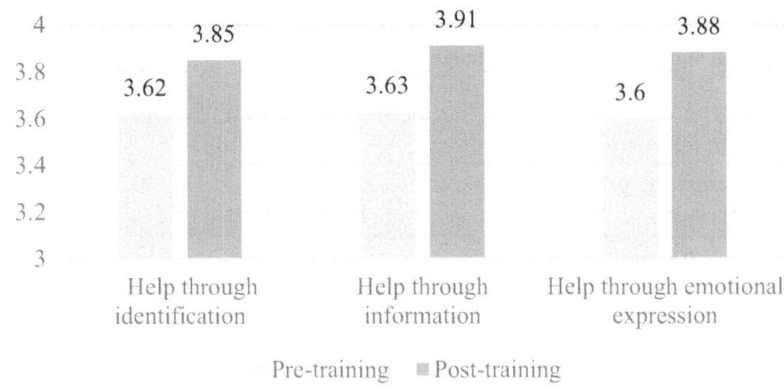

Fig. 3 Development of changes on the help axes

Increasing awareness of how behaviors affect others; Facilitating effective interpersonal exchanges within the group; Learning from feedback received from others, as well as from the feedback given to others; and Working effectively in the "here and now", linking external content with experiences, relationships, and feelings within the group.

The remaining group facilitation competencies also improved, but the differences were marginal for *Developing counseling and group facilitation skills* ($p = 0.070$), and did not reach statistical significance for the following competencies: *Increasing confidence and comfort with group work; and Exploring personal issues relevant to group work* ($p > 0.100$).

6.2.4 Self-Awareness to Facilitate a Group

According to the results in Table 4, in the post-training condition, participants were more aware of the *need for specific training to lead a group* than in the pre-training condition. There was a parallel decrease in *the perception of one's own abilities to guide a group*, with the differences being statistically marginal.

7 Discussion

In alignment with the goals of Agenda 2020, which aim to foster a more just, inclusive, peaceful, and sustainable future, higher education plays a pivotal role as a transformative agent. The need to reflect on the future and to train professionals capable of addressing current challenges underscores the importance of incorporating active educational methodologies that foster the development of complex skills.

This study describes the experiences of 41 s-year undergraduate Psychology students in Spain who participated in an experiential training group as part of their Group Psychology course, following a concentric circles approach. Experiential

Table 3 Comparison of means between pre-training and post-training situations for subject competencies, with standard deviation in parentheses

Variables	Pre	Post	*t*	*p*	*d*
General competencies in group psychology					
Knowing and interpreting group interaction and development processes	2.88 (1.14)	3.88 (0.72)	−6.407	0.000	−1.048
Identifying the group structure and its most relevant elements	3.08 (1.12)	3.80 (0.65)	−4.134	0.000	−0.786
Identifying and understanding the processes of influence occurring within the group	2.98 (1.05)	3.75 (0.81)	−4.669	0.000	−0.821
Observing and analyzing group dynamics and structure, and being able to identify group and intergroup problems and needs	3.13 (1.16)	3.90 (0.63)	−4.206	0.000	−0.825
Observing and analyzing interaction processes and social influence	3.08 (1.21)	3.73 (0.68)	−3.523	0.001	−0.622
Being able to identify conflicts. needs. and group objectives, and choosing appropriate intervention strategies	3.03 (1.07)	3.30 (0.94)	−1.602	0.117	−0.268
Developing personal skills to work effectively with groups	3.00 (1.20)	3.80 (0.88)	−4.066	0.000	−0.760
Developing skills to coordinate group work, manage conflicts, and improve performance	3.08 (1.19)	3.43 (0.93)	−1.595	0.119	−0.327
Specific group facilitation competencies					
Increasing confidence and comfort with group work	4.12 (0.81)	4.19 (0.71)	−0.534	0.596	−0.092
Encouraging self-development and personal growth, as well as a deeper understanding of interactions with others	3.88 (0.84)	4.20 (0.68)	−2.239	0.031	−0.418
Exploring personal issues relevant to group work	3.58 (0.98)	3.75 (0.81)	−0.909	0.369	−0.189
Increasing awareness of how behaviors impact others	4.02 (0.69)	4.32 (0.65)	−2.080	0.044	−0.447
Facilitating effective interpersonal exchanges within the group	4.12 (0.81)	4.41 (0.63)	−2.147	0.038	−0.399
Learning from the feedback we receive from others and the comments we give to others	4.07 (0.69)	4.49 (0.60)	−2.800	0.008	−0.649
Ability to work in the "here and now", linking external content with experiences, relationships, and feelings within the group	3.80 (0.84)	4.12 (0.78)	−2.311	0.026	−0.061
Developing counseling and group facilitation skills	3.90 (0.97)	4.17 (0.70)	−1.862	0.070	−0.319

Table 4 Comparison of means between pre-training and post-training situations for training awareness, with standard deviation in parentheses

Variables	Pre	Post	*t*	*p*	*d*
Self-awareness to facilitate a group					
I feel that I have the abilities to lead a group	3.47 (0.85)	3.17 (0.87)	1.91	0.063	0.349
I consider that leading a group is difficult	3.77 (0.73)	3.85 (0.83)	−0.52	0.608	−0.102
I consider that specific training is necessary to lead a group	3.75 (0.93)	4.43 (0.74)	−4.28	0.000	−0.809

group training programs have scarcely been evaluated in the university setting, and their effectiveness remains largely unknown, particularly at the undergraduate level (Arnoso & Gonzalo, 2022; Pamukçu, 2022). This study contributes to the scientific literature by rigorously assessing the impact of a brief T-group training experience on group work competencies. Specifically, we examine group work readiness, group change processes, the development of facilitation skills, and awareness of the importance of group training. To the best of our knowledge, this is the first study conducted in the Spanish context to analyze the effectiveness of this type of training in undergraduate students using a pre-post quantitative methodology to assess its impact following a brief four-session intervention, with a variety of standardized measures. Its design seeks to provide evidence on the feasibility and benefits of integrating such experiences into university education, taking into account the time constraints and academic workload inherent to the ECTS credit system.

The high level of participation in the experience, with 81.36% of students involved in six training groups, suggests that this activity is both engaging and appealing. This finding is particularly significant considering that two-thirds of the participants (66.67%) completed the training program in its entirety, demonstrating sustained commitment to the initiative. However, the overall dropout rate of 14.58% and the limited participation in the evaluation protocol, with only half of the participants completing both the pre- and post- questionnaires, highlight certain challenges that warrant further analysis. As suggested by Anderson and Price (2001), aligning participants' expectations with the objectives of educational interventions is essential to reduce dropout rates and resistance, which are critical factors for the success of experiential group training. Additionally, it is crucial to address the management of participants' boundaries, as they engage both as learners and as group members. Research has documented concerns related to dual relationships, confidentiality risks, and emotional distress, which are significant considerations in ensuring the effectiveness and well-being of participants (Doshi et al., 2019; Haber & Deaton, 2019).

The results obtained regarding the impact of the training program are promising, as evidence from the post-training phase indicates the achievement of the established objectives at various levels: readiness for group work, group change processes, specific competencies in group facilitation, and awareness of the importance of group training.

The results reflect that, after the training, participants reported higher expectations for group work, suggesting that the experience contributed to increasing their confidence in group dynamics. However, changes in both positive and critical skills did not reach statistical significance, which may indicate that these dimensions require a longer training period to observe significant improvements, or that participants already had high scores and motivation for group work at the outset.

In the post-training condition, participants also reported improved group functioning, with significant increases in some group change processes such as universality, resocialization, and interpersonal learning. According to Yalom and Leszcz (2020), these factors are essential components for facilitating change in group and educational settings, as they promote connection among members and meaningful learning through interpersonal interaction. Additionally, significant improvements were observed in the perception of the help received through identification, information, and expression.

Group facilitation also improved significantly with the training. Participants reported an increase in their ability to foster self-development and personal growth, as well as a deeper understanding of their interactions with others. This is consistent with the literature on T-Groups, which highlights their potential to promote self-reflection and interpersonal learning (Corey, 2016; Yalom & Leszcz, 2020).

Additionally, progress was observed in the awareness of the impact of behaviors on others, the facilitation of effective interpersonal exchanges, and learning through feedback. These competencies are essential for creating a participatory and cooperative learning environment, aligning with the literature that highlights group interaction as a driving force for the development of social skills (Chang et al., 2017; Corey & Corey, 2016) and conflict resolution within the group. This reinforces the importance of experiential learning in the training of facilitators, in line with the recommendations of leading group psychotherapy organizations (Hahn et al., 2022; Whittingham et al., 2021).

Another significant improvement was the ability to work in the 'here and now,' connecting external content with experiences, relationships, and emotions within the group. In this way, the social becomes personal, and the external becomes internal (Foulkes, 1948). This skill allows participants to focus their attention on the present (Yalom & Leszcz, 2020) and helps, among other things, to reflect on the emotions that arise within the group and transform them into the necessary knowledge for change to occur.

These findings underscore the effectiveness of T-Groups as a pedagogical tool in higher education, aligning with the educational model of UPV/EHU and the Sustainable Development Goals (UNESCO, 2017). In particular, group training promotes systemic thinking by allowing participants to understand the interconnection of group processes and their impact on social dynamics. Furthermore, the development of collaboration and effective communication skills reinforces the importance of teamwork in solving complex problems, a key aspect of education for sustainability. In addition, self-awareness and emotional self-regulation, enhanced through group interaction, are fundamental for the training of professionals with a reflective and ethical perspective. These findings are in line with SDG 4 (Quality Education), which highlights the importance of innovative and participatory methodologies to prepare students for the challenges of the twenty-first century. In this regard, the observed effects demonstrate that the use of experiential methodologies promotes active, student-centered learning approaches, fostering peer learning and self-directed learning. Moreover, the development of skills such as communication, self-awareness, active listening, and collaboration is directly related to SDG 5 (Gender Equality) and SDG 10 (Reduced Inequalities), as these competencies promote equitable group dynamics and more egalitarian relationships. Furthermore, the increase in emotional self-regulation and awareness of the impact of one's behavior on others reinforces the importance of emotional intelligence in educational and professional contexts, which aligns with SDG 3 (Good Health and Wellbeing). The implementation of experiential methodologies in universities also reflects a commitment to building inclusive and strong institutions, in line with SDG 16 (Peace, Justice, and Strong Institutions). Finally, the type of relationship established with students and the institutional culture promoted through these methodologies reinforce the importance of cooperation and community learning, essential elements of SDG 17 (Partnerships for the Goals).

Although improvements were perceived in various competencies related to Group Psychology and group facilitation, some areas did not show statistically significant changes, such as the ability to identify conflicts, coordinate group work, and manage group performance. Similarly, in specific group facilitation competencies, no significant improvements were observed in areas such as confidence and comfort with group work, and exploring personal issues relevant to group work. These results show that, while T-Group training promotes the development of interpersonal and reflective competencies, certain group facilitation skills require additional formative experiences beyond T-Groups. In any case, it is considered a success that brief and easily adaptable experiences to the higher education context enable students to acquire interpersonal and reflective competencies.

There is no universally accepted framework for structuring experiential group training in higher education institutions, resulting in programs with variations in design, approach, and duration (Feiner, 1998; Ieva et al., 2009; Pamukçu, 2022). In this context, competency-based training models, which assess the acquisition of facilitation skills beyond mere participation time, are gaining relevance (Ni et al., 2023; Urkmez & Singhani, 2023). From this perspective, a training program like the one proposed in this study can be easily adaptable, as it relies on the use of standardized instruments and validated measures. This approach ensures consistency in data comparison and facilitates the application of the model in diverse contexts. However, future studies could benefit from integrating both quantitative and qualitative methodologies, including participant reflective journals, to gain a deeper understanding of participants' experiences and the factors influencing their effectiveness.

The transferability of experiential training to different cultural contexts is a key aspect. In addition, disciplinary contexts such as counselor education, social work, psychology, and teacher training are among the fields where experiential methods like

T-Groups have been most widely applied and studied. However, studies exploring the adaptation of experiential methods to specific cultural settings (Chang et al., 2017; Pamukçu, 2022; Terrazas-Carrillo et al., 2022) typically focus on specific populations, such as Taiwanese, Turkish, or Latin American students, without comprehensively addressing their applicability in more diverse contexts. These works highlight the need to expand beyond predominant Western frameworks by incorporating collectivist group healing traditions and alternative group structures where relevant (Jakubkaitė & Kočiūnas, 2013; Terrazas-Carrillo et al., 2022).

Additionally, recent works have explored the digital adaptation of experiential training, demonstrating an effective transfer of learning to virtual environments (Urkmez et al., 2021; Weinberg, 2023). They suggest that these formats can be comparable to in-person teaching in developing key competencies such as leadership self-efficacy and group facilitation. However, further research is needed to understand how to optimize these methodologies in different institutional contexts.

Considering the above, the described experience is easily generalizable to other settings. Evidence suggests that, within the university context, promoting experiences that provide students with the opportunity to debate and reflect on topics of their interest, in a peer-interaction space, fosters both learning and preparedness. This approach, where students are given a voice and provided with time for reflection and collaboration, with the professor acting as a guide, can contribute to the development of key competencies for addressing future challenges. Thus, in other faculties, universities, and countries, similar experiences, such as debate groups or activities focused on challenges and questions, could be equally effective, without the need to replicate the exact same methodology.

Finally, participants demonstrated a greater awareness of the need for specific training to lead groups, indicating a more realistic and nuanced understanding of the required competencies. This increased awareness was accompanied by a tendentially lower perception of their own abilities to lead a group, which could be interpreted as an increase in self-demand or a greater recognition of the challenges involved in group facilitation. This result is particularly interesting because, following the experience, there was a heightened sensitivity to the difficulty of leading a group and the responsibility it entails, which is crucial when working with people. In the context of higher education, brief experiences addressing emotional and relational aspects should be promoted throughout undergraduate studies, and more intensive training should be encouraged in postgraduate specialization courses or doctoral programs.

To increase the practical value of this study, several recommendations are provided for university administrators, faculty, and curriculum designers. First, it is emphasized that facilitators should have a strong theoretical foundation and practical experience in group dynamics. Second, it is essential to ensure the voluntariness of participation, highlighting the importance of attendance, confidentiality, and self-disclosure within the group. Next, the use of concentric circles, where participants alternate between the roles of participant and observer, is recommended as an effective strategy to foster reflection and learning from different perspectives. Finally, complementing experiential training with standardized measures and participant reflective journals contributes to a better understanding of the lived experience and the factors influencing its effectiveness. These recommendations can serve as a practical guide or framework for implementing experiential training in different academic settings.

Funding This study has been partially funded by a Basque Government grant to Research Groups ("LOTUNEAK: Human Bonds for Change" Consolidated Research Group, IT1797-26). This work is also part of the Campus Bizia Lab Project, "Experiential Training Groups as a Tool for Interpersonal Learning and Social Transformation," coordinated by the Vice-Rectorate for Scientific-Social Development and Transfer at the University of the Basque Country (UPV/EHU), in collaboration with the Sustainability Directorate.

References

Anderson, R. D., & Price, G. E. (2001). Experiential groups in counselor education: Student attitudes and instructor participation. *Counselor Education and Supervision, 41*(2), 111–119. https://doi.org/10.1002/j.1556-6978.2001.tb01275.x

ANECA, National Agency for Quality Assessment and Accreditation. (2005). *Libro Blanco del Título del Grado en Psicología [White Paper on the Qualification of the Degree in Psychology]*. http://www.aneca.es/var/media/150356/libroblanco_psicologia_def.pdf

Arnoso, A., & Gonzalo, A. (2022). Experiential group psychotherapy training with undergraduate students in Spain. *International Journal of Group Psychotherapy, 72*(4), 358–396. https://doi.org/10.1080/00207284.2022.2126844

Baker, E., Burlingame, G. M., Cox, J. C., Beecher, M. E., & Gleave, R. L. (2013). The Group Readiness Questionnaire: A convergent validity analysis. *Group Dynamics: Theory, Research, and Practice, 17*(4), 299–314. https://doi.org/10.1037/a0034477

Bloom, B. S., Engelhart, M. D., Furst, E. J., Hill, W. H., & Krathwohl, D. R. (1956). *Handbook I: Cognitive domain* (pp. 483–498). David McKay.

Chang, S. H., Bhat, C. S., & Chen, Y. A. (2017). Experiential group training and group leadership development in Taiwanese school counsellor trainees. *Asia Pacific Journal of Counselling and Psychotherapy, 8*(1), 66–85. https://doi.org/ https://doi.org/10.1080/21507686.2016.1264435

Cohen, J. (1977). *Statistical power analysis for the behavioral sciences*. Academic Press.

Corey, G. (2016). *Theory and practice of group counseling*. Cengage Learning.

Corey, G., & Corey, M. S. (2016). Group psychotherapy. In J. C. Norcross, G. R. VandenBos, D. K. Freedheim, & R. Krishnamurthy (Eds.), *APA Handbook of clinical psychology: Applications and methods* (pp. 289–306). American Psychological Association.

Council for Accreditation of Counseling and Related Programs. (2016). *2016 CACREP standards*. https://www.cacrep.org/for-programs/2016-cacrep-standards/

Doshi, J., Patel, T., & Kumar Bharti, S. (2019). Smart Farming using IoT, a solution for optimally monitoring farming conditions. *Procedia Computer Science, 160*, 746–751. https://doi.org/10.1016/j.procs.2019.11.016

Falgares, G., Venza, G., & Guarnaccia, C. (2017). Learning psychology and becoming psychologists: Developing professional identity through group experiential learning. *Psychology Learning & Teaching, 16*(2), 232–247. https://doi.org/10.1177/1475725717695148

Feiner, S. E. (1998). Course design: An integration of didactic and experiential approaches to graduate training of group therapy. *International Journal of Group Psychotherapy, 48*(4), 439–460. https://doi.org/10.1080/00207284.1998.11491566

Foulkes, S. H. (1948). *Introduction to group-analytic psychotherapy*. Karnac.

Foulkes, S. H., & Antony, E. J. (1965). *Group psychotherapy: The psychoanalytic approach*. Karnac.

Fuhriman, A., & Burlingame, G. M. (2001). Group psychotherapy training and effectiveness. *International Journal of Group Psychotherapy, 51*(3), 399–416. https://doi.org/10.1521/ijgp.51.3.399.49889

Haber, R., & Deaton, J. D. (2019). Facilitating an experiential group in an educational environment: Managing dual relationships. *International Journal of Group Psychotherapy, 69*(4), 434–458. https://doi.org/10.1080/00207284.2019.1656078

Hahn, A., Paquin, J. D., Glean, E., McQuillan, K., & Hamilton, D. (2022). Developing into a group therapist: An empirical investigation of expert group therapists' training experiences. *American Psychologist*. https://doi.org/10.1037/amp0000956

Ieva, K. P., Ohrt, J. H., Swank, J. M., & Young, T. (2009). The impact of experiential groups on master students' counselor and personal development: A qualitative investigation. *The Journal for Specialists in Group Work, 34*(4), 351–368. https://doi.org/10.1080/01933920903219078

Jakubkaitė, B., & Kočiūnas, R. A. (2013). Experiential groups in group counselors training: a view through the multiplicity of contexts. *Problems of Psychology in the 21st Century, 5*, 12–22.

Kolb, D. A. (1984). *Experiential learning: Experience as the source of learning and development*. Prentice Hall.

Kolb, A. Y., & Kolb, D. A. (2005). Learning styles and learning spaces: Enhancing experiential learning in higher education. *Academy of Management Learning & Education, 4*(2), 193–212. https://doi.org/10.5465/amle.2005.17268566

Kolb, A. Y., & Kolb, D. A. (2009). The learning way: Meta-cognitive aspects of experiential learning. *Simulation & Gaming, 40*(3), 297–327. https://doi.org/10.1177/1046878108325713

Leszcz, M. (2018). The evidence-based group psychotherapist. *Psychoanalytic Inquiry, 38*(4), 285–298. https://doi.org/10.1080/07351690.2018.1444853

Lewin, K. (1946). Action research and minority problems. *Journal of Social Issues, 2*, 34–46.

McClelland, D. C. (1973). Testing for competence rather than for "intelligence." *American Psychologist, 28*(1), 1–14. https://doi.org/10.1037/h0034092

McCowan, T. (2023). The crosscutting impact of higher education on the sustainable development goals. *International Journal of Educational Development, 103*, Article 102945. https://doi.org/10.1016/j.ijedudev.2023.102945

Merta, R., Johnson, P., & McNeil, K. (1995). Updated research on group work: Educators, course work, theory, and teaching methods. *The Journal for Specialists in Group Work, 20*(3), 132–142. https://doi.org/10.1080/01933929508411337

Merta, R. J., Wolfgang, L., & McNeil, K. (1993). Five models for using the experiential group in the preparation of group counselors. *The Journal for Specialists in Group Work, 18*, 200–207. https://doi.org/10.1080/01933929308413755

Nathan, V., & Poulsen, S. (2004). Group-analytic training groups for psychology students: A qualitative study. *Group Analysis, 37*(2), 163–177. https://doi.org/10.1177/0533316404042855

Ni, C. F., Lin, C. C., & Dykeman, C. (2023). Exploring multimodality with online peer-facilitated experiential learning in group work training. *The Journal for Specialists in Group Work, 48*(4), 299–316. https://doi.org/10.1080/01933922.2023.2251141

Pamukçu, B. (2022). An exploration of Turkish counselor trainees' development in experiential training groups: An uncontrolled case study. *International Journal of Group Psychotherapy, 72*(1), 64–92. https://doi.org/10.1080/00207284.2021.2015602

Pierre, B. K. (2014). Student attitudes and instructor participation in experiential groups. *The Journal for Specialists in Group Work, 39*(3), 194–211. https://doi.org/10.1080/01933922.2014.919048

Ramos, D. I. (2021). Contribución de la educación superior a los Objetivos de Desarrollo Sostenible desde la docencia [Higher education's contribution to the sustainable development goals from the teaching perspective]. *Revista Española De Educación Comparada, 37*, 89–110. https://doi.org/10.5944/reec.37.2021.27763

Shumaker, C., Ortiz, C., & Brenninkmeyer, L. (2011). Revisiting experiential group training in counselor education: A survey of master's-level programs. *The Journal for Specialists in Group Work, 36*(2), 111–128. https://doi.org//https://doi.org/10.1080/01933922.2011.562742

Terrazas-Carrillo, E., Garcia, E., & Vera, A. (2022). Experiential groups: Narratives of Latino master's level students in counseling psychology during the Covid-19 pandemic. *The Journal for Specialists in Group Work, 47*(3–4), 188–207. https://doi.org/10.1080/01933922.2022.2126049

Tourón, J., & Martín, D. (2019). *Aprender y enseñar en la universidad hoy. Una guía práctica para profesores [Learning and teaching at university today. A practical guide for teachers].* UNIR editorial.

UNESCO. (2015). *Education 2030: Incheon declaration and framework for action for the implementation of sustainable development goal 4: Ensure inclusive and equitable quality education and promote lifelong learning opportunities for all.* https://uis.unesco.org/sites/default/files/documents/education-2030-incheon-framework-for-action-implementation-of-sdg4-2016-en_2.pdf

UNESCO. (2017). *Educación para los Objetivos de Desarrollo Sostenible. Objetivos de aprendizaje [Education for the Sustainable Development Goals. Learning Objectives].* https://unesdoc.unesco.org/ark:/48223/pf0000247444

UPV/EHU. (2019). *EHUagenda for sustainable development.* https://www.ehu.eus/es/web/iraunkortasuna/ehuagenda-2030

Urkmez, B., Pinkney, C., Amparbeng, D. B., Gunawan, N., Isiko, J. O., Tomlinson, B., & Bhat, C. S. (2021). Experience of graduate counseling students during COVID-19: Application for group counseling training. *Professional Counselor, 11*(4), 475–492. https://doi.org/ https://doi.org/10.15241/bu.11.4.475

Urkmez, B., & Singhani, S. (2023). Effectiveness of experiential group training in developing leadership self-efficacy of counselors-in-training. *The Journal for Specialists in Group Work, 48*(4), 335–346. https://doi.org/10.1080/01933922.2023.2246538

Weinberg, H. (2023). Online training process groups for therapists: A proposed model. *International Journal of Group Psychotherapy, 73*(2), 141–165. https://doi.org/10.1080/00207284.2023.2170236

Whittingham, M., Lefforge, N. L., & Marmarosh, C. (2021). Group psychotherapy as a specialty: An inconvenient truth. *American Journal of Psychotherapy, 74*(2), 60–66. https://doi.org/10.1176/appi.psychotherapy.20200037

Yalom, I. D. (1995). *The theory and practice of group psychotherapy* (4th ed.). Basic Books.

Yalom, I. D., & Leszcz, M. (2020). *The theory and practice of group psychotherapy.* Basic Books.

Zukerfeld, R. (1999). Salud mental y enfermedad crónica: Dispositivos grupales y autogestión [Mental health and chronic illness: Group devices and self-management]. *Vertex, Revista Argentina De Psiquiatría, 10*(36), 91–99.

Non-financial Reporting and SDGs: Advancing Sustainability in Higher Education Institutions Through Measuring and Reporting

Maria Serena Chiucchi, Angelica Sardini, and Roberta Ciccola

Abstract Higher Education Institutions (HEIs) are increasingly acknowledged as relevant for the achievement of Sustainable Development Goals (SDGs), primarily through teaching, research, and third mission activities. Non-financial reporting has emerged as a communication tool for providing a detailed disclosure of HEIs' efforts to achieve the SDGs. Existing literature underscores how its adoption not only enhances legitimacy and stakeholder engagement, but also supports decision-making, promoting future actions aimed at achieving the SDGs. However, the extent to which HEIs may leverage non-financial reporting to integrate SDGs into their core organizational processes and strategies remains not adequately explored. The aim of this study is to explore if and how the adoption of non-financial reporting may support the integration of SDGs within HEIs. To achieve this aim, the research relies on an in-depth longitudinal case study of an Italian university which has been adopting a non-financial reporting—and more specifically an Integrated Report (IFRS Foundation, 2022)—since 2020. Findings suggest that the adoption of the non-financial reporting influences the integration of SDGs within HEIs, as it facilitates the shift from mere disclosure to the effective definition of actions and indicators through the explicit formalization of the link between the University's actions and their impact on SDGs. The study contributes to the ongoing debate on the role of HEIs in sustainable development, by exploring the role of non-financial reporting in mobilizing efforts to integrate SDGs into strategic and organizational practices. The non-financial reporting emerges as a tool to bridge the gap between disclosure and the execution of SDGs-oriented initiatives.

M. S. Chiucchi · A. Sardini (✉) · R. Ciccola
Department of Management, Università Politecnica delle Marche, Ancona, Italy
e-mail: a.sardini@pm.univpm.it

M. S. Chiucchi
e-mail: m.s.chiucchi@univpm.it

R. Ciccola
e-mail: r.ciccola@univpm.it

M. S. Chiucchi
IRCCS INRCA, Ancona, Italy

W. Leal Filho et al. (eds.), *Introducing the UN SDGs in Higher Education Institutions*, World Sustainability Series, https://doi.org/10.1007/978-3-032-23017-1_39

Keywords Sustainable Development Goals · SDGs · Non-financial reporting · Integrated reporting · Higher education institutions

1 Introduction

In 2015, the United Nations (UN) established a global agenda aimed at fostering sustainability and driving transformative change by 2030. This agenda is structured around the 17 Sustainable Development Goals (SDGs), which collectively address critical issues such as poverty, inequality, and climate change, while promoting global peace, justice, and well-being. The SDGs represent a universal call to action, encouraging to work collaboratively against these pressing challenges.

Higher Education Institutions (HEIs) have emerged as critical actors in achieving these goals, given their role in fostering sustainable development through teaching, research, and third mission activities (Ceulemans et al., 2015a; Lozano et al., 2015), as well as their significant responsibility in educating future generations and preparing society for urgent challenges. Additionally, HEIs are increasingly expected to broaden their impact by reducing negative effects while enhancing the positive economic, social, and environmental outcomes of their activities. In response to these expectations, a growing number of HEIs are adopting internal policies and initiatives, such as waste management programs and student welfare strategies, to align their strategy and operations with sustainability principles (Sepasi et al., 2019; Serafini et al., 2022).

In this context, non-financial reporting has emerged as a key process and tool to support these efforts, providing a structured approach for organizations, including HEIs, to communicate their SDG-related initiatives, track progress, and foster stakeholder engagement. Non-financial reporting allows HEIs to demonstrate their critical role in advancing sustainable development by transparently sharing their values, initiatives, and actions to address global sustainability challenges (Brusca et al., 2018).

Given their relevance, these themes have garnered increasing attention within academic literature. On the one hand, scholars emphasize how HEIs integrate SDGs into their activities. This includes exploring how HEIs contribute to sustainable development, analyzing the factors that facilitate or constrain their contributions to the SDGs (Chankseliani & McCowan, 2021; De Iorio et al., 2022; Ferguson & Roofe, 2020), investigating their engagement in implementing the SDGs (Leal Filho et al., 2023), and assessing the extent of SDG dissemination (Caputo et al., 2021). On the other hand, the research focuses also on the adoption of non-financial reporting and its associated effects within HEIs. Specifically, some studies suggest that non-financial reporting enhances competitiveness and reputation, promotes stakeholder engagement, and increases sustainability awareness in HEIs (Hassan et al., 2019; Monteiro et al., 2024b). Other articles examine the characteristics, advantages, and limitations of non-financial reporting in HEIs (Siboni et al., 2013), including the motivations for its implementation (Brusca et al., 2018) and the comprehensiveness of non-financial reporting within HEIs (Hassan et al., 2019; Sepasi et al., 2019).

While existing studies explore how HEIs integrate the SDGs as well as the non-financial reporting implementation within HEIs, a substantial gap remains in understanding if and how non-financial reporting can support the integration of SDGs within HEIs (Guthrie & Neumann, 2007). Given the significant role of HEIs in advancing sustainable development, examining whether and how they can leverage non-financial reporting to embed sustainability into their core organizational processes and strategies is crucial. This research addresses this gap by investigating if and how the adoption of non-financial reporting may support the integration of SDGs within HEIs. Drawing on a longitudinal case study of an Italian university, which has been employing a non-financial reporting—and more specifically an Integrated Report (IFRS Foundation, 2022)—since 2020, this research examines whether and how the implementation of non-financial reporting influences the development and implementation of SDG-related objectives, actions, and initiatives.

This chapter is structured as follows. The next section provides a review of the existing literature on non-financial reporting in the context of HEIs and their role in advancing SDGs, while the third section describes the research method. The fourth section presents the findings, and it is followed by the discussion and conclusions of the chapter.

2 Literature Review

2.1 The Role of Higher Education Institutions for Sustainable Development Goals

In recent years, climate change, air and water pollution, resource waste and socioeconomic inequalities have exerted growing pressure on organizations to prioritzse and commit to sustainable development. The SDGs have been introduced as viable solutions to tackle these challenges and achieve higher levels of sustainability (Bebbington & Unerman, 2018; Mio et al., 2020). Formulated by the United Nations General Assembly in the context of the UN 2030 Agenda, the SDGs aim to "stimulate action over the next 15 years in areas of critical importance for humanity and the planet" (UN, 2015, p. 3). They include 17 goals focused on economic, social and environmental issues directed at the improvement of well-being, the equitable access to education, the reduction of inequalities and the promotion of peace and prosperity.

The attainment of the SDGs requires coordination and collaboration on multiple levels—national, organizational, and societal. With the advancement of the UN 2030 Agenda, HEIs are called upon to play an active role in sustainable development, contributing through their efforts in educating individuals, advancing knowledge, and promoting innovation (Chankseliani & McCowan, 2021).

Scholars have recently recognized that the issue of the SDGs has generated a new area of research in the HEIs field and have begun to explore the role of these institutions in pursuing the 17 SDGs (Albareda-Tiana et al., 2018; Blasco et al., 2020; Caputo et al., 2021; Raimo et al., 2024). HEIs, as "centres of creative thinking and innovation" (Hassan et al., 2019, p. 865), have been acknowledged as key actors in the implementation of SDGs in several key areas: teaching, research, governance and campus operations, and community engagement (Amaral et al., 2015; Leal Filho et al., 2018, 2023; Lozano et al., 2015). Following this perspective, HEIs can contribute to the attainment of the SDGs both directly, through teaching and research, and indirectly, by engaging in areas such as institutional governance, campus management, and community involvement (Caputo et al., 2021).

Regarding teaching, given their significant role in educating future generations, HEIs have a responsibility to transmit knowledge and influence students' attitudes toward sustainability (Dobson, 2007; Weybrecht, 2017). Several studies show that HEIs are actively working to align scientific literacy with SDGs, equipping students with the critical thinking abilities needed to be a "dynamic agent of change" for sustainable development (Avelar et al., 2019; Roque de Rosa et al., 2024; Manes-Rossi et al., 2020). Specifically, HEIs are progressively embedding sustainability-related themes aligned with the SDGs into their educational programs and curricula (Avelar et al., 2019; Palma et al., 2023; Roque de Rosa et al., 2024). HEIs adopt different strategies to incorporate sustainability into their curricula, such as designing a foundational course on sustainability, creating a specialized pathway or program focused on sustainable development, or embedding sustainability topics throughout various existing disciplinary courses (Ceulemans & De Prins, 2010; Lozano, 2010; Lozano & Young, 2013).

Regarding research, within HEIs the creation of inter- and transdisciplinary research projects plays a key role in producing knowledge, designing evidence-based solutions, and fostering innovation which can facilitate the implementation of the 2030 Agenda at both local and national levels (Monteiro et al., 2024a). Specifically, HEIs can capitalize on insights gained from their research activities to introduce innovative solutions and technologies that foster the social and economic progress of their surrounding community (Blasco et al., 2020).

HEIs play a critical role for the achievement of SDGs also through governance and campus operations, community engagement, collaboration and partnerships with public and private institutions (Menon & Suresh, 2020). Initiatives implemented by HEIs include awareness campaigns, solid waste management, and renewable energy programs (Duram & Williams, 2015; Ghosh, 2011; Leal Filho et al., 2021). These efforts are often complemented by measures for water and energy conservation, the establishment of campus gardens, green building projects and sustainability audits (Murray, 2018). By "walking the talk" (Leal Filho et al., 2021, p. 3) and embedding SDGs into their initiatives and operational policies, HEIs have the potential to serve as role model for other public and private institutions.

2.2 Non-financial Reporting in Higher Education Institutions

Considering the efforts of HEIs to achieve sustainable development and the SDGs through teaching, research, and community engagement initiatives, the adoption of non-financial reporting emerges as a significant instrument to support these efforts. Non-financial reporting is a broad and evolving concept (Stolowy & Paugam, 2018) which encompasses various reporting practices disclosing information on an organization's environmental, social, governance and generally non-financial performance (Ahunov, 2023; Dimes & Molinari, 2024; Lisi et al., 2023). Within the university context, it includes different forms of disclosure such as sustainability reporting, social reporting and integrated reporting (Manes Rossi et al., 2020; Iacuzzi et al., 2020).

Non-financial reporting allows HEIs to systematically communicate their contributions to sustainability, providing a detailed disclosure of their alignment with the SDGs and emphasizing their commitment in supporting these sustainability objectives (Caputo et al., 2021).

The process of reporting, managing, addressing, and shaping strategies related to sustainable development issues through non-financial reporting holds significant potential for HEIs to drive impactful contributions toward achieving the SDGs (Adams, 2018; Alonso-Almeida et al., 2015; Aversano et al., 2020). The development of non-financial reporting enables the integration of economic, social, and environmental risks and opportunities, facilitating evaluation and performance improvement (Blasco et al., 2020; Caputo et al., 2021; Ceulemans et al., 2015a, 2015b).

Specifically, the "processual" dimension of non-financial reporting is of significant importance in this context, as its development process entails the involvement of various actors in data collection, analysis, and interpretation, thus influencing internal decision-making (Lueg & Radlach, 2016; Traxler et al., 2020).

Several studies have explored the motivations and the drivers for its implementation within HEIs (Aras et al., 2022; Bice & Coates, 2016; Ceulemans et al., 2015b; Manes Rossi et al., 2018). A proactive leadership, the presence of research groups, dedicated coordinating units, and scholars specialized in sustainability are considered as key influential internal factors that contribute to the development of non-financial reporting (Adams, 2013; Blanco-Portela et al., 2017; Fonseca et al., 2011; Larran Jorge et al., 2019). However, the literature highlights that one of the main motivations for its adoption lies in the ability of non-financial reporting to meet stakeholder expectations (Fonseca et al., 2011; Lopatta & Jaeschke, 2014; Manes Rossi et al., 2018). Non-financial reporting, indeed, is regarded as a means to increase the reputation of HEIs, as it allows them to benefit from enhancing transparency in communicating their value creation activities to stakeholders (Brusca et al., 2018; De Villiers et al., 2025; Hassan et al., 2019). Therefore, it has been considered not only as a tool for disclosing HEIs' efforts toward the achievement of the SDGs, but also for enhancing their legitimacy, enabling stakeholders to assess the tangible outputs and broader impacts of their activities (Aras et al., 2022; Bice & Coates, 2016; Monteiro et al.,

2024b). In addition to its role in evaluating activities and communicating sustainability performance, non-financial reporting also represents a relevant instrument to identify areas for improvement and inform the strategic planning of future actions aimed at sustainable development (Chatelain-Ponroy & Morin-Delerm, 2016; Moggi, 2023; Sepasi et al., 2019). The scope and the influence of non-financial reporting extend beyond disclosure, as highlighted by several scholars who have investigated its internal relevance (e.g., De Villiers et al., 2025; Stubbs & Higgins, 2014). Specifically, research has emphasized the importance of "mobilizing" the aspects that are measured and disclosed, thus moving from measurement and reporting to action (Catasús & Gröjer, 2006; Catasús et al., 2007; Chiucchi, 2013). At the same time, the literature has also emphasized that, when adopting external reporting tools, their benefits fully materialize when reporting not only addresses external stakeholders but also actively supports internal decision-making and performance measurement, according to an outside-in approach (Maas et al., 2016).

Furthermore, non-financial reporting plays a crucial role in influencing HEIs' ranking positions by facilitating cross-institutional comparisons and providing verifiable data for accreditation bodies (Ceulemans et al., 2015a). In this context, the study of Hansen et al. (2021) highlights the relevance of coordinating SDG assessments and information to streamline data collection for sustainability rankings.

Another stream of research has instead examined the current state of non-financial reporting in HEIs, focusing on its characteristics, comprehensiveness, and quality (Alonso-Almeida et al., 2015; Caputo et al., 2021; Lozano et al., 2013). However, these studies emphasize that the adoption of non-financial reporting in HEIs is in its infancy (Ceulemans et al., 2015a; Fonseca et al., 2011; Hassan et al., 2019).

Overall, the studies reviewed reveal the role of HEIs in contributing to SDGs and highlight the growing importance of non-financial reporting in supporting their sustainability efforts. However, the literature review reveals a gap in understanding whether and how non-financial reporting effectively supports the integration of SDGs within HEIs (Ceulemans et al., 2015a; Siboni et al., 2013; Yanez et al., 2019). The SDGs represent a significant opportunity for HEIs to act as catalysts for change toward sustainable development (Weybrecht, 2017) since they have the possibility to influence many of them directly or indirectly. Non-financial reporting is a tool for disclosing sustainability and generally non-financial performance and for engaging with stakeholders. At the same time, literature on non-financial reporting has underlined that it may also promote actions and enactment of the dimensions that are measured. Moving from these premises, this study investigates if and how non-financial reporting effectively influences the integration of the SDGs within HEIs.

3 Research Method

To achieve the aim of this study, this paper adopts a single, longitudinal case study approach (Ahrens & Dent, 1998; Denzin & Lincoln, 2017). The case study method has been chosen for several reasons. First, the case study method has been widely recognized as particularly suitable for the in-depth investigation of contemporary phenomena that are yet unexplored (Yin, 2014). Second, this method is considered appropriate for answering 'how' questions by providing holistic and systemic thinking. Third, case studies allow for gathering rich information, making them suitable for examining the complexities involved in non-financial reporting and evaluating the associated impacts on the integration of SDGs.

The case under scrutiny is the Università Politecnica delle Marche, a medium Italian university which has been adopting a non-financial report—and more specifically an Integrated Report (IFRS Foundation, 2022)—since 2020.

This University has been selected since it is among the first Italian universities that have adopted an Integrated Report. Second, in recent years, the Università Politecnica delle Marche has placed a stronger focus on sustainability, as evidenced by its participation in the Italian Sustainable University Network (RUS) and in various international sustainability-related rankings in which this University has progressively improved its results. Moreover, the selection of the case study was guided also by considerations of access to relevant and internal information.

The case period under investigation is 2020–2024; the adoption of a longitudinal perspective allowed to examine the changes and developments regarding the University's approach towards SDGs and non-financial reporting. Data collection relied primarily on participant observations, semi-structured interviews and document analysis. Specifically, the Authors were part of the team that designed and prepared the Integrated Report and one of them is the Rector's Delegate for Performance Measurement and Integrated Reporting. Semi-structured interviews with academics and administrative staff members both involved in integrated reporting and in sustainable development initiatives within the University such as, among others, the General Director, the Rector's Delegate for Sustainability, and the Head of the Evaluation Unit and Quality Assurance Office. The interviews were tape-recorded and immediately transcribed.

The data collected with the interviews were supplemented with document analysis of both internal and publicly available documents including Integrated Reports, strategic plans, the official University website, meeting minutes, documents related to the University's participation in international sustainability rankings and other administrative documents relevant to sustainability initiatives and reporting. This extensive document analysis helped to triangulate the data collected from the participant observations and interviews, enhancing the validity and reliability of collected data.

Finally, the data retrieved from the data collection were progressively coded according to the insights provided by the literature reviewed in Sect. 2. The data analysis was conducted in a back-and-forth movement between data and theory, within an iterative interplay with data collection by the three authors. The data items were analyzed separately and then regularly discussed, ensuring consistency and internal validity. The results of this analysis are presented in the next section.

4 Findings

The Università Politecnica delle Marche (hereinafter, UnivPM) is a public Italian university founded in 1969. With more than 17,500 students, UnivPM is considered a medium-sized university in Italy; it offers educational programs at different levels in Agriculture, Economics, Engineering, Medicine and Science besides doing research and promoting third mission activities related to those areas.

The University is actively committed to promoting sustainability, as demonstrated by several initiatives launched in recent years, including the adoption of non-financial reporting practices. In 2016, the University began to publish a Social Report, and starting from 2020, it introduced an Integrated Report (hereinafter referred to as the "IR"), prepared according to the Integrated Reporting Framework (IFRS Foundation) and the 2021 version of the Global Reporting Initiative (GRI) Sustainability Reporting Guidelines.

The Rector and the General Director are keen on underlining that the reasons for adopting an IR are various. It has been adopted with the goal of implementing a performance measurement and reporting system aimed at ensuring transparency and accountability of the University towards the community and the territory. Additionally, it represents a tool for facilitating ongoing dialogue with the University's main stakeholders such as students, companies, suppliers, institutional bodies and the local communities. At the same time, the IR has been introduced to achieve a relevant internal purpose: it is aimed at diffusing the culture of economic, social and environmental sustainability and supporting the planning and control process within the University.

The IR illustrates the University's business model, its activities, the goals it has set, and the results achieved, showing how the University generates "shared value" to society. The IR is organized into four main sections. The first outlines the University's vision, mission, strategy, stakeholders, and governance. The second section provides a detailed description of the University's Business Model and its value creation process. According to the Integrated Reporting framework, the third section shows the University's main "capitals" (i.e., human, economic, relational, physical and environmental), which represent the inputs of its value creation process. The most important section of the document focuses on the University's core missions, i.e. teaching, research, and third mission, and finally the last part is dedicated to improvement objectives and alignment tables (with GRI and SDGs). The Integrated Reporting process is carried out by a dedicated Working Group composed of academics and administrative staff members. At the beginning, the project was co-coordinated by the Rector's Delegate for Performance Measurement and Integrated Report and by the current General Director. A relevant role in the Group was attributed to the Head of the

Office "Financial and Non-Financial Reporting and Organizational Performance". The assignment of project coordination to those involved, in various functions, in the University's management accounting activities reflects the broad scope and foresight with which the project has been approached. The IR aims to enhance communication and transparency externally while also fostering synergies with management accounting to support the governance decision-making process. The Personnel of the above-mentioned offices as well as other University Researchers and Post Doc have participated over the years in the design and implementation of the Report. The Rector's Delegate for Gender Issues and the Rector's Delegate for Sustainability have contributed to the reporting process for their areas of responsibilities. The composition of the Working Group for the IR was entirely designed to facilitate the integration between sustainability reporting, financial reporting, and management accounting. It is important to underline that the reporting process involves the entire organization, particularly the various administrative offices which actively participate in data collection, providing the necessary information and also suggesting some improvements thus contributing to the relevance and accuracy of the document.

Within the IR, significant attention is dedicated to sustainability, highlighting initiatives, events, teaching programs, and research projects referred to it. In 2020, UnivPM started its first steps towards integrating the SDGs into the IR. However, from 2021, a more structured approach to SDGs integration has been promoted. Specifically, in the section dedicated to the Business Model (Fig. 1), where its capitals and the three core missions are described, the primary actions undertaken to achieve specific SDGs are identified for each mission. For example, in the realm of teaching, the University implements actions to support students who lack financial resources by enacting policies that help them both financially and in terms of services. This approach fosters the cultural growth for students who need this support thus contributing to SDG1 and SDG4. Additionally, by promoting high quality interdisciplinary education, UnivPM seeks to enhance the employability of its graduates, addressing SDG 8. Besides using narratives, another significant step toward the awareness of the University's contribution towards SDGs has been the development of an SDGs alignment table, which identifies where information related to each SDG target can be found. This adds clarity and transparency to the actions' impact over SDGs which is consistent with the increasing attention paid towards them.

The IR is used by the University Governance and the Heads of Administrative Offices as a key reference for obtaining reliable data about the activities carried out in the University, sustainability issues and the impact on SDGs. This role is particularly emphasized by the General Director who considers the IR the University's most important Report that discloses all the relevant information referred to the activities carried out by UnivPM. As such, he considers it and promotes its use as the main and most complete source of information across University offices and for the University Governance.

It is important to underline that the IR has represented a powerful stimulus and a source of information for the integration of SDGs into the University Strategic Plan. The work carried out during the Integrated Reporting process to explicitly map

Fig. 1 The UnivPM Business Model—(UnivPM Integrated Report—y 2023, p. 44)

the impacts of the University's activities on the SDGs within the University's business model has also been included into the Strategic Plan. Moreover, recently, these aspects, along with the impact materiality analysis conducted during the Integrated Reporting process, have been used as the inputs for the definition of specific indicators within the Strategic Plan, aimed at monitoring the University's contribution to SDGs.

In line with this perspective, considering the experience and the awareness gained in monitoring sustainability within the IR, the UnivPM General Director has recently established a working group for the implementation of a Sustainability Plan in which long term objectives and strategy referred to Sustainability will be devised. In this context, the materiality analysis will be considered an important foundation for the development of this document. The materiality analysis enables a series of analyses and decisions regarding both the key themes to prioritize and the relevant stakeholders to engage with, representing the premise for the definition of objectives and actions. Beyond reflecting the University's attention to stakeholder interests, the materiality analysis also acted as a key instrument to foster stakeholder engagement, providing an opportunity for stakeholders to actively contribute to the identification and prioritization of the most relevant issues. This is particularly evident in the 2023 IR, where the materiality analysis—conducted through questionnaires—broadened its scope by involving a broad range of stakeholders in identifying the material topics. While in previous years the questionnaire was addressed to UnivPM's internal stakeholders, primarily students, technical, administrative and academic staff, in 2023 it was extended also to external stakeholders. The process included several significant external stakeholders of the University, such as other universities, research institutions, educational institutions, business associations, for-profit, non-profit, and public organizations, regional healthcare service entities, suppliers, national institutional bodies, student welfare agencies, regional authorities, local governments, and University spin-offs. This broader engagement aimed to enhance their involvement in the process and ensure that the identification of material topics captured not only the internal perspective but also the expectations, needs, and priorities of the broader socio-economic and institutional context in which UnivPM operates. The identification of stakeholders' areas of interest, their impact, and relevance, also enabled by the inclusion of external stakeholders in the materiality analysis, strengthens the role of the IR as a tool to support strategic planning, as the insights generated by the analysis represent an essential input to inform the strategic planning process.

Therefore, the IR is not used only to measure and disclose the impact of the activities of UnivPM toward the sustainable development, but it is also a starting point for decision-making aligned with SDGs and sustainability, enabling the evaluation of the strengths and weaknesses of the University's activities. As explained by the Rector's Delegate for Sustainability:

> The IR provides the data needed to identify potential actions to propose and assess the strengths and weaknesses of our University in terms of sustainability. Based on this analysis, discussions can then focus on areas or aspects that need to be improved to fill this gap.—Rector's Delegate for Sustainability.

In this regard, it is worth highlighting that, gender data included in the IR are an important basis for devising some of the actions included in the Gender Equality Plan and are used also to monitor them. As highlighted by the Rector's Delegate for Performance Measurement and Integrated Report:

> The IR (and before) the Social Report have been for years the document where gender issues were illustrated and monitored. From 2022 a Gender Equality Plan has been produced by UnivPM, nevertheless, it has been decided to keep all the information (also those referred to gender issues) referred to the Capitals and the core missions all together, within the IR, to highlight all relevant aspects and have an overall picture of the phenomena.—Rector's Delegate for Performance Measurement and Integrated Report.

Another example of the role of IR as a "data source" is the support provided for participation in international rankings. UnivPM takes part in three international sustainability-related rankings (i.e. Green Metric, Times Higher Education (THE) Impact, QS Sustainability). These rankings have a similar structure, as they primarily require universities to provide evidence of their sustainability efforts, evaluating aspects such as publications on sustainability-related topics, academic courses on sustainability, engagement with local communities to promote sustainable development, etc. Specifically, they assess what actions the University undertakes and how these actions are implemented according to specific SDGs, particularly the THE impact ranking, which evaluates universities based on their contributions to the SDGs through research, teaching, and institutional policies. The participation in these rankings is relevant also to witness and demonstrate the impact of the actions undertaken towards SDGs.

> The progress achieved so far acknowledges the commitment of the entire University community, which is increasingly aware that the transition toward a more sustainable and inclusive path is possible and it is a shared responsibility.—Rector's Delegates for Research and International Rankings and Rector's Delegate for Sustainability.

Also, the IR represents a key resource in this process, offering a comprehensive dataset used to extract the necessary information required for these assessments. As highlighted by the Head of the Evaluation Unit and Quality Assurance Office, without this document, participation in these rankings would have been significantly more challenging:

> The IR is an extremely valuable source of information because it allows to address a relevant amount of all requests. It provides both qualitative data – what we do and how we do it – and quantitative data, such as waste, emissions, and water consumption. We specifically rely on those data, making it a crucial data source. It is the "cornerstone document" fundamental to provide evidence.—Head of the Evaluation Unit and Quality Assurance Office.

Thus, the IR acts not only as a structured repository of data, providing both quantitative and qualitative data, but also represents documentary evidence that supports the accuracy of the information reported. Therefore, this dual role makes it a crucial validation tool that UnivPM can rely on to substantiate its claims, supporting the participation in sustainability rankings.

The structured presentation of information within the IR significantly streamlines the data collection process for sustainability rankings. By organizing data in a clear and systematic way, the Report ensures that relevant information is readily accessible and aligned with ranking criteria. Specifically, the presence in the IR of a table that illustrates the alignment between SDGs and the actions and initiatives of UnivPM has provided crucial support in streamlining the data collection process, making it easier for UnivPM to respond comprehensively to ranking requests.

By structuring the information in a way that visually conveys how and through which activities UnivPM contributes to the achievement of the SDGs, the IR not only aligns with the format and categorization about SDGs required by ranking bodies but also enhances the University's ability to efficiently respond to diverse external requests, including those from governmental and ministerial bodies (e.g. report the initiatives undertaken towards development countries to improve specific SDGs), reinforcing its commitment to accountability and sustainable development.

Standardizing the information related to the SDGs and integrating it into a single comprehensive report makes the IR the first explicit and structured formalization of the link between the University's activities and the SDGs as it systematically connects all SDGs-related data within a unified framework, ensuring coherence and accessibility. As highlighted by the Rector's Delegate for Sustainability:

> It is becoming clear how important it is to have a document that is as a reference point for everyone, where the same information can be found, including a lot of important details regarding, for example, sustainability projects […]It is a document in which these discussions (about sustainability and SDGs) are becoming more common…which helps us communicate more easily with each other.—Rector's Delegate for Sustainability.

Moreover, it explicitly articulates the SDGs impacted and the specific actions through which these contributions are realized, making these connections visible, more transparent and structured. Therefore, the IR provides a sustainability "*narrative*" that has been spread throughout the University Personnel both through the Document but also through "Apostles" such as the General Director and the Delegates and Heads of Offices involved who have not only contributed to the preparation of the document but who have also used it in official and internal meetings and, in general, in mobilizing the Report and the information included. This dissemination has enhanced awareness within the institution, not only of sustainability issues and the SDGs themselves but, more importantly, of UnivPM's contributions to them, thus promoting a diffusion of an SDGs literacy. As highlighted by the Rector's Delegate for Performance Measurement and Integrated Reporting:

> The IR has helped clearly identify the parameters and categories for observing certain activities we implement. I believe it has contributed to establishing shared ways of viewing information and interpreting reality within a common framework.—Rector's Delegate for Performance Measurement and Integrated Reporting.

The structured reporting process has also played a crucial role in this context, shaping the University's organizational practices, alongside initiatives for plastic reduction, sustainable mobility, and training workshops. The Rector's Delegate for Sustainability clarified:

> Over time, I have come to fully appreciate the usefulness of the Integrated Report, not only for the information it provides for participation in rankings but also for enhancing awareness of sustainability initiatives within UnivPM and for understanding how we approach various activities and initiatives. —Rector's Delegate for Sustainability.

An initiative that reflects the growing diffusion of a sustainability culture among the Personnel is the training course "Sustainable UnivPM" which is aimed at illustrating all the activities the University does towards sustainability as well as the Planning, Control and Reporting System referred to it. The IR is illustrated during the course in order to show how sustainability is embedded in the University activities (e.g., reflecting on how Research, Teaching and Third mission may be 'sustainable') and how it can be measured. Personnel are then engaged in defining objectives, indicators and actions for achieving them specifically referred to their working activities that can show their impact on economic, social and environmental performance. This resonates with one of the ideas that lies behind the IR and which is expressed by the Rector and the General Director in their preface to the last IR "*Reading the pages of this document, everyone can recognize themselves and gain awareness of how their work contributes to the common vision*". Thus, engaging the Personnel in the abovementioned activity reinforces the awareness and a common sustainability culture.

The dissemination of this sustainability orientation within the University is also reflected in its educational offerings, with more than forty percent of the courses at UnivPM addressing sustainability-related topics. UnivPM offers a wide range of bachelor's and master's degree programs which cover areas such as industrial sustainability and green engineering, environmental sciences and protection, agricultural, forestry and landscape sciences, sustainability management and circular economy, as well as marine biology and ecosystem studies. Furthermore, UnivPM has participated in the National PhD Programme in Sustainable Development and Climate Change and, in 2023, provided additional funding for scholarships within this programme.

5 Discussion and Conclusions

This study investigates whether and how non-financial reporting influences the integration of SDGs within HEIs. Using the longitudinal case study of UnivPM, which has employed IR since 2020, we explore the extent to which non-financial reporting facilitates the incorporation of SDGs into organizational and strategic practices. The findings suggest that non-financial reporting plays a key role in advancing sustainability efforts within HEIs, enhancing transparency and accountability. Over the years, UnivPM has implemented a more systematic approach to SDG integration within the IR, explicitly linking the SDGs impacted and the specific actions through which these contributions are realized in many ways: through its Business Model, through narratives, and through tables showing the correspondence between SDGs and the indicators/information illustrated within the Report. As a consequence, the IR at UnivPM represents the first explicit formalization of the connection between the University's actions and their impact on SDGs.

Formally expressing this link has contributed to enhancing SDGs literacy across the University Personnel. The influence that the IR has had at the organizational and strategic level lies in its role in harmonizing and codifying the approaches, processes, and tangible effects of the University's actions on the SDGs, establishing a coherent and structured framework that describes how the University actively contributes to their advancement. Specifically, in strategic planning, this clear articulation of the UnivPM's actions and the SDGs impacted has represented the premise and the starting point for the definition of objectives and actions. Therefore, this tool is not only used to disclose the impact of activities on the SDGs but also serves as a resource for informing decision-making processes aimed at improving the University's future actions. The expanded approach in the materiality analysis provides further evidence of the role of this tool. The inclusion of external stakeholders relevant to the University in the materiality analysis ensured that external perspectives were systematically considered. Through the explicit mapping of stakeholders' expectations, the University is able to better align its actions with the SDGs and ensure that its initiatives and plans also reflect the priorities of the broader socio-economic and institutional context in which it operates. In this way, the IR contributes to mobilizing the integration of SDGs, ensuring that both internal and external stakeholders influence the direction of future initiatives aimed at sustainable development.

The dissemination of the sustainability "narrative" throughout the University is also the result of the non-financial reporting process, which involves multiple actors in the design of the Report and in the collection, analysis, and interpretation of SDGs-related information.

Moreover, the IR has become a primary source of reliable data for UnivPM's participation in international sustainability rankings and for responding ministerial requests. It has been considered as the primary accountability document to access reliable and structured SDGs-related data. For sustainability rankings, its role extends beyond that of a mere data repository that optimizes the data collection process, ultimately embodying essential documentary evidence to attest to the University's commitment to sustainability initiatives.

A key contribution of this study is to provide a more in-depth comprehension of the role of the non-financial reporting for the integration of SDGs into the university's operations. Although the literature highlights the potential of non-financial reporting in communicating HEIs' contributions to sustainability by disclosing their alignment with the SDGs and informing the strategic planning of future actions aimed at sustainable development (Adams, 2018; Caputo et al., 2021; Moggi, 2023), this study explores in greater depth its effects on the organizational and strategic practices of HEIs. This analysis provides further insight on how the non-financial reporting process enables HEIs to make meaningful contributions toward achieving the SDGs (Adams, 2018; Alonso-Almeida et al., 2015; Aversano et al., 2020). Specifically, the influence of non-financial reporting extends beyond enhancing the reputation and legitimacy of HEIs (Brusca et al., 2018; De Villiers et al., 2025; Hassan et al., 2019) as it is able to "mobilize" actions towards managing sustainability and SDGs. This study offers a deeper understanding of how non-financial reporting contributes to the formulation of objectives and action lines (Yáñez et al., 2019), specifically emphasizing its role as a key reference document that articulates the link between the impact of HEIs' activities and the SDGs. More than a tool for compliance and disclosure, non-financial reporting constitutes a strategic resource that supports the definition of strategic objectives and indicators used in the internal decision-making processes (Chatelain-Ponroy & Morin-Delerm, 2016; Moggi, 2023). Through the harmonization and codification of the approaches, mechanisms, and tangible outcomes of institutional actions on the SDGs, non-financial reporting underscores its strategic relevance. It represents the reference document to inform decision-making, providing a clear link between the university's actions and the SDGs impacted, thereby ensuring that sustainability commitments translate into concrete plans and actions (De Villiers et al. 2025). The increased stakeholder engagement resulting from the integration of external stakeholders into the non-financial reporting process has further strengthened this role, turning non-financial reporting into a means that may be used to rethink priorities and identify strategic areas where the university could contribute more effectively to the SDGs.

Moreover, this research illustrates also an outside-in approach (Maas et al., 2016), in which a tool designed to disclose HEIs' efforts to contribute to the SDGs to stakeholders also functions as a key resource and reference document for internal decision-making. Furthermore, the non-financial reporting is not just a tool used to disclose and assess the efforts of HEIs to contribute to SDGs, but it represents a means to spread the SDGs literacy and awareness of their contribution throughout the university.

Finally, this study provides an in-depth understanding of how non-financial reporting influences HEIs' ranking positions (Ceulemans et al., 2015a). The non-financial reporting represents a key document in structuring and consolidating the data required for rankings, ensuring a systematic and coherent representation of the HEIs' sustainability efforts. By guaranteeing the reliability and accessibility of information, it provides a structured narrative that not only facilitates data collection for ranking but also validates the University's contributions that are essential for participating in these rankings. Although grounded in a single case, this research offers insights that may hold theoretical relevance beyond the specific context analysed. While HEIs may differ in terms of size, governance, culture, and context, the findings shed light on broader aspects of how a reporting tool can generate effects and trigger organizational change. In particular, the case illustrates how non-financial reporting, rather than functioning solely as a tool for disclosure, can act as a catalyst for the integration of SDGs within the university. Therefore, this study may contribute to a deeper understanding of the underlying mechanisms through which non-financial reporting can "mobilize" action within organizations (Catasús & Gröjer, 2006; Catasús et al., 2007), supporting a shift from mere measurement and reporting toward strategies.

A similar perspective can be found also in the study by Iacuzzi et al. (2020) on another Italian university, in which non-financial reporting is described as a tool that supports incremental change. In that case, however, the strategic plan served as the foundation upon which non-financial reporting was structured, acting as a "pull" mechanism for its implementation. In contrast, the present study reveals that, in the case of UnivPM, non-financial reporting played an active role, contributing to integrate the SDGs orientation into the strategic plan and representing a key source of information to support its formulation. This divergence highlights the need for further exploration of the conditions under which non-financial reporting can function as a "push" mechanism, influencing and shaping strategic planning within HEIs. Moreover, in addition to the mobilizing mechanism toward strategic planning, the possibility of non-financial reporting to foster also specific actions needs a more in-depth investigation and can be a future development path of this study.

This study has also practical implications that may support HEIs in enhancing non-financial reporting and SDG integration. First, our findings suggest that HEIs could benefit from establishing a dedicated working group that includes both academic and administrative staff to coordinate non-financial reporting process, thus facilitating the integration between sustainability reporting, financial reporting, and management accounting and contributing to the diffusion of a sustainability culture across all organizational levels. Second, our findings indicate that the integration of SDG-related performance indicators into strategic planning processes can encourage universities to move beyond disclosure toward actionable commitments. Third, our analysis reveals that initiatives aimed at promoting awareness and engagement among university personnel, such as training courses, can facilitate a shared understanding of sustainability-related activities and their impact throughout the organization. Finally, the findings suggest that developing frameworks for mapping university activities to specific SDGs can improve the accessibility and strategic use of non-financial reporting.

A more comprehensive understanding of the role of non-financial reporting in HEIs would require investigating also how it drives the formulation of strategic objectives aligned with the SDGs. Future research could extend this analysis by investigating more in-depth whether and how non-financial reporting contributes to embedding SDGs into HEIs' strategic planning and the development of concrete actions. Finally, as this study provides evidence into a single case study, future research could adopt a comparative approach across multiple HEIs to understand differences in non-financial reporting implementation and its effectiveness in integrating SDGs within the university context.

References

Adams, C. A. (2013). Sustainability reporting and performance management in universities: Challenges and benefits. *Sustainability Accounting, Management and Policy Journal, 4*(3), 384–392. https://doi.org/10.1108/SAMPJ-12-2012-0044

Adams, C. A. (2018). Debate: Integrated reporting and accounting for sustainable development across generations by universities. *Public Money & Management, 38*(5), 332–334. https://doi.org/10.1080/09540962.2018.1477580

Ahrens, T., & Dent, J. F. (1998). Accounting and organizations: Realizing the richness of field research. *Journal of Management Accounting Research, 10*.

Ahunov, H. (2023). Non-financial reporting in hybrid organizations—A systematic literature review. *Meditari Accountancy Research, 31*(6), 1757–1797. https://doi.org/10.1108/MEDAR-01-2022-1558

Albareda-Tiana, S., Vidal-Raméntol, S., & Fernández-Morilla, M. (2018). Implementing the sustainable development goals at University level. *International Journal of Sustainability in Higher Education, 19*(3), 473–497. https://doi.org/10.1108/IJSHE-05-2017-0069

Alonso-Almeida, D. M., et al. (2015). Diffusion of sustainability reporting in universities: Current situation and future perspectives. *Journal of Cleaner Production, 106*, 144–154. https://doi.org/10.1016/j.jclepro.2014.09.048

Amaral, L. P., Martins, N., & Gouveia, J. B. (2015). Quest for a sustainable university: A review. *International Journal of Sustainability in Higher Education, 16*(2), 155–172. https://doi.org/10.1108/IJSHE-02-2013-0017

Aras, G., Furtuna, K. O., & Kazak, H. E. (2022). The nexus between stakeholders' materiality and sustainable development goals: Evidence from higher education institutions. *International Journal of Sustainability in Higher Education, 23*(1), 114–134. https://doi.org/10.1108/IJSHE-11-2020-0439

Avelar, A. B. A., da Silva-Oliveira, K. D., & da Silva, P. R. (2019). Education for advancing the implementation of the Sustainable Development Goals: A systematic approach. *The International Journal of Management Education, 17*(3), Article 100322. https://doi.org/10.1016/j.ijme.2019.100322

Aversano, N., Di Carlo, F., Sannino, G., Tartaglia Polcini, P., & Lombardi, R. (2020). Corporate social responsibility, stakeholder engagement, and universities: New evidence from the Italian scenario. *Corporate Social Responsibility and Environmental Management, 27*(4), 1892–1899. https://doi.org/10.1002/csr.1934

Bebbington, J., & Unerman, J. (2018). Achieving the United Nations Sustainable Development Goals: An enabling role for accounting research. *Accounting, Auditing & Accountability Journal, 31*(1), 2–24.

Bice, S., & Coates, H. (2016). University sustainability reporting: Taking stock of transparency. *Tertiary Education and Management, 22*(1), 1–18. https://doi.org/10.1080/13583883.2015.1115545

Blanco-Portela, N., Benayas, J., Pertierra, L. R., & Lozano, R. (2017). Towards the integration of sustainability in Higher Education Institutions: A review of drivers of and barriers to organisational change and their comparison against those found of companies. *Journal of Cleaner Production, 166*, 563–578. https://doi.org/10.1016/j.jclepro.2017.07.252

Blasco, N., Brusca, I., & Labrador, M. (2020). Drivers for universities' contribution to the sustainable development goals: An analysis of Spanish public universities. *Sustainability, 13*(1), 89. https://doi.org/10.3390/su13010089

Brusca, I., Labrador, M., & Larran, M. (2018). The challenge of sustainability and integrated reporting at universities: A case study. *Journal of Cleaner Production, 188*, 347–354. https://doi.org/10.1016/j.jclepro.2018.03.292

Caputo, F., Ligorio, L., & Pizzi, S. (2021). The contribution of higher education institutions to the SDGs—An evaluation of sustainability reporting practices. *Administrative Sciences, 11*(3), 97. https://doi.org/10.3390/admsci11030097

Catasús, B., & Gröjer, J. E. (2006). Indicators: On visualizing, classifying and dramatizing. *Journal of Intellectual Capital, 7*(2), 187–203. https://doi.org/10.1108/14691930610661854

Catasús, B., Ersson, S., Gröjer, J. E., & Yang Wallentin, F. (2007). What gets measured gets… on indicating, mobilizing and acting. *Accounting, Auditing & Accountability Journal, 20*(4), 505–521. https://doi.org/10.1108/09513570710762566

Ceulemans, K., & De Prins, M. (2010). Teacher's manual and method for SD integration in curricula. *Journal of Cleaner Production, 18*(7), 645–651. https://doi.org/10.1016/j.jclepro.2009.09.014

Ceulemans, K., Lozano, R., & Alonso-Almeida, M. D. M. (2015a). Sustainability reporting in higher education: Interconnecting the reporting process and organisational change management for sustainability. *Sustainability, 7*(7), 8881–8903. https://doi.org/10.3390/su7078881

Ceulemans, K., Molderez, I., & Van Liedekerke, L. (2015b). Sustainability reporting in higher education: A comprehensive review of the recent literature and paths for further research. *Journal of Cleaner Production, 106*, 127–143. https://doi.org/10.1016/j.jclepro.2014.09.052

Chankseliani, M., & McCowan, T. (2021). Higher education and the sustainable development goals. *Higher Education, 81*(1), 1–8. https://doi.org/10.1007/s10734-020-00652-w

Chatelain-Ponroy, S., & Morin-Delerm, S. (2016). Adoption of sustainable development reporting by universities: An analysis of French first-time reporters. *Accounting, Auditing & Accountability Journal, 29*(5), 887–918. https://doi.org/10.1108/AAAJ-06-2014-1720

Chiucchi, M. S. (2013). Intellectual capital accounting in action: Enhancing learning through interventionist research. *Journal of Intellectual Capital, 14*(1), 48–68. https://doi.org/10.1108/14691931311289011

De Iorio, S., Zampone, G., & Piccolo, A. (2022). Determinant factors of SDG disclosure in the university context. *Administrative Sciences, 12*(1), 21. https://doi.org/10.3390/admsci12010021

De Villiers, C., Dimes, R., & Molinari, M. (2025). Determinants, mechanisms and consequences of UN SDGs reporting by universities: Conceptual framework and avenues for future research. *Journal of Public Budgeting, Accounting & Financial Management, 37*(2), 329–349. https://doi.org/10.1108/JPBAFM-07-2023-0125

Denzin, N. K., & Lincoln, Y. S. (Eds.). (2017). *The Sage handbook of qualitative research.* Sage Publication.

Dimes, R., & Molinari, M. (2024). Non-financial reporting and corporate governance: A conceptual framework. *Sustainability Accounting, Management and Policy Journal, 15*(5), 1067–1093. https://doi.org/10.1108/SAMPJ-04-2022-0212

Dobson, A. (2007). Environmental citizenship: Towards sustainable development. *Sustainable Development, 15*(5), 276–285. https://doi.org/10.1002/sd.344

Duram, L. A., & Williams, L. L. (2015). Growing a student organic garden within the context of university sustainability initiatives. *International Journal of Sustainability in Higher Education, 16*(1), 3–15. https://doi.org/10.1108/IJSHE-03-2013-0026

Ferguson, T., & Roofe, C. G. (2020). SDG 4 in higher education: Challenges and opportunities. *International Journal of Sustainability in Higher Education, 21*(5), 959–975. https://doi.org/10.1108/IJSHE-12-2019-0353

Fonseca, A., Macdonald, A., Dandy, E., & Valenti, P. (2011). The state of sustainability reporting at Canadian universities. *International Journal of Sustainability in Higher Education, 12*(1), 22–40. https://doi.org/10.1108/14676371111098285

Ghosh, S. (2011). Participation in the green power partnership: An analysis of higher education institutions as partners in the program. *International Journal of Sustainability in Higher Education, 12*(4), 306–321. https://doi.org/10.1108/14676371111168241

Guthrie, J., & Neumann, R. (2007). Economic and non-financial performance indicators in universities: The establishment of a performance-driven system for Australian higher education. *Public Management Review, 9*(2), 231–252. https://doi.org/10.1080/14719030701340390

Hansen, B., Stiling, P., & Uy, W. F. (2021). Innovations and challenges in SDG integration and reporting in higher education: A case study from the University of South Florida. *International Journal of Sustainability in Higher Education, 22*(5), 1002–1021. https://doi.org/10.1108/IJSHE-08-2020-0310

Hassan, A., Adhikariparajuli, M., Fletcher, M., & Elamer, A. (2019). Integrated reporting in UK higher education institutions. *Sustainability Accounting, Management and Policy Journal, 10*(5), 844–876. https://doi.org/10.1108/SAMPJ-03-2018-0093

Iacuzzi, S., Garlatti, A., Fedele, P., & Lombrano, A. (2020). Integrated reporting and change: Evidence from public universities. *Journal of Public Budgeting, Accounting & Financial Management, 32*(2), 291–310. https://doi.org/10.1108/JPBAFM-08-2019-0120

IFRS Foundation. (2022). *International framework January 2021.* Available at: https://integratedreportingsa.org/ircsa/wp-content/uploads/2021/01/International-IR-Framework-2021-with-changes-highlighted.pdf

Larrán Jorge, M., Andrades Peña, F. J., & Herrera Madueño, J. (2019). An analysis of university sustainability reports from the GRI database: An examination of influential variables. *Journal of Environmental Planning and Management, 62*(6), 1019–1044. https://doi.org/10.1080/09640568.2018.1457952

Leal Filho, W., Raath, S., Lazzarini, B., Vargas, V. R., de Souza, L., Anholon, R., & Orlovic, V. L. (2018). The role of transformation in learning and education for sustainability. *Journal of Cleaner Production, 199*, 286–295. https://doi.org/10.1016/j.jclepro.2018.07.017

Leal Filho, W., Salvia, A. L., & Eustachio, J. H. P. P. (2023). An overview of the engagement of higher education institutions in the implementation of the UN Sustainable Development

Goals. *Journal of Cleaner Production, 386*, Article 135694. https://doi.org/10.1016/j.jclepro.2022.135694

Leal Filho, W., Frankenberger, F., Salvia, A. L., Azeiteiro, U., Alves, F., Castro, P., & Ávila, L. V. (2021). A framework for the implementation of the Sustainable Development Goals in university programmes. *Journal of Cleaner Production, 299*, Article 126915. https://doi.org/10.1016/j.jclepro.2021.126915

Lisi, S., Mignacca, B., & Grimaldi, M. (2023). Non-financial reporting and SMEs: A systematic review, research agenda, and novel conceptualization. *Journal of Management & Organization, 1–23*,. https://doi.org/10.1017/jmo.2023.43

Lopatta, K., & Jaeschke, R. (2014). Sustainability reporting at German and Austrian universities. *International Journal of Education Economics and Development, 5*(1), 66–90. https://doi.org/10.1504/IJEED.2014.059868

Lozano, R., & Young, W. (2013). Assessing sustainability in university curricula: Exploring the influence of student numbers and course credits. *Journal of Cleaner Production, 49*, 134–141. https://doi.org/10.1016/j.jclepro.2012.07.032

Lozano, R., Ceulemans, K., Alonso-Almeida, M., Huisingh, D., Lozano, F. J., Waas, T., & Hugé, J. (2015). A review of commitment and implementation of sustainable development in higher education: Results from a worldwide survey. *Journal of Cleaner Production, 108*, 1–18. https://doi.org/10.1016/j.jclepro.2014.09.048

Lozano, R. (2010). Diffusion of sustainable development in universities' curricula: An empirical example from Cardiff University. *Journal of Cleaner Production, 18*(7), 637–644. https://doi.org/10.1016/j.jclepro.2009.07.005

Lueg, R., & Radlach, R. (2016). Managing sustainable development with management control systems: A literature review. *European Management Journal, 34*(2), 158–171. https://doi.org/10.1016/j.emj.2015.11.005

Maas, K., Schaltegger, S., & Crutzen, N. (2016). Integrating corporate sustainability assessment, management accounting, control, and reporting. *Journal of Cleaner Production, 133*, 859–862. https://doi.org/10.1016/j.jclepro.2016.06.006

Manes-Rossi, F., Tiron-Tudor, A., Nicolò, G., & Zanellato, G. (2018). Ensuring more sustainable reporting in Europe using non-financial disclosure—De facto and de jure evidence. *Sustainability, 10*(4), 1162. https://doi.org/10.3390/su10041162

Manes-Rossi, F., Nicolò, G., & Argento, D. (2020). Non-financial reporting formats in public sector organizations: A structured literature review. *Journal of Public Budgeting, Accounting & Financial Management, 32*(4), 639–669. https://doi.org/10.1108/JPBAFM-03-2020-0037

Menon, S., & Suresh, M. (2020). Synergizing education, research, campus operations, and community engagements towards sustainability in higher education: A literature review. *International Journal of Sustainability in Higher Education, 21*(5), 1015–1051. https://doi.org/10.1108/IJSHE-03-2020-0089

Mio, C., Panfilo, S., & Blundo, B. (2020). Sustainable development goals and the strategic role of business: A systematic literature review. *Business Strategy and the Environment, 29*(8), 3220–3245. https://doi.org/10.1002/bse.2568

Moggi, S. (2023). Sustainability reporting, universities and global reporting initiative applicability: A still open issue. *Sustainability Accounting, Management and Policy Journal, 14*(4), 699–742. https://doi.org/10.1108/SAMPJ-05-2022-0257

Monteiro, S., Ribeiro, V., & Molho, C. (2024a). A 5 pillars approach to the sustainable development goals performance and reporting in Portuguese higher education institutions. Proposal for an applied framework. *International Journal of Sustainability in Higher Education, 25*(1):104–123. https://doi.org/10.1108/IJSHE-02-2023-0047

Monteiro, S., Ribeiro, V., Vilhena, E., Lemos, K., & Molho, C. (2024b). Determinants of online-reporting on sustainable development goals: The case of Portuguese Higher Education Institutions. *Social Responsibility Journal, 20*(3), 462–484. https://doi.org/10.1108/SRJ-01-2023-0044

Murray, J. (2018). Student-led action for sustainability in higher education: A literature review. *International Journal of Sustainability in Higher Education, 19*(6), 1095–1110. https://doi.org/10.1108/IJSHE-09-2017-0164

Palma, L. C., De Oliveira, L. M., Alves, N. B., & Figueiró, P. S. (2023). Sustainability in Business Administration programs in Brazil: What curricula changes have taken place in the past ten years? *International Journal of Sustainability in Higher Education, 24*(6), 1347–1363. https://doi.org/10.1108/IJSHE-11-2021-0482

Raimo, N., Nicolò, G., L'Abate, V., & Vitolla, F. (2024). Analyzing the factors affecting university contributions to achieving the sustainable development goals in European Union countries. *Sustainable Development*. https://doi.org/10.1002/sd.3013

Rosa, M. R. D., Moggi, S., Boscarioli, C., & Freitas Zara, K. R. D. (2024). Examining the implementation of UN sustainable development goals in Brazilian universities. *International Journal of Sustainability in Higher Education Ahead-of-Print*. https://doi.org/10.1108/IJSHE-02-2024-0117

Sepasi, S., Braendle, U., & Rahdari, A. H. (2019). Comprehensive sustainability reporting in higher education institutions. *Social Responsibility Journal, 15*(2), 155–170. https://doi.org/10.1108/SRJ-01-2018-0009

Serafini, P. G., de Moura, J. M., de Almeida, M. R., & de Rezende, J. F. D. (2022). Sustainable development goals in higher education institutions: A systematic literature review. *Journal of Cleaner Production, 370*, Article 133473. https://doi.org/10.1016/j.jclepro.2022.133473

Siboni, B., Del Sordo, C., & Pazzi, S. (2013). Sustainability reporting in state universities: An investigation of Italian pioneering practices. *International Journal of Social Ecology and Sustainable Development (IJSESD), 4*(2), 1–15.

Stolowy, H., & Paugam, L. (2018). The expansion of non-financial reporting: An exploratory study. *Accounting and Business Research, 48*(5), 525–548. https://doi.org/10.1080/00014788.2018.1470141

Stubbs, W., & Higgins, C. (2014). Integrated reporting and internal mechanisms of change. *Accounting, Auditing & Accountability Journal, 27*(7), 1068–1089. https://doi.org/10.1108/AAAJ-03-2013-1279

Traxler, A. A., Schrack, D., & Greiling, D. (2020). Sustainability reporting and management control—A systematic exploratory literature review. *Journal of Cleaner Production, 276*, Article 122725. https://doi.org/10.1016/j.jclepro.2020.122725

United Nations. (2015). *Transforming our world: The 2030 Agenda for Sustainable Development*. Available at: https://sdgs.un.org/2030agenda

Weybrecht, G. (2017). From challenge to opportunity–Management education's crucial role in sustainability and the Sustainable Development Goals—An overview and framework. *The International Journal of Management Education, 15*(2), 84–92. https://doi.org/10.1016/j.ijme.2017.02.008

Yáñez, S., Uruburu, Á., Moreno, A., & Lumbreras, J. (2019). The sustainability report as an essential tool for the holistic and strategic vision of higher education institutions. *Journal of Cleaner Production, 207*, 57–66. https://doi.org/10.1016/j.jclepro.2018.09.171

Yin, R. K. (2014). *Case study research design and methods* (5th ed., p. 282). Sage.

Re-Imagining Capacity Building in Higher Education to Promote Learning and Teaching for Sustainability

Niamh Guiry, John Barimo, Catherine O'Mahony, and Maria Kirrane

Abstract This chapter explores the short course 'Learning & Teaching with the SDGs', developed at University College Cork to enable teaching staff to modify or (re) design their modules and courses to align with and embed the United Nations Sustainable Development Goals (SDGs). This was achieved by providing staff with an understanding of the theoretical background and values of Education for Sustainable Development (ESD) alongside practical examples of effective SDG integration via teaching and assessment methods. This digital badge promotes action-oriented ESD as part of a broader initiative to upskill and empower teaching staff in the area of sustainability whilst aligning with institutional and governmental strategies. This chapter illustrates the potential for digital badges (micro-credentials) to be adopted as pedagogical tools in Higher Education and uses key ESD competencies as an analytical framework to assess how well-grounded this short course is within these established sustainability proficiencies. ESD is a rapidly evolving area that must embody adaptability in response to ongoing development challenges, curricular innovation, and the overarching polycrisis. Drawing upon institutional resources and examples of global good practice, this course demonstrates the potential for ESD

N. Guiry (✉)
School of Law, University College Cork, Cork, Ireland
e-mail: nguiry@ucc.ie

Jointly funded by Taighde Éireann-Research Ireland and the Environmental Protection Agency Government of Ireland Postgraduate Scholarship under Grant number GOIPG/2023/4774, Cork, Ireland

J. Barimo
Sustainable Development Solutions Network (SDSN) Ireland, University College Cork, Ireland.
Sustainability Institute, University College Cork, Cork, Ireland
e-mail: john.barimo@ucc.ie

C. O'Mahony
Centre for the Integration of Research, Teaching and Learning, University College Cork, Cork, Ireland
e-mail: catherine.omahony@ucc.ie

M. Kirrane
Office of Sustainability and Climate Action, University College Cork, Cork, Ireland
e-mail: m.kirrane@ucc.ie

W. Leal Filho et al. (eds.), *Introducing the UN SDGs in Higher Education Institutions*, World Sustainability Series, https://doi.org/10.1007/978-3-032-23017-1_40

to be an interactive and collaborative process which enhances teaching and learning in Higher Education by integrating sustainability into class materials, content, and assessments for the benefit of learners and educators alike.

Keywords Education for sustainable development · Digital badge · Micro-credentials · Sustainable development goals (SDGs) · Continuing professional development (CPD) · Curriculum development

1 Introduction

Higher Education is seen as a crucial partner and enabler in achieving the United Nations (UN) Sustainable Development Goals (SDGs). Moreso, it plays a key role in ensuring that learners acquire knowledge and skills through approaches such as Education for Sustainable Development (ESD) and sustainable lifestyles. Yet, Higher Education staff often lack sufficient time, and the expertise needed to incorporate an ESD focus into their teaching (Guiry et al., 2022). Few Higher Education Institutions (HEIs) offer 'professional development in sustainability for their staff'(Bonilla-Jurado et al., 2024). For those that do, ESD is often viewed as an advanced staff development topic rather than a core requirement (Zhang 2025). This chapter outlines the approach taken at University College Cork (UCC) to support staff in ESD and encourage engagement with the SDGs through the provision of online, self-directed learning recognised via a digital badge. Digital badges are electronic micro-credentials that represent a valuable skill or proficiency gained by learners in a particular area (Besser & Newby 2019). In the context of UCC, they are used to recognise non-accredited learning and sit within a robust and transparent quality assurance structure and are both valued and utilised by staff. Indeed, Ireland is the first European country to develop a national framework for micro-credentials (Irish Universities Association, 2023). The recognition approach offered in this course thus sits within good practice in badge design, being 'internally valued' alongside robust offerings (Hickey et al., 2015) and has proven popular for recognising engagement in staff professional development activities. The purpose of the online course 'Learning & Teaching with the SDGs' is to provide an ESD pedagogical tool which enables UCC teaching staff to enrich teaching and learning activities across all disciplines. In this chapter, we uncover the rationale guiding the development of the short course and provide details of its design and enactment. In doing so we hope to inspire others in their ESD journeys. We use the core ESD competencies to reflectively explore the course and illustrate the degree to which it aligns with or departs from these skills and values. The outcomes, impacts, limitations, and future opportunities of this online course are subsequently discussed. Finally, this chapter concludes with reflections on the role of Higher Education and educators in the integration of ESD, through which educators and learners can be empowered with the essential skills and adaptive understandings to respond to and tackle emergent and ongoing SDG-related challenges.

2 Policy and Institutional Context

Unanimously ratified by 193 UN member states in 2015 after 3 years of formulation, consultation, and negotiations (Dodds et al., 2017), the 2030 Agenda and its SDGs represent a globally defined 'plan of action for people, planet and prosperity' (UN, 2015). With 17 goals and 169 targets to be achieved by 2030, the SDGs channel the three pillars of sustainable development (social, environmental, and economic), and their success is contingent on the collaborative mobilisation of 'Governments as well as parliaments, the United Nations system and other international institutions, local authorities, indigenous peoples, civil society, business and the private sector, the scientific and academic community—and all people' (UN 2015, Sec. 52).

The goals of the framework include ensuring the health and well-being of all (SDG 3), providing universal access to quality education (SDG 4), reducing social inequalities (SDG 10), and promoting peaceful, just, and inclusive societies (SDG 16). Whilst being subject to a range of criticism (*e.g.*, Adelman 2018); Spangenberg 2017; Kopnina 2015), it is arguable that this 'near-universal acceptance of and reporting on a non-binding framework by states has created a newfound global paradigm', one which is grounded in voluntary commitments and a whole-of-society approach to sustainable development (Guiry 2024). Indeed, HEIs play a vital role in furthering the realisation of sustainable development, and in this case, the aspirations of the SDGs, through teaching, learning, research, and reporting activities (Leal Filho et al., 2024). Evidence of robust engagement with the SDGs and issues of sustainability can be found in many institutions, including UCC. The 2030 Agenda, SDGs, and SDG targets, including the means of implementation, are 'universal, indivisible and interlinked' (UN 2015, Sec. 71), whilst the need for ESD to evolve beyond ontological reductionism has been advocated by Sterling (2024). UCC has been 'ranked highly' by the UI Green Metric World University Rankings for institutional sustainability since 2011, ranking fourth globally in 2024, and was the first University in the world to be awarded a Green Flag by the Foundation for Environmental Education (Kirrane et al., 2020; UI Green Metric (UI 2024)). The Green-Campus programme, pioneered by UCC students in partnership with staff in 2007 (Reidy et al., 2015), provides a seven-step framework that promotes awareness of key sustainable development challenges and facilitates the cultivation of active citizenship. Step 5 of this framework, *Linking to Learning on Campus*, requires the broad embedding of environmental education across a range of disciplines and courses (An Taisce 2021). Given that its origins lie in student activism, student leadership is one of the building blocks of the UCC Green Campus programme and guides its approach to sustainability initiatives and UCC policy. Expanding on previous internal policies, UCC launched its first Sustainability Strategy in 2016, using the SDGs as a framework to formulate its overarching strategic aims across nine key sustainability areas (UCC 2016). Integrating sustainability across academic courses and increasing engagement with issues of sustainability is a priority of this programme and UCC's overall Academic Strategy (UCC 2023). Furthermore, these internal documents aligned with strategic priorities of the Government of Ireland (Department of Education 2014 and 2022) and the UN (UNESCO 2017 and 2020), especially with regards to ESD capacity building amongst teaching staff. While the Government of Ireland clearly recognised the role of education in SDG implementation, policy initially focused on issues of social equality and inclusion but also gave regard to climate literacy (Government of Ireland 2018). 'Learning & Teaching with the SDGs' was developed in accordance with the University's strategic goals and as a follow-on from several sustainability initiatives, including UCC's Connected Curriculum, the SDG Toolkit for Teaching and Learning, and the 2021 Green-Campus Network Consultation. The Connected Curriculum proposed a new framing of the educational offerings at UCC and included sustainability as one of its six foci. This was followed by the development of the SDG Toolkit to create a pan-disciplinary framework applicable to all ESD skill levels so to increase capacity amongst teaching staff internally and globally as an open-source web-based resource grounded in transformative sustainability learning (TSL) and critical pedagogy (Barimo et al., 2026). The web-based toolkit was coupled with continuous professional development (CPD), initially offered as 90-minute workshops, and evolving into 5-hour curriculum 'Design Sprints'. While the toolkit and associated workshops promoted transdisciplinarity, to address the holistic nature of the 2030 Agenda, they also provided users with the disciplinary knowledge needed to take action in their respective fields and the wider community (Barimo et al., 2026). The toolkit's initial mapping exercises allows users to identify the connections between their curriculum and the SDGs, and subsequently their research to promote research-based teaching practices (UCC SDG Toolkit 2021). 'Learning & Teaching with the SDGs' is a reimagining of this toolkit based on feedback from staff that they needed more time and content to explore these ideas. The 2021 Green-Campus Network Consultation brought together students and staff from 15 different Irish HEIs to identify common opportunities and challenges in relation to ESD in Higher Education in Ireland (Guiry et al., 2022). The online course sought to address some of these identified challenges by strengthening ESD competencies amongst staff and facilitating the effective integration of the SDGs into academic programmes.

3 An Integrative Approach to Course Development

The short course is a 30-hour asynchronous offering recognised via a digital badge and enables teaching staff to review and reconstruct their curriculum and incorporate student-centric, problem-based approaches to ESD. Course development followed an integrative approach with Students as Partners (SaP), the TSL framework, engaged teaching, and research, and the competencies approach (see Sect. 4). It is composed of four modules that learners must complete to receive their digital badge recognition (Fig. 1). This section provides an overview of the course. It outlines the content and key materials covered in each of the four modules so to demonstrate the relevance and approach taken towards systemic ESD integration by 'Learning & Teaching with the SDGs'.

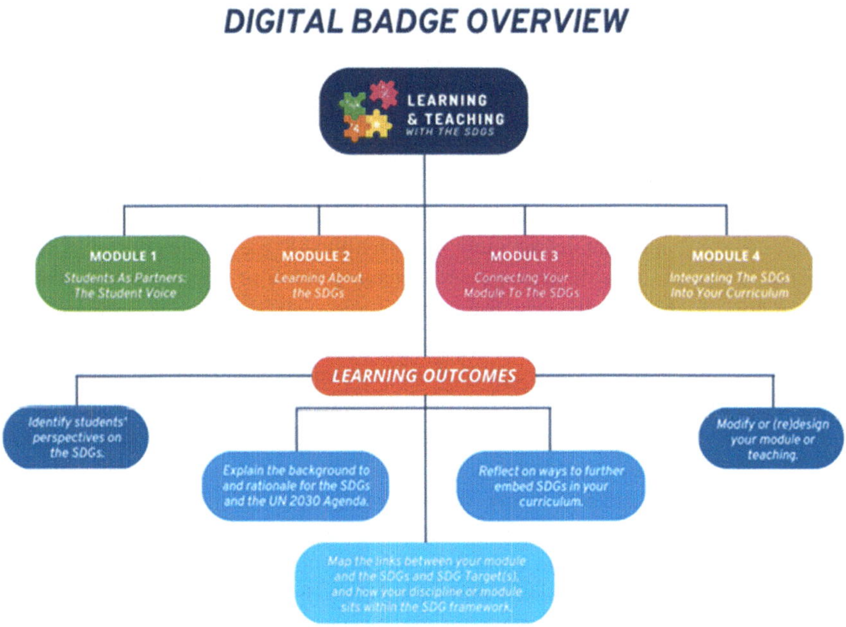

Fig. 1 Overview of course modules and learning outcomes of the short course 'Learning & Teaching with the SDGs'

This course adopts a stepwise approach, with participants completing formative assessments tasks that accrue to a module review and design that explicitly highlight alignments between their existing or future curriculum and the SDGs. Using this approach, participants are provided with a foundational knowledge of key ESD theory and policy in addition to practical outputs that will inform their own future teaching, learning, and assessment practices. The course is an opt-in offering for staff and participants can decide which section of the course to begin with and whether or not they complete the assignments. A digital badge is only awarded for those who complete all the assignments. This design aligns with self-determination theory by providing supports to grow participants' competence yet ensuring that staff autonomy and academic freedom are respected. *Module 1: Students As Partners—The Student Voice* begins by highlighting the importance of incorporating a SaP co-creative approach to curriculum design. Using SaP as the starting point for this course was a deliberate design choice, as it ensures the student voice frames the subsequent course content. It also provides that staff see the relatedness of the topic, which can be a powerful motivator for action. SaP is a participatory pedagogical approach that works to foster a fundamental academic relationship from which both students and staff can benefit (Healey et al., 2014). This approach confronts 'traditional hierarchies' and offers students a holistic opportunity to collaborate with staff and co-design curricula (Cook-Sather & Alter 2011). The student partnership conceptual model developed by Healey et al., (2014) is highlighted as a useful framework in the course, as it allows staff to explore student partnership. Previous CPD offerings called curriculum 'Design Sprints' brought UCC staff and students together to consider the redesign of modules to integrate the SDGs (Barimo et al., 2026). This early engagement recognises the advice of Whelehan (2020), that any 'tokenistic' or late-stage inclusion of students in curricula development should be avoided. Instead, teaching staff should strive to be active listeners and use innovative methods to cultivate impactful student engagement opportunities. In the online course, staff are encouraged to partner with students when making sense of the SDGs and linking their teaching to the global goals. Some staff have taken this further, with students embarking on lengthy projects to map entire programmes to the SDGs as part of SaP research initiatives. As Vermeulen & Emplit (2024, p. 2) note, the 'co-creation of course contents and learning activities, for and with students, may be implemented to sustain a student-centred evolution of teaching' in alignment with the Bologna Process. While engagement with the SaP approach when designing for ESD is encouraged, different levels of participation may be suitable depending on the context (Bovill & Bulley 2011). For instance, first-year undergraduate students may require more guidance to contribute to curriculum design in comparison to postgraduate students. However, by embracing open dialogue and facilitating meaningful participation, teaching staff can work to increase student confidence and build a sense of curricula ownership amongst their students. This partnership approach may also help to alleviate staff anxiety related to being less expert in this area as it recognises that this work is emergent and requires collective insights and reflection. The authors view transdisciplinary practices and collaboration as key components of effective and engaging ESD content, and both play a central role in this course. *Module 2: Learning About the SDGs* introduces learners to the holistic interdependent nature of the 2030 Agenda, which states that '[t] he Sustainable Development Goals and targets are integrated and indivisible, global in nature and universally applicable, taking into account different national realities, capacities and levels of development and respecting national policies and priorities' (UN 2015, Sec. 55). Learners are also presented with some of the limitations of the framework. For example, the online course highlights the lack of explicit inclusion of the LGBTQI+ community in the SDG framework, something that arguably creates tensions with the 'inclusive ideology' of the SDGs and the general ethos of the framework, that 'no one will be left behind' (Mills 2015; Guiry 2023). Another example of a criticism of the SDGs explored in this module is the inadequate incorporation of the rights, perspectives, and traditional knowledge of Indigenous Peoples (see Yap & Watene 2019). As Magni (2017) elucidates, recognition and integration of such considerations into the 2030 Agenda is critical to effectively address environmental crises. It would also align with international recognition of indigenous rights and the role of indigenous and traditional communities in conservation (*e.g.,* UN 2007; UN 2024). Additionally, SDG targets to reduce greenhouse gas emissions appear woefully inadequate, however, parallel negotiations for the legally binding Paris Agreement at the UN Climate Change Conference (COP21) allowed for the two UN frameworks to develop in tandem (Dodds et al., 2017). These examples highlight the need for critical thinking in ESD, which was emphasised in this CPD course. While the SDGs provide useful transdisciplinary language that can be used to navigate shared challenges, it is important to constructively analyse this framework to identify potential tensions and gaps, so that a more effective and representative sustainability framework can be established post-2030. SDG 17 (Partnership for the Goals) and the importance of civic engagement are also detailed in this module. Civic engagement through community collaboration is a UCC strategic priority (UCC 2017) and a national benchmark in Higher Education in Ireland (Higher Education Authority 2011, 2018). Long-term strategic partnerships bring HEIs and communities together to collaborate and address shared development challenges. As El-Jardali et al., (2018) comment, HEIs are 'uniquely placed to lead the cross-sectoral implementation of the SDGs, providing an invaluable source of expertise in research and education on all sectors of the SDGs'. Learners are also introduced to engaged teaching and research approaches, which can be integrated across the curricula to create impactful community actions. Reference is also made to existing UCC resources, including the Civic Engagement Toolkit (Hally et al., 2021). It should be noted that in Ireland, and increasingly across Europe, engaged research is an umbrella term for co-created or co-produced research conducted 'with' society and not 'for' society and includes methodologies such as participatory action research, citizen science, or public patient involvement (Irish Universities Association and Irish Research Council 2016). *Module 3: Connecting Your Modules to the SDGs* allows staff to map the connections between their respective discipline and the SDGs using the SDG Toolkit for Teaching and Learning (see Barimo et al., 2026). Learners identify which SDG Targets align with their discipline and reflect on their teachings through 'formal, informal, or hidden curriculum'. These activities facilitate intentional and proactive SDG integration into curriculum, which is further developed in Module 4. The course recognises that SDG curriculum integration is likely to evolve as teaching staff become more familiar with the SDG framework in the context of guiding student development. Digital badge learners are given a stepwise guide on how curriculum modification or redesign may progress over time. Using the UCC Connected Curriculum self-evaluation tool (Connected Curriculum 2018), learners are given examples of how ESD can begin as an introductory class element with disciplinary perspectives on sustainability before progressing to a stage where the SDGs are central to assessments, learning outcomes, and other core design features. A key feature throughout this digital badge is the video interviews of expert teaching staff who share their insights on successful alignment of respective disciplines with the SDGs. A broad array of disciplinary backgrounds was featured in these videos including law, business, engineering, nursing, and humanities. These real-life, peer-to-peer examples of good practice provide learners with insights into innovative teaching and learning methods being implemented locally and allow staff to reflect on how they could best integrate the SDGs across UCC via their teaching and assessment practices. In the final module, *Module 4: Integrating the SDGs into Your Curriculum,* participants are introduced to the TSL framework which aims to localise the learner experience between universities and local communities so to 'find or create necessary common ground to minimize conflicts of interests' and rebuild relationships between universities and local communities to create collaborative research, which lies at the heart of engaged teaching and research (Sipos et al., 2008). Furthermore, as TSL promotes the 'co-creation of knowledge', it also prompts educators to be 'self-reflective and adaptive in response to rapidly changing environments'. It also invokes the Head, Heart, and Hands model, which 're-imagines the cognitive, psychomotor, and affective learning domains' outlined in Bloom's Taxonomy and emphasises the importance of the 'translation of passion and values' into proactive behaviour (Sipos et al., 2008; Bloom et al., 1964). Learners discover the significance of action-oriented teaching, an ESD pedagogy that empowers students to make informed decisions and equips them with skills to effectively respond to sustainability challenges in their personal and professional lives (Sinakou et al., 2019). Action-oriented teaching approaches 'encourage self-directed learning' and foster collaboration, 'often with an inter- or transdisciplinary focus'. The course developers recognised that some teaching programmes will have a more applied focus in response to sustainable development challenges. However, there is also scope for theoretical and conceptual modules to incorporate concepts of sustainability in a broader sense.

4 Aligning with ESD Competencies

To strategically design and deliver effective ESD in Higher Education, one must be equipped with practical skills alongside an understanding of the theoretical underpinnings and core competencies relevant to sustainability education. ESD competencies focus on learning outputs thereby enabling educators to use problem-solving strategies to equip learners with the ability to 'modify and shape the future of society, and to guide its social, economic, technological, and ecological changes along the lines of sustainable development' (de Haan 2010). As illustrated by Rieckmann (2018), there are eight principal competencies found in international ESD discourse that are highlighted as essential for advancing sustainability knowledge:

- Systems thinking competency
- Collaboration competency
- Integrated problem-solving competency
- Critical thinking competency
- Strategic competency
- Normative competency
- Self-awareness competency
- Anticipatory competency

This competency framework described by Rieckmann (2018), which also served as a foundation part of GreenComp—The European Sustainability Competence Framework (Bianchi et al., 2022), was applied to this online course.

Systems thinking competency is the ability to recognise the 'characteristics', 'interconnections', and 'function' of a system (Arnold & Wade 2015). The SDGs deal with complex, interlinked issues and this course encourages learners to adopt a holistic approach to sustainable development. Throughout the four modules, learners are taken through the origins and nature of the SDGs, the connections between sustainable development and their discipline, and the importance of achieving these goals for the well-being of people and the planet. Systems thinking competency is also promoted through the SDG Toolkit activities, particularly the Mapping Tool, as it enables an in-depth investigation, identification, and understanding of the connections between the SDGs and their curriculum. Furthermore, the 2030 Agenda's own propensity for integrative and interconnected approaches to sustainable development suggests an evolution of thought beyond Cartesian/Newtonian reductionism. Collaboration competency is a key feature of this badge. Long-term partnership and collaborative synergies are key 'drivers of change' and are critical if all SDGs are to be achieved (Leal Filho et al., 2022). This course promotes collaboration competency in three different ways. Firstly, the student-staff partnership is highlighted as a significant component for the creation of high-quality, relevant, and impactful ESD experiences. The potential partnership opportunities are particularly emphasised across Module 1. Secondly, the value of interdisciplinary academic collaboration is demonstrated, drawing upon local examples to emphasise the importance of sharing experiences and group problem-solving. Lastly, the importance of community collaboration through engaged teaching and research is stressed throughout. Cultivating collaborative partnerships with societal stakeholders adds a real-world dimension to student learning and allows sustainable development challenges to be tackled collectively, which may increase student motivation as opposed to the 'chalk and board' lecture style. Arguably, there is an overlap between this and integrated problem-solving competency, engaging with the other 'problem-solving frameworks' to tackle and overcome sustainable development issues (Wiek et al., 2015). Through inter- and transdisciplinary collaborative activities, effective solutions can be explored and applied. These competencies could have been strengthened by incorporating interactive workshop sessions into the course which would enhance the relatedness aspect of 'Learning & Teaching with the SDGs'. Critical thinking allows learners to 'apply higher-order, rational thinking skills such as analysis, synthesis, problem recognition, and problem-solving, inference, and evaluation' (Taimur & Sattar 2019). This competency is promoted in Module 2 when discussing the limitations of the SDG framework. While the SDGs can be regarded as a starting point for considering less anthropocentric dimensions of sustainability, the heuristic value of the SDGs cannot be understated for critical exploration of relational connections between students and their local surroundings (Barimo & Mullally 2023). Strategic competency can particularly be seen in Modules 3 and 4 when learners are given practice-based examples of how to connect their discipline with the SDGs. There are some interconnections between critical and normative competency in this course. Normative competence allows learners to understand, consider, and apply the values embedded in their belief systems and navigate sustainability norms and conflicts (Rieckmann 2018, Sá et al., 2022). Using institutional and local examples to encourage learners to challenge their established perceptions of the SDG paradigm, deep transformative learning experiences can take place (Giangrande et al., 2019). Self-awareness competency is the ability to be aware of one's role in society and how it relates to and is impacted by others (Jaakkola et al., 2022). HEI teaching staff have a critical part to play in educating current and future generations on sustainability issues, including helping learners navigate these complex challenges and giving them the confidence to take action. From the focus on empowering students and understanding how teaching practices relate to SDGs, to the significance of ESD overall, this competency occurs throughout the digital badge. All eight competencies are mentioned in Module 4 to spark ideas related to ESD integration, as some or indeed all may fit with disciplinary teaching practices and learning objectives. By understanding these competencies, teaching staff can design curriculum to be supportive of these sustainability proficiencies (Brundiers et al., 2021). Arguably, the least emphasised ESD competency is anticipatory, which fosters futures thinking. While the evolutionary process of redesigning modules is highlighted, a weakness of this badge is that it does not address how teaching staff could deal with potential difficulties pertaining to curriculum changes and the use of experimental assessment techniques. A lack of understanding or confidence amongst educators may inhibit the adoption of experimental practices and thus present additional challenges to ESD integration. As Serdyukov (2017) wrote '[t] o create innovations, we need innovators'. In these cases, staff would require supportive environments and experienced peers to guide them through challenging aspects of curriculum design to ensure authentic transformative learning experiences are created. To truly 'transform capacities', learners require the above-outlined sustainability competencies in addition to 'corresponding values and motivational drivers' (Rieckmann 2018). These eight competencies offer a useful parameter to gauge the degree to which this online course corresponds to core ESD skills and values, and by extension, empowers learners throughout their sustainability journey. Overall, 'Learning & Teaching with the SDGs' is well-aligned with these internationally supported ESD competencies, with weak alignment with the anticipatory element as discussed. These competencies provide learners with a well-rounded skillset that can be employed when navigating sustainability challenges and opportunities. Not only relevant to CPD resources, this exercise of self-evaluation could also be applied to teaching and learning activities. This would allow educators to reflect on existing methods and practices and identify any existing gaps or weak competency alignments, which could subsequently be addressed through incremental adjustments or more robust curriculum changes as needed.

5 Outcomes, Limitations, and Future Opportunities

To date, a total of 53 learners (including students and teaching and professional services staff) were awarded digital badges. This number does not include individuals who may have removed themselves from the course or those who are auditing without completing assessments for badge attainment. The highest levels of engagement can be seen in the College of Business and Law (Table 1). The College of Medicine and Health participated the least in this course, despite significant topical relevance to the SDGs. Whilst full completion of the badge has been limited, the course has been successful in addressing concerns raised in the 2021 Green-Campus Network Consultation (Guiry et al., 2022) and providing resources complementary to the SDG Toolkit. For instance, the digital badge helps to address some of the main identified challenges facing teaching staff in Irish HEIs (limited skills, knowledge, and resources related to ESD) by offering accessible and practical step-by-step guidance on how to effectively incorporate issues of sustainability into respective modules and programmes. The need to develop peer-informed and collaborative CPD with a preference for training that is delivered in a 'flexible and non-prescriptive manner' was emphasised in this report (Guiry et al., 2022). Thus, in this regard, the course was effective in producing a timely, adaptable, and interdisciplinary resource from which both students and staff can benefit.

Reflecting upon the future of this initiative, there is scope to expand this badge on a national scale. This expansion would allow for the pooling of resources, avoid unnecessary course duplication, accelerate the integration of ESD across Higher Education curricula, and ultimately align with the growing interest in establishing an ESD standard in Higher Education (Guiry et al., 2022). Unfortunately, competitiveness within and amongst HEIs still confounds the mainstreaming of ESD into the curriculum (Byrne et al., 2017). There are several limitations related to this digital badge to acknowledge. Firstly, this course lacks an interactive element that would facilitate peer learning and the development of soft skills such as effective communication, active listening, and teamwork. The benefits of peer-to-peer learning are well-established, and incorporating interactive workshop sessions into the course would strengthen its impact. Limited staffing and resources prohibited active workshop sessions. Secondly, 'Learning & Teaching with the SDGs' is not a mandatory course for teaching staff as institutional culture respects academic freedom. Comprehensive integration of the SDGs across UCC curricula cannot be achieved through this course alone and would require a variety of modalities, *e.g.*, sharing best practice though a horizontal community of practice and dedicated ESD seminars. Institutional and national incentives and release time are needed to compel teaching staff to dedicate their time and attention to ESD and curriculum (re)design.

Table 1 Breakdown of course completion across UCC

College	Total	Lecturer	Professional Services	Student
College of arts, Celtic studies, and social sciences	9	7	1	1
Central (cross-cutting units not designated within a college)	13	1	12	0
College of business and law	14	13	0	1
College of medicine and health	5	4	1	0
College of science, engineering, and food science	10	5	5	0
External to UCC	2	2	0	0

Thirdly, while a gender balance was achieved among the video interviewees, racial and ethnic diversity is lacking. A diverse range of interviewee videos would have provided a broader perspective on ESD pedagogical approaches whilst aligning with the inclusive nature of the ESD ethos and institutional equality, diversity, and inclusion practices. Fourthly, ESD and innovative pedagogies are rapidly evolving and any CPD course, resource, or digital badge needs to be dynamic and updated frequently to reflect global sustainability challenges. This would enable other disciplinary approaches and innovations to be referenced. Finally, HEI teaching staff are under significant time constraints and may not have the time to complete a 30-hour course. The existing asynchronous delivery of content and lack of direct human interaction may be a factor in the low completion rate. However, it is likely that many participants utilise this short course as a reference library given its ungated format. A once-off workshop session providing staff with an overview of the SDGs and guiding participants through the SDG Toolkit's Mapping Tools may initially suffice while a community of practice could recommend bespoke workshops according to collective need. While the rise in popularity of digital badges in recognition of discrete learning could result in greater support for these pedagogical tools (Ahn et al., 2014), the widespread development of digital badges risks compromising their authority as an 'effective assessment tool', unless credibility and course authenticity can be established (Farmer & West 2016). The quality assurance and governance structure in place in UCC and the nascent national framework for micro-credentials ensures that these flexible, discrete forms of learning hold greater trust and respect across the academic community. The national recognition also ensures greater portability and recognition of the digital badge outside of UCC, but some work is needed to ensure recognition outside the Irish Higher Education sector. Key stakeholders, such as future employers and national bodies, must recognise the value and flexibility of badges to increase learner engagement (Davis & Singh 2015). Additionally, some learners may be unfamiliar with the concept of digital badges, lack digital connectivity, or have limited technological skills, all of which could act as a barrier to the widespread use of online courses and digital badges as a tool for full and effective ESD implementation and SDG integration within and across HEIs.

6 Conclusion

As this case study illustrates, the online course 'Learning & Teaching with the SDGs' offers a useful opportunity to further ESD efforts by integrating sustainability into teaching and learning activities across UCC. The course has noted limitations in relation to its non-mandatory nature and wholly online delivery. Nevertheless, as this chapter illustrates, 'Learning & Teaching with the SDGs' has the potential to enhance the teaching content being delivered on campus whilst also allowing teaching staff to become members of an emerging and active community committed to enhancing the topicality and impact of their courses, modules, and programmes. Thus, this course helps to ensure that students and staff across all disciplines understand the complexities of local, regional, and global sustainable development and are equipped with the essential skills, values, and knowledge to think critically, challenge, and tackle existing unsustainable norms, institutional practices, and local issues, and have their efforts recognised via a digital badge. Furthermore, the promotion of engaged teaching and research grounds action-oriented pedagogies in the local context. HEIs are a key partner of and contributor to SDG implementation. In the face of ever-growing and complex global ecological, social, and economic challenges, HEIs have a duty to amalgamate sustainability and the values, ethos, and teachings of ESD into the foundation of all university practices. In this context, the authors also see the utility of incorporating emerging UN conventions advancing sustainability and/or sustainable development into ESD activities and sustainability integration efforts. For example, the Kunming-Montreal Global Biodiversity Framework (UNEP 2022) begins to steer global governance towards a less exploitive view of natural ecosystems while recognising 'contributions of indigenous peoples and local communities as custodians of biodiversity and as partners in its conservation, restoration and sustainable use'. Additionally, the UN Pact for the Future (UN 2024) could serve as a useful heuristic for anticipatory competency development as it states that 'we must conserve, restore, and sustainably use our planet's ecosystems and natural resources' as a basic right for future generations. Protocols such as these will invariably influence the establishment of the agenda for post-2030 sustainable development. Going forward, such perspectives could be used to enhance current and emerging ESD approaches in HEIs. By taking inspiration from institutional, local, and global best practices and upskilling teaching staff through flexible and peer-informed trainings and resources, educators can guide learners to have voice and agency in matters of sustainability so to manifest the imaginary and better shape the societies and planet that they will inherit.

7 Acknowledgements

This project was funded by the National Forum for the Enhancement of Teaching and Learning under the 2020 SATLE funding. We sincerely thank the members of the SDG Toolkit Advisory Committee and the following students, staff, and alumni who contributed to this course: Dr. Gerald Mullally, Asha Woodhouse, Áine Ní Dháiligh, Mark Falvey, Rebecca Doocey, Caoimhe Flynn, Prof Edmond Byrne, Dr. Angela Flynn, Prof Owen McIntyre, Dr. Briony Supple, Dr. F. Jerry Reen, Prof Caitríona Ní Dhúill, Dr. Noreen Byrne, and William O'Halloran.

References

Adelman, S. (2018). The sustainable development goals, anthropocentrism and neoliberalism. In D. French & L. J. Kotzé (Eds.), *Sustainable development goals: Law, theory and implementation* (pp. 15–40). Edward Elgar. https://doi.org/10.4337/9781786438768.00008

Ahn, J., Pellicone, A., & Butler, B. S. (2014). Open badges for education: What are the implications at the intersection of open systems and badging? *Research in Learning Technology, 22*, 1–13. https://doi.org/10.3402/rlt.v22.23563

Arnold, R. D., & Wade, J. P. (2015). A definition of systems thinking: A systems approach. *Procedia Computer Science, 44*, 669–678. https://doi.org/10.1016/j.procs.2015.03.050

Barimo, J., & Mullally, G. (2023). Walking at the margins of the Anthropocene: A novel curriculum to enhance sustainability learning and teaching. In *International Conference on Sustainable Development Conference*. https://ic-sd.org/wp-content/uploads/2023/10/2023-submission_319.pdf

Barimo, J.F, O'Mahony, C., Mullally, G., Byrne, E., O'Halloran, J., Reidy, D., & Kirrane, M. (2026). Integrating Sustainability Across University Curricula: Creating a Dynamic Pan-Disciplinary Framework and Open-Source Materials. In W. Leal Filho, R.F Sari, and L. Viera Trevisan (Eds.), *Green Metric in Higher Education - Measuring and Reporting on Sustainability Initiatives at Higher Education Institutions, World Sustainability Series,* Springer Nature. Pp 1031-1050. https://doi.org/10.1007/978-3-032-00361-4_56

Besser, E. D., & Newby, T. J. (2019). Exploring the role of feedback and its impact within a digital badge system from student perspectives. *TechTrends, 63*, 485–495. https://doi.org/10.1007/s11528-019-00386-2

Bianchi, G., Pisiotis, U., & Cabrera Giraldez, M. (2022). GreenComp—The European sustainability competence framework. In M. Bacigalupo & Y. Punie (Eds.), *EUR 30955 EN*. Publications Office of the European Union. https://doi.org/10.2760/13286

Bloom, B. S., Masia, B. B., & Krathwohl, D. R. (1964). *Taxonomy of educational objectives (two volumes: The affective domain & the cognitive domain).* David McKay & Co.

Bonilla-Jurado, D., Zumba, E., Lucio-Quintana, A., Yerbabuena-Torres, C., Ramírez-Casco, A., & Guevara, C. (2024). Advancing university education: Exploring the benefits of education for sustainable development. *Sustainability, 16*(17), 1–27. https://doi.org/10.3390/su16177847

Bovill C & Bulley CJ (2011). A model of active student participation in curriculum design: Exploring desirability and possibility. In Rust C (Ed.) *Improving student learning.* The Oxford Centre for Staff and educational development. (pp. 176–188). https://eprints.gla.ac.uk/57709/1/57709.pdf

Brundiers, K., Barth, M., Cebrián, G., Cohen, M., Diaz, L., Doucette-Remington, S., Dripps, W., Habron, G., Harré, N., Jarchow, M., Losch, K., Michel, J., Mochizuki, Y., Rieckmann, M., Parnell, R., Walker, P., & Zint, M. (2021). Key competencies in sustainability in higher education—Toward an agreed-upon reference framework. *Sustainability Science, 16*, 13–29. https://doi.org/10.1007/s11625-020-00838-2

Byrne, E., Sage, C., & Mullally, G. (2017). Transdisciplinarity within the university: Emergent possibilities, opportunities, challenges and constraints. In E. Byrne, G. Mullally, & C. Sage (Eds.), *Transdisciplinary perspectives on transitions to sustainability*. Routledge. https://doi.org/10.4324/9781315550206

Connected Curriculum. (2018). *Centre for the Integration of research, teaching and learning.* University College Cork. https://www.ucc.ie/en/cirtl/resources/connectedcurriculum/

Cook-Sather, A., & Alter, Z. (2011). What is and what can be: How a liminal position can change learning and teaching in higher education. *Anthropology and Education Quarterly, 42*(1), 37–53. https://doi.org/10.1111/j.1548-1492.2010.01109.x

Davis, K., & Singh, S. (2015). Digital badges in afterschool learning: Documenting the perspectives and experiences of students and educators. *Computers & Education, 88*, 72–83. https://doi.org/10.1016/j.compedu.2015.04.011

de Haan, G. (2010). The development of ESD-related competencies in supportive institutional frameworks. *International Review of Education, 56*, 315–328. https://doi.org/10.1007/s11159-010-9157-9

Department of Education. (2014). *'Education for sustainability' the national strategy on education for sustainable development in Ireland, 2014–2020.* https://assets.gov.ie/70926/7e8f9ccf9069441e960d7f32a38dc7b9.pdf

Department of Education (2022). *ESD to 2030: second national strategy on education for sustainable development.* https://assets.gov.ie/228330/c69895a6-88f0-4132-b6d1-9085a9c31996.pdf

Dodds, F., Donoghue, D., & Roesch, J. L. (2017). *Negotiating the sustainable development goals: A transformational agenda for an insecure world.* Routledge. https://doi.org/10.4324/9781315527093

El-Jardali, F., Ataya, N., & Fadlallah, R. (2018). Changing roles of universities in the era of SDGs: Rising up to the global challenge through institutionalising partnerships with governments and communities. *Health Research Policy and Systems, 16*(38), 1–5. https://doi.org/10.1186/s12961-018-0318-9

Farmer, T., & West, R. E. (2016). Opportunities and challenges with digital open badges. *Educational Technology, 56*(5), 45–48. https://edtechbooks.org/lidtfoundations/opportunities_and_challenges_with_digital_open_badges

Giangrande, N., White, R. M., East, M., Jackson, R., Clarke, T., Coste, M. S., & Penha-Lopes, G. (2019). A competency framework to assess and activate education for sustainable development: Addressing the UN sustainable development goals 4.7 challenge. *Sustainability, 11*(10), 1–16. https://doi.org/10.3390/su11102832

Government of Ireland. (2018). *The sustainable development goals national implementation plan 2018–2020.* https://assets.gov.ie/19344/32f9bdd2aae2464caae37760edd1da04.pdf

Guiry, N. (2023). *No one will be left behind: The SDG framework and LGBTQ+ rights. The roundtable of the Cork online law review.* https://www.corkonlinelawreview.com/single-post/no-one-will-be-left-behind-the-sdg-framework-and-lgbtq-rights

Guiry, N. (2024). Why do states adhere to the sustainable development goals? *Transnational Environmental Law, 13*(2), 400–427. https://doi.org/10.1017/S2047102524000190

Guiry, N., Barimo, J., Byrne, E., O'Mahony, C., Reidy, D., Dever, D., Mullally, G., Kirrane, M., & O'Mahony, M. J. (2022). *Education for sustainable development: Co-creating common areas of need and concern.* The National Forum for the Enhancement of Teaching and Learning in Higher Education. https://doi.org/10.13140/RG.2.2.12453.58080

Hally, R., O'Mahony, C., O'Halloran, C., & Galvin, M. (2021). *Civic and community engagement in the curriculum: Connecting the curriculum for societal impact. Civic engagement toolkit.* University College Cork. https://www.ucc.ie/en/cirtl/resources/cetoolkit/

Healey, M., Flint, A., & Harrington, K. (2014). *Engagement through partnership: Students as partners in learning and teaching in higher education.* Higher Education Academy. https://s3.eu-west-2.amazonaws.com/assets.creode.advancehe-document-manager/documents/hea/private/resources/engagement_through_partnership_1568036621.pdf

Hickey, D. T., Willis, J. E., & Quick, J. D. (2015). *Where badges work better.* ELI Brief.

Higher Education Authority (2011). *National strategy for higher education to 2030.* : Department of Education and Skills. ISBN—9781406425420. https://assets.gov.ie/24558/c90f9fae0a70444cbe20feeff7b55558.pdf.

Higher Education Authority. (2018). *Higher education system performance framework 2018–2020.* Higher Education Authority. https://hea.ie/assets/uploads/2018/01/higher-education-system-performance-framework-2018-2020.pdf

Irish Universities Association (2023, July 12) *New Micro-credentials Platform Launch Gives Major Boost to Upskilling and Reskilling in Ireland.* https://www.iua.ie/press-releases/launch-of-microcreds-ie/

Irish Universities Association, Irish Research Council. (2016). *Engaged research: Society and higher education addressing grand societal challenges together.* Campus Engage. https://research.ie/assets/uploads/2017/04/Engaged-Research.pdf

Jaakkola, N., Karvinen, M., Hakio, K., Wolff, L., Mattelmäki, T., & Friman, M. (2022). Becoming self-aware—How do self-awareness and transformative learning fit in the sustainability competency discourse? *Frontiers in Education, 7*, 1–13. https://doi.org/10.3389/feduc.2022.855583

Kirrane, M. J., Pelton, C., Mehigan, P., Poland, M., Mullally, G., & O'Halloran, J. (2020). "Reaching for the STARs": A collaborative approach to transparent sustainability reporting in higher education, the experience of a European university in achieving STARS gold. In W. Leal Filho, P. R. Borges de Brito, & F. Frankenberger (Eds.), *International business, trade and institutional sustainability (World Sustainability Series)* (pp. 991–1008). Springer Nature. https://doi.org/10.1007/978-3-030-26759-9

Kopnina, H. (2015). The victims of unsustainability: A challenge to sustainable development goals. *International Journal of Sustainable Development & World Ecology, 23*, 113–121. https://doi.org/10.1080/13504509.2015.1111269

Leal Filho, W., Sierra, J., Price, E., Eustachio, J. H. P. P., Novikau, A., Kirrane, M., Dinis, M. A. P., & Salvia, A. L. (2024). The role of universities in accelerating the sustainable development goals in Europe. *Scientific Reports, 14*(15464), 1–14. https://doi.org/10.1038/s41598-024-65820-9

Leal Filho, W., Wall, T., Barbir, J., Alverio, G. N., Dinis, M. A. P., & Ramirez, J. (2022). Relevance of international partnerships in the implementation of the UN sustainable development goals. *Nature Communications, 13*(613), 1–4. https://doi.org/10.1038/s41467-022-28230-x

Magni, G. (2017). Indigenous knowledge and implications for the sustainable development agenda. *European Journal of Education, 52*(4), 437–447. https://doi.org/10.1111/ejed.12238

Mills E (2015). *'Leave no one behind': Gender, sexuality and the sustainable development goals.* IDS Evidence Report 154, : IDS. https://opendocs.ids.ac.uk/articles/report/_Leave_No_One_Behind_Gender_Sexuality_and_the_Sustainable_Development_Goals/26470093

Reidy, D., Kirrane, M. J., Curley, B., Brosnan, D., Koch, S., Bolger, P., Dunphy, N., McCarthy, M., Poland, M., Ryan-Fogarty, Y., & O'Halloran, J. (2015). A journey in sustainable development in an urban campus. In W. Leal Filho, L. Brandli, O. Kuznetsova, & A. Paço (Eds.), *Integrative approaches to sustainable development at university level. World sustainability series* (pp. 599–613). https://doi.org/10.1007/978-3-319-10690-8_41

Rieckmann, M. (2018). Learning to transform the world: Key competencies in ESD. In A. Leicht, J. Heiss, & W. J. Byun (Eds.), *Issues and trends in education for sustainable development* (pp. 39–59). UNESCO. https://unesdoc.unesco.org/ark:/48223/pf0000261802

Sá, P., Lourenço, M., & Carlos, V. (2022). Sustainability competencies in higher education research: An analysis of doctoral theses in Portugal. *European Journal of Investigation in Health, Psychology and Education, 12*(4), 387–399. https://doi.org/10.3390/ejihpe12040028

Serdyukov, P. (2017). Innovation in education: What works, what doesn't, and what to do about it? *Journal of Research in Innovative Teaching & Learning, 10*(1), 4–33. https://doi.org/10.1108/JRIT-10-2016-0007

Sinakou, E., Donche, V., Boeve-de Pauw, J., & Van Petegem, P. (2019). Designing powerful learning environments in education for sustainable development: A conceptual framework. *Sustainability, 11*(21), 1–23. https://doi.org/10.3390/su11215994

Sipos, Y., Battisti, B., & Grimm, K. (2008). Achieving transformative sustainability learning: Engaging head, hands and heart. *International Journal of Sustainability in Higher Education, 9*(1), 68–86. https://doi.org/10.1108/14676370810842193

Spangenberg, J. H. (2017). Hot air or comprehensive progress? A critical assessment of the SDGs. *Sustainable Development, 25*(4), 311–321. https://doi.org/10.1002/sd.1657

Sterling SR (2024). *Learning and sustainability in dangerous times: The Stephen Sterling reader.* Agenda Publishing: ISBN: 9781788216913.

Taimur, S., & Sattar, H. (2019). Education for sustainable development and critical thinking competency. In W. Leal Filho, A. M. Azul, L. Brandli, P. G. Özuyar, & T. Wall (Eds.), *Quality education: Encyclopaedia of the UN sustainable development goals* (pp. 238–248). https://doi.org/10.1007/978-3-319-95870-5_64

An Taisce. (2021). *Smarter sustainable campus communities: A guide for campuses embarking on the Green campus Programme*. Dublin. https://www.greencampusireland.org/wp-content/uploads/2018/11/Green-Campus-Guidebook-2021-2022.pdf

UCC. (2016). *UCC sustainability strategy*. University College Cork.

UCC. (2017). *UCC civic engagement plan 2017–2022. Together with and for community*. Cork. https://www.ucc.ie/en/media/centralmedia/UCC_Civic_Engage_2017a.pdf

UCC. (2023). *Securing Our Future: University College Cork Strategic Plan 2023–2028*. https://www.ucc.ie/en/media/aboutucc/presidentx27soffice/ucc2028/ucc_securing_our_future_strategic_plan_23_28.pdf

UCC SDG Toolkit. (2021). *About the project*. University College Cork. https://www.ucc.ie/en/sdg-toolkit/about/

UI Green Metric. (2024). https://greenmetric.ui.ac.id/

UN. (2007). *General Assembly Resolution A/RES/61/295*. United Nations declaration on the rights of indigenous peoples. 2 October 2007. https://www.un.org/en/genocideprevention/documents/atrocity-crimes/Doc.18_declaration%20rights%20indigenous%20peoples.pdf

UN. (2015) *Transforming Our World: The 2030 Agenda for sustainable development*. Resolution adopted by the General Assembly on 25 September 2015. 42809:1–13. https://sdgs.un.org/2030agenda

UN. (2024) *Pact for the future, global digital compact and declaration on future generations*. A/RES/79/1. https://www.un.org/sites/un2.un.org/files/sotf-pact_for_the_future_adopted.pdf

UNEP. (2022). *The Kunming-Montreal global biodiversity framework*. The 15th conference of the parties to the convention on biological diversity. https://www.cbd.int/doc/decisions/cop-15/cop-15-dec-04-en.pdf

UNESCO. (2017). *Education for sustainable development goals: Learning objectives*. UNESCO. https://unesdoc.unesco.org/ark:/48223/pf0000247444

UNESCO. (2020). *Education for sustainable development: A roadmap*. UNESCO. https://unesdoc.unesco.org/ark:/48223/pf0000374802

Vermeulen V & Emplit P (2024). Students as teacher whisperers: A case study of course co-creation for and with students. , *2024 European Learning & Teaching Forum Learning for in and with Society*. ISSN: 2593–9602.

Whelehan, D. F. (2020). Students as partners: A model to promote student engagement in post-COVID-19 teaching and learning. *AISHE Journal, 12*(3), 1–10. https://ojs.aishe.org/index.php/aishe-j/article/view/479

Wiek, A., Bernstein, M. J., Foley, R. W., Cohen, M., Forrest, N., Kuzdas, C., Kay, B., & Withycombe Keeler, L. (2015). Operationalising competencies in higher education for sustainable development. In M. Barth, G. Michelsen, M. Rieckmann & I. Thomas (Eds.), *Routledge Handbook of higher education for sustainable development* (pp. 241–260). Routledge.

Yap, M. L., & Watene, K. (2019). The sustainable development goals (SDGs) and indigenous peoples: Another missed opportunity? *Journal of Human Development and Capabilities, 20*(4), 451–467. https://doi.org/10.1080/19452829.2019.1574725

Zhang, T. (2025). *Staff development in learning and teaching at European Universities*. EUA.

Enhancing SDG 17: A Case Study of Co-Creation at a Norwegian University in a European University Alliance

Tove Holm, Tanja Ellingsen, Umair Najeeb Mughal, Ingunn Skjesol, Amsale Kassahun Temesgen, Vivi Marie Lademoe Storsletten, and Kaspar Bredahl Rasmussen

Abstract In 2024, there are 65 European university alliances, each comprising several universities across Europe. These alliances aim to enhance cooperation among European higher education institutions. One of the European university alliances is SEA-EU, which consists of nine universities. In May 2024, each of the nine universities was allocated one Sustainable Development Goal (SDG) based on published research related to the SDGs. Nord University was allocated SDG 17, which focuses on global partnerships for sustainable development. In October 2024, the universities involved in the alliance showcased how they enhanced the SDGs at an SEA-EU event. Additionally, the forum explored the idea of expanding collaborations to include partners from diverse regions. This chapter will explore how the activities of

T. Holm (✉)
Department of Academic and Student Affairs, Nord University, Bodø, Norway
e-mail: tove.holm@nord.no

T. Ellingsen
Faculty of Social Science, Department of History, Geography and International Relations, Nord University, Bodø, Norway
e-mail: tanja.ellingsen@nord.no

U. N. Mughal
Department of Research and Innovation, Nord University, Bodø, Norway
e-mail: umair.n.mughal@nord.no

I. Skjesol
Faculty for Nursing and Health Sciences, Nord University, Namsos, Norway
e-mail: ingunn.skjesol@nord.no

A. K. Temesgen · V. M. L. Storsletten
Department of Marketing, Organization and Management, Nord University Business School, Bodø, Norway
e-mail: amsale.k.temesgen@nord.no

V. M. L. Storsletten
e-mail: vivi.m.storsletten@nord.no

K. B. Rasmussen
Department of Academic and Student Affairs, Nord University, Bodø, Norway
e-mail: kaspar.b.rasmussen@nord.no

W. Leal Filho et al. (eds.), *Introducing the UN SDGs in Higher Education Institutions*, World Sustainability Series, https://doi.org/10.1007/978-3-032-23017-1_41

universities can make a vital contribution to international educational collaborations and can enhance the role of universities in fostering partnerships for sustainable development. In other words, we are looking at transdisciplinarity in action. The authors showcase Nord University's commitment to SDG 17, detailing examples of how it focuses on collaboration and co-creation as part of its day-to-day activities and work in a European university alliance. We argue that enhancing collaboration or working in a multi-, inter-, and transdisciplinary manner is a great first step, but have identified that it is important to start working at the next step, with more focus on co-creation.

Keywords SDG17 · European University Alliances · Co-creation · Partnership

1 Introduction

As of 2025, there are 65 European university alliances, each comprising several universities across Europe. These alliances aim to enhance cooperation among European higher education institutions, enhance the quality and competitiveness of European education, promote innovation and research, and facilitate student and staff mobility across Europe. They aim to create a more integrated and cohesive European educational arena The alliances are co-funded by the European Union (European Commission, 2024a, 2024b).

One of the alliances is SEA-EU, European University of the Seas, consisting of nine partner universities: Nord University in Norway, the University of Algarve in Portugal, Brest in France, Cadiz in Spain, Gdansk in Poland, Kiel in Germany, Malta, Split in Croatia, and Parthenope in Italy (SEA-EU, 2025).

In October 2024, the European University Alliance SEA-EU arranged the Future Leadership Forum in Naples, focusing on integrating the Sustainable Development Goals (SDGs) within higher education. One SDG was assigned to each SEA-EU partner university in May 2024, based on published research related to the SDGs, summarised from Scopus (SEA-EU, 2025). In Scopus, this data is accessible on each university's profile page in a section related to the SDGs. Scopus uses a combination of keyword queries and machine learning to map documents relating to specific SDGs (Scopus, 2024). Each of the nine SEA-EU partners contributed to the forum, expressing their perspectives on its SDGs. This event showcased how European universities promote Open Science and shared values, these being essential for fostering international cooperation toward sustainable development. Additionally, the forum explored expanding collaborations in order to include partners from diverse other regions such as North Carolina, Angola, Senegal, Mozambique, Colombia, and Morocco, thereby enhancing the intercontinental dialogue and cooperation between the European Union (EU), African Union (AU), and the United States of America (USA). These efforts reflect the university's dedication to its educational mission and contribution to broader sustainable development goals (SEA-EU, 2025).

Nord University was allocated SDG 17. Nord University has published the most research on SDG 14, followed by SDG 3, and then SDG 17. However, since other universities in the alliance had more research on SDG 14 and SDG 4, Nord University was allocated SDG 17 (SEA-EU, 2025). SDG 17 emphasizes the need for partnerships between governments, private sectors, and civil societies in order to achieve the ambitious Agenda 2030 (United Nations, 2024).

This chapter will explore how Nord University's activities can be vital contributors to international educational collaborations and the role of universities in fostering partnerships for sustainable development. In other words, we are looking at transdisciplinarity in action. The authors from Nord University showcase its commitment to SDG 17, with examples of how it focuses on collaboration and co-creation as part of its day-to-day activities and work in a European university alliance.

2 Methods

Action research was chosen as the methodology for this chapter because we are studying transdisciplinarity in action. Multidisciplinary work involves collaboration across different disciplines. Interdisciplinary cooperation builds on multidisciplinary efforts to achieve a holistic or joint understanding. Transdisciplinary approaches go a step further by working in an interdisciplinary manner and including stakeholders such as academia, politicians, communities, and the private sector (Didham et al., 2024). Transdisciplinarity is characterized by its focus on: "wicked problems" that need creative solutions, its reliance on stakeholder involvement, and its focus on engaged, socially responsible science. In simultaneously studying multiple levels of, and angles on, reality, transdisciplinary work provides the potential to invigorate scholarly and scientific inquiry both in and outside the academy" (Bernstein, 2015, p. 1).

Since there are no existing approaches to enhancing SDG 17 in universities; the only way to study this is by implementing it ourselves. The results are intended for practical application, so they must be based on practice, not just theory. We have therefore included both theoretical background and practical examples from Nord University. We do it by showcasing how Nord University enhances co-creation and collaboration as part of its day-to-day activities and work in the European University Alliance SEA-EU. We share our findings on enhancing SDG 17 at Nord University. Our current focus is on academic and administrative roles within Nord University. Through this, we contribute not only to scholarly work on transdisciplinary approaches to enhancing SDGs, specifically SDG 17, but also to how this knowledge can be accessed through this type of action-based study. Another reason for choosing action research is the necessity of involving students, faculty members, and stakeholders in the process of enhancing SDG 17. This methodology involves collaboration among researchers and practitioners to diagnose a problem and develop a solution, or at least suggest future steps based on the findings.

Action research is often criticized for its lack of repeatability, precision, and for focusing too much on action at the expense of theory (Bryman & Bell, 2011). We have addressed these concerns by describing each step and presenting the results at seminars.

3 Results

In this section, we present our results of the literature review on SDG 17, the role of universities in SDG 17, and Nord University's contribution to SDG 17.

3.1 The Role of Universities in SDG 17

The SDGs are 17 global objectives set by the UN in the 2030 Agenda to promote economic, social, and environmental sustainability (United Nations, 2024). Universities play a crucial role in advancing the SDGs, acting as key catalysts for progress (Raimo et al., 2024). In addition, developing university sustainability competencies can lead to more sustainability-literate decision-makers, leaders, and scientists (Lozano et al., 2022).

3.1.1 SDG 17

SDG 17 calls for strengthening the means of implementing Agenda 2030. It calls for investing in global partnerships to realize sustainable development worldwide.

SDG 17 has 19 targets. The targets focus on mobilizing resources, enhancing global partnerships, promoting sustainable technologies, supporting developing countries, improving trade, and fostering policy coherence to achieve the SDGs. The target that aligns best with the objectives of universities is target 17.6, that aims to "enhance the global partnership for sustainable development, complemented by multi-stakeholder partnerships that mobilize and share knowledge, expertise, technologies, and financial resources to support the achievement of sustainable development goals in all countries, particularly in developing countries" (United Nations, 2024).

Maltais et al. (2018) have studied the research needs for SDG 17. They have found that achieving the ambitions of Agenda 2030 will require mobilizing political will and deepening partnerships between the government, the private sector, and civil society. It will also require increasing international cooperation and improving the fit and international coherence between policies and initiatives. SDG 17 addresses these needs and calls for actions to increase capacity for implementing the SDGs at all levels. As such, meeting the targets and ambitions of SDG 17 is essential for the successful advancement of the entire SDG agenda.

3.1.2 Why is SDG 17 Relevant for Universities?

Universities play a crucial role in fostering partnerships through research, innovation, knowledge sharing, and community engagement (Bowser et al., 2024; Leal Filho et al., 2021, 2024). They drive technological advances, promote sustainable development, and build capacity by training future generations. As a result, universities are essential players in achieving global progress. Chankseliani and McCowan (2021) argue that universities have evolved from educational institutions for elites to centres of research and societal engagement. They now play a crucial role in societal development by educating professionals like teachers, doctors, and engineers, and conducting research to enhance our understanding of life.

The actions for SDG 17 are aligned with the recommendations found in the most recent reports on the development of European higher education: the "Draghi Report on The Future of European Competitiveness" by Mario Draghi (Draghi, 2024) and "Align, Act, Accelerate" by Manuel Heitor (Heitor, 2023). In both reports, the importance of collaboration in order to enhance global educational partnerships for a leading Europe is emphasized. The Draghi Report outlines strategies to boost Europe's competitiveness and sustainable development through global partnerships. The report highlights the need for enhanced international collaboration and resource mobilization, and the role of technology and innovation in driving sustainable development to strengthen developing countries' capacities and align national policies with global ones (Draghi, 2024). The "Align, Act, Accelerate" report promotes science, technology, and innovation to drive sustainable development and global progress (Heitor, 2023).

At the same time, we acknowledge the critical role of academia in advancing international cooperation, education, and innovation to address complex global challenges, simultaneously recognizing that enhancing the traditional economy, like competitiveness, does not align with Agenda 2030.

3.1.3 Why is Cooperation with Universities Important for Enhancing the Global Sustainable Development Agenda?

As one concrete example, universities play a crucial role in aligning the needs of migrants with the demands of the workforce. In the European Union, the workforce is projected to decrease from 265 million in 2022 to 258 million by 2030 (European Commission, 2023). Additionally, by 2030, an estimated 21 million people, most of whom lack tertiary education, may need to change occupations (McKinsey, 2024). Therefore, we argue that higher education should focus on providing solutions for upskilling and reskilling those already in the workforce. Developing lifelong learning opportunities is essential for competitiveness in higher education across Europe.

Universities are uniquely positioned to prepare new postdoctoral researchers and PhD candidates to tackle global challenges by integrating real-world problems into their research agendas. This involves not only equipping them with the latest scientific and technological knowledge, but also fostering generic skills like problem-solving, ethical decision-making, and collaborative teamwork across disciplines and cultures. By embedding these capabilities, universities can contribute to a workforce that is adaptable and equipped to deal with the complexities associated with global challenges such as climate change, healthcare disparities, food security, and energy sustainability.

Disruptive technologies such as artificial intelligence (AI), blockchain, renewable energy technologies, and advanced biotechnologies are transforming industries and societies. AI is revolutionizing how data is processed and interpreted, enabling breakthroughs in everything from climate modelling to personalized medicine. AI also has opposing sides, like there are risks that it reduces critical thinking, might spread bias and/or misinformation, and can lead to cheating and leak of personal data. Blockchain technology promises greater transparency and security in transactions, potentially reshaping financial systems and supply chains. Renewable energy technologies are critical in the transition towards sustainable energy systems, reducing dependency on fossil fuels and mitigating climate change. Meanwhile, biotechnologies are at the forefront of medical innovations, agricultural improvements, and environmental conservation.

PhD researchers and postdoctoral fellows play a pivotal role in advancing these technologies, contributing not only to scientific and technological progress but also ensuring that these advancements align with the SDGs. Their work often leads to innovative solutions that address complex societal challenges, making them invaluable for promoting a sustainable future. By integrating cutting-edge research with a deep understanding of ethical, social, and environmental considerations, they help ensure that technological progress benefits society broadly and sustainably.

During the Industrial Revolution, rapid industrialization led to significant economic growth and societal transformation. However, this period also marked the beginning of significant environmental degradation. Industries expanded with little regard for their environmental impact, leading to pollution, resource depletion, and considerable ecological damage. The consequences of these actions are still being managed today, highlighting the need for a balanced approach to economic development.

Academia can provide a comprehensive perspective on the dual nature of business growth, emphasizing both the benefits and the potential drawbacks. Through rigorous research and analysis, universities can offer insights into the complexities of sustainable development and help devise strategies that minimize negative impacts while maximizing positive outcomes. This role is crucial not only in mitigating the effects of past actions but also in guiding future decisions to ensure that development is truly sustainable.

Markauskaite et al. (2023) discuss how we should not forget the human aspect as we go through the digital revolution. They argue that "Teaching in a sustainable university requires a fluency to embrace different ways of knowing and a collective awareness of how the digital is intertwined with human practices across the educational ecosystem." (p. 181).

Incorporating digital solutions into the educational and research missions of universities strengthens their role as key contributors to society. By fostering a deeper understanding and developing solutions that consider both the advantages and disadvantages of technological and industrial advancements, academia can help steer the world toward a more sustainable and equitable future. Moreover, universities can expand their impact by intensifying community engagement with stakeholders such as NGOs, governments, and the private sector. This collaborative approach can enhance the relevance of research and increase the practical application of innovations, ensuring that academic work contributes directly to societal needs and sustainable development. Through such partnerships, universities can play a pivotal role in building a bridge between theoretical knowledge and practical implementation, thus strengthening their contribution to SDG 17.

For succeeding with co creation, transdisciplinary cooperation is needed, meaning not just cooperation or working in a multidisciplinary manner, but co-creating by involving partners in the development work. Working in a transdisciplinary manner always means not restricting activity to any general field. This might be a reason why it is challenging; there is a lack of a joint methodology and language within the group. Bursztyn and Drummond (2013) argue that interdisciplinarity should not be regarded as anti-disciplinary but rather as a way to apply or merge solutions from different fields for tackling complex challenges. Another critical point is that educators and students should be involved in developing sustainability competencies (Lozano et al., 2025).

3.2 *Nord University's Contribution to SDG 17*

According to Strategy 2030, Nord University will "contribute to sustainable development and value creation through research, education, cooperation with society and business, and as a member of the international academic community" (Nord University, 2024).

Nord University joined the European university alliance SEA-EU in 2022. The mission statement of SEA-EU emphasizes the alliance's commitment to "living sustainably by, from, and with the sea." It highlights the partner universities' social responsibilities and their role in implementing the SDGs. The vision includes fostering sustainable cooperation across various levels of the organization and ensuring that educational and research activities are aligned with the broader context of sustainable development (SEA-EU, 2025).

The SEA-EU alliance project 2.0, co-funded by the European Union, is divided into several work packages, each focusing on different aspects of the alliance's objectives. Each work package is further divided into tasks that address detailed objectives and deliverables, ensuring a comprehensive approach to achieving the alliance's goals. SEA-EU has 75 associated partners, such as research centers, businesses, city and provincial authorities, public entities, NGOs, non-profit organizations, international clusters of excellence, and ports. Nord University is associated with nine partners. In 2025, a total of 65 European university alliances are cooperating with over 2,200 associated partners (European Commission, 2025). In January 2025, the SEA-EU Alliance participated in the 4EU + 1CORE Workshop for European university alliances, sharing insights on best practices for governance models. At the workshop, the challenges of involving associated partners were also discussed. It was emphasized that co-creation would go more smoothly if the alliances could define clear roles for all partners and keep up continuous monitoring to sustain involvement.

In the following section, we will showcase Nord University's work on SDG 17, with examples of how the university is using collaboration and co-creation in its day-to-day activities and work in the European university alliance SEA-EU. Research collaboration is taking place, albeit we do not focus on it in this chapter. We instead present examples of community engagement, cooperation within the SEA-EU alliance on education, and global partnership.

3.2.1 Cooperation and Co-Creation with Partners at Nord University

According to Holand and Ellingsen (2024), co-creation is a recommended working method for succeeding with Agenda 2020 in Norway. Nord University has recognized this, and so here are four examples of such cooperation: with industry (3.2.1.1); across borders in the Arctic (3.2.1.2); with the global research community (3.2.1.3); and with local actors handling incidents and crises (3.2.1.4).

The Business Mentor Program: Bridging Academia and Industry

The "Næringslivsmentor" ("Business Mentor" in English) Program connects experienced business mentors with academic courses and research projects at Nord University. This initiative strengthens collaboration between industry and academia, ensuring that students gain practical, real-world insights. By fostering knowledge transfer, the program equips students with skills to address contemporary challenges while bridging the gap between academic knowledge and needs of industry (Nord University, 2025b).

Nordic Bridge: Cross-Border Collaboration with Global Impact

The Nordic Bridge Project underscores Nord University's commitment to fostering collaboration in the Northern Periphery and Arctic regions. The project enhances student mobility and cross-border cooperation by creating a digital matchmaking portal that matches societal challenges with academic expertise. It addresses regional challenges with global implications, exemplifying SDG 17 by leveraging partnerships in order to drive sustainable development (Nord University, 2025c).

EU Funding Success and Marie Skłodowska-Curie Actions

Nord University has secured EU funding to advance its research and educational goals. The university is increasingly active in Marie Skłodowska-Curie Actions

(MSCA), such as Postdoctoral Fellowships and Doctoral Networks. These programs enable Nord University to attract researchers from around the world, providing them with cutting-edge education and training. These researchers, in turn, contribute to their home regions by enhancing education, research, and capacity for innovation (Nord University, 2025d).

Exercise Nord—a Full-Scale Preparedness Exercise

Exercise Nord is another prominent example of bridging the gap between the university sector and society, and thus SDG 17 at Nord University. This is a full-scale preparedness exercise in which actors within preparedness train to handle significant incidents and crises. Linked to the exercise, numerous other activities are taking place in the same week including seminars and professional forums directed at several different professions: business, maritime sector, journalists, security and defence, police, etc. In addition, public meetings and several activities both on campus and in the city are taking place (Nord University, 2025e).

The exercise is the Nordic region's most significant civil mainland exercise and is the largest of its kind in Europe under the auspices of a university. During the exercise, Nord University collaborates with many agencies, including the main rescue centre, the police, the fire service, municipalities, the county council, emergency services, hospitals, the Home Guard, voluntary organizations, and the Coast Guard. In addition, several hundred students at bachelor's and master's levels participated in the exercise across numerous disciplines and faculties. This is an excellent opportunity for professionals to meet across sectors in order to exercise their skills and for students to become better prepared for their future work. It is also a good way to ensure that everyone, is familiar with the latest research and with policy discussions within the field, which various studies stress is of great importance (Torani et al., 2019).

3.2.2 Cooperation and Co-Creation Within the European University Alliance SEA-EU on Education

One example of a task that enhances the SDGs in SEA-EU is task 4.4, "SEA-EU Goes Greener, which focuses on promoting sustainability and green practices within the SEA-EU alliance. The main objectives of the task are to: (1) establish a network of individuals committed to sustainability across partner universities; (2) implement small steps towards making university campuses more sustainable; (3) develop a Green Charter that outlines good practices and ambitions within SEA-EU. These efforts aim to support the universities' contributions to Agenda 2030 and foster a culture of sustainability within the academic community (SEA-EU, 2025).

Work package 2 of the SEA-EU project deals with education and training. Co-creation and transdisciplinarity are key among the tasks that are included in this work package. One of the tasks in work package 2 is task 2.2. The task aims to encourage innovative training activities that foster inclusive digital and green interdisciplinary and creative skills. This involves creating training pathways for academic teachers and staff so that they acquire essential, inclusive, digital, green, interdisciplinary, and innovative skills for the future. This will be achieved by providing diverse courses within the alliance and fostering the exchange of experiences among academic teachers and staff. The process of creating these courses and tools started with assessing the level of competence and needs within the alliance. This assessment was done through the means of a survey. The survey was designed by means of co-creation in a core academic team and with input from an extended interdisciplinary panel of experts and other representatives in the alliance. In addition, we took inspiration from research articles, previous surveys, and the SEA-EU's mission statements.

The survey data presents diverse responses from partner universities involved in the SEA-EU survey. In total, we received 674 responses from the nine universities. We gathered information that could assist in achieving the objectives of task 2.2, for example, the level of skills for teaching sustainability and the use of digital technologies in teaching and learning activities. We also identified which skills respondents wished to acquire through the alliance.

The exchange of knowledge and skills that task 2.2 envisions aligns well with SDG 17. It will enable educators to collaborate. Those with more experience on a topic could transfer their knowledge to their colleagues in the alliance. In contrast, those who lack experience could learn from their colleagues. This would enhance the universities toward sustainable development. In addition to indirectly contributing to SDG 17, educators in the SEA-EU alliance also address SDG 17 directly in their teaching activities. Figures 1 and 2 illustrate that this varies between disciplinary fields. SDG 17 is most frequently discussed in business and administration, social sciences, education, and health and welfare.

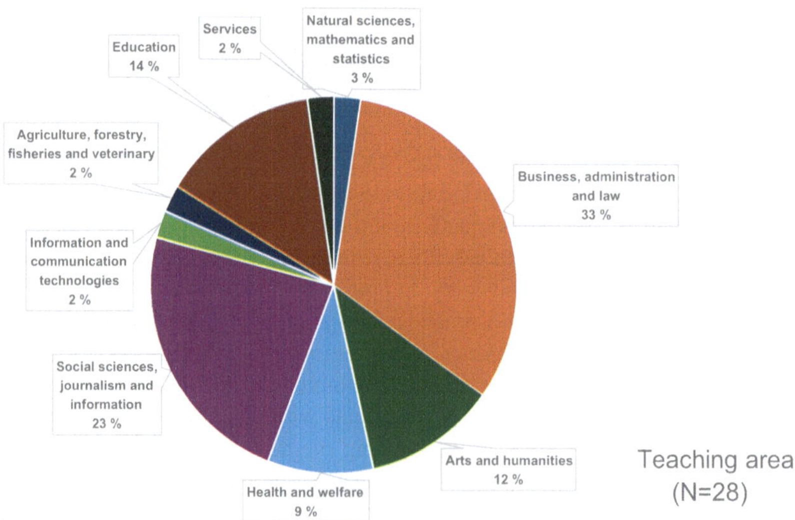

Fig. 1 Teaching areas of those whose teaching addresses SDG 17 at Nord University

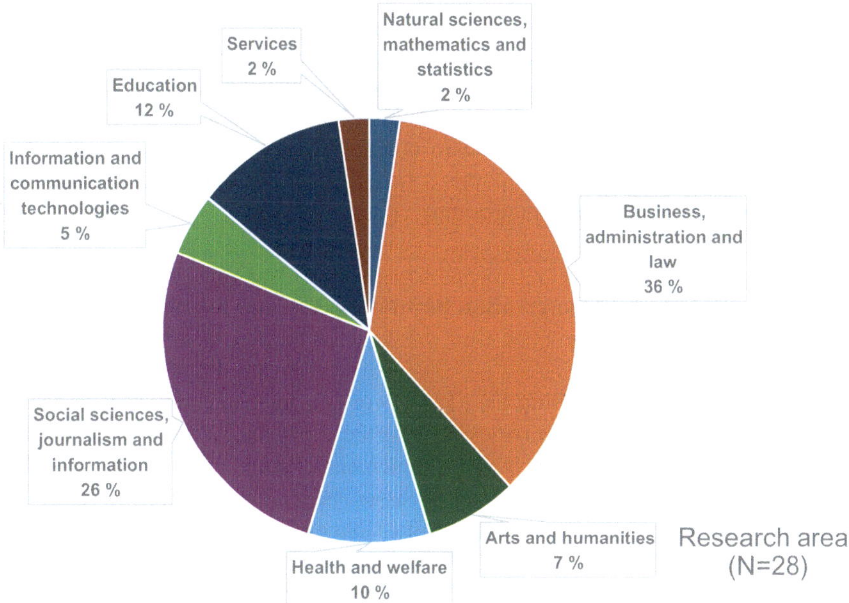

Fig. 2 Research areas of those whose teaching addresses SDG 17 at Nord University

We are now in an ongoing process to narrow down the list of courses that will be available digitally to academic staff within the alliance. This is done through a number of workshops where task representatives engage in discussions with each other and the coordinators of the course responsible for identifying courses that match the visions and objectives of task 2.2 and the alliance as a whole.

Another task under work package 2 is task 2.4, focusing on joint educational programs. European Joint Education Programs are collaborative initiatives mounted by different European universities with the aim of enhancing educational quality, promoting staff and student mobility, and fostering European cooperation (European Commission, 2024a, 2024b). As of the writing of this chapter, three of the four joint programs in which Nord University is involved achieved accreditation.

We offer here the joint Master Programme in Sustainability Transition in Organisations and Resilience Management (STORM) as an example of transdisciplinarity and co-creation in the development of education programs. Early in the process of designing the Master's program, a committee was established with representatives from the participating six universities with a view to bringing together the unique insights and experiences from their respective universities and countries.

This committee adopted a bottom-up approach to align the programme with labour market needs. Each of the partner universities organized workshops to gather input from research experts, industry, and different organizations. By involving a diverse range of stakeholders, a comprehensive, multi-perspective, and transdisciplinary design was assured.

These consultations highlighted the essential skills future managers need in order to address environmental and societal challenges. In addition, they emphasized the importance of a curriculum focused on environmental awareness, resilience, adaptable management skills, a holistic understanding of the natural world, and ethical leadership. These insights were used to create a stakeholder-driven programme and five Programme Intended Learning Outcomes (PILOs), which guided the development of the full course programme.

3.2.3 Cooperation and Co-creation Beyond the European University Alliance SEA-EU

One of the work packages within SEA EU is work package five, which specifically points to the importance of universities bridging the gap between the university sector and society and, in this sense, is a direct implementation of SDG 17. This includes tighter collaboration and co-creation with associated partners from the maritime industry, the ports, and the cities/municipalities that each of the nine universities has brought into the alliance (SEA-EU, 2025). As part of this work, Nord University has been involved in everything from offering summer school courses and blended intensive programs, to participating in research projects and promoting more collaboration between the cities and ports involved in the alliance. Through service learning (Bringle & Hatcher, 1996), students within the alliance are eventually expected to be part of a global society hub, connecting with various associated partners from industry and municipality based on the nine different university need analyses.

In addition to stronger collaboration within the alliance and with associated partners in the industry or municipalities, the work package also emphasizes the importance of an extraterritorial strategy—"Stronger Partnerships for Common Global Goals." Under the SEA-EU Alliance, Nord University collaborates with global partners, including the African Union, to address wicked challenges such as climate change, poverty, and democratic governance. By sharing its expertise and fostering educational exchanges, Nord University actively contributes to capacity-building in Africa, Asia, and South America. This engagement supports co-creation solutions that are locally relevant and globally impactful.

With this in mind, and as a result of various strategies and plans presented by the European Commission over the last four to five years, Nord University initiated a workshop within the SEA-EU alliance directed at "Strengthening Ties to African Universities." As stated in the EU Commission Report in 2020, "Towards a comprehensive strategy with Africa," "Africa is Europe's closest neighbour, and the two continents have numerous ties that bind Africa and the European Union (EU) together, both as a result of history, proximity and shared interests" (European Commission, 2020). This was later followed up in several meetings in Brussels, and in 2022 the EU-AU Summit flagged up the idea of "Two Unions, a joint vision", and declared that "the aims of the partnership are solidarity, security, peace and sustainable and sustained economic development and prosperity for the citizens of the African Union and the European Union, now and in the future, bringing together people, regions and organizations" (European Council, 2022). The initiative started with Nord surveying all universities within the alliance of the scientific staff's experiences collaborating with African universities in research or education, and their future interest. Based on the results, we were invited to a first workshop within the alliance in October 2024, and a more detailed set of workshops is was planned at each of the nine universities to be conducted in the spring of 2025. Ideas and experiences have been shared across the alliance, and Nord also participated in Southern African-Nordic (SANORD) meetings and conferences in 2024, presenting some of the findings that have so far been well received. SANORD is working towards advancing collaboration between institutions in the South African and Nordic Regions (SANORD, 2025). This type of collaboration will also be directed towards more African universities and the SEA EU Alliance. Further concretization, collaboration, and co-creation is needed, but the first foundations have now been laid.

4 Discussion

In this chapter, we discuss the way further.

Much collaboration and co-creation is already ongoing, but the question that follows is how can we engage more faculty and staff in this work. The main author started to explore what initiatives were in place and how she could facilitate a broader involvement in initiatives connected to SDG 17.

Nord University represents northern Europe within the SEA-EU alliance, with campuses in nine locations in Northern and Central Norway (Nord University, 2025a). The High North faces unique challenges, including climate change, resource management, demographic shifts, and economic diversification (European Parliament, 2021). These challenges require targeted research and innovative solutions to ensure sustainable development. Nord University, located in a region rich in natural resources such as those associated with the aquaculture industry, is well-positioned to contribute to these efforts. The university fosters resilience and development in the Arctic by researching blue and green growth, indigenous community support, and innovation and entrepreneurship. Nord's contributions to the public sector are deeply rooted in its educational programs that focus on public security, health and welfare, art and culture, and education. These programs prepare graduates to tackle societal challenges like climate change, disruptive technologies, and democratic governance (Nord University, 2024). This multidisciplinary approach enables Nord University to holistically address regional and global challenges, creating a ripple effect of knowledge and innovation. The presence of the university within various districts in peripheral areas allows for a stronger connection to stakeholders in the various districts, and this in turn leads to collaboration.

4.1 New Steps for Developing Initiatives Connected to SDG 17 at Nord University

As mentioned in Sect. 3.2.1., Holand and Ellingsen (2024) have, in their recent explorative study on co-creation, identified that co-creation is a suggested working method for succeeding with Agenda 2020 in Norway. This is partly because cooperation between partners is historically more typical in Norway than in similar countries. Røiseland et al. (2024) have conducted a study that has focused on the issue of advancing co-creation in local governance. According to them, "One of the important findings of the research presented here is the insight that the context and content of co-creation initiatives matter. Therefore, extending further research to other fields and countries is a logical next step". The interviews done by Røiseland et al. focused on: (1) what co-creation meant for those who were interviewed; (2) motivation for co-creation practices; (3) the consequences of initiatives of co-creation; (4) questions about challenges; (5) how the actors coped with them. Inspired by Røiseland et al., two of the authors tried to apply the same questions to the staff at Nord University at large but failed to receive sufficient responses.

Looking for a different approach, the main author made contact with the faculty that has been the least involved in the initiatives described in this chapter, the Faculty of Nursing and Health Science.

Even though sustainability is highly relevant within the field of health education, as there is not enough access to health personnel to uphold the Nordic Welfare Model (Norway's Government, 2023), sustainability within the health and welfare services has not been central to the initiatives connected to SDG 17 at Nord University. Dossey et al. (2019) argue that the SDG framework should be embedded in the professional role of nurses as it "can help us contribute to global health through local means, empowering us to nurture healthy people living on a healthy planet."

The official Norwegian report, "Time to Act: The Personnel in a Sustainable Health and Care Service" (Norway's Government, 2023) underlines the severity of the recruitment issue by pointing out that it is not a national problem but an international one. "Like other countries, Norway faces major challenges with recruiting sufficient personnel. The situation will become even more serious towards 2040." Policy debates across Europe stress how social welfare policies are becoming more aligned in the face of similar social, economic, and political problems (Garcés et al., 2003). Social sustainability has become a rising concept in the debate centering on how we should respond to these issues.

In short, the principle of social sustainability implies a conscious responsibility towards our successors and obliges us to take account of all future societies when we judge our actions from a moral standpoint. From this position, policies cannot be considered to be morally acceptable if they either favor the few, or favor all now but prejudice others in the future (Garcés et al., 2003).

The health professions cannot reach sustainable development in isolation; jointly favorable partnerships must be promoted across disciplines and sectors in order to initiate engagement and foster long-term allegiance from various stakeholders. As mentioned earlier, interdisciplinary research is vital for achieving the SDGs. This also concerns research on health topics. Keynejad et al. (2021) argue that early career researchers are well positioned to find innovative solutions to essential research questions. International cooperation and interdisciplinary education for early career researchers can lead to the fostering of new perspectives and the incorporation of broader health factors into future research. Universities ought to welcome the need for an extreme reimagining of borders and connections.

4.2 Co-Creation Workshop

As mentioned earlier, Nord University aims to "contribute to sustainable development and value creation through research, education, cooperation with society and business, and as a member of the international academic community." (Nord University, 2024). Despite significant activity in these four areas across Nord University, there is limited evidence of coordination between them. Further, there is little clarity on how linkages between these areas can contribute to sustainable development and value creation for society.

At the Faculty of Nursing and Health Science, many staff members collaborate within and outside academia, both nationally and internationally, but such collaboration is rarely integral to the faculty's activities. Therefore, collaboration that reaches outside the immediate area served by the university tends to be dependent on individual relationships rather than faculty structures or initiatives. A more systematic approach would generate greater benefits both within the university and for associated stakeholders.

To build on the existing networks, a workshop was planned and conducted for all faculty members in March 2025. At the workshop, existing successful collaborations were identified and awareness about them raised in order to support the process of co-creating ways to develop further collaborative practices within and across all four areas of core activity.

At the workshop, the participants identified existing local, regional, and international links with individuals and institutions and framed these in terms of particular activities or research areas. The participants were asked to identify the different types of collaborative links and activities and describe their experience with specific types of teaching, study programs, or research topics. This will be explored further in the upcoming future by examining the extent to which existing collaborations are documented through formal agreements or are primarily founded in informal relationships.

Following the workshop, a report summarizing existing collaborations and networks was produced, along with a process for prospectively reporting new and emerging collaborations to ensure a continually updated database. Secondly, we will propose a method for staff to draw on the database to identify potential collaborators for other projects to extend and intensify existing connections.

5 Conclusion and Future Ambitions

In this chapter, we have explored how the activities of universities are, and can be, essential contributors to international collaborations. We have also looked at the role of universities have in fostering partnerships for sustainable development, SDG 17, or in other words, transdisciplinarity in action. We argue that universities have a key role in promoting partnerships through knowledge sharing, innovation, research, and community engagement. The authors showcase Nord University's commitment to SDG 17, with examples of how it enhances collaboration and co-creation as part of the day-to-day activities at the university and in a European university alliance, SEA-EU. Beyond the work already done, we share steps taken to facilitate a broader involvement in initiatives connected to SDG 17 at the university.

The university shapes a sustainable and inclusive future through education, collaboration, and innovation. Through its educational programs, Nord University equips students with the tools that allow them to understand and address complex social challenges. Programs focusing on democracy, societal readiness, nursing, and education prepare graduates to become leaders in their respective fields of expertise. By fostering community engagement through partnerships at the university, through the European university alliance, through the securing EU funding, through participation in Marie Skłodowska-Curie Actions (MSCA) initiatives, and through the conducting of impactful research, Nord University reinforces its commitment to SDG 17.

We argue that enhancing collaboration or working in a multi-, inter-, and transdisciplinary manner is a great first step, but we have also identified that it is important to start working at the next step, working more with co-creation. We are sharing some steps that have been taken, but there is huge potential for more collaboration, meaning that we see this as an ongoing developmental process.

The associated partners we have in SEA-EU are from, among others, research centers, businesses, city and provincial authorities, and ports, and all are mostly involved in activities that enhance economic and ecological aspects of sustainability. The core for enhancing sustainability is not about emphasizing different aspects of sustainability, but rather about taking into consideration all aspects, ecological, economic, and social, and in finding new ways of thinking about such things. Therefore, we think that it will be crucial to include NGOs and companies that enhance social sustainability as associated partners in the alliance.

In this chapter, we have presented how SDG 17 can be enhanced through the incorporation of the perspective of educators and staff. We also acknowledge that, so far, the voices of the stakeholders and the students are missing. Currently, one of the goals in the alliance involves establishing a Global Society Hub. In the hub students can adopt the method of service learning and become involved in community service. It would therefore be interesting and important to hear both from students and stakeholders regarding their thoughts about co-creation and SDG 17. In upcoming research, the focus should be on how SDG 17 relates to students, like the courses offered, student exchange within and beyond the alliance, but also collaboration across cities, involving everything from internships, housing, community work, etc.

There are a lot of changes going on in society, like demographic changes, increased need for preparedness, and climate change, which require extensive adaptation. Universities play a crucial role in future societal development, which is why we argue that they should take a more active role by fostering more critical reflection and discussion. The European university alliances have a key role in this; they are the catalysts of transdisciplinary sustainable change in higher education in Europe.

We encourage other universities and stakeholders to engage in similar work and initiatives. In this chapter we have shown how a higher education institution can foster partnership and enhance its contributions to SDG 17 through collaborating with diverse institutions at an international level.

References

Bernstein, J. H. (2015). Transdisciplinarity: A review of its origins, development, and current issues. *Journal of Research Practice, 11*(1).

Bowser, G., Ho, S. S., Ziebell, A., & Lazendic-Galloway, J. (2024). Networking and collaborating: The role of partnerships across sectors to achieve educational goals in sustainability. *Sustainable Earth Reviews, 7*(1), 17.

Bringle, R. G., & Hatcher, J. A. (1996). Implementing service learning in higher education. *The Journal of Higher Education, 67*(2), 221–239.

Bryman, A., & Bell, E. (2011). *Business research methods*. Oxford University Press.

Bursztyn, M., & Drummond, J. (2013). Sustainability science and the university: Pitfalls and bridges to interdisciplinarity. *Environmental Education Research, 20*(3), 313–332.

Chankseliani, M., & McCowan, T. (2021). Higher education and the sustainable development goals. *Higher Education, 81*(1), 1–8.

Didham, R. J., Fujii, H., & Torkar, G. (2024). Exploring Interdisciplinary Approaches to Education for Sustainable Development. *Nordic Journal of Comparative and International Education, 8*(2). https://doi.org/10.7577/njcie.5877

Dossey, Barbara M., PhD, RN; Rosa, William E., MS, RN; Beck, Deva-Marie, PhD, RN. (2019) Nursing and the sustainable development goals: From Nightingale to now. AJN, *American Journal of Nursing, 119*(5), 44–49. https://doi.org/10.1097/01.NAJ.0000557912.35398.8f

Draghi, M. (2024). The Future of European Competitiveness: A Competitiveness Strategy for Europe. European Commission, https://commission.europa.eu/topics/eu-good, thank yiucompetitiveness/draghi-report_en, retrieved 31.1.2025

European Commission. (2020). "Towards a comprehensive strategy with Africa," Joint Communication to the European Parliament and Council. Brussels, EU.

European Commission. (2023). Directorate-General for Employment, Social Affairs and Inclusion. Employment and Social Developments in Europe 2023. Publications Office of the European Union (2023). https://doi.org/10.2767/089698

European Commission. (2024). European Universities initiative. https://education.ec.europa.eu/education-levels/higher-education/european-universities-initiative, Retrieved 10.1.2025

European Commission. (2024). Commission rolls out plans for a European degree. https://commission.europa.eu/news/commission-rolls-out-plans-european-degree-2024-03-27_en. Retrieved 28.3.2025

European Commission. (2025). Directorate-General for Education, Youth, Sport and Culture, Grumbinaitė, I., Colus, F. and Buitrago Carvajal, H., Report on the outcomes and transformational potential of the European Universities initiative, Publications Office of the European Union, 2025, https://doi.org/10.2766/32313, Retrieved 13.2.2025.

European Council. (2022). Sixth European Union—African Union Summit: A Joint Vision for 2030.

European Parliament. (2021). Report on the Arctic: opportunities, concerns, and security challenges. A9-0239/2021.

Garcés, J., Ródenas, F., & Sanjosé, V. (2003). Towards a new welfare state: The social sustainability principle and health care strategies. *Health Policy, 65*(3), 201–215. https://doi.org/10.1016/S0168-8510(02)00200-2

Heitor, M. (2023). *Align, act, accelerate: Science, technology, and innovation for sustainable development*. Springer.

Holand, A. M, & Ellingsen, T. (2024). Never time to meet? Time as a constraint in cross-sectoral co-creation (Experiences from an Explorative Study of Co-Creation). In *Time in Our Times. Stretching Contemporary Understandings of Time*, pp. 127–153. https://doi.org/10.1515/9783111428970-006

Keynejad, R. C., Yapa, H. M., & Ganguli, P. (2021). Achieving the sustainable development goals: Investing in early career interdisciplinarity. *Humanities Social Science Communication, 8*, 153. https://doi.org/10.1057/s41599-021-00834-6

Leal Filho, W., Frankenberger, F., A. L., Salvia, U., Azeiteiro, F., Alves, P., Castro, M., Will, J., Platje, V. O., Lovren, L., Brandli, E., Price, F., Doni, M., Mifsud., L. V. Ávila (2021). A framework for the implementation of the Sustainable Development Goals in university programmes. *Journal of Cleaner Production, 299*, 126915.

Leal Filho, W., Dibbern, T., Dinis, M. A. P., Cristofoletti, E. C., Mbah, M. F., Mishra, A., Clarke, A., Samuel, N., Apraiz, J., C., Abubakar, I., R., & Aina, Y. A. (2024). The added value of partnerships in implementing the UN sustainable development goals. *Journal of Cleaner Production, 438*, 140794.

Lozano, R., Bautista-Puig, N., & Barreiro-Gen, M. (2022). Developing a sustainability competencies paradigm in Higher Education or a White Elephant? *Sustainable Development, 30*(5), 870–883. https://doi.org/10.1002/sd.2286

Lozano, R., Barreiro-Gen, M., Ortiz-Martínez, G., Mendoza, A., Segalàs, J., Caeiro, S., Simão, J., Favi, C., Gładysz, B., Alló, M., Gago-Cortés, C., Tortato, U., Mónus, F., Pietikäinen, J., Södervik, I., Ratvio, R., Bostanci, S., Tóth, M. D., Benayas, J., Munguía, M., T., J., Cervantes, D., X., F., Díaz, R., A., L., Corona, N., R., Desha, C., Boddy, R., H., J., Djekic, I., Caldera, S., Sáenz, O., Azizi, L:, & Sassen, R. (2025). Students' Perspectives on sustainability competences and pedagogical approaches: Groupings and interrelations results from 17 international case studies. *Sustainable Development*. https://doi.org/10.1002/sd.3368

Maltais, A., Weitz, N., & Persson, Å. (2018). SDG 17: Partnerships for the Goals. A Review of Research Needs. Technical annex to the Formas report Forskning för Agenda 2030: Översikt av forskningsbehov och vägar framåt. Stockholm Environment Institute, Stockholm.

Markauskaite, L., Carvalho, L., & Fawns, T. (2023). The role of teachers in a sustainable university: From digital competencies to postdigital capabilities. *Educational Technology Research and Development, 71*(1), 181–198.

McKinsey. (2024). Some 21 million people may need to change occupations. https://www.mckinsey.com/featured-insights/future-of-work/explore-the-future-of-work-in-europe?page=/map/employment/outlook, Retrieved 2.10.2024

Nord University. (2024). Strategy 2030, https://www.nord.no/en/about/strategy-2030, Retrieved 4.10.2024

Nord University. (2025a). About Faculties and Centres. https://www.nord.no/en/about/faculties-and-centres, Retrieved 10.1.2025.

Nord University. (2025b). Business Mentor. Shaping Tomorrow's Workforce: Join our programme for visionary business leaders! https://www.nord.no/forskning/naeringslivsmentor, Retrieved 12.2.2025.

Nord University. (2025c). Nordic Bridge. Empowering Connections: Making Academia-Society Collaboration Accessible. https://www.nord.no/en/research/nordic-bridge, Retrieved 12.2.2025.

Nord University. (2025d). Unlock your research horizons with MSCA Postdoctoral Fellowships at Nord University. https://www.nord.no/en/research/for-researchers/research-funding/msca, Retrieved 12.2.2025.

Nord University (2025e). Øvelse Nord. https://www.nord.no/om/kontakt-oss/krise-og-beredskap/ovelse-nord, Retrieved 12.2.2025.

Norway's Government. (2023). Time to Act. The Personnel in a Sustainable Health and Care Service. Norwegian Official Report (NOU) 2023: 4.

Raimo, N., Nicolò, G., L'Abate, V., & Vitolla, F. (2024). Analyzing the factors affecting university contributions to achieving the Sustainable Development Goals in European Union countries. *Sustainable Development, 32*(6), 6033–6044.

Røiseland, A., Sørensen, E., & Torfing, J. (Eds.). (2024). *Advancing co-creation in local governance.* The Role of Coping Strategies and Constructive Hybridization Policy.

SANORD. (2025). Southern African-Nordic Center (SANORD). https://sanord.uwc.ac.za/, Retrieved 12.2.2025.

Scopus. (2024). Blog.scopus.com: Sustainable Development Goals (SDGs) are now available on Scopus. https://blog.scopus.com/posts/sustainable-development-goals-now-on-scopus-author-profiles, Retrieved 31.1.2025

SEA-EU. (2025). https://sea-eu.org/, Retrieved 10.1.2025.

Torani, S., Majd, P. M., Maroufi, S. S., Dowlati, M., & Sheikhi, R. A. (2019). The importance of education on disasters and emergencies: A review article. *Journal of Education and Health Promotion, 8*(1), 85. https://doi.org/10.4103/jehp.jehp_262_18

United Nations. (2024). https://sdgs.un.org/goals/goal17, Retrieved 5.7.2024.

Challenges and Opportunities in Advancing Sustainability: The Case of Green Campus Nord

Tanja Ellingsen and Benedicte Forsland

Abstract As providers of research, education and innovation as well as their inherited position for taking social responsibility, higher education institutions (HEIs) are vital in the quest to achieve the Sustainable Development Goals (SDGs) and are key to their accomplishment. As a result, many universities and HEIs across the world have integrated the SDGs within their organization, often as part of their overall university strategy. This paper focuses on Nord University and its pilot-project "Green campus Nord" initiated in the period 2019–2021. The project focused on six main aspects: own climate footprint, green areas and activities, research, education, entrepreneurship and innovation and social responsibility. A large N-survey was conducted among students and staff concerning their knowledge, attitudes and behavior towards environmental issues and the SDGs ($N = 2{,}858$). In this paper we share some of the results, achievements, opportunities and challenges we met when implementing and embedding this within the organization of Nord University. Our findings indicate that the project received high interest and attention from both students and staff, as well as from other universities in Norway and external partners. The project had the potential to become a common denominator bridging administration, education and research. However, the lack of further funding, Covid-19, and other issues, halted the project. In this sense, Green campus Nord offers valuable knowledge and lessons on how sustainability-initiatives can and ought to be integrated and adapted within higher education institutions.

Keywords Sustainability-transformation · Higher education institutions · SDGs · Barriers · Holistic approach

T. Ellingsen (✉) · B. Forsland
Faculty of Social Science, Nord University, Bodø, Norway
e-mail: tanja.ellingsen@nord.no

B. Forsland
e-mail: benedicte.forsland@nord.no

W. Leal Filho et al. (eds.), *Introducing the UN SDGs in Higher Education Institutions*, World Sustainability Series, https://doi.org/10.1007/978-3-032-23017-1_42

1 Introduction

In the last decade, the Paris Agreement, and the international spotlight on Agenda 2030 and its 17 Sustainable Development Goals (SDGs), have swiftly trickled down to influence national and local agendas, demanding reforms in various sectors, including higher education and research (Leal Filho, 2023).

In fact, higher education institutions (HEIs) are often acknowledged for their significant role in combating the global environmental crisis and advancing the Sustainable Development Goals (SDGs). Being uniquely positioned not only as knowledge producers and educators, but also as societal actors with a responsibility to take a lead in sustainable transformation, HEIs are considered critical actors in bridging local and global sustainability efforts (Sterling, 2021; Parr et al., 2022). In 2017, the United Nations thus published a manual for how universities could integrate the SDGs in education as well as within the organization (SDSN Australia/Pacific, 2017). Against this background, a growing number of higher education institutions (HEIs) have adopted policies to decrease their carbon footprint and are striving to achieve the Sustainable Development Goals (SDGs) (Alcántara-Rubio et al., 2022). Annually, *Times Higher Education* releases the University Impact Rankings, highlighting and honoring universities that score highly in multiple United Nations Sustainable Development Goals (SDGs)[1]. Moreover, an expert report from UNESCO underlines that sustainability should become a core activity and should shape objectives for universities and colleges (UNESCO, 2022). Although some scholars (Kohl et al., 2022) argue that the higher education (HE) sector's explicit inclusion in the Agenda and the SDGs is limited, it has become something that both external stakeholders and the HEIs themselves consider as part of their role within society. This is true also for HEIs in Norway, although this very much still is a work in progress.

This chapter focuses on a sustainability initiative at Nord University in the Northern part of Norway. It is called "Green Campus Nord" and was initiated in October 2019. It started off as a bottom-up initiative by sustainability researchers at Nord University and Nordland Research Center (an associated research-institute) aiming to lower the carbon footprint at Nord University and working towards Agenda 2030 and the 17 SDGs. The initiative received funding for approximately 6 months and was considered a pilot project on the main campus of Nord University—campus Bodø. Despite receiving a lot of interest and enthusiasm both from within and outside the institution, the project never received additional funding, and after an additional ten months of unpaid work, the project initiators unfortunately had to leave the project and their ideas and ambitions behind. Today, only a limited part of the initiative still exists, despite Agenda 2030 and the SDGs being central to the strategy of Nord University. Based on theories of organizational and institutional change, this chapter discusses advantages and disadvantages of bottom-up versus top-down approaches to promoting and advancing sustainability within the university. We then present some of the initiatives that were taken and their key results, including a survey conducted among students and employees about the university and its role with regards to climate change and the SDGs. Finally, we delve into some of the successes of the project, while also identifying some of the barriers to institutional transformation that were experienced through the work with Green Campus Nord.

2 Social Transformation and Higher Education Institutions

The interconnection between societal structures and environmental challenges is emphasized within environmental sociology. As Boström and Lidskog (2024) point out, environmental issues are not merely ecological phenomena but are deeply rooted in social systems. These systems are shaped by patterns of consumption, economic priorities, and institutional norms that often perpetuate unsustainable practices. Addressing these "social roots" necessitates a focus on systemic change that moves beyond technical or policy-oriented solutions, engaging with the deeper cultural and structural dimensions of environmental problems. The same could be said about all the SDG's.

Relating this to HEIs and sustainability transformation, Owens and Halfacre-Hitchcock (2006) highlight that universities can be considered "microcosms" of society, and, as such, their experiences may guide efforts for change at the societal level. Expanding on this, Sporn and Godonoga (2024) emphasize that HEIs are not only shaped by societal transformations but also actively contribute to them. In addition, within the literature on social transformation and environmental sociology, two concepts seem essential: Beck's (2021) concept of the cosmopolitan society, and Rittel and Webber's (1973) concept of "wicked problems". The argument is that in an increasingly interconnected world, traditional nation-state boundaries are considered inadequate for tackling issues like climate change, or other environmental, social and economic challenges such as the SDGs, as they are "wicked problems"[1] that require transnational cooperation and collective solutions. This in turn shifts the focus from national self-interest to a recognition of shared global responsibilities (Beck, 2021). By adopting such an approach, HEIs can thus position themselves as critical actors in bridging local and global sustainability efforts. However, wicked problems may also manifest in more localized, but no less complex forms. Universities, as microcosms of society, face their own version of these dilemmas when implementing sustainability initiatives/the SDGs. For example, encouraging a shift from air travel to train travel as part of emission reduction strategies may align with environmental goals but clash with practical concerns such as time efficiency and financial constraints. In this sense, each of the two concepts offers crucial insights into the role of HEIs as transformers of sustainability, and the complexities they, as a result, encounter. We elaborate further on the latter in the following section.

[1] Wicked problems are defined by their interconnectedness, contradictions, and resistance to straightforward solutions (Rittel & Webber, 1973).

2.1 Barriers to Transformation

Like the difficulties and complexities with social transformation of society, the path to sustainable transformation within institutions is also beset with barriers that hinder their ability to enact meaningful change. According to Boström and Lidskog (2024) some of the key challenges include institutional inertia, conflicting interests among stakeholders, and fragmented policy responses. These barriers often result in a piecemeal approach to sustainability, where actions address symptoms rather than systemic issues. This "problem-solving mentality" can limit the potential for transformative change, as it fails to engage with the broader structures and cultural norms that sustain environmental problems. Leal Filho et al. (2022) build on this by examining barriers associated with social sustainability implementation. They argue that entrenched power dynamics and institutionalized practices often resist change, even when there is a clear mandate for action. Overcoming these barriers requires more than just policy adjustments; it demands a cultural shift that redefines institutional priorities and values. Such shifts are not merely desirable but necessary to lay the groundwork for transformative initiatives. This perspective provides a foundation for understanding the challenges HEIs face as they navigate the complexities of sustainability.

In response to these barriers, Boström and Lidskog (2024) propose the concept of "realistic enabling", which bridges the gap between ambitious sustainability goals and the practical realities of implementation. This approach emphasizes incremental steps that align with institutional capacities while maintaining a focus on long-term, systemic change. Realistic enabling prioritizes actions that are both achievable and transformative, leveraging small-scale initiatives to build momentum for larger shifts.

Central to this approach are participatory processes, as they foster engagement and ownership among stakeholders. As Boström and Lidskog (2024) highlight, involving diverse voices in decision-making ensures that sustainability initiatives are contextually relevant and inclusive. Thus, Beck's (2021) vision of "a cosmopolitan society" also applies to micro-societies as to HEIs. In this sense, examining the approaches used to develop sustainable campuses becomes valuable. Institutional changes in universities are particularly intriguing, as variations in governance structures pose unique challenges for achieving campus sustainability. Although there probably are numerous ways and a variety of approaches that HEIs can take in order to reach these goals. The most common changes, though, are those initiated by the administration, students, and faculty, or a blend of these, and can be referred to as top-down and bottom-up approaches (Brinkhurst et al., 2011). Hence, most universities would choose either a top-down approach to sustainability transformation within their own institution or a bottom-up one. The former may focus on expectations and demands for the institutional change coming from top leadership or administration, while the latter centres on an expectation or demand from students or staff for institutional change. See Fig. 1.

The top-down approach is taken by the top university management and administrative leaders, and is channelled mainly through official statements and policy

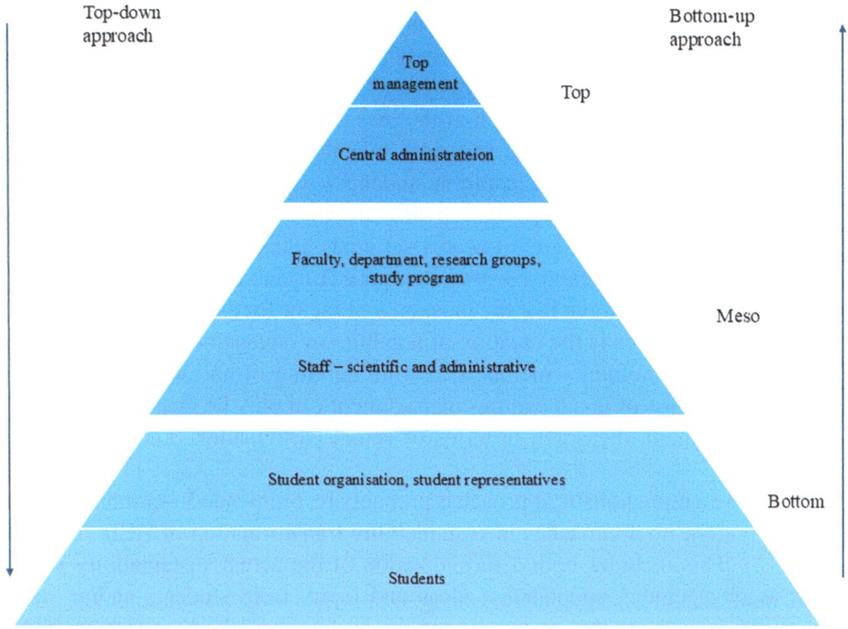

Fig. 1 Top-down and Bottom-up approach used in higher education for sustainable development

documents, which then are communicated down to the lower levels for implementation (Wang et al., 2020) The bottom-up approach on the other hand focuses on empowerment and student and faculty participation, which according to Disterheft et al. (2012) allows for greater integration and application of sustainability in universities. Although the top-down approach offers the benefits of clearly defined goals, efficient decision-making, and resource allocation, it still presents several challenges that must be addressed (Leal Filho et al., 2015). Bottom-up approaches in the context of sustainability in higher education institutions (HEIs) have the advantages of increased engagement, being open to diverse ideas and input from stakeholders, and having the possibility of tailoring it to local realities. However, without institutional support (support from leaders at the top or from a policy document) it will be difficult because governance for transformational change and leadership are considered necessary prerequisites (Purcell et al., 2019; Leal Filho et al., 2023; Ssekamatte, 2023). Lozano et al. (2015) highlight that the absence of a proper institutional framework explicitly stating the mission, vision, and policies of many higher education institutions (HEIs) hampers their progress towards sustainability. Moreover, a large N-study conducted by Avila et al. (2017) concluded that many top leaders may not grasp the importance of sustainability, making it challenging for them to prioritize sustainability initiatives on campus, especially as sustainability practices have typically not been economically recognized in annual performance reports of HEIs (Ávila et al., 2017).

Ferrer-Balas et al. (2008), on the other hand, argue that the main barrier to sustainable transformation of HEIs, is not in leadership, but the difficulty of involving the users of the campus in sustainable initiatives. This thus favors more participatory and bottom-up approaches. Although at the meso-level, and not the bottom level, a study by Mazon et al. (2020) for instance, indicates that academics have significant influence when it comes to inspiring students to accept sustainability-focused changes. Furthermore, it suggests that regular and thorough evaluation of campus sustainability efforts and progress towards set goals helps unite people while also inspiring them and giving them a sense of purpose (Lozano et al., 2015). Similarly, Leal Filho et al. (2021) argue that the most crucial factor for any implementation of green campus initiatives is the creation of a culture of engagement and acceptance among academic community members and the building of an awareness that the improvement process of green campus management can only be guaranteed if there is social engagement, measured in terms of actual participation (León-Fernández et al., 2018).

In this sense a more holistic approach is probably recommended—combining top-down and bottom-up approaches in sustainability transformation of HEIs (Lozano et al, 2015). This includes both a formalization of the actual sustainability work, as well as engagement, spontaneity, ideas and input, from students and/or staff. Some might even argue that co-creation is the preferred method for sustainability transformation in HEIs (Mbah & Ezegwu, 2024). However, research indicates that many universities have not yet developed effective relationships with internal and external parties (Blanco-Portela et al., 2017), making co-creation a less realistic option. Finally, regardless of the chosen approach for institutional change that has been chosen, a common obstacle faced by faculty and staff is the fact that they are not receiving the financial support required to carry out campus sustainability initiatives (Jorge et al., 2015). For instance, Dahle and Neumayer (2001) demonstrated that the most significant hindrance was budget constraints, which were partly due to a lack of awareness of how green initiatives can lower costs. Institutional reluctance to change was also a key factor. Moreover, Brandli et al (2015) also point to factors such as the degree of importance attributed to sustainability, where the crucial obstacle is a lack of collaboration between universities, coupled with inadequate governmental policies to support sustainability. We will return to this discussion and various aspects later in our analysis of success and barriers for Green Campus Nord. First, however, a few words on the SDGs and higher education in Norway more generally.

2.2 Norway, Higher Education and the SDGs

In the 2024 Sustainable Development Report, Norway has a score of 82.2, and ranks number 7 out of 167 countries (Sachs et al., 2024), leaving it as one of the most sustainable countries in the world. And this is despite Norway being one of the world's leading fossil fuel and emissions exporter. As a country Norway has a long history of being concerned with environmental issues and sustainability, the origins of this tracing back to the Brundtland commission and report in 1987 (Brundtland, 1987). As with other United Nations member states, Norway has dedicated itself to the attainment of the SDGs. This was stated as early as 2016, one year after the Paris agreement and Agenda 2030 and the 17 SDGs had been signed, and in 2017 they started integrating sustainability as part of their national strategy. However, three years later, in 2020, the Norwegian government was criticized for its lack of cohesion and measureable progress. Thus, in 2021, the Norwegian government launched an action plan intended to localize the global goals within Norway and demonstrate how diverse actors, such as companies, can incorporate and implement the SDGs in their work (Regjeringen, 2021). It is highlighted in the action plan from 2021 as well as in previous national documents, that to achieve the goals by 2030, all parts of society need to contribute.

In the context of Norwegian higher education institutions (HEIs), sustainability has been discussed for at least 15 years, but it did not truly gather momentum until 2016/2017, with the University of Bergen taking a lead. Soon other Norwegian universities followed, and in 2018 all the larger Norwegian universities had incorporated some sort of plan for this within the institution. Smaller universities, such as Nord (and others) were, however, lagging. The Minister of Research and Higher Education at the time, Iselin Nybø, also highlighted at the contact conference in 2019 that "green transformation in the knowledge sector was important" and in January 2020 she even launched the idea of a national green competition between universities. It is at this moment in time that the Green Campus Nord initiative started.

2.3 *Nord University and the Green Campus Nord Initiative*

Nord University, located in the northern parts of Norway, is a young university, and consists of five different faculties: the Faculty of Biosciences and Aquaculture; the Faculty of Education and Arts; the Faculty of Social Sciences; the Faculty of Nursing and Health Sciences, and Nord University Business School. Its main focus-areas are said to be blue and green growth, innovation and entrepreneurship, societal security and welfare, health and education. The university strives to be globally relevant but with a strong regional profile and focus. With around 11,000 students and approximately 1,200 employees, Nord university is one of the smaller universities in Norway. Despite being a small institution, Nord University is however, like other HEIs in Norway, held to high standards for environmental performance. With nine campuses stretching 884 km from the northern to middle parts of Norway, and the two largest campuses—Bodø and Levanger—being 11 h by train apart, reducing the carbon footprint is not an easy task (see Fig. 2).

The ideas and thoughts of a green campus initiative at Nord University were first discussed at a research meeting in late 2018. At that point in time Nord university was part of the Eco-Lighthouse Foundation, but did not have any strategy or action plan towards reducing its carbon footprint or working towards sustainability. There were, however, signals from minister Iselin Nybø that she would want to call for a

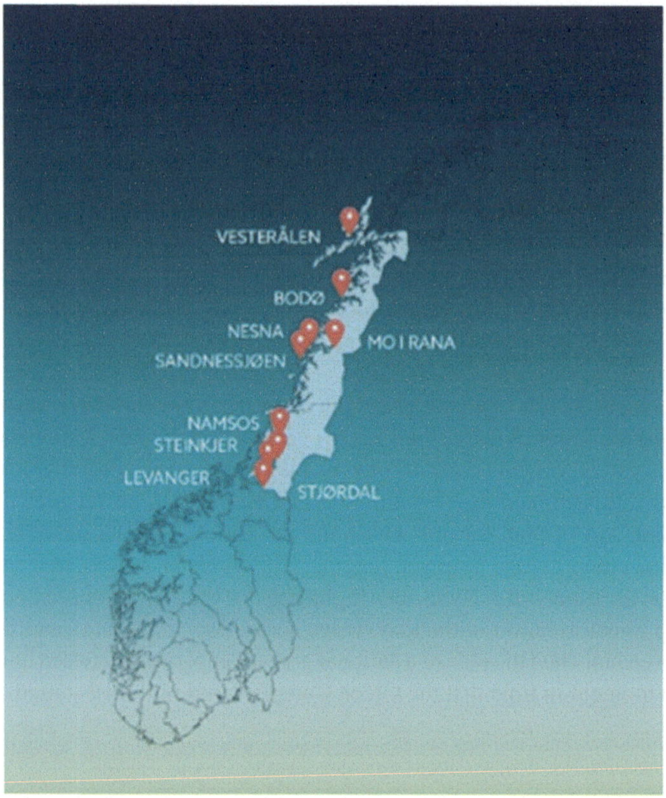

Fig. 2 Map over the nine campuses of Nord university

sustainability competition between Norwegian universities. After first applying and receiving some funding from the municipality for the project in March 2019, the rector of Nord University decided to chip in on the funding, and a pilot study of Campus Bodø—the largest of Nord's nine campuses, was initiated, with its starting point in September 2019. The funding would cover expenses for a research assistant, as well as some of the hours that were allotted to the researchers allocated to the project. It would also cover the expenses related to the planned activities for a six-month period. The pilot project was to have a focus on the Bodø campus, but with transfer value to the other eight campuses of Nord university as well.

From the beginning, the Green Campus project at Nord University, aimed to facilitate a green transition at the Bodø campus through a structured and strategic approach. Guided by six overarching focus areas—(1) reducing the campus climate footprint, (2) enhancing green spaces and sustainable activities, (3) integrating sustainability into research, (4) promoting sustainability in education, (5) strengthening green entrepreneurship and (6) social responsibility. In this regard, the project from the beginning relied on a holistic approach connecting different functions by means of teaching, research, campus activities, and strategies aimed at engaging and involving communities and stakeholders (Sterling, 2021).

The project was organized into three interrelated work packages. The first work package focused on *identifying key emission sources and areas for improvement*, both by conducting a carbon footprint mapping (energy, travel, procurement), as well as a survey on knowledge, attitudes and behaviours towards environmental issues and the SDGs among staff and students. Finally, this work-package also aimed at mapping courses and study-programs as well as the research at Nord University that addressed the 17 SDGs. The second work package translated the findings from the footprint analysis into *actionable sustainability measures*. These initiatives included energy efficiency improvements, upgrading waste-sorting infrastructure and increasing accessibility to recycling facilities, installing additional bike racks and incentivizing the use of public transportation, expanding and maintaining green spaces to enhance biodiversity and create a more inviting campus environment, and testing and developing a climate budgeting system to track progress and prioritize sustainability actions effectively. The third work package emphasized *outreach and collaboration*, ensuring long-term engagement among stakeholders. Key activities included community engagement through hosting events such as the "Grønsj" (Green lunch) seminars and debate nights to facilitate discussions on sustainability challenges and solutions, showcasing academic contributions by highlighting student-led enterprises and academic programs and courses focused on sustainability to inspire participation and innovation, co-creation with stakeholders and dissemination of results and updates through newsletters, a dedicated project website, and social media channels to maximize visibility and impact.

So, to what extent did the Green Campus Nord project then succeed in sustainability transformation of Nord University?

3 Methodology

To evaluate the successes and challenges of the Green Campus Project, we use two primary methodological approaches: a survey of students and staff, and participatory observation as active project members. These methods provide both quantitative and qualitative insights into the project's processes, outcomes, and obstacles.

The survey assessed knowledge, attitudes, and behaviors related to sustainability at Nord University, covering environmental awareness, transportation habits, waste management, energy use, and barriers to sustainable practices such as cost and institutional constraints. It was distributed to all students and employees, promoted via campus screens, posters, informational stands, and the university's intranet, with reminders sent via email. A total of 2,858 respondents participated, yielding a 21.72% response rate. A 21.72% response rate is generally considered good for surveys and response rates for online surveys typically range from 20 to 30% (Rea & Parker, 2014). Responses came from all campuses, with the main campuses Bodø and Levanger having the highest participation. The majority (71%) were students, while academics (15%) and administrative staff (9%) were also represented. In all 2,041 of the respondents were students, while 817 were staff. This means that we have a higher response rate from the staff (68.8%) than from the students (18.8%). Like other surveys on sustainability, it is possible that those individuals most concerned with climate change and sustainability were more likely to answer than those who are less concerned. However, given the high number of respondents both across ages, genders, faculties and campuses, the overall findings are valid. Further, our main objective in this chapter is to look at the institution and Nord university as a whole; however, we present differences between students and staff when applicable.

Participatory observation provided deeper insights into project implementation and stakeholder interactions. As active members, the authors participated in planning meetings, workshops, and sustainability initiatives, documenting real-time decision-making processes and institutional challenges. According to Kimber and Maertens (2023), participant observation offers an insider's perspective, revealing organizational constraints and everyday sustainability practices. However, Fangen (2010: 72) warns of the risk of losing objectivity when deeply embedded in a setting. To counter this, systematic notes were recorded during meetings and events, ensuring analytical rigor. Balancing the dual role of researcher and practitioner required careful reflexivity to minimize bias while maintaining a nuanced understanding of institutional sustainability efforts (Fangen, 2010).

4 Accomplishments and Barriers: Experiences from Green Campus Nord

As mentioned, the Green Campus Nord initiative received funding from the rector at Nord University, as a *pilot*-project running for approximately 6 months. After six months the funding ran out, and although there was hope for additional funding, the decision seemed to take time, and one of the key persons involved ended up having to apply for another job outside the university, as she had been hired to work specifically on that sole project. The same week, in March 2020, the Covid-19 pandemic reached Norway and the university as well as most of the Norwegian society closed. It was gradually reopened from June 2020, but the activities and teaching were highly regulated, and staff members were also recommended to do home-officing even until summer 2021. During this period the project leader continued to be involved in the Green Campus initiative, hoping for additional funding from the top management. Without allocated time or financial resources and with restrictions on campus, the initiative faced significant setbacks and was ultimately halted.

In 2021, the Green Campus Nord was then replaced with a full-time administrative position in the administration, in addition to a project named "Sustainability at Nord", with the HR-manager as the project-leader, and two researchers were assigned the role as a reference-group. Despite putting Agenda 2030 in the university strategy in 2021, a large extent of the Green Campus initiative and spirit fizzled out, although some parts of it remain to this day. When discussing the achievements of and barriers to the Green Campus initiative, below we focus our analysis on the project's achievements from September 2019-June 2021, with a main emphasis on the first 6 months. We base our evaluation of success based on four main criterias: (1) The extent of anchoring sustainability transformation within top leadership and within the organization, (2) The actual organizational sustainability changes that occurred related to (a) own carbon footprint and (b) enhancing education, research and innovation of the SDGs, (3) The ability to bridge with society towards sustainability transformation, and (4) The level of engagement and enthusiasm for the sustainability transformation initiative. Finally, as the status of the sustainability transition of Nord today is important also for evaluating the success of Green Campus Nord, we also touch upon this as a fifth element at the very end of the chapter.

4.1 Anchoring Within Top Leadership and Within the Organization

As stated by Purcell (2019), sustainability transition is highly dependent on support from top management or from a policy document. The ideas and ambitions for the Green Campus project were presented at a strategic leader meeting and received a lot of praise. However, a concrete Action Plan or a Strategic Vision for Nord University related to the SDGs was not in place at the time. However, very early on the initiative was able to put down a network of researchers across all five faculties, one person from campus services as well as a few students from various student organizations and a student entrepreneurial pop-up. The project was also presented to the central administration at an internal seminar and was invited to two (out of five) of the faculty's strategy seminars, as well as to the Student Council. With some help from communication and the graphic design office, Green Campus Nord also got its own logo and an official webpage at Nord. The project also administrated a dedicated Facebook account.

Green campus Nord was able to gain a lot of attention and raise awareness around SDGs on campus relatively quickly, both among students and staff. This was partly due to various informational campaigns and small videos of students being interviewed about their environmental behaviour, or tested on their waste-sorting skills, or promotion of the Green Campus Nord survey etc.

Through the survey, staff and students were asked to rate their level of concern for the environment and sustainability on a scale from 1 to 10. As shown in Fig. 3a, b, the majority of respondents rated their concern between 7 and 8. However, staff generally reported slightly higher levels of concern compared to students. When assessing the severity of climate change, most respondents rated it between 8 and 10, with staff again demonstrating a stronger sense of urgency. This indicates a high level of environmental awareness across the university community, particularly among employees. These results highlight a widespread recognition of the urgency of

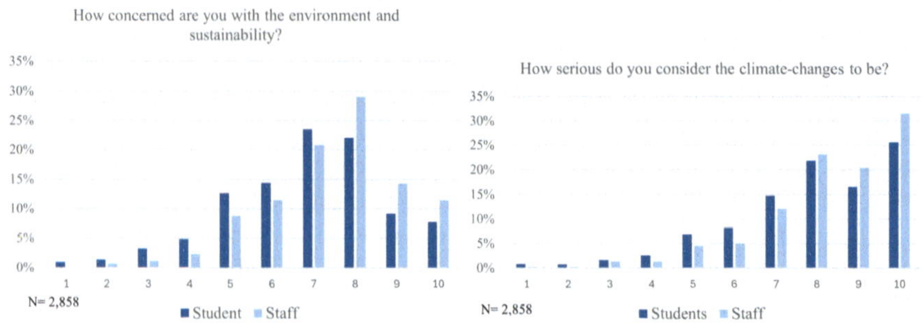

a) Concern for the environment *b) Level of severity of global climate change*

Fig. 3 **a** Concern for the environment, **b** Level of severity of global climate change

climate issues within the institution, with staff displaying a greater degree of concern compared to students.

The survey also explored perceptions of responsibility in addressing climate change. When asked who they believed held the greatest responsibility for solving climate-related challenges, responses varied. A significant portion of respondents stated that responsibility should be shared equally among individuals, businesses, and policymakers. However, when evaluating whether key actors were doing enough to mitigate climate change, responses were more divided. Governments and businesses were frequently perceived as not having taken sufficient action, whereas educational institutions, including Nord University, were generally viewed more positively. However, 49.1% still answered that Nord could do more, and interestingly this percentage was slightly higher for members of staff than students (Fig. 4).

a) Perceptions of Actors responsibility *b) Evaluation of Actors*

Fig. 4 **a** Perceptions of Actors responsibility, **b** Evaluation of Actors

4.2 *Organizational Sustainable Transformations*

When evaluating the level of success of the Green Campus Nord project it is based on to what extent the project led to organizational changes with regards to sustainability, either in terms of reducing Nord's own carbon footprint, or in terms of enhancing education, innovation and research relating to the SDGs.

4.2.1 Reduction of Nord's Own Carbon Footprint

The most important practical environmental outcome was the delivery of a carbon footprint report of the university. It included areas such as procurement, energy, transport and waste. However, data on travel and other aspects were still missing. Apart from the report, the Green Campus project was also able to enhance recycling on campus by improving waste management systems and reducing some of the electricity consumption in university buildings, by installing better lighting systems and temperature control. The student canteen decided to offer more vegetarian alternatives, as well as leftovers for low prices, and to restrict their use of plastic containers and glasses. The project looked at ways to promote public transport, parking for electric vehicles, and installing bicycle sheds. While the latter remained at the planning stage when the project ended, the actual installation did take place later, and as planned. Finally, the representative from campus services became part of a Norwegian university network for environmental advisors.

4.2.2 Education, Research and Innovation and the SDGs

An important ambition of Green Campus Nord was to do a mapping of the study programs and courses, as well as research in terms of each of the SDG Goals. However, parts of top management were reluctant to spend time on this. As a result, Green Campus Nord failed with its intended mapping of education and research and the SDGs at Nord. Nevertheless, a course named "Sustainability in Practice" was developed, and some of the larger research projects were mentioned as being at the core of Nord's research within sustainability. The potential for more change was huge though.

Our survey indicated, for instance, first a strong demand for more education on sustainability topics (over 40%), and secondly that many respondents reported that environmental and climate-related issues were insufficiently integrated into their study programs. The question was limited to students. While some students indicated that sustainability was frequently addressed in their courses, a considerable proportion reported limited exposure to these topics. This gap between interest and curriculum content suggests an opportunity for the university to further strengthen the inclusion of sustainability within the academic programs (Fig. 5).

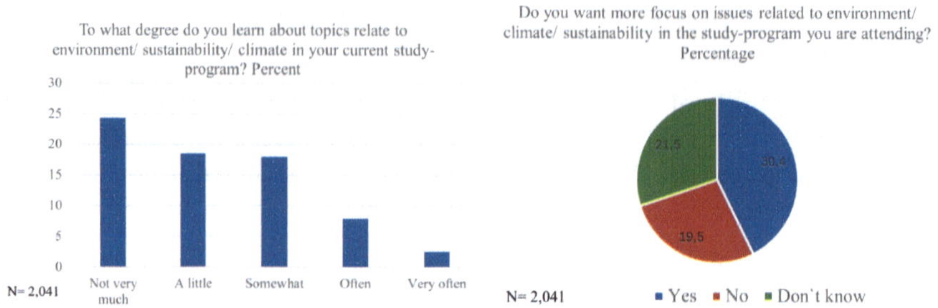

a) Sustainability-topics in Study-Program *b) Percentage wanting more focus on sustainability*

Fig. 5 **a** Sustainability-topics in Study-Program, **b** Percentage wanting more focus on sustainability

A small incremental step in that direction was when the Green Campus initiative was written into the Erasmus + application for 2021–2027, offering SDGs days for international students. A SDG Quiz and debate on the student pub and social arena "Samfunnet" directed at students also took place. In addition, examples of green entrepreneurship and innovation at Nord were highlighted and displayed at seminars for students and staff. Related to research, Green Campus Nord arranged regular internal seminars (green lunches) for sustainability researchers to meet and discuss some of the things that they had been working on, or other hot topics within the field.

4.3 Bridging with Society in Sustainable Transformation

From the beginning, Green Campus Nord successfully brought external stakeholders into the reference-group. This included external stakeholders that where in charge of the university buildings (Statsbygg), the waste management of the municipality (IRIS), the welfare-organization of –student housing, kindergardens and canteens (Studentinord), as well as Bodø municipality and Nordland County. As the project progressed it also collaborated with several other local sustainability networks, including one for businesses run by the local sports club Bodø Glimt playing in the European league.

We were also invited to meetings with a local shopping center in relation to recycling, and one of the larger civil society organizations involved in rehabilitating former drug addicts through giving them jobs involving recycling, bicycle-repair etc. Finally, the project also participated in meetings with architects involved in the urban planning of the part of the city were Nord University and campus Bodø is situated, and with Bodø 2024, as the city was to become a capital of culture in 2024. The university kindergartens, as well as one of the local schools, also joined in on occasions.

The project and researchers from Nord also initiated numerous events with a sustainability theme. Some of them were done with external parties, such as the Bodø-situated Norwegian Aviation Museum on "Flightshame", or "Dougnut economy discussion" for Bodø with Bodø municipality. We also arranged an Innovation Day for students in connection with the yearly television campaign against plastic in the ocean, and a retail event on campus. Students from across faculties also joined in on a national beach-cleaning event, and eventually a whole week was celebrated on campus and within town. named Gjenbruksuka (Recycling Week). This was a collaboration between the university, the waste company and the municipality, and various other organizations. The week included everything from re-design, to cooking, to public debates and presentation of research etc., taking place downtown, on campus, as well as at the student-pub and social arena "Samfunnet". Finally in 2021, Green Campus Nord and Bodø municipality had the honour to arrange a National Sustainability Workshop with all local actors involved in sustainability transformation together with the Norwegian Minister of Municipality.

4.4 A Culture of Engagement for Sustainable Transformation

Green Campus Nord came about in many ways as a bottom-up initiative, although it was initiated by academic staff rather than students. In this sense it could be argued that the project was initiated at the meso-level which Mazon et al. (2020) found performed particularly well in terms of student engagement. Whether for that reason or for another, the initiative was met with a relatively high level of engagement and enthusiasm from start to finish, and this even increased throughout the project as new partners became involved etc. Thus, the important culture of engagement that Leal Filho et al. (2021) stressed seemed to be present. When asked if they would like to be actively involved in the Green Campus Nord project, about 30% wanted to contribute more directly. Preferred forms of engagement included assisting in developing visions for Nord's sustainability strategy, helping launch new initiatives, and participating in cultural or social events related to sustainability. When asked to rate the importance of the Green Campus Project on a scale from 1–10, 53.8% of respondents scored it 8 or higher, thus indicating broad support for the initiative and its objectives. While both students and staff acknowledged its importance, staff showed a significantly higher willingness to engage in sustainable transformation at Nord University. 40.9% of staff expressed a desire to participate directly, compared to 21.6% of students. Yet, even for students, 1 out of 5 wanted to commit to direct participation (Fig. 6).

This, combined with a rather holistic approach in terms of partners both internally and externally, as well as several activities directed towards students and staff, and within the city, resulted in the initiative receiving local and national media attention. Other Norwegian universities also contacted the project in order to participate in the sharing of knowledge. In sum, the network laid grounds for the important cultural shift that Boström and Lidskog (2024) stress is needed in order for societal transformation to take place. See Fig. 7.

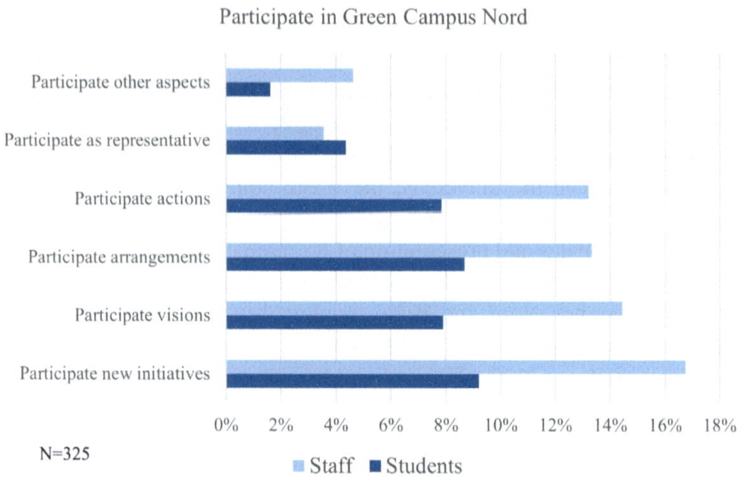

Fig. 6 Percentage of students and staff wanting to participate in the sustainability transformation of Nord University

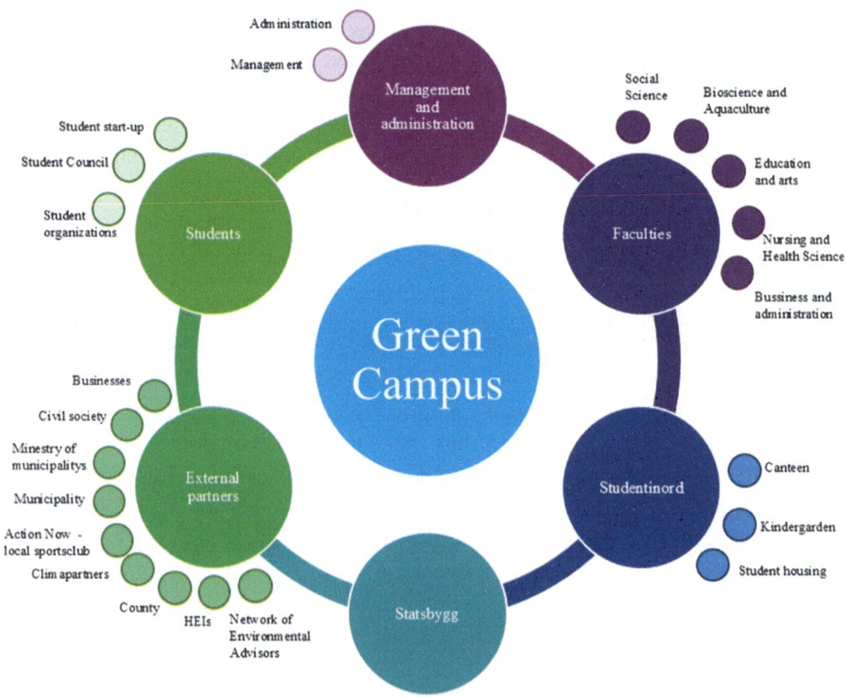

Fig. 7 Green campus network

4.5 Sustainability at Nord University Today

Green Campus Nord emerged when the Norwegian government urged HEIs and businesses to take greater SDG responsibility. A proposed university competition was quickly shelved in the face of criticism, leaving no concrete carbon-cut or SDG requirements for academia. Shortly afterwards, the Covid-19 pandemic shut down campuses, reducing engagement in Green Campus Nord among leadership, staff, and students.

Nevertheless, as the national strategy highlighted the importance of the SDGs, once the pandemic seemed to be under control, the focus on the SDGs continued. In the revision of the Norwegian university and college act of 2021, the importance of HEIs as key contributors to sustainable restructuring is emphasized. The amendment to the law implies that the social mission of the educational institutions—teaching, research and innovation/communication—has been expanded to include a fourth aspect: "The purpose of universities and colleges is to contribute to environmental, socially and economically sustainable development' (University and Colleges Act, 2005, § 1–1). Since then, sustainable development has also become part of the government's allocation letter to the HEIs. That same year Nord University decided to insert sustainability and Agenda 2030 as part of the 2030-strategy of the university. The university replaced the Green Campus Nord project with "Sustainability at Nord," with the latter being run by a project leader from the central administration. This reduced the project's bottom-up approach and connection with students and stakeholders (see Figs. 1, and 7). Many sustainability activities ceased, among them special events and external collaborations. However, some initiatives, like Recycling Week and the Carbon Footprint Report, continued. Since 2021, all study programs in Norway must integrate the 17 SDGs into learning outcomes. In 2022, Nord University joined the SEA EU alliance, shifting its sustainability focus towards European collaboration rather than internal and local sustainable transformation. Despite sustainability being part of Nord's and SEA-EU's strategic commitments, it is still very top down in its ethos, and most students and staff do not know much about SEA EU or how it contributes to the SDGs. Nord still even lacks an Action Plan for carbon reduction or SDGs.

5 Conclusion

The Green Campus Nord project experienced both successes and challenges, offering insights into sustainability initiatives in HEIs. Considering that the funding and allocation of resources were limited, the overall conclusion is that the project did achieve a lot in a short time. However, it failed to receive long-term funding. This ultimately led to its demise, this being further accelerated by the Covid-19 pandemic, which restricted campus activities. The top leadership also seemed to lack a deep understanding of the importance of the SDGs for all campus activities. Sustainability cannot relate solely to a single part of the institution. It must be integrated into every facet of HEIs, influencing decisions from procurement to teaching, to stakeholder and student engagement. This is what a holistic approach entails. In addition, the project's replacement by "Sustainability at Nord" transferred the responsibility for sustainability from the meso (scientific staff) to the top level of the organization (administration). This soon led to a reduction in activity and engagement within the organization. Staff and students—that according to the survey were eager to participate in a green transition of the university (see Fig. 6)—were no longer contacted. Most of the networking-activities and contacts across faculties at Nord University and with external parties (see Fig. 7) were also abandoned.

Obviously, transforming societies is not an easy task. The same could be said about HEIs. Achieving sustainability transformation in higher education institutions (HEIs) necessitates a combination of top-down and bottom-up strategies, dedicated resources, and strong policy support. While Green Campus Nord fostered strong engagement both within and beyond campus, and management praised the project, their ambitions at the time did not align with those of the initiators. Hence, the possible cultural shift ended in the taking of incremental steps, illustrated by the limited aspect of the initiative active today. However, sustainability and SDGs remain on Nord University's agenda, now within the SEA-EU framework. This has resulted in three new joint programs (bachelor and master-level) and several blended intensive programs and courses focusing on sustainability. A green charter for the alliance is also currently in progress. Managing a European alliance is very top-down driven, and although this has the advantage of the funding and the leadership being secured and in place, without a larger engagement and involvement of staff and students and local stakeholders, there is however a risk that the focus on sustainability within SEA-EU will also end up being implemented in an incremental manner. Hopefully, due to the Green Campus Nord project, the survey and experiences, SEA-EU and the top management at Nord are, however, able to see the importance and value of a more bottom-up or a "whole institutional approach". Given the actual outreach of a European University Alliance such as SEA-EU, this certainly has the potential of bringing something to pass that is of a more transformative nature.

References

Alcántara-Rubio, L., Valderrama-Hernández, R., Solís-Espallargas, C., & Ruiz-Morales, J. (2022). The implementation of the SDGs in universities: A systematic review. *Environmental Education Research, 28*(11), 1585–1615.

Ávila, L. V., Leal Filho, W., Brandli, L., Macgregor, C. J., Molthan-Hill, P., Özuyar, P. G., & Moreira, R. M. (2017). Barriers to innovation and sustainability at universities around the world. *Journal of Cleaner Production, 164*, 1268–1278.

Beck, U. (2021). The World Risk Society as a Cosmopolitan Society. In K. Rasborg (Ed.), *Ulrich Beck: Theorising World Risk Society and Cosmopolitanism* (pp. 160–163). Springer.

Blanco-Portela, N., Benayas, J., Pertierra, L. R., & Lozano, R. (2017). Towards the integration of sustainability in Higher Education Institutions: A review of drivers of and barriers to organisational change and their comparison against those found of companies. *Journal of Cleaner Production, 166*, 563–578.

Brandli, L. L., Leal Filho, W., Frandoloso, M. A. L., Korf, E. P., & Daris, D. (2015). The environmental sustainability of Brazilian Universities: Barriers and pre-conditions. *Integrating sustainability thinking in science and engineering curricula: Innovative approaches, methods and tools*, 63–74.

Boström, M., & Lidskog, R. (2024). *Environmental sociology and social transformation*. Routledge.

Brinkhurst, M., Rose, P., Maurice, G., & Ackerman, J. D. (2011). Achieving campus sustainability: Top-down, bottom-up, or neither? *International Journal of Sustainability in Higher Education, 12*(4), 338–354. https://doi.org/10.1108/14676371111168269

Brundtland, G. H. (1987). *Report of the World Commission on Environment and Development: Our Common Future.*

Dahle, M., & Neumayer, E. (2001). Overcoming barriers to campus greening: A survey among higher educational institutions in London, UK. *International Journal of Sustainability in Higher Education, 2*(2), 139–160.

Disterheft, A., da Silva Caeiro, S. S. F., de Miranda Azeiteiro, U. M., & Leal Filho, W. (2012). *Implementing Sustainability at the Campus–Towards a Better Understanding of Participation Processes within Sustainability Initiatives.* na.

Fangen, K. (2010). *Deltakende observasjon.* Fagbokforlaget.

Ferrer-Balas, D. J., Adachi, S., Banas, C. I., Davidson, A., Hoshikoshi, A., Mishra, Y., Motodoa, M., & Onga, M. O. (2008). An international comparative analysis of sustainability transformation across seven universities. *International Journal of Sustainability in Higher Education, 9*(3), 295–316.

Jorge, M. L., Madueno, J. H., Cejas, M. Y. C., & Peña, F. J. A. (2015). An approach to the implementation of sustainability practices in Spanish universities. *Journal of Cleaner Production, 106*, 34–44.

Kimber, L. R., & Maertens, L. (2023). Participant Observation. In L. R. Kimber, L. Maertens, & F. Badache (Eds.), *International Organizations and Research Methods: An Introduction* (pp. 33–41). University of Michigan Press.

Kohl, K., Hopkins, C., Barth, M., Michelsen, G., Dlouhá, J., Razak, D. A.,... & Toman, I. (2022). A whole-institution approach towards sustainability: a crucial aspect of higher education's individual and collective engagement with the SDGs and beyond. *International Journal of Sustainability in Higher Education, 23*(2), 218–236.

Leal Filho, W., Aina, Y. A., Dinis, M. A., Pimenta, Purcell, W., & Nagy, G. J. (2023). Climate change: Why higher education matters?, *Science of The Total Environment, 892*, September 2023, 164819. https://doi.org/10.1016/j.scitotenv.2023.164819.

Leal Filho, W., Salvia, A. L., Vasconcelos, C. R. P., Anholon, R., Rampasso, I. S., Eustachio, J. H. P. P.,... & Sharifi, A. (2022). Barriers to institutional social sustainability. *Sustainability Science, 17*(6), 2615–2630.

Leal Filho, W., Frankenberger, F., Salvia, A. L., Azeiteiro, U., Alves, F., Castro, P.,... & Ávila, L. V. (2021). A framework for the implementation of the Sustainable Development Goals in university programmes. *Journal of Cleaner Production, 299*, 126915.

Leal Filho, W., Shiel, C., & Paço, A. D. (2015). Integrative approaches to environmental sustainability at universities: An overview of challenges and priorities. *Journal of Integrative Environmental Sciences, 12*(1), 1–14.

León-Fernández, Y., Gomera, A., Antúnez, M., Martínez-Escrich, B., Villamandos, F., & Vaquero, M. (2018). Enhancing environmental management in universities through participation: The case of the University of Córdoba. *Journal of Cleaner Production, 172*, 4328–4337.

Lozano, R., Ceulemans, K., Alonso-Almeida, M., Huisingh, D., Lozano, F. J., Waas, T.,... & Hugé, J. (2015). A review of commitment and implementation of sustainable development in higher education: results from a worldwide survey. *Journal of Cleaner Production, 108*, 1–18.

Mazon, G., Ribeiro, J. M. P., de Lima, C. R. M., Castro, B. C. G., & de Andrade, J. B. S. O. (2020). The promotion of sustainable development in higher education institutions: Top-down bottom-up or neither? *International Journal of Sustainability in Higher Education, 21*(7), 1429–1450.
Mbah, M. F., & Ezegwu, C. (2024). Impactful methodological considerations for knowledge co-creation in sustainable development research. *Sustainability, 17*(1), 52.
Owens, K. A., & Halfacre-Hitchcock, A. (2006). As green as we think? The case of the College of Charleston green building initiative. *International Journal of Sustainability in Higher Education, 7*(2), 114–128.
Parr, A. (2022). *Knowledge-driven actions: transforming higher education for global sustainability* (pp. 1–100). UNESCO.
Purcell, W. M., Henriksen, H., & Spengler, J. D. (2019). Universities as the engine of transformational sustainability toward delivering the sustainable development goals:"Living labs" for sustainability. *International Journal of Sustainability in Higher Education, 20*(8), 1343–1357.
Rea, L. M., & Parker, R. A. (2014). *Designing and conducting survey research: A comprehensive guide.* John Wiley & Sons.
Regjeringen (2021). Mål med mening—Norges handlingsplan for å nå bærekraftsmålene innen 2030. Stortingsmelding 40 (2020–2021). Regjeringen.
Rittel, H. W., & Webber, M. M. (1973). Dilemmas in a general theory of planning. *Policy Sciences, 4*(2), 155–169.
Sachs, J. D., Lafortune, G., & Fuller, G. (2024). *The SDGs and the UN Summit of the Future. Sustainable Development Report 2024.* Paris: SDSN, Dublin University Press. https://doi.org/10.25546/108572
SDSN (Sustainable Development Solutions Network) Australia/Pacific. (2017). Getting started with the SDGs in universities: a guide for universities, higher educationinstitutions, and the academic sector. Australia, New Zealand and Pacific Edition, Sustainable Development Solutions Network—Australia/Pacific, Melbourne, https://ap-unsdsn.org/wp-content/uploads/University-SDG-Guide_web.pdf
Sporn, B., & Godonoga, A. (2024). Higher education institutions as change agents in society: Perspectives on adaptation and impact. *European Journal of Higher Education, 14*(sup1), 1–9. https://doi.org/10.1080/21568235.2024.2412764
Ssekamatte, D. (2023). The role of the university and institutional support for climate change education interventions at two African universities. *Higher Education, 85*(1), 187–201.
Sterling, S. (2021, May). Educating for the future we want. In GTI Forum Opening Essay. Cambridge: Tellus Institute. https://greattransition.org/gti-forum/pedagogy-transition-sterling
UNESCO. (2022). *Education for sustainable development: a roadmap.* United Nations: Unesco.

Sowing Seeds for Regenerative Futures: Transdisciplinary Approaches in Higher Education

A. Kent, Jeanine Gregersen-Hermans, Guillermo Palau Salvador, Azadeh Babaghaderi, Eundeok Kim, Halima Begum, Alice Annelin, Kshamta Hunter, Tara Ivanochko, and Jan Brown

> The classroom remains the most radical space of possibility in the academy.
> –bell hooks (2014).

Abstract In the face of escalating global challenges such as climate change, social inequality, and ecological degradation, higher education institutions (HEIs) must evolve to effectively support the Sustainable Development Goals (SDGs). Transdisciplinary approaches, which foster innovative educational practices for positive change, have gained increasing attention from educators and scholars. By integrating

A. Kent (✉)
Williams, Dalhousie University, Halifax, Canada
e-mail: kn830866@dal.ca

J. Gregersen-Hermans
Zuyd University of Applied Sciences, Maastricht, Netherlands

G. P. Salvador
Universitat Politècnica de València, Valencia, Spain

A. Babaghaderi
Univeristy of Windsor, Windsor, Canada

E. Kim
Florida State University, Tallahassee, USA

H. Begum
Universiti Utara Malaysia, Kuala Lumpur, Malaysia

A. Annelin
Umeå University, Umea, Sweden

K. Hunter · T. Ivanochko
University of British Columbia, Vancouver, Canada

J. Brown
Liverpool John Moore University, Liverpool, UK

W. Leal Filho et al. (eds.), *Introducing the UN SDGs in Higher Education Institutions*, World Sustainability Series, https://doi.org/10.1007/978-3-032-23017-1_43

diverse perspectives and collective thinking, these approaches break down traditional academic silos, encouraging collaboration among scholars, practitioners, and community 'heartholders' to drive impactful action toward achieving the SDGs. This commentary reflects on the critical role of transdisciplinary approaches in higher education, highlighting examples of successful implementations from universities around the world. These models have had significant impacts on curriculum development, student engagement, and community partnerships, demonstrating the potential for transformative educational environments that embrace complexity and promote inclusivity. We explore the challenges and opportunities that arise when adopting transdisciplinary methods and call for higher education institutions to integrate these approaches to contribute meaningfully to the achievement of the SDGs. This commentary aims to expand the understanding of regenerative education, offering insights and practical strategies to inspire HEIs to adopt transdisciplinary practices as a pathway toward a more equitable, resilient, and sustainable future.

Keywords Climate change · Community partnership · Higher education · Sustainable development goals · Transdisciplinary approach

1 Introduction

This paper is presented as a reflective transdisciplinary commentary, drawing on collaborative engagements, documented gatherings, and co-created experiences across global institutions. The insights and goals emerged through iterative feedback loops, grounded in lived practice and shared values. This narrative methodology reflects the action-oriented nature of regenerative learning and is appropriate for exploring emerging pathways in higher education transformation.

In the face of escalating global crises such as climate change, social inequality, and ecological degradation, higher education institutions (HEIs) are increasingly called upon to assume a transformative role (Wals et al., 2022). The United Nations' 2030 Agenda emphasizes that HEIs hold a critical responsibility to contribute to sustainable development by embedding the Sustainable Development Goals (SDGs) into their core missions and functions. Particularly, SDG 4.7 urges education systems to foster the knowledge, skills, and values necessary to promote sustainable development, global citizenship, and respect for human rights, cultural diversity, and environmental stewardship (Giangrande et al., 2019). Yet despite this imperative, many universities remain entrenched in traditional, siloed structures, limiting their capacity to address the complex, interconnected challenges of the twenty-first century (Filho et al., 2018; Pereira & Žalėnienė, 2021). At this critical juncture, a transdisciplinary approach—one that integrates diverse perspectives and collective, relational thinking—offers a promising pathway for regenerating higher education and fostering flourishing futures (Horn et al., 2022; Gazo & Tasdemir, 2020). Such approaches not only encourage collaboration across academic disciplines but also bring universities, practitioners, and communities together, creating fertile ground for innovative educational models that inspire transformative change.

This commentary reflects on the work of the *Impact Network for Regenerative Futures* (IN4RF), a transdisciplinary, global initiative committed to co-creating pathways for regenerative sustainable futures. Our network engages students, faculty, and community "heartholders" in reimagining education through the lens of regeneration, encouraging a shift away from linear, disciplinary thinking toward a more holistic, systemic approach. We aim to explore how these practices are already transforming educational landscapes and call on others to join us in this vital work. By sharing insights and lessons from our collective efforts, we invite educators, scholars, and communities to consider how transdisciplinary education can empower the next generation of leaders and changemakers. This commentary not only serves as a reflection on our journey but also as a call to action for others who share our vision: to transform higher education into a force for ecological stewardship, social justice, and global citizenship. Together, we can contribute to a more equitable, sustainable, and resilient future for all. Positioned alongside global initiatives such as the UN's Principles for Responsible Management Education (PRME), UNESCO's Education for Sustainable Development (ESD) roadmap, and the Inner Development Goals (IDGs), IN4RF contributes to a growing international movement reimagining education as a catalyst for systemic, regenerative change.

2 Literature Synthesis-Transdisciplinary Educational Models for Sustainability

Higher education institutions (HEIs) have increasingly adopted transdisciplinary approaches to address the multifaceted challenges posed by sustainability (Sterling, 2010). This paradigm shift is driven by the necessity to integrate diverse knowledge systems and foster collaboration between academic and non-academic actors, thus facilitating the co-creation of solutions to real-world problems (Rudhumbu et al., 2017; Strachan et al., 2019). Transdisciplinary education models involve the active engagement of external partners, including community stakeholders, in the learning process, enhancing the relevance and applicability of academic knowledge (Aneas, 2014; Hoinle et al., 2021).

Transdisciplinary learning encourages communication across disciplinary boundaries, facilitating knowledge exchange between academic and non-academic participants. This co-creation of knowledge is essential for fostering the competencies required for responsible global citizenship, particularly in the context of sustainability challenges (Hilger & Keil, 2022). Moreover, these educational models promote hands-on learning and community engagement, allowing students to apply theoretical concepts to practical challenges, thereby enhancing their problem-solving skills (Tejedor & Segals, 2018). Transdisciplinary education is also characterized by collaborative processes, including co-design, co-production, and co-dissemination of knowledge (Wiek et al., 2011). These processes help bridge the gap between academia and society, ensuring that educational models are inclusive and reflect the diverse perspectives of all stakeholders (Horn et al., 2023). The active participation of non-academic partners, such as community members and industry experts, enriches the educational experience by bringing real-world relevance to theoretical frameworks (Hoinle et al., 2021).

Implementing transdisciplinary models in HEIs requires substantial institutional support. Universities must prioritize interdisciplinary collaboration by incentivizing cross-disciplinary teaching and research (Dieleman & Juárez-Nájera, 2015; Rocha et al., 2020). Additionally, aligning academic strategies with global sustainability frameworks, such as the United Nations Sustainable Development Goals (SDGs), can enhance the impact of transdisciplinary education (Fortune et al., 2021; Kubisch et al., 2021). Without institutional backing, these initiatives may face challenges due to resource constraints and resistance to change (Holzer et al., 2019). Curriculum redesign is another critical strategy for embedding sustainability into higher education. Integrating sustainability concepts across all academic disciplines, particularly in business courses, can foster a holistic understanding of the planetary boundary and Indigenous knowledge frameworks, which weaves social-ecological and relational systems into learning and action (Escobar, 2020; Findler et al., 2019; Rockström et al., 2024; Gazo & Tasdemir, 2020). Creating transdisciplinary minor programs that focus on sustainable development offers an additional pathway for promoting sustainability in higher education (Bootsma et al., 2013; Jia et al., 2019).

Stakeholder involvement is a core component of transdisciplinary educational models. Designing education for sustainability requires embracing real-world challenges that extend far beyond traditional academic exercises. This shift calls for a collaborative and integrated model of learning that fosters connections across disciplines and communities (Bootsma et al., 2013). Aneas (2014) describes sustainable transdisciplinary education as a dynamic framework of maps and socio-cultural perspectives, allowing diverse viewpoints to illuminate complex issues from multiple angles. Additionally, the success of such programs hinges on the depth and quality of collaboration with stakeholders (Meyer et al., 2017). Engaging with community stakeholders allows for the co-creation of knowledge that is both academically rigorous and socially relevant (Baumber et al., 2019). Long-term engagement with community partners can help HEIs address complex social-ecological challenges, although this process may be hindered by power imbalances and institutional inertia (Lake et al., 2022). Nonetheless, community-involved pedagogies foster deeper understanding and innovation by incorporating diverse perspectives into the learning process (Steger et al., 2021). The creation of transdisciplinary communities of practice, which bring together students, academics, and community stakeholders, further enhances the learning experience. These communities provide a platform for social learning and collaborative problem-solving, which are crucial for addressing sustainability challenges (Cundill et al., 2015). However, achieving successful community engagement requires careful management of power dynamics to ensure that all voices are heard and valued (Schmidt et al., 2020).

Challenges and Opportunities.

Despite the significant potential of transdisciplinary education to address global sustainability challenges, HEIs face numerous barriers to adopting these models. Resistance to change remains a persistent challenge, often rooted in the entrenched culture of static and didactic teaching approaches that dominate traditional higher education (Holzer et al., 2019; Hooks, 2014). These traditional methods, while familiar and straightforward, often lack the dynamism required to prepare students for the complexities of twenty-first-century challenges. Moreover, the absence of clear motivation and incentives for faculty and administrators to embrace transdisciplinary approaches further compounds this inertia. Faculty members may perceive these approaches as an additional burden on their already stretched resources, particularly in environments where tenure and promotion systems prioritize disciplinary outputs over collaborative, transdisciplinary work.

External political and policy factors further exacerbate these challenges. A global shift toward nationalism and the rise of governments deprioritizing sustainability—such as potential policy changes under an administration hostile to climate action—can undermine HEIs' efforts to pursue regenerative education. Reduced funding for sustainability research and education, combined with policies promoting short-term economic gains over long-term environmental stewardship, can create an unfavorable environment for transdisciplinary initiatives. This broader context underscores the vulnerability of higher education to political and policy shifts, particularly in regions where sustainability is framed as secondary to nationalistic goals. Financial constraints within institutions compound the issue. Sustainable education research and initiatives often receive less funding compared to traditional academic programs, limiting the resources available for innovation. Institutions operating within tight budgets may deprioritize transdisciplinary education in favor of more cost-efficient, siloed models. Additionally, administrative goals and institutional priorities may not align with the long-term, resource-intensive nature of transdisciplinary programs, further complicating their adoption and integration into existing structures.

The diversity of needs and expectations among stakeholders also presents challenges. Faculty, students, and external partners often have differing priorities, making alignment around shared objectives difficult. Unequal power dynamics within participatory processes can exacerbate these challenges, sometimes marginalizing non-academic participants such as community members or Indigenous knowledge holders (Adelle, 2021; Schmidt et al., 2020). Addressing these barriers requires a paradigm shift within HEIs. Beyond fostering open communication and inclusivity, it is crucial to align heartholders and stakeholders around shared values and a common purpose, as articulated through the concept of shared axiology (Antonacopoulou 2022). Institutional leadership must actively support transdisciplinary initiatives by providing funding, recognizing collaborative work in tenure and promotion decisions, and investing in faculty development programs to build capacity for new teaching, research models, and co-creating impact (Williams et al., 2024). HEIs must also advocate for sustainability at the policy level, using their influence on champion climate action and resist the erosion of progress in sustainability education.

While the challenges are considerable, they are not insurmountable. By addressing institutional inertia, providing adequate incentives and resources, and embracing innovative teaching methods, HEIs can create inclusive, sustainable, and impactful transdisciplinary education models. Strong stakeholder engagement, institutional support, and policy advocacy are critical to overcoming these barriers, enabling HEIs to harness the transformative potential of transdisciplinary approaches and contribute meaningfully to a more sustainable future.

3 Methodology Framing

This paper adopts a **reflective transdisciplinary commentary** approach grounded in the lived practices of the Impact Network for Regenerative Futures (IN4RF). Rather than following a single empirical method, the work is shaped by principles of collaborative inquiry, narrative reflection, and systems thinking. The three goals and associated initiatives emerged organically through iterative dialogues among network members, informed by documented events, shared teaching practices, student engagement, and co-creative gatherings held between 2022 and 2024.

Our methodology reflects the **transdisciplinary ethos** of the network—relational, emergent, and context-sensitive. While no formal data collection tools were used, insights were validated through collective sense-making, shared writing, and peer feedback. We acknowledge the limitations of this narrative approach, including the absence of traditional measurement; however, we argue that this framing enables a deeper exploration of regenerative educational practices. The aim is not to generalize, but to offer situated insights that inspire others navigating similar transitions in higher education.

4 The Network—Who we Are

The Impact Network for Regenerative futures is a global collective of around 70 educators, researchers, students, and professionals dedicated to advancing sustainability through education, research, and community engagement. This diverse group, spanning continents and disciplines, shares a commitment to fostering a more inclusive, equitable, and sustainable (even more–regenerative) future by integrating sustainability into teaching and leadership practices.

Network members come from various locations around the world, including countries such as Canada, United States, Spain, Malaysia, the United Kingdom, the Netherlands, Germany, Sweden, Bosnia and Herzegovina, and Chile, Zimbabwe. Their work is influenced by the local contexts in which they work, ranging from rural agricultural systems in Southern Europe to indigenous communities in Southeast Asia, and from bustling urban centers in North America to the riverine regions of Northern Europe. A common thread uniting network members is their shared focus on justice, equity, regeneration and sustainability. Through their teaching and research, they seek to challenge traditional models of education and instead foster environments where students and community members can co-create knowledge. By incorporating diverse perspectives, they foster learning spaces where innovation and creativity are nurtured and where the principles of empathy, collaboration, and inclusivity are deeply embedded.

In addition to teaching, network members are actively engaged in interdisciplinary projects and partnerships that bridge local and global contexts. They often use approaches such as transformative learning, systemic innovation, and design and relational thinking to address issues ranging from climate change to social inequality. Their methodologies also emphasize the integration of indigenous knowledge and the co-production of solutions with marginalized communities.

5 Facing a Polycrisis: The Urgent Call for Transformative Change

We are navigating an era of immense global challenges, often described as a polycrisis, where the interlocking crises of climate change, biodiversity loss, and growing social inequality require urgent attention and action. These crises are not isolated; they are deeply interconnected, amplifying one another and demanding a comprehensive rethinking of how we approach education, sustainability, and societal resilience.

Anthropocene and Environmental Crisis.

The environmental crises of our time are unparalleled in scale and severity. Climate change and ecological degradation have accelerated to the point where entire ecosystems are collapsing, with devastating consequences for life on Earth (Steffen et al., 2018; Wunderling et al., 2022). This epoch, often referred to as the Anthropocene, reflects a turning point in human history—one in which human activity has become a dominant force shaping the planet's environment (Steffen et al., 2015). However, it is critical to acknowledge that the drivers of these crises are not evenly distributed; they are largely fueled by the actions of the wealthiest and most industrialized nations, whose patterns of economic consumption disproportionately contribute to environmental harm (Altvater et al., 2016). Meanwhile, the effects are felt globally, with marginalized communities, particularly those in the Global South, Indigenous peoples, and youth, bearing the brunt of these environmental disasters (Islam & Winkel, 2017). As the planet warms, biodiversity declines, and natural resources become increasingly strained, it becomes clear that "business as usual" approaches in education, development, and policy are no longer sufficient (Hoffman, 2021).

Economic and Social Inequality.

Compounding the environmental crisis is the deepening economic and social inequality. Disparities in wealth, access to education, and resource distribution continue to widen, reinforcing cycles of poverty, social injustice, and environmental degradation (Kamande et al., 2024). These issues are not just symptoms of a broken system; they are core drivers of the current polycrisis. Environmental destruction disproportionately affects the world's poorest communities, exacerbating existing inequalities. To address these interconnected crises, approaches must be holistic, recognizing the inseparability of social and ecological systems (Escobar, 2020; O'Brien et al., 2023). A failure to do so will only deepen the divisions and destruction we face.

The Urgent Need for Educational Transformation.

Our current educational systems, shaped by linear, mechanistic thinking, are ill-equipped to address the complexity and interconnectedness of the global challenges before us (O'Brien et al., 2023). Traditional approaches to education often emphasize siloed disciplines, overlooking the necessity of collaborative, transdisciplinary approaches that reflect the realities of the twenty-first century. The scope of the polycrisis demands more than incremental changes—it requires a paradigm shift in how we educate future generations (Singer-Brodowski, 2023). This is where the IN4RF steps in. Our network aims to transform educational practices by fostering transdisciplinary, creative, and regenerative approaches that break down traditional academic silos and encourage collective problem-solving. We believe education must evolve to prepare students not just to survive but to thrive in a rapidly changing world. At the same time, it is crucial to address the growing eco-anxiety among emerging leaders, as pessimism about the future can paralyze action. Research highlights the role of hope in enabling young people to face social-ecological challenges constructively; fostering "critical hope" can empower students to navigate uncertainty, confront complexity, and maintain trust in their ability to drive change (Ojala, 2017). By integrating diverse perspectives and prioritizing systems and relational thinking, we aim to empower learners to become agents of transformation, capable of shaping regenerative futures for both people and the planet.

The urgency of the current polycrisis calls for a radical rethinking of education. The IN4RF is committed to pioneering new pathways of education that respond to the complexities of our world and foster the capacity for students, educators, and communities to drive meaningful change. Now more than ever, we need educational systems that prepare us not just to cope with the crises at hand, but to actively co-inspire and co-create flourishing, resilient futures.

6 Approach: Regenerative Futures through Transdisciplinary Education

The three goals outlined below were co-developed through iterative dialogue across IN4RF's global network, rooted in shared values of justice, regeneration, and community engagement. These goals crystallized over time through reflective synthesis of our collective experiences during student-led workshops, international conference gatherings, and local engagement initiatives. Rather than being predefined, they emerged as thematic anchors that guided our evolving practice toward regenerative education. To prepare students to address the interconnected complex global challenges and achieve regenerative futures, we must transform our traditional education system. IN4RF has developed the following three goals to successfully renew the focus of HEIs: enabling transformational learning, building transdisciplinary approaches, and supporting sustainable-as-flourishing futures.

Goal 1: Enabling Transformational Learning.

To enable transformational learning, fostering a new paradigm in higher education is essential. IN4RF is committed to promoting transformational learning by moving beyond traditional teaching methods to embrace holistic, experiential, and participatory approaches. We aim to cultivate future leaders capable of addressing complex, interrelated global challenges. The specific approaches to meet this goal are as follows: first, integrating sustainability into curricula (SDG target 4.7) is urgently required. We are actively working to embed sustainability education into core curricula across disciplines. By integrating the principles of sustainability, regeneration, and social justice, we empower students to sense the interconnections and reciprocity between economics, society, and the biosphere. Second, our network emphasizes experiential and participatory learning, where students engage in real-world challenges through community projects, sustainability workshops, and fieldwork. This hands-on approach deepens their understanding of complex systems and fosters critical thinking, collaboration, and innovation. Third, we promote a transdisciplinary approach to knowledge in learning environments that break down traditional academic silos. By bringing together students, faculty, and practitioners from diverse fields and cultures, we enable nascent ways of thinking and problem-solving that reflect the complex nature of sustainability challenges. Fourth, we empower students to become proactive agents of change. Through our programs, students develop the leadership capacities, resilience, and creativity needed to engage with the polycrises of climate change, social inequality, and ecological degradation. Fifth, we establish communities of practice where students and educators co create knowledge and pathways that embrace possible futures. These learning communities provide an inclusive space for dialogue, reflexivity, and collective action around sustainability issues.

Goal 2: Building Transdisciplinary Approaches.

To build transdisciplinary approaches that bridge disciplines and tackle complex global challenges is imperative. IN4RF is committed to advancing transdisciplinary approaches that go beyond and transform traditional disciplinary boundaries, enabling a deeper understanding of interconnected global issues such as climate change, loss of biodiversity, and social inequality. Transdisciplinarity can be understood as fostering greater coherence, unity, and integration of knowledge, where traditional disciplinary divisions become less relevant or are fundamentally redefined (Reinecke et al., 2024; Hibbert & Sieldok 2014). In line with Meadows' (1999) systems framework, we view transdisciplinary education as a critical leverage point—an intervention that can shift the deep structure of higher education systems toward more regenerative and resilient pathways. Guided by this vision, we have developed the following four approaches to advance our mission:

First, to break down silos in education, we aim to restructure educational frameworks by encouraging cross-disciplinary collaboration between departments and faculties. This holistic approach ensures that students and educators explore sustainability challenges from multiple perspectives, fostering an integrated and comprehensive understanding of real-world problems. Second, collaborative knowledge production, including diverse, transdisciplinary players, is central to our mission. We hold space for collaboration not only between academic disciplines but also with external stakeholders, such as community organizations, industry leaders, and Indigenous knowledge holders. This transcending and transformative approach leverages diverse insights and experiences to co-create innovative, practical solutions for sustainable futures.

Third, through the network, we actively support and encourage interdisciplinary research and projects that connect environmental sciences, social sciences, business, and policymaking. These projects address the Sustainable Development Goals (SDGs) by promoting systems thinking and creative solutions that integrate social-ecological dimensions. Finally, focusing on engagement beyond academia, we actively involve non-academic stakeholders in our research and educational practices. By engaging with local communities, NGOs, businesses, and policymakers, we foster meaningful dialogue and ensure that the knowledge we produce is grounded in practical realities and geared toward creating tangible impacts.

Goal 3: Supporting Sustainable-as-Flourishing Futures.

Sustainability in Regenerative Futures goes beyond harm mitigation to actively promote human, ecological, and economic flourishing. To create futures where individuals, communities, and ecosystems thrive, this method integrates regenerative sustainability into educational programs. Initially, we need to shift the focus of sustainability from minimizing damage (World Commission on Environment and Development, 1987) to promoting regeneration and flourishing (Müller, 2020). This perspective inspires students to sense and design systems that support and regenerate all forms of life through the teaching and practice of sustainability. By reframing sustainability as a positive and regenerative force, students are empowered to prevent environmental degradation and cultivate conditions that enhance life for everyone. Environmental degradation, social inequity, and economic instability demand a transdisciplinary approach to problem-solving. At IN4RF, we foster systemic and relational thinking by integrating multiple fields of study, equipping students to tackle complex challenges from diverse perspectives. This approach enables the creation of comprehensive pathways that address ecological, social, and economic issues holistically. Recognizing that sustainable futures require diverse pathways; we cultivate a new generation of thinkers capable of collaborating across disciplines to drive lasting change.

Central to our mission is the prioritization of regenerative practices. These strategies focus on building resilience, fostering innovation, and encouraging adaptation in curricula, research, and community engagement. By inspiring students and educators to think creatively, we prepare them to support communities and ecosystems in dynamic and evolving contexts. This adaptive mindset empowers students to apply regenerative principles across various fields, contributing to the sustainability and well-being of both communities and the environment. Our goal is to motivate students not only to understand sustainability but to live it. Through leadership development, systems thinking, and ethical decision-making, we empower them to initiate regenerative practices in their careers and communities. By equipping them with the skills to create sustainable solutions, we ensure that the next generation will play a pivotal role in advancing global regeneration efforts.

Finally, we emphasize the importance of translating theory into action. By collaborating with community partners, policymakers, and businesses, our educational programs make sustainability a tangible and actionable concept. This integration of theory and practice ensures that students can actively contribute to building a future rooted in regenerative principles, enhancing the resilience and vitality of communities and ecosystems. These efforts align with the core pillars of Regenerative Futures, which envision sustainability as flourishing—embracing growth, adaptability, and resilience. Together, we strive to create a regenerative world where people, communities, and ecosystems thrive in harmony, supporting all life.

7 Key Initiatives for Regenerative Futures

- Seeds and Scenario Development

  The "Seeds and Scenario Development" program fosters creative thinking and collaborative exploration of pathways toward sustainable futures. This endeavor brings together diverse participants to envision the future of educational practices, community projects, and research innovations, focusing on scenarios that are scalable for broader societal change. Participants employ systemic thinking and foresight techniques to investigate multiple pathways, considering diverse environmental, social, and economic contexts. The program embodies a transdisciplinary learning model that integrates both academic experts and community partners, adding real-world relevance to the educational process (Horn et al., 2023; Hoinle et al., 2021). By bringing together individuals from academia, community organizations, and industry, the initiative emphasizes cross-sector and interdisciplinary collaboration. This diverse engagement enriches scenario building, ensuring that potential outcomes are inclusive, adaptable, and resilient across varying settings.

For higher education institutions (HEIs) to maximize their societal impact, it is essential to move beyond disciplinary silos and foster partnerships that extend within and outside academia. These partnerships should bridge disciplines, institutions, and national boundaries, promoting a more integrated and comprehensive approach to sustainability. For instance, collaborations between HEIs in developed and developing countries offer notable potential for enriching curricula, supporting cross-institutional internships, and enhancing diversity within educational programs (Ferguson & Roofe, 2020). The program not only encourages envisioning future challenges and opportunities for sustainability but also supports the creation of detailed action plans, strategic partnerships, and a "seed bank" of transformative projects. This repository of outcomes is designed for dissemination within and beyond the network, forming a foundation for future research, policy initiatives, and educational reforms aimed at advancing regenerative systems. By promoting such integrative and forward-thinking practices, the effort contributes to the development of sustainable, resilient, and equitable systems that can adapt to diverse contexts and scale effectively for lasting societal impact.

- Transdisciplinary Student Network (TSN)

  The Transdisciplinary Sustainability Network (TSN) is home to a global community of passionate student leaders committed to sustainability. Through a variety of immersive learning experiences—such as hands-on workshops, project-based learning, and collaborative research—students gain insights into the complex, interconnected nature of global sustainability issues. The network serves as a dynamic platform, bringing together students from fields like environmental science, business, engineering, social sciences, and the arts to collaboratively tackle sustainability challenges. TSN encourages students to actively engage in shaping sustainable futures by building cross-disciplinary skills, fostering teamwork, and connecting academic knowledge to real-world issues (Burandt & Dlouhá, 2015; Fonseca & Molderez, 2018). Their involvement in community-based projects alongside local organizations, businesses, and policymakers helps them develop the skills and insight needed to drive tangible, lasting change. Guided by a transformative sustainability framework (McCullough & Pelcher, 2021), each project directly targets pressing concerns like climate change, loss of biodiversity, and social inequality, ensuring their work contributes meaningfully to society. By connecting academic work with real-world stakeholders, students can address specific Sustainable Development Goals (SDGs) and the global partnership among students, faculty mentors, and community members fosters an environment where innovative and practical solutions are co-created and applied.
- Hybrid Learning Approaches

  With the intention of incubating creativity and innovation, encouraging a broad and diverse range of dialogues and empowering the wider community, a hybrid approach to lifelong learning has been established at the heart of all activities in the IN4RF collaboration. Based on equity and justice principles it is acknowledged that, due to a wide range of factors, not all colleagues will be able to participate in the various projects at the same time and in the same places. However, since participation—whether active or passive—is considered essential to fostering a flourishing environment, an open and welcoming approach to integrating sustainable practices is being embraced.

Recognizing that in the post-digital world, where there are "...no sharp contrasts between digital, online and distributed learning environments and physical, onsite or co-located environments" (Nørgård 2021, p.1712) the Impacts Network for Regenerative Futures website has been constructed (https://advancingesd.com/in4rf/) to provide a real time portal that promotes and celebrates the wide variety of hybrid approaches being undertaken globally.

Using cathedral thinking (Krznaric, 2020) to develop a sustainable community of practice it is acknowledged that long-term visioning, commitment and intergenerational thinking must be part of the initial project. This means that those initiating the projects do so with the hope that future generations will benefit from the activities and outputs of those projects. In this post-digital world, new long-term opportunities emerge to develop innovative and complex learning environments, where institutions, society, learners, and professionals collaborate in partnerships and engage in closer dialogues on more equal and inclusive terms (Nørgård et al., 2019).

- Knowledge Mobilization

  To transform diverse knowledge into practical applications and actionable strategies that foster regenerative practices, the IN4RF community has established a robust management framework to support its long-term development. This framework, outlined in the original funding documentation and detailed on the community website (https://advancingesd.com/in4rf/), reflects the community's primary goal: to build a global network for knowledge sharing on transformative learning, driving positive societal impact with a clear and sustained focus on the future.

The IN4RF community fosters an inclusive culture of patience and perseverance, encouraging members to actively participate in activities they are passionate about or feel comfortable engaging in. These activities range from strategic planning and development to case study presentations and active co-creation. They take place in various formats, including face-to-face gatherings such as symposiums, conference presentations (e.g., the Transatlantic Symposium on Sustainable Development in Hasselt, Belgium), on-site visits, and online events like strategic development meetings, case study sharing, and local student-focused workshops. All gatherings are digitally documented, with reflections encouraged and shared on the IN4RF website. To ensure authenticity and inclusivity, creative data capture methods are promoted, allowing community members to communicate in ways that align with their unique perspectives and positionalities.

Creating collaborative networks, investing in education, and building capacity are essential for shaping regenerative futures. Within the IN4RF community, every voice matters. To uphold this principle, members from diverse global backgrounds and roles—including educators, learners, professionals, and community representatives—are encouraged to lead and actively participate in knowledge mobilization activities. This inclusive approach fosters spaces for reflexive practice, enabling members to listen, learn, and draw inspiration from a wide range of experiences. By placing reflective practices at the heart of the IN4RF community, individual and collective stories can be celebrated, shared, and further developed collaboratively.

These adaptive practices allow the IN4RF to evolve in response to the community's needs, ensuring that regenerative visioning remains central to its mission, both now and in the future.

8 Critical Perspectives and Challenges

In an interconnected world, no decision exists in isolation. Every local or community action reverberates across broader social, ecological, and political systems. The Impact Network for Regenerative Futures (IN4RF) was founded on this very understanding—that collective decision-making and collaboration are essential in creating resilient futures. In a time when, as Zizek (2011) observes, it is "easier to imagine the end of the world than the end of capitalism," the network serves as a space for reimagining our shared realities and challenging the systemic constraints that shape how we perceive and engage with the world.

The capitalist patterns that dominate much of our thinking, and seep into institutional learning, condition us to see the world through narrow, materialistic lenses, often obscuring the true complexity and interconnectedness of our global systems. These patterns lock us into familiar ways of knowing, being, and doing, making it difficult to sense alternative realities or envision futures beyond the capitalist framework. However, the IN4RF serves as a catalyst for awakening—a space where network dialogue allows the invisible to become visible, as Merleau-Ponty (1968) suggests. Through the network's collaborative platforms, participants shed their blindfolds, revealing the deeper connections that exist between individuals, communities, and ecosystems.

The IN4RF facilitates the development of resilient competencies—tools and modes of thinking that disrupt patterned assumptions. As Bohm and Biederman (1999) argue, we must resist the temptation to accept fixed conclusions and instead engage in a process of continuous questioning and reflexivity. Within the network, this disruption of assumptions is encouraged through transdisciplinary collaboration, where diverse perspectives and insights come together to co-create approaches that challenge the status quo. This process of awakening to new realities can be disorienting, as Plato's *Allegory of the Cave* reminds us. When network members step out of familiar ways of thinking and engage in collective inquiry, they confront a world that is brighter, more complex, and often more uncomfortable than they previously imagined. But this discomfort is necessary. It is through this resilience thinking—the ability to test, reflect, and explore—that network members unlock new possibilities for collective action and transformative change.

The IN4RF serves as a crucible for this kind of resilience. It offers a platform for members to engage in dialogue, experimentation, and critical reflection on the assumptions and systems that shape our world. In doing so, the network fosters a radically different understanding of the world around us, one that is more attuned to the complexity of our interconnected global systems and more capable of fostering regenerative, equitable futures. By co-creating spaces for resilience thinking and action, the IN4RF co-inspires and empowers participants to not only respond to crises but also to reimagine the very foundations of our social-ecological system. The network's emphasis on systems and relational thinking, collaborative learning, and community co-creation enables participants to engage with the 'being of being'—the essential nature of existence—and challenge the limitations of current structures. Through these collaborative, reflective processes, the network offers a pathway toward a regenerative future that is deeply attuned to the interdependencies between people, place, and planet.

Despite the promise of transdisciplinary collaboration, our journey has not been without challenges. Sustaining long-term engagement across diverse global contexts requires ongoing commitment, and disparities in access—particularly digital access—have at times limited participation. We are also aware of potential power imbalances in partnerships, including the risk of privileging academic voices over community knowledge holders. Furthermore, we recognize the tension between meaningful engagement and the danger of performative transdisciplinarity, where collaboration is more symbolic than substantive. To counter this, the network continually reflects on its practices, prioritizes co-creation over representation, and creates spaces for dissent, uncertainty, and learning from failure. These moments of friction have deepened our awareness of the care, humility, and time required to build genuinely regenerative and inclusive educational futures.

9 Discussion

There is a clear call for international network groups to collaborate and co-create relevant and flexible curricula that promote transdisciplinary learning in higher education. This could involve integrating courses from various disciplines and encouraging students to explore real-world problems that require diverse perspectives. However, deeper knowledge comes from taking the teacher, student and community practitioner out of the traditional structure of a learning environment and instead facilitating an experiential space where new knowledge can be created. Learning and sharing ideas from innovative and creative mindsets can help to provide such opportunities and creating a safe space for participants to test new methods of learning is paramount. Like SDG Labs, Living Labs, and other European and global sustainability education initiatives, IN4RF extends these models by weaving together transdisciplinary action, community-based knowledge co-production, and regenerative principles rooted in place-based and relational learning.

HEI must adapt to the changing global demands, where students are looking for customized learning, for example through micro-certificates and modular learning. Demand for critical thinking, adaptability, and interdisciplinary knowledge is more valuable than ever. The polycrisis is also putting resource constraints and increasing competition for HEI to collaborate through global partnerships to share knowledge and expertise through joint research initiatives, such as IN4RF.

Through curricular reform and pedagogical innovations, IN4RF will foster connections and cultivate spaces and relationships for deeper explorations of twenty-first century teaching and learning approaches. Fostering collaboration among students from different disciplines can enhance creativity and innovation and provide learning environments that facilitate communication across fields, which are essential. Transforming education towards transdisciplinary learning environments entails applying activities that help prepare students with the capacity building towards regenerative future mindsets (Backman et al., 2019). Institutions can provide hands-on learning experiences through projects that focus on regenerative practices, such as urban farming, community resilience projects, and clean energy initiatives. Experiential learning promotes active engagement and practical application of transdisciplinary knowledge (Scholz & Steiner, 2015). We need to prepare graduates for varied career paths that may not fit into a single discipline. Regenerative futures necessitate strong leadership, empathy, communication, and collaboration skills (Edwards et al., 2024).

10 Call to Action

Beyond disciplines, HEIs also need to emerge out of the traditional hierarchies and silos of roles and positions. Professors, researchers, and staff must be encouraged to collaborate across disciplines, and be supported through joint appointments, interdisciplinary research teams, and shared funding for transdisciplinary initiatives. Higher education can act as a hub for innovative thinking where diverse fields converge to create impactful research. Strengthening connections between academia and local communities can facilitate the sharing of knowledge (Scholz & Steiner, 2015) and co-create solutions that enhance local resilience and promote regenerative practices.

The IN4RF is a timely and emergent space for scholars, students and community stakeholders to experiment and challenge our notions of education, to innovate and collaborate on relevant models and frameworks for action and ultimately share our learning with the broader global community. By prioritizing fairness, social justice, and equity we incorporate diverse global perspectives in all our teaching and research efforts and can broaden students' understanding of how regenerative practices vary across cultures and environments. The IN4RF encourages lifelong learning around regenerative practices that can help individuals adapt to changing circumstances and contribute to ongoing community and societal transformation. By embracing these implications, higher education can better prepare students for the complexities of the modern world and foster a culture of innovation and collaboration.

11 Looking Ahead: Scaling Initiatives, Future, and Long-Term Vision

As the IN4RF continues to evolve, our long-term vision centers on expanding the reach and depth of our initiatives to foster regenerative education and societal transformation. Scaling our impact involves both enhancing current projects and developing innovative strategies to ensure sustainability, inclusivity, and resilience across diverse global contexts.

Scaling Initiatives.

A key focus is scaling the "Seeds and Scenario Development" program, which brings together transdisciplinary collaborators to envision and prototype pathways toward sustainable futures. By expanding this initiative, we aim to create a global repository of "transformative seeds" that can be adapted and implemented in varied local contexts. This effort will involve the development of partnerships with additional HEIs, community organizations, and policymakers to co-create regionally specific strategies while maintaining a shared commitment to regenerative principles.

Future.

We are also dedicated to deepening the integration of transdisciplinary practices in educational curricula. This includes establishing formal partnerships between HEIs and local communities to offer experiential learning opportunities that address pressing challenges like biodiversity loss, climate change, and social inequities. Additionally, we aim to grow the Transdisciplinary Sustainability Network (TSN), enabling students worldwide to collaborate on joint projects, share knowledge, and develop the skills needed to implement regenerative solutions. These partnerships will help bridge academic knowledge with real-world applications, fostering lasting social and environmental impacts.

Long-Term Vision.

Our long-term vision is to cultivate a global community of practice where education, research, and societal engagement converge to create sustainable, flourishing futures. Central to this vision is embedding regenerative principles into every facet of our work, from institutional policies and academic frameworks to community-driven projects. Through creative data capture, inclusive dialogue, and adaptive management frameworks, we aim to foster a culture of continuous learning and innovation. By building capacity within educational systems and equipping future leaders with the skills to navigate complexity, IN4RF aspires to be a catalyst for global regeneration efforts.

Together, we envision a future where collaborative, transdisciplinary approaches empower individuals, institutions, and communities to thrive in harmony with the planet, ensuring a regenerative world for generations to come.

12 Conclusion: IN4RF's Potential for Change

The Impact Network for Regenerative Futures (IN4RF) stands as a transformative force, redefining how education, research, and community engagement intersect to address complex global challenges. Through its commitment to transdisciplinary approaches, IN4RF is fostering innovative practices that break down traditional silos and empower participants to envision and co-create sustainable, regenerative futures. By emphasizing collaborative knowledge production and experiential learning, IN4RF equips individuals with the skills, values, and tools necessary to navigate the interconnected social, ecological, and economic challenges of the twenty-first century. Programs such as the "Seeds and Scenario Development" initiative and the Transdisciplinary Sustainability Network exemplify how the network brings diverse voices together to generate practical solutions, foster resilience, and inspire systemic change. These efforts underscore IN4RF's potential to not only influence higher education but also transform societal structures, contributing to the achievement of the SDGs.

Looking ahead, IN4RF's inclusive, adaptive practices ensure it can evolve in response to emerging needs while staying true to its mission of advancing regenerative systems. Its focus on bridging global and local contexts, engaging community "heartholders," and cultivating reflective practices positions the network as a model for impactful, sustainable change. In a time of unprecedented global crises, IN4RF offers a beacon of hope and innovation. By nurturing a generation of leaders who are equipped to think systemically, act ethically, and collaborate inclusively, the network has the potential to catalyze meaningful transformation. Together, we can reimagine education as a force for regeneration, building a future where people, communities, and ecosystems thrive in harmony. We have the elements to co-create the world anew.

Acknowledgments The authors wish to thank the members of the Impact Network for Regenerative Futures (IN4RF) for their invaluable contributions and ongoing commitment to advancing sustainability through education, research, and community engagement. We are deeply grateful to the institutions, community partners, and students who have supported and inspired this work. Special thanks to the Canadian Social Sciences and Humanities Research Council (SSHRC) for their support through the Connection Grant, which has been instrumental in enabling this initiative. Your generosity and vision have made this collaborative effort possible.

Conflict of Interest Statement The authors declare that there are no conflicts of interest related to the content or funding of this paper. All contributions were made in the spirit of advancing knowledge and promoting the shared mission of the Impact Network for Regenerative Futures.

References

Adelle, C., Görgens, T., Kroll, F., & Losch, B. (2021). Co-production of knowledge in transdisciplinary communities of practice: Experiences from food governance in South Africa. *Science and Public Policy*. https://doi.org/10.1093/scipol/scaa069

Altvater, E., Crist, E. C., Haraway, D. J., Hartley, D., Parenti, C., & McBrien, J. (2016). *Anthropocene or capitalocene?: Nature, history, and the crisis of capitalism*: *Pm press*.

Aneas, A. (2014). Transdisciplinary technology education: A characterisation and some ideas for implementation in the university. *Studies in Higher Education, 40*(9), 1715–1728. https://doi.org/10.1080/03075079.2014.899341

Antonacopoulou, E. P. (2022). Partnering for impact: A grand challenge and design for co creating a just, resilient and flourishing society. *The Journal of Applied Behavioral Science, 58*(4), 571–594.

Backman, M., Pitt, H., Marsden, T., Mehmood, A., & Mathijs, E. (2019). Experiential approaches to sustainability education: Towards learning landscapes. *International Journal of Sustainability in Higher Education, 20*(1), 139–156. https://doi.org/10.1108/IJSHE-06-2018-0109

Baumber, A., Kligyte, G., Bijl-Brouwer, M., & Pratt, S. (2019). Learning together: A transdisciplinary approach to student–staff partnerships in higher education. *Higher Education Research and Development, 39*, 395–410. https://doi.org/10.1080/07294360.2019.1684454

Bohm, D., & Biederman, C. (1999). *Bohm–Bierderman Correspondence: Volume one-Creative and science*. Routledge.

Bootsma, M., Vermeulen, W. J., Dijk, J. V., & Schot, P. (2013). Added value and constraints of transdisciplinary case studies in environmental science curricula. *Corporate Social Responsibility and Environmental Management, 21*(3), 155–166. https://doi.org/10.1002/csr.1314

Cundill, G., Roux, D., & Parker, J. (2015). Nurturing communities of practice for transdisciplinary research. *Ecology and Society, 20*(2), 22. https://doi.org/10.5751/ES-07580-200222

Dieleman, H., & Juárez-Nájera, M. (2015). Introducing a transdisciplinary curriculum to foster student citizenship: A challenge beyond curricula reform. *International Journal of Higher Education, 1*(1), 3–18. https://doi.org/10.1504/ijhes.2015.073463

Dlouhá, J., & Burandt, S. (2015). Design and evaluation of learning processes in an international sustainability-oriented study program. In search of a new educational quality and assessment method. *Journal of Cleaner Production, 106*, 247–258.

Edwards, M., Lindberg, A., Larsson, M., & Angle, J. (2024). *Regenerative business voices: Values-based entrepreneurship for sustainable enterprises*. Routledge, Taylor & Francis Group. https://www.routledge.com/Regenerative-Business-Voices-Values-based-Entrepreneurship-for-Sustainable-Enterprises/Edwards-Lindberg-Larsson-Angel/p/book/9781032343273

Escobar, A. (2020). *Pluriversal politics: The real and the possible*. Duke University Press.

Ferguson, T., & Roofe, C. G. (2020). SDG 4 in higher education: Challenges and opportunities. *International Journal of Sustainability in Higher Education, 21*(5), 959–975.

Filho, W., Raath, S., Lazzarini, B., Vargas, V., Souza, L., Anholo, R., Quelhas, O., Haddad, R., Kļaviņš, M., & Orlovic, V. (2018). The role of transformation in learning and education for sustainability. *Journal of Cleaner Production, 199*, 286–295. https://doi.org/10.1016/J.JCLEPRO.2018.07.017

Findler, F., Schönherr, N., Lozano, R., Reider, D., & Martinuzzi, A. (2019). The impacts of higher education institutions on sustainable development. *International Journal of Sustainability in Higher Education*. https://doi.org/10.1108/IJSHE-07-2017-0114

Fortune, T., Nicolacopoulos, T., & Horey, D. (2021). Conceptualising and educating for global citizenship: The experiences of academics in an Australian university. *Higher Education Research and Development, 41*, 1089–1103. https://doi.org/10.1080/07294360.2021.1882960

Giangrande, N., White, R. M., East, M., Jackson, R., Clarke, T., Saloff Coste, M., & Penha-Lopes, G. (2019). A competency framework to assess and activate education for sustainable development: Addressing the UN sustainable development goals 4.7 challenge. *Sustainability, 11*(10), 2832.

Hilger, A., & Keil, A. (2022). Education for sustainable development with transdisciplinary-oriented courses—Experiences and recommendations for future collaborations in higher education teaching. *Journal of Geography in Higher Education, 46*(3), 427–446. https://doi.org/10.1080/03098265.2021.1946765

Hoffman, A. J. (2021). *Management as a calling: Leading business, serving society*. Stanford University Press.

Hoinle B, Roose I & Shekhar H (2021). Creating Transdisciplinary teaching spaces. *Sustainability*, 13(7), article 7 https://doi.org/10.3390/su13073680.

Holzer, J., Adamescu, C., Cazacu, C., Ríaz-Delgado, R., Dick, J., Méndez, P., Santamaría, L., & Orenstein, D. (2019). Evaluating transdisciplinary science to open research-implementation spaces in European social-ecological systems. *Biological Conservation, 238*. https://doi.org/10.1016/j.biocon.2019.108228

hooks, b. (2014). *Teaching to transgress: Education as a practice of freedom*: Routledge.

Horn, A., Scheffelaar, A., Urias, E., & Zweekhorst, M. (2022). Training students for complex sustainability issues: A literature review on the design of inter- and transdisciplinary higher education. *International Journal of Sustainability in Higher Education*. https://doi.org/10.1108/ijshe-03-2021-0111

Horn, A., Scheffelaar, A., Urias, E., & Zweekhorst, M. (2023). Training students for complex sustainability issues: A literature review on the design of inter- and transdisciplinary higher education. *International Journal of Sustainability in Higher Education, 24*(1), 1–27. https://doi.org/10.1108/IJSHE-03-2021-0111

Islam, S. N., & Winkel, J. (2017). Climate change and social inequality. *United Nations Department of economic and social affairs (DESA)*. https://www.un.org/esa/desa/papers/2017/wp152_2017.pdf

Jia, Q., Wang, Y., & Fengting, L. (2019). Establishing transdisciplinary minor programme as a way to embed sustainable development into higher education system. *International Journal of Sustainability in Higher Education, 20*(1), 157–169. https://doi.org/10.1108/IJSHE-05-2018-0095

Kamande, A., Walker, J., Martin, M., & Lawson, M. (2024). *The commitment to reducing inequality index 2024*. Oxfam International. https://policy-practice.oxfam.org/resources/the-commitment-to-reducing-inequality-index-2024-621653/.

Krznaric, R. (2020). *The good ancestor: How to think long term in a short-term world*. Penguin Random House.

Kubisch, S., Parth, S., Deisenrieder, V., Oberauer, K., Stötter, J., & Keller, L. (2021). From Transdisciplinary research to Transdisciplinary education—The role of schools in contributing to community Well-being and sustainable development. *Sustainability*. https://doi.org/10.3390/su13010306

Lake, D., Motley, P. M., & Moner, W. (2022). Completing the CiCLE: Long-term assessment of community-involved collaborative learning ecosystems for social innovation in higher education. *Social Enterprise Journal, 18*(1), 28–50.

McCullough, B. P., & Pelcher, J. A. (2021). Instructor–student mentoring: Strengths of transformative sustainability learning and its direct application to impact industry and curricular refinement. *Sustainability, 13*(19), 10768.

Meadows, D. H. (1999). *Leverage points: Places to intervene in a system* (Vol. No. 994). The Sustainability Institute.

Merleau-Ponty, M. (1968). *The visible and the invisible: Followed by working notes*. Northwestern University Press.

Meyer, J., Mader, M., Zimmermann, F. M., & Çabiri, K. (2017). Training sessions fostering transdisciplinary collaboration for sustainable development. *International Journal of Sustainability in Higher Education, 18*(5), 738–757. https://doi.org/10.1108/ijshe-02-2016-0032

Molderez, I., & Fonseca, E. (2018). The efficacy of real-world experiences and service learning for fostering competences for sustainable development in higher education. *Journal of Cleaner Production, 172*, 4397–4410.

Müller, E. (2020). Regenerative development as natural solution for sustainability. In *The Elgar companion to geography, transdisciplinarity and sustainability* (pp. 201–218). Edward Elgar Publishing.

Nørgård, R. T. (2021). Theorising hybrid lifelong learning. *British Journal of Educational Technology, 52*, 1709–1723.

Nørgård, R. T., Mor, Y., & Bengtsen, S. (2019). Networked learning in, for, and with the world. In A. Littlejohn, J. Jaldemark, E. Vrieling-Teunter, & F. Nijland (Eds.), *Networked professional learning: Emerging and equitable discourses for professional development: Research in networked learning* (pp. 71–88). Springer.

O'Brien, K., Carmona, R., Gram-Hanssen, I., Hochachka, G., Sygna, L., & Rosenberg, M. (2023). Fractal approaches to scaling transformations to sustainability. *Ambio, 52*(9), 1448–1461.

Ojala, M. (2017). Hope and anticipation in education for a sustainable future. *Futures, 94*, 76–84.

Reinecke, J., Little, L. M., Simons, T., Bliese, P., Dencker, J., Roberson, Q., & Gruber, M. (2024). Advancing management theory through interdisciplinary research: Challenges and opportunities. *Academy of Management Journal, 67*(6), 1421–1427. https://doi.org/10.5465/amj.2024.4006

Rocha, P. L. B. D., Pardini, R., Viana, B. F., & El-Hani, C. N. (2020). Fostering inter-and transdisciplinarity in discipline-oriented universities to improve sustainability science and practice. *Sustainability Science, 15*(3), 717–728.

Rockström, J., Donges, J. F., Fetzer, I., Martin, M. A., Wang-Erlandsson, L., & Richardson, K. (2024). Planetary boundaries guide humanity's future on earth. *Nature Reviews Earth & Environment, 5*(11), 773–788. https://doi.org/10.1038/s43017-024-00597-z

Rudhumbu, N., Zhou, L., & Nhundu, K. (2017). Transdisciplinary research in higher education: Towards a paradigm for sustainable development. *IOSR Journal of Business and Management, 19*(1), 13–19.

Schmidt, L., Falk, T., Siegmund-Schultze, M., & Spangenberg, J. (2020). The objectives of stakeholder involvement in Transdisciplinary research. *Ecological Economics, 176*, 106751. https://doi.org/10.1016/j.ecolecon.2020.106751

Scholz, R., & Steiner, G. (2015). The real type and ideal type of transdisciplinary processes: Part I—Theoretical foundations. *Sustainability Science, 10*(4), 527–544. https://doi.org/10.1007/s11625-015-0326-4

Sieldok, F., & Hibbert, P. (2014). The organization of interdisciplinary work: Modes, drivers, and barriers. *International Journal of Management Reviews, 16*, 194–210.

Singer-Brodowski, M. (2023). *The potential of transformative learning for sustainability transitions: Moving beyond formal learning environments* (pp. 1–19). Environment.

Steffen, W., Broadgate, W., Deutsch, L., Gaffney, O., & Ludwig, C. (2015). The trajectory of the Anthropocene: The great acceleration. *The Anthropocene Review, 2*(1), 81–98.

Steffen, W., Rockström, J., Richardson, K., Lenton, T. M., Folke, C., Liverman, D., & Crucifix, M. (2018). Trajectories of the earth system in the Anthropocene. *Proceedings of the National Academy of Sciences, 115*(33), 8252–8259.

Steger, C., Klein, J., Reid, R., Lavorel, S., Tucker, C., Hopping, K., Marchant, R., Teel, T., Cuni-Sanchez, A., Dorji, T., Greenwood, G., Huber, R., Kassam, K.-A., Kreuer, D., Nolin, A., Russell, A., Sharp, J., Šmid Hribar, M., Thorn, J., Grant, G., & Waiswa, D. (2021). Science with society: Evidence-based guidance for best practices in environmental transdisciplinary work. *Global Environmental Change, 68*, 102240. https://doi.org/10.1016/J.GLOENVCHA.2021.102240

Sterling, S. (2010). Transformative learning and sustainability: Sketching the conceptual ground. *Learning and Teaching in Higher Education, 5*, 17–33.

Strachan, S., Marshall, S., Murray, P., Coyle, E., & Sonnenberg-Klein, J. (2019). Using vertically integrated projects to embed research-based education for sustainable development in undergraduate curricula. *International Journal of Sustainability in Higher Education, 20*(8), 1313–1328. https://doi.org/10.1108/IJSHE-10-2018-0198

Tasdemir, C., & Gazo, R. (2020). Integrating sustainability into higher education curriculum through a transdisciplinary perspective. *Journal of Cleaner Production, 265*, 121759. https://doi.org/10.1016/j.jclepro.2020.121759

Tejedor, G., & Segals, J. (2018). Action research workshop for transdisciplinary sustainability science. *Sustainability Science, 13*(2), 493–502.

Wals, A. E. J., Pinar, W. F., Macintyre, T., Chakraborty, A., Johnson, D., Waghid, Y., Tusiime, M., Grange, L. L. L., Razak, D. A., Accioly, I., Xu, Y., Humphrey, N., Iyengar, R., Chaves,

M., Herring, E., Vickers, E. A., Santamaria, R. D. P., Korostelina, K. V., & Pherali, T. (2022). Curriculum and Pedagogy in a Changing World. https://doi.org/10.56383/yudj7139

Wiek, A., Withycombe, L., & Redman, C. L. (2011). Key competencies in sustainability: A reference framework for academic program development. *Sustainability Science, 6*(2), 203–218. https://doi.org/10.1007/s11625-011-0132-6

Williams, K. A., Baidoo, L., Walker, T. R., Sundararajan, B., & Davis, A. (2024). The quest for impact research: Position, strategies and future directions. In T. Wall, L. Viera Trevisan, W. Leal Filho, & A. Shore (Eds.), *Sustainability in business education, research and practices*. World Sustainability Series. Springer. https://doi.org/10.1007/978-3-031-55996-9_8

World Commission on Environment and Development. (1987). *Our common future*. Oxford University Press.

Wunderling, N., Winkelmann, R., Rockström, J., Loriani, S., Armstrong McKay, D. I., Ritchie, P. D., & Donges, J. F. (2022). Global warming overshoots increase risks of climate tipping cascades in a network model. *Nature Climate Change*, 1–8.

Žalėnienė, I., & Pereira, P. (2021). Higher education for sustainability: A global perspective. *Geography and Sustainability*. https://doi.org/10.1016/J.GEOSUS.2021.05.001

Zizek, S. (2011). *Living in the end times*. Verso Books.

Advancing the SDGs Through Higher Education: Opportunities and Challenges in the Triple Helix Model

Tracey Dodd, Chanvi Singh, Gamithri Karunasena, and Daniel Prior

Abstract This chapter explores the role of Higher Education Institutions (HEIs) in advancing the United Nations Sustainable Development Goals (SDGs) with a focus on the Triple Helix model of university-industry-government interactions. We examine existing research to summarise the current state of knowledge on key success factors that enable successful university-industry-government collaborations. Building on this emerging framework, we explore and articulate how HEIs, through the Triple Helix model, can influence action to advance the SDGs. Our systematic review of 103 papers reveals that successful university-industry-government collaborations share four critical elements regardless of regional context: stable policy frameworks, entrepreneurial institutional cultures, effective knowledge management systems, and boundary-spanning capabilities. These findings have significant implications for HEIs globally, particularly as they navigate the complex demands of implementing SDGs within resource constraints and competing institutional priorities. Based on these insights, our review and emerging model offer important insights for university-industry-government collaborations aimed at advancing the SDGs and points to promising areas for future research. By integrating macro-level policy frameworks, organisational capacity, team-level dynamics, and individual motivations within the Triple Helix model, we provide a comprehensive approach to fostering innovation and sustainability. The model underscores the need for alignment at all levels of collaboration, recognising that successful outcomes depend not only on structural factors but also on relational dynamics and shared commitments to sustainable development. By addressing these factors, universities, industries, and governments can create a robust foundation for advancing the SDGs and contributing to sustainable development.

T. Dodd (✉) · C. Singh
College of Business and Law, Adelaide University, Adelaide, Australia
e-mail: tracey.dodd@adelaide.edu.au

G. Karunasena · D. Prior
UNSW Canberra School of Business, UNSW Canberra, Canberra , Australia

W. Leal Filho et al. (eds.), *Introducing the UN SDGs in Higher Education Institutions*, World Sustainability Series, https://doi.org/10.1007/978-3-032-23017-1_44

Keywords Higher Education Institutions (HEIs) · United Nations Sustainable Development Goals (SDGs) · Triple helix model · University-industry-government collaboration · Sustainability

1 Introduction

Higher Education Institutions (HEIs) hold a unique position in society as hubs of knowledge creation, innovation, and social influence (Bellandi et al., 2023). In the context of global challenges such as climate change, social inequality, and economic instability, HEIs are increasingly being called upon to play a transformative role in advancing sustainable development (Dodd et al., 2022). The adoption of the United Nations (2015) Sustainable Development Goals (SDGs) in 2015 provided a framework for addressing these challenges, outlining 17 interconnected objectives designed to guide global efforts toward a more equitable, inclusive, and sustainable future. HEIs are not only well-positioned to advance these goals but are also uniquely capable of aligning their teaching, research, and community engagement activities with the SDG agenda (Dodd, 2024; Tereshchenko et al., 2024).

However, despite widespread agreement about HEIs' potential to advance the SDGs, many institutions struggle to effectively implement their contributions (Hou et al., 2023; Prior et al., 2024). This challenge manifests across multiple levels—from institutional policies and systems to departmental practices and individual engagement (Dodd et al., 2022; Tereshchenko et al., 2024). As Prior et al. (2024) note, even basic tasks like measuring SDG content in educational offerings can present significant resource burdens, particularly for larger institutions.

HEIs face increasing pressure to demonstrate meaningful contributions to the SDGs while navigating resource constraints and competing priorities. The Triple Helix model (Etzkowitz & Leydesdorff, 2000; Leydesdorff & Etzkowitz, 1998) provides a useful framework for understanding these interactions, emphasising the dynamic relationships among universities, businesses, and policymakers. This model conceptualizes innovation as emerging from the dynamic, non-linear relationships among three institutional spheres: universities (focusing on knowledge creation and education), industry (emphasizing commercial application and market-based innovation), and government (providing regulatory frameworks and public funding). Unlike traditional linear models of knowledge transfer, the Triple Helix emphasizes the overlapping and increasingly interdependent nature of these sectors, where each adopts some functions traditionally associated with the others. For example, universities engage in commercialization activities, firms develop academic partnerships for basic research, and governments create innovation-focused policy frameworks. This interactive model suggests that innovation and societal progress emerge most effectively when these three sectors engage in reciprocal partnerships, leveraging their unique strengths to drive knowledge creation, technological advancement, and policy development.

The Triple Helix model has evolved differently across international contexts, reflecting regional economic priorities, cultural factors, and governance structures. For example, Zheng and Hu (2018) demonstrates how universities in China have adapted the model to address development challenges through stronger state direction, functioning as institutional entrepreneurs in regional innovation systems. In Europe, studies by Galán-Muros et al. (2015) show that universities employ various mechanisms to support business collaboration, with significant differences between Eastern and Western European approaches. Meanwhile, Wakkee et al. (2019) highlight how universities across different regions have integrated sustainable development into their entrepreneurial activities, with varying emphases on economic, social, and environmental priorities. These variations significantly influence how HEIs approach sustainability initiatives and integrate SDG priorities into their collaborative activities across different global contexts.

By embracing the Triple Helix approach, HEIs can move beyond isolated efforts and integrate their teaching, research, and community engagement activities with industry expertise and governmental support. This alignment not only enhances institutional capacity for implementing and measuring SDG-related initiatives but also fosters more strategic and scalable solutions that can meaningfully advance global sustainability goals. However, achieving this integration presents considerable challenges. HEIs must navigate differences in organizational cultures, align diverse stakeholder interests, and overcome resource limitations—all while maintaining academic integrity and addressing regional socioeconomic realities that influence their sustainability priorities.

Despite the growing recognition of the importance of HEI-industry-government collaboration, significant gaps remain in understanding the key factors that drive successful partnerships (Leišytė & Sigl, 2018). Many initiatives struggle with misaligned incentives, bureaucratic barriers, and difficulties in measuring impact, limiting their effectiveness in advancing the SDGs (Dodd et al., 2022; Marullo et al., 2021). While the Triple Helix model offers a conceptual foundation for these interactions, there is a need for a more detailed examination of the structural, relational, and operational elements that contribute to successful collaboration (Marques et al., 2024). To address this gap, we conduct a comprehensive review of the existing literature on HEI-industry-government partnerships. By examining these interconnected dimensions, we provide practical guidance for HEIs seeking to enhance their SDG contributions while highlighting critical tensions and implementation challenges that must be addressed.

This chapter begins by exploring the role of HEIs in advancing the SDGs and reviewing existing research on HEI-industry-government interactions to identify key themes. We then present a multilevel framework and discuss its implications for practice, offering insights for policymakers, institutional leaders, and educators on how to position HEIs as effective agents of sustainable development.

2 The Role of Higher Education Institutions in Advancing the SDGs

As noted above, HEIs play a critical role in advancing the SDGs through teaching, research, and collaborative efforts with industry and government (Maietta, 2015). By embedding sustainability across curricula, HEIs prepare future leaders to tackle global challenges (Dodd, 2024). For instance, Arizona State University requires all incoming students to complete a sustainability course (Inside Higher Ed, 2024), while business schools worldwide are increasingly integrating environmental and social governance principles into their programmes (Yang et al., 2021). These initiatives reflect a growing commitment to equipping students with the knowledge and skills necessary to address pressing societal and environmental issues.

HEIs also contribute significantly to sustainable development through research that influences policy and industry practices (Cottafava et al., 2022). For example, the University of Edinburgh Business School developed a climate risk framework that informs financial models at Standard Chartered Bank, showcasing the practical applications of academic research in driving corporate sustainability strategies (University of Edinburgh Business School, 2024). Other institutions, such as the University of Adelaide in Australia, have established sustainability as a focus area, fostering networks to address global challenges such as energy transitions and climate change, encouraging interdisciplinary collaboration, and disseminating innovative solutions (University of Adelaide, 2024).

The intersection of defence and sustainable development highlights another critical area where HEIs can make a significant impact, especially given the overarching goal of the SDGs to support world peace. For instance, the defence sector, often perceived as detached from sustainability efforts, holds considerable potential to contribute to the SDGs, particularly SDG 16, which focuses on peace, justice, and strong institutions (United Nations, 2024). Transparency International (2024) emphasises the importance of good governance and transparency in defence spending to prevent the misallocation of resources away from essential services like health and education. Furthermore, research commissioned by the UK Ministry of Defence, conducted by RAND Europe (2020) and the University of Exeter, explores the implications of climate change for defence strategies. This highlights the need for industries central to the SDGs to adapt to emerging environmental challenges and foster collaboration between industry and government.

However, many sustainability initiatives in higher education face significant challenges (Dodd et al., 2022). While statistics on failure rates are limited, studies have identified several common barriers. Prior et al. (2024) highlight that even basic tasks like measuring SDG content in educational offerings can present substantial resource burdens for institutions. Additionally, many HEIs rely on subjective interpretations and manual approaches to assess their SDG performance, leading to potential bias and inefficiencies in reporting (Gutiérrez-Mijares et al., 2023). Understanding these barriers is crucial for developing effective frameworks that foster successful sustainability initiatives in higher education.

Given the extensive body of work that has examined these challenges, we conducted a literature review as a first step. We examined research that has explored the factors that drive successful HEI-industry-government partnerships and those that contribute to their failure. Based on these findings, we then synthesise existing research to provide an emerging framework and research agenda for addressing the complexity and interconnectedness of global sustainability challenges associated with the SDGs.

We argue that using the literature is a crucial step in strengthening the evidence base for more effective sustainability initiatives in higher education that drive meaningful change. By analysing the current body of work, this chapter aims to develop a nuanced understanding of how HEIs contribute to sustainable development and how collaborations across sectors can be optimised to achieve the SDGs.

3 Method

We conducted a review of research published between 2013 and 2024. This timeframe captures a significant evolution in how Triple Helix models are understood and operationalised, with research shifting from traditional, linear models of collaboration to more dynamic, multidimensional frameworks that consider cognitive, social, and technological dimensions of partnerships (Maietta, 2015; O'Dwyer et al., 2022). Our search encompassed three search strings (listed in the Appendix) and four major academic databases (Scopus, Web of Science, ProQuest, and EBSCOhost). This search yielded 2,033 papers. After removing duplicates, papers underwent three rounds of screening: initial title and abstract review, date and peer review screening, and detailed qualitative assessment.

We selected a systematic review approach to provide a comprehensive, structured, and transparent synthesis of the diverse literature on Triple Helix collaborations. This approach allowed us to identify patterns and themes across multiple studies and minimize bias in study selection. To ensure methodological rigor, we used Covidence, a systematic review management tool, to streamline the screening process. Two independent reviewers assessed titles and abstracts against predefined inclusion and exclusion criteria, focusing on empirical research and theoretical contributions examining success factors in university-industry-government collaborations related to the SDGs. Disagreements were resolved through team discussions until consensus was reached.

Studies were excluded if they lacked clear research methodology, failed to provide empirical evidence or theoretical contributions, or did not explicitly examine the dynamics of university-industry-government interactions. Opinion pieces, editorials, descriptive case studies without analytical frameworks, and publications without rigorous peer review were also excluded to maintain the analytical focus of the review. This process yielded 103 papers for final analysis (see Fig. 1). The final articles included in the review are marked with an asterisk in the reference list.

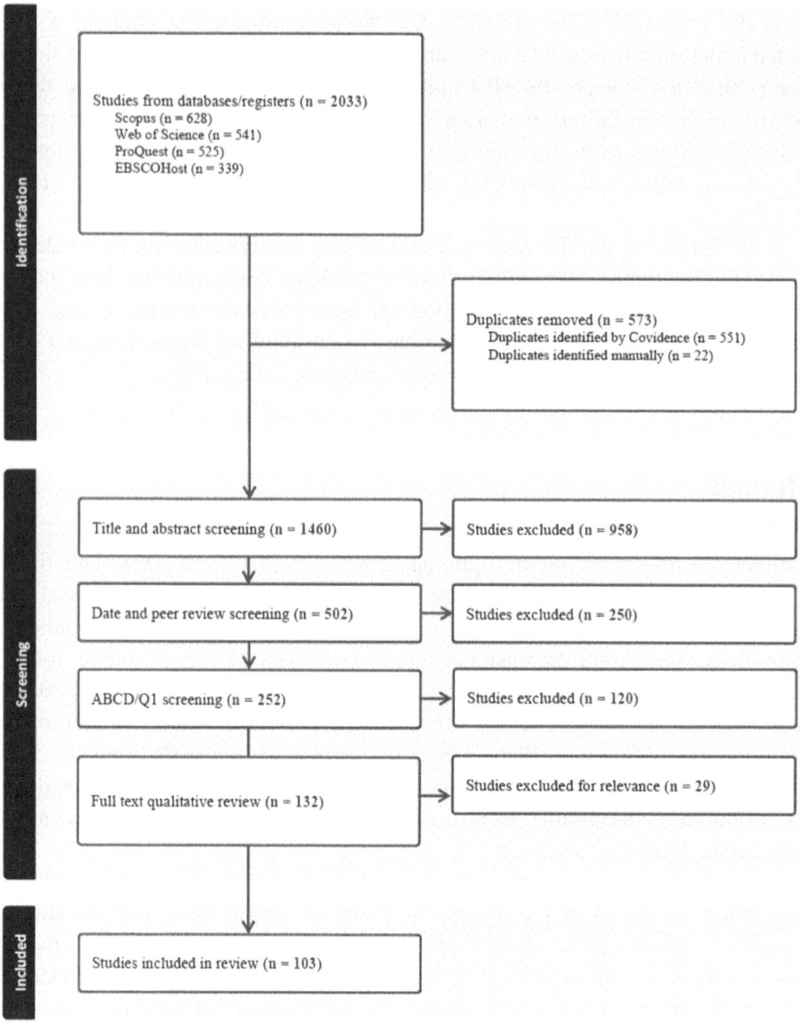

Fig. 1 Systematic review process: study selection and screening

Using NVivo qualitative analysis software, we systematically coded each paper to capture key themes, including focus areas, methodological approaches, barriers, enablers, and outcomes. Regular research team meetings ensured consistency in interpretation, with disagreements resolved through discussion until consensus was reached. This process resulted in a final data structure comprising three dimensions (macro, organisational, and team and individual-level dynamics) and 13 overarching themes, summarised in Table 1 and discussed below.

Geographical Distribution and Scope.

Table 1 Data structure and themes

Dimension	Theme	Codes
Macro-level dynamics	Policy stability and funding	Long-term funding, multi-year commitments, predictable support, government subsidies, SDG 9
	Intellectual Property (IP) frameworks	IP protection, technology transfer, knowledge sharing, legal enforcement
	Sustainable innovation ecosystems	Collaborative networks, trust-building, structured partnerships, evolving interactions
	Proximity and collaboration	Geographic, cognitive, and institutional proximity, knowledge exchange, shared frameworks
	Digital transformation	R&D networks, data governance, technological infrastructure, IP challenges
Organisational-level dynamics	Culture and leadership	Entrepreneurial mindset, institutional policies, leadership commitment, cultural change
	Entrepreneurial orientation	Third mission, commercialisation, business incubators, Technology Transfer Offices
	Knowledge management systems	Knowledge capture, sharing routines, IP management
	Organisational processes	Bureaucracy, administrative barriers, agility, decision-making complexity
	Trust and governance	Shared governance, flexible IP policies, transparency, strategic leadership
Team and individual-Level dynamics	Team composition and expertise	Academic-industry mix, interdisciplinary collaboration, knowledge diversity
	Leadership in cross-sector collaboration	Project management, bridging institutional logics, conflict resolution
	Departmental culture and social norms	Institutional attitudes, internal collaboration culture, support for partnerships

(continued)

Table 1 (continued)

Dimension	Theme	Codes
	Capability & Boundary spanning	Cross-sector communication skills; adaptability; network development; open communication
	Gender & Systemic barriers	Gender-based access factors; commercialization motivations
	Motivation & Prior experience	Academic versus industry motivations; career advancement; entrepreneurial mindset; networks; time and work schedules, career pathways; industry collaboration history

Notably, in addition to the themes summarised in Table 1, our analysis of the 103 papers included in the final review revealed a distinct geographical concentration within the existing literature. A significant portion of the studies originated from or focused on European contexts, particularly Western European nations such as the UK (Gilman & Serbanica, 2014), Italy (Cesaroni & Piccaluga, 2015; Marullo et al., 2021), Germany (Leišytė & Sigl, 2018), and the Netherlands (Bronneberg et al., 2023).

Studies from North America predominantly focused on the United States (Cohen et al., 2020; Libaers, 2015), revealing well-established institutional frameworks for technology transfer and stronger commercialization orientations. East Asian research, largely centered on China, Japan, and South Korea (Li et al., 2018; Son et al., 2019; Yoshioka-Kobayashi, 2019), demonstrated more state-directed innovation policies and different governance structures in collaborative relationships.

Research from Australia/Oceania, the Middle East, Latin America, and Africa was notably less prevalent. The limited studies from these regions, such as those by Danquah et al. (2024) on Ghana, Jonbekova et al. (2020) on Kazakhstan, and Dalmarco et al. (2018) on Brazil, indicate important contextual differences that influence how Triple Helix collaborations function in diverse economic environments.

These geographical differences raise important considerations about the transferability of findings across diverse contexts. Regional variations in institutional frameworks, resource availability, and cultural factors appear to significantly shape collaboration dynamics. For example, Hojeij's (2024) work on university-industry collaboration in the Arab world reveals distinct patterns shaped by unique cultural and institutional environments.

This geographical imbalance in the literature points to a valuable opportunity for future research examining Triple Helix collaborations in more diverse global contexts, particularly where different approaches to advancing the SDGs may be required due to varying economic and institutional conditions.

These geographical differences raise interesting considerations about the transferability of findings across diverse economic contexts. For example, the limited representation of studies from less developed countries raises questions about whether the identified factors and dynamics are equally applicable in regions with differing institutional arrangements, resource availability, and policy frameworks.

Studies by Danquah et al. (2024) on university-industry collaboration in Ghana and Jonbekova et al. (2020) on partnerships in Kazakhstan, for example, demonstrate how regional economic and institutional differences significantly shape collaboration dynamics. Similarly, Dalmarco et al. (2018) highlight the distinct constraints faced by entrepreneurial universities in Brazil compared to those in developed economies, while Liu et al. (2024) and Li et al. (2018) illustrate how university engagement in China is influenced by state policies and regional development priorities.

In interpreting our findings, we, therefore, acknowledge that the predominance of studies from more developed economies limits our ability to make definitive claims about how these factors operate in other contexts. While many organizational and team-level factors may have universal relevance, their manifestations likely vary significantly across different economic and infrastructural environments. For instance, Hojeij's (2024) overview of university-industry collaboration in the Arab world reveals how cultural and institutional factors create distinct collaboration patterns compared to Western contexts.

This limitation underscores the need for more research examining Triple Helix collaborations in diverse global contexts, particularly in Africa, parts of Asia, and Latin America, where institutional arrangements, resource availability, and policy frameworks may necessitate different approaches to advancing the SDGs through university-industry-government partnerships.

Several other methodological limitations should also be acknowledged. Our focus on English-language, peer-reviewed publications potentially excluded valuable insights from non-English and grey literature sources. The 2013–2024 timeframe, while capturing contemporary practices, excluded historical perspectives. Additionally, the subjective elements inherent in qualitative coding may introduce interpretive bias despite our systematic approach. Finally, our emphasis on identifying broad patterns across diverse contexts necessarily sacrificed some depth in specific institutional or regional settings. These limitations should be considered when interpreting our findings and framework.

4 Findings

4.1 Macro-Level Dynamics

Our analysis revealed several critical macro-level factors that drive successful university-industry-government collaborations. At the policy level, two elements emerged as particularly important for advancing SDG 9 (Industry, Innovation, and

Infrastructure): consistent long-term funding commitments and clear IP frameworks. Government funding models providing predictable, multi-year support allow universities and industry partners to plan and execute longer-term projects to address complex sustainability challenges. For instance, Wang and Liu (2022) found that sustained industrial funding, backed by strong IP enforcement mechanisms, led to more successful technology transfer between universities and industry. This is supported by Fan et al., (2019, p. 1377), who observe that when organisations can rely on stable funding, they are more likely to share knowledge and commit resources to joint projects that advance innovation goals:

> [W]e found that governmental funding positively affects the UIC [university-industry collaboration] management mechanism and UIC regulation implementation, which subsequently influenced innovation performance of universities. The results are consistent with existing research, which finds that there is a positive relationship between governmental subsidised funding and innovation outcomes of universities.

The importance of policy stability for sustainable development is further reinforced by O'Dwyer et al. (2022), who demonstrated that network-specific government investments fostered greater knowledge sharing between academics and industry partners. Rather than one-off grants, these sustained investments helped build ongoing collaborative relationships necessary for tackling long-term sustainability challenges.

The broader sustainable innovation ecosystem also emerged as a fundamental success factor. The findings suggest well-supported innovation through partnerships is essential for economic development, job creation, and systemic innovation. Donati et al. (2023) highlight how these ecosystems support collaborative networks by creating platforms that enable multiple actors to work together through "evolving helical patterns". Tereshchenko et al. (2024) emphasise the importance of a comprehensive approach that strengthens trust and simplifies organisational interactions, noting that effective ecosystems require nuanced co-operation between different institutional actors. The literature consistently points to these ecosystems as critical infrastructures that bridge different organisational logics, enabling knowledge exchange and collaborative innovation by creating structured yet adaptable environments where universities, industries, and government entities can interact and generate new insights:

> [G]overnments began to acknowledge the economic potential of UI collaborations and to establish new models, platforms, and policy initiatives for supporting such collaborations on a national level. Similarly, the notion of knowledge communities began to be relevant for knowledge management, the development of social capital, and the facilitation of new knowledge creation and innovation. (Oliver, 2022, p. 1615)

The analysis also revealed that proximity—geographic, cognitive, and institutional—plays a crucial role in shaping university-industry-government collaborations, with particular relevance to SDG 9 (Industry, Innovation, and Infrastructure). Maietta (2015) found that geographical proximity fundamentally influences university-firm research collaborations, with location crucial for accessing specialised research capabilities. Cognitive proximity emerged as a critical factor in knowledge transfer, with studies showing that partners must develop a shared understanding to effectively collaborate. As O'Dwyer et al. (2022) demonstrate, successful collaborations depend on partners' ability to bridge different expertise while maintaining a common conceptual framework. The research highlights that proximity is also about creating meaningful connections across organisational boundaries. These proximity dynamics are important in complex collaborative projects, where partners must navigate different institutional logics, research approaches, and organisational cultures to generate innovative solutions.

Finally, the literature highlights that digital technologies are reshaping how collaborations occur. For instance, research by Xing et al. (2024) demonstrates that digital technologies are reshaping R&D networks, offering new mechanisms for knowledge exchange and institutional adaptation. As highlighted in studies by Wirsich et al. (2016), these technological transformations create new opportunities for cross-sector collaboration while introducing complex challenges around data governance and IP management. The literature emphasises that successful digital co-operation requires more than technological infrastructure—it demands carefully negotiated institutional frameworks that balance openness with protection (Jabbar et al., 2024). Research by Wirsich et al., (2016, p. 719) suggests that digital platforms are not simply tools but dynamic environments that fundamentally reconfigure how universities, industries, and governments interact and generate knowledge:

> UIC increases the ability to explore. Scientific research further extends the knowledge base of companies and helps them to identify complementary technologies for recombination. Additionally, knowledge spillovers from universities to companies occur and may enhance the firm's ability to understand and merge existing technologies.

4.2 *Organisational-Level Dynamics*

At the organisational level, our systematic analysis reveals several critical factors that shape university-industry-government collaboration effectiveness. First, culture and leadership emerged as key enablers of innovation (SDG 9). Yoshioka-Kobayashi (2019) demonstrates how universities that successfully foster entrepreneurial mindsets achieve more effective industry engagement for knowledge transfer, highlighting the importance of institutional policies in strengthening the entrepreneurial culture. Based on a qualitative study of barriers and enablers of successful UICs, O'Dwyer et al. (2022) found that leadership commitment to cultural change proves particularly crucial, with senior management support significantly influencing collaboration outcomes. However, balancing entrepreneurial activities with traditional academic values requires careful management; as Yoshioka-Kobayashi (2019, p. 426) notes, "An entire organisation cannot become an entrepreneurial body by only introducing new policies; the shift requires both resources and culture". Adegbile et al., (2021, p. 7) point out that.

> Collaboration opportunities tend to be higher in universities which have a strong entrepreneurial orientation towards the commercialization of academic knowledge, where

there exists an environment (such as university structure, culture, policies, and governance) which facilitates collaboration.

Entrepreneurial orientation emerged as central to achieving effective collaborations. Universities increasingly embrace what scholars term the 'third mission'—contributing to economic and social development alongside traditional teaching and research roles (Alexander et al., 2015). The research shows an entrepreneurial orientation significantly enhances the ability of universities to advance sustainable development through industry partnerships. Dalmarco et al. (2018) observe how successful entrepreneurial universities establish comprehensive support systems through business incubators and technology transfer offices (TTOs)—specialised units within universities that manage the institution's IP and facilitate the transfer of research discoveries to industry:

> The TTO acts as an intermediary between the technology producers (academic researchers) and companies and entrepreneurs... allowing for specialized services in mapping potential industrial partners, intellectual property management and new business development… in terms of organizational structure, the creation of a specialized TTO inside a university can be essential to the development of partnerships with industry. (Dalmarco et al., 2018, p. 102)

This structural support enables universities to more effectively translate research into practical applications while maintaining academic integrity. Studies show that universities with well-established TTOs achieve higher rates of knowledge transfer and industry collaboration compared to those without such dedicated support structures (Dalmarco et al., 2018; Marullo et al., 2021; Son et al., 2019).

Effective knowledge management systems also emerged as critical for successful collaboration. Research by Tootell et al. (2020) highlights that organisations must develop robust mechanisms for capturing, sharing, and protecting knowledge across institutional boundaries. TTOs play a pivotal role in translating research into sustainability solutions, though their success varies considerably based on available resources and institutional support. For example, Ates et al. (2024) found that strong knowledge sharing routines enhance learning between academic and industry partners, with academics gaining a deeper understanding of industry challenges while industry partners learn about revolutionary processes and technologies. As Tootell et al. note (2020, p.743).

> Knowledge creation is the driver of sustained performance... and the ability for firms to form cooperative relationships with other organizations is vital for their competitive success.

These relationships require both formal knowledge management systems and the development of shared understanding between partners to effectively translate research into practical sustainability solutions (Alexander et al., 2015; Ates et al., 2024).

Organisational processes emerged as a critical success factor in university-industry collaborations (O'Dwyer et al., 2022; Temel et al., 2021). The literature highlights that university bureaucracy and complex administrative processes often create barriers to effective industry partnerships (Jonbekova et al., 2020). Research shows that bureaucratic processes and administrative requirements create significant obstacles, particularly around transaction-oriented factors. As a study found, university bureaucracy, when combined with lack of time, resources, and rewards, represents the most frequently cited barrier by academics engaging in industry partnerships (Perkmann et al., 2021). These institutional processes often conflict with industry's need for more agile and responsive decision-making (Jonbekova et al., 2020).

Finally, trust emerges as a critical enabler in university-industry collaborations, transcending mere contractual arrangements (O'Dwyer et al., 2022). While initially rooted in individual relationships, trust evolves into an organisational capability, fundamentally shaped by institutional policies, governance structures, and collaborative practices (Ulhøi et al., 2012). Kunttu and Neuvo, (2019) demonstrate that flexible IP policies, shared governance mechanisms, and strategic leadership are instrumental in building organisational trust. As interactions progress, trust becomes a dynamic organisational resource, with integrity and transparency increasingly critical as partners share complex, tacit knowledge (Ulhøi et al., 2012). Government policies and funding mechanisms can further cultivate this organisational trust by creating collaborative platforms that facilitate knowledge exchange (O'Dwyer et al., 2022; Perkmann et al., 2021). Ultimately, trust becomes a strategic organisational asset that reduces transactional safeguards and enables more open, innovative partnerships across institutional boundaries (Jabbar et al., 2024).

4.3 Team and Individual-Level Dynamics

Our analysis also revealed that the interplay between team composition and individual characteristics is crucial in advancing SDGs through more effective cross-sector partnerships. Teams that effectively combine academic and commercial expertise demonstrate enhanced capabilities for managing complex projects. Research shows that teams effectively combining academic and commercial expertise demonstrate enhanced capabilities for managing complex projects (Tootell et al., 2020). For instance, Messeni Petruzzelli and Murgia (2021) found that university-SME collaborations with diverse team composition generated more innovative technological outputs, particularly when teams balanced different knowledge domains. This finding aligns with Ting et al. (2019), who demonstrate that researcher competence and team expertise diversity significantly influence successful university-industry partnerships. Adegbile et al., (2021, p. 7) emphasize the importance of team composition and institutional environment:

> Collaboration opportunities tend to be higher in universities which have a strong entrepreneurial orientation towards the commercialization of academic knowledge, where there exists an environment (such as university structure, culture, policies, and governance) which facilitates collaboration. The effectiveness of interdisciplinary teams is particularly enhanced when supported by organizational structures that enable knowledge integration across institutional boundaries and promote entrepreneurial approaches to problem-solving.

Studies by Jones and Coates (2020) and Tootell et al. (2020) reinforce this, showing that teams incorporating both academic insights and industry knowledge enhance project outcomes through complementary skills and perspectives, leading to more practical and implementable solutions for sustainability challenges (Wirsich et al., 2016).

Leadership and departmental culture were also seen as critical in shaping team performance. Strong leadership proves essential for navigating the complexities of cross-sector collaborations while maintaining focus on team cohesion and broader societal goals. Research, such as that in O'Dwyer et al., (2022, p. 920), underscores the importance of leaders who can bridge different institutional logics, fostering mutual understanding between academic and commercial partners:

> Much of the success achieved by this UIC is attributed to the positive impact of the project manager. The project manager was vital in harmonizing different objectives, needs, perspectives, and modes of operation of academic and industrial organisations.

Within departments, social norms and attitudes significantly influence collaboration outcomes. Sharifi et al. (2013) illustrate that supportive departmental cultures valuing engagement with real-world problems create environments more conducive to substantive and productive collaboration. Peer support and positive departmental attitudes enhance teams' ability to tackle complex societal challenges by promoting collaborative approaches to problem-solving.

At the individual level, boundary-spanning capabilities emerged as crucial for achieving SDGs 17 (Partnerships for the Goals). Research highlights that individuals with strong cross-sector communication skills, adaptability, and prior industry experience are more likely to succeed in bridging academic research with real-world impact (Abramo & D'Angelo, 2021). These boundary-spanning capabilities include cultural awareness, networking proficiency, and the ability to navigate differing institutional priorities. Open communication and the development of professional networks also play a pivotal role, enabling individuals to build trust and effectively manage collaboration complexities and achieve broader societal targets.

Abramo and D'Angelo (2021, p. 1862) point to the critical role individuals play:

> The determinants of variety and frequency of public-private interactions lie above all in the characteristics of the individual researcher, much more than those of their home organizations.

Gender equity (SDG 5) also emerged as a significant consideration. Studies by Abramo and D'Angelo (2021) reveal gender-based disparities in collaboration participation, with research groups led by women exhibiting a lower likelihood of "showing interest in R&D co-operation with firms" (Abramo & D'Angelo, 2021, p. 1864). Finally, individual motivation and prior experience strongly shape collaboration success in advancing SDGs. Studies indicate that motivations extend beyond financial incentives to include career advancement, professional development, and the desire to contribute to meaningful societal impact (Cohen et al., 2020; Olmos-Peñuela et al., 2016). Research by Cohen et al. (2020) reveals that different academic fields show distinct motivational patterns, with life sciences academics being particularly motivated by social impact potential. Prior industry experience and collaboration history also significantly influence individual success in university-industry partnerships, with studies showing that early exposure to collaborative projects fosters both the skills and confidence necessary for managing cross-sector partnerships (Plantec et al., 2023).

5 Emerging Model

Building on the findings of our literature review, we present an emerging model that highlights how much is beyond the control of HEIs in advancing the SDGs. This section introduces an integrated framework that conceptualizes how the Triple Helix model can effectively support SDG implementation in higher education. While HEIs engage in solid strategies to advance the SDGs, as highlighted in Sect. 2, we argue in Sect. 1 that one of the many reasons SDG advancement stalls in HEIs is their reliance on industry and government. Our systematic review reveals that although the Triple Helix model—comprising HEI, government, and industry collaboration—offers conceptual promise, action is still stalling. The framework we propose addresses these challenges by mapping the complex interrelationships between key stakeholders while identifying critical success factors at multiple levels.

To address challenges in HEI, government, and industry collaborations as they relate to sustainable development, we propose a Triple Helix model to advance the SDGs. The following framework visually represents the dynamic interplay between universities, industry, and government in pursuing sustainable development goals. As illustrated in Fig. 2, our model emphasizes that to achieve meaningful progress, HEIs must consider and seek to influence macro, organizational, team, and individual-level factors critical for cross-sectoral efforts.

The Triple Helix Framework for SDG Advancement illustrates the dynamic interplay between universities, industry, and government in pursuing sustainable development. At its core, the framework demonstrates how these three realms overlap and interact around SDG advancement, supported by key enabling factors. Each brings distinct contributions to the collaborative effort: universities provide research and teaching expertise, as well as innovation capacity; industry contributes market application capabilities, resource mobilisation, and technology transfer; while government delivers the essential policy framework, funding stability, and regulatory support needed for sustainable development.

The framework positions SDG advancement at the centre where all three sectors overlap, emphasising that sustainable development requires a coordinated effort from all stakeholders and that no single element can achieve the SDGs in isolation. This integration is supported by three critical enabling factors: digital transformation, which facilitates communication and knowledge sharing across sectors; trust and governance, which ensures accountability and effective partnership management; and sustainable innovation, which drives creative solutions aligned with SDG objectives.

The framework operates across multiple levels, beginning with macro-level conditions that shape collaboration potential, including policy frameworks, ecosystem

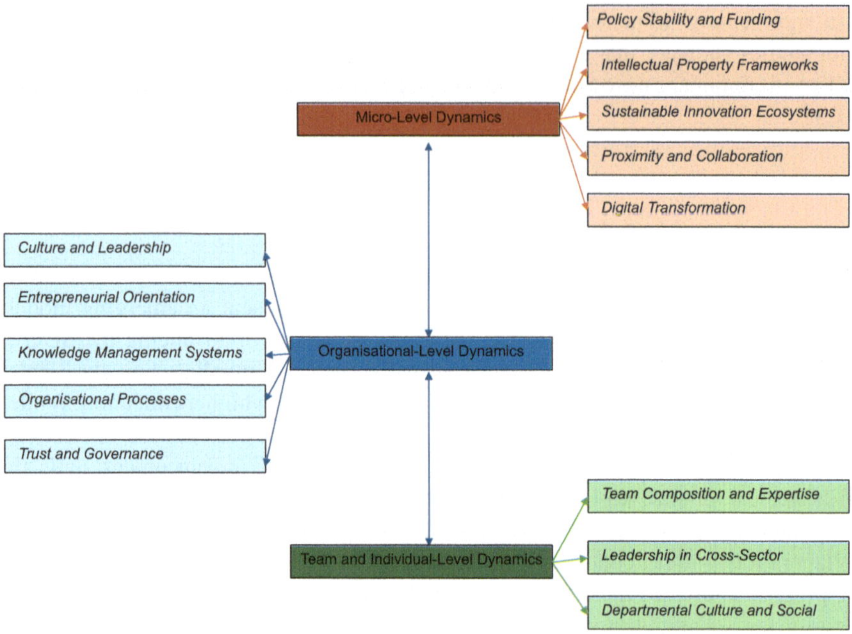

Fig. 2 Emerging model

dynamics, proximity considerations, and the impact of digital transformation. At the organisational level, internal factors within an organisation's control become crucial, such as culture and leadership alignment, entrepreneurial mindset cultivation, knowledge management systems, strategic resource allocation, and cultural transformation approaches. The team and individual level focus on operational elements that drive success, including interdisciplinary team composition, cross-sector expertise integration, project management capabilities, conflict resolution mechanisms, and supportive departmental culture.

For effective implementation, organisations seeking to engage in the triple helix model to advance the SDGs must first leverage the enabling factors while building strong institutional foundations through organisational alignment. They need to simultaneously develop team capabilities for cross-sector collaboration and nurture a culture that supports sustainable development initiatives. This framework emphasises that while external factors significantly influence collaboration success, institutions maintain considerable control over internal organisational and team dynamics that can enhance their ability to contribute to SDG advancement through cross-sector partnerships. The framework ultimately suggests that successful SDG advancement requires a holistic approach that considers and integrates all these elements while recognising the unique role and contribution of each in the collaborative process.

Moving from conceptual understanding to practical application, we now examine how organizations can operationalize this framework. Effective implementation requires institutions to navigate both external constraints and internal opportunities, focusing on elements within their control while developing strategies to influence broader ecosystem factors. The following considerations provide guidance for HEIs seeking to leverage the Triple Helix model in their sustainability initiatives.

While the framework proposed in this chapter offers valuable insights into how HEIs, industry, and government can collaborate to advance the SDGs, several limitations, not yet explored, must be acknowledged. Notably, the studies reviewed did not specifically mention the SDGs in their analyses, indicating a significant gap in the existing literature. Although many of these studies addressed issues related to sustainability, governance, and cross-sector collaboration, they did not explicitly link their findings to the SDGs or consider how HEI-industry-government partnerships can contribute to achieving these global goals.

This gap underscores the need for more targeted research that explicitly focuses on the role of HEIs in advancing the SDGs. For example, Prior et al. (2024) highlight that the SDGs, when considered collectively, present inherent tensions and paradoxes. These contradictions arise because the goals encompass a broad spectrum of social, economic, and environmental priorities that do not always align. For instance, economic growth (SDG 8) can conflict with climate action (SDG 13), while industrial development (SDG 9) may have unintended consequences for biodiversity (SDG 15). Addressing these challenges requires a nuanced analytical approach, which is where our framework may be valuable.

Our framework allows for a structured analysis of SDG tensions by considering interactions across different levels of governance (global, national, local), sectors (public, private, civil society), and temporal scales (short-term vs. long-term sustainability impacts). This perspective could help to unpack trade-offs and synergies by identifying where conflicts emerge and how they might be mitigated through governance mechanisms, policy coordination, or institutional design. For example, at a global level, international agreements can provide overarching guidelines to balance economic and environmental goals, while national governments can tailor policies to address local socio-economic conditions. Similarly, through industry-academic partnerships, businesses and universities can contribute to practical solutions that integrate multiple SDG objectives. By applying this multilevel lens, policymakers, researchers, and practitioners can develop more adaptive and context-specific strategies to navigate SDG paradoxes. Rather than viewing tensions as barriers to progress, this framework suggests that these contradictions can be productive sites of innovation, where new governance models and institutional arrangements emerge to reconcile competing priorities.

More nuanced analysis of these varying degrees of alignment would be useful in critically unpacking the claims in Prior et al. (2024), particularly in relation to how HEIs navigate SDG-related paradoxes. This would help differentiate between cases where universities are genuinely driving progress on SDGs and where they are simply aligning with goals that fit within their existing structures. Furthermore, it would illuminate where institutional and strategic shifts are necessary to move beyond business as usual, ensuring that universities do not merely adopt SDG rhetoric but actively contribute to systemic change. Applying this lens would deepen understanding of how HEIs can reconcile the tensions between their conventional roles and the broader ambitions of the SDGs, highlighting opportunities for innovation, cross-sector collaboration, and new governance mechanisms.

Further, our framework does not incorporate the Quadruple Helix model, which extends the traditional Triple Helix by including civil society as a fourth pillar. McAdam et al. (2016) highlight the importance of integrating societal actors into university-industry-government collaborations to enhance technology commercialisation and knowledge diffusion. By not explicitly considering the role of civil society, our framework may overlook critical contributions from community engagement, public discourse, and social innovation in advancing SDGs. Future refinements of our model could incorporate this dimension, acknowledging how non-governmental organisations, grassroots movements, and the public shape sustainable development initiatives. Expanding our framework to a Quadruple Helix approach would provide a more comprehensive understanding of SDG advancement by recognising the reciprocal influence of civil society in fostering inclusive and impactful cross-sector collaborations.

Another limitation of our research is its reliance primarily on peer-reviewed studies, which, while valuable, represent only a portion of the available evidence on HEI-industry-government collaborations and SDG advancement. Peer-reviewed publications often focus on specific academic or theoretical perspectives, potentially overlooking practical insights and emerging trends in the field. To address this, future research could expand beyond traditional academic sources and include grey literature such as reports from international organisations, governmental bodies, industry associations, and non-governmental organisations. These sources may offer a more comprehensive understanding of how HEIs are engaging with the SDGs in real-world settings, capturing practical challenges, innovations, and institutional strategies that may not be fully represented in scholarly publications. Incorporating grey literature would enrich the research by providing a broader, more diverse range of perspectives, which could lead to a deeper understanding of the opportunities and barriers faced by HEIs in their efforts to contribute to sustainable development.

6 Conclusion

In conclusion, while this chapter offers valuable insights into how HEIs, industry, and government can collaborate to advance the SDGs, several limitations highlight the need for further research. Our emerging framework provides a structured approach to understanding the dynamics of cross-sector collaboration and institutional innovation, but more specific and in-depth studies on HEIs' direct engagement with the SDGs are required to fully capture the complexities of these relationships.

The framework presented in this chapter has significant implications for advancing SDGs through higher education on a global scale. While our findings emerged primarily from research in developed economies, they point to universal principles that can be adapted across diverse international contexts. For instance, the emphasis on sustainable innovation ecosystems provides a template that can be customized to address region-specific sustainability challenges, from water security in arid regions to coastal resilience in island nations. Furthermore, our findings suggest that international organizations and funding bodies should prioritize support for developing robust policy frameworks and institutional capacity in regions where Triple Helix collaborations are less established. Future research should particularly focus on how these collaborative models can be effectively implemented in resource-constrained environments typical of many Global South countries, where HEIs may lack the infrastructure and support systems common in more developed regions.

Future research should extend our emerging framework to offer a more comprehensive view of the real-world challenges and innovations driving HEI contributions to sustainable development. By expanding the scope of inquiry and embracing diverse sources of knowledge, researchers can develop a more nuanced understanding of how HEIs can effectively navigate SDG-related tensions and drive meaningful, systemic change. This approach will not only enhance our understanding of the role of HEIs in advancing the SDGs but also inform practical strategies and policy recommendations for creating more effective and consequential collaborations among academia, industry, and government.

Appendix: Search Strings

(1) Search String: TS = ((universit* NEAR/5 industr*) AND (collaborat* OR partner*) AND (evolution OR trend* OR chang*) AND (effective* OR success* OR impact*) AND (commerciali* OR "technology transfer" OR innovat* OR entrepren*))

(2) Search String: TS = ((universit* OR academi* OR "higher education") AND (cultur* NEAR/5 (chang* OR transform* OR adapt*)) AND (driver* OR barrier* OR enabler* OR challenge* OR factor*) AND (industr* NEAR/5 (engage* OR collaborat* OR partner*)) AND (commerciali* OR "technology transfer" OR "knowledge transfer" OR innovat*) OR ("institutional change" AND (universit* OR academi*) AND (industr* OR business*))

(3) Search String: TS = ((universit* OR academi*) AND (polic* OR initiative*) AND (industr* NEAR/5 (engag* OR collaborat*)) AND (commerciali* OR "technology transfer") AND (impact* OR outcome*)). Illustrative results: Acworth (2008), Al-Tabbaa and Ankrah (2016), Arshed et al. (2021), Azagra-Caro et al. (2019), Azman et al. (2018), Björklund et al. (2019), Blankesteijn et al. (2020), Bojko et al. (2020), Cantner et al. (2023), Carvalho De Mello et al. (2016), Chai and Shih (2016), Chais et al. (2018), Chang (2017), Chen and Zhang (2023), Chen et al. (2024), Dalmarco et al. (2019), Davey and Galan-Muros (2020), D'este et al. (2019), Dolmans and a. M., Walrave, B., Read, S., Van Stijn, N. (2021), Dwivedi et al. (2024), Edgar and Kharazmi (2022), Fernández-Pérez De La Lastra et al. (2023), Filieri et al. (2014), Filippetti and Savona (2017), Gaspar Pacheco et al. (2024), Gerbin and Drnovsek (2020), Guerrero et al. (2021), Hao and Yunlong (2014), Hasche and Linton

(2021), Hoppmann (2021), Jonsson et al. (2015), Kuzior et al. (2024), Lee and Miozzo (2014), Li and Tang (2021), Ma et al. (2019), Mahdad et al. (2020), Mascarenhas et al. (2018), Mcadam et al. (2017), Murga Pinillos (2024), Muscio and Vallanti (2014), Noble et al. (2018), Okamuro and Nishimura (2018), Pitt et al. (2020), Plata (2024a), Plata (2024b), Pujotomo et al. (2023), Rosendo Rios et al. (2016), Rybnicek and Königsgruber (2018), Schaeffer et al. (2021), Shin et al. (2023), Sjöö and Hellström (2019), Song et al. (2019), Striukova et al. (2015), Taxt (2023), Troath (2023), Veletanlić and Sá (2019), Weerasinghe and Dedunu (2020), Zhao et al. (2018)

References

Abramo, G., & D'Angelo, C. A. (2021). Drivers of academic engagement in public–private research collaboration: An empirical study. *The Journal of Technology Transfer, 47*(6), 1861–1884.*

Acworth, E. B. (2008). University–industry engagement: The formation of the Knowledge Integration Community (KIC) model at the Cambridge-MIT institute. *Research Policy, 37* (8), 1241–1254.*

Adegbile, A. S., Sarpong, D., & Kolade, O. (2021). Environments for Joint University-Industry Laboratories (JUIL): Micro-level dimensions and research implications. *Technological Forecasting and Social Change, 170*, 120888.*

Alexander, A. T., Miller, K., & Fielding, S. (2015). Open for business: Universities, entrepreneurial academics and open innovation. *International Journal of Innovation Management, 19*(6), 1540013–1540037.*

Al-Tabbaa, O., & Ankrah, S. (2016). Social capital to facilitate 'engineered' university–industry collaboration for technology transfer: A dynamic perspective. *Technological Forecasting and Social Change, 104*, 1–15.*

Arshed, N., Ahmad, W., & Hanif, U. (2021). A spatial temporal exploration of factors motivating academia-industry collaboration. *Journal of the Knowledge Economy, 13*(1), 521–540.*

Ates, A., Paton, S., Bititci, U., & Kemal Konyalıoğlu, A. (2024). From transfer to co-creation: action research perspectives in knowledge transfer partnership (KTP) projects. *Production Planning & Control, 36*(2), 236–249.*

Azagra-Caro, J. M., Tijssen, R. J. W., Tur, E. M., & Yegros-Yegros, A. (2019). University-industry scientific production and the great recession. *Technological Forecasting and Social Change, 139*, 210–220.*

Azman, N., Sirat, M., Pang, V., Lai, Y. M., Govindasamy, A. R., & Din, W. A. (2018). Promoting university–industry collaboration in Malaysia: Stakeholders' perspectives on expectations and impediments. *Journal of Higher Education Policy and Management, 41*(1), 86–103.*

Bellandi, M., Corpakis, D., & Donati, L. (2023). Governing new and traditional partnerships for innovation and development in the post-pandemic world. *Triple Helix, 10*, 1–11.*

Björklund, T. A., Keipi, T., Celik, S., & Ekman, K. (2019). Learning across silos: Design Factories as hubs for co-creation. *European Journal of Education, 54*(4), 552–565.*

Blankesteijn, M., Bossink, B., & Van Der Sijde, P. (2020). Science-based entrepreneurship education as a means for university-industry technology transfer. *International Entrepreneurship and Management Journal, 17*(2), 779–808.*

Bojko, M. M., Knapińska, A., & Tomczyńska, A. (2020). Academic entrepreneurship and the research productivity in Poland. *Industry and Innovation, 28*(4), 486–506.*

Bronneberg, M., Pieterse, J., & Post, G. (2023). Brainport Eindhoven: born from crisis—25 years as a triple helix governed ecosystem. *Journal of Innovation Management, 11*(1), 36–67.*

Cantner, U., Kalthaus, M., & Yarullina, I. (2023). Outcomes of science-industry collaboration: Factors and interdependencies. *The Journal of Technology Transfer, 49*(2), 542–580.*

Carvalho De Mello, J. M., De Fuentes, C., & Iacobucci, D. (2016). Introduction to the special issue: Universities as interactive partners. *Science and Public Policy, 43*(5), 581–584.*

Cesaroni, F., & Piccaluga, A. (2015). The activities of university knowledge transfer offices: Towards the third mission in Italy. *The Journal of Technology Transfer, 41*(4), 753–777.*

Chai, S., & Shih, W. (2016). Bridging science and technology through academic–industry partnerships. *Research Policy, 45*(1), 148–158.*

Chais, C., Patrícia Ganzer, P., & Munhoz Olea, P. (2018). Technology transfer between universities and companies. *Innovation & Management Review, 15*(1), 20–40.*

Chang, S.-H. (2017). The technology networks and development trends of university-industry collaborative patents. *Technological Forecasting and Social Change, 118*, 107–113.*

Chen, X., & Zhang, G. (2023). Mapping the research on university-industry collaborative innovation of individuals: A scientometric analysis. *IEEE Access, 11*, 86318–86334.*

Chen, Z., Little, V. J., & Thuan, N. H. (2024). The evolving role of technology transfer offices in the entrepreneurial university: Go-betweens or playmakers? *The Journal of Technology Transfer*. https://doi.org/10.1007/s10961-024-10123-4*

Cohen, W. M., Sauermann, H., & Stephan, P. (2020). Not in the job description: The commercial activities of academic scientists and engineers. *Management Science, 66*(9), 4108–4117.*

Cottafava, D., Ascione, G. S., Corazza, L., & Dhir, A. (2022). Sustainable development goals research in higher education institutions: An interdisciplinarity assessment through an entropy-based indicator. *Journal of Business Research, 151*, 138–155.

Dalmarco, G., Hulsink, W., & Blois, G. V. (2018). Creating entrepreneurial universities in an emerging economy: Evidence from Brazil. *Technological Forecasting and Social Change, 135*, 99–111.*

Dalmarco, G., Hulsink, W., & Zawislak, P. A. (2019). New perspectives on university-industry relations: an analysis of the knowledge flow within two sectors and two countries. *Technology Analysis & Strategic Management, 31*(11), 1314–1326.*

Danquah, M. M., Onyancha, O. B., & Avuglah, B. K. (2024). Patterns and trends of university-industry research collaboration in Ghana between 2011 and 2020. *Information Discovery and Delivery, 53*(1), 90–108.*

Davey, T., & Galan-Muros, V. (2020). Understanding entrepreneurial academics - how they perceive their environment differently. *Journal of Management Development, 39*(5), 599–617.*

D'este, P., Llopis, O., Rentocchini, F., & Yegros, A. (2019). The relationship between interdisciplinarity and distinct modes of university-industry interaction. *Research Policy, 48*(9), 103829.*

Dodd, T. (2024). Conceptualising and evaluating the impact of business training programmes delivered to marginalized individuals, education for sustainable development: The contribution of universities. World Sustainability Series (in press), pp. 1–20. https://www.springer.com/series/13384

Dodd, T., Graves, C., & Hentzen, J. (2022). Impact and university business training courses delivered to the marginalized: A systematic review. *Academy of Management Learning & Education, 21*(3), 449–469.

Dolmans, S. A. M., Walrave, B., Read, S., & Van Stijn, N. (2021). Knowledge transfer to industry: How academic researchers learn to become boundary spanners during academic engagement. *The Journal of Technology Transfer, 47*(5), 1422–1450.*

Donati, L., Bellandi, M., & Corpakis, D. (2023). Evolving helical patterns supporting systemic innovation for sustainability transitions. *Triple Helix, 10*, 1–35.

Dwivedi, Y. K., Jeyaraj, A., Hughes, L., Davies, G. H., Ahuja, M., Albashrawi, M. A., Al-Busaidi, A. S., Al-Sharhan, S., Al-Sulaiti, K. I., Altinay, L., Amalaya, S., Archak, S., Ballestar, M. T., Bhagwat, S. A., Bharadwaj, A., Bhushan, A., Bose, I., Budhwar, P., Bunker, D., Walton, P. (2024).

"Real impact": Challenges and opportunities in bridging the gap between research and practice–Making a difference in industry, policy, and society. *International Journal of Information Management, 78*, 102750.*

Edgar, G., & Kharazmi, O. A. (2022). Systems evaluation of university-industry collaboration efficiency in Iran: Current situation and proposed policy framework. *Journal of the Knowledge Economy, 14*(2), 645–675.*

Etzkowitz, H., & Leydesdorff, L. (2000). The dynamics of innovation: From national systems and "Mode 2" to a Triple Helix of university–industry–government relations. *Research Policy, 29*(2), 109–123.

Fan, H. -L., Huang, M. -H., & Chen, D. -Z. (2019). Do funding sources matter? The impact of university-industry collaboration funding sources on innovation performance of universities. *Technology Analysis & Strategic Management, 31*(11), 1368–1380.*

Fernández-Pérez De La Lastra, S., Foncubierta-Rodríguez, M. -J., & Sánchez-Gardey, G. (2023). Toward classification of transfer research: an exploratory analysis based on indicators of academic engagement of knowledge transfer in academia–industry. *The Journal of Technology Transfer, 49*(3), 1043–1061.*

Filieri, R., Mcnally, R. C., O'dwyer, M., & O'malley, L. (2014). Structural social capital evolution and knowledge transfer: Evidence from an Irish pharmaceutical network. *Industrial Marketing Management, 43*(3), 429–440.*

Filippetti, A., & Savona, M. (2017). University–industry linkages and academic engagements: Individual behaviours and firms' barriers. Introduction to the special section. *The Journal of Technology Transfer, 42*(4), 719–729.*

Galán-Muros, V., Van Der Sijde, P., Groenewegen, P., & Baaken, T. (2015). Nurture over nature: How do European universities support their collaboration with business? *The Journal of Technology Transfer, 42*(1), 184–205.*

Gaspar Pacheco, A. I., Ferreira, J., Simoes, J., Mota Veiga, P., & Dabic, M. (2024). Mechanisms for facilitating academic entrepreneurship in higher education. *International Journal of Entrepreneurial Behavior & Research, 30*(6), 1448–1479.*

Gerbin, A., & Drnovsek, M. (2020). Knowledge-sharing restrictions in the life sciences: Personal and context-specific factors in academia–industry knowledge transfer. *Journal of Knowledge Management, 24*(7), 1533–1557.*

Gilman, M., & Serbanica, C. (2014). University–industry linkages in the UK: Emerging themes and 'unanswered' questions. *Prometheus, 32*(4), 322–343.*

Guerrero, M., Herrera, F., & Urbano, D. (2021). Does policy enhance collaborative-opportunistic behaviours? Looking into the intellectual capital dynamics of subsidized industry–university partnerships. *Journal of Intellectual Capital, 22*(6), 1055–1081.*

Gutierrez-Mijares, M. E., Josa, I., Casanovas-Rubio, M. D. M., & Aguado, A. (2023). Methods for assessing sustainability performance at higher education institutions: A review. *Studies in Higher Education, 48*(8), 1137–1158.

Hao, Z., & Yunlong, D. (2014). Research on the relationship of institutional innovation, organizational learning and synergistic effect: An empirical study of Chineses university spin-offs. *Journal of Industrial Engineering and Management, 7*(3), 645–659.*

Hasche, N., & Linton, G. (2021). University–industry collaboration: constructing a business model lab for student venture creation. *International Journal of Entrepreneurial Behavior & Research, 27*(5), 1241–1263.*

Hojeij, Z. (2024). An overview of university-industry collaboration in the Arab world. *Journal of Innovation and Entrepreneurship, 13*(1), 1–24.*

Hoppmann, J. (2021). Hand in hand to Nowhereland? How the resource dependence of research institutes influences their co-evolution with industry. *Research Policy, 50*(2), 104145.*

Hou, X., Gao, X., Yin, S., & Li, J. (2023). A three evolutionary game model for driving mechanism of industry-university-research collaborative innovation in agricultural innovation ecosystems. *PLoS One, 18*(8), e0289408.*

Inside Higher Ed. (2024, August 1). Sustainability course as major requirement at ASU. *Inside Higher Ed*. https://www.insidehighered.com/news/student-success/academic-life/2024/08/01/sustainability-course-major-requirement-asu

Jabbar, A., Apostolidis, C., Baines, N., Devine, A., Christofi, M., & Trivedi, S. (2024). Help those helping others—the role of universities in facilitating digitalisation and virtualisation in non-profit organisations. *The Journal of Technology Transfer*. https://doi.org/10.1007/s10961-024-10109-2*

Jonbekova, D., Sparks, J., Hartley, M., & Kuchumova, G. (2020). Development of university–industry partnerships in Kazakhstan: Innovation under constraint. *International Journal of Educational Development, 79*, 102291.*

Jones, S. E., & Coates, N. (2020). A micro-level view on knowledge co-creation through university-industry collaboration in a multi-national corporation. *Journal of Management Development*, 39(5), 723–738.*

Jonsson, L., Baraldi, E., Larsson, L.-E., Forsberg, P., & Severinsson, K. (2015). Targeting academic engagement in open innovation: Tools, effects and challenges for university management. *Journal of the Knowledge Economy, 6*(3), 522–550.*

Kunttu, L., & Neuvo, Y. (2019). Balancing learning and knowledge protection in university-industry collaborations. *The Learning Organization, 26*(2), 190–204.*

Kuzior, A., Samoilikova, A., Kossek, W., Krišková, P., & Vasylieva, T. (2024). Is digital development a factor of university-industry R&D collaboration and vice versa? *Production Engineering Archives, 30*(2), 204–213.*

Lee, H.-F., & Miozzo, M. (2014). How does working on university–industry collaborative projects affect science and engineering doctorates' careers? Evidence from a UK research-based university. *The Journal of Technology Transfer, 40*(2), 293–317.*

Leišytė, L., & Sigl, L. (2018). Academic institutional entrepreneurs in Germany: navigating and shaping multilevel research commercialization governance. *Triple Helix, 5*(1), 1–23.*

Leydesdorff, L., & Etzkowitz, H. (1998). Triple Helix of innovation: Introduction. *Science and Public Policy, 25*(6), 358–364.

Li, F., Chen, J., & Su, Y. S. (2018). Managing the university-industry collaborative innovation in China. *Journal of Organizational Change Management, 31*(1), 62–82.*

Li, Y., & Tang, Y. (2021). A dynamic capabilities perspective on pro-market reforms and university technology transfer in a transition economy. *Technovation, 103*, 102224.*

Libaers, D. (2015). Time allocations across collaborations of academic scientists and their impact on efforts to commercialize novel technologies: is more always better? *R&D Management, 47*(2), 180–197.*

Liu, Y., Fu, W., & Schiller, D. (2024). University–industry linkages in China from the agency perspective of university engagement: A systematic literature review. *Science and Public Policy, 51*(3), 435–449.*

Ma, L., Liu, Z., Huang, X., & Li, T. (2019). The impact of local government policy on innovation ecosystem in knowledge resource scarce region: Case study of Changzhou, China. *Science, Technology and Society, 24*(1), 29–52.*

Mahdad, M., De Marco, C. E., Piccaluga, A., & Di Minin, A. (2020). Why open innovation is easier said than done: An organizational identity perspective. *Journal of Innovation Management, 8*(2), 47–67.*

Maietta, O. W. (2015). Determinants of university–firm R&D collaboration and its impact on innovation: A perspective from a low-tech industry. *Research Policy, 44*(7), 1341–1359.*

Marques, C. S., Ferreira, J. J., Marques, C. P., & Leal, C. (2024). Understanding university-industry collaboration: a social exchange perspective on the entrepreneurial university. *The Journal of Technology Transfer*. https://doi.org/10.1007/s10961-024-10084-8*

Marullo, C., Piccaluga, A., & Cesaroni, F. (2021). From knowledge to impact. An investigation of the commercial outcomes of academic engagement with industry. *Technology Analysis & Strategic Management, 34*(9), 1065–1080.*

Mascarenhas, C., Ferreira, J. J., & Marques, C. (2018). University–industry cooperation: A systematic literature review and research agenda. *Science and Public Policy, 45*(5), 708–718.*

Mcadam, M., Miller, K., & Mcadam, R. (2016). Understanding Quadruple Helix relationships of university technology commercialisation: A micro-level approach. *Studies in Higher Education, 43*(6), 1058–1073.*

Mcadam, M., Miller, K., & Mcadam, R. (2017). University business models in disequilibrium–engaging industry and end users within university technology transfer processes. *R&D Management, 47*(3), 458–472.*

Messeni Petruzzelli, A., & Murgia, G. (2021). A multilevel analysis of the technological impact of university-SME joint innovations. *Journal of Small Business Management, 61*(4), 1896–1928.*

Murga Pinillos, A. Y. (2024). University-industry open innovation: Main enablers and practices based on a scoping review. *International Journal of Innovation Science*. https://doi.org/10.1108/ijis-08-2023-0194*

Muscio, A., & Vallanti, G. (2014). Perceived obstacles to university–industry collaboration: Results from a qualitative survey of Italian academic departments. *Industry and Innovation, 21*(5), 410–429.*

Noble, D., Charles, M. B., & Keast, R. (2018). The research collaboration paradox: A tale of two governance narratives in an Australian innovation setting. *Australian Journal of Public Administration, 77*(4), 597–603.*

O'Dwyer, M., Filieri, R., & O'Malley, L. (2022). Establishing successful university–industry collaborations: Barriers and enablers deconstructed. *The Journal of Technology Transfer, 48*(3), 900–931.*

Okamuro, H., & Nishimura, J. (2018). Whose business is your project? A comparative study of different subsidy policy schemes for collaborative R&D. *Technological Forecasting and Social Change*, 127, 85–96.*

Oliver, A. L. (2022). Holistic ecosystems for enhancing innovative collaborations in university–industry consortia. *The Journal of Technology Transfer, 47*(5), 1612–1628.*

Olmos-Peñuela, J., Benneworth, P., & Castro-Martínez, E. (2016). Does it take two to tango? Factors related to the ease of societal uptake of scientific knowledge. *Science and Public Policy, 43*(6), 751–764.*

Perkmann, M., Salandra, R., Tartari, V., McKelvey, M., & Hughes, A. (2021). Academic engagement: A review of the literature 2011–2019. *Research Policy, 50*(1), 104114.

Pitt, R. N., Satcher, L. A., & Drew, A. M. (2020). Optimism, innovativeness, and competitiveness: The relationship between entrepreneurial orientations and the development of science identity in scientists. *Social Currents, 7*(2), 155–172.*

Plantec, Q., Cabanes, B., Le Masson, P., & Weil, B. (2023). Early-career academic engagement in university–industry collaborative PhDs: Research orientation and project performance. *Research Policy, 52*(9), 104856*

Plata, C. (2024). Innovation investment and performance: the moderating role of university collaborations. *Journal of Management History, 31*(1), 204–227.*

Plata, C. (2024). University-Industry Relationships beyond technology transfer: The role of intangible knowledge. *Triple Helix, 10*(1), 1–35.*

Prior, D. D., Mysore Seshadrinath, S., Zhang, M., & McCormack, M. (2024). Measuring sustainable development goals (SDGs) in higher education through semantic matching. *Studies in Higher Education*. https://doi.org/10.1080/03075079.2024.2386625

Pujotomo, D., Syed Hassan, S. a. H., Ma'aram, A., & Sutopo, W. (2023). University–industry collaboration in the technology development and technology commercialization stage: A systematic literature review. *Journal of Applied Research in Higher Education, 15*(5), 1276–1306.*

RAND Europe. (2020). Climate change implications for UK defence. *RAND Europe*. https://www.rand.org/randeurope/research/projects/2020/climate-change-implications-for-uk-defence.html

Rosendo Rios, V., Zhang, Y., & Ghauri, P. N. (2016). Empirical analysis of the key factors that can contribute to university-industry cooperational success from a relationship marketing approach. *European Journal of International Management, 10*(6), 527–547.*

Rybnicek, R., & Königsgruber, R. (2018). What makes industry–university collaboration succeed? A systematic review of the literature. *Journal of Business Economics, 89*(2), 221–250.*
Schaeffer, P. R., Guerrero, M., & Fischer, B. B. (2021). Mutualism in ecosystems of innovation and entrepreneurship: A bidirectional perspective on universities' linkages. *Journal of Business Research, 134*, 184–197.*
Sharifi, H., Liu, W., & Ismail, H. S. (2013). Higher education system and the 'open' knowledge transfer: a view from perception of senior managers at university knowledge transfer offices. *Studies in Higher Education, 39*(10), 1860–1884.*
Shin, B., Rask, M., & Kahma, N. (2023). Measuring the Quadruple Helix in social media: A case study of university–industry collaboration. *Technological Forecasting and Social Change, 194*, 122711.*
Sjöö, K., & Hellström, T. (2019). University–industry collaboration: A literature review and synthesis. *Industry and Higher Education, 33*(4), 275–285.*
Son, H., Chung, Y., & Hwang, H. (2019). Do technology entrepreneurship and external relationships always promote technology transfer? Evidence from Korean public research organizations. *Technovation*, 82–83, 1–15.*
Song, Y., Elsner, W., Zhang, Z., & Berger, R. (2019). Collaborative innovation and policy support: the emergence of trilateral networks. *Applied Economics, 52*(34), 3651–3668.*
Striukova, L., Carl, A., Professor James A. Powell, D., & Rayna, T. (2015). University-industry knowledge exchange. *European Journal of Innovation Management, 18*(4), 471–492.*
Taxt, R. E. (2023). Motivations for academic engagement and commercialisation: A case study of actors' collaboration in third mission activities from three European universities. *Industry and Higher Education, 38*(2), 136–149.*
Temel, S., Dabić, M., Murat Ar, I., Howells, J., Ali, M., & Yesilay, R. B. (2021). Exploring the relationship between university innovation intermediaries and patenting performance. *Technology in Society, 66*, 101665*
Tereshchenko, D., Demchenko, V., & Khomenko, O. (2024). Emerging best strategies and capabilities for university-industry cooperation: A literature review 2000–2023. *Journal of Innovation and Entrepreneurship, 13*(28), 1–45.*
Ting, S. H., Yahya, S., & Tan, C. L. (2019). The influence of researcher competence on university-industry collaboration. *Journal of Entrepreneurship in Emerging Economies, 11*(2), 277–303.*
Tootell, A., Kyriazis, E., Billsberry, J., Ambrosini, V., Garrett-Jones, S., & Wallace, G. (2020). Knowledge creation in complex inter-organizational arrangements: understanding the barriers and enablers of university-industry knowledge creation in science-based cooperation. *Journal of Knowledge Management, 25*(4), 743–769.*
Transparency International. (2024). Defense and sustainable development. *Transparency International*. https://ti-defence.org/really-care-sustainable-development-better-think-defense/
Troath, S. (2023). The political economy of Australian militarism: On the emergent military–industrial–academic complex. *Journal of Global Security Studies, 8*(4), ogad018*
Ulhøi, J., Neergaard, H., & Bjerregaard, T. (2012). Beyond unidirectional knowledge transfer. *The International Journal of Entrepreneurship and Innovation, 13*(4), 287–299.*
United Nations. (2015). *Transforming our world: The 2030 agenda for sustainable development*. United Nations. https://sdgs.un.org/2030agenda
United Nations. (2024). Linking good security sector governance and SDG16. *United Nations*. https://sdgs.un.org/partnerships/linking-good-security-sector-governance-and-sdg16
University of Adelaide. (2024). FAME sustainability initiative. *University of Adelaide*. https://www.adelaide.edu.au/iser/fame-sustainability
University of Edinburgh Business School. (2024). Climate change risk in finance executive education. *University of Edinburgh Business School*. https://www.business-school.ed.ac.uk/executive-education/climate-change-risk-in-finance
Veletanlić, E., & Sá, C. (2019). Government programs for university–industry partnerships: Logics, design, and implications for academic science. *Research Evaluation, 28*(2), 109–122.*

Wakkee, I., Van Der Sijde, P., Vaupell, C., & Ghuman, K. (2019). The university's role in sustainable development: Activating entrepreneurial scholars as agents of change. *Technological Forecasting and Social Change, 141*, 195–205.*

Wang, W., & Liu, Y. (2022). Industrial funding and university technology transfer: The moderating role of intellectual property rights enforcement. *The Journal of Technology Transfer, 47*(5), 1549–1572.*

Weerasinghe, I. M. S., & Dedunu, H. H. (2020). Contribution of academics to university–industry knowledge exchange: A study of open innovation in Sri Lankan universities. *Industry and Higher Education, 35*(3), 233–243.*

Wirsich, A., Kock, A., Strumann, C., & Schultz, C. (2016). Effects of university–industry collaboration on technological newness of firms. *Journal of Product Innovation Management, 33*(6), 708–725.*

Xing, Z., Fang, D., Wang, J., & Zhang, L. (2024). How does institutional theory illuminate the influence of the digital economy on R&D networks? *European Journal of Innovation Management*. https://doi.org/10.1108/ejim-11-2023-0958*

Yang, Z., Chen, H., Du, L., Lin, C., & Lu, W. (2021). How does alliance-based government-university-industry foster cleantech innovation in a green innovation ecosystem? *Journal of Cleaner Production, 283*, 124559.*

Yoshioka-Kobayashi, T. (2019). Institutional factors for academic entrepreneurship in publicly owned universities in Japan: Transition from a conservative anti-industry university collaboration culture to a leading entrepreneurial university. *Science, Technology and Society, 24*(3), 423–445.*

Zhao, Z., Broström, A., & Cai, J. (2018). Promoting academic engagement: university context and individual characteristics. *The Journal of Technology Transfer, 45*(1), 304–337.*

Zheng, C., & Hu, M. C. (2018). An exploration of the application of universities as artificial institutional entrepreneurs: The case of China. *Journal of Public Affairs, 18*(1), e1697.*

The manufacturer's authorised representative in the EU is Springer Nature Customer Service Centre GmbH, Europaplatz 3, 69115 Heidelberg, Germany. If you have any concerns regarding our products, please contact ProductSafety@springernature.com

Printed and bound by CPI Group (UK) Ltd, Croydon, CR0 4YY
17/09/2026
02218456-0020